Advances in Experimental Medicine and Biology

Volume 1349

Advances in Experimental Medicine and Biology provides a platform for scientific contributions in the main disciplines of the biomedicine and the life sciences. This series publishes thematic volumes on contemporary research in the areas of microbiology, immunology, neurosciences, biochemistry, biomedical engineering, genetics, physiology, and cancer research. Covering emerging topics and techniques in basic and clinical science, it brings together clinicians and researchers from various fields.

Advances in Experimental Medicine and Biology has been publishing exceptional works in the field for over 40 years, and is indexed in SCOPUS, Medline (PubMed), Journal Citation Reports/Science Edition, Science Citation Index Expanded (SciSearch, Web of Science), EMBASE, BIOSIS, Reaxys, EMBiology, the Chemical Abstracts Service (CAS), and Pathway Studio.

2020 Impact Factor: 2.622

More information about this series at http://www.springer.com/series/5584

Lei Zhou
Editor

Ion Channels in Biophysics and Physiology

Springer

Editor
Lei Zhou
Shenzhen Bay Laboratory, Institute of Molecular
Physiology / Virginia Commonwealth
University School of Medicine
Shenzhen, Guangdong Province, P. R. China

ISSN 0065-2598 ISSN 2214-8019 (electronic)
Advances in Experimental Medicine and Biology
ISBN 978-981-16-4256-2 ISBN 978-981-16-4254-8 (eBook)
https://doi.org/10.1007/978-981-16-4254-8

This Springer imprint is published by the registered company Springer Nature Singapore Pte Ltd.
The registered company address is: 152 Beach Road, #21-01/04 Gateway East, Singapore 189721, Singapore

Contents

Part II Physiological Function

Part I

Biophysical Mechanism

Venom-Derived Peptides Inhibiting Voltage-Gated Sodium and Calcium Channels in Mammalian Sensory Neurons

1

Arsalan Yousuf, Mahsa Sadeghi, and David J. Adams

Abstract

Pain management is a serious worldwide problem that affects the physical and mental health of all affected humans. As an alternative to opioids, pharmaceutical companies are seeking other sources of potential analgesics that have fewer adverse side effects. Animal venoms are a natural cocktail of a complex mixture of salts, peptides, and proteins. Most animals that produce venoms release them for the purpose of prey capture and/or defense against other vertebrates. Over the last 30 years, many venom-derived peptides have been shown to be active against numerous voltage-gated ion channels in the mammalian somatosensory nervous system. Voltage-gated ion channels and in particular sodium, potassium, and calcium channels are fundamental to the transmission of all somatosensory information from the periphery to the central nervous system. This information can be chemical, mechanical, or thermal sensation that can result from touch to a more painful sensation of tissue injury. These voltage-gated ion channels open or close in response to changes in membrane potential to permit ion movement across the cell membrane. In this chapter, we screened the scientific literature characterizing venom-derived peptides that target voltage-gated sodium and calcium channels and exhibit analgesic properties. Depending on peptide activity, these can either inhibit voltage-gated sodium or calcium channels completely by binding to the pore of the channel or modulate the activity by binding to other regions such as the voltage sensor of the channel.

Keywords

Sodium channels · Calcium channels · Voltage sensor · Pore blockers · Gating modifiers · Pain

1.1 Introduction

1.1.1 Sensory Neurons and Pain Signaling

Pain is an alert message and is experienced as an unpleasant feeling by the brain that something is wrong in the body. It can result from tissue injury in the periphery or from damage to internal organs. Pain can be either acute or chronic, depending on whether the recovery from an injury also stops the pain perception or whether

A. Yousuf · D. J. Adams (✉)
Illawarra Health and Medical Research Institute (IHMRI), University of Wollongong, Wollongong, NSW, Australia
e-mail: arsalan@uow.edu.au; djadams@uow.edu.au

M. Sadeghi
Department of Anatomy, University of California San Francisco, San Francisco, CA, USA
e-mail: mahsa.sadeghi@ucsf.edu

L. Zhou (ed.), *Ion Channels in Biophysics and Physiology*, Advances in Experimental Medicine and Biology 1349, https://doi.org/10.1007/978-981-16-4254-8_1

the brain still keeps on perceiving pain even after the injury gets healed. The human somatosensory nervous system is designed in a delicate way to transmit noxious information from the periphery to the central nervous system [1]. Specialized primary sensory neurons called nociceptors innervate the skin and muscles and detect injurious stimuli, such as high temperature, mechanical pressure, or chemicals, which could be endogenous mediators, such as low pH or other irritants. These sensory neurons express a vast network of ligand-gated and voltage-gated ion channels and G-protein-coupled receptors to transduce these stimuli. The nociceptors synapse with the dorsal horn neurons of the spinal cord, and from here, the sensory information is relayed to the brain stem and thalamus and finally to the cortical networks [1].

1.1.2 Venom-Derived Peptides

Treatment of chronic pain is a serious economic problem worldwide, and treatments with opioids have given rise to enormous side effects. Scientists, hence, therefore have focused on developing new alternatives.

Venoms are a complex mixture of organic and inorganic molecules, salts, proteins, and peptides that are released from animals for a purpose of defense or prey. The composition of these venoms can range from relatively simple to complex and can contain thousands of individual components with distinct pharmacological activity. Animal venoms have been used as medicines, for centuries, in different civilizations [2, 3]. Broadly, venoms can be categorized as cytotoxic or neurotoxic; either mechanism can induce pain on envenomation, although many venoms are either non-noxious or even analgesic [4]. The latter includes venoms that open or block voltage-gated ion channels, either directly or indirectly, via G-protein-coupled receptors [5]. The interaction of venom-derived peptides with ion channels and GPCRs have constantly led to the development of medicinal therapeutics.

Venoms-derived peptides can be categorized into two different types as to how they influence voltage-gated ion channels. They can be either (i) pore blockers that can directly inhibit the flow of cations blocking the ion conduction pore or (ii) gating modifiers that means they can interact with regions of the channel that modifies the opening or closing of the channel [6–8] (Fig. 1.1). As the venom composition is so huge and complex, scientists have derived small peptide sequences (20–30 amino acid long) that can have distinct pharmacological profiles. A sum of these peptides has made to clinical trials to treat chronic pain.

The most common and diverse voltage-gated ion channels in the human body are sodium channels, potassium channels, and calcium channels with different subtypes. Here we highlight the structure and importance of voltage-gated sodium channels and voltage-gated calcium channels and specifically enlist all the venom-derived peptides discovered so far that target these channels in the mammalian sensory nervous system. Most of these peptides are promiscuous, meaning they hit different ion channels at equally strong potency.

1.2 Voltage-Gated Sodium Channels

VGSC's have a major role in transmitting sensory signals from the periphery to the central nervous system and neurotransmitter release. They are the main components of membrane action potential initiation, generation, and propagation. A total of 10 sodium channel subtypes are currently known (Nav1.1–1.9 and Nax) [9, 10]. Using RNA transcriptomics and gene expression studies, it is known that Na_V subtypes $Na_V1.1$, 1.2, 1.3, and 1.6 are distributed both in the central (CNS) and in the peripheral (PNS) nervous systems, while $Na_V1.7$, 1.8, and 1.9 are located exclusively in the PNS, and 1.4 and 1.5 are exclusively expressed in the skeletal muscle and heart muscle, respectively [11, 12].

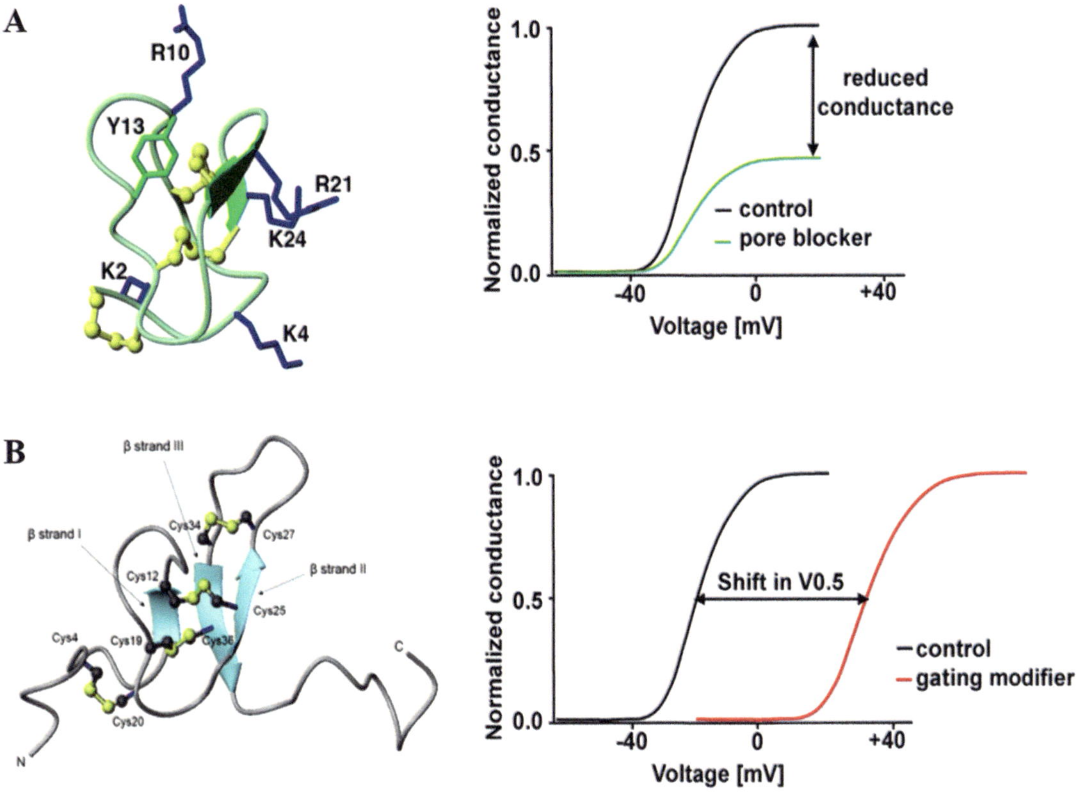

Fig. 1.1 ω-conotoxin MVIIA (pore blocker) and ω-agatoxin IVA (gating modifier) affecting two different modes of ion channel conductances [6–8]. Adapted and reproduced with permission from Elsevier *Neuropharmacology*

The structure of the sodium channel is universally similar for all subtypes. The human Na_V α-subunit consists of four pore-forming domains (DI-DIV), with each domain made up of six transmembrane segments (S1–S6). S1–S4 segment of each domain contains the voltage sensor, whereas S5/S6 forms the channel pore [13–15]. In addition to α-subunits, there are four β-subunits (β1–β4) that are believed to modify the gating and expression of Na_Vs [14, 16]. As venom peptides can act either as pore blockers or as gating modifiers, an example of peptide toxin KIIIA on $Na_V1.2$ and Huwentoxin on $Na_V1.7$ is shown in Fig. 1.2 to explain how such toxins interact with different domains of the sodium channels using cryo-electron microscopy [17]. A detailed list of venom-derived peptides targeting different Na_V subtypes in sensory neurons is further detailed in Table 1.1.

1.3 Voltage-Gated Calcium Channels

Voltage-gated calcium channels (VGCC) are also pivotal in the normal functioning of neurons and other excitable cells. They mediate depolarization-induced calcium influx into neurons and other excitable cells in response to an action potential. These channels are involved in many physiological functions such as contraction, secretion, gene transcription, and cell proliferation. Therefore, they initiate many cellular signaling events like neurotransmitter release and activation of calcium-dependent kinases. VGCCs also serve as potential drug targets in the therapy of neurological and cardiovascular diseases [18–20] and also in pain management due to their regulatory role in neurotransmitter release at nociceptive synapses. A list of venom-

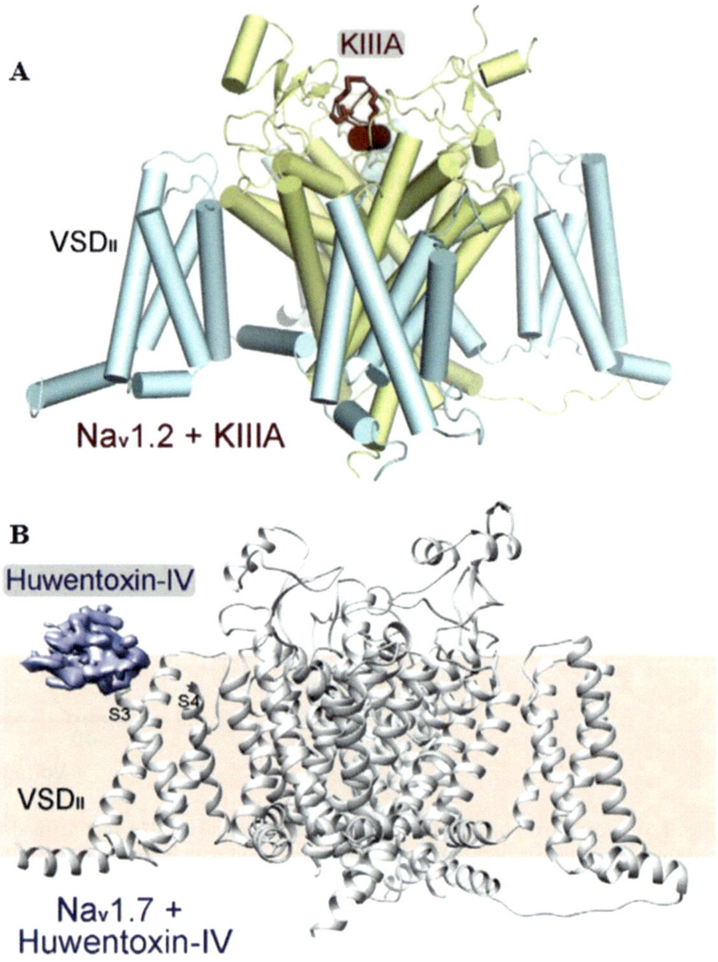

Fig. 1.2 Different modes of binding between peptidic toxins and Nav channels. Shown here is a summary of available structures between Nav channels and peptidic toxins [17]. Adapted and reproduced with permission from The American Association for the Advancement of Science

derived peptides targeting different Ca_V subtypes in sensory neurons is highlighted in Table 1.2.

Voltage-gated calcium channels are broadly classified into two categories: low voltage-activated (LVA) channels requiring moderate membrane depolarization to open and high voltage-activated (HVA) channels that require stronger depolarization for activation [21]. The low voltage-activated (LVA) channels are also termed as T-type (transient) channels, whereas high voltage-activated (HVA) channels are further characterized into L-type, N-, P/Q-, and R-type channels. As the $Ca_V\alpha1$ is the main subunit component forming the structural pore of the channel, 10 different $Ca_V\alpha1$ subunits are identified in the mammalian genome. The low voltage-activated (LVA) T-type channels comprise three types of $Ca_V\alpha1$ subtypes ($Ca_V3.1$, $Ca_V3.2$, and $Ca_V3.3$), and the high voltage-activated (HVA) channels are further subdivided into Cav1 family encoding four different types of L-type channels (Cav1.1–Cav1.4) and Cav2 family encompassing P/Q-type (Cav2.1), N-type (Cav2.2), and R-type (Cav2.3) channels [21]. The L-type channels are the major Cav channels in cardiac, smooth muscle, and skeletal muscle, also broadly expressed in neurons, auditory hair cells, pancreas cells, and retinae [22]. N-, P/Q-, and R-type channels are expressed both in the central and in the peripheral nervous systems.

Apart from a principal functional pore-forming α1 subunit, calcium channels are also associated with a disulfide-linked α2δ dimer and

Table 1.1 Venom-derived peptides inhibiting Nav channels expressed in mammalian sensory neurons

Peptide	Peptide source	Peptide activity	IC_{50}	Reference
Na_V1.1				
μ-conotoxin GIIIA	*C. geographus*	Pore blocker	0.3 μM	[26]
μ-conotoxin PIIIA	*C. purpurascens*	Pore blocker	0.1 μM	[26]
μ-conotoxin SxIIIA	*C. striolatus*	Pore blocker	0.4 μM	[26]
μ-conotoxin BuIIIA	*C. bullatus*	Pore blocker	0.35 μM	[26, 27]
μ-conotoxin BuIIIB	*C. bullatus*	Pore blocker	0.36 μM	[26, 27]
μ-conotoxin KIIIA	*C. kinoshitai*	Pore blocker	290 nM	[26, 28]
μ-conotoxin TIIIA	*C. tulipa*	Pore blocker	0.9 μM	[26]
μ-conotoxin SmIIIA	*C. stercusmuscarum*	Pore blocker	4 nM	[26, 28]
PnCS1	Hybrid peptide from PnTX1 and μ-KIIIA	Pore blocker	0.8 μM	[29]
PnCS1[W4K]	Hybrid peptide from PnTX1 and μ-KIIIA	Pore blocker	0.4 μM	[29]
PnCS1[W7Y]	Hybrid peptide from PnTX1 and μ-KIIIA	Pore blocker	2.2 μM	[29]
PhlTx1	*Phlogiellus* genus	Gating modifier	280 nM	[12]
MeuNaTxa-12	*Mesobuthus eupeus*	Gating modifier	0.91 μM	[27, 30]
Cyrtx-1a	*Cyriopagopus schioedtei*	Gating modifier	72 nM	[31]
Cd1a	*Ceratogyrus darlingi*	Gating modifier Cav2.2 off target (pore blocker)	2.2 μM	[28, 32]
MeuNaTxa-13	*Mesobuthus eupeus*	Gating modifier	2.5 μM	[27, 30]
ATX-II	*Anemonia sulcata*	Gating modifier	6 nM	[27, 33, 34]
CGTX-II	*Bunodosoma cangicum*	Gating modifier	0.16 μM	[27, 35]
Bc-III	*Bunodosoma caissarum*	Gating modifier	300 nM	[34]
AFT-II	*Anthopleura fuscoviridis*	Gating modifier	391 nM	[27, 34]
$GVIIJ_{SSG}$	*C. geographus*	Pore blocker	11 nM	[27, 36]
Pn3a	*Pamphobeteus nigricolor*	Gating modifier	37 nM	[37]
μ-TRTX-Df1a	*Davus fasciatus*	Gating modifier (also inhibits Cav3)	14 nM	[28, 38]
ProTx III	*Thrixopelma pruriens*	Not known	500 nM	[28, 39]
m3-HwTx IV	*Haplopelma schmitdi*	Gating modifier	8.4 nM	[28, 40]
μ-SLPTX-Ssm6a	*Scolopendra subspinipes mutilans*	Gating modifier	4.1 μM	[28, 41]
HnTx III	*Ornithoctonus hainana*	Gating modifier	1.27 μM	[28, 42]
Pre1a	*Psalmopoeus reduncus*	Gating modifier	57 nM	[43]
PaurTx3	*Phrixotrichus auratus*	Gating modifier	610 nM	[44, 45]
GsAFI	*Grammostola rosea*	Multi modal action (Pore blocker and gating modifier)	360 nM	[28, 46]
GsAFII	*Grammostola rosea*	Multi modal action (pore blocker and gating modifier)	5.7 μM	[28, 46]
GrTx1	*Grammostola rosea*	Multi modal action (pore blocker and gating modifier)	0.63 μM	[27, 46]

(continued)

Table 1.1 (continued)

Peptide	Peptide source	Peptide activity	IC_{50}	Reference
GsMTx4	*Grammostola rosea*	Multi modal action (pore blocker and gating modifier)	12 μM	[46]
Nav1.2				
ProTx-II	*Thrixopelma pruriens*	Gating modifier	136 nM	[28, 47]
ProTx-III	*Thrixopelma pruriens*	Gating modifier	300 nM	[28, 39]
Pn3a	*Pamphobeteus nigricolor*	Gating modifier	124 nM	[28, 37]
μ-TRTX-Df1a	*Davus fasciatus*	Gating modifier (also inhibits Cav3)	1.9 nM	[28, 38]
m3-HwTx IV	*Haplopelma schmitdi*	Gating modifier	11.9 nM	[28, 40]
Cd1a	*Ceratogyrus darlingi*	Gating modifier Cav2.2 off target (pore blocker)	130 nM	[28, 32]
HnTx IV	*Ornithoctonus hainana*	Gating modifier	1.7 nM	[28, 48]
μ-SLPTX-Ssm6a	*Scolopendra subspinipes mutilans*	Gating modifier	813 nM	[28, 41]
HwTx-IV	*Haplopelma schmitdi*	Gating modifier	5.8 nM	[28, 49]
GsAFI	*Grammostola rosea*	Multi modal action (pore block and gating modifier)	600 nM	[28, 46]
CcoTx I	*Ceratogyrus cornuatus*	Gating modifier	6.1 nM	[28, 50]
HnTx III	*Ornithoctonus hainana*	Gating modifier	275 nM	[28, 42]
PhlTx1	*Phlogiellus* genus	Gating modifier	73 nM	[28, 51]
μ-conotoxin KIIIA	*C. kinoshitai*	Pore blocker	5 nM	[26, 28, 52]
μ-conotoxin SmIIIA	*C. stercusmuscarum*	Pore blocker	1.3 nM	[26, 28]
ATX-II	*Anemonia sulcata*	Gating modifier	41 nM	[27, 33, 34]
AFT-II	*Anthopleura fuscoviridis*	Gating modifier	2 μM	[27, 34]
Lqh-2	*Leiurus quinquestriatus hebraeus*	Gating modifier	1.8 nM	[53, 54]
PaurTx3	*Phrixotrichus auratus*	Gating modifier	0.6 nM	[27, 44]
GrTx1	*Grammostola rosea*	Multi modal action (pore blocker and gating modifier)	0.23 μM	[27, 46]
Bc-III	*Bunodosoma caissarum*	Not known	1.5 μM	[27, 34]
PnTx1	*Phoneutria nigriventer*	Pore blocker	34 nM	[27, 55]
$GVIIJ_{SSG}$	*C. geographus*	Pore blocker	41 nM	[27, 36]
μ-conotoxin TIIIA	*C. tulipa*	Pore blocker	45 nM	[26, 27]
μ-conotoxin SIIIA	*C. striatus*	Pore blocker	50 nM	[26, 27]
μ-conotoxin BuIIIA	*C. bullatus*	Pore blocker	12 nM	[26, 27]
μ-conotoxin MIIIA	*C. magus*	Pore blocker	450 nM	[26, 27]
μ-Conotoxin PIIIA	*C. purpurascens*	Pore blocker	0.6 μM	[26]
Nav1.3				
ProTx-II	*Thrixopelma pruriens*	Gating modifier	343 nM	[28, 47]
ProTx-III	*Thrixopelma pruriens*	Gating modifier	900 nM	[28, 39]
Pn3a	*Pamphobeteus nigricolor*	Gating modifier	210 nM	[28, 37]
μ-TRTX-Df1	*Davus fasciatus*	Gating modifier Inhibits Cav3 currents	3 nM	[28, 38]
m3-HwTx IV	*Haplopelma schmitdi*	Gating modifier	7.2 nM	[28, 40]

(continued)

Table 1.1 (continued)

Peptide	Peptide source	Peptide activity	IC_{50}	Reference
HnTx IV	*Ornithoctonus hainana*	Gating modifier	18 nM	[28, 48]
HwTx-IV	*Haplopelma schmitdi*	Gating modifier	13 nM	[28, 49]
GsAFI	*Grammostola rosea*	Multi modal action (pore blocker and gating modifier)	1.3 μM	[28, 46]
HnTx III	*Ornithoctonus hainana*	Gating modifier	491 nM	[28, 42]
PhlTx1	*Phlogiellus* genus	Gating modifier	201 nM	[28, 51]
μ-conotoxin KIIIA	*C. kinoshitai*	Pore blocker	8 μM	[26, 28]
μ-conotoxin SmIIIA	*C. stercusmuscarum*	Pore blocker	35 nM	[26, 28]
GpTx-1	*Grammostola portei*	Gating modifier	20 nM	[28, 56]
ATX-II	*Anemonia sulcata*	Gating modifier	759 nM	[27, 33, 34]
AFT-II	*Anthopleura fuscoviridis*	Gating modifier	460 nM	[27, 34]
Bc-III	*Bunodosoma caissarum*	Not known	1.5 μM	[27, 34]
GrTx1	*Grammostola rosea*	Multi modal action (pore blocker and gating modifier)	0.77 μM	[27, 46]
$GVIIJ_{SSG}$	*C. geographus*	Pore blocker	15 nM	[27, 36]
μ-conotoxin BuIIIA	*C. bullatus*	Pore blocker	0.35 μM	[26, 27]
PnCS1	Hybrid peptide from PnTX1 and μ-KIIIA	Pore blocker	1.1 μM	[29]
PnCS1[W4K]	Hybrid peptide from PnTX1 and μ-KIIIA	Pore blocker	1.3 μM	[29]
PnCS1[W7Y]	Hybrid peptide from PnTX1 and μ-KIIIA	Pore blocker	4.1 μM	[29]
Nav1.6				
μ-conotoxin SmIIIA	*C. stercusmuscarum*	Pore blocker	160 nM	[26, 28]
μ-conotoxin SIIIA	*C. striatus*	Pore blocker	0.8 μM	[26]
μ-conotoxin CnIIIA	*C. consors*	Pore blocker	7.1 μM	[26]
μ-conotoxin GIIIA	*C. geographus*	Pore blocker	0.7 μM	[26]
μ-conotoxin PIIIA	*C. purpurascens*	Pore blocker	100 nM	[26]
μ-conotoxin SxIIIA	*C. striolatus*	Pore blocker	0.6 μM	[26]
μ-conotoxin BuIIIA	*C. bullatus*	Pore blocker	4.4 μM	[26, 27]
μ-conotoxin BuIIIB	*C. bullatus*	Pore blocker	1.8 μM	[26]
μ-conotoxin KIIIA	*C. kinoshitai*	Pore blocker	240 nM	[24, 28]
PnCS1	Hybrid peptide from PnTX1 and μ-KIIIA	Pore blocker	0.7 μM	[29]
PnCS1[W4K]	Hybrid peptide from PnTX1 and μ-KIIIA	Pore blocker	0.7 μM	[29]
PnCS1[W7Y]	Hybrid peptide from PnTX1 and μ-KIIIA	Pore blocker	0.7 μM	[29]

(continued)

Table 1.1 (continued)

Peptide	Peptide source	Peptide activity	IC_{50}	Reference
ATX-II	*Anemonia sulcata*	Gating modifier	180 nM	[27, 34]
AFT-II	*Anthopleura fuscoviridis*	Gating modifier	300 nM	[27, 34]
BC-III	*Bunodosoma caissarum*	Gating modifier	900 nM	[27, 34]
ProTx-II	*Thrixopelma pruriens*	Gating modifier Cav3.1 off target	47 nM, 86 nM	[27, 28, 47, 57]
PhlTx1	*Phlogiellus* genus	Gating modifier	491 nM	[12, 31]
CGTX-II	*Bunodosoma cangicum*	Gating modifier	50 nM	[27, 35]
AM-6120	*Chilobrachys jingzhao*	Gating modifier	604 nM	[28, 58]
Pn3a	*Pamphobeteus nigricolor*	Gating modifier	129 nM	[28, 37]
μ-TRTX-Df1a	*Davus fasciatus*	Gating modifier Inhibits Cav3 currents	7.6 nM	[28, 38]
m3-HwTx IV	*Haplopelma schmitdi*	Gating modifier	6.8 nM	[28, 40]
CcoTx I	*Ceratogyrus cornuatus*	Gating modifier	50 nM	[28, 50]
Pre1a	*Psalmopoeus reduncus*	Gating modifier	221 nM	[43]
Cyrtx-1a	*Cyriopagopus schioedtei*	Gating modifier	115 nM	[31]
JzTx-14	*Chilobrachys jingzhao*	Gating modifier	159 nM	[45]
Nav1.7				
CnIIIC	*Conus consors*	Pore blocker	489 nM	[59, 60]
μ-Conotoxin SmIIIA	*Conus stercusmuscarum*	Pore blocker	1.3 μM	[26, 28]
μ-Conotoxin KIIIA	*Conus kinoshitai*	Pore blocker	290 nM	[26, 28]
PnCS1	Hybrid peptide from PnTX1 and μ-KIIIA	Pore blocker	0.9 μM	[29]
PnCS1[W4K]	Hybrid peptide from PnTX1 and μ-KIIIA	Pore blocker	7.4 μM	[29]
PnCS1[W7Y]	Hybrid peptide from PnTX1 and μ-KIIIA	Pore blocker	8.2 μM	[29]
ProTx-I	*Thrixopelma pruriens*	Gating modifier	51 nM	[27, 61]
ProTx-II	*Thrixopelma pruriens*	Gating modifier	300 nM	[47]
ProTx-III	*Thrixopelma pruriens*	Gating modifier	2.1 nM	[28, 39]
Lqh-2	*Leiurus quinquestriatus hebraeus*	Gating modifier	32 nM	[53, 54]
Lqh-3	*Leiurus quinquestriatus hebraeus*	Gating modifier	13.6 nM	[53, 54]
GpTx-1	*Grammostola portei*	Gating modifier	10 nM,	[56]
VsTx-3	*Grammostola rosea*	Gating modifier	430 nM	[12, 62]
GpTx-1-71	*Grammostola porteri*	Gating modifier	1.6 nM	[56]
μ-SLPTX-Ssm6a	*Scolopendra subspinipes mutilans*	Gating modifier	25 nM	[41]
Pn3a	*Pamphobeteus nigricolor*	Gating modifier	(hNav = 0.9 nM) (rNav = 1.5 nM) (mNav = 4.4 nM)	[28, 37]
$GVIIJ_{SSG}$	*C. geographus*	Pore blocker	41 nM	[36]
JzTX V	*Chilobrachys jingzhao*	Gating modifier	0.6 nM	[28, 63, 64]
JzTX-1	*Chilobrachys jingzhao*	Gating modifier	348 nM	[45]
AM-8145	*Chilobrachys jingzhao*	Gating modifier	0.5 nM	[28, 63, 64]
AM-0422	*Chilobrachys jingzhao*	Gating modifier	0.8 nM	[28, 63, 64]

(continued)

Table 1.1 (continued)

Peptide	Peptide source	Peptide activity	IC_{50}	Reference
AM-6120	*Chilobrachys jingzhao*	Gating modifier	0.8 nM (hNav1.7) 5.4 nM (mNav1.7)	[58]
JzTx-14	*Chilobrachys jingzhao*	Gating modifier	189 nM	[45]
JzTX-34	*Chilobrachys jingzhao*	Gating modifier	610 nM	[28, 65]
μ-TRTX-Df1a	*Davus fasciatus*	Gating modifier (also inhibits Cav3)	1.9 nM	[28, 38]
Cd1a	*Ceratogyrus darlingi*	Gating modifier Cav2.2 off target (pore blocker)	3 μM	[28, 32]
HnTx III	*Ornithoctonus hainana*	Gating modifier	232 nM	[28, 42]
HnTx IV	*Ornithoctonus hainana*	Gating modifier	21 nM	[28, 48]
MrVIA	*C. marmoreus*	Not known	345 nM	[59]
MrVIB	*C. marmoreus*	Not known	345 nM	[59]
HwTx-IV	*Haplopelma schmitdi*	Gating modifier	26 nM	[29, 49]
m3-HwTx-IV	*Haplopelma schmitdi*	Gating modifier	3.3 nM	[28, 40]
CcoTx I	*Ceratogyrus cornuatus*	Gating modifier	129 nM	[28, 50]
PhlTx1	*Phlogiellus* genus	Gating modifier	254 nM	[28, 51]
Phlo1a	*Phlogius sp.*	Gating modifier	459 nM	[45, 66]
Phlo1b	*Phlogius sp.*	Gating modifier	360 nM	[45, 66]
Phlo2a	*Phlogius sp.*	Gating modifier	333 nM	[45, 66]
HnTx III	*Ornithoctonus hainana*	Gating modifier	232 nM	[42]
Pre1a	*Psalmopoeus reduncus*	Gating modifier	15 nM	[43]
Cyrtx-1a	*Cyriopagopus schioedtei*	Gating modifier	130 nM	[31]
GsAFI	*Grammostola rosea*	Multi modal action (pore block and gating modifier)	40 nM	[28, 46]
GsAFII	*Grammostola rosea*	Multi modal action (pore blocker and gating modifier)	1.03 μM	[28, 46]
GsMTx4	*Grammostola rosea*	Multi modal action (pore blocker and gating modifier)	7.4 μM	[46]
Nav1.8				
μO-MrVIB	*C. marmoreus*	Not known	326 nM	[59]
ProTx-I	*Thrixopelma pruriens*	Gating modifier	27 nM	[27, 61]
ProTx-II	*Thrixopelma pruriens*	Gating modifier	19 nM, 146 nM, 486 nM	[28, 47]
MrVIB (μO-conotoxin	*C. marmoreus*	Gating modifier	102 nM	[27, 67]
MfVIA (μO-conotoxin)	*C. magnificus*	Gating modifier	529 nM	[27, 68]
HSTX-I	*Haemadipsa sylvestris*	Not known	2.44 μM	[27, 69]
Cd1a	*Ceratogyrus darlingi*	Gating modifier Cav2.2 off target (pore blocker)	7 μM	[28, 32]
JzTx-14	*Chilobrachys jingzhao*	Gating modifier	824 nM	[45]
VsTx-3	*Grammostola rosea*	Gating modifier	770 nM	[12, 62]
Nav1.9				
HSTX-I	*Haemadipsa sylvestris*	Gating modifier	3.3 μM	[27, 69]

Table 1.2 Venom-derived peptides inhibiting Cav channels expressed in mammalian sensory neurons

Peptide	Peptide source	Peptide activity	IC$_{50}$	Reference
Cav1 (L-type calcium channels)				
Calciseptine	*Dendroaspis polylepis*	Pore blocker	430 nM	[70]
Calcicludine	*Dendroaspis angusticeps*	Pore blocker	88 nM	[71, 72, 73]
ω-TRTX-Cc1a (Cc1a)	*Pelinobius muticus*	Pore blocker	Ca$_V$1.2: 825 nM Ca$_V$1.3: 2.24 μM	[74]
Glacontryphan-M	*Conus marmoreus*	Not known	~20 nM	[75]
PnTx3–6	*Phoneutria nigriventer*	Pore blocker	607 nM	[76]
Kurtoxin	*Parabuthus transvaalicus*	Gating modifier	70 nM	[77]
ω-filistatoxin-Kh1a	*Filista hibernalis*	Pore blocker	26.8 nM	[78]
ω-agatoxin IIIA	*Agelenopsis aperta*	Pore blocker	0.35 nM	[79, 80]
ω-agatoxin IIIB	*Agelenopsis aperta*	Pore blocker	47 nM	[80]
ω-agatoxin IIID	*Agelenopsis aperta*	Pore blocker	130 nM	[80]
Cav2 (N-, P/Q-, R-type calcium channels)				
Peptide	Peptide source	Peptide activity	IC$_{50}$/K$_d$	Reference
Cav2.1 (P/Q-type Calcium channels)				
ω-agatoxin IVA	*Agelenopsis aperta*	Gating modifier	3 nM	[81, 82]
ω-agatoxin IVB	*Agelenopsis aperta*	Gating modifier	3 nM	[83, 84]
ω-conotoxin GVIA	*Conus geographus*	Pore blocker	1 μM	[85, 86]
ω-conotoxin MVIIA	*Conus catus*	Pore blocker	156 nM	[85, 87]
ω-conotoxin MVIIC	*Conus magus*	Pore blocker	0.6 nM	[85, 86]
ω-conotoxin CVIA	*Conus catus*	Pore blocker	850 nM	[85, 86]
ω-conotoxin CVIB	*Conus catus*	Pore blocker	11 nM	[85, 86]
ω-conotoxin CVIC	*Conus catus*	Pore blocker	31 nM	[85, 86]
ω-conotoxin CVID	*Conus catus*	Pore blocker	55 μM	[85, 86]
ω -Lsp-IA	*Geolycosa sp*	Pore blocker and Gating modifier.	10 nM	[86, 88]
ω-conotoxin CnVIIA	*Conus consors*	Pore blocker	179 nM	[87]
PnTx3-6	*Phoneutria nigriventer*	Pore blocker	263 nM	[76]
Kurtoxin	*Parabuthus transvaalicus*	Gating modifier	15 nM	[77]
ω-filistatoxin-Kh1a	*Filista hibernalis*	Pore blocker	4.3 nM	[78]
ω-Phonetoxin IIA	*Phoneutria nigriventer*	Not known	0.16 nM	[89, 90]
Cav2.2 (N-type calcium channels)				
Kurtoxin	*Parabuthus transvaalicus*	Gating modifier	460 nM	[77]
ω-filistatoxin-Kh1a	*Filista hibernalis*	Pore blocker	2.3 nM	[78]
ω-conotoxin GVIA	*Conus geographus*	Pore blocker	2 pM	[85, 87]
ω-conotoxin GVIIA	*Conus geographus*	Pore blocker	22.9 nM	[85, 91]
ω-conotoxin MVIIA	*Conus catus*	Pore blocker	8 nM	[85, 92]
ω-conotoxin MVIIB	*Conus catus*	Pore blocker	101 pM	[85, 93]
ω-conotoxin MVIIC	*Conus magus*	Pore blocker	7 nM	[85, 86]
ω-conotoxin SVIA	*Conus. striatus*	Pore blocker	1.46 μM	[85, 93]
ω-conotoxin SVIB	*Conus striatus*	Pore blocker	1.09 nM	[85, 93, 126]
ω-conotoxin MVIIA/SO-3	*Conus. striatus*	Pore blocker	160 nM	[94]
ω-conotoxin CVIA	*Conus catus*	Pore blocker	0.6 nM	[85, 86]
ω-conotoxin CVIB	*Conus catus*	Pore blocker	8 nM	[85, 86]

(continued)

Table 1.2 (continued)

Cav2 (N-, P/Q-, R-type calcium channels)				
Peptide	Peptide source	Peptide activity	IC_{50}/K_d	Reference
ω-conotoxin CVIC	*Conus catus*	Pore blocker	7.6 nM	[85, 86]
ω-conotoxin CVID	*Conus catus*	Pore blocker	0.07 nM	[85, 86]
ω-conotoxin CVIE	*Conus catus*	Pore blocker	2.6 nM, 0.12 nM	[85, 95]
ω-conotoxin CVIF	*Conus catus*	Pore blocker	19.9 nM	[85, 95]
ω-conotoxin FVIA	*Conus fulman*	Pore blocker	11.5 nM	[85, 92]
ω-conotoxin RVIA	*Conus. radiatus*	Pore blocker	229 nM	[85, 93]
ω-conotoxin TVIA	*Conus. tulipa*	Pore blocker	228 pM	[85, 93]
ω-conotoxin CnVIIA	*Conus consors*	Pore blocker	2.3–3.7 pM	[85, 87]
ω-agatoxin IIA	*Agelenopsis aperta*	Pore blocker	10 nM	[96–98]
ω-agatoxin IIIA	*Agelenopsis aperta*	Pore blocker	1.4 nM	[79, 96, 99]
ω-agatoxin IIIB	*Agelenopsis aperta*	Pore blocker	140 nM	[96, 99, 100]
ω-agatoxin IIID	*Agelenopsis aperta*	Pore blocker	35 nM	[96, 99]
glycerotoxin (GLTx)	*Glycera tridactyla*	Activate N-type Ca_V channel	50 pM	[101, 102]
α-conotoxin Eu1.6	*Conus eburneus*	Not known	1 nM	[103]
SNX-325	*Segestria florentina*	Pore blocker	3-30 nM	[104]
Huwentoxin-1	*Selenocosmia huwena*	Pore blocker	~100 nM; also blocks L-type Ca_V	[105]
ω-ctenitoxin-Pr1a (PRTx3-7)	*Phoneutria reidyi*	Gating modifier	436 nM; also partially blocks P/Q, R-types.	[96, 106]
PnTx3-6	*Phoneutria nigriventer*	Pore blocker	122 nM	[107]
ω-Phonetoxin IIA	*Phoneutria nigriventer*	Not known	0.16 nM	[89, 90]
Cav2.3 (R-type calcium channels)				
ω-theraphotoxin-Hg1a (SNX-482)	*Hysterocrates gigas*	Gating modifier	30 nM	[108, 109]
ω-filistatoxin-Kh1a	*Filista hibernalis*	Pore blocker	96.4 nM	[78]
Peptide	Peptide source	Peptide activity	IC_{50}/K_d	Reference
Cav 3 (T-type calcium channels)				
Protoxin I	Tarantula *Thrixoppelma pruriens*	Gating modifier	Cav3.1: 0.2 μM Cav3.2: 32 μM Cav3.3: 5.4 μM	[110–112]
Protoxin II	*Thrixoppelma pruriens*	Gating modifier	Cav3.1: 0.8 μM Cav3.2: ~1 μM	[110, 111, 113, 114]
Kurtoxin	*Parabuthus transvaalicus*	Gating modifier	15 nM	[115]
PnTx3-6	*Phoneutria nigriventer*	Pore blocker	136 nM	[107]
Other Ca_V blockers (IC_{50} not determined)				
ω-agatoxin IA	*Agelenopsis aperta*	Not known	Inhibition of K^+-induced Ca^{2+} influx	79
ω-grammotoxin SIA	*Grammostola spatulata*	Not known	P/Q- and N-type	[116–118]
PsPTx3	*Parabuthus transvaalicus*	Not known	Preferentially Cav3.2	[78]

(continued)

Table 1.2 (continued)

Other Ca_V blockers (IC_{50} not determined)				
PnTx3-3	*Phoneutria nigriventer*	Pore blocker	L-, P/Q- and R-type	[119]
ω-PnTx3-2	*Phoneutria nigriventer*	Pore blocker	L-type	[120–122]
ω-PnTx3-5	*Phoneutria nigriventer*	Pore blocker	L-type channels	[120, 122, 123]
ω-oxotoxin (OxyTx1 and OxyTx2)	*Oxyopes lineatus*	Pore blocker	P/Q-, N- or L-type	[124]
Funnel-web spider toxin (FTX)	*Agelenopsis aperta*	Gating modifier	N-type channels	[125]

intracellular phosphorylated β subunit transmembrane γ subunit possibly present in some VGCC families [23, 24]. The amino acid sequence of Ca_V channel α1 subunit is similar to that of the sodium channel, consisting of four repeated domains (I–IV), each of which contains six transmembrane segments (S1–S6) and a membrane-associated loop between transmembrane segments S5 and S6 (P-loop), the first four segment of each domain is believed to control voltage-dependent activation [25]. The intracellular loop connecting domains I and II in the α1 subunit contains the binding site for the β subunit, known as the α interaction domain (AID).

References

1. Finnerup NB, Kuner R, Jensen TS (2021) Neuropathic pain: from mechanisms to treatment. Physiol Rev 101(1):259–301
2. Bhattacharjee P, Bhattacharyya D (2014) Therapeutic use of snake venom components: a voyage from ancient to modern India. Mini-Rev Org Chem 11:45–54
3. Utkin YN (2015) Animal venom studies: current benefits and future developments. World J Biol Chem 6:28–33
4. Panagides N, Jackson TN, Ikonomopoulou MP, Arbuckle K, Pretzler R, Yang DC, et al (2017) How the cobra got its flesh-eating venom: cytotoxicity as a defensive innovation and its co-evolution with hooding, aposematic marking, and spitting. Toxins 9:103
5. Jami S, Erickson A, Brierley SM, Vetter I (2017) Pain causing venom peptides: insights into sensory neuron pharmacology. Toxins 10(1):15
6. Bourinet E, Zamponi GW (2017) Block of voltage-gated calcium channels by peptide toxins. Neuropharmacology 127:109–115
7. Ryu JH, Jung HJ, Konishi S, Kim HH, Park ZY, Kim JI (2017) Structure-activity relationships of ω-Agatoxin IVA in lipid membranes. Biochem Biophys Res Comm 482(1):170–175
8. Sousa SR, McArthur JR, Brust A, Bhola RF, Rosengren KJ, Ragnarsson L, Dutertre S, Alewood PF, Christie MJ, Adams DJ, Vetter I, Lewis RJ (2018) Novel analgesic ω-conotoxins from the vermivorous cone snail *Conus moncuri* provide new insights into the evolution of conopeptides. Sci Rep 8 (1):13397
9. Eijkelkamp N, Linley JE, Baker MD, Minett MS, Cregg R, Werdehausen R, Rugiero F, Wood JN (2012) Neurological perspectives on voltage-gated sodium channels. Brain 135(Pt. 9):2585–2612
10. Goldin AL, Barchi RL, Caldwell JH, Hofmann F, Howe JR, Hunter JC, Kallen RG, Mandel G, Meisler MH, Netter YB, Noda M, Tamkun MM, Waxman SG, Wood JN, Catterall WA (2000) Nomenclature of voltage-gated sodium channels. Neuron 28:365–368
11. Ray P, Torck A, Quigley L, Wangzhou A, Neiman M, Rao C, Lam T, Kim JY, Kim TH, Zhang MQ, Dussor G, Price TJ (2018) Comparative transcriptome profiling of the human and mouse dorsal root ganglia: an RNA-seq-based resource for pain and sensory neuroscience research. Pain 159 (7):1325–1345
12. Cardoso FC (2020) Multi-targeting sodium and calcium channels using venom peptides for the treatment of complex ion channels-related diseases. Biochem Pharmacol 181:114107. https://doi.org/10.1016/j.bcp.2020.114107
13. Yu FH, Catterall WA (2003) Overview of voltage-gated sodium channel family. Genome Biol 4 (207):1–7
14. Ahern CA, Payandeh J, Bosmans F, Chanda B (2016) The hitchhiker's guide to the voltage-gated sodium channel galaxy. J Gen Physiol 147:1–24

15. Dib-Hajj SD, Yang Y, Black JA, Waxman SG (2013) The Na_V1.7 sodium channel: from molecule to man. Nat Rev Neurosci 14:49–62
16. Körner J, Lampert A (2020) Sodium channels. In: Fritzsch B (ed) The senses: a comprehensive reference, 2nd edn. Academic Press, Cambridge, MA, pp 120–141
17. Pan X, Li Z, Huang X, Huang G, Gao S, Shen H, Liu L, Lei J, Yan N (2019) Molecular basis for pore blockade of human Na^+ channel Na_v1.2 by the μ-conotoxin KIIIA. Science 363(6433):1309–1313
18. Cain SM, Snutch TP (2011) Voltage-gated calcium channels and disease. Biofactors 37(3):197–205. https://doi.org/10.1002/biof.158
19. Bourinet E, Altier C, Hildebrand ME, Trang T, Salter MW, Zamponi GW (2014) Calcium-permeable ion channels in pain signaling. Physiol Rev 94 (1):81–140. https://doi.org/10.1152/physrev.00023.2013
20. Zamponi GW (2015) Targeting voltage-gated calcium channels in neurological and psychiatric diseases. Nat Rev Drug Discov 15:19–34
21. Catterall WA, Perez-Reyes E, Snutch TP, Striessnig J (2005) International union of pharmacology. XLVIII Nomenclature and structure-function relationships of voltage-gated calcium channels. Pharmacol Rev 57 (4):411–425
22. Namkung Y, Skrypnyk N, Jeong MJ, Lee T, Lee MS, Kim HL, Chin H, Suh PG, Kim SS, Shin HS (2001) Requirement for the L-type calcium channel α1D subunit in postnatal pancreatic beta cell generation. J Clin Invest 108:1015–1022
23. Curtis BM, Catterall WA (1984) Purification of the calcium antagonist receptor of the voltage-sensitive calcium channel from skeletal muscle transverse tubules. Biochemistry 23(10):2113–2118. https://doi.org/10.1021/bi00305a001
24. Takahashi M, Seagar MJ, Jones JF, Reber BF, Catterall WA (1987) Subunit structure of dihydropyridine-sensitive calcium channels from skeletal muscle. Proc Natl Acad Sci U S A 84:5478–5482
25. Catterall WA (2010) Ion channel voltage sensors: structure, function, and pathophysiology. Neuron 67 (6):915–928. https://doi.org/10.1016/j.neuron.2010.08.021
26. Wilson MJ, Yoshikami D, Azam L, Gajewiak J, Olivera BM, Bulaj G et al (2011) Conotoxins that differentially block sodium channels NaV1.1 through 1.8 identify those responsible for action potentials in sciatic nerve. Proc Natl Acad Sci U S A 108:10302–10307
27. Bajaj S, Han J (2019) Venom-derived peptide modulators of cation-selective channels: friend, foe or frenemy. Front Pharmacol 10:58
28. Eagles DA, Chow CY, King GF (2020) Fifteen years of Na_V 1.7 channels as an analgesic target: why has excellent in vitro pharmacology not translated into in vivo analgesic efficacy? Br J Pharmacol. https://doi.org/10.1111/bph.15327
29. Peigneur S, da Costa Oliveira C, de Sousa Fonseca FC, McMahon KL, Mueller A, Cheneval O, et al (2021) Small cyclic sodium channel inhibitors. Biochem Pharmacol 183:114291
30. Zhu S, Peigneur S, Gao B, Lu X, Cao C, Tytgat J (2012) Evolutionary diversification of Mesobuthus α-scorpion toxins affecting sodium channels. Mol Cell Proteomics 11(1):M111.012054. https://doi.org/10.1074/mcp.M111.012054
31. Goncalves TC, Benoit E, Kurz M, Lucarain L, Fouconnier S, Combemale S et al (2019) From identification to functional characterization of cyriotoxin-1a, an antinociceptive toxin from the spider *Cyriopagopus schioedtei*. Br J Pharmacol 176 (9):1298–1314
32. Sousa SR, Wingerd JS, Brust A, Bladen C, Ragnarsson L, Herzig V et al (2017) Discovery and mode of action of a novel analgesic β-toxin from the African spider *Ceratogyrus darling*. PLoS One 12: e0182848
33. Chahine M, Plante E, Kallen RG (1996) Sea anemone toxin (ATX II) modulation of heart and skeletal muscle sodium channel a-subunits expressed in tsA201 cells. J Membr Biol 152:39–48
34. Oliveira JS, Redaelli E, Zaharenko AJ, Cassulini RR, Konno K, Pimenta DC et al (2004) Binding specificity of sea anemone toxins to Nav 1.1-1.6 sodium channels. Unexpected contributions from differences in the IV/S3-S4 outer loop. J Biol Chem 279:33323–33335. https://doi.org/10.1074/jbc.M404344200
35. Zaharenko AJ, Schiavon E, Ferreira WA, Lecchi M, Freitas JC, De Richardson M et al (2012) Characterization of selectivity and pharmacophores of type 1 sea anemone toxins by screening seven Na_V sodium channel isoforms. Peptides 34:158–167. https://doi.org/10.1016/j.peptides.2011.07.008
36. Gajewiak J, Azam L, Imperial J, Walewska A, Green BR, Bandyopadhyay PK et al (2014) A disulfide tether stabilizes the block of sodium channels by the conotoxin O-GVIIJ. Proc Natl Acad Sci U S A 111:2758–2763
37. Deuis JR, Dekan Z, Wingerd JS, Smith JJ, Munasinghe NR, Bhola RF, et al (2017) Pharmacological characterisation of the highly Na_V1.7 selective spider venom peptide Pn3a. Sci Rep 7:40883. https://doi.org/10.1038/srep40883
38. Cardoso FC, Dekan Z, Smith JJ, Deuis JR, Vetter I, Herzig V et al (2017) Modulatory features of the novel spider toxin μ-TRTX-Df1a isolated from the venom of the spider *Davus fasciatus*. Br J Pharmacol 174:2528–2544
39. Cardoso F, Dekan Z, Rosengren K, Erickson A, Vetter I, Deuis J (2015) Identification and characterization of ProTx-III [μ-TRTX-Tp1a], a new voltage-gated sodium channel inhibitor from venom of the tarantula *Thrixopelma pruriens*. Mol Pharmacol

88:291–303. https://doi.org/10.1124/mol.115.098178
40. Rahnama S, Deuis JR, Cardoso FC, Ramanujam V, Lewis RJ, Rash LD et al (2017) The structure, dynamics and selectivity profile of a NaV1.7 potency-optimised huwentoxin-IV variant. PLoS ONE 12:e0173551
41. Yang S, Xiao Y, Kang D, Liu J, Li Y, Undheim EAB (2013) Discovery of a selective $Na_V1.7$ inhibitor from centipede venom with analgesic efficacy exceeding morphine in rodent pain models. Proc Natl Acad Sci U S A 110:17534–17539
42. Liu Z, Cai T, Zhu Q, Deng M, Li J, Zhou X et al (2013) Structure and function of hainantoxin-III, a selective antagonist of neuronal tetrodotoxin-sensitive voltage-gated sodium channels isolated from the chinese bird spider *Ornithoctonus hainana*. J Biol Chem 288:20392–20403. https://doi.org/10.1074/jbc.M112
43. Wingerd JS, Mozar CA, Ussing CA, Murali SS, Chin YK, Cristofori-Armstrong B, Durek T, Gilchrist J, Vaughan CW, Bosmans F, Adams DJ, Lewis RJ, Alewood PF, Mobli M, Christie MJ, Rash LD (2017) The tarantula toxin β/δ-TRTX-Pre1a highlights the importance of the S1-S2 voltage-sensor region for sodium channel subtype selectivity. Sci Rep 7(1):974
44. Bossmans F, Rash L, Zhu S, Diochot S, Lazdunski M, Escoubas P, Tytgat J (2006) Four novel tarantula toxins as selective modulators of voltage-gated sodium channel subtypes. Mol Pharmacol 69(2):419–429
45. Dongol Y, Cardoso FC, Lewis R (2019) Spider knottin pharmacology at voltage-gated sodium channels and their potential to modulate pain pathways. Toxins 11(11):626
46. Redaelli E, Cassulini RR, Silva DF, Clement H, Schiavon E, Zamudio FZ et al (2010) Target promiscuity and heterogeneous effects of tarantula venom peptides affecting Na^+ and K^+ ion channels. J Biol Chem 285:4130–4142
47. Schmalhofer WA, Calhoun J, Burrows R, Bailey T, Kohler MG, Weinglass AB et al (2008) ProTx-II, a selective inhibitor of $Na_V1.7$ sodium channels, blocks action potential propagation in nociceptors. Mol Pharmacol 74:1476–1484
48. Cai T, Luo J, Meng E, Ding J, Liang S, Wang S, Liu Z (2015) Mapping the interaction site for the tarantula toxin hainantoxin-IV (β-TRTX-Hn2a) in the voltage sensor module of domain II of voltage gated sodium channels. Peptides 68:148–156
49. Xiao Y, Bingham JP, Zhu W, Moczydlowski E, Liang S, Cummins T (2008) Tarantula huwentoxin-IV inhibits neuronal sodium channels by binding to receptor site 4 and trapping the domain II voltage sensor in the closed configuration. J Biol Chem 283:27300–27313
50. Agwa A, Peigneur S, Chow C, Lawrence N, Craik D, Tytgat J et al (2018) Gating modifier toxins isolated from spider venom: modulation of voltage-gated sodium channels and the role of lipid membranes. J Biol Chem 293:9041–9052
51. Nicolas S, Zoukimian C, Bosmans F, Montnach J, Diochot S, Cuypers E et al (2019) Chemical synthesis, proper folding, Na_V channel selectivity profile and analgesic properties of the spider peptide phlotoxin 1. Toxins 11:367
52. Pan X, Li Z, Huang X, Huang G, Gao S, Shen H, Liu L, Lei J, Yan N (2019) Molecular basis for pore blockade of human Na^+ channel $Na_v1.2$ by the μ-conotoxin KIIIA. Science 363(6433):1309–1313
53. Chen H, Lu S, Leipold E, Gordon D, Hansel A, Heinemann SH (2002) Differential sensitivity of sodium channels from the central and peripheral nervous system to the scorpion toxins Lqh-2 and Lqh-3. Eur J Neurosci 16(4):767–770
54. Leipold E, Lu S, Gordon D, Hansel A, Heinemann SH (2004) Combinatorial interaction of scorpion toxins Lqh-2, Lqh-3, and Lqh alpha IT with sodium channel receptor sites-3. Mol Pharmacol 65 (3):685–691
55. Silva AO, Peigneur S, Diniz MRV, Tytgat J, Beirão PSL (2012) Inhibitory effect of the recombinant *Phoneutria nigriventer* Tx1 toxin on voltage-gated sodium channels. Biochimie 94:2756–2763
56. Murray JK, Ligutti J, Liu D, Zou A, Poppe L, Li H et al (2015) Engineering potent and selective analogues of GpTx-1, a tarantula venom peptide antagonist of the $Na_V1.7$ sodium channel. J Med Chem 58:2299–2314
57. Maertens C, Cuypers E, Amininasab M, Jalali A, Vatanpour H, Tytgat J (2006) Potent modulation of the voltage-gated sodium channel Nav1.7 by OD1, a toxin from the scorpion *Odonthobuthus doriae*. Mol Pharmacol 70:405–414
58. Wu B, Murray JK, Andrews KL, Sham K, Long J et al (2018) Discovery of Tarantula venom-derived Na_V 1.7-inhibitory JzTx-V peptide 5-Br-Trp24 analogue AM-6120 with systemic block of histamine-induced Pruritis. J Med Chem 61(21):9500–9512
59. Knapp O, McArthur JR, Adams DJ (2012) Conotoxins targeting voltage gated sodium channel subtypes: potential analgesics? Toxins 4 (11):1236–1260
60. Markgraf R, Leipold E, Schirmeyer J, Paolini-Bertrand M, Hartley O, Heinemann SH (2012) Mechanism and molecular basis for the sodium channel subtype specificity of μ-conopeptide CnIIIC. Br J Pharmacol 167:576–586
61. Middleton RE, Warren VA, Kraus RL, Hwang JC, Liu CJ, Dai G et al (2002) Two tarantula peptides inhibit activation of multiple sodium channels. Biochemist 41:14734–14747
62. Cherki RS, Kolb E, Langut Y, Tsveyer L, Bajayo N, Meir A (2014) Two tarantula venom peptides as potent and differential Na_V channels blockers. Toxicon 77:58–67

63. Moyer BD, Murray JK, Ligutti J, Andrews K, Favreau P, Jordan JB, et al (2018) Pharmacological characterization of potent and selective $Na_V1.7$ inhibitors engineered from *Chilobrachys jingzhao* tarantula venom peptide JzTx-V. PLoS ONE 13(5): e0196791
64. Murray JK, Wu B, Tegley CM, Nixey TE, Falsey JR, Herberich B et al (2019) Engineering $Na_V1.7$ inhibitory JzTx-V peptides with a potency and basicity profile suitable for antibody conjugation to enhance pharmacokinetics. ACS Chem Biol 14:806–818
65. Zeng X, Li P, Chen B, Huang J, Lai R, Liu J, Rong M (2018) Selective closed-state Nav1.7 blocker JZTX-34 exhibits analgesic effects against pain. Toxins 10 (2):64
66. Chow CY, Cristofori-Armstrong B, Undheim EA, King GF, Rash LD (2015) Three peptide modulators of the human voltage-gated sodium channel 1.7, an important analgesic target, from the venom of an Australian tarantula. Toxins 7:2494–2513
67. Ekberg J, Jayamanne A, Vaughan CW, Aslan S, Thomas L, Mould J et al (2006) μO-conotoxin MrVIB selectively blocks Nav1.8 sensory neuron specific sodium channels and chronic pain behavior without motor deficits. Proc Natl Acad Sci U S A 103:17030–17035
68. Vetter I, Dekan Z, Knapp O, Adams DJ, Alewood PF, Lewis RJ (2012) Isolation, characterization and total regioselective synthesis of the novel μO-conotoxin MfVIA from *Conus magnificus* that targets voltage-gated sodium channels. Biochem Pharmacol 84:540–548
69. Wang G, Long C, Liu W, Xu C, Zhang M, Li Q et al (2018) Novel sodium channel inhibitor from leeches. Front Pharmacol 9:186
70. De Weille JR, Schweitz H, Maest P, Tartart A, Lazdunski M (1991) Calciseptine, a peptide isolated from black mamba venom, is a specific blocker of the L-type calcium channel. Proc Natl Acad Sci U S A 88 (6):2437–2440
71. Schweitz H, Heurteaux C, Bois P, Moinier D, Romey G, Lazdunskit M (1994) Calcicludine, a venom peptide of the Kunitz-type protease inhibitor family, is a potent blocker of high-threshold Ca^{2+} channels with a high affinity for L-type channels in cerebellar granule neurons. Proc Natl Acad Sci U S A 91(3):878–82
72. Stotz SC, Spaetgens RL, Zamponi GW (2000) Block of voltage-dependent calcium channel by the green mamba toxin calcicludine. J Membr Biol 174 (2):157–165
73. Wang X, Du L, Peterson BZ (2007) Calcicludine binding to the outer pore of L-type calcium channels is allosterically coupled to dihydropyridine binding. Biochemistry 46(25):7590–7598. https://doi.org/10.1021/bi7001696
74. Klint JK, Berecki G, Durek T, Mobli M, Knapp O, King GF, Adams DJ, Alewood PF, Rash LD (2014) Isolation, synthesis and characterization of ω-TRTX-Cc1a, a novel tarantula venom peptide that selectively targets L-type Ca_V channels. Biochem Pharmacol 89(2):276–286. https://doi.org/10.1016/j.bcp.2014.02.008
75. Hansson K, Ma X, Eliasson L, Czerwiec E, Furie B, Furie BC, Rorsman P, Stenflo J (2004) The first γ-carboxyglutamic acid-containing contryphan. A selective L-type calcium ion channel blocker isolated from the venom of *Conus marmoreus*. J Biol Chem 279(31):32453–32463
76. Vieira LB, Kushmerick C, Hildebrand ME, Garcia E, Stea A, Cordeiro MN, Richardson M, Gomez MV, Snutch TP (2005) Inhibition of high voltage-activated calcium channels by spider toxin PnTx3-6. J Pharmacol Exp Therap 314(3):1370–1377. https://doi.org/10.1124/jpet.105.087023
77. Sidach SS, Mintz IM (2002) Kurtoxin, a gating modifier of neuronal high- and low-threshold Ca^{2+} channels. J Neurosci 22(6):2023–2034. https://doi.org/10.1523/jneurosci.22-06-02023.2002
78. Pringos E, Vignes M, Martinez J, Rolland V (2011) Peptide neurotoxins that affect voltage-gated calcium channels: a close-up on ω-agatoxins. Toxins 3 (1):17–42. https://doi.org/10.3390/toxins3010017
79. Olivera BM, Miljanich GP, Ramachandran J, Adams ME (1994) Calcium channel diversity and neurotransmitter release: the ω-conotoxins and ω-agatoxins. Annu Rev Biochem 63:823–867
80. Ertel EA, Warren VA, Adams ME, Griffin PR, Cohen CJ, Smith MM (1994) Type III ω-agatoxins: a family of probes for similar binding sites on L- and N-type calcium channels. Biochemistry 33:5098–5108
81. Mintz IM, Venemat VJ, Swiderek KM, Lee TD, Bean BP, Adamst ME (1992) P type calcium channels blocked by the spider toxin ω-Aga-IVA. Nature 355:827–829
82. Fisher TE, Bourque CW (1995) Distinct w-agatoxin-sensitive calcium currents in somata and axon terminals of rat supraoptic neurones. J Physiol 489 (2):383–388
83. Adams ME, Mintz IM, Reily MD, Thanabal V, Bean BP (1993). Structure and properties of ω-agatoxin IVB, a new antagonist of P-type calcium channels. Mol Pharmacol 44(4): 681–688
84. Sitges M, Galindo CA (2005) ω-Agatoxin-TK is a useful tool to study P-type Ca^{2+} channel-mediated changes in internal Ca^{2+} and glutamate release in depolarised brain nerve terminals. Neurochem Int 46(1):53–60. https://doi.org/10.1016/j.neuint.2004.07.004
85. Ramírez D, Gonzalez W, Fissore RA, Carvacho I (2017) Conotoxins as tools to understand the physiological function of voltage-gated calcium (Ca_V) channels. Mar Drugs 15(10):313
86. Lewis RJ, Nielsen KJ, Craik DJ, Loughnan ML, Adams DA, Sharpe IA, Luchian T, Adams DJ, Bond T, Thomas L et al (2000) Novel ω-conotoxins from *Conus catus* discriminate among neuronal

calcium channel subtypes. J Biol Chem 275:35335–35344
87. Favreau P, Gilles N, Lamthanh H, Bournaud R, Shimahara T, Bouet F, et al (2001) A new ω-conotoxin that targets N-type voltage-sensitive calcium channels with unusual specificity. Biochemistry 40(48):14567–14575
88. Pluzhnikova K, Vassilevskia A, Korolkovaa Y, Fisyunovb A, Iegorovab O, Krishtalb O, Grishin E (2007) ω-Lsp-IA, a novel modulator of P-type Ca^{2+} channels. Toxicon 50:993–1004
89. Cassola AC, Jaffe H, Fales HM, Afeche SC, Magnoli F, Cipolla-Neto J (1998) ω-Phonetoxin-IIA: a calcium channel blockerfrom the spider *Phoneutria nigriventer*. Pflügers Arch 436 (4):545–552
90. Dos Santos RG, Van Renterghem C, Martin-Moutot-N, Mansuelle P, Cordeiro MN, Diniz CR, et al (2002) *Phoneutria nigriventer* ω-phonetoxin IIA blocks the Cav2 family of calcium channels and interacts with ω-conotoxin-binding sites. J Biol Chem 277(16): 13856–13862
91. Miljanich GP, Bitner RS, Bowersox SS, Fox JA, Valentino KL, Yamashiro DH, Tsubokawa M (1993) Screening Method for Neuroprotective Compounds. U.S. Patent 5,424,218 A
92. Lee S, Kim Y, Keun Back S, Choi HW, Yeon Lee J, Ho Jung H, Ha Ryu J, Suh HW, Sik Na H, Jeong Kim H, Rhim H, Il Kim J (2010) Analgesic effect of highly reversible ω-conotoxin FVIA on N type Ca^{2+} channels. Mol Pain 6:97–109. https://doi.org/10.1186/1744-8069-6-97
93. Miljanich GP, Bowersox SS, Fox JA, Valentino KL, Bitner RS, Yamashiro DH (1993). Compositions for delayed treatment of ischemia-related neuronal damage. WO Patent 1993010145 A1
94. Wang F, Yan Z, Liu Z, Wang S, Wu Q, Yu S, Ding J, Dai Q (2016) Molecular basis of toxicity of N-type calcium channel inhibitor MVIIA. Neuropharmacology 101:137–145. https://doi.org/10.1016/j.neuropharm.2015.08.047
95. Berecki G, Motin L, Haythornthwaite A, Vink S, Bansal P, Drinkwater R, Wang CI, Moretta M, Lewis RJ, Alewood PF, Christie MJ, Adams DJ (2010) Analgesic ω-conotoxins CVIE and CVIF selectively and voltage-dependently block recombinant and Native N-type calcium channels. Mol Pharmacol 77(2):139–148. https://doi.org/10.1124/mol.109.058834
96. Sousa SR, Vetter I, Lewis RJ (2013) Venom peptides as a rich source of $Ca_V2.2$ channel blockers. Toxins 5 (2):286–314
97. Bindokas VP, Adams ME (1989) ω-Aga-I: a presynaptic calcium channel antagonist from venom of the funnel web spider, *Agelenopsis aperta*. J Neurobiol 20(4):171–188
98. Adams ME, Bindokas VP, Hasegawa L, Venema VJ (1990) ω-Agatoxins: novel calcium channel antagonists of two subtypes from funnel web spider (*Agelenopsis aperta*) venom. J Biol Chem 265 (2):861–867. https://doi.org/10.1016/s0021-9258(19)40129-4
99. Ertel EA, Warren VA, Adams ME, Griffin PR, Cohen CJ, Smith MM (1994) Type III ω-agatoxins: a family of probes for similar binding sites on L- and N-type calcium channels. Biochemistry 33(17):5098–5108
100. Yan L, Adams ME (2000) The spider toxin ω-Aga IIIA defines a high affinity site on neuronal high voltage-activated calcium channels. J Biol Chem 275:21309–21316
101. Meunier FA, Feng ZP, Molgo J, Zamponi GW, Schiavo G, Schenning MP, Proctor DT, Ragnarsson L, Barbier J, Lavidis NA, Molgo JJ, Schiavo G (2002) Glycerotoxin synchronises spontaneous quantal neurotransmitter release independently of action potentials. Proc Aust Physiol Soc 21:6733–6743. http://www.aups.org.au/Proceedings/37/98P
102. Richter S, Helm C, Meunier FA, Hering L, Campbell LI, Drukewitz SH, Undheim EAB, Jenner RA, Schiavo G, Bleidorn C (2017) Comparative analyses of glycerotoxin expression unveil a novel structural organization of the bloodworm venom system. BMC Evol Biol 17(1):64. https://doi.org/10.1186/s12862-017-0904-4
103. Liu Z, Bartels P, Sadeghi M, Du T, Dai Q, Zhu C, Yu S et al (2018) A novel α-conopeptide Eu1.6 inhibits N-type (Ca_V 2.2) calcium channels and exhibits potent analgesic activity. Sci Rep 8(1):1004
104. Newcomb R, Palma A, Fox J, Gaur S, Lau K, Chung D, Cong R, Bell JR, Home B, Nadasdi L, Ramachandran J (1995) SNX-325, A novel calcium antagonist from the spider *Segestria florentina*. Biochemistry 34(26):8341–8347. https://doi.org/10.1021/bi00026a015
105. Peng K, Chen XD, Liang SP (2001) The effect of Huwentoxin-I on Ca^{2+} channels in differentiated NG108-15 cells, a patch-clamp study. Toxicon 39 (4):491–498. https://doi.org/10.1016/S0041-0101(00)00150-1
106. Vieira LB, Pimenta AM, Richardson M, Bemquerer MP, Reis HJ, Cruz JS, Gomez MV, Santoro MM, Ferreira-de-Oliveira R, Figueiredo SG et al (2007) Leftward shift in the voltage-dependence for Ca^{2+} currents activation induced by a new toxin from *Phoneutria reidyi* (Aranae, Ctenidae) venom. Cell Mol Neurobiol 27:129–146
107. Vieira LB, Kushmerick C, Hildebrand ME, Garcia E, Stea A, Cordeiro MN, Richardson M, Gomez MV, Snutch TP (2005) Inhibition of high voltage-activated calcium channels by spider toxin PnTx3-6. J Pharmacol Exp Therap 314(3):1370–1377. https://doi.org/10.1124/jpet.105.087023
108. Bourinet E, Stotz SC, Spaetgens RL, Dayanithi G, Lemos J, Nargeot J, Zamponi GW (2001) Interaction of SNX482 with domains III and IV inhibits activation gating of α1E (CaV2.3) calcium channels.

Biophys J 81(1):79–88. https://doi.org/10.1016/S0006-3495(01)75681-0
109. Kimm T, Bean BP (2014) Inhibition of A-type potassium current by the peptide toxin SNX-482. J Neurosci 34(28):9182–9189. https://doi.org/10.1523/JNEUROSCI.0339-14.2014
110. Ohkubo T, Yamazaki J, Kitamura K (2010) Tarantula toxin ProTx-I differentiates between human T-type voltage-gated Ca^{2+} channels Cav3.1 and Cav3.2. J Pharm Sci 112(4):452–458. https://doi.org/10.1254/jphs.09356FP
111. Bladen C, Hamid J, Souza IA, Zamponi GW (2014) Block of T-type calcium channels by protoxins I and II. Mol Brain 7(1):36. https://doi.org/10.1186/1756-6606-7-36
112. Salari A, Vega BS, Milescu LS, Milescu M (2016) Molecular interactions between tarantula toxins and low-voltage-activated calcium channels. Sci Rep 6 (1):1–12. https://doi.org/10.1038/srep23894
113. Edgerton BM, Blumenthal KM, Hanck DA (2010) Inhibition of the activation pathway of the T-type calcium channel $Ca_V3.1$ by ProTxII. Toxicon 56 (4):624–636
114. Olamendi-Portugal T, Inés García B, López-González I, Van der Walt J, Dyason K, Ulens C, Tytgat J, Felix R, Darszon A, Possani LD (2002) Two new scorpion toxins that target voltage-gated Ca^{2+} and Na^{+} channels. Biochem Biophy Res Comm 299(4):562–568. https://doi.org/10.1016/S0006-291X(02)02706-7
115. Chuang RSI, Jaffe H, Cribbs L, Perez-Reyes E, Swartz KJ (1998) Inhibition of T-type voltage-gated calcium channels by a new scorpion toxin. Nat Neurosci 1(8):668–674. https://doi.org/10.1038/3669
116. Lampe RA, Defeo PA, Davison MD, Young J, Herman JL, Spreen RC, Horn MB, Mangano TJ, Keith RA (1993) Isolation and pharmacological characterization of omega-grammotoxin SIA, a novel peptide inhibitor of neuronal voltage-sensitive calcium channel responses. Mol Pharmacol 44 (2):451–460. http://molpharm.aspetjournals.org/content/44/2/451.abstract
117. McDonough SI, Lampe RA, Keith RA, Bean BP (1997) Voltage-dependent inhibition of N-and P-type calcium channels by the peptide toxin ω-grammotoxin-SIA. Mol Pharmacol 52 (6):1095–1104
118. Piser TM, Lampe RA, Keith RA, Thayer SA (1995) Omega-grammotoxin SIA blocks multiple, voltage-gated, Ca^{2+} channel subtypes in cultured rat hippocampal neurons. Mol Pharmacol 48 (1):131–139. http://molpharm.aspetjournals.org/content/48/1/131.abstract
119. Dalmolin GD, Silva CR, Rigo FK, Gomes GM, do Nascimento Cordeiro M, Richardson M, Silva MAR, Prado MAM, Gomez MV, Ferreira J (2011) Antinociceptive effect of Brazilian armed spider venom toxin Tx3–3 in animal models of neuropathic pain. Pain 152(10):2224–2232
120. Cordeiro MdN, de Figueiredo SG, Valentim AdC, Diniz CR, von Eickstedt VR, Gilroy J, Richardson M (1993) Purification and amino acid sequences of six Tx3 type neurotoxins from the venom of the Brazilian 'armed' spider *Phoneutria nigriventer* (keys). Toxicon 31(1):35–42
121. Oliveira SM, Silva CR, Trevisan G, Villarinho JG, Cordeiro MN, Richardson M, Borges MH, Castro CJ Jr, Gomez MV, Ferreira J, Ezequiel Dias F, Horizonte B (2016) Antinociceptive effect of a novel armed spider peptide Tx3-5 in pathological pain models in mice. Pflügers Arch Eur J Phys 468:881–894. https://doi.org/10.1007/s00424-016-1801-1
122. Peigneur S, de Lima ME, Tytgat J (2018) *Phoneutria nigriventer* venom: a pharmacological treasure. Toxicon 151:96–110. https://doi.org/10.1016/j.toxicon.2018.07.008
123. Maria E, Pereira R, Souza JM, Carobin NV, Figueira Silva J, Santos DC, Antonio C, Júnior S, Scardua Binda N, Borges MH, Alves R, Nagem P, Kushmerick C, Ferreira J, Castro Junior J, Ribeiro M, Vinicius Gomez M (2020) Phoneutria toxin PnTx3-5 inhibits TRPV1 channel with antinociceptive action in an orofacial pain model. Neuropharmacology 162:107826–107835. https://doi.org/10.1016/j.neuropharm.2019.107826
124. Villegas E, Adachi-Akahane S, Bosmans F, Tytgat J, Nakajima T, Corzo G (2008) Biochemical characterization of cysteine-rich peptides from Oxyopes sp. venom that block calcium ion channels. Toxicon 52(2):228–236. https://doi.org/10.1016/j.toxicon.2008.05.019
125. Wang G, Lemos JR (1994) Effects of funnel web spider toxin on Ca^{2+} currents in neurohypophysial terminals. Brain Res 663(2):215–222
126. Ramilo CA, Zafaralla GC, Nadasdij L, Hammerland LG, Yoshikami D, Gray WR, Kristipatij R, Ramachandran J, Miljanichj G, Olivera BM, Cruz LJ (1992) Novel α-and ω-conotoxins from *Conus striatus* venom. Biochemistry 31(41):9919–9926

Advancing Ion Channel Research with Automated Patch Clamp (APC) Electrophysiology Platforms

2

Damian C. Bell and Mark L. Dallas

Abstract

Since its development on the cusp of the new millennium, automated patch clamp (APC) technology has matured over the last two decades. The increased throughput it afforded promised a new paradigm in ion channel recordings: It offered the potential to overcome the time-consuming, low-throughput bottleneck arising from manual patch clamp (MPC) investigations. This chapter highlights the advances in technology, showing how APC platforms have 'democratised' ion channel recordings, lowering the technical bar whilst substantially raising throughput. It will describe the background of the seminal first-generation and updates on advances in second-generation platforms. Furthermore, the chapter summarises the advances APC has made in ion channel studies, including finding new tool compounds and medicines. New functionality and applications on APC platforms give ion channel researchers flexible tools to study ion channels with high quality and high throughput.

D. C. Bell (✉)
Sophion Bioscience A/S, Ballerup, Denmark
e-mail: dbe@sophion.com

M. L. Dallas
School of Pharmacy, University of Reading, Reading, UK

Keywords

Electrophysiology · Ion channels · Automated patch clamp · Manual patch clamp

Abbreviations

APC	Automated patch clamp
CHO	Chinese hamster ovary
CRO	Contract research organisation
d.p.	Data points
HEK293	Human embryonic kidney 293
hERG	Human ether-a-go-go-related gene
HTS	High throughput screen
MPC	Manual patch clamp
MTS	Medium throughput screen

2.1 Introduction

The biggest advances in recording ion channel currents occurred in the 1970–1980s, a decade in which the godfathers of electrophysiology formalized the methods and recording hardware that allowed measurement of ion currents from single cells via a technique dubbed whole-cell patch clamp [1, 2], subsequently referred to as manual patch clamp (MPC). MPC has maintained its place in the electrophysiologists' armoury and continues to contribute to our understanding of ion channels but does have downsides. It requires substantial technical training and teaching

L. Zhou (ed.), *Ion Channels in Biophysics and Physiology*, Advances in Experimental Medicine and Biology 1349, https://doi.org/10.1007/978-981-16-4254-8_2

resources to become competent, a growing problem in the competitive and time-starved modern-day academia [3]. As an highly labour-intensive and very low-throughput technique (~10–30 data points, d.p./day, where a d.p. is defined as a current recording under different test conditions e.g. a four-point consecutive drug concentration response would be five d.p., a control current, plus four drug concentration tested currents), MPC is not an efficient method and thus hamstrings ion channel research and drug development.

Since 1999, automated patch clamp (APC) technologies were developed to make ion channel recordings easier and more efficient by lowering the technical requirements whilst increasing throughput (for reviews on the early developments of APC technology see [4–7]). The main breakthrough in making ion channel recordings easier and faster was planar patch clamp: whilst MPC used a microscope and micromanipulator to locate and place a recording glass electrode pipette on an adherent cell, APC used robots to apply suspensions of ion channel expressing cells onto planar arrays of recording sites (or planar recording chips; see [8]) – see Fig. 2.1. In APC, a combination of gravity and negative pressure enables high electrical resistance seals (100 s MΩ – several GΩ) to form between the cell membrane and the recording site hole.

This chapter will discuss a number of APC platforms, describing key developments that have deepened our understanding of ion channels and advanced our ability to discover novel ion channel therapeutics.

2.2 APC Platforms: Key Developments Over Two Decades

Developments of APC technologies have grown apace since the first Apatchi-1 (Sophion Bioscience) and AutoPatch (CeNes/Xention) were made in 1999 – see Fig. 2.2. In brief, the initial technologies in APC developments used three main types of recording:

1. A robotic visualisation and movement of recording electrode to an adherent cell, essentially replicating what a researcher would do in MPC;
2. Inverted pipettes that were 'backfilled with cell suspensions', allowing trapping and sealing of a single cell in the pipette tip; and
3. Planar patch clamp where robotics applies a cell suspension to a planar array of recording sites.

After these initial format developments, the planar patch clamp quickly evolved to be the clear winner. Planar arrays of recording sites, or planar recording chips, gave the simplest and most consistent method of robotic cell application and sealing and vastly increased the recording capacity of APC by as much as >100-fold over MPC. In gaining the highest throughput capability of thousands of d.p./day, compromises were made; these compromises, it is argued, may have reduced the early adoption of the technology. One such compromise was that the early APC platforms used a 'loose-seal' (typically ~100 MΩ) configuration, which differed from the more accurate and giga-ohm (GΩ) seals routinely attained in MPC recordings [1]. The PatchXpress compromised on having lower throughput (limited to 16 simultaneous recordings) but was the first APC with the ability to achieve the favoured, accurate GΩ seal formation. Another compromise that allowed higher throughput platforms to achieve GΩ seals was the use of 'seal enhancer' in recording solutions (e.g. PatchLiner [10] and the SyncroPatch96 [11], Nanion). The high Ca^{2+} content of this 'seal enchancer' (e.g. 40 mM) has an impact on recordings that are sensitive and/or modulated by Ca^{2+}. The need to use such high divalent ion 'seal enhancer' supplements has been reduced over time, with developments in cell lines, culturing and understanding of the chemistry of glass/lipid interactions. Finally, the fixed well format used on a number of platforms (see Table 2.1) can also cause problems: the lack of wash through of cells or test compounds can have knock-on effects to later elements of the recording (e.g. desensitisation of ligand-gated ion channels

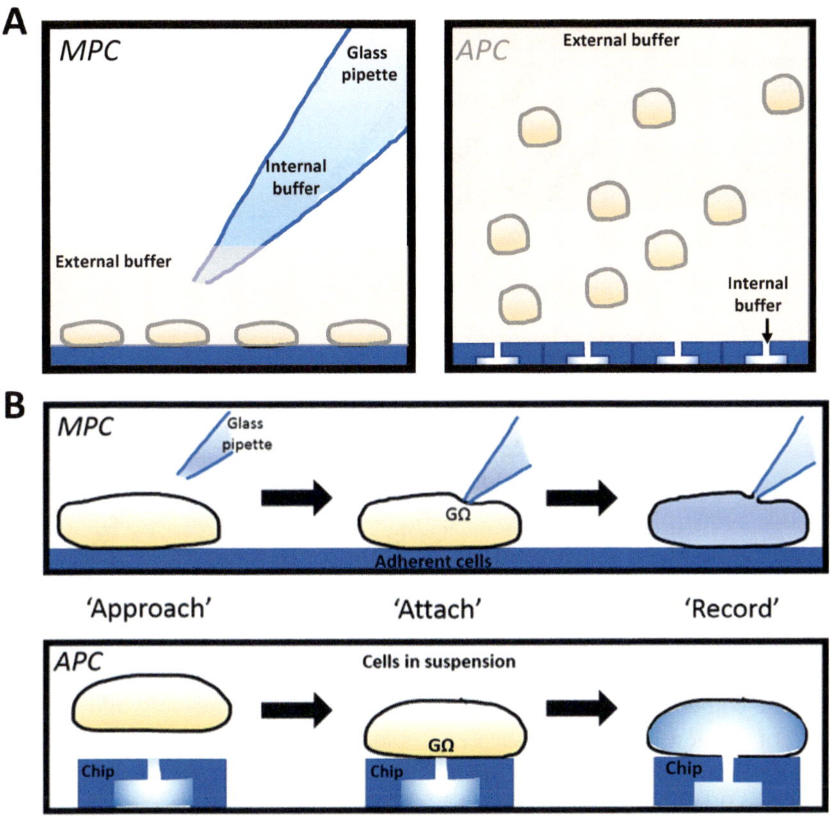

Fig. 2.1 Manual and automated patch clamp methodologies. (**a**) (*Left panel*) Manual patch clamp (MPC) involves the use of a glass pipette, filled with an internal recording buffer, containing a silver-chloride electrode in circuit with a patch clamp amplifier. Adherent cells on coverslips are maintained in an external recording buffer. (*Right panel*) Automated patch clamp (APC; also referred to as planar patch clamp) involves the use of recording chips with planar arrays of recording sites (four recording sites are shown); each site contains an internal buffer and an electrode in circuit with an amplifier. A suspension of cells in an external buffer is applied to the recording sites via an automated pipettor. (**b**) (*Top panel*) To obtain recordings in MPC the glass pipette is manoeuvred using a micro-manipulator onto an adherent cell ('Approach'). The glass pipette then makes contact with the cell and then through the application of negative pressure forms an electrical seal, in the order of GΩ magnitude ('Attach'). Via the application of negative pressure a hole is made in the patch of membrane patch under the glass pipette: this allows the electrode, the recording solution (shaded blue) in the pipette and cell to become one compartment and electrically contiguous ('Record'). (*Bottom panel*) APC, uses cells in suspension that are applied to planar arrays of recording sites on the recording chip (for clarity only a single recording site is shown; 'Approach'). Under gravity and negative pressure a GΩ electrical seal forms between cell membrane and the recording site ('Attach'). Further negative pressure is applied to form a hole in the cell membrane to give the whole cell configuration, whereby the electrode and internal recording solution (shaded blue) situated beneath the recording site are made electrically contiguous with the cytosol ('Record'). Adapted from Perkel et al. [8] and Bell and Dallas [9]

due to incomplete or inefficient washout of applied ligand). Another documented example was that when cell suspension applications were made to a fixed recording well: Surplus cells (i.e. all the cells added but not sealed on the recording site or sites of a recording well) acted as 'lipid sinks', whereby lipophilic compounds (e.g. terfenadine) were absorbed into the surplus cell lipid membranes. This lipid absorption lowered the free concentration of the compounds

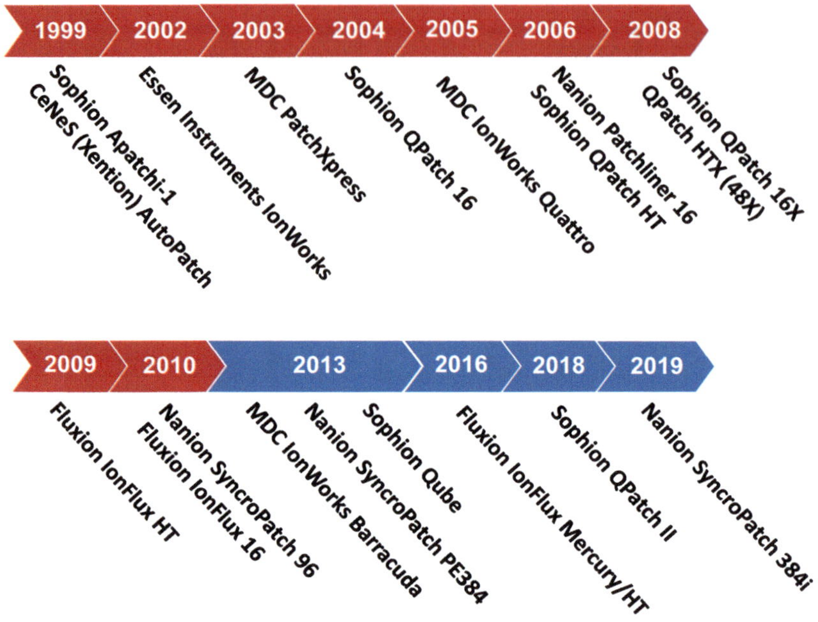

Fig. 2.2 Two decades of automated patch clamp platform development**.** A timeline showing the main automated patch clamp (APC) platforms. First-generation development years are shaded red and second generation are shaded blue

and resulted in lower potencies being measured for these lipophilic compounds [12].

A number of reviews describe the key developments between the first and the second generation of APC platforms [9, 13–15]. Table 2.1 summarises the main capabilities of the second generation of APC platforms. For example, IonWorks, the first 384-recording capable APC platform, only allowed two compound additions, whilst its improved, later sibling, the IonWorks Barracuda, had multiple additions by using an add-dilute-remove solution addition cycle [16, 17]. Further improvements in fluidics followed in other platforms, Nanion's

Table 2.1 A comparison of key features of the second generation of APC platforms (2013 onwards)

Feature		IonWorks Barracuda (MDC, USA)	Qube (Sophion, Denmark)	IonFlux HT/Mercury/Ultra (Fluxion, USA)	QPatch II (Sophion, Denmark)	SyncroPatch 384PE/384i (Nanion, Germany)
Recording sites		384	384	16/64/256	48	384 (768) (+2nd module)
Throughput (d.p./day)		~8 k	~16 k	~1.5 k/5 k/18 k	~3–5 k	~18 k (36 k)
Substrate		Planar, plastic	Planar, polymer	Lateral, PDMS	Planar, silicon/glass	Planar, glass
Seal resistances		~100–300 MΩ	>1 GΩ	>1 GΩ	> 1 GΩ	>1 GΩ
Amplifier compensation		No	Yes	Yes	Yes	Yes
Internal perfusion		No	Yes (offline)	No	No	Yes (online)
External solution exchange	Format	Fixed well	Microfluidics	Microfluidics	Microfluidics	Fixed well
	Cycle	Add-dilute-remove	Add-displace-replace	Continuous	Add-displace-replace	Add-dilute-remove
Current clamp		No	Yes	Yes	Yes	Yes
Temperature control		No	Yes – BoN	Yes – ambient	Yes - BoN	Yes - ambient

d.p./day data points per day, *PDMS* poly-dimethyl-siloxane, *BoN* bed-of-nails. Copied with permission from Bell & Fermini, J. Pharm. Tox. Methods, 2021

SyncroPatch 384 PatchEngine (384PE) and 384i [15, 18], use low-volume pipettor additions and recording site wells, whilst Sophion's QPatch II [19], Qube [20, 21] and Fluxion's HT, Mercury and Ultra platforms all have planar patch clamp recording plates with built-in micro-fluidics channels allowing low-volume, efficient and rapid exchange of solutions into and through the cell recording sites. The second-generation APC platforms maintain high throughput whilst achieving high-quality GΩ seals. With the exception of the IonWorks Barracuda, all of the second-generation APC platforms have current-clamp and temperature control modes, adding further functionality. Finally, developments made from first to second generation have seen the estimated cost per d.p. fall some 3- to 10-fold [9].

2.3 Early APC Adoption: Ion Channels in Drug Discovery

With such automation and vastly increased throughput, a clear application was in drug discovery safety pharmacology. To this end, cardiac ion channel safety testing researchers were early adopters and advanced the assay capabilities of APC platforms. For example, in some of the earliest APC screening campaigns, drug libraries were tested against the hERG (Kv11.1) ion channel using IonWorks HT [22, 23]. More recently, a wide range of industry and academic cardiac safety researchers have worked together to define a standardized hERG liability definition [24]. Although much of the focus of cardiac safety APC assays has been on the chemotype promiscuous hERG channel, other critical cardiac sodium, potassium, and calcium ion channel currents have seen increasing attention using these platforms [25–28]. This work has culminated in a broader scope of cardiac safety pharmacology: The comprehensive in vitro pro-arrhythmia assay (CiPA) is a strategy aimed at standardising assays on key cardiac ionic currents (I_{Kr}, I_{Ks}, I_{K1}, I_{NaFast}, I_{NaLate}, I_{CaL}), across different labs and various APC platform assays [29].

Alongside the early adoption in cardiac safety pharmacology, an early application of APC technology was the secondary screening of high-throughput primary screening, checking that 'hits' arising in the primary screen were actually modulating the ion channel target and not an artefactual response observed in the primary screen (e.g. fluorescent screens can give false positive hits due to compound auto-fluorescence). After using a large, first (primary) fluorescence-based HTS, drug discovery researchers might follow-up with a secondary, smaller medium-throughput screen (MTS) to validate the primary HTS compound 'hits' (e.g. [30]); [7], provide a good summary of such a screening cascade). A number of drug discovery ion channel programmes have followed this screening cascade format, with APC routinely used to determine and verify compound activity in a 'hit' focused, limited secondary screen: for example, HCN channels [31]; Nav channels [32–35], Cav3 (T-type) calcium channels [36] and Cav2 (N-type) calcium channels [37].

However, more recently, the high-throughput and high-quality recordings afforded by the second generation of APC (see Table 2.1) have given drug discovery researchers the tools to make APC-driven HTS possible earlier in the cascade. A screening model being adopted is that the HTS primary and hit-validation secondary screens are being combined into a single mid-to-high throughput screen with the rationale that the improved, data-rich read-out of APC gives stronger, more robust data earlier. One such single APC 158 k compound screen was made by Chambers et al. [20] in a drug discovery programme versus the Nav1.7 ion channel, a chronic pain target, using the Qube. Along these lines, Danahay et al. [38] adopted a mixed fluorescence-based assay and mid-throughput QPatch APC screening model, running the two assay formats in parallel to find compounds that potentiated TMEM16A chloride currents for the development to treat cystic fibrosis [38]. In these APC screens. the traditional primary and secondary screens are driven by a single APC HTS screen, making ion channel drug discovery faster, more efficient and consequently cheaper.

2.4 APC: Advancing Ion Channel Research

Apart from the higher throughput and lowered user technical hurdles, the most obvious advantage that APC brings is in the name: automation. Unlike MPC, APC allows full walk-away recording capability: Most APC platforms have fully automated cell preparation (onboard centrifugation of cell suspensions and resuspension in external recording solution); membrane seal and whole cell formation; and voltage (or current) clamp and test/drug solution exchange protocols. Consequently, as long as sufficient cell suspension volume, solutions, recording and compound plates (both often queued in automated elevator stacks) are provided, several hours (e.g. overnight) of unattended recordings are readily achievable. Add to this the capability that the second-generation APC platforms (see Table 2.1) all have remote access and control which allows monitoring of experimental progress, writing and execution of voltage and test solution application protocols and analysis of completed experimental data. With the growing trend of working remotely – accelerated by the Covid-19 pandemic – such remote access and control of APC platforms will become increasingly advantageous and routinely adopted in ion channel labs.

Earlier, when describing the development of the first generation of APC platforms, the sacrifices that APC R&D adopted were defined that allowed a broader user base by lowering the technical ability needed to make ion channel recordings whilst vastly increasing throughput. However, the most efficient and widely adopted solution, that of planar patch clamp (i.e. arrays of recording sites in a planar recording plate), in the design and development of APC technology provided advantages, giving researchers additional capabilities as standard, not readily achievable in MPC.

Table 2.1 summarises these additional capabilities that APC technology allows, further described here:

2.4.1 Control of Internal Cell Solution Dialysis or Exchange

SyncroPatch 96/384PE/384i APC platforms allow the exchange of the internal cell solution: On these platforms, internal solution exchange can be performed 'online' during ongoing recordings, with the experimental protocol set to exchange the internal solution, i.e. allowing continuous recording throughout and following exchange. Using the Patchliner and Qube, the internal solution can be changed 'off-line': recordings can be paused and internal solution changed manually before continuing with recordings. A number of labs have used APC platforms to study internal solution exchange: TREK-1 channel modulation via intracellular pH was demonstrated using the 384PE [39]; direct PIP2 modulation of human EAG channels was shown on the Patchliner [40]; activation of TRPC5 or TMEM16A channels was shown on the SyncroPatch 384PE via Ca^{2+} containing internal solutions [41].

In MPC recordings, the internal solution exchange is limited to the internal pipette solution dialyzing the intracellular environment: Consequently, the timing of cell dialysis is diffusion dependent and a one-time event after achieving whole-cell access to the recording cell.

2.4.2 Control of Experimental Temperature

APC platforms use two formats to control experimental temperature (see Table 2.1):

(a) control by the ambient environment, whereby the platform cabinet temperature and/or recording solutions can be heated or cooled [e.g. TRPV1 and TPRV3 studies on the Patchliner platform [11, 42].
(b) control by bed-of-nails (BoN; an array of amplifier electrode connectors), whereby temperature measured at the BoN controls

the recording site temperature by heating/cooling fluid that flows via tubes that snake in and out of the BoN. This format takes into account the significant heat generated by the BoN, which can alter the recording site temperature by over +3 °C [43].

Of particular importance in ion channel, safety testing is the biophysical and pharmacological effects of temperature [44]. The cardiac safety implications of temperature on compound modulation of hERG (Kv11.1) ion channels have been studied using IonFlux [45, 46]. An examination of the effects of temperature on kinetics was performed across 40 different Kv channels using the Patchliner [47].

Temperature-controlled experiments though possible on MPC would require specialised add-on temperature control equipment.

2.4.3 Recording Site Fluid Applications: Microfluidics and Fixed-Well

The use of microfluidic channels (e.g. IonFlux HT, Mercury and Ultra; QPatch, QPatch II and Qube) or fixed-well formats (IonWorks Barracuda, Patchliner, SyncroPatch PE384 and 384i; see Table 2.1) allows expensive, limited quantity molecules (e.g. crude venom fraction purifications, synthesized peptides or antibodies) to be tested in low-volume (2–20 μl) applications [48–56].

Microfluidic channels or pipettor robot 'stacking' (where a liquid column 'stacks' different test solutions in the pipettor to be applied sequentially) also provide rapid external solution exchange rates (with complete solution exchange in 10–50 ms), fast enough to allow fast desensitising ligand-gated ion channel recordings [57–63].

Such perfusion is possible in MPC; however, it requires specific add-on fast perfusion microinjection units.

2.4.4 Planar Patch Clamp Recording Plates

In MPC, using a long, glass micropipette as the recording electrode introduces mechanical noise to the recording. Although this mechanical noise is abrogated by the use of vibration-dampening air tables, mechanical noise is still picked up by the glass electrode acting as a mechanical 'antenna'. Planar patch clamp does not have this problem, so anti-vibrational air tables are not needed on APC platforms. Furthermore, this reduced mechanical noise in APC is best seen in the greater success rates for longer recordings (>30 min) that are possible on APC platforms [5].

2.4.5 Multi-Hole Patch Clamp

APC recording plates can employ more than one recording site per recording well. With multiple recording sites per well, multiple cells can seal and be recorded from in a single recording well. This multi-hole patch clamp (also known as population or ensemble) allows multiple whole-cell currents to be averaged in a single recording well. Consequently, by recording currents across multiple cells, multi-hole patch clamp gives improved recording success rates and averages variable ion channel expression across several cells [64, 65].

2.4.6 Pressure Control

To position, seal and attain whole-cell electrical access on cells landing on recording sites in planar arrays, APC platforms have the ability to control pressure at each recording site. This pressure control has been extended beyond its original role in seal and whole-cell formation to apply different pressures to cells during recordings, allowing pressure protocols to be applied to mechanosensitive channels. For instance, QPatch was used for cell volume and membrane stretch pressure studies on cells expressing the mechanosensitive BK and KCNQ channels [66].

2.4.7 Optogenetics and Optical Stimulation

Both Sophion and Nanion have adapted their Qube and SyncroPatch 384PE APC platforms to have a fully integrated LED-based optogenetics unit allowing simultaneous optical and voltage stimulation. Using these specially modified APC platforms, they have shown that optogenetically modified cell lines (i.e. cell lines expressing optically activated channelrhodopsins – for an optogenetics review see [67]) can be simultaneously optically stimulated and voltage-clamped with the resulting currents recorded [68, 69]. Additional functionality is possible with these LED-capable APC, such as releasing caged compounds and in light-based actuation studies of intracellular events [68] and investigating photoswitchable compounds in the modulation of ion channels [69]. These LED-modified APC platforms to date are in-house R&D proof-of-concept platforms and as such are currently not on the market. However, researchers interested in using the optogenetically capable Qube 384-Opto or SyncroPatch 384 PE should contact the manufacturers (*personal communications* Dr. Sandra Wilson, Sophion or Dr. Alison Obergrussberger, Nanion).

Optopatch spiking HEK cells, an optogenetics assay-ready cell line has been developed by researchers at Harvard University [70, 71], allowing an optogenetics approach to high-throughput screening of Nav1.7 ion channel modulators. Recently, this optogenetics screening cell line was used to generate data in a pilot screen of Nav1.7 blockers: these optogenetics screen data were compared to data generated on the IonWorks Barracuda APC (MDC) with the two sets of data showing good correlation [72]. Another example where optogentically modified ion channel cell lines have been used to good effect in combination with APC are in human-induced pluripotent stem cell-derived cardiomyocytes (Cor.4U cells, NCardia) expressing the light-activated channel channelrhodopsin-2 (ChR2). Using the LED capable SyncroPatch 384PE action potentials in Cor.4U-ChR2 cells can be elicited via 1 ms pulses of blue light [69]. These examples of optogenetically modified ion channel cell lines provide further evidence that combining APC with optogenetic screens and experimentation is likely to become increasingly viable assay formats in the future.

These advancements in APC were initially based on what was capable in MPC recordings. However, APC R&D has made significant improvements in ion channel recordings, and in many respects surpasses existing MPC capabilities (e.g. microfluidics channels, internal perfusion and planar patch clamp). In other capabilities where MPC has traditionally held a clear advantage, in recent years, APC developments are making good progress to catch up, allowing far higher throughput in these capabilities typically dominated by MPC. For instance, due to low numbers and potentially mixed, heterogeneous populations of native cells in acutely isolated cell preparations, ion channel recordings on primary cells have often been challenging on APC platforms. However, there are now several labs that have successfully employed APC to make recordings on primary cells [5], and specific examples include pancreatic cells [73], T-cells [74, 75], red blood cells [76] and cortical neurons [77]. An area of research that better lends itself to APC is induced pluripotent stem cells (iPSC), which allow greater numbers and homogeneity of these pseudo-native cells [11, 73, 78]. A potential improvement in human iPSC cardiomyocyte APC recordings is 'dynamic clamp': In Patchliner recordings, an electronic adjustment of the resting membrane potential is made by adding an I_{K1}-modelled current, resulting in a more negative resting membrane potential, redolent of native and mature cardiomyocytes [25, 26, 79].

2.5 Concluding Remarks

APC technology has made ion channel recordings easier, faster and more efficient, widening the potential user base for a previously technically challenging technique. The 'democratisation' of ion channel recordings by APC is arguably the biggest advancement in our ability to study ion channels, since Neher and Sakmann's [80] seminal publication ushered a golden era in patch clamp techniques [1, 2, 80]. Understandably, biotechs, pharmas and CROs were early adopters of APC technology, but more academic labs are accessing and adopting the technology via collaborations, consortia and shared, core facilities. For example, a recent publication highlights the democratizing advantages of APC (fostering collaborations, higher throughput, lower technical hurdles and rapid advancement of the field): in a collaboration between Washington University and Genentech, Wandi Zhu, a summer internship student, completed an impressively thorough biophysical and pharmacological analysis of 39 compounds against the bacterial NaChBac and human Nav1.7 ion channels [81].

With APC technology becoming commonplace, the future of ion channel research looks to be in fine fettle. Indeed, this APC-enabled explosion in ion channel research is likely to lead to significant breakthroughs in our understanding of ion channel physiology and pathophysiology. For instance, the improved capabilities provide the potential for greater definition of selected patient cohorts, channelopathies and rare diseases, giving more scope to treat individuals with increasingly personalised (or stratified) medicinal regimes. For instance, the application of APC allowed researchers to test gain or loss of function mutations in Nav and Cav channels – these mutant channel data will inform potential future precision medicine treatments in patients carrying these channelopathies [82]. Consequently, on the back of this accelerated research, ion channel treatments will be found helping patients across a wide range of diseases, and the first blockbuster drug is imminent.

Acknowledgements We would like to thank Dr. Alison Obergrussberger (Nanion), Drs. Sandra Wilson and Göran Mattsson (Sophion) and Dr. Ali Yehia (Fluxion) for their help with publications and clarifications. MD acknowledges the support received to carry out ion channel research in both MPC and APC formats from various sources.

Competing Interests The authors declare no competing interests. Since writing this chapter DCB has joined Sophion Biosciences A/S; however, before he joined the manuscript was written and completed as an objective review of the APC field.

References

1. Hamill OP, Marty A, Neher E, Sakmann B, Sigworth FJ (1981) Improved patch-clamp techniques for high-resolution current recording from cells and cell-free membrane patches. Pflügers Arch: Eur J Physiol 391:85–100
2. Sakmann B, Neher E (1984) Patch clamp techniques for studying ionic channels in excitable membranes. Annu Rev Physiol 46:455–472
3. Nurse, P. (2015). Ensuring a successful UK research endeavour: a review of the UK Research Councils
4. Dunlop J, Bowlby M, Peri R, Vasilyev D, Arias R (2008) High-throughput electrophysiology: an emerging paradigm for ion-channel screening and physiology. Nat Rev Drug Discov 7:358–368
5. Milligan CJCJ, Li J, Sukumar P, Majeed Y, Dallas MLML, English A et al (2009) Robotic multiwell planar patch-clamp for native and primary mammalian cells. Nat Protoc 4:244–255
6. Priest, B.T., Cerne, R., Krambis, M.J., Schmalhofer, W.A., Wakulchik, M., Wilenkin, B., et al. (2004). Automated electrophysiology assays (Eli Lilly & Company and the National Center for Advancing Translational Sciences)
7. Terstappen GC, Roncarati R, Dunlop J, Peri R (2010) Screening technologies for ion channel drug discovery. Future Med Chem 2:715–730
8. Perkel JM (2010) High-throughput ion channel screening: a 'patch'-work solution. BioTechniques 48:25–29
9. Bell DC, Dallas ML (2018) Using automated patch clamp electrophysiology platforms in pain-related ion channel research: insights from industry and academia. Br J Pharmacol 175:2312–2321
10. Brüggemann A, Stoelzle S, George M, Behrends JC, Fertig N (2006) Microchip technology for automated and parallel patch-clamp recording. Small 2:840–846
11. Stoelzle S, Obergrussberger A, Brüggemann A, Haarmann C, George M, Kettenhofen R et al (2011) State-of-the-art automated patch clamp devices: heat activation, action potentials, and high throughput in ion channel screening. Front Pharmacol 2:76
12. Bridgland-Taylor MH, Hargreaves AC, Easter A, Orme A, Henthorn DC, Ding M et al (2006)

Optimisation and validation of a medium-throughput electrophysiology-based hERG assay using IonWorks™ HT. J Pharmacol Toxicol Methods 54:189–199
13. Liu C, Li T, Chen J (2019) Role of high-throughput electrophysiology in drug discovery. Curr Protoc Pharmacol 87:e69
14. McGivern JG, Ding M (2020) Ion channels and relevant drug screening approaches. SLAS Discov 25:413–419
15. Obergrussberger A, Friis S, Brüggemann A, Fertig N (2020) Automated patch clamp in drug discovery: major breakthroughs and innovation in the last decade. Expert Opin Drug Discov:1–5
16. Gillie DJ, Novick SJ, Donovan BT, Payne LA, Townsend C (2013) Development of a high-throughput electrophysiological assay for the human ether-à-go-go related potassium channel hERG. J Pharmacol Toxicol Methods 67:33–44
17. Kuryshev YA, Brown AM, Duzic E, Kirsch GE (2014) Evaluating state dependence and subtype selectivity of calcium channel modulators in automated electrophysiology assays. Assay Drug Dev Technol 12:110–119
18. Obergrussberger A, Brüggemann A, Goetze TA, Rapedius M, Haarmann C, Rinke I et al (2016) Automated patch clamp meets high-throughput screening: 384 cells recorded in parallel on a planar patch clamp module. J Lab Autom 21:779–793
19. Schupp M, Park SH, Qian B, Yu W (2020) Electrophysiological studies of GABAA receptors using QPatch II, the next generation of automated patch-clamp instruments. Curr Protoc Pharmacol 89:e75
20. Chambers C, Witton I, Adams C, Marrington L, Kammonen J (2016) High-throughput screening of NaV1.7 modulators using a giga-seal automated patch clamp instrument. Assay Drug Dev Technol 14:93–108
21. Qian B, Park SH, Yu W (2020) Screening assay protocols targeting the Nav1.7 channel using qube high-throughput automated patch-clamp system. Curr Protoc Pharmacol 89(1):e74
22. Danker T, Möller C (2014) Early identification of hERG liability in drug discovery programs by automated patch clamp. Front Pharmacol 5:203
23. Sorota S, Zhang XS, Margulis M, Tucker K, Priestley T (2005) Characterization of a hERG screen using the IonWorks HT: comparison to a hERG rubidium efflux screen. Assay Drug Dev Technol 3:47–57
24. Windley MJ, Abi-Gerges N, Fermini B, Hancox JC, Vandenberg JI, Hill AP (2016) Measuring kinetics and potency of hERG block for CiPA. J Pharmacol Toxicol Methods
25. Becker N, Horváth A, De Boer T, Fabbri A, Grad C, Fertig N et al (2020) Automated dynamic clamp for simulation of IK1 in human induced pluripotent stem cell–derived cardiomyocytes in real time using patchliner dynamite8. Curr Protoc Pharmacol 88:e70
26. Goversen B, Becker N, Stoelzle-Feix S, Obergrussberger A, Vos MA, van Veen TAB et al (2018) A hybrid model for safety pharmacology on an automated patch clamp platform: using dynamic clamp to join iPSC-derived cardiomyocytes and simulations of Ik1 ion channels in real-time. Front Physiol 8
27. Guo D, Jenkinson S (2019) Simultaneous assessment of compound activity on cardiac Nav1.5 peak and late currents in an automated patch clamp platform. J Pharmacol Toxicol Methods 99
28. Sanson C, Schombert B, Filoche-Rommé B, Partiseti M, Bohme GA (2019) Electrophysiological and pharmacological characterization of human inwardly rectifying Kir 2.1 channels on an automated patch-clamp platform. Assay Drug Dev Technol 17:89–99
29. Brinkwirth N, Takasuna K, Doi M, Becker N, Obergrussberger A, Friis S et al (2020) Reliable identification of cardiac liability in drug discovery using automated patch clamp: benchmarking best practices and calibration standards for improved proarrhythmic assessment. J Pharmacol Toxicol Methods
30. Gilbert DF, Islam R, Lynagh T, Lynch JW, Webb TI (2009) High throughput techniques for discovering new glycine receptor modulators and their binding sites. Front Mol Neurosci 2:17
31. Vasilyev DV, Shan QJ, Lee YT, Soloveva V, Nawoschik SP, Kaftan EJ et al (2009) A novel high-throughput screening assay for HCN channel blocker using membrane potential-sensitive dye and FLIPR. J Biomol Screen 14:1119–1128
32. Bagal SK, Chapman ML, Marron BE, Prime R, Storer RI, Swain NA (2014) Recent progress in sodium channel modulators for pain. Bioorg Med Chem Lett 24:3690–3699
33. Klement G, Babich O, Larsson O, Lund P-E, Malmberg A, Sandberg L et al (2012) Identification of novel NaV1.7 antagonists using high throughput screening platforms. Comb. Chem High Throughput Screen 15:713–720
34. Kornecook TJ, Yin R, Altmann S, Be X, Berry V, Ilch CP et al (2017) Pharmacologic characterization of AMG8379, a potent and selective small molecule sulfonamide antagonist of the voltage-gated sodium channel NaV1.7. J Pharmacol Exp Ther
35. Trivedi S, Dekermendjian K, Julien R, Huang J, Lund PE, Krupp J et al (2008) Cellular HTS assays for pharmacological characterization of Na V1.7 modulators. Assay Drug Dev Technol 6:167–179
36. Xie XS, Van Deusen AL, Vitko I, Babu DA, Davies LA, Huynh N et al (2007) Validation of high throughput screening assays against three subtypes of Cav3 T-type channels using molecular and pharmacologic approaches. Assay Drug Dev Technol 5:191–203
37. Swensen AM, Herrington J, Bugianesi RM, Dai G, Haedo RJ, Ratliff KS et al (2012) Characterization of the substituted N-triazole oxindole TROX-1, a small-molecule, state-dependent inhibitor of Ca(V)2 calcium channels. Mol Pharmacol 81:488–497

38. Danahay HL, Lilley S, Fox R, Charlton H, Sabater J, Button B et al (2020) TMEM16A potentiation: a novel therapeutic approach for the treatment of cystic fibrosis. Am J Respir Crit Care Med 201:946–954
39. Sauter DRP, Sørensen CE, Rapedius M, Brüggemann A, Novak I (2016) pH-sensitive K+ channel TREK-1 is a novel target in pancreatic cancer. Biochim Biophys Acta Mol Basis Dis 1862:1994–2003
40. Han B, He K, Cai C, Tang Y, Yang L, Heinemann SH et al (2016) Human EAG channels are directly modulated by PIP 2 as revealed by electrophysiological and optical interference investigations. Sci Rep 6:1–13
41. Brinkwirth N, Friis S, Goetze T, Rapedius M, Costantin J, Brüggemann A et al (2017) Investigation of the ion channels TMEM16A and TRPC5 and their modulation by intracellular calcium. Biophys J 112:413a
42. Papakosta M, Dalle C, Haythornthwaite A, Cao L, Stevens EB, Burgess G et al (2011) The chimeric approach reveals that differences in the TRPV1 pore domain determine species-specific sensitivity to block of heat activation. J Biol Chem 286:39663–39762
43. Lei CL, Clerx M, Beattie KA, Melgari D, Hancox JC, Gavaghan DJ et al (2019) Rapid characterization of hERG channel kinetics II: temperature dependence. Biophys J 117:2455–2470
44. Guo J, Zhan S, Lees-Miller JP, Teng GQ, Duff HJ (2005) Exaggerated block of hERG (KCNH2) and prolongation of action potential duration by erythromycin at temperatures between 37 °C and 42 °C. Hear Rhythm 2:860–866
45. Golden AP, Li N, Chen Q, Lee T, Nevill T, Cao X et al (2011) IonFlux: a microfluidic patch clamp system evaluated with human ether-à-go-go related gene channel physiology and pharmacology. Assay Drug Dev Technol 9:608–619
46. Kauthale RR, Dadarkar SS, Husain R, Karande VV, Gatne MM (2015) Assessment of temperature-induced hERG channel blockade variation by drugs. J Appl Toxicol 35:799–805
47. Ranjan R, Logette E, Marani M, Herzog M, Tâche V, Scantamburlo E et al (2019) A kinetic map of the homomeric voltage-gated potassium channel (Kv) family. Front Cell Neurosci 13:358
48. Chow CY, Chin YK-Y, Walker AA, Guo S, Blomster LV, Ward MJ et al (2020) Venom peptides with dual modulatory activity on the voltage-gated sodium channel Na V 1.1 provide novel leads for development of antiepileptic drugs. ACS Pharmacol Transl Sci 3:119–134
49. Gonçalves TC, Benoit E, Kurz M, Lucarain L, Fouconnier S, Combemale S et al (2019) From identification to functional characterization of cyriotoxin-1a, an antinociceptive toxin from the spider *Cyriopagopus schioedtei*. Br J Pharmacol 176:1298–1314
50. Israel MR, Dash TS, Bothe SN, Robinson SD, Deuis JR, Craik DJ et al (2020) Characterization of synthetic Tf2 as a NaV1.3 selective pharmacological probe. Biomedicines 8:155
51. Nicolas S, Zoukimian C, Belgium FB, Montnach J, Diochot S, Cuypers E et al (2019) Chemical synthesis, proper folding, nav channel selectivity profile and analgesic properties of the spider peptide phlotoxin 1. Toxins (Basel) 11:367
52. Sharma G, Deuis JR, Jia X, Mueller A (2020) Recombinant production, bioconjugation and membrane binding studies of Pn3a, a selective Na V 1. 7 inhibitor. Biochem Pharmacol:114148
53. Tzakoniati F, Xu H, Li T, Garcia N, Kugel C, Payandeh J et al (2020) Development of photocrosslinking probes based on Huwentoxin-IV to map the site of interaction on Nav1.7. Cell Chem Biol 27:306–313. e4
54. Williams WA, Linley JE, Jones CA, Shibata Y, Snijder A, Button J et al (2019) Antibodies binding the head domain of P2X4 inhibit channel function and reverse neuropathic pain. Pain 160:1989–2003
55. Xu H, Li T, Rohou A, Arthur CP, Tzakoniati F, Wong E et al (2019) Structural basis of Nav1.7 inhibition by a gating-modifier spider toxin. Cell 176:702–715. e14
56. Zoukimian C, Meudal H, De Waard S, Ouares KA, Nicolas S, Canepari M et al (2019) Synthesis by native chemical ligation and characterization of the scorpion toxin AmmTx3. Bioorganic Med Chem 27:247–253
57. Alijevic O, McHugh D, Rufener L, Mazurov A, Hoeng J, Peitsch M (2020) An electrophysiological characterization of naturally occurring tobacco alkaloids and their action on human $\alpha 4\beta 2$ and $\alpha 7$ nicotinic acetylcholine receptors. Phytochemistry 170:112187
58. Gilbert DF, Islam R, Lynagh T, Lynch JW, Webb TI (2013) High throughput techniques for discovering new glycine receptor modulators and their binding sites. Front Mol Neurosci 2
59. Harvey AJ, Avery TD, Schaeffer L, Joseph C, Huff BC, Singh R et al (2019) Discovery of BNC375, a potent, selective, and orally available Type i positive allosteric modulator of $\alpha 7$ nAChRs. ACS Med Chem Lett 10:754–760
60. Skarratt KK, Gu BJ, Lovelace MD, Milligan CJ, Stokes L, Glover R et al (2020) A P2RX7 single nucleotide polymorphism haplotype promotes exon 7 and 8 skipping and disrupts receptor function. FASEB J 34:3884–3901
61. Sophocleous RA, Berg T, Finol-Urdaneta RK, Sluyter V, Keshiya S, Bell L et al (2020) Pharmacological and genetic characterisation of the canine P2X4 receptor. Br J Pharmacol 177:2812–2829
62. Stead C, Brown A, Adams C, Nickolls SJ, Young G, Kammonen J et al (2016) Identification of positive allosteric modulators of glycine receptors from a high-throughput screen using a fluorescent membrane potential assay. J Biomol Screen 21:1042–1053
63. Yehia A, Wei H (2020) Studying nicotinic acetylcholine receptors using the IonFlux™ microfluidic-based automated patch-clamp system with continuous

perfusion and fast solution exchange. Curr Protoc Pharmacol 88
64. Dale TJ, Townsend C, Hollands EC, Trezise DJ (2007) Population patch clamp electrophysiology: a breakthrough technology for ion channel screening. Mol BioSyst 3:714–722
65. Finkel A, Wittel A, Yang N, Handran S, Hughes J, Costantin J (2006) Population patch clamp improves data consistency and success rates in the measurement of ionic currents. J Biomol Screen 11:488–496
66. Hammami S, Willumsen NJ, Olsen HL, Morera FJ, Latorre R, Klaerke DA (2009) Cell volume and membrane stretch independently control K^+ channel activity. J Physiol 587:2225–2231
67. Fenno L, Yizhar O, Deisseroth K (2011) The development and application of optogenetics. Annu Rev Neurosci 34:389–412
68. Boddum, K., Skafte-Pedersen, P., Rolland, J.F., and Wilson, S. (2021). Optogenetics and optical tools in automated patch clamping. In: Methods in Molecular Biology.
69. Obergrussberger A, Goetze TA, Brinkwirth N, Becker N, Friis S, Rapedius M et al (2018) An update on the advancing high-throughput screening techniques for patch clamp-based ion channel screens: implications for drug discovery. Expert Opin Drug Discov 13:269–277
70. McNamara HM, Zhang H, Werley CA, Cohen AE (2016) Optically controlled oscillators in an engineered bioelectric tissue. Phys Rev X 6:031001
71. Zhang H, Reichert E, Cohen AE (2016) Optical electrophysiology for probing function and pharmacology of voltagegated ion channels. Elife:5
72. Zhang H, Moyer BD, Yu V, McGivern JG, Jarosh M, Werley CA et al (2020) Correlation of optical and automated patch clamp electrophysiology for identification of NaV1.7 inhibitors. SLAS Discov 25:434–446
73. Becker N, Stoelzle S, Göpel S, Guinot D, Mumm P, Haarmann C et al (2013) Minimized cell usage for stem cell-derived and primary cells on an automated patch clamp system. J Pharmacol Toxicol Methods 68:82–87
74. Li T, Lu G, Chiang EY, Chernov-Rogan T, Grogan JL, Chen J (2017) High-throughput electrophysiological assays for voltage gated ion channels using SyncroPatch 768PE. PLoS One 12:e0180154
75. Ong ST, Bajaj S, Tanner MR, Chang SC, Krishnarjuna B, Ng XR et al (2020) Modulation of lymphocyte potassium channel K V 1.3 by membrane-penetrating, joint-targeting immunomodulatory plant defensin. ACS Pharmacol Transl Sci
76. Rotordam MG, Fermo E, Becker N, Barcellini W, Brüggemann A, Fertig N et al (2019) A novel gain-of-function mutation of piezo1 is functionally affirmed in red blood cells by high-throughput patch clamp. Haematologica 104:e179–e183
77. Toh MF, Brooks JM, Strassmaier T, Haedo RJ, Puryear CB, Roth BL et al (2020) Application of high-throughput automated patch-clamp electrophysiology to study voltage-gated ion channel function in primary cortical cultures. SLAS Discov 25:447–457
78. Haythornthwaite A, Stoelzle S, Hasler A, Kiss A, Mosbacher J, George M et al (2012) Characterizing human ion channels in induced pluripotent stem cell-derived neurons. J Biomol Screen 17:1264–1272
79. Li W, Luo X, Ulbricht Y, Wagner M, Piorkowski C, El-Armouche A et al (2019) Establishment of an automated patch-clamp platform for electrophysiological and pharmacological evaluation of hiPSC-CMs. Stem Cell Res 41:101662
80. Neher E, Sakmann B (1976) Single-channel currents recorded from membrane of denervated frog muscle fibres. Nature 260:799–802
81. Zhu W, Li T, Silva JR, Chen J (2020) Conservation and divergence in NaChBac and NaV1.7 pharmacology reveals novel drug interaction mechanisms. Sci Rep 10:10730
82. Heyne HO, Baez-Nieto D, Iqbal S, Palmer D, Brunklaus A, Collaborative E et al (2019) Predicting functional effects of missense variants in voltage-gated sodium and calcium channels. BioRxiv:671453

Ion Channels in Biophysics and Physiology: Methods & Challenges to Study Mechanosensitive Ion Channels

3

Yun Lyna Luo and Jerome Lacroix

...one cannot dismiss the possibility of a specially adapted molecular mechanism, for example some piezo-electric substance, being responsible for the conversion of stretch into a spindle potential.

Bernhard Katz, J. Physiol. (1950) III, 261–282

Abstract

In a seminal work published in 1950, Sir B. Katz showed that the electrical response of the frog muscle spindle varies directly with the rate and amplitude of muscle stretch. This observation led him to propose the existence of a piezoelectric substance in this organ, setting the stage for the field of mechanobiology (Katz, J Physiol 111, 261–282, 1950). Despite this early work, the identity of the molecules responsible for the conversion of mechanical stimuli into biological signals has remained hidden for decades. This delay is often attributed to the inherent difficulty to precisely quantify the mechanical deformations of biological samples. In contrast to other forms of stimuli such as ligand concentration and membrane potential, quantifying mechanical deformations of cell membranes is not trivial. Mechanical forces produce a complex array of membrane deformations including bending, thinning, compression, expansion, and shear, and thus, have components in many strain dimensions. In addition, due to the viscoelastic nature of cells, these deformations may have linear and nonlinear components. In spite of these experimental challenges, Sukharev et al. cloned the first mechanosensitive ion channel from the bacteria *E. coli* in the mid-1990s (Sukharev et al. Nature, 265–268, 1994). Two decades later, several protein families encompassing dozens of eukaryotic mechanosensitive ion channels have been identified, depicting an astonishing diversity of force-activated molecular machines. In this chapter, we intend to provide an overview of the current state of knowledge and technical challenges to study how cell membranes deform upon mechanical stress and how ion channel proteins detect these deformations to engage homeostatic cellular responses.

Y. L. Luo
College of Pharmacy, Western University of Health Sciences, Pomona, CA, USA
e-mail: luoy@westernu.edu

J. Lacroix (✉)
College of Biomedical Sciences, Western University of Health Sciences, Pomona, CA, USA
e-mail: jlacroix@westernu.edu

L. Zhou (ed.), *Ion Channels in Biophysics and Physiology*, Advances in Experimental Medicine and Biology 1349, https://doi.org/10.1007/978-981-16-4254-8_3

Keywords

Molecular mechanotransduction · Ion channels · Mechanical stress · Membrane mechanical properties · Force-from-lipids

3.1 Mechanical Properties of Biological Materials

Any material, including solids, liquids, and gases can sustain three general forms of physical deformations depending on the direction of applied forces: compression, tension (expansion), or shear (Fig. 3.1).

In biological systems, compressional forces are produced when tumors grow, shear forces are produced when blood flow creates frictions against the vascular wall, and tensional forces are produced during development when cells pull on each other as the embryo grows. The amount of physical deformation produced by mechanical forces largely depends on the mechanical properties of the material under stress. For instance, elastic materials deform instantly and reversibly upon application of mechanical stress. In these materials, the mechanical deformation (strain) is proportional to the amplitude of the mechanical stimulus (stress). In linear elastic materials, the ratio of the stress over strain is a constant called the elastic Young's modulus or stiffness:

$$\text{stiffness} = \frac{\text{Stress}}{\text{Strain}} \tag{3.1}$$

The strain is a dimensionless number that represents the relative change in the space occupied by the material (i.e., its length, volume, or curvature). Thus, stress and stiffness have the same dimension, which is that of a force divided by a surface area, that is, a pressure usually expressed in Newton per square meter (N m^{-2}) or Pascal (Pa). Because the amplitude of the strain in a linearly elastic material is proportional to the amplitude of the applied stress, the material will not deform further, even if the stress is maintained over time.

Viscous materials behave very differently from elastic materials in that the amount of deformation depends not only upon the amplitude of the stress but also on the amount of time the stress is being applied. Hence, there exists a linear relationship between the applied stress and the *rate* of mechanical deformation, also called strain rate. In linear viscous materials (also called Newtonian fluids), the ratio of the stress over strain rate is a constant called viscosity:

$$\text{viscosity} = \frac{\text{Stress}}{\text{Strain rate}} \tag{3.2}$$

Like many naturally occurring materials, biological tissues possess both an elastic and a viscous component: they exhibit viscoelastic properties (Fig. 3.2). In addition, the viscosity and stiffness may change as a function of the amplitude or the nature of applied stress (nonlinearity). Because of these complex mechanical properties, it is inherently difficult to characterize the strain experienced by cells and tissues upon given mechanical stress.

3.2 Detection of Mechanical Forces at the Cellular Level

The cell membrane is a relatively fragile material that cannot bear elastic deformations above 3–4%. To prevent mechanical disruption of their membrane, most cells adapt to mechanical deformations of the lipid bilayer, for instance, by supplying the membrane with more lipids from intracellular reservoirs or by changing cytosolic ionic concentrations to reduce the osmotic flow of water. The lytic tension of a cell membrane, that is, the mechanical stress producing its rupture can, however, be increased by specific biological adaptations. For instance, plant and bacteria protect their cells by building thick extracellular walls made of long and highly branched polymeric chains of cellulose (plant) or peptidoglycan (bacteria). In contrast, animal cells build an intracellular cytoskeleton made of networks of actin filaments. The actin cytoskeleton of animal cells is highly dynamic, that is, it is constantly

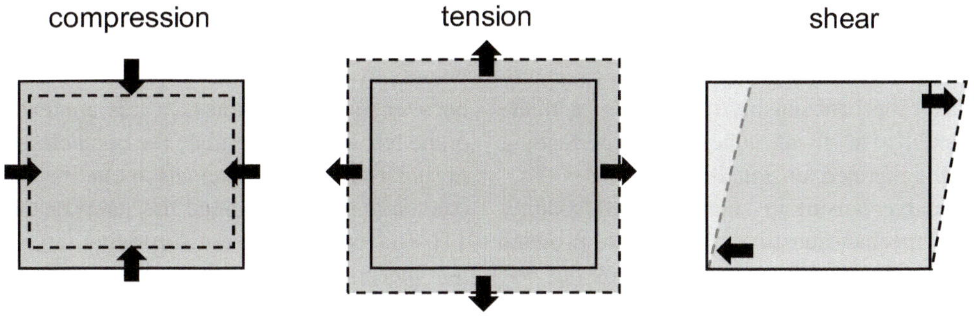

Fig. 3.1 Three possible mechanical deformations of materials

being deconstructed and rebuilt, allowing cells to change their morphology and migrate. These additional molecular structures, which are attached to the lipid bilayer by protein–protein or protein–lipid interactions, resist tensional deformations, and thus, increase the apparent stiffness of the cell membrane.

Besides external mechanical stress, cells themselves can generate mechanical forces to interrogate the mechanical properties of their surrounding microenvironment. The existence of cell-generated and environment-generated mechanical forces divides mechanotransduction signaling into two groups: **inside-out** signaling, where cells used force to gauge the mechanical properties of their local microenvironment, and **outside-in** signaling, where cells sense external mechanical stress produced by their environment. To enable inside-out signaling, animal cells employ actin filaments connected via adapter proteins to extracellular matrix (EM) proteins. This network of cables is physically pulled toward the interior of the cell by the activity of molecular motor proteins called myosin. Similar to how the tension of a fishing line increases when a fish is hooked, the tension in the actin cytoskeleton and its associated proteins increases when the stiffness of the extracellular environment surrounding the cell increases [3, 4]. The molecular mechanisms by which changes in tension in this web of molecular cables is sensed and interpreted by cells are not completely understood. However, recent studies indicate that specific adaptor proteins change their conformation under tension, unveiling binding sites that can be recognized by specific signaling proteins,

Fig. 3.2 Examples of time-dependent deformations in different materials

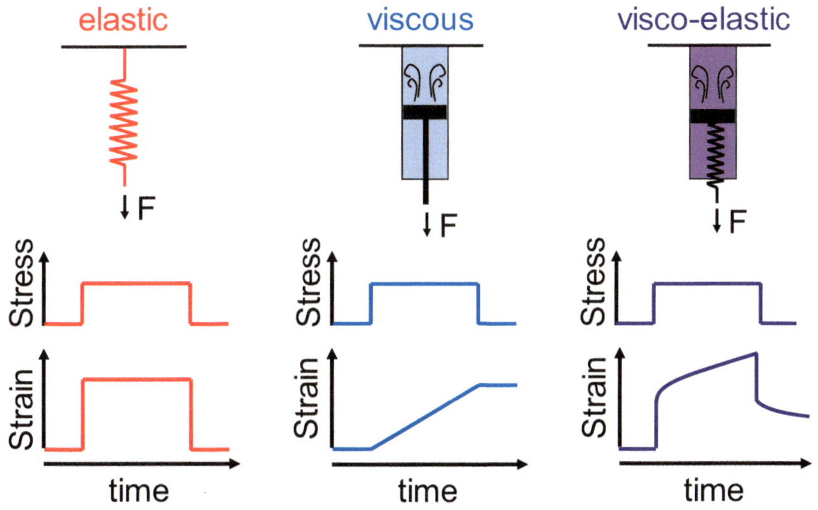

effectively transducing tensional forces into biological responses. The cell membrane itself seems to deform during this process, as recently identified mechanosensitive channels activate preferentially at focal adhesions experiencing high cell-generated tensional forces [5].

To enable outside-in signaling, cells mainly rely on mechanosensitive membrane proteins that sense viscoelastic deformations of the cell membrane induced by external mechanical forces. Among known mechanosensitive membrane proteins are mechanosensitive ion channels and G-protein coupled receptors. Activation of mechanosensitive ion channels enables faster transduction of mechanical stimuli into biological signals as compared to metabotropic receptors and/or G-proteins [6–8]. Indeed, mechanosensitive ion channels rapidly increase ionic permeabilities of the cell membrane upon mechanical deformations of the lipid bilayer and/or its associated structures including cytoskeletal elements and extracellular filaments tethered to membrane proteins. This increased permeability typically enables intracellular uptake of calcium ions and/or instant changes in the membrane potential.

3.3 Principles of Mechano-Electrical Transduction in Mechanosensitive Ion Channels

Most ion channels are not permanently open and rather populate discrete structural conformations that can either be nonconducting (closed or inactivated) or conducting (open). The relative free energy difference between these conformations determines the likeliness of channels to populate these states given a set of physicochemical conditions. The free energy of a protein conformation depends not only on interaction energies between protein residues but also upon interaction energies between protein residues and the surrounding lipids, solvent molecules, and associated protein(s). Changes in the surrounding molecular environment of a membrane protein are, therefore, likely to change the free energy associated with its conformations, potentially changing the free energy difference between functional states. In this context, membrane tension can modulate the open/closed equilibrium of many ion channels, including those not conventionally considered mechanosensitive [9–11]. But this is not very surprising, considering that mechanical deformations of the bilayer have profound effects on lipid–protein interactions, and thus, free energies differences between functional states. On the other hand, a much lower number of ion channels are regulated over *a large fraction of their activity profile* by mechanical stimuli. This means that their open probability can be tuned down near its lowest physiological value up to near its highest physiological value solely by varying the amplitude of mechanical stimuli. In this chapter, we shall use the acronym MSCs (mechanosensitive channels) to refer to these specialized mechanosensitive ion channels. At scientific meetings, MSCs are often called "professional mechanosensitive ion channels" by experts in the field.

In contrast to other ion channels partially modulated by mechanical stimuli, MSCs possess highly sophisticated molecular mechanosensory mechanisms that amplify the effects of mechanical stress on the open/closed thermodynamic equilibrium. Understanding how these molecular mechanisms operate currently represents one of the greatest and most difficult challenges in the field of ion channels. Several factors explain this difficulty, such as the heterogeneity of physiological mechanical stresses applied to the cell membrane (tension, compression, and shear), the complexity of viscoelastic membrane deformations (bending, thinning, expansion, changes in lipid organization, and lipid composition, etc.), and the astonishing diversity of molecular architectures by which MSCs sense mechanical stimuli.

A growing body of evidence supports the existence of two broad mechanisms by which mechanical deformations of the cell membrane and its associated filaments are transmitted to MSCs. The first one is the *force-from-lipid* (FFL) paradigm, where the protein detects changes in membrane lipid organization, composition, and/or changes in membrane thickness, curvature, or lateral tension. The second one is the *force-from-filament* (FFF) principle, where mechanical forces are transmitted *via* molecular tethers such as cytoskeletal or EM filaments that are physically attached to the protein [12–14].

To date, the FFL paradigm is by far the most common and the best understood mode of MSC activation. Within the FFL paradigm, membrane stretch (increase bilayer tension) is the most common and best understood mechanical stimulus. The amplitude and direction of mechanical forces exerted onto a membrane protein embedded in a lipid bilayer under tension vary dramatically along the vertical (z-axis) of the membrane. Indeed, moderate repulsive interactions (positive pressure) are produced at two places, first at the level of the polar headgroups due to electrostatic repulsions and second at the level of the hydrophobic core due to lipid tail entropy (Fig. 3.3a, b). On the other hand, stronger attractive interactions (negative pressure) are produced at the boundary between the hydrophilic headgroups the hydrophobic aliphatic core to prevent the entry of water molecules into the bilayer [15].

Membrane stretch is a tensional elastic strain of the lipid bilayer. When the membrane stretches, lipids are pulled further apart despite attractive hydrophobic forces that tend to keep them close together. Hence, membrane stretch induces an increase in the area per lipid (or a decrease in lipid density), leading to (1) a reduction of lipid–protein repulsive interactions and (2) an increase of lipid–protein attractive interactions (Fig. 3.3c). The net negative pressure imposes a thermodynamic penalty (increased free energy) on more compact conformations, thus favoring protein conformations with a larger membrane surface area (membrane footprint). Hence, membrane stretch can modulate thermodynamic equilibria between protein conformations providing that these conformations occupy different membrane footprints.

The Gibbs free energy difference between open and closed states ($G_{\text{open}} - G_{\text{closed}} = \Delta G$) can be calculated by the following formula:

$$\Delta G = -k_{\mathrm{B}}T \ln \left[\frac{P_{\text{open}}}{P_{\text{closed}}}\right] = -\gamma \Delta A + \Delta G_{\text{protein}} + \Delta G_{\text{membrane}} \quad (3.3)$$

With k_{B} is the Boltzmann constant, T the temperature, P_{open} and P_{closed}, respectively the probability of the channel being open and closed, γ the membrane tension (in N m^{-1}), ΔA the relative change in the membrane surface footprint associated with channel opening, $\Delta G_{\text{protein}}$ the free energy of channel opening in absence of tension and $\Delta G_{\text{membrane}}$ the free energy of membrane deformations in absence of tension. From this equation, it is evident that the larger the surface change between open and closed states, the lower the tension needed to open a stretch-activated ion channel.

Membrane stretch does not only reduce lipid density. A reduced lipid density facilitates the penetration of lipid tails from one leaflet into the other. This effect tends to reduce the average thickness of the bilayer. An immediate consequence of membrane thinning is the exposure of hydrophobic protein regions to the aqueous solvent (hydrophobic mismatch). This hydrophobic mismatch may generate a thermodynamic penalty that could shift the closed/open equilibrium of the channel, acting as a genuine physicochemical stimulus [13].

3.4 Families of Mechanosensitive Ion Channels

To date, MSCs have been identified at least eight evolutionary-unrelated families spanning all domains of life. This strongly suggests that MSCs have emerged independently from different molecular ancestors during evolution. This also suggests that mechanosensitivity is a property of ion channels that can be achieved by a variety of molecular mechanisms. This may not appear very surprising to some readers since, as we discussed earlier, the cell membrane can be mechanically deformed in many ways. However, even the same mechanical stimulus—such as a membrane stretch—can be detected by ion

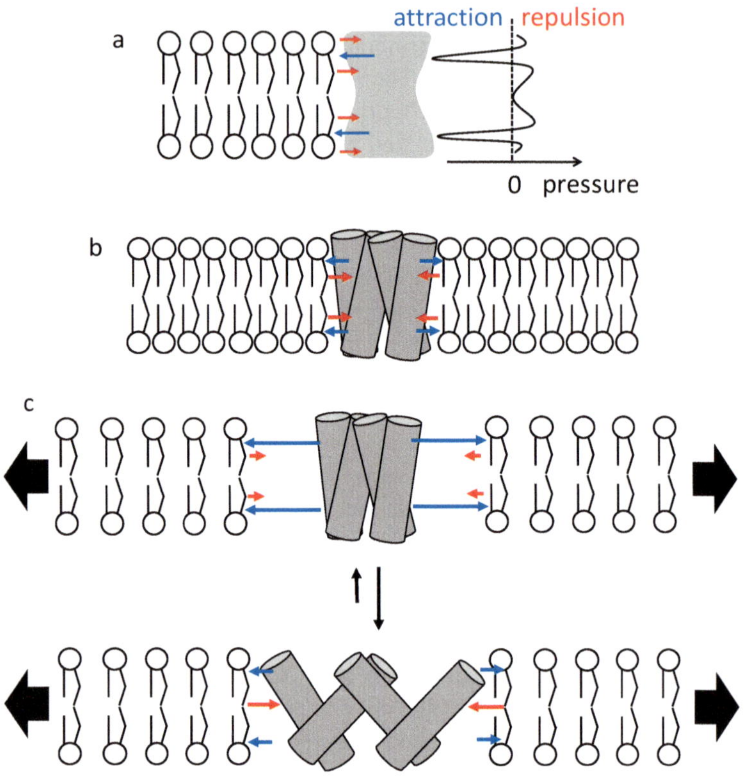

Fig. 3.3 Opening of a mechanosensitive ion channel by membrane stretch. (**a**) Interplay between attractive and repulsive lipid–protein interactions along the normal axis of the bilayer. (**b**) Balance of forces in a protein–lipid system at equilibrium. (**c**) Membrane stretch imposes a larger negative net pressure on compact versus extended conformations.

channels exhibiting strikingly different structures (Fig. 3.4 and Table 3.1). This suggests that different molecular mechanisms are able to sense common mechanical cues.

In prokaryotes, MSCs are mainly formed by mechanosensitive ion channels of small, intermediate, and large conductance (MscS, MscI, and MscL). In eukaryotes, MSCs are found in some members of the Transient Receptor Potential (TRP) and Two-pore-domain K^+ (K_2P) ion channels families and all members of the PIEZO [16], OSCA/TMEM63 [17, 18], DEG/ENaC channels [19], transmembrane channel-like protein (TMC), and LRRC8/SWELL (i.e., Volume Regulated Anion Channels, or VRAC) [20–22] ion channels families. Table 3.1 summarizes mechanical modalities known to stimulate specific MSCs.

Molecular insights into the conformational changes underlying mechanosensitive gating are available for some MSCs. Bacterial MSCs are formed by the assembly of five to seven transmembrane subunits that arrange like the iris of a camera around a large central pore. It has been proposed that stretch and membrane thinning produce a tilt of transmembrane helices surrounding the central pore, effectively supporting an iris-like gating mechanism [44, 45]. K_2P channels are dimeric polymodal MSCs activated by a variety of physicochemical stimuli besides mechanical cues. These stimuli include pH, temperature, and lipids. Mechanical activation of K_2P channels is thought to be mediated by the stretch-dependent rotation of the fourth transmembrane helix, preventing occlusion of the ion permeation pathway by surrounding lipid molecules [46].

PIEZO channels possess a three-bladed propeller structure that creates a local curvature (inverted dome or bowl-like shape) in the lipid bilayer [47–49]. Membrane stretch is expected to flatten the curved propeller domains of PIEZOs, ultimately leading to opening of their pore [50, 51]. Interestingly, this dome-like mechanism produces a large change of the membrane-projected surface area of the channel [51–53]. By imposing a local membrane curvature,

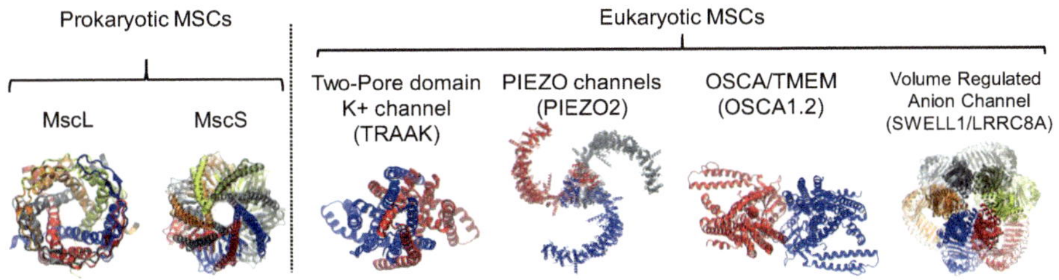

Fig. 3.4 Structural diversity of mechanosensitive ion channels (structures are viewed from above the membrane plane)

PIEZO channels would amplify the ΔA parameter in Eq. (3.3), thus, increasing the intrinsic mechanosensitivity of the protein. PIEZOs are responsible for a large number of force-regulated physiological functions including touch and pain sensation, proprioception, vascular, and neuronal development, blood pressure regulation, blood flow sensing, cell volume regulation, epithelial homeostasis, and bone formation [54–58]. It is, thus, not surprising that PIEZOs respond to virtually every tested mechanical stimulus including membrane stretching, cell poking, fluid shear stress, substrate displacement, hypotonic shocks, vibration, and even ultrasound simulation.

SWELL/LRRC8 are volume-regulated channels and their gating mechanisms remain unclear, as conflicting results have been reported with respect to their sensitivity to lipids or ionic strength [59]. TRP channels are a large family of polymodal ion channels, some of them exhibiting mechanosensitivity. However, recent evidence suggests these isoforms do not obey the force-from lipid paradigm, implying mechanosensitivity of TRPs is mediated by the FFF paradigm [60]. ENaCs are thought to sense shear stress by the flow-induced movement of their extracellular domain although no direct evidence has been shown to support this gating mechanism.

Table 3.1 Effective mechanical stimuli for known mechanosensitive ion channels

MSC family members	Selectivity	Gating paradigm	Effective mechanical stimuli
Bacterial MSCs (MscS, MscL)	No selectivity	FFL	Hypotonic shock [23] Membrane stretch [2] Hydrostatic pressure [24]
K_2P channels (TRAAK and TREK1/2)	Potassium	FFL	Hypotonic shock [25] Fluid shear stress [26] Membrane stretch [25, 26] Indentation [25]
PIEZOs	Cationic	FFL (FFF?)	Hypotonic shock [27–29] Fluid shear stress [6, 30–33] Membrane stretch [16] Hydrostatic pressure [34] Indentation [16] Substrate motion [35]
OSCA/TMEM	Cationic	FFL	Membrane stretch [18] Indentation [18]
SWELL/LRRC8 (VRAC)	Anionic	FFL	Hypotonic shock
TRP	Cationic	FFF	Hypotonic shock [36, 37] Fluid shear stress [38] Substrate motion [39]
ENaC/Degenerin	Sodium	?	Shear stress [40–42]
TMC	Cationic (calcium)	FFF	Hair cell bundle displacement [43]

Gating mechanisms in MSCs

OSCA/TMEM and TMCs are relatively new families of mechanosensitive ion channels and their gating mechanisms are unclear.

3.5 Experimental Methods to Stimulate Mechanosensitive Ion Channels

We shall describe some of the numerous techniques recently developed to study MSCs—starting with the easiest ones and ending with the most challenging ones—to inform readers interested in pursuing these experiments themselves about the level of technical expertise needed for implementation in the lab.

Osmotic shocks: The easiest experiment to induce a mechanical stretch of the plasma membrane is to create an osmotic flow of water directed toward the interior of the cell. Although the concept of osmosis is commonly explained by the idea that water diffuses less freely when they interact with solute molecules, this idea has been challenged many times as more complex chemical theories of osmosis have recently emerged [61]. For practical purposes, creating an osmotic gradient is simply done by immersing cells in a hypotonic solution, that is, with an osmolality lower than that of the cytosol compartment, which is typically around 280–320 mOsmol L^{-1} kg^{-1}. Except for some cells adapted for survival in a hypotonic environment—such as *Xenopus laevis* oocytes, which are normally laid in river streams—most cell membranes are quite permeable to water because they possess specialized transmembrane water-permeable transmembrane proteins called aquaporins. Reducing the concentration of solutes outside the cell hence creates an inwardly directed osmotic flow of water molecules entering the cell through aquaporins, leading the cell to swell. It is naturally difficult to quantify the amount of membrane stretch induced by a hypotonic shock because cells may express a variable amount of aquaporins, affecting the rate of water diffusion into the cell. In addition, the resting tension (i.e., the tension before the osmotic shock) may fluctuate widely due to the presence of invaginations (e.g., caveolae) or evaginations (e.g., membrane blebs). As a net flow of water molecules enter the cell, these membrane domains may provide additional lipid areas to buffer the mechanical stretch of the lipid bilayer. Finally, the presence of a sub-membrane (cortical) actin cytoskeleton provides additional mechanical resistance to deformations of the plasma membrane, which will oppose hydrostatic forces pushing on the cell membrane from the interior of the cell.

Fluid shear stress: A second method consists of immersing cells into a viscous fluid (any physiological saline solution) set in motion. The movement of fluid molecules near the cell surface creates a frictional force (shear stress, τ) that applies a force against the cell surface of area A:

$$\tau = \frac{F}{A} \tag{3.4}$$

When cells are physically attached to a substrate (i.e., the bottom of a petri dish or surrounding cells), a force equal to the shear and opposite to the direction of flow appears at the points of attachment (e.g., focal adhesions), producing mechanical deformations of the cell (shear strain). The mechanical effects induced by shear stress are difficult to model because the cell membrane itself behaves as a two-dimensional fluid. Lateral diffusion of individual lipid molecules increases in the presence of shear stress. In addition, studies have indicated that shear stress tends to reduce lipid order. Reproducible shear stress stimulations are now possible for cells, which can be seeded inside commercial flow chambers and microfluidics systems. For those interested in mimicking physiological flow conditions, most of these commercial systems enable different modes of shear stress stimulation, including laminar, pulsatile, or turbulent. Alternatively, shear stress can be produced by flowing solution in the vicinity of cells of interest using microperfusion systems. Although this second approach is compatible with in situ studies (e.g., brain slices) and with electrophysiological interrogations using patch-clamp methods, it is more difficult to predict the exact amount of shear stress experienced by the cell because it is influenced not only by the rate of flow through the perfusing pipette and the viscosity of the flowing solution but also by the distance and angle of the pipette with respect to the cell. Variants of this technique have recently been used to highlight the functional role of PIEZO1 channels present in tissues experiencing shear stress including the vascular and lymphatic endothelium [30, 32, 33, 62–64].

Substrate displacement: Many tissues experience large-scale mechanical deformations. During development, young tissues constantly experience pulling and pushing forces from the growing embryo. Some tissues continue to experience mechanical stretch after development: the bladder epithelium expands and contracts several times per day while the lung epithelium expands and contracts 12–18 times per minute. To mimic large-scale tissue mechanics, cells of interest can be cultured onto elastic biocompatible materials such as matrix bonded silicone rubber. These materials can be stretched in a precise direction (axial, biaxial, or isotropic) using various commercially available systems.

Pressure-clamp (gigaseal pressurization): This method is one of the two gold standard methods to study mechanosensitive ion channels. Like a conventional patch-clamp electrophysiology experiment, an experimentalist carefully approaches a fire-polished glass micropipette toward the surface of a cell of interest. A gigaohm electrical seal is then created by applying brief episodes of negative air pressure at the backside of the patch pipette while monitoring a voltage-induced current between electrodes placed in the bath and the pipette solution. The area of membrane trapped by the pipette (the "patch") is now electrically isolated from the bath solution. Ion channel-mediated ionic currents can be readily recorded in a cell-attached mode or after the patch has been excised from the rest of the cell membrane (excised patch mode). Mechanical activation of ion channels can be controlled by applying subsequent pulses of positive or negative pressure in the pipette. The changes in air pressure either pull or push the column of fluid directly above the membrane patch. This creates a stretch in either the convex or concave membrane configuration.

Mechanical indentation (cell poking): This is the second gold standard method to study MSCs. In this technique, the cell is mechanically stimulated by a round-shaped probe (usually a blunt fire-polished pipette) of 2–5 μm tip diameter. The speed and extent of the probe displacement are precisely controlled by a commercially available piezoelectric microstage to which the poking probe is physically attached. The angle of the probe with respect to the cell surface at the point of impact is determined by the micromanipulator holding the piezo-electric microstage. In most studies, mechanical indentations are produced incrementally: the minimal probe displacement enabling the probe to touch the cell surface is subtracted to the minimal probe displacement producing mechanically activated ionic current. This subtracted poking distance represents the mechanical threshold for MSC activation and can be compared between different MSCs, cell types, or experimental conditions.

Micropillar arrays: This method is perhaps one of the most innovative method recently developed. Cells are directly grown on an array of elastomeric micropillars [65]. Each individual pillar can be displaced using a micropipette mounted onto a nano-positioning device or micro-manipulator, enabling local mechanical perturbations on desired membrane microdomains such as neuronal processes, filopodia, and so on. A major drawback of this technique is the fact that these elastomeric arrays require advanced fabrication that may be costly, precluding general use of the technique.

Atomic force microscopy (AFM): AFM is a form of scanning probe microscopy that enables the determination of the stiffness, imaging, and mechanical manipulation of a sample. This is accomplished by oscillating a probe (or tip) *via* a piezo-electrically actuated cantilever. This technique is evolving rapidly in the field of mechanobiology, as several groups have reported the use of AFM to mechanically stimulate mechanosensitive channels embedded into a flat lipid bilayer [51] as well as in living cells [66].

Magnetic and acoustic tweezers: Mechanical actuation of ion channels is possible using several noninvasive approaches. In magnetic tweezing, magnetic beads or magnetic nanoparticles are physically attached to the cell or to the channel protein itself [67]. An electromagnetic field can be applied from a permanent or magnet or electromagnet positioned near the cell while the activity of the channel protein is being recorded using patch-clamp electrophysiology or fluorescence imaging.

Acoustic tweezing consists of delivering pulses of ultrasound, typically in the 100 kHz–100 MHz range, to a sample. When sound waves propagate in a homogenous medium, their amplitude progressively attenuates due to frictional forces between molecules as they vibrate around their resting positions. When sound waves reach an interface between two mediums of different acoustic impedance (the acoustic impedance of the medium depends on its density and stiffness), part of the wave is reflected backward, producing an impulse directed forward. The resulting acoustic radiation force, or acoustic pressure, can be tuned to mechanically actuate cells or tissues. However, the acoustic impedance mismatch produced by a single cell may be too small, as the mechanical properties of cells (which are mostly made of water) are not too different from that of their aqueous environment. Hence, to increase the amplitude of the acoustic radiation force, it is possible to attach microbubbles or micron-size polystyrene beads to the cells of interest. These particles produce strong acoustic impedance mismatches and thus are easily stirred by an acoustic beam [23]. Focused ultrasound transducers with higher frequencies (>1MHz) can focus acoustic energy into sub-millimeter spatial regions, and thus, are preferable for molecular and cellular investigations which often require the use of small-size samples [68]. A major trade-off for using high-frequency ultrasound is the more rapid attenuation of the amplitude of the acoustic wave due to increasing molecular friction at higher vibration frequency.

3.6 Computational Approaches to Study Gating Mechanisms in Mechanosensitive Ion Channels

Advances in X-ray crystallography and cryo-electron microscopy (cryo-EM) have driven generations of scientists on a path to understanding the fundamental mechanisms of membrane proteins at the atomic level. However, static protein structures do not explain protein function, which is based on the transitions between functional states. One of today's challenges is to make use of these atomic-level structures using computational methods that can provide new insights on the structure–dynamic–function relationships and generate testable predictions or hypotheses. Molecular dynamics (MD) simulations is a powerful tool for investigating the dynamics of membrane proteins embedded in a solvated membrane bilayer model. Typical MD simulation engines, CHARMM, AMBER, NAMD, GROMACS, OpenMM, LAMMPS, TINKER, and DESMOND, compute the numerical solutions of Newton's equation of motion ($\boldsymbol{F} = m\boldsymbol{a}$) using time steps of few femtoseconds. The forces acting on every atom—actual atoms in all-atom (AA) systems, or pseudo-atoms (beads) in coarse-grained (CG) systems—are calculated from the potential energy function defined by the force field. Because the probability of a certain conformation's occurrence exponentially depends on energetics, small improvements in force field accuracy can lead to considerable improvement in predicting conformation ensembles. It is thus not surprising that decades of efforts have been devoted to improving various versions of force fields, including polarizable force fields.

MD simulations can serve several purposes. The most straightforward use is to generate multiple replicas of the time evolution of protein motions, that is, a short movie (usually up to sub-millisecond) of the microscopic protein movement, if the motion at the available time scale is of interest [69]. The second and most common purpose is to sample Boltzmann-distributed equilibrium configurations to calculate the thermodynamic and kinetic quantities from ensemble averages that control protein function and can be related to experimental observables. The third is to bias the system to follow a process of interest. In all cases, the main challenge is to ensure:

- A minimum model system that allows sufficient sampling without compromising the protein function under investigation. Decisions regarding what types of lipids to use and how many protein domains/subunits need to be included in the model have to be made based on known experimental evidence and the question of interest.
- An accurate description of molecular mechanical force fields that balance interactions between protein and its environment including water, lipids, ions, and small molecules. Extra validation of force fields is usually necessary when the system includes exotic lipid types, multivalence ions, or flexible molecular ions.
- A rigorous statistical mechanical framework within which we can produce meaningful predictions.

Mechanical stimuli in silico: For 30 years, our understanding of direct force transmission from lipids to proteins has been limited to the FFL paradigm. The force-from-lipid is influenced by both (1) the intrinsic mechanical properties of the lipid bilayer and (2) lipid–protein interactions. The former involves understanding the effect of the bilayer mechanical properties—such as bilayer stiffness, compressibility, bending rigidity—on protein energetics, whereas the latter focuses on specific chemical interactions between lipid molecules and protein residues. Both mechanisms have been partially explored using in vitro or cell-based experimental techniques, however, a comprehensive physical description of the FFL paradigm is lacking. This is because experimental measurements of interactions between lipids and proteins are mostly indirect. In addition, many challenges remain for precisely measuring the direction and amplitude of mechanical forces, as described in the previous section of this book chapter. Therefore, experimental measurements and molecular simulations need each other to solve these challenges. Below we describe few case-by-case simulation approaches that have been—or can be—used to mimic various physiological mechanical stimuli. We briefly discuss the pros and cons of each approach.

Mechanical property of membrane bilayer: The mechanical properties of the membrane bilayer are directly related to its lipid composition. Changing the lipid composition experimentally is nontrivial and often affects multiple membrane properties, making interpretations difficult. The beauty of computer modeling is that the lipid composition can be precisely controlled and fine-tuned. At equilibrium, bilayer area compressibility (K_A) and bilayer bending constants (K_C) are critical mechanical properties that determine the membrane's ability to compress, expand, or bend. While there is considerable uncertainty in the experimental measurement of these properties, those intrinsic properties of the membrane can be calculated directly from lipid-only simulations (without protein).

Bilayer area compressibility: The area compressibility K_A quantifies the response of membrane area to tension, which under physiological conditions may arise from various perturbations, such as osmotic swelling, addition or extraction of lipids, or other amphipathic molecules from the bilayer, or changes in surface tension. Assuming the undulations are small so the difference in projected areas and local areas is negligible, the area compressibility K_A can be calculated from [70]:

$$K_{\mathrm{A}} = \frac{k_{\mathrm{B}} T \langle A \rangle}{\langle \delta A^2 \rangle} = A \left(\frac{d\gamma}{dA} \right) \tag{3.5}$$

$$-\frac{2k_{\mathrm{B}}T}{\langle A \rangle} \ln p \left(\frac{\delta A}{\langle A \rangle} \right) = K_{\mathrm{A}} \left(\frac{\delta A}{\langle A \rangle} \right)^2 + C' \tag{3.6}$$

where $\langle A \rangle$ is the average total area of the bilayer, $\langle \delta A^2 \rangle$ is the mean square fluctuation, k_B and T are the Boltzmann's constant and the temperature, γ

is surface tension. According to Eq. (3.5), bilayer K_A can be calculated either using a series of constant tension simulations using NPγT ensemble or using the equilibrium thermal fluctuations of the bilayer at zero tension. Equation (3.6) is an alternative form of Eq. (3.5), which uses the probability distribution of the area change around the mean instead of the mean-square area fluctuations, thus reducing the sensitivity to the outliers and deviations from the elastic regime [71]. When dealing with asymmetric bilayer, the K_A values for each bilayer leaflet are needed. To address this problem, a novel approach was recently developed to estimate K_A based on local thermal fluctuations of the leaflet thickness. In this approach, each leaflet is viewed as a collection of more than one parallel elastic blocks that have the same average area but different instantaneous areas. The interleaflet coupling is shown to be equivalent to $\sigma^2(A)$. The local area fluctuation in Eq. (3.6) is converted to the local thickness fluctuation assuming volume conservation [71].

Bilayer bending rigidity: The classic Helfrich–Canham expression for binding free energy of a symmetric and homogeneous bilayer normal to the z-axis is given by:

$$F_{\text{bend}} = \int_{\text{membrane}} \frac{1}{2} K_C (c_x + c_y)^2 dxdy \quad (3.7)$$

where c_x and c_y are the local values of the membrane curvature along x and y axes defining the bilayer plane. Assuming the membrane behaves as a structureless thin sheet, the bending rigidity K_c can be estimated from the spectral analysis of the bilayer thermal fluctuation during the MD simulations. In this approach, the power spectrum of the local height deviation can be related to $k_B T/(K_c q^4)$ in which q is the wave vector in the limit of small q values. This approach necessitates sampling of bilayer undulations of a relatively large membrane system containing at least 1000 lipids. A modification of this approach is to compute lipid tilt fluctuations instead of local height fluctuations, which can be obtained using a smaller membrane size of ~400 lipids [72].

In addition to the Fourier space methodology described above, a real-space fluctuation analysis has been recently developed that requires an even smaller membrane size and has been tested on inhomogeneous multicomponent lipid bilayers. The bilayer deformation free energy related to the bending can be expressed as the quadratic approximation for the free energies associated with monolayer lipid splay S. Splay can be expressed by the divergence of two local vectors, lipid director vector $\boldsymbol{n}$ and the local vector normal to the lipid–water interface $\boldsymbol{N}$ (Eq. (3.8) and Fig. 3.5). Since the probability of lipid splays can be expressed as Boltzmann distribution, by constructing the distributions of lipid splays in the MD simulations and fitting a quadratic function in Eq. (3.9), K_c is estimated as the coefficients of the quadratic term.

$$-\frac{2k_B T}{A} \ln P(S) = K_C \left(\nabla \vec{n} - \nabla \vec{N} \right)^2 + C'_S \quad (3.8)$$

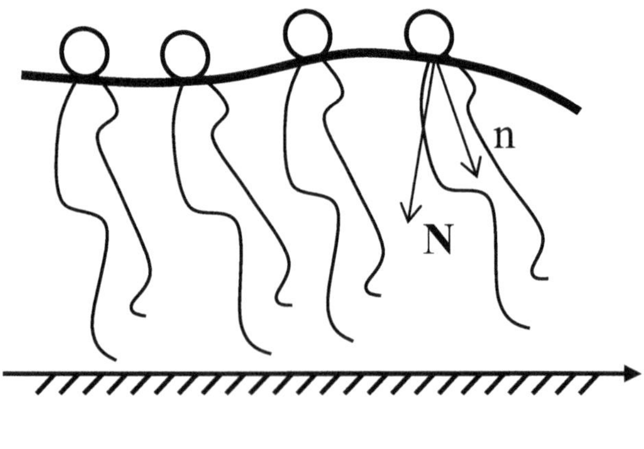

Fig. 3.5 Illustration of lipid director vector (n) and the local vector normal to the lipid–water interface (N)

$$f_{\text{bending}} = \frac{1}{2}K_C\langle S\rangle^2 = \frac{1}{2}K_C\left\langle \nabla\vec{n} - \nabla\vec{N}\right\rangle^2 \quad (3.9)$$

where A represents the area per lipid and C_S is normalized constant [73].

Free energy of membrane bending can also be calculated using enhanced sampling MD simulations. Most of the enhanced sampling methods based on bias potential need predefined coordinates, also called, collective variables, reaction coordinates, or order parameters. Collective variables are defined as a low dimensional function of the atomistic coordinate describing the slow motion in the process of interest. In the membrane deformation process, a collective variable defined from the instantaneous atomic curvature can be used to explore the free energy landscape associated with membrane bending [74]. Recently, a grid-based collective variable was used to calculate the free energy landscape of a predefined membrane deformation [75]. Such approaches have great potential as they are advantageous for studying inhomogeneous and asymmetric bilayers with a length scale below 100 Å.

Challenges in simulating asymmetric bilayers and activation of mechanosensitive channels: A challenge in simulating asymmetric bilayers is due to the use of periodic boundary conditions, in which the area of the top and bottom leaflets is constrained to be identical. Hence, in MD simulations, the change of the leaflet area due to leaflet tension depends on the compressibility modulus of the bilayer, not on the compressibility modulus of each leaflet. To avoid leaflet mismatch, Doktorova et al. proposed that one must first construct each asymmetric bilayer with zero leaflet tension by adjusting the number of lipids in each leaflet to satisfy $\frac{N_t}{N_b} = \frac{N_t'(T_t+K_A)}{N_b'(T_b+K_A)}$, in which T_t and T_b are tension in each leaflet, where subscripts "t" and "b" denote "top" and "bottom" leaflet [71]. N and N' denote the ideal and current number of lipids in each leaflet, respectively. The zero leaflet tension bilayer can then be used to compare the mechanical properties of bilayers. Following this requirement, Jiang et al. recently computed key mechanical properties (bilayer binding rigidity, area compressibility, and surface shear viscosity) of several asymmetric atomistic bilayer models with anionic phosphatidylinositol bisphosphate (PIP) or phosphatidylserine (PS) in the inner leaflet [76]. The energetic cost of the mechanical deformation of membrane, such as the short-ranged hydrophobic mismatch and long-ranged membrane bending, has been found to regulate mechanosensitive channels at the single-channel level and also multi-channel localization and cooperativity. Thus, it is necessary to quantify the key mechanical properties of membrane when simulating mechanosensitive channels in complex bilayer. Mechanical activation of mechanosensitive ion channels can be stimulated via a change in membrane tension, membrane topology, or other mechanical forces. While it is easier to manipulate force in MD simulations than in experiments, caution is needed to avoid simulation artifacts. When large tension has to be applied to accelerate the opening of a channel within simulated timescale, membrane integrity needs to be checked. Other approaches to accelerate the channel opening while maintaining the membrane integrity include manipulating membrane curvature [77], directly applying forces to protein atoms, protein-bound lipids, or distributing the applied force based on the lateral distance between lipids and protein using steered MD [78]. In all cases, the open conformation of the channel generated from such nonequilibrium process need to be validated against experimental data, such as ionic conductance, selectivity, and mutant phenotypes

References

1. Katz B (1950) Depolarization of sensory terminals and the initiation of impulses in the muscle spindle. J Physiol 111:261–282. https://doi.org/10.1113/jphysiol.1950.sp004479. PMC1392822
2. Sukharev SI, Blount P, Martinac B, Blattner FR, Kung C (1994) A large-conductance mechanosensitive channel in *E. coli* encoded by mscL alone. Nature 368:265–268. https://doi.org/10.1038/368265a0
3. Morimatsu M, Mekhdjian AH, Chang AC, Tan SJ, Dunn AR (2015) Visualizing the interior architecture of focal adhesions with high-resolution traction maps.

Nano Lett 15:2220–2228. https://doi.org/10.1021/nl5047335
4. Morimatsu M, Mekhdjian AH, Adhikari AS, Dunn AR (2013) Molecular tension sensors report forces generated by single integrin molecules in living cells. Nano Lett 13:3985–3989. https://doi.org/10.1021/nl4005145. PMC3815579
5. Ellefsen KL, Holt JR, Chang AC, Nourse JL, Arulmoli J, Mekhdjian AH, Abuwarda H, Tombola F, Flanagan LA, Dunn AR, Parker I, Pathak MM (2019) Myosin-II mediated traction forces evoke localized Piezo1-dependent Ca(2+) flickers. Commun Biol 2:298. https://doi.org/10.1038/s42003-019-0514-3. PMC6685976
6. Xu J, Mathur J, Vessieres E, Hammack S, Nonomura K, Favre J, Grimaud L, Petrus M, Francisco A, Li J, Lee V, Xiang FL, Mainquist JK, Cahalan SM, Orth AP, Walker JR, Ma S, Lukacs V, Bordone L, Bandell M, Laffitte B, Xu Y, Chien S, Henrion D, Patapoutian A (2018) GPR68 Senses flow and is essential for vascular physiology. Cell 173(762–775):e716. https://doi.org/10.1016/j.cell.2018.03.076. PMC5951615
7. Dela Paz NG, Melchior B, Frangos JA (2017) Shear stress induces Galphaq/11 activation independently of G protein-coupled receptor activation in endothelial cells. Am J Physiol Cell Physiol 312:C428–C437. https://doi.org/10.1152/ajpcell.00148.2016. PMC5407018
8. Chachisvilis M, Zhang YL, Frangos JA (2006) G protein-coupled receptors sense fluid shear stress in endothelial cells. Proc Natl Acad Sci U S A 103: 15463–15468. https://doi.org/10.1073/pnas.0607224103. PMC1622845
9. Gu CX, Juranka PF, Morris CE (2001) Stretch-activation and stretch-inactivation of Shaker-IR, a voltage-gated K+ channel. Biophys J 80:2678–2693. https://doi.org/10.1016/S0006-3495(01)76237-6. PMC1301455
10. Laitko U, Juranka PF, Morris CE (2006) Membrane stretch slows the concerted step prior to opening in a Kv channel. J Gen Physiol 127:687–701. https://doi.org/10.1085/jgp.200509394. PMC2151533
11. Lin W, Laitko U, Juranka PF, Morris CE (2007) Dual stretch responses of mHCN2 pacemaker channels: accelerated activation, accelerated deactivation. Biophys J 92:1559–1572. https://doi.org/10.1529/biophysj.106.092478. PMC1796836
12. Bavi N, Cox CD, Perozo E, Martinac B (2017) Toward a structural blueprint for bilayer-mediated channel mechanosensitivity. Channels (Austin) 11:91–93. https://doi.org/10.1080/19336950.2016.1224624. PMC5398584
13. Perozo E, Kloda A, Cortes DM, Martinac B (2002) Physical principles underlying the transduction of bilayer deformation forces during mechanosensitive channel gating. Nat Struct Biol 9:696–703. https://doi.org/10.1038/nsb827
14. Cox CD, Bavi N, Martinac B (2019) Biophysical principles of ion-channel-mediated mechanosensory transduction. Cell Rep 29:1–12. https://doi.org/10.1016/j.celrep.2019.08.075
15. Martinac B, Bavi N, Ridone P, Nikolaev YA, Martinac AD, Nakayama Y, Rohde PR, Bavi O (2018) Tuning ion channel mechanosensitivity by asymmetry of the transbilayer pressure profile. Biophys Rev 10:1377–1384. https://doi.org/10.1007/s12551-018-0450-3. PMC6233343
16. Coste B, Mathur J, Schmidt M, Earley TJ, Ranade S, Petrus MJ, Dubin AE, Patapoutian A (2010) Piezo1 and Piezo2 are essential components of distinct mechanically activated cation channels. Science 330: 55–60. https://doi.org/10.1126/science.1193270. PMC3062430
17. Jojoa-Cruz S, Saotome K, Murthy SE, Tsui CCA, Sansom MS, Patapoutian A, Ward AB (2018) Cryo-EM structure of the mechanically activated ion channel OSCA1.2. Elife 7:e41845. https://doi.org/10.7554/eLife.41845. PMC6235563
18. Murthy SE, Dubin AE, Whitwam T, Jojoa-Cruz S, Cahalan SM, Mousavi SAR, Ward AB, Patapoutian A (2018) OSCA/TMEM63 are an evolutionarily conserved family of mechanically activated ion channels. Elife 7:e41844. https://doi.org/10.7554/eLife.41844. PMC6235560
19. Geffeney SL, Cueva JG, Glauser DA, Doll JC, Lee TH, Montoya M, Karania S, Garakani AM, Pruitt BL, Goodman MB (2011) DEG/ENaC but not TRP channels are the major mechanoelectrical transduction channels in a *C. elegans* nociceptor. Neuron 71:845–857. https://doi.org/10.1016/j.neuron.2011.06.038. PMC3170654
20. Kefauver JM, Saotome K, Dubin AE, Pallesen J, Cottrell CA, Cahalan SM, Qiu Z, Hong G, Crowley CS, Whitwam T, Lee WH, Ward AB, Patapoutian A (2018) Structure of the human volume regulated anion channel. Elife 7:e38461. https://doi.org/10.7554/eLife.38461. PMC6086657
21. Syeda R, Qiu Z, Dubin AE, Murthy SE, Florendo MN, Mason DE, Mathur J, Cahalan SM, Peters EC, Montal M, Patapoutian A (2016) LRRC8 Proteins form volume-regulated anion channels that sense ionic strength. Cell 164:499–511. https://doi.org/10.1016/j.cell.2015.12.031. PMC4733249
22. Qiu Z, Dubin AE, Mathur J, Tu B, Reddy K, Miraglia LJ, Reinhardt J, Orth AP, Patapoutian A (2014) SWELL1, a plasma membrane protein, is an essential component of volume-regulated anion channel. Cell 157:447–458. https://doi.org/10.1016/j.cell.2014.03.024. PMC4023864
23. Heureaux J, Chen D, Murray VL, Deng CX, Liu AP (2014) Activation of a bacterial mechanosensitive channel in mammalian cells by cytoskeletal stress. Cell Mol Bioeng 7:307–319. https://doi.org/10.1007/s12195-014-0337-8. PMC4297646
24. Petrov E, Rohde PR, Martinac B (2011) Flying-patch patch-clamp study of G22E-MscL mutant under high hydrostatic pressure. Biophys J 100:1635–1641. https://doi.org/10.1016/j.bpj.2011.02.016. PMC3072613

25. Del Marmol J, Rietmeijer RA, Brohawn SG (2018) Studying mechanosensitivity of two-pore domain K(+) channels in cellular and reconstituted proteoliposome membranes. Methods Mol Biol 1684:129–150. https://doi.org/10.1007/978-1-4939-7362-0_11. PMC6202064
26. Patel AJ, Honore E, Maingret F, Lesage F, Fink M, Duprat F, Lazdunski M (1998) A mammalian two pore domain mechano-gated S-like K+ channel. EMBO J 17:4283–4290. https://doi.org/10.1093/emboj/17.15.4283. PMC1170762
27. Lacroix JJ, Botello-Smith WM, Luo Y (2018) Probing the gating mechanism of the mechanosensitive channel Piezo1 with the small molecule Yoda1. Nat Commun 9:2029. https://doi.org/10.1038/s41467-018-04405-3
28. Syeda R, Florendo MN, Cox CD, Kefauver JM, Santos JS, Martinac B, Patapoutian A (2016) Piezo1 channels are inherently mechanosensitive. Cell Rep 17:1739–1746. https://doi.org/10.1016/j.celrep.2016.10.033. PMC5129625
29. Gottlieb PA, Bae C, Sachs F (2012) Gating the mechanical channel Piezo1: a comparison between whole-cell and patch recording. Channels (Austin) 6: 282–289. https://doi.org/10.4161/chan.21064. PMC3508907
30. Jetta D, Gottlieb PA, Verma D, Sachs F, Hua SZ (2019) Shear stress-induced nuclear shrinkage through activation of Piezo1 channels in epithelial cells. J Cell Sci 132:jcs226076. https://doi.org/10.1242/jcs.226076
31. Maneshi MM, Ziegler L, Sachs F, Hua SZ, Gottlieb PA (2018) Enantiomeric Abeta peptides inhibit the fluid shear stress response of PIEZO1. Sci Rep 8: 14267. https://doi.org/10.1038/s41598-018-32572-2. PMC6155315
32. Wang S, Chennupati R, Kaur H, Iring A, Wettschureck N, Offermanns S Endothelial cation channel PIEZO1 controls blood pressure by mediating flow-induced ATP release. J Clin Invest 126:4527–4536. https://doi.org/10.1172/JCI87343. PMC5127677
33. Ranade SS, Qiu Z, Woo SH, Hur SS, Murthy SE, Cahalan SM, Xu J, Mathur J, Bandell M, Coste B, Li YS, Chien S, Patapoutian A (2014) Piezo1, a mechanically activated ion channel, is required for vascular development in mice. Proc Natl Acad Sci U S A 111: 10347–10352. https://doi.org/10.1073/pnas.1409233111. PMC4104881
34. Luo M, Ho KKY, Tong Z, Deng L, Liu AP (2019) Compressive stress enhances invasive phenotype of cancer cells via piezo1 activation. BioRxiv. https://doi.org/10.1101/513218
35. Servin-Vences MR, Moroni M, Lewin GR, Poole K (2017) Direct measurement of TRPV4 and PIEZO1 activity reveals multiple mechanotransduction pathways in chondrocytes. Elife 6. https://doi.org/10.7554/eLife.21074. PMC5279942
36. Liedtke W, Choe Y, Marti-Renom MA, Bell AM, Denis CS, Sali A, Hudspeth AJ, Friedman JM, Heller S (2000) Vanilloid receptor-related osmotically activated channel (VR-OAC), a candidate vertebrate osmoreceptor. Cell 103:525–535. https://doi.org/10.1016/s0092-8674(00)00143-4. PMC2211528
37. Strotmann R, Harteneck C, Nunnenmacher K, Schultz G, Plant TD (2000) OTRPC4, a nonselective cation channel that confers sensitivity to extracellular osmolarity. Nat Cell Biol 2:695–702. https://doi.org/10.1038/35036318
38. Gao X, Wu L, O'Neil RG (2003) Temperature-modulated diversity of TRPV4 channel gating: activation by physical stresses and phorbol ester derivatives through protein kinase C-dependent and -independent pathways. J Biol Chem 278:27129–27137. https://doi.org/10.1074/jbc.M302517200
39. Mochizuki T, Sokabe T, Araki I, Fujishita K, Shibasaki K, Uchida K, Naruse K, Koizumi S, Takeda M, Tominaga M (2009) The TRPV4 cation channel mediates stretch-evoked Ca2+ influx and ATP release in primary urothelial cell cultures. J Biol Chem 284:21257–21264. https://doi.org/10.1074/jbc.M109.020206. PMC2755849
40. Shi S, Carattino MD, Hughey RP, Kleyman TR (2013) ENaC regulation by proteases and shear stress. Curr Mol Pharmacol 6:28–34. https://doi.org/10.2174/18744672112059990027. PMC3697921
41. Satlin LM, Sheng S, Woda CB, Kleyman TR (2001) Epithelial Na(+) channels are regulated by flow. Am J Physiol Renal Physiol 280:F1010–F1018. https://doi.org/10.1152/ajprenal.2001.280.6.F1010
42. Carattino MD, Sheng S, Kleyman TR (2004) Epithelial Na+ channels are activated by laminar shear stress. J Biol Chem 279:4120–4126. https://doi.org/10.1074/jbc.M311783200
43. Jia Y, Zhao Y, Kusakizako T, Wang Y, Pan C, Zhang Y, Nureki O, Hattori M, Yan Z (2019) TMC1 and TMC2 proteins are pore-forming subunits of mechanosensitive ion channels. Neuron. https://doi.org/10.1016/j.neuron.2019.10.017
44. Yefimov S, van der Giessen E, Onck PR, Marrink SJ (2008) Mechanosensitive membrane channels in action. Biophys J 94:2994–3002. https://doi.org/10.1529/biophysj.107.119966. PMC2275678
45. Betanzos M, Chiang CS, Guy HR, Sukharev S (2002) A large iris-like expansion of a mechanosensitive channel protein induced by membrane tension. Nat Struct Biol 9:704–710. https://doi.org/10.1038/nsb828
46. Brohawn SG, Campbell EB, MacKinnon R (2014) Physical mechanism for gating and mechanosensitivity of the human TRAAK K+ channel. Nature 516:126–130. https://doi.org/10.1038/nature14013. PMC4682367
47. Wang L, Zhou H, Zhang M, Liu W, Deng T, Zhao Q, Li Y, Lei J, Li X, Xiao B (2019) Structure and mechanogating of the mammalian tactile channel PIEZO2. Nature 573:225–229. https://doi.org/10.1038/s41586-019-1505-8
48. Zhao Q, Zhou H, Chi S, Wang Y, Wang J, Geng J, Wu K, Liu W, Zhang T, Dong M-Q, Wang J, Li X, Xiao B (2018) Structure and mechanogating

mechanism of the Piezo1 channel. Nature. https://doi.org/10.1038/nature25743
49. Saotome K, Murthy SE, Kefauver JM, Whitwam T, Patapoutian A, Ward AB (2018) Structure of the mechanically activated ion channel Piezo1. Nature 554:481–486. https://doi.org/10.1038/nature25453. PMC6010196
50. Botello-Smith WM, Jiang W, Zhang H, Ozkan AD, Lin YC, Pham CN, Lacroix JJ, Luo Y (2019) A mechanism for the activation of the mechanosensitive Piezo1 channel by the small molecule Yoda1. Nat Commun 10:4503. https://doi.org/10.1038/s41467-019-12501-1. PMC6776524
51. Lin YC, Guo YR, Miyagi A, Levring J, MacKinnon R, Scheuring S (2019) Force-induced conformational changes in PIEZO1. Nature 573:230–234. https://doi.org/10.1038/s41586-019-1499-2
52. Guo YR, MacKinnon R (2017) Structure-based membrane dome mechanism for Piezo mechanosensitivity. Elife 6. https://doi.org/10.7554/eLife.33660
53. Haselwandter CA, MacKinnon R (2018) Piezo's membrane footprint and its contribution to mechanosensitivity. Elife 7. https://doi.org/10.7554/eLife.41968. PMC6317911
54. Zhao Q, Zhou H, Li X, Xiao B (2019) The mechanosensitive Piezo1 channel: a three-bladed propeller-like structure and a lever-like mechanogating mechanism. FEBS J 286:2461–2470. https://doi.org/10.1111/febs.14711
55. Wu J, Lewis AH, Grandl J (2017) Touch, tension, and transduction: the function and regulation of Piezo ion channels. Trends Biochem Sci 42:57–71. https://doi.org/10.1016/j.tibs.2016.09.004
56. Murthy SE, Dubin AE, Patapoutian A (2017) Piezos thrive under pressure: mechanically activated ion channels in health and disease. Nat Rev Mol Cell Biol. https://doi.org/10.1038/nrm.2017.92
57. Geng J, Zhao Q, Zhang T, Xiao B (2017) In touch with the mechanosensitive Piezo channels: structure, ion permeation, and mechanotransduction. Curr Top Membr 79:159–195. https://doi.org/10.1016/bs.ctm.2016.11.006
58. Douguet D, Honore E (2019) Mammalian mechanoelectrical transduction: structure and function of force-gated ion channels. Cell 179:340–354. https://doi.org/10.1016/j.cell.2019.08.049
59. Konig B, Hao Y, Schwartz S, Plested AJ, Stauber T A FRET sensor of C-terminal movement reveals VRAC activation by plasma membrane DAG signaling rather than ionic strength. Elife 8. https://doi.org/10.7554/eLife.45421. PMC6597245
60. Nikolaev YA, Cox CD, Ridone P, Rohde PR, Cordero-Morales JF, Vasquez V, Laver DR, Martinac B (2019) Mammalian TRP ion channels are insensitive to membrane stretch. J Cell Sci 132. https://doi.org/10.1242/jcs.238360
61. Borg FG (2003) What is osmosis? Explanation and understanding of a physical phenomenon. arXiv
62. Choi D, Park E, Jung E, Cha B, Lee S, Yu J, Kim PM, Lee S, Hong YJ, Koh CJ, Cho CW, Wu Y, Li Jeon N, Wong AK, Shin L, Kumar SR, Bermejo-Moreno I, Srinivasan RS, Cho IT, Hong YK Piezo1 incorporates mechanical force signals into the genetic program that governs lymphatic valve development and maintenance. JCI Insight 4. https://doi.org/10.1172/jci.insight.125068. PMC6483520
63. Nonomura K, Lukacs V, Sweet DT, Goddard LM, Kanie A, Whitwam T, Ranade SS, Fujimori T, Kahn ML, Patapoutian A (2018) Mechanically activated ion channel PIEZO1 is required for lymphatic valve formation. Proc Natl Acad Sci U S A. https://doi.org/10.1073/pnas.1817070115
64. Li J, Hou B, Tumova S, Muraki K, Bruns A, Ludlow MJ, Sedo A, Hyman AJ, McKeown L, Young RS, Yuldasheva NY, Majeed Y, Wilson LA, Rode B, Bailey MA, Kim HR, Fu Z, Carter DA, Bilton J, Imrie H, Ajuh P, Dear TN, Cubbon RM, Kearney MT, Prasad KR, Evans PC, Ainscough JF, Beech DJ (2014) Piezo1 integration of vascular architecture with physiological force. Nature 515:279–282. https://doi.org/10.1038/nature13701. PMC4230887
65. Poole K, Herget R, Lapatsina L, Ngo HD, Lewin GR (2014) Tuning Piezo ion channels to detect molecular-scale movements relevant for fine touch. Nat Commun 5:3520. https://doi.org/10.1038/ncomms4520. PMC3973071
66. Gaub BM, Muller DJ (2017) Mechanical stimulation of Piezo1 receptors depends on extracellular matrix proteins and directionality of force. Nano Lett. https://doi.org/10.1021/acs.nanolett.7b00177
67. Wu J, Goyal R, Grandl J Localized force application reveals mechanically sensitive domains of Piezo1. Nat Commun 7:12939. https://doi.org/10.1038/ncomms12939. PMC 5063965
68. Lacroix JJ, Ozkan AD (2019) Multiplexing focused ultrasound stimulation with fluorescence microscopy. J Vis Exp. https://doi.org/10.3791/58781
69. McNulty R, Ulmschneider JP, Luecke H, Ulmschneider MB Mechanisms of molecular transport through the urea channel of *Helicobacter pylori*. Nat Commun 4:2900. https://doi.org/10.1038/ncomms3900
70. Feller SE, Pastor RW (1999) Constant surface tension simulations of lipid bilayers: The sensitivity of surface areas and compressibilities. J Chem Phys 111:1281–1287. https://doi.org/10.1063/1.479313
71. Doktorova M, LeVine MV, Khelashvili G, Weinstein H (2019) A new computational method for membrane compressibility: bilayer mechanical thickness revisited. Biophys J 116:487–502. https://doi.org/10.1016/j.bpj.2018.12.016. PMC6369663
72. Venable RM, Brown FLH, Pastor RW (2015) Mechanical properties of lipid bilayers from molecular dynamics simulation. Chem Phys Lipids 192:60–74. https://doi.org/10.1016/j.chemphyslip.2015.07.014. PMC4684433

73. Doktorova M, Harries D, Khelashvili G (2017) Determination of bending rigidity and tilt modulus of lipid membranes from real-space fluctuation analysis of molecular dynamics simulations. Phys Chem Chem Phys 19:16806–16818. https://doi.org/10.1039/c7cp01921a. PMC5538590
74. Bouvier B (2019) Curvature as a collective coordinate in enhanced sampling membrane simulations. J Chem Theory Comput 15:6551–6561. https://doi.org/10.1021/acs.jctc.9b00716
75. Fiorin G, Marinelli F, Faraldo-Gomez JD (2019) Direct derivation of free energies of membrane deformation and other solvent density variations from enhanced sampling molecular dynamics. J Comput Chem. https://doi.org/10.1002/jcc.26075
76. Jiang W, Lin YC, Luo YL (2021) Mechanical properties of anionic asymmetric bilayers from atomistic simulations. J Chem Phys 154(22):224701. https://doi.org/10.1063/5.0048232
77. Jiang W, Del Rosario JS, Botello-Smith W, Zhao S, Lin YC, Zhang H, Lacroix J, Rohacs T, Luo YL (2021) Crowding-induced opening of the mechanosensitive Piezo1 channel in silico. Commun Biol 4(1):84. https://doi.org/10.1038/s42003-020-01600-1
78. Rajeshwar TR, Anishkin A, Sukharev S, Vanegas JM (2021) Mechanical activation of MscL revealed by a locally distributed tension molecular dynamics approach. Biophys J 120(2):232-242. https://doi.org/10.1016/j.bpj.2020.11.2274

The Polysite Pharmacology of TREK K_{2P} Channels

4

Lianne Pope and Daniel L. Minor Jr

Abstract

K_{2P} (KCNK) potassium channels form "background" or "leak" currents that have critical roles in cell excitability control in the brain, cardiovascular system, and somatosensory neurons. Similar to many ion channel families, studies of K_{2P}s have been limited by poor pharmacology. Of six K_{2P} subfamilies, the thermo- and mechanosensitive TREK subfamily comprising K_{2P}2.1 (TREK-1), K_{2P}4.1 (TRAAK), and K_{2P}10.1 (TREK-2) are the first to have structures determined for each subfamily member. These structural studies have revealed key architectural features that underlie K_{2P} function and have uncovered sites residing at every level of the channel structure with respect to the membrane where small molecules or lipids can control channel function. This polysite pharmacology within a relatively small (~70 kDa) ion channel comprises four structurally defined modulator binding sites that occur above (Keystone inhibitor site), at the level of (K_{2P} modulator pocket), and below (Fenestration and Modulatory lipid sites) the C-type selectivity filter gate that is at the heart of K_{2P} function. Uncovering this rich structural landscape provides the framework for understanding and developing subtype-selective modulators to probe K_{2P} function that may provide leads for drugs for anesthesia, pain, arrhythmia, ischemia, and migraine.

Keywords

K_{2P} channel · TREK subfamily · Ruthenium red · ML335/ML402 · Fluoxetine · PIP_2

4.1 Introduction

Ion channel proteins facilitate the flow of bioelectricity that underlies the physiology of thought, movement, mood, and sensation [1]. The K_{2P} (KCNK) potassium channel family comprises a set of 15 members (Fig. 4.1a) of the voltage-gated ion channel (VGIC) superfamily [2] that have central roles in controlling cell excitability by

L. Pope
Cardiovascular Research Institute, University of California San Francisco, CA, US

D. L. Minor (✉)
Cardiovascular Research Institute, University of California San Francisco, CA, US

Departments of Biochemistry and Biophysics, and Cellular and Molecular Pharmacology, University of California, San Francisco, CA, USA

California Institute for Quantitative Biomedical Research, University of California, San Francisco, CA, USA

Kavli Institute for Fundamental Neuroscience, University of California, San Francisco, CA, USA

Molecular Biophysics and Integrated Bio-imaging Division, Lawrence Berkeley National Laboratory, Berkeley, CA, USA
e-mail: daniel.minor@ucsf.edu

L. Zhou (ed.), *Ion Channels in Biophysics and Physiology*, Advances in Experimental Medicine and Biology 1349, https://doi.org/10.1007/978-981-16-4254-8_4

producing "leak" potassium currents that are largely time and voltage-independent [3–5]. The 15 K_{2P} subtypes comprise six subfamilies (TREK, TWIK, TASK, TALK, THIK, and TRESK) that within each subfamily are related by sequence similarities and regulation by shared types of signals [6] (Fig. 4.1a). A diverse range of stimuli affect K_{2P}s depending on the subfamily type and include physical gating commands such as pressure and temperature (TREK) [4, 7], external and internal pH (TREK, TASK, TALK, and TWIK) [4, 8], chemicals such as volatile anesthetics (TREK, TASK, THIK, TRESK) and antidepressants (TREK) [4, 9], lipids and polyunsaturated fatty acids (PUFAs) (TREK, THIK) [4, 9], and protein–protein interactions with partners such as 14–3-3, G-proteins, Protein Kinase A, and Protein Kinase C (TREK, TASK, TALK, TWIK, TRESK) [4, 9].

K_{2P}s are named for their unique architecture. Each subunit bears two pore-forming domains, PD1 and PD2, each comprising two transmembrane helices (M1–M2 and M3–M4) bridged by a pore helix (P1 and P2) and selectivity filter (SF1 and SF2) (Fig. 4.1b). K_{2P} subunits dimerize to create a channel in which the pore is intrinsically heterotetrameric due to sequence differences between PD1 and PD2. Structures have been determined for exemplars of five of the fifteen K_{2P}s subtypes (Fig. 4.1a), K_{2P}1.1 (TWIK-1), [10] K_{2P}2.1 (TREK-1), [11] K_{2P}3.1 (TASK-1), [12] K_{2P}4.1 (TRAAK), [13] and K_{2P}10.1 (TREK-2) [14] (Fig. 4.1c, d). These structures reveal a common protein scaffold that defines the K_{2P} family.

Similar to other VGICs [15, 16], the two transmembrane segments of the pore domains form outer (M1 and M3) and inner (M2 and M4) helices that define the pore and support the pore helices and selectivity filter [10–14]. The pore helices and selectivity filter coordinate a set of four potassium ions on the channel central axis. K_{2P}s have a unique structural feature, the CAP domain (Fig. 4.1b, c). This extracellular structural element forms an arch directly over the channel pore and creates the bifurcated extracellular ion pathway (EIP) from which the ions exit the channel after passing through the selectivity filter [10, 13]. The M1 helix is domain-swapped between the two subunits, but how this structural intertwining of the subunits impacts function, assembly, or biogenesis is not clear. Each subunit also bears sequences at both the N- and C-termini that are likely to be unstructured on their own but that provide sites for protein–protein interactions that impact function [4, 9].

Unlike other potassium channels, the K_{2P} principal gate is the selectivity filter "C-type" gate [11, 17–22]. In line with this mechanism of control, the structures of K_{2P}1.1 (TWIK-1), [10] K_{2P}2.1 (TREK-1) [11], K_{2P}4.1 (TRAAK), [13] and K_{2P}10.1 (TREK-2) [14] show that these channels lack the inner gate that is present in most other VGIC superfamily members (Fig. 4.1d). The K_{2P} structures have shown that the M4 helix is mobile and can adopt conformations that range between an "up" state and a "down" state [13, 14, 23–25]. The "down" state creates a fenestration just below the P2 pore helix that is open to the center of the membrane bilayer [10, 13, 24, 25]. These M4 conformational changes are linked to C-type gate control [17, 18, 25–27] but do not impede access from the intracellular side [26]. Intriguingly, the recent K_{2P}3.1 (TASK-1) [12] structure shows a "down" state in which M4 creates an intracellular barrier, termed the "X-gate" that appears to be a special feature of the TASK subfamily [12] and that highlights structural diversification of the M4 segment within the K_{2P} family.

K_{2P}s have roles in a multitude of physiological responses and pathological conditions such as action potential propagation [28, 29], anesthetic responses [30, 31], microglial surveillance [32], sleep duration [33], pain, [34–36] arrhythmia [37], ischemia [30, 38, 39], cardiac fibrosis [40], depression [41], migraine [42], intraocular pressure regulation [43], pulmonary hypertension [44], acute respiratory distress syndrome [45], and cancer [46]. Despite these clear physiological roles, the pharmacology of K_{2P}s is generally poor [9, 47] and has been a barrier to understanding K_{2P} function. The paucity of reagents to probe K_{2P} activity has motivated multiple efforts that have begun to define new K_{2P} modulators [11, 36,

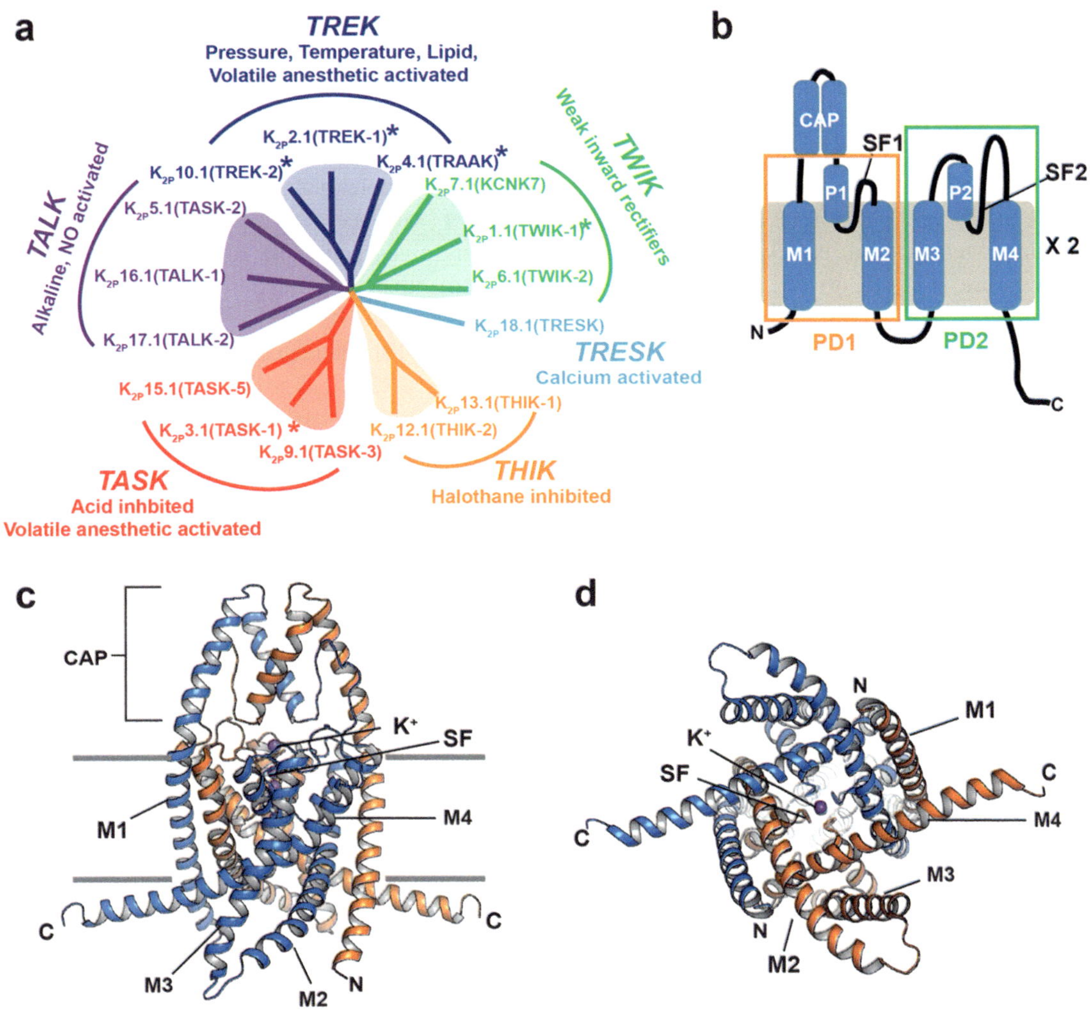

Fig. 4.1 K_{2P} channel relationships and architecture (**a**) K_{2P} channel dendrogram. Subfamilies are indicated. Asterisks indicate structurally characterized K_{2P}s. (**b**) K_{2P} subunit diagram. Pore domains 1 and 2 (PD1 and PD2), transmembrane helices (M1–M4), pore helices (P1 and P2), selectivity filters (SF1 and SF2), and CAP domain are indicated. (**c** and **d**) Cartoon diagram of the K_{2P}2.1 (TREK-1) structure (PDB:6CQ6) [11]. Chains are colored marine and orange. Potassium ions are purple. (**c**), side view, (**d**), cytoplasmic view. Channel elements are labeled as in "b"

48–52] and key structural aspects of K_{2P} channel pharmacology [11, 14, 53, 54].

K_{2P}s are thought to be potential therapeutic targets for pain [47, 55, 56], anesthesia [9, 47], arrhythmia [57, 58], ischemia [59], depression [60], and migraine [9, 61]. Although there are no approved drugs that target K_{2P}s specifically, recent advances in the discovery of new classes of a variety of K_{2P} modulators should enable elaboration of a suite of new pharmacological tools directed at this channel family [11, 12, 36, 48, 49, 51, 62–64].

4.2 The TREK Subfamily: Model Polymodal Ion Channels

The TREK subfamily comprising K_{2P}2.1 (TREK-1), K_{2P}10.1 (TREK-2), and K_{2P}4.1 (TRAAK) is both the most extensively studied K_{2P} family and the only subfamily for which structures of each subtype are known [11, 13, 14] (Fig. 4.1a). TREK channels are polymodal ion channels that respond to diverse physical and chemical gating cues including temperature, pressure, pH, and modulatory lipids [7, 65]. The sensors for these signals reside in different parts of the channel. The intracellular C-terminal tail is key to modulation by temperature [17, 18, 66–68], pressure [18, 69, 70], intracellular pH [70, 71], responses to lipids such as phosphoinositol [4, 5] bis-phosphate (PIP_2) [68, 72, 73], and control by phosphorylation [17, 74, 75]. The sensor for extracellular pH is a histidine [19, 76] located in the loop that connects the P1 helix to the CAP domain [11]. Gating cues from the extracellular pH sensor [19, 76] and the C-terminal tail [17, 18] converge on the selectivity filter C-type gate and make this channel element the nexus of signal integration and functional control [11, 17, 18, 21, 22].

TREK subfamily channels are found throughout the central and peripheral nervous system [4, 28, 29, 34, 47, 77], the eye [43], and the heart [58]. Since their discovery, TREK subfamily channels have been implicated as therapeutic targets for pain, ischemia, and depression [47, 60, 65, 78]. $K_{2P}2.1$ (TREK-1) and $K_{2P}10.1$ (TREK-2) share ~65% sequence similarity, whereas $K_{2P}2.1$ (TREK-1) and $K_{2P}4.1$ (TRAAK) share only ~40% similarity [79–81]. Most of this sequence divergence is embedded in the N- and C-terminal cytoplasmic regions. In line with the high degree of conservation in the core elements of the channel, the structures of homodimers of each of the TREK subfamily members are similar to each other [11]. The three subtypes also have been shown to heterodimerize and provide further functional diversity from this subfamily [42, 82–84]. An important consequence of heterodimerization is that it creates a channel in which the PD1 and PD2 domains are all different from each other. How these differences manifest in functional diversification remains to be defined. Understanding how the similarities and differences within the TREK subfamily contribute to function is of critical importance for developing pharmacological tools and potential therapeutics targeted toward this complex subfamily.

Both activation and inhibition of TREK channels have proposed therapeutic benefits. The ability of TREK channels to stabilize the membrane potential and reduce excitability together with their high expression in sensory neurons gives activators of this K_{2P} subfamily the potential to function as analgesics or anesthetics [35, 47, 85] and as agents against migraine. [42] Interestingly, the original studies of $K_{2P}2.1$ (TREK-1) knockout mice indicated that inhibition of this channel might have a role in mitigating depression [41]. Pharmacologically relevant antipsychotics have also been reported to inhibit $K_{2P}2.1$ (TREK-1) [86]. Consequently, there has been an effort to explore $K_{2P}2.1$ (TREK-1) inhibitors, such as the peptide spadin [87, 88] and small-molecule norfluoxetine [60] as new directions for treating depression. Inhibiting $K_{2P}2.1$ (TREK-1) with such agents would facilitate membrane depolarization, but how such effects could result in modulation of the mental disease remains unclear and harder to understand than the links between TREK channel activation and pain suppression.

Because TREK subfamily channels are readily studied in a variety of experimental systems from potassium transport deficient yeast [48] to *Xenopus* oocytes [11, 22, 52, 54, 75, 80], to transfected mammalian cells [11, 22, 75, 80, 81, 89], to reconstitution assays using purified channels [13, 27, 51, 90], this K_{2P} subfamily has been a key model for understanding K_{2P} channel biophysics in general and is leading the way in demonstrating the potential of K_{2P}s as pharmacological targets [47].

4.3 The Polysite Pharmacology of TREK Channels

Structural studies of TREK K_{2P}s have revealed a strikingly rich structural landscape for functional control, especially given their modest size (~70 kDa). Binding sites for small molecules are found at every layer of the protein starting from its extracellular side through the portion that interacts with the membrane bilayer inner leaflet (Fig. 4.2). This polysite pharmacology comprises four defined binding sites for small molecules or lipids: the Keystone inhibitor site [54], the K_{2P} modulator pocket, [11] the Fenestration site [14, 53], and the Modulatory lipid site [11]. Each offers a distinct structural environment and mechanism for controlling K_{2P} function.

4.3.1 The Keystone Inhibitor Site: Block by Polynuclear Ruthenium Amines

The trinuclear oxo-bridged ruthenium amine ruthenium red (RuR) [91] is a polycation with many biological applications [92], including a ~50 year legacy of use as an inhibitor of diverse ion channels. RuR has been shown to inhibit three K_{2P} channels, two from the TREK subfamily, $K_{2P}4.1$ (TRAAK) [93, 94] and $K_{2P}10.1$ (TREK-2) [93], as well as $K_{2P}9.1$ (TASK-3) [95–97]. Functional studies showed that a negatively charged residue at the base of the K_{2P} CAP domain comprises a key RuR sensitivity determinant in the natively RuR sensitive channels $K_{2P}9.1$ (TASK-3) [95–97] and $K_{2P}10.1$ (TREK-2). Further, placing a negatively charged residue at the CAP base is sufficient for rendering a non-RuR sensitive K_{2P} responsive to RuR inhibition [54, 93]. Hence, this negative residue is both necessary and sufficient for RuR sensitivity in the context of a K_{2P} channel.

Structural studies of a RuR-sensitive $K_{2P}2.1$ (TREK-1) mutant, I110D [54], revealed that RuR inhibits K_{2P}s in a 1:1 stoichiometry matching functional studies [93, 94] and places one ruthenium amine moiety directly over the channel pore while the remainder of the RuR molecule occupies one of the two branches of the extracellular ion pathway (EIP). This "finger in the dam" mechanism provides both, and even an electrostatic and physical barrier that prevents the flow of potassium ions through the selectivity filter.

RuR interacts directly with the negatively charged residues that form the RuR-sensitivity determinant and that constitute the "Keystone inhibitor site" at the base of the CAP domain. The principal mode of binding is through a multi-pronged interaction made by the two acidic residues at the Keystone inhibitor site with multiple RuR elements. This sort of direct engagement of RuR by multiple acidic sidechains is likely to contribute to RuR block of other classes of RuR-sensitive channels where the binding site is thought to be rich in acidic residues such as TRP channels [98–105], the mitochondrial calcium uniporter (MCU) [106–109], CALHM calcium channels [110–112], ryanodine receptors [113, 114], and Piezo channels [115, 116]. The dinuclear ruthenium amine, Ru360 [117], an inhibitor of the mitochondrial calcium uniporter [106, 118, 119] not previously known to affect potassium channels also binds to the Keystone inhibitor site in a similar way, although due to its reduced electrostatic interactions relative to RuR, Ru360 is a weaker blocker (IC_{50} = 0.287 vs. 11.3 μM, for RuR and Ru360, respectively) [54] (Fig. 4.2).

Once sites of modulator action are known, it is possible to use the structural information to alter the protein or the ligand to create molecules having new properties. Using a structure-based protein engineering approach, our lab-created RuR super-responder $K_{2P}2.1$ (TREK-1) mutants having IC_{50}s in the low nanomolar range by placing acidic resides at Asn147 site at the external mouth of the selectivity filter in conjunction with the I110D mutation (IC_{50} = 12.7 nM) [54]. Because of the shared pore architecture among K_{2P}s, this strategy is generalizable to other K_{2P} members to create subtypes endowed with a high-affinity RuR sensitivity and could provide a means for exploring their functions. The demonstration that compounds such as RuR and Ru360 can block K_{2P} function by reaching through the EIP raises the possibility of identifying other classes of molecules that could work similarly. Two interesting directions for making subtype-selective modulators directed at the Keystone inhibitor site would be to capitalize on the renewed interest in synthesizing novel polyruthenium amine derivatives [120] or to design compounds having moieties that interact with the Keystone inhibitor

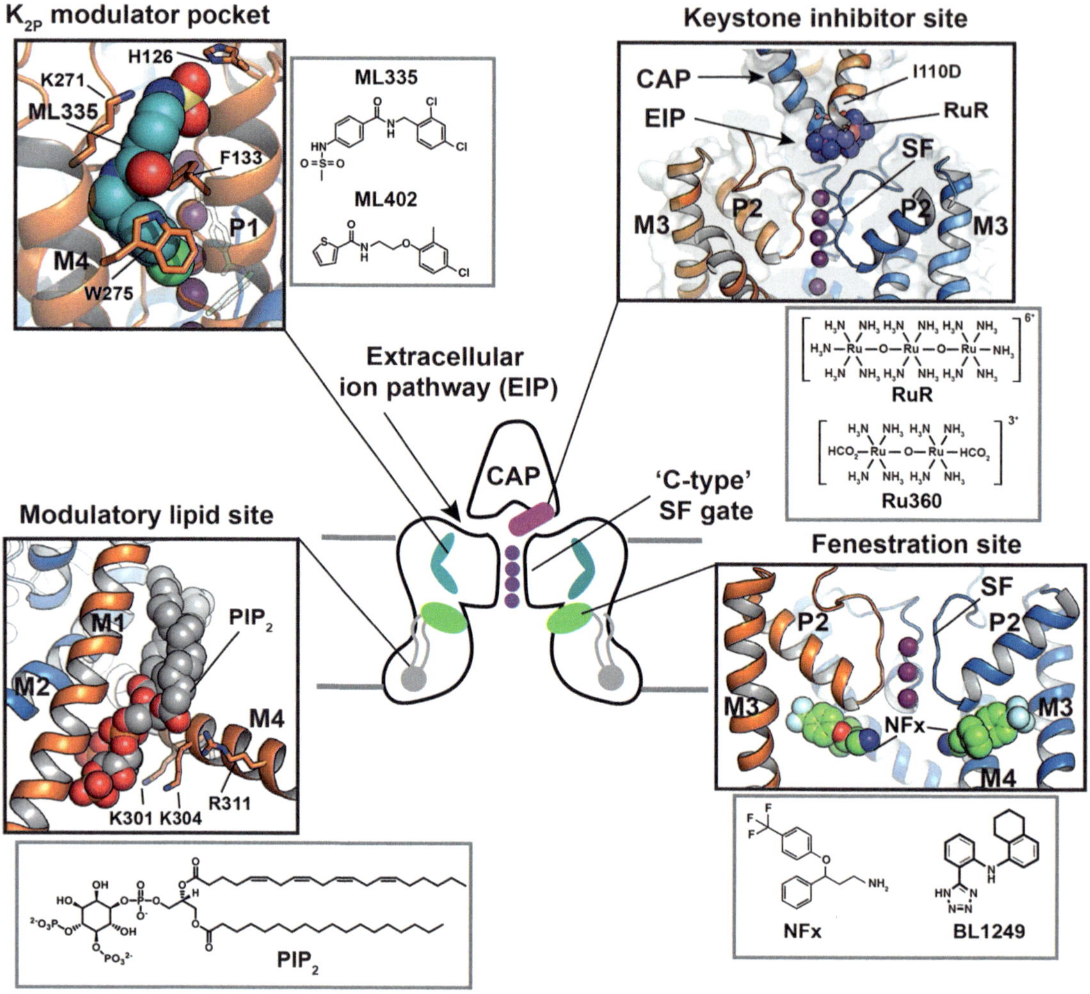

Fig. 4.2 Polysite model of TREK subfamily modulation. Central cartoon shows the locations of structurally defined K_{2P} small molecule binding sites including the Keystone inhibitor site (magenta), K_{2P} modulator pocket (cyan), fenestration site (green), and modulatory lipid site (grey). CAP and "C-type" SF gates are indicated. Potassium ions are shown (purple). Grey lines denote the lipid bilayer. Black boxes show the details of the individual sites. Grey boxes show modulator chemical structures

site but that also make specific contacts to non-conserved features of CAP exterior. Biologics, such as nanobodies, may be particularly suited to this type of molecular recognition mode as one can envision that a long variable loop from the nanobody could reach through the EIP to block the pore while other parts of the protein recognize subtype-specific features of the CAP exterior and EIP entryway.

4.3.2 The K_{2P} Modulator Pocket: A Cryptic Small Molecule Binding Site for K_{2P} Control

The K_{2P} modulator pocket (Fig. 4.2) is unrelated to any previously known small molecule binding pocket in the VGIC superfamily and was discovered in structural studies of $K_{2P}2.1$ (TREK-1) with two novel activators, ML335

(N-[(2,4-dichlorophenyl)methyl]-4-(methanesulfonamido) benzamide) and ML402 (N-[2-(4-chloro-2-methylphenoxy)ethyl]thiophene-2-carboxamide (Fig. 4.2) [11]. This L-shaped pocket is found in the P1–M4 interface, an intersubunit interface involved in C-type gating [17, 18]. Both compounds bind in similar ways and act as molecular wedges that stabilize the P1–M4 interface and directly activate the channel selectivity filter C-type gate [11, 22]. In the unliganded structure, the K_{2P} modulator pocket is occluded by P1–M4 interface interactions that require small movements of few residues to open, making this pocket a cryptic site that relies on conformational change similar to cryptic sites described for soluble proteins [121]. Rigidification of the P1–M4 interface is central to channel activation [11, 22]. The observation that these two compounds stabilize this intersubunit interface highlights the general importance of intersubunit interfaces as sites of channel control.

ML335 and ML402 are remarkably selective, activating $K_{2P}2.1$ (TREK-1) and $K_{2P}10.1$ (TREK-2) but not $K_{2P}4.1$ (TRAAK) [11] (Fig. 4.3). This strong subtype selectivity originates from a single lysine residue on the N-terminal end of M4 that engages in a cation–π interaction with the upper ring of each of the activators (Fig. 4.2). The equivalent residue in $K_{2P}4.1$ (TRAAK) is glutamine and exchanging K→Q in $K_{2P}2.1$ (TREK-1) and Q→K in $K_{2P}4.1$ (TRAAK) at this position is sufficient for rendering the former insensitive to ML335 and ML402 activation and the latter sensitive to activation by both compounds [11]. The importance of a single amino acid difference in an otherwise conserved small molecule binding pocket underscores the potential for exploiting local differences and structural knowledge to develop subtype-selective K_{2P} modulators.

The K_{2P} modulator pocket is unique to K_{2P}s [11]. Currently, there are no known natural ligands for this site, but it seems unlikely that such a well-defined binding site is only recognized by two unnatural small molecules. Stabilization of the P1–M4 interface is central for integrating gating cues that arise in other parts of the protein, particularly the C-terminal tail [11, 17, 18, 22]. Because the K_{2P} modulator pocket is in the center of this interface, it seems very likely that there are natural compounds such as lipids, metabolites, signaling molecules, or regulatory proteins that target this site. Hence, understanding the normal function of this part of the channel and whether Nature has exploited natural compounds to affect K_{2P} activity through the K_{2P} modulator pocket remains an important direction for future study.

4.3.3 The Fenestration Site: A Binding Site for Activators and Inhibitors

The K_{2P} M4 transmembrane helix is a key moving element and serves as a means to convey gating cues from temperature [17, 18, 25, 26], pressure [18, 26, 27], and phosphorylation [17] to the C-type gate. Structural studies of TREK subfamily channels have defined two extreme positions of the M4 helix termed "up" and "down" [13, 14, 23–25]. The "down " state creates a fenestration just below the P2 pore helix that is open to the center of the membrane bilayer [10, 13, 24, 25], the "Fenestration site" (Fig. 4.2). Structural studies of $K_{2P}10.1$ (TREK-2) have shown that this site binds to the K_{2P} inhibitors fluoxetine and norfluoxetine [14]. These compounds bind to a site defined by the lower part of the P2 pore helix and M4 (Fig. 4.2) and require the M4 helix to adopt the "down" position.

Remarkably, crystal structures of a $K_{2P}10.1$ (TREK-2) complex with a brominated version of an activator, the fenamate BL-1249, although not defining the entire compound, strongly indicate that this molecule and perhaps other activators bind to the Fenestration site created by the "down" M4 position [53]. How can the binding of a small molecule to the same site yield opposite functional outcomes of inhibition and activation? Clearly, the answer cannot be in the stabilization of the M4 "down" state over the "up" state as the binding of both inhibitors and activators to the Fenestration site requires an M4

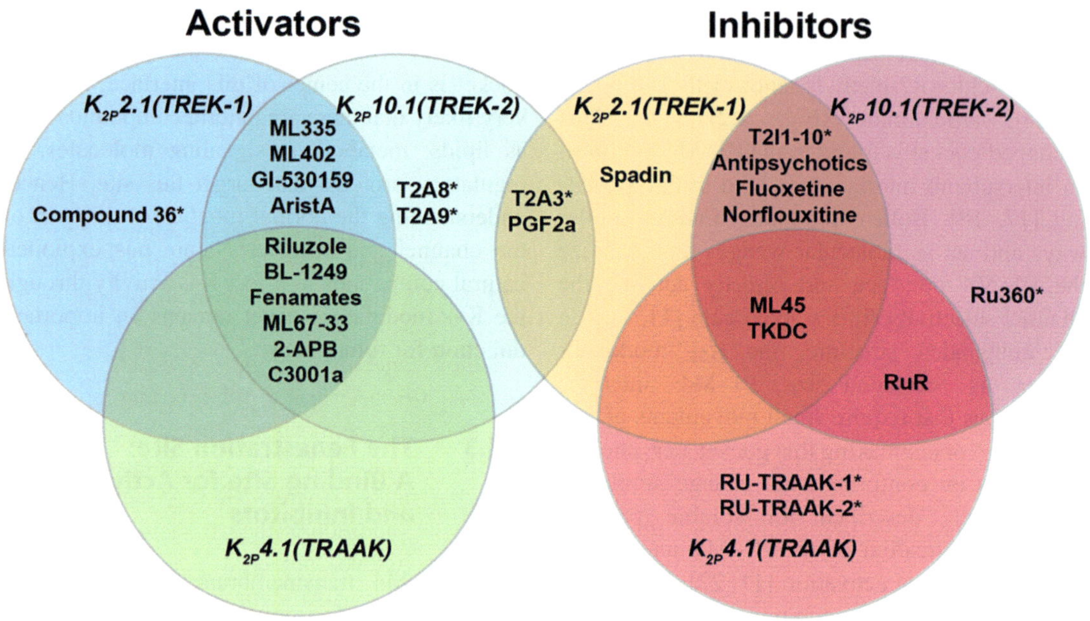

Fig. 4.3 Selectivity profiles of TREK subfamily small molecule activators and inhibitors. Asterisks indicate modulators lacking a complete profile of subtype selectivity

"down" conformation [14, 53]. Interestingly, it is suggested that activators such as BL-1249 use their tetrazole moiety to create a binding site for potassium ions in the central cavity and thereby stabilize the selectivity filter C-type gate [53]. Given this type of mechanism, it is notable that the norfluoxetine structure places the norfluoxetine amine just below the selectivity filter where its expected positive charge could provide an unfavorable modification to the potassium ion conduction pathway that would lead to channel inhibition (Fig. 4.2). The fenestration site is commonly found in the VGIC superfamily of which K_{2P}s are members and serves as the site of action for multiple types of activators of different classes of potassium channels [53]. Understanding the relationship between the occupation of this site, effects on the selectivity filter C-type gate, and the relationship between the properties of activators and inhibitors that can inhabit this site is an important challenge for further development of K_{2P} modulators.

4.3.4 The Modulatory Lipid Site: PIP_2 and the C-Terminal Tail

PIP_2 is an important modulatory lipid for TREK subfamily channels [68, 72, 73]. The likely site of PIP_2 action has been located in a series of $K_{2P}2.1$ (TREK-1) structures [11, 22]. These show the presence of a phospholipid that co-purified with the channel and that was bound to a groove at the M1/M2/M4 interface (Fig. 4.2). The phosphoinositol headgroup contacts an electropositive patch on the C-tail comprising five residues implicated in PIP_2 modulation (Arg297, Lys301, Lys302, Lys304, and R311) [68, 72] (Fig. 4.2). This same stretch of the C-terminal tail also contains the intracellular proton sensor site, Glu306 [71], and inhibitory phosphorylation site, Ser300 [74]. The key PIP_2-interacting residues are in a portion of the channel that is most affected by movements of M4 between the "up" and "down" positions. Hence, it seems likely that regulatory impacts of the modulatory lipid, intracellular pH sensor, and phosphorylation site on the C-terminal tail are all tightly intertwined with M4 motions [122]. Further study is needed to unravel these interactions, to understand whether other lipids reported to impact TREK channel function, such as phosphatidyl serine and phosphatidic acid [68, 123], compete with PIP_2 at this site, whether this site can be targeted by small molecules, and to define how changes in this lower part of the channel impact the dynamics and function of the C-type gate.

4.4 Subtype Specific Modulators in the TREK Subfamily

The growing progress in developing modulators for the TREK subfamily has been well-reviewed recently elsewhere [9, 47] and is, therefore, not reiterated here. As new modulators are uncovered, one key question is whether it will be possible to create subtype-specific modulators that can not only distinguish among the various K_{2P} subfamilies but also among the different subtypes within a subfamily. Such a high level of target selectivity would provide powerful tools for delineating the biological functions of these channels and could provide starting points for the development of subtype-selective pharmacology for these therapeutically relevant targets.

With respect to subtype selectivity, currently characterized TREK subfamily activators mostly fall into two categories. There are many examples of molecules that affect all three members of the TREK subfamily such as Riluzole [89, 124], BL-1249 [52], fenamates [36], ML67–33 [48], 2-Aminoethoxydiphenyl Borate (2-APB) [125, 126], and C3001a [85]. Some of these, such as BL-1249 [52] show limited selectivity for $K_{2P}2.1$ (TREK-1) and $K_{2P}10.1$ (TREK-2) over $K_{2P}4.1$ (TRAAK). The second category of compounds activate $K_{2P}2.1$ (TREK-1) and $K_{2P}10.1$ (TREK-2) but not $K_{2P}4.1$ (TRAAK) and include ML335 [11], ML402 [11], GI-530159 [127], and aristolochic acid (AristA) [128]. Given that $K_{2P}2.1$ (TREK-1) and $K_{2P}10.1$ (TREK-2) have sequences that are more similar to each other than they are to $K_{2P}4.1$ (TRAAK) (Fig. 4.1a), it is not surprising that $K_{2P}4.1$ (TRAAK) exhibits different responses to some modulators. There is a report of a $K_{2P}2.1$ (TREK-1) selective activator, Compound 36 [36], based on studies of less selective caffeic acid ester activators [36, 56, 63], but a detailed understanding of this high degree of selectivity remains to be defined. The compounds T2A8 and T2A9 are also reported to activate $K_{2P}10.1$ (TREK-1) with some selectivity [62]. The fact that there are already compounds showing some degree of selectivity provides an encouraging sign that developing activators having better subtype selectivity is a goal that can be reached.

TREK subfamily inhibitors show slightly more selectivity than activators and follow the same pattern having broadly acting inhibitors and inhibitors showing some selectivity. ML45 [48] and TKDC [129] inhibit all three TREK subfamily channels. There are a set of molecules that inhibit $K_{2P}2.1$ (TREK-1) and $K_{2P}10.1$ (TREK-2) but not $K_{2P}4.1$ (TRAAK), namely, antipsychotics [86], fluoxetine [84], and norfluoxetine [130]. A series of inhibitors (T2I1–10) unable to discriminate between $K_{2P}2.1$ (TREK-1) and $K_{2P}10.1$ (TREK-2) have also been reported [62], but whether these compounds affect $K_{2P}4.1$ (TRAAK) remains to be established. The polyruthenium blockers RuR and Ru360 show an unusual inhibition profile. RuR and Ru360 inhibit $K_{2P}10.1$ (TREK-2) but not $K_{2P}2.1$ (TREK-1) [54, 93, 94]. $K_{2P}4.1$ (TRAAK) is sensitive to RuR but the mechanism of inhibition must be different from the "finger in the dam" mechanism as $K_{2P}4.1$ (TRAAK) lacks the defining acidic residue in the Keystone inhibitor site [54]. Ru-TRAAK-1 and Ru-TRAAK-2 inhibit $K_{2P}4.1$ (TRAAK) as well as K_{2P}s from other subfamilies, such as $K_{2P}1.1$ (TWIK-1), $K_{2P}3.1$ (TASK-1), and $K_{2P}18.1$ (TRESK) [51]. The action of these compounds on other TREK subfamily members has not been reported, but given their ability to inhibit K_{2P}s outside of the TREK subfamily, it would be surprising if they did not also show some activity against the more closely related $K_{2P}2.1$ (TREK-1) or $K_{2P}10.1$ (TREK-2). The peptide spadin has been reported to act as a $K_{2P}2.1$ (TREK-1)-selective inhibitor [87, 88], but its mechanism of action remains unclear. T2A3, T2A8, T2A9, and the bioactive lipid 11-deoxyprostaglandin-F2α form an unusual class K_{2P}2.1 (TREK-1) inhibitors that are reported to also act as K_{2P}10.1 (TREK-2) activators [62]. How such dual action occurs is not known, but has been proposed to involve a short part of the P2–M4 loop [62]. As with the activators, the growing examples of subtype-selective inhibitors indicate that developing better and more selective inhibitors of the TREK subfamily should also be feasible, especially as more structural data about how such molecules interact with K_{2P}s becomes available.

So far, the origins of subtype selectivity for TREK modulators are understood for only two sites at the level of the atomic interactions, the Keystone inhibitor site, and the K_{2P} modulator pocket. For the Keystone inhibitor site, the negative charge at the Keystone inhibitor site is the principal determinant controlling RuR and Ru360 inhibition [54] (see Sect. 1.3.1). How RuR affects K_{2P}4.1 (TRAAK) remains unknown, as this channel lacks the negative residue at the Keystone inhibitor site and is inhibited with a stoichiometry higher than the 1:1 interaction of the Keystone inhibitor site [93, 94]. The basis of the subtype selectivity of the ML335 and ML402 activators [11] arises from a single lysine in the K_{2P} modulator pocket that controls subtype selectivity (see Sect. 1.3.2). Although not yet mapped in atomic detail, BL-1249 shows a ~10-fold selectivity for K_{2P}2.1 (TREK-1) and K_{2P}10.1 (TREK-2) over K_{2P}4.1 (TRAAK) that originates from differences in the M2/M3 helix interface [52]. Understanding the molecular origins of the subtype specificity for this compound as well as GI-530159 [127], aristolochic acid (AristA) [128], and C3001a [85] will require further studies that combines both structural and functional approaches.

4.5 Perspectives on K_{2P} Channel Polysite Pharmacology

Structural data for complexes of K_{2P}2.1 (TREK-1) and K_{2P}10.1 (TREK-2) with various modulators have uncovered a surprisingly large number of unique sites for small molecule control present at every level of the channel structure with respect to the membrane. These sites are arranged above (Keystone inhibitor site [54]), at the level of (K_{2P} modulator pocket [11]), and below (fenestration [14] and modulatory lipid [11] sites) the structure that is at the heart of channel function, the C-type gate [17, 18, 20–22]. It seems likely that there is a fifth site within the channel cavity, as there is good functional evidence that this architectural feature is targeted in TREK channels by alkylammonium pore blockers [9, 53] and this shared K_{2P} architectural element is the site of crystallographically defined small molecule block of K_{2P}3.1 (TASK-1) [12]. All of these sites have functions that are crucial for the normal functioning and modulation of K_{2P}s. The intersection of small molecule modulators and key sites of channel modulation emphasizes the importance of building an integrated understanding of modulator action and the basic mechanisms of channel function.

The structural pharmacology of the Keystone inhibitor site, K_{2P} modulator pocket, and Fenestration site has been defined by crystal structures of K_{2P}s complexed with nonnatural compounds that exert powerful effects on channel function. These findings highlight two key outstanding questions: To what extent has Nature has capitalized on these control points to influence TREK channel activity? and What are the naturally occurring modulators that target these sites? Answering such questions will be important for developing a better understanding of the roles of K_{2P}s in physiological responses. One known natural modulator of great physiological interest whose site remains to be defined on the TREK subfamily is the site of action of the activator arachidonic acid [75, 131]. From a structural perspective, although K_{2P}4.1 (TRAAK) was the first member of the TREK subfamily structurally characterized [13] and received its name due to its sensitivity to arachidonic acid [131], it remains the only channel in the TREK subfamily lacking any modulator-bound structures. Given the fact that this channel stands apart with respect to the selectivity of many modulators (Fig. 4.3), defining the site of arachidonic acid action in the TREK subfamily along with obtaining structural data for small molecule:K_{2P}4.1 (TRAAK) complexes will provide important guides for unraveling natural mechanisms of channel modulation and better templates for subtype-selective modulator discovery.

The clear crosstalk between various K_{2P} moving parts and the C-type gate [17, 18, 52] raises the question of whether the action of modulators at the various sites influence each other. The strength of polyruthenium amine block at the Keystone inhibitor site is not influenced by C-type gate stabilization by compounds occupying the K_{2P} modulator pocket [54], but the extent to which there may be crosstalk among the other sites remains to be evaluated. Understanding such interactions could provide insight into how to boost the efficacy of current modulators and will refine our understanding of how signals from various parts of the channel impinge on the C-type gate.

The current structural knowledge of modulator sites provides a framework to discover a new chemical matter that can affect K_{2P} function in novel ways. Such molecules may be engineered to block or enhance the consequences of various physical and chemical stimuli or to modify the channel chemically so that its biogenesis, distribution, and interaction with other cellular components can be followed in complex cell types and tissues. Besides further crystallographic studies of new K_{2P}: modulator complexes, an obvious key advance will be to understand structural consequences of heterodimer formation [42, 82–84] and to image K_{2P}s in lipid membrane environments using cryo-electron microscopy (cryo-EM) so that interactions between the channel and bilayer can be better understood. The initial burst of activity surrounding K_{2P} structural pharmacology is the first of many waves that will fill out our understanding of this important ion channel family and should lead to new and novel therapeutic strategies for a host of brain, cardiac, and nervous system diseases.

Acknowledgments We thank F. C. Chatelain for comments on the manuscript and P. Deal for help with figure preparation. This work was supported by grant NIH-R01-MH093603 to D.L.M.

References

1. Hille B (2001) Ion channels of excitable membranes, 3rd edn. Sinauer Associates, Inc., Sunderland, MA
2. Yu FH, Yarov-Yarovoy V, Gutman GA, Catterall WA (2005) Overview of molecular relationships in the voltage-gated ion channel superfamily. Pharmacol Rev 57:387–395
3. Enyedi P, Czirjak G (2010) Molecular background of leak K+ currents: two-pore domain potassium channels. Physiol Rev 90:559–605
4. Feliciangeli S, Chatelain FC, Bichet D, Lesage F (2014) The family of K2P channels: salient structural and functional properties. J Physiol. https://doi.org/10.1113/jphysiol.2014.287268
5. Renigunta V, Schlichthorl G, Daut J (2015) Much more than a leak: structure and function of K(2)p-channels. Pflugers Arch 467:867–894
6. Goldstein SA et al (2005) International Union of Pharmacology. LV Nomenclature and molecular relationships of two-P potassium channels. Pharmacol Rev 57:527–540
7. Douguet D, Honore E (2019) Mammalian mechanoelectrical transduction: structure and function of force-gated ion channels. Cell 179:340–354
8. Sepulveda FV, Pablo Cid L, Teulon J, Niemeyer MI (2015) Molecular aspects of structure, gating, and physiology of pH-sensitive background K2P and Kir K+-transport channels. Physiol Rev 95:179–217
9. Sterbuleac D (2019) Molecular determinants of chemical modulation of two-pore domain potassium channels. Chem Biol Drug Des 94:1596–1614
10. Miller AN, Long SB (2012) Crystal structure of the human two-pore domain potassium channel K2P1. Science 335:432–436
11. Lolicato M et al (2017) K2P2.1 (TREK-1)-activator complexes reveal a cryptic selectivity filter binding site. Nature 547:364–368
12. Rödström KEJ et al (2020) A lower X-gate in TASK channels traps inhibitors within the vestibule. Nature 582:443–447
13. Brohawn SG, del Marmol J, MacKinnon R (2012) Crystal structure of the human K2P TRAAK, a lipid- and mechano-sensitive K+ ion channel. Science 335:436–441
14. Dong YY et al (2015) K2P channel gating mechanisms revealed by structures of TREK-2 and a complex with Prozac. Science 347:1256–1259
15. Payandeh J, Minor DL Jr (2014) Bacterial Voltage-Gated Sodium Channels (BacNas) from the soil, sea, and salt lakes enlighten molecular mechanisms of electrical signaling and pharmacology in the brain and heart. J Mol Biol. https://doi.org/10.1016/j.jmb.2014.08.010

16. Catterall WA, Wisedchaisri G, Zheng N (2017) The chemical basis for electrical signaling. Nat Chem Biol 13:455–463
17. Bagriantsev SN, Clark KA, Minor DL Jr (2012) Metabolic and thermal stimuli control K(2P)2.1 (TREK-1) through modular sensory and gating domains. EMBO J 31:3297–3308
18. Bagriantsev SN, Peyronnet R, Clark KA, Honore E, Minor DL Jr (2011) Multiple modalities converge on a common gate to control K2P channel function. EMBO J 30:3594–3606
19. Cohen A, Ben-Abu Y, Hen S, Zilberberg N (2008) A novel mechanism for human K2P2.1 channel gating. Facilitation of C-type gating by protonation of extracellular histidine residues. J Biol Chem 283:19448–19455
20. Piechotta PL et al (2011) The pore structure and gating mechanism of K2P channels. EMBO J 30:3607–3619
21. Schewe M et al (2016) A non-canonical voltage-sensing mechanism controls gating in K2P K(+) channels. Cell 164:937–949
22. Lolicato M et al (2020) K2P channel C-type gating involves asymmetric selectivity filter order-disorder transitions. bioRxiv. https://doi.org/10.1101/2020.03.20.000893
23. Brohawn SG, Campbell EB, MacKinnon R (2013) Domain-swapped chain connectivity and gated membrane access in a Fab-mediated crystal of the human TRAAK K+ channel. Proc Natl Acad Sci U S A 110:2129–2134
24. Brohawn SG, Campbell EB, MacKinnon R (2014) Physical mechanism for gating and mechanosensitivity of the human TRAAK K+ channel. Nature 516:126–130
25. Lolicato M, Riegelhaupt PM, Arrigoni C, Clark KA, Minor DL Jr (2014) Transmembrane helix straightening and buckling underlies activation of mechanosensitive and thermosensitive K (2P) channels. Neuron 84:1198–1212
26. McClenaghan C et al (2016) Polymodal activation of the TREK-2 K2P channel produces structurally distinct open states. J Gen Physiol 147:497–505
27. Aryal P et al (2017) Bilayer-mediated structural transitions control mechanosensitivity of the TREK-2 K2P channel. Structure 25:708–718. e702
28. Kanda H et al (2019) TREK-1 and TRAAK are principal K(+) channels at the nodes of ranvier for rapid action potential conduction on mammalian myelinated afferent nerves. Neuron. https://doi.org/10.1016/j.neuron.2019.08.042
29. Brohawn SG et al (2019) The mechanosensitive ion channel TRAAK is localized to the mammalian node of Ranvier. eLife 8
30. Heurteaux C et al (2004) TREK-1, a K+ channel involved in neuroprotection and general anesthesia. EMBO J 23:2684–2695
31. Lazarenko RM et al (2010) Anesthetic activation of central respiratory chemoreceptor neurons involves inhibition of a THIK-1-like background K(+) current. J Neurosci 30:9324–9334
32. Madry C et al (2018) Microglial ramification, surveillance, and interleukin-1beta release are regulated by the two-pore domain K(+) channel THIK-1. Neuron 97:299–312. e296
33. Yoshida K et al (2018) Leak potassium channels regulate sleep duration. Proc Natl Acad Sci U S A 115:E9459–E9468
34. Alloui A et al (2006) TREK-1, a K+ channel involved in polymodal pain perception. EMBO J 25:2368–2376
35. Devilliers M et al (2013) Activation of TREK-1 by morphine results in analgesia without adverse side effects. Nat Commun 4:2941
36. Vivier D et al (2017) Development of the first two-pore domain potassium channel TREK-1 (TWIK-Related K+ Channel 1)-selective agonist possessing in vivo anti-nociceptive activity. J Med Chem. https://doi.org/10.1021/acs.jmedchem.6b01285
37. Decher N et al (2017) Sodium permeable and "hypersensitive" TREK-1 channels cause ventricular tachycardia. EMBO Mol Med 9:403–414
38. Laigle C, Confort-Gouny S, Le Fur Y, Cozzone PJ, Viola A (2012) Deletion of TRAAK potassium channel affects brain metabolism and protects against ischemia. PLoS One 7:e53266
39. Wu X et al (2013) Involvement of TREK-1 activity in astrocyte function and neuroprotection under simulated ischemia conditions. J Mol Neurosci 49:499–506
40. Abraham DM et al (2018) The two-pore domain potassium channel TREK-1 mediates cardiac fibrosis and diastolic dysfunction. J Clin Invest 128:4843–4855
41. Heurteaux C et al (2006) Deletion of the background potassium channel TREK-1 results in a depression-resistant phenotype. Nat Neurosci 9:1134–1141
42. Royal P et al (2019) Migraine-associated TRESK mutations increase neuronal excitability through alternative translation initiation and inhibition of TREK. Neuron 101:232–245. e236
43. Yarishkin O et al (2018) TREK-1 channels regulate pressure sensitivity and calcium signaling in trabecular meshwork cells. J Gen Physiol 150:1660–1675
44. Lambert M et al (2018) Loss of KCNK3 is a hallmark of RV hypertrophy/dysfunction associated with pulmonary hypertension. Cardiovasc Res 114:880–893
45. Schwingshackl A (2016) The role of stretch-activated ion channels in acute respiratory distress syndrome: finally a new target? Am J Physiol Lung Cell Mol Physiol 311:L639–L652
46. Petho Z, Najder K, Bulk E, Schwab A (2019) Mechanosensitive ion channels push cancer progression. Cell Calcium 80:79–90
47. Mathie A, Veale EL, Cunningham KP, Holden RG, Wright PD (2020) Two-pore domain potassium

channels as drug targets: anesthesia and beyond. Annu Rev Pharmacol Toxicol. https://doi.org/10.1146/annurev-pharmtox-030920-111536
48. Bagriantsev SN et al (2013) A high-throughput functional screen identifies small molecule regulators of temperature- and mechano-sensitive K2P channels. ACS Chem Biol 8:1841–1851
49. Tian F et al (2019) A small-molecule compound selectively activates K2P Channel TASK-3 by acting at two distant clusters of residues. Mol Pharmacol 96:26–35
50. Wright PD et al (2019) Pranlukast is a novel small molecule activator of the two-pore domain potassium channel TREK2. Biochem Biophys Res Commun. https://doi.org/10.1016/j.bbrc.2019.09.093
51. Su ZW, Brown EC, Wang WW, MacKinnon R (2016) Novel cell-free high-throughput screening method for pharmacological tools targeting K+ channels. Proc Natl Acad Sci USA 113:5748–5753
52. Pope L et al (2018) Protein and chemical determinants of BL-1249 action and selectivity for K2P channels. ACS Chem Neurosci 9:3153–3165
53. Schewe M et al (2019) A pharmacological master key mechanism that unlocks the selectivity filter gate in K (+) channels. Science 363:875–880
54. Pope L, Lolicato M, Minor DL Jr (2020) Polynuclear ruthenium amines inhibit K2P channels via a "Finger in the Dam" mechanism. Cell Chem Biol. https://doi.org/10.1016/j.chembiol.2020.01.011
55. Gada K, Plant LD (2019) Two-pore domain potassium channels: emerging targets for novel analgesic drugs: IUPHAR Review 26. Br J Pharmacol 176:256–266
56. Vivier D, Bennis K, Lesage F, Ducki S (2016) Perspectives on the two-pore domain potassium channel TREK-1 (TWIK-Related K(+) Channel 1). A novel therapeutic target? J Med Chem 59:5149–5157
57. Hancox JC, James AF, Marrion NV, Zhang H, Thomas D (2016) Novel ion channel targets in atrial fibrillation. Expert Opin Ther Targets 20:947–958
58. Decher N, Kiper AK, Rinne S (2017) Stretch-activated potassium currents in the heart: focus on TREK-1 and arrhythmias. Prog Biophys Mol Biol 130:223–232
59. Bagal SK et al (2013) Ion channels as therapeutic targets: a drug discovery perspective. J Med Chem 56:593–624
60. Borsotto M et al (2015) Targeting two-pore domain K(+) channels TREK-1 and TASK-3 for the treatment of depression: a new therapeutic concept. Br J Pharmacol 172:771–784
61. Enyedi P, Czirjak G (2015) Properties, regulation, pharmacology, and functions of the K(2)p channel, TRESK. Pflugers Arch 467:945–958
62. Dadi PK et al (2016) Selective small molecule activators of TREK-2 channels stimulate DRG c-fiber nociceptor K2P currents and limit calcium influx. ACS Chem Neurosci. https://doi.org/10.1021/acschemneuro.6b00301
63. Rodrigues N et al (2014) Synthesis and structure-activity relationship study of substituted caffeate esters as antinociceptive agents modulating the TREK-1 channel. Eur J Med Chem 75:391–402
64. Liao P et al (2019) Selective activation of TWIK-related acid-sensitive K(+) 3 subunit-containing channels is analgesic in rodent models. Sci Transl Med 11
65. Honore E (2007) The neuronal background K2P channels: focus on TREK1. Nat Rev Neurosci 8:251–261
66. Maingret F et al (2000) TREK-1 is a heat-activated background K(+) channel. EMBO J 19:2483–2491
67. Kang D, Choe C, Kim D (2005) Thermosensitivity of the two-pore domain K+ channels TREK-2 and TRAAK. J Physiol 564:103–116
68. Chemin J et al (2005) A phospholipid sensor controls mechanogating of the K+ channel TREK-1. EMBO J 24:44–53
69. Maingret F, Fosset M, Lesage F, Lazdunski M, Honoré E (1999) TRAAK is a mammalian neuronal mechano-gated K^+ channel. J Biol Chem 274:1381–1387
70. Maingret F, Patel AJ, Lesage F, Lazdunski M, Honore E (1999) Mechano- or acid stimulation, two interactive modes of activation of the TREK-1 potassium channel. J Biol Chem 274:26691–26696
71. Honore E, Maingret F, Lazdunski M, Patel AJ (2002) An intracellular proton sensor commands lipid- and mechano-gating of the K(+) channel TREK-1. EMBO J 21:2968–2976
72. Chemin J et al (2007) Up- and down-regulation of the mechano-gated K(2P) channel TREK-1 by PIP (2) and other membrane phospholipids. Pflugers Arch 455:97–103
73. Lopes CM et al (2005) PIP2 hydrolysis underlies agonist-induced inhibition and regulates voltage gating of two-pore domain K+ channels. J Physiol 564:117–129
74. Murbartian J, Lei Q, Sando JJ, Bayliss DA (2005) Sequential phosphorylation mediates receptor- and kinase-induced inhibition of TREK-1 background potassium channels. J Biol Chem 280:30175–30184
75. Patel AJ et al (1998) A mammalian two pore domain mechano-gated S-like K+ channel. EMBO J 17:4283–4290
76. Sandoz G, Douguet D, Chatelain F, Lazdunski M, Lesage F (2009) Extracellular acidification exerts opposite actions on TREK1 and TREK2 potassium channels via a single conserved histidine residue. Proc Natl Acad Sci U S A 106:14628–14633
77. Acosta C et al (2014) TREK2 expressed selectively in IB4-binding C-fiber nociceptors hyperpolarizes their membrane potentials and limits spontaneous pain. J Neurosci 34:1494–1509

78. Waxman SG, Zamponi GW (2014) Regulating excitability of peripheral afferents: emerging ion channel targets. Nat Neurosci 17:153–163
79. Bang H, Kim Y, Kim D (2000) TREK-2, a new member of the mechanosensitive tandem-pore K+ channel family. J Biol Chem 275:17412–17419
80. Fink M et al (1996) Cloning, functional expression and brain localization of a novel unconventional outward rectifier K+ channel. EMBO J 15:6854–6862
81. Lesage F, Maingret F, Lazdunski M (2000) Cloning and expression of human TRAAK, a polyunsaturated fatty acids-activated and mechano-sensitive K(+) channel. FEBS Lett 471:137–140
82. Lengyel M, Czirjak G, Enyedi P (2016) Formation of functional heterodimers by TREK-1 and TREK-2-two-pore domain potassium channel subunits. J Biol Chem 291:13649–13661
83. Levitz J et al (2016) Heterodimerization within the TREK channel subfamily produces a diverse family of highly regulated potassium channels. Proc Natl Acad Sci U S A 113:4194–4199
84. Blin S et al (2016) Mixing and matching TREK/TRAAK subunits generate heterodimeric K2P channels with unique properties. Proc Natl Acad Sci U S A 113:4200–4205
85. Qiu Y et al (2020) TREK Channel family activator with a well-defined structure-activation relationship for pain and neurogenic inflammation. J Med Chem 63:3665–3677
86. Thummler S, Duprat F, Lazdunski M (2007) Antipsychotics inhibit TREK but not TRAAK channels. Biochem Biophys Res Commun 354:284–289
87. Maati HMO et al (2012) Spadin as a new antidepressant: absence of TREK-1-related side effects. Neuropharmacology 62:278–288
88. Mazella J et al (2010) Spadin, a sortilin-derived peptide, targeting rodent TREK-1 channels: a new concept in the antidepressant drug design. PLoS Biol 8: e1000355
89. Lesage F, Terrenoire C, Romey G, Lazdunski M (2000) Human TREK2, a 2P domain mechano-sensitive K+ channel with multiple regulations by polyunsaturated fatty acids, lysophospholipids, and Gs, Gi, and Gq protein-coupled receptors. J Biol Chem 275:28398–28405
90. Brohawn SG, Su Z, MacKinnon R (2014) Mechanosensitivity is mediated directly by the lipid membrane in TRAAK and TREK1 K+ channels. Proc Natl Acad Sci U S A 111:3614–3619
91. Fletcher JM, Greenfield BF, Hardy CJ, Scargill D, Woodhead JL (1961) Ruthenium red. J Chem Soc:2000–2006
92. Clarke MJ (2002) Ruthenium metallopharmaceuticals. Coordin Chem Rev 232:69–93
93. Braun G, Lengyel M, Enyedi P, Czirjak G (2015) Differential sensitivity of TREK-1, TREK-2 and TRAAK background potassium channels to the polycationic dye ruthenium red. Br J Pharmacol 172:1728–1738
94. Czirjak G, Enyedi P (2002) Formation of functional heterodimers between the TASK-1 and TASK-3 two-pore domain potassium channel subunits. J Biol Chem 277:5426–5432
95. Musset B et al (2006) Effects of divalent cations and spermine on the K+ channel TASK-3 and on the outward current in thalamic neurons. J Physiol 572:639–657
96. Czirjak G, Enyedi P (2003) Ruthenium red inhibits TASK-3 potassium channel by interconnecting glutamate 70 of the two subunits. Mol Pharmacol 63:646–652
97. Gonzalez W et al (2013) An extracellular ion pathway plays a central role in the cooperative gating of a K(2P) K+ channel by extracellular pH. J Biol Chem 288:5984–5991
98. Caterina MJ et al (1997) The capsaicin receptor: a heat-activated ion channel in the pain pathway. Nature 389:816–824
99. Strotmann R, Harteneck C, Nunnenmacher K, Schultz G, Plant TD (2000) OTRPC4, a nonselective cation channel that confers sensitivity to extracellular osmolarity. Nat Cell Biol 2:695–702
100. Guler AD et al (2002) Heat-evoked activation of the ion channel, TRPV4. J Neurosci 22:6408–6414
101. Caterina MJ, Rosen TA, Tominaga M, Brake AJ, Julius D (1999) A capsaicin-receptor homologue with a high threshold for noxious heat. Nature 398:436–441
102. Voets T et al (2002) Molecular determinants of permeation through the cation channel TRPV4. J Biol Chem 277:33704–33710
103. Arif Pavel M et al (2016) Function and regulation of TRPP2 ion channel revealed by a gain-of-function mutant. Proc Natl Acad Sci U S A 113:E2363–E2372
104. Voets T et al (2004) TRPM6 forms the Mg^{2+} influx channel involved in intestinal and renal Mg^{2+} absorption. J Biol Chem 279:19–25
105. Story GM et al (2003) ANKTM1, a TRP-like channel expressed in nociceptive neurons, is activated by cold temperatures. Cell 112:819–829
106. Kirichok Y, Krapivinsky G, Clapham DE (2004) The mitochondrial calcium uniporter is a highly selective ion channel. Nature 427:360–364
107. Chaudhuri D, Sancak Y, Mootha VK, Clapham DE (2013) MCU encodes the pore conducting mitochondrial calcium currents. elife 2:e00704
108. Rahamimoff R, Alnaes E (1973) Inhibitory action of Ruthenium red on neuromuscular transmission. Proc Natl Acad Sci U S A 70:3613–3616
109. Moore CL (1971) Specific inhibition of mitochondrial Ca++ transport by ruthenium red. Biochem Biophys Res Commun 42:298–305
110. Ma Z et al (2012) Calcium homeostasis modulator 1 (CALHM1) is the pore-forming subunit of an ion channel that mediates extracellular Ca^{2+} regulation of

neuronal excitability. Proc Natl Acad Sci U S A 109: E1963–E1971
111. Dreses-Werringloer U et al (2013) CALHM1 controls the Ca(2)(+)-dependent MEK, ERK, RSK and MSK signaling cascade in neurons. J Cell Sci 126:1199–1206
112. Choi W, Clemente N, Sun W, Du J, Lu W (2019) The structures and gating mechanism of human calcium homeostasis modulator 2. Nature. https://doi.org/10.1038/s41586-019-1781-3
113. Ma J (1993) Block by ruthenium red of the ryanodine-activated calcium release channel of skeletal muscle. J Gen Physiol 102:1031–1056
114. Smith JS et al (1988) Purified ryanodine receptor from rabbit skeletal muscle is the calcium-release channel of sarcoplasmic reticulum. J Gen Physiol 92:1–26
115. Coste B et al (2012) Piezo proteins are pore-forming subunits of mechanically activated channels. Nature 483:176–181
116. Zhao QC et al (2016) Ion permeation and mechanotransduction mechanisms of mechanosensitive piezo channels. Neuron 89:1248–1263
117. Ying WL, Emerson J, Clarke MJ, Sanadi DR (1991) Inhibition of mitochondrial calcium ion transport by an oxo-bridged dinuclear ruthenium ammine complex. Biochemistry 30:4949–4952
118. Baughman JM et al (2011) Integrative genomics identifies MCU as an essential component of the mitochondrial calcium uniporter. Nature 476:341–345
119. Oxenoid K et al (2016) Architecture of the mitochondrial calcium uniporter. Nature 533:269–273
120. Woods JJ, Wilson JJ (2019) Inhibitors of the mitochondrial calcium uniporter for the treatment of disease. Curr Opin Chem Biol 55:9–18
121. Hardy JA, Wells JA (2004) Searching for new allosteric sites in enzymes. Curr Opin Struct Biol 14:706–715
122. Soussia IB et al (2018) Antagonistic effect of a cytoplasmic domain on the basal activity of polymodal potassium channels. Front Mol Neurosci 11:301
123. Chemin J et al (2005) Lysophosphatidic acid-operated K+ channels. J Biol Chem 280:4415–4421
124. Duprat F et al (2000) The neuroprotective agent riluzole activates the two P domain K(+) channels TREK-1 and TRAAK. Mol Pharmacol 57:906–912
125. Zhuo RG et al (2016) Allosteric coupling between proximal C-terminus and selectivity filter is facilitated by the movement of transmembrane segment 4 in TREK-2 channel. Sci Rep 6:21248
126. Beltran L, Beltran M, Aguado A, Gisselmann G, Hatt H (2013) 2-Aminoethoxydiphenyl borate activates the mechanically gated human KCNK channels KCNK 2 (TREK-1), KCNK 4 (TRAAK), and KCNK 10 (TREK-2). Front Pharmacol 4:63
127. Loucif AJC et al (2017) GI-530159, a novel, selective, mechanosensitive two-pore-domain potassium (K2P) channel opener, reduces rat dorsal root ganglion neuron excitability. Br J Pharmacol. https://doi.org/10.1111/bph.14098
128. Veale EL, Mathie A (2016) Aristolochic acid, a plant extract used in the treatment of pain and linked to Balkan endemic nephropathy, is a regulator of K2P channels. Br J Pharmacol 173:1639–1652
129. Luo Q et al (2017) An allosteric ligand-binding site in the extracellular cap of K2P channels. Nat Commun 8:378
130. Kennard LE et al (2005) Inhibition of the human two-pore domain potassium channel, TREK-1, by fluoxetine and its metabolite norfluoxetine. Br J Pharmacol 144:821–829
131. Fink M et al (1998) A neuronal two P domain K^+ channel stimulated by arachidonic acid and polyunsaturated fatty acids. EMBO J 17:3297–3308

Calcium Channel Splice Variants and Their Effects in Brain and Cardiovascular Function

5

Yeow Sean Qing Zhang, Kelvin Wei Zhern Loh, and Tuck Wah Soong

Abstract

Calcium ions serve as an important intracellular messenger in many diverse pathways, ranging from excitation coupling in muscles to neurotransmitter release in neurons. Physiologically, the concentration of free intracellular Ca^{2+} is up to 10,000 times less than that of the extracellular concentration, and increases of 10- to 100-fold in intracellular Ca^{2+} are observed during signaling events. Voltage-gated calcium channels (VGCCs) located on the plasma membrane serve as one of the main ways in which Ca^{2+} is able to enter the cell. Given that Ca^{2+} functions as a ubiquitous intracellular messenger, it is imperative that VGCCs are under tight regulation to ensure that intracellular Ca^{2+} concentration remains within the physiological range. In this chapter, we explore VGCCs' inherent control of Ca^{2+} entry as well as the effects of alternative splicing in $Ca_V2.1$ and posttranslational modifications of $Ca_V1.2$/$Ca_V1.3$ such as phosphorylation and ubiquitination. Deviation from this physiological range will result in deleterious effects known as calcium channelopathies, some of which will be explored in this chapter.

Keywords

Voltage-gated calcium channels · CACNA1C · CACNA1D · CACNA1A · Cav1.2 · Cav1.3 · Cav2.1 · Channelopathies · Splicing · Modulation

Yeow S. Q. Z. · K. W. Z. Loh
Department of Physiology, Yong Loo Lin School of Medicine, National University of Singapore, Singapore, Singapore

NUS Graduate School for Integrative Sciences and Engineering, Singapore, Singapore

T. W. Soong (✉)
Department of Physiology, Yong Loo Lin School of Medicine, National University of Singapore, Singapore, Singapore

NUS Graduate School for Integrative Sciences and Engineering, Singapore, Singapore

Healthy Longevity Translational Research Programme and Cardiovascular Disease Translational Research Programme, National University of Singapore, Singapore, Singapore

Neurobiology Programme, Life Science Institute, National University of Singapore, Singapore, Singapore
e-mail: phsstw@nus.edu.sg

5.1 Ion Channels in Biophysics and Physiology

5.1.1 Introduction to LTCC

L-type calcium channels (LTCC) are a subset of voltage-gated calcium channels (VGCC). LTCCs comprise calcium channel (Ca_v) isoforms $Ca_v1.1$–$Ca_v1.4$ and are encoded by the human genes *CACNA1S*, *CACNA1C*, *CACNA1D*, and *CACNA1F*, respectively. They are called

L. Zhou (ed.), *Ion Channels in Biophysics and Physiology*, Advances in Experimental Medicine and Biology 1349, https://doi.org/10.1007/978-981-16-4254-8_5

"L-type" because of their long-lasting Ba^{2+} conductance, which is attributed to the slow inactivation kinetics of LTCCs (~500 ms). Unlike the other VGCCs, LTCCs are the only group that can bind to various 1,4-dihydropyridines (DHP) and are thus also called dihydropyridine receptors (DHPR).

LTCCs are transmembrane proteins found on the plasma membrane of excitable cells such as neurons, skeletal, cardiac, and smooth muscles [1, 2]. Similar to other VGCCs, LTCCs are voltage-gated and they open in response to membrane depolarization, thereby allowing for the selective influx of calcium ions (Ca^{2+}) into the cytoplasm. This Ca^{2+} influx regulates various downstream functions like synaptic vesicle anchorage and release, gene expression, and excitation–contraction (EC) coupling [3]. While the topological structure of LTCCs is generally similar across the tissues they are expressed in, their functions differ according to their tissue localization and biophysical properties. Further discussions will be focused on the cardiovascular system for the purpose of this review.

5.1.2 Structure and Localization of LTCC

LTCCs are transmembrane protein complexes made up of four to five subunits depending on where they are expressed—α_1-, β-, $a_2\delta$-, and γ-subunits [2]. In skeletal and cardiac muscles, the γ-subunit is not associated with $Ca_v1.2$ and $Ca_v1.3$ isoforms. The ~250 kDa α_1-subunit forms the main pore of the channel complex and is comprised of four repeat domains (DI–IV), with each domain made up of six transmembrane segments (S1–S6) that are linked by intracellular loops. To date, there are ten known mammalian isoforms of α_1-subunit of which 4 are LTCCs. The S4 of each domain acts as the voltage sensor for the channel due to the presence of positively charged amino acids, while the S5 and S6 segments form the activation gate of the α_1-subunit and the S5–S6 reentrant loop lines the selectivity filter. Upon membrane depolarization, the change in membrane potential is "sensed" by the four S4 segments, resulting in a conformational change from close to open state, thereby facilitating the flux of Ca^{2+} ions across the membrane.

The auxiliary β- and $a_2\delta$-subunits are necessary for the proper expression and insertion of the α_1-subunit into the plasma membrane. It was reported that the overexpression of either the β- or $a_2\delta$-subunits enhanced the current density of the α_1-subunit [4]. Conversely, the current density is reduced when either of these subunits was transiently knocked down. These data suggest that the function of the α_1-subunit is regulated by the coexpression of the auxiliary subunits.

Co-immunoprecipitation experiments revealed that the β-subunit physically associates with the α_1-subunit. More importantly, this association occurs within a region between DI and DII linker (I–II loop), named as the α_1-interacting domain (AID) [5]. Interestingly, in vitro experiments revealed a potential mechanism by which the binding of the β-subunit to the AID is important in regulating the expression of the α_1-subunit. Of note, disruption to this interaction was reported to have resulted in the poly-ubiquitination of the α_1-subunit, leading to proteasomal degradation and thus decreased trafficking of the α_1-subunit to the plasma membrane [6].

5.1.3 LTCC in the Cardiovascular System—Function

In the cardiovascular system (CVS), LTCC isoforms $Ca_v1.2$ and $Ca_v1.3$ are found to be expressed. $Ca_v1.2$ is the predominant isoform expressed in excitable cells [1]. Conversely, $Ca_v1.3$ is found to be dominantly expressed in the sinoatrial node (SAN) and the atrioventricular node (AVN) and is important for pacemaking activity [7].

$Ca_v1.2$ and $Ca_v1.3$ are responsible for the initiation of the excitation–contraction (EC) coupling [3], a process whereby an electrical signal is converted into mechanical signals. Macroscopically, this can be seen simply as "contraction" of the muscle cells. Microscopically, the depolarizing electrical signals are sensed by the

S4 "voltage sensors," and the ensuing protein conformational change leads to channel activation. The resultant Ca^{2+} influx causes a transient increase in intracellular Ca^{2+} concentration. The binding of Ca^{2+} by the ryanodine receptors (RyR) localized to the membrane of sarcoplasmic reticulum (SR) activates the RyRs to open to release Ca^{2+} within the intracellular store of the SR, and this process further increases cytoplasmic Ca^{2+} concentration. This process, also known as calcium-induced calcium release (CICR), triggers a cascade of downstream biochemical signals, ultimately allowing for the contraction of muscle fibers via activation of tropomyosin. Although the sliding motion of a few tropomyosin proteins may seem insignificant, nevertheless, the coordinated, simultaneous sliding of tropomyosin across the entire muscle fiber is sufficient to generate much mechanical force for various processes such as vasocontraction, contraction of the myocardium during systole, and concentric movements such as weightlifting.

In the heart, this coordinated EC coupling process and pumping action of the myocardium are essential for maintaining a constant blood supply throughout the body. In the vessels, $Ca_v1.2$ is essential for maintaining the basal vascular tone of blood vessels, contributing to the maintenance of blood pressure. In both events, mutations of $Ca_v1.2$ or $Ca_v1.3$ result in channelopathies and could lead to serious health complications such as those discussed subsequently.

5.1.4 Channelopathies in the CVS—$Ca_v1.2$

5.1.4.1 Timothy Syndrome

Timothy Syndrome (TS) was first discovered by Katherine W. Timothy and is a rare multi-organ disorder that affects not only the myocardium but also the brain. TS is associated with a gain-of-function mutation in the α_1-subunit of $Ca_v1.2$. There are two unique mutations categorized, G406R and G402S, which occur in the mutually exclusive exons 8 or 8a of the α_1-subunit [8, 9]. As such, TS can be further categorized as a classical type I TS or an atypical type II TS depending on which exon is affected. Of note, exon 8 and 8a are both found in the S6 segment of DI, which forms part of the activation gate. In type I TS, G406R mutation was observed in exon 8a, while in type II TS, either G406R or G402S can be found in exon 8. In both events, this gain-of-function mutation results in reduced voltage-dependent inactivation (VDI) of the $Ca_v1.2$ channel, thereby resulting in a sustained increase in Ca^{2+} influx, long QT syndrome (LQTS), and thus arrhythmias. In most cases, arrhythmias remain the primary underlying cause of death for patients presenting with TS. As TS is a multisystem disorder, it was observed that in addition to cardiac symptoms, patients with TS also present with extracardiac phenotypes. These typically include syndactyly, dysmorphic facial features, and mental retardation.

5.1.4.2 Brugada Syndrome

In contrast to TS, Brugada Syndrome (BS) occurs due to loss-of-function mutations in the α_1-subunit of $Ca_v1.2$. This results in shorter QT intervals and contributes to the manifestation of arrhythmias [10, 11]. While gain-of-function mutations within the α_1-subunit are responsible for TS, loss-of-function mutations in the β-subunit also contribute to the progression of BS. It has been reported that in Chinese hamster ovary (CHO) cells, the loss-of-function mutations in both α_1- and β-subunits led to decreased current density. Despite this, however, it is interesting to note the surface expression of $Ca_v1.2$ in BS was not affected when investigated in neuronal rat brain [11].

Despite the differences in mutations, these results, when taken together, strongly suggest that the function of $Ca_v1.2$ within the myocardium needs to be tightly regulated. As seen earlier, dysregulation of either the main pore-forming α_1-subunit or the auxiliary β-subunit could result in serious health complications.

5.1.5 Channelopathies in the CVS—$Ca_v1.3$

5.1.5.1 Cardiac Dysfunction/Arrhythmia

The $Ca_v1.3$ isoform is encoded by the *CACNA1D* gene, and like $Ca_v1.2$, channelopathies due to mutations in *CACNA1D* could result in abnormal heart functions. $Ca_v1.3$ is activated at a more hyperpolarizing potential relative to $Ca_v1.2$, enabling its contribution to pacemaking in the SAN and AVN [7]. In a study done to better understand the contributions of $Ca_v1.3$ mutations to cardiac symptoms, a family presenting with $Ca_v1.3$ mutations was recruited and their electrocardiogram (ECG) studied [12]. It was then reported that severe bradycardia was observed at rest between a 12- to 24-h period in individuals with $Ca_v1.3$ mutations but not in the healthy individuals. Yet, it is interesting to note that unlike $Ca_v1.2$ mutations, these affected individuals did not present with any abnormalities to either the QRS complex or the QT intervals, thereby indicating that this bradycardia was associated with the conductance of the nodes. Further screening and characterizing of mutations in this family revealed the insertion of a glycine residue in various positions of the α_1-subunit. Of note, the insertion of a glycine residue at position 404 was reported to be found only in the mutually exclusive exon 8b. Again, this exon is located in the S6 of DI, which is part of the activation gate. Functional studies using patch-clamp experiments into the differences of biophysical properties in tsA-203 cells between wild-type and mutant $Ca_v1.3$ revealed that the insertion of the glycine residue results in a nonconducting variant of $Ca_v1.3$. Moreover, reverse transcriptase-polymerase chain reaction (RT-PCR) results showed that exon 8b is predominant in the SAN and not in the ventricles of the myocardium. Thus, the cardiac symptom presented due to this $Ca_v1.3$ mutation is due to SAN dysfunction and arrhythmia and not due to ventricular dysfunctions.

5.1.6 Regulation of $Ca_v1.2$

As discussed earlier, it is imperative that the activity of Ca_v in the heart is tightly regulated to function within a specific narrow range, the deviation of which could lead to serious health implications. Therefore, it is only natural that various mechanisms, for example, posttranslational modifications (PTMs), are in place to properly regulate the function of these calcium channels. An in-depth review of $Ca_v1.2$ post-translation modification has been recently published [13].

$Ca_v1.2$ is known to be phosphorylated by various protein kinases, namely, serine/threonine protein kinases A (PKA), C (PKC), G (PKG), and Ca2+/Calmodulin-dependent protein kinase II (CaMKII) [14]. It has been proposed that phosphorylation of serine residues in Ca_v is a mechanism that regulates protein–protein interactions. Of note, PKA phosphorylation of serine 1928 (S1928) was reported to transiently disrupt the interaction between $Ca_v1.2$ and the β_2-adrenergic receptor (β_2AR). Yet, despite being the most abundant phosphorylation site of $Ca_v1.2$ in the myocardium, S1928 was not involved in the β-adrenergic upregulation of $Ca_v1.2$ in the heart. Interestingly, S1928 can also be phosphorylated by PKC. However, unlike the prior PKA-dependent phosphorylation event, S1928 phosphorylation by PKC was reported to increase $Ca_v1.2$ current density. Furthermore, PKG was also reported to phosphorylate S1928. In this case, however, a reduced calcium current was reported in the transfected human embryonic kidney (HEK) 293 cells. Despite phosphorylation of the same serine residue within the full-length $Ca_v1.2$ channel, it is intriguing that phosphorylation by different protein kinases resulted in different downstream effects on the channel. From this, it can be seen that S1928 phosphorylation, while important, was not solely responsible for the changes in the function of $Ca_v1.2$. Nevertheless, it is imperative that S1928 phosphorylation be considered as part of the possible mechanisms for regulating $Ca_v1.2$ channel function.

Ubiquitination is another common PTM known to regulate protein expression and turnover rates. Ret finger protein 2 (Rfp2), or tripartite motif-containing 13 (TRIM13), poly-ubiquitinates lysine residues located within the intracellular II–III loop, leading to proteasomal degradation of $Ca_v1.2$ [15]. However, it remains to be proven if Rfp2 physically associates with $Ca_v1.2$. Yet, when the β-subunit is co-expressed with the α_{1C}-subunit, there is a significant reduction in ubiquitination of the α_{1C}-subunit. Therefore, it can be hypothesized that the auxiliary β-subunit is necessary for the expression of $Ca_v1.2$. A recent study has shown that disruption of the interaction between the α_{1C}-subunit and the β-subunit by a novel binding partner, Galectin-1 (Gal-1), led to increased ubiquitination and turnover of $Ca_v1.2$ α_1-subunit protein [6]. More importantly, the study also revealed that overexpression of Gal-1 is associated with hypotension. Conversely, knockdown of Gal-1 protein via siRNAs led to increased blood pressure. It can thus be said that although the α_1-subunit is the main pore-forming subunit, its expression at the protein level is regulated tightly by auxiliary subunits like the β-subunit. Of interest, Gal-1 binding to the I–II loop is hugely reduced in the presence of alternative exon 9*, an exon that is selectively expressed in smooth but not cardiac muscle.

5.1.7 Conclusion

All in all, this section has discussed the importance of LTCC $Ca_v1.2$ and $Ca_v1.3$ in the CVS. Modulation of Ca_v function has to be tightly regulated such that its activity remains within a physiological range, outside of which could lead to the pathogenesis of cardiovascular diseases. Both de novo or familial genetic mutations of the *CACNA1C* gene are associated with a variety of diseases comprising cardiac and extracardiac symptoms. Thus, further studies to understand the regulation of $Ca_v1.2$ in the context of cardiovascular diseases are necessary to better treat cardiac diseases and improve the quality of life of patients.

5.2 $Ca_V2.1$

Much like the previous Ca_V1 channel family, channels in the Ca_V2 channel family comprise hetero-multimetric assemblies of a pore-forming $Ca_V\alpha_1$ subunit, auxiliary $Ca_V\beta$, and $Ca_V\alpha_2\delta$ subunits, with the α_1-subunit determining the channel subtype. There are three-channel subtypes in the Ca_V2 family, $Ca_V2.1$, $Ca_V2.2$, and $Ca_V2.3$, that are encoded by three genes *CACNA1A*, *CACNA1B*, and *CACNA1E*, respectively [16–18]. In this chapter, we will focus mainly on the history of $Ca_V2.1$ as well as the structure and function of this channel along with the various mechanisms that govern its modulation.

5.2.1 History of $Ca_V2.1$ Channel

$Ca_V2.1$ gives rise to both the P-type Ca^{2+} currents, aptly named P-type as it was first recorded in Purkinje neurons [19, 20], and Q-type Ca^{2+} currents that were first recorded in cerebellar granule cells [21]. Although both currents displayed very similar electrophysiological properties, they differed in their inactivation kinetics as well as their sensitivity to ω-Agatoxin IVA, with P-type Ca^{2+} channels exhibiting a higher sensitivity to ω-Agatoxin IVA [21]. Hence, it was initially thought that P-type and Q-type Ca^{2+} currents were generated by different α1 subunits [21], although both P-type and Q-type currents were found to be required for neurotransmitter release [22–25]. What eventually resolved this mystery was a combination of molecular cloning and electrophysiological methods.

The amino acid sequence of $Ca_V2.1$ was first obtained in 1991 from the cloned cDNA sequence of a rabbit brain calcium channel, then referred to as BI [16], or also as class A calcium channel by Snutch's group [26]. Northern blot analysis of the BI voltage-gated calcium (Ca_V) channel showed that it was highly expressed in the brain, particularly in the cerebellum, and at a far lower level in the heart. Using mutant mice models with different patterns of cerebellar degeneration, it was deduced that BI calcium channels were highly expressed in both Purkinje cells and granule cells [16]. This was later confirmed in a later study where the spatial distribution of BI mRNA in the rat cerebellum (rbA for rat variant) showed a high expression of rbA mRNA in Purkinje cells and a lower expression of rbA mRNA in granule cells [27]. The electrophysiological and pharmacological properties as well as expression patterns were also similar to that observed of P-type channels in Purkinje cells, leading to postulations that BI/rbA Ca^{2+} channels correspond to the P-type Ca^{2+} channels [28], although there was initial caution to concluding that the BI channel was the same as the P-type Ca^{2+} channel due to certain differences in sensitivity to ω-Agatoxin IVA and inactivation kinetics [29].

Subsequently, there was also standardization of the calcium channel nomenclature, with class A calcium channel subunit (α_{1A}) gradually being the accepted name. Evidence for α_{1A} corresponding to P-type Ca^{2+} channel became more compelling when a study co-expressed different calcium channel β subunits with rat α_{1A} in *Xenopus* oocytes and was able to generate a current waveform similar to the P-type Ca^{2+} current observed in Purkinje cells when α_{1A} was expressed with β_{2a}. Interestingly, when α_{1A} was expressed with β_{1b} or β_3, current waveforms similar to the Q-type Ca^{2+} current observed in cerebellar granule cells were produced. A more native expression system involving rat neostriatal and cortical neurons further confirmed the effect of the β-subunit expression on α_{1A} channel kinetics [30]. Coupled with the in situ localization results showing that α_{1A} transcripts were also found to be highly expressed in Purkinje cells and granule cells, cells where P-type channels are highly expressed, it was highly conceivable that α_{1A} transcripts could correspond to P and Q-type Ca^{2+} channels [27]. The most convincing piece of evidence for α_{1A} giving rise to P and Q-type Ca^{2+} channels finally came when screening of a rat brain cDNA library was performed [31]. Three alternatively spliced α_{1A} isoforms were identified, with a valine insertion in the domain I–II linker (Val_{421}), an insertion of Asn and Pro (N_{1605}–P_{1606}), due to the use of an alternative 5′ splice donor site, in the extracellular linker between transmembrane segments S3 and S4 of domain IV, and 10 different amino acid residues in the carboxyl tail adjacent to domain IV S6 (also identified as EF hand-like motif). Expression of the different splice variants in *Xenopus* oocytes revealed alterations in channel properties. When α_{1A} was co-expressed with β_{2a}, there was slower inactivation kinetics similar to that observed in P-type Ca^{2+} channels. On the other hand, when α_{1A} was co-expressed with β_1, β_3, or β_4 subunits, faster inactivation kinetics similar to those observed in Q-type Ca^{2+} channels was seen. Sensitivity to ω-Aga-IVA was also found to be affected by alternative splicing. When the α_{1A} splice variant lacking N_{1605}–P_{1606} was expressed in HEK 293 cells, the sensitivity of α_{1A} to ω-Aga IVA was reduced, similar to native P-type channels. α_{1A} splice variants containing N_{1605}–P_{1606} was shown to be more sensitive to ω-Aga IVA, resembling Q-type channels. It was thus proposed that both P-type and Q-type are phenotypic variants arising from the alternative splicing of the α_{1A} transcript [31], hence solving the issue of the molecular origins of the P/Q-type Ca^{2+} channel.

In the following sections, we will continue to explore alternative splicing and its possible physiological roles.

5.2.2 $Ca_V2.1$ Channel Diversity

Following the discovery of the three alternatively spliced loci in the human *CACNA1A*, further studies have revealed a total of seven and nine alternative splice sites in human (Fig. 5.1) and rat cerebellum, respectively [32, 33]. However, between human and rat cerebellum, there are only five common splice loci, some of which we will examine in greater detail.

5.2.3 Exon 31

Exon 31 encodes for the NP insert in the extracellular loop of $Ca_V2.1$ domain IV S3–S4 and

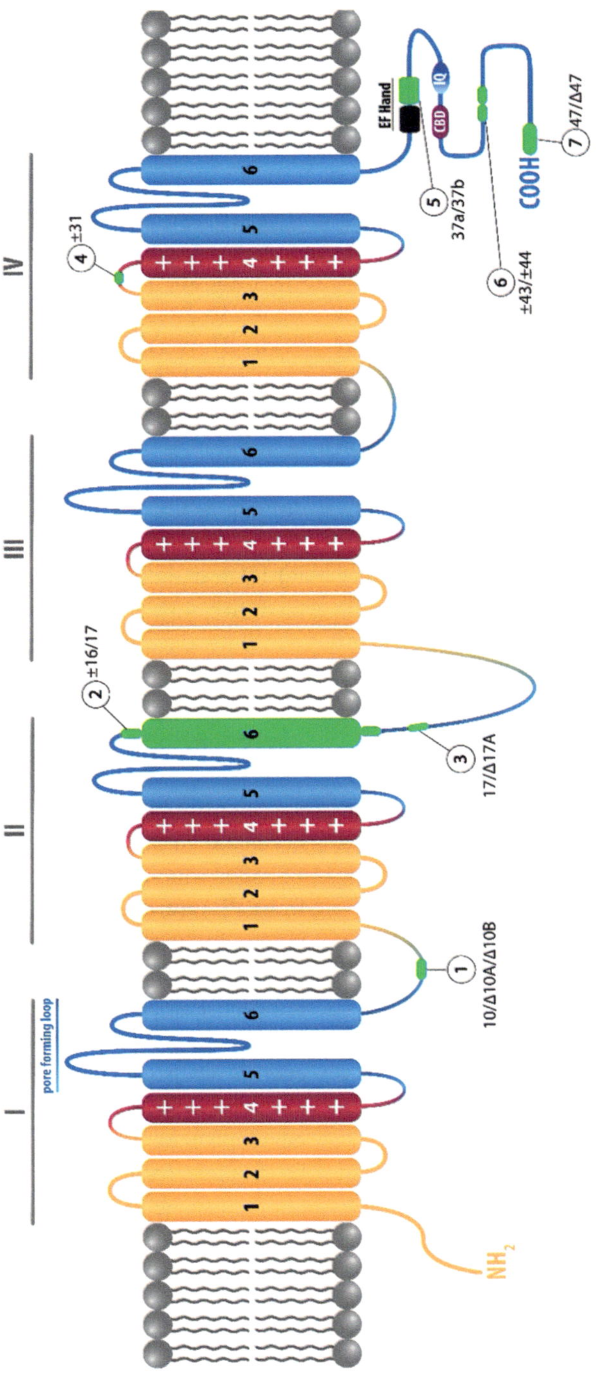

Fig. 5.1 Seven splice loci of human α_{1A} identified by transcript scanning

inclusion of NP decreases $Ca_V2.1$ affinity for ω-Aga-IVA as well as slowing channel activation and deactivation kinetics [31, 34]. It is worth noting that NP is encoded by a 6 nucleotide mini-exon (AATCCG) flanked by GT/AG acceptor-donor sites which are nested within intron 31 of humans, which is different from that observed in the rat variant where inclusion of NP was due to an alternative 5′ splice donor site located at the 3′ end of exon 31 [31]. When examining splice variant distribution in a human cerebellum cDNA library, 95% of the splice variants examined contained the +NP variant [31]. However, when single-cell RT-PCR of rat Purkinje cells was performed, the majority of the $Ca_V2.1$ transcripts were found to lack exon 31, which agrees with the required concentration of ω-Aga-IVA needed to almost completely block Ca^{2+} channel currents in rat Purkinje cells [24]. One possible explanation as to why the results from the human cerebellar cDNA library and the single-cell RT PCR of rat Purkinje cells do not match could be due to the proportion of Purkinje cells to granule cells in the cerebellum, where the granule cells can outnumber the Purkinje cells by 100–300 times [35, 36]. A human cDNA library constructed from bulk cerebellar tissue would hence have a higher proportion of granule cell transcripts, possibly masking the exact transcript expression profile in Purkinje cells.

5.2.4 Exon 37

Exon 37, together with exon 36, encodes for a Ca^{2+} binding site, known as an EF-hand-like motif, comprising 29 amino acids [31, 37, 38]. The general EF-hand motif consists of an α-helix E, a linker region, and an α-helix F [39]. Exon 37 encodes for part of the linker region as well as the α-helix F region. Mutually exclusive splicing of exon 37 results in variants EFa and EFb [31, 33]. Both splice variants are highly similar, with only a substitution of 10 amino acids out of a region of 30 amino acids [31]. Electrophysiological characterization involving transient overexpression of splice variants in HEK 293 cells revealed that both splice variants possessed almost similar properties, with the exception of EFa being able to undergo Ca^{2+}-dependent facilitation (CDF) and Ca^{2+}-dependent inactivation (CDI) (both processes will be explored in greater detail later), while EFb was only able to undergo CDI [40]. Subsequent experiments involving single-channel recordings of the two splice variants showed that EFa, when in the CDF mode, had an increase in channel open probability as well as an increase in channel open duration. On the other hand, EFb was found to be locked in a normal gating mode [41]. How the differences in amino acid sequences between the two splice variants determine whether the channel is able to undergo CDF is still unknown, but chimeric channel analysis involving the EF-hand region and the pre-IQ-IQ domain of $Ca_V2.1$ and $Ca_V1.3$ as well as $Ca_V2.2$ showed that both the EF-hand region and the pre-IQ-IQ domain of $Ca_V2.1$ are required for eliciting CDF [42, 43].

While the physiological significance for the splice variants is not known, an unequal distribution of splice variants has been observed in the cerebellum, with EFa being the dominant splice variant expressed in adulthood [44, 45]. However, it was also observed in mice where EFb was the dominant variant from the embryonic stage up till P5, after which there was a developmental switch from P7 onward to an EFa dominant expression pattern in the cerebellum [45]. Coincidentally, the developmental switch observed in the cerebellum between P5 and P7 coincides with a time period where a climbing fiber (CF) is selectively strengthened, a process that is critical for proper cerebellar circuitry formation [46], hence suggesting that the developmental switch of the splice variants could be required for CF strengthening and proper formation of cerebellar circuitry.

Chronic reduction of network activity in rat primary hippocampal neuron cultures via toxin application resulted in the upregulation of EFa but not EFb mRNA, with the greatest increase observed at 24 hr. An increase in $Ca_V2.1$ protein expression was also observed following network deprivation, which was blocked when EFa was knocked down with miRNA, suggesting that neurons are able to control EFa and EFb splice variant levels in a homeostatic manner [47].

While the verdict is still open on the physiological significance of exon 37 splice variants, the evidence presented earlier seems to suggest that modulation of splice variant expression is activity dependent. To date, the splicing factors governing the expression of exon 37 splice variants are still unknown.

5.2.5 Exon 43/44

Alternative splicing of exons 43 and 44 gives rise to combinations where both exons can be present or absent in four combinations (±43/±44). Transcript scanning of a human cerebellum cDNA library showed that $43^+/44^+$ splice variant was the most abundant (80–90% of total colonies counted), followed by $43^+/44^-$, $43^-/44^+$ and $43^-/44^-$ [33, 48]. Overexpression of the different splice variants in HEK 293 cells followed by electrophysiological characterization revealed that alternative splicing at exons 43 and 44 did not affect CDF. However, Ca^{2+} current amplitude and CDI were shown to be affected. CDI was the greatest in $43^-/44^-$, while the other splice variant combinations had similar CDI values. $43^-/44^-$ splice variant had a Ca^{2+} current amplitude that was twofold larger than that of the $43^+/44^+$ splice variant [33]. Given the increases in current amplitude and CDI, with CDI being affected by global increases in Ca^{2+}, it was postulated that the various splice variants at this particular locus affect surface channel expression. Furthermore, the presence of exon 44 was required for binding to Rab3-interacting molecules (RIM), scaffolding proteins important for the anchoring of $Ca_V2.1$ to synaptic vesicles at the active zone. Disrupting binding of $Ca_V2.1$ to α-RIMs was found to result in a release of suppression of voltage-dependent inactivation (VDI), and hence, $Ca_V2.1$ splice variants lacking exon 44 were found to have a stronger VDI [48].

5.2.6 Exon 47

Addition of a pentanucleotide GGCAG at the start of exon 47 results in an in-frame translation of exon 47 to produce a long version of the C terminus (termed as 47), allowing for the insertion of a polyglutamine (polyQ) tract. On the other hand, omission of GGCAG in splice variants causes a frameshift, resulting in the generation of a stop codon near the beginning of exon 47, thus producing a shorter variant of the channel (Δ47) [33, 49]. Studies involving knock-in mouse models of the human splice variants at this particular locus show no change in the intrinsic electrophysiological properties of the calcium channel [50–52], although studies utilizing heterologous transient overexpression systems provided conflicting evidence [53–55]. This further highlights the need to study the effects of calcium channel mutations, specifically polyQ expansion discussed here, in a more physiological setting. While the basic properties of the calcium channel are not altered by alternative splicing, it must be noted that the longer splice variant containing an expanded polyglutamine tract is an underlying cause for spinocerebellar ataxia 6 (SCA6), which will be covered in greater detail in the later section subsequently concerning $Ca_V2.1$ channelopathies.

5.2.7 Modulation of Calcium Channels

5.2.7.1 Voltage-Dependent Inactivation (VDI)

VGCCs open in response to membrane depolarization, allowing for the influx of Ca^{2+} ions. However, as Ca^{2+} is an important second messenger for many molecular pathways in the cell, there exist feedback mechanisms such as VDI and Ca^{2+}-dependent inactivation (CDI) to prevent unregulated and excessive Ca^{2+} entry.

VDI is common across all VGCC subtypes, but the degree of VDI varies for individual subtypes and is greatly modulated in high voltage-activated (HVA) channels, such as Ca_V1

and Ca_V2, by $Ca_V\beta$ subunits [56, 57]. VDI has both fast and slow components to it, with fast inactivation occurring on a timescale of milliseconds, while slow inactivation is generally observed over a more prolonged membrane depolarization time frame (~1 min) and is not as well understood [58, 59]. Hence for this section, we will be focusing on fast VDI.

Generally, when HVA channels are co-expressed with β_{1b} or β_3 subunits, VDI is observed to be accelerated, while β_4 is observed to not affect VDI by much and the inactivation kinetics are like that of α_1 alone. However, when expressed with β_{2a}, inactivation kinetics is significantly slowed down [27, 60].

To study VDI on its own, Ba^{2+} is used as the charge carrier instead of Ca^{2+} to avoid evoking Ca^{2+}-dependent inactivation [61]. The earliest attempt to uncover structural determinants of calcium channel inactivation involved chimeric channels produced from rat $Ca_V2.1$ and marine ray $Ca_V2.3$. Subsequent electrophysiological analysis of the chimeric channel showed that differences in domain IS6 of $Ca_V2.1$ and $Ca_V2.3$ were the reason for the difference in inactivation kinetics between the two Ca_V2 subtypes [62]. Further evidence for the importance of S6 segment to VDI came from studies where the S6 segments of domain II or III of slowly inactivating $Ca_V1.2$ were swapped with the corresponding S6 segments of rapidly inactivating $Ca_V2.3$, transforming $Ca_V1.2$ to a rapidly inactivating phenotype [63]. Additionally, mutations in IIS6 and IVS6 regions of $Ca_V2.1$ in familial hemiplegic migraine 1 (FHM1) patients resulted in altered inactivation kinetics [64, 65]. Introduced point mutations in IIS6, IIIS6, and IVS6 of L-type calcium channels were also found to underlie an increase in inactivation kinetics [66–68]. One other compelling piece of evidence for S6 to be critical for VDI comes from the G406R mutation in TS patients, as mentioned in the earlier section. G406 is at the end of IS6 segment, and mutation to the G406R results in a near-total abolishment of VDI [69]. These studies thus support the notion that all four S6 regions of HVA channels are involved in VDI.

Apart from S6 segments of HVA channels being implicated in VDI, the intracellular loop linking domain I–II has been implicated in VDI [64, 68, 70, 71], leading some to propose the hinged-lid model as a molecular basis for VDI [72]. An alternative theory based on current kinetics and steady-state activation of $Ca_V1.2$ proposed that the calcium channel can exist in four different states, a closed resting state where the pore is closed and the voltage sensors lock the pore (R), an activated closed state where the pore is closed but the voltage sensors are in their "up" position (A), an activated open state where the pore is open and the voltage sensors are in their "up" position (O), and a deactivated open state where the pore is open but the voltage sensors are in their down position [73]. A recent review by Hering et al. [74] discusses this model in greater detail along with the recently discovered crystal structure of $Ca_V1.1$ [75, 76].

5.2.7.2 Ca^{2+}-Dependent Inactivation (CDI)

CDI serves as a form of negative feedback regulation for VGCCs and was first discovered in Paramecium where it was observed that inactivation of VGCCs occurred at a faster rate in solutions containing Ca^{2+} than solutions containing Ba^{2+}, and increased buffering of intracellular Ca^{2+} resulted in decreased CDI, suggesting that accumulation of intracellular Ca^{2+} is required for CDI [77, 78]. This was also supported by the observation that CDI has a U-shaped dependency on voltage, a very telling sign of a Ca^{2+}-regulated process [79]. However, the extent to which Ca^{2+} chelators affect CDI differs between $Ca_V\alpha$ subunits and will be explored further subsequently.

To study CDI, a double-voltage pulse protocol is used where a prepulse voltage step to various potentials is delivered, and a test pulse of a fixed voltage is delivered after a brief pause [77]. The difference between the remaining current of Ba^{2+} and Ca^{2+} after a stipulated duration is taken to be a measure of pure CDI. [80] It must however be noted that for channels exhibiting faster inactivation kinetics such as Ca_V1 channels, shorter depolarizing pulses, and consequently a shorter index r will be used [81].

Advances in experimental techniques allowed for the isolation and recording of Ca^{2+} currents in both multicellular and single myocytes, revealing the existence of CDI [82–84]. Single-channel recordings of an L-type calcium channel from an adult rat cardiac myocyte revealed that Ca^{2+} influx of a single channel was sufficient for CDI to occur [80]. However, while it was agreed that Ca^{2+} influx through the calcium channel was responsible for providing negative feedback, the mechanism by which it exerted its effects was then still largely unknown. Identification of an EF-hand Ca^{2+} binding motif in the C-terminal region of $Ca_V1.2$ [38] gave rise to the possibility that the binding of Ca^{2+} to this region could be responsible for the initiation of CDI. Chimeric channel studies involving $Ca_V1.2$ (observed to have strong CDI) and $Ca_V2.3$ (observed to have little CDI) helped to provide the first evidence for the EF-hand motif to be involved in CDI. When the C-terminal region of $Ca_V2.3$ and $Ca_V1.2$ were swapped, it resulted in $Ca_V1.2$ exhibiting weak CDI and $Ca_V2.3$ exhibiting strong CDI [85]. These results implied that the proximal third of the carboxyl terminal (CI region) as well as the EF-hand motif within this region were integral to CDI. Subsequent studies utilizing point mutations in the critical residues within the EF-hand region required for Ca^{2+} binding were unable to completely eliminate CDI, suggesting that the Ca^{2+} sensor may reside somewhere else within the CI region [86, 87].

Calmodulin (CaM), a Ca^{2+} binding protein, is a ubiquitously expressed cytoplasmic protein and is involved in many Ca^{2+}-related processes. Downstream of the EF-hand region is an IQ CaM-binding motif that was shown to be essential for CDI [88]. Coexpression of mutant CaM that was unable to bind Ca^{2+} along with $Ca_V1.2$ resulted in ablation of CDI, confirming the role of CaM as a Ca^{2+}-sensor in initiating CDI of L-type calcium channels. Glutathione S-transferase (GST) pulldown experiments involving GST fusion proteins with various regions of $Ca_V1.2$ CI region confirmed that the IQ domain in $Ca_V1.2$ is the site for CaM interaction [89]. GST pulldown experiments also demonstrated that the IQ domains of Ca_V2 family were able to interact with CaM in a Ca^{2+}-dependent manner. Further evidence for CaM being the Ca^{2+} sensor for VGCCs came from studies where Ca^{2+}-free CaM (apoCaM) being pre-associated with the α_1-subunit were reported [90–92].

Using the C-terminus of the rat $Ca_V2.1$ α_1-subunit as bait, a yeast two-hybrid screen of a rat brain cDNA library identified CaM to be interacting with the C-terminus. When low Ca^{2+} buffering conditions were used (0.5 mM EGTA), CDI of $Ca_V2.1$ channels was observed [93]. On the other hand, under high Ca^{2+} buffering conditions (10 mM EGTA), CDI was not observed in transfected cells [81, 94], suggesting that CDI in $Ca_V2.1$ is dependent on global increases in Ca^{2+}. This hypothesis was later confirmed with recombinant mutant CaM co-expression with $Ca_V2.1$ in HEK 293 cells, where it was observed that Ca^{2+} binding to the N-lobe of CaM (sensitive to global increases in Ca^{2+}) was responsible for CDI [81, 95]. This is in contrast to CDI of $Ca_V1.2$, where CDI was observed even in high Ca^{2+} buffering conditions and even at single-channel level [96], demonstrating that CDI in $Ca_V1.2$ is dependent on local increases in Ca^{2+} near the channel pore. In this initial study demonstrating the importance of CaM to Ca^{2+} regulation of $Ca_V2.1$ channels, a site downstream of the IQ-like motif, calmodulin-binding domain (CBD), was thought to be vital for CaM binding and consequently Ca^{2+} regulation of the channel [93, 94, 97]. However, subsequent studies have called the importance of this binding site into question, where deletion of CBD was found to have no effect on CDI [44, 81]. However, there is consensus for the importance of the IQ-like motif for CDI of $Ca_V2.1$ channels to occur [43, 81, 97, 98].

While the abovementioned lines of evidence present a common mechanism by which VGCCs are Ca^{2+} regulated, the IQ domain, discovery of another site for CaM binding in $Ca_V1.2/1.3$ but not Ca_V2 channels adds another layer of control to Ca^{2+} regulation of VGCCs. Given that C-lobe of CaM has a stronger binding affinity to Ca^{2+}, it has always been traditionally viewed that local increases in Ca^{2+} drive CaM C-lobe mediated processes, while global increases in Ca^{2+} were thought to only drive CaM N-lobe mediated processes [95, 99]. A chimeric channel consisting of the N-terminus of $Ca_V1.2$ and $Ca_V2.2$ was found to demonstrate strong CDI even in the presence of strong Ca^{2+} buffering. Likewise, when the N-terminus of $Ca_V1.2$ was substituted in $Ca_V2.1$, CDI was observed under strong buffering conditions, demonstrating CDI driven by local Ca^{2+} increase, unlike the usual CDI driven by the global increase in Ca^{2+} in Ca_V2 channels [81, 100]. Subsequent alanine point mutations of the N-terminus combined with in situ FRET as well as electrophysiological experiments confirmed the existence of a CaM-binding element in the N-terminus of $Ca_V1.2/1.3$ which was subsequently named as NSCaTE [100]. NSCaTE was shown to bind to the N-lobe of Ca^{2+}/CaM with a local Ca^{2+} sensitivity, resulting in CDI.

5.2.7.3 Ca^{2+}-Dependent Facilitation (CDF)

On the flipside of CDI is Ca^{2+}-dependent facilitation (CDF), a positive feedback regulatory mechanism of VGCCs where further Ca^{2+} entry is increased due to enhanced channel opening. CDF has been observed in $Ca_V1.2$, $Ca_V1.3$, $Ca_V2.1$, and Ca_V3 channels.

The common mechanism underlying CDF in $Ca_V1.2/1.3$ is through the actions of CaMKII [101, 102]. In $Ca_V1.2$, CaMKII becomes tethered to the $\alpha 1$ subunit near the IQ domain, where it is then able to be activated by Ca^{2+}/CaM and subsequently phosphorylating the channel and inducing CDF [103, 104]. CaMKII also induces CDF in $Ca_V1.2$ by phosphorylating β_{2A}, which when bound to $Ca_V1.2$ increases channel open probability, resulting in CDF [105–107].

Like $Ca_V1.2$, CDF of $Ca_V1.3$ is also dependent on CaMKII. However, when $Ca_V1.3$ and CaMKII were co-expressed in HEK 293T cells, no CDF was observed. Only when $Ca_V1.3$, CaMKII, and densin-180, a scaffolding protein that is highly expressed at excitatory synapses and CaMKII-interacting, was CDF elicited [108, 109]. In hippocampal neurons, a long duration form of CDF was found to be dependent on CaMKII and densin-180, and also functionally coupled to $K_{Ca}3.1$, an intermediate-conductance Ca^{2+}-gated potassium channel that is thought to be important in regulating neuronal excitability [110].

CDF of $Ca_V2.1$ is widely studied in the Calyx of Held synapse, where Ca^{2+} currents are largely mediated by $Ca_V2.1$ channels [111]. At this synapse, when high-frequency depolarizations are applied, CDF followed by CDI is observed in $Ca_V2.1$ channel recordings [112]. Efforts to elucidate the molecular mechanisms underlying CDF came in the form of studies overexpressing recombinant $Ca_V2.1$ where it was discovered that like CDI, CDF is also dependent on CaM binding to the IQ domain of $Ca_V2.1$ [81, 93, 94]. As mentioned earlier, the N-lobe of CaM detects increases in global Ca^{2+} and induces CDI in $Ca_V2.1$ channels. The binding of Ca^{2+} to the C-lobe of CaM is largely triggered by local Ca^{2+} increase and results in CDF [81], with single-channel recordings confirming the local Ca^{2+} sensitivity of CaM in inducing CDF [41]. In L-type channels, binding of Ca^{2+} to the C-lobe of CaM results in CDI while in $Ca_V2.1$, CDF is evoked. How is it possible that binding to the same C-lobe results in opposite Ca^{2+} regulation? Crystal structures of Ca^{2+}/CaM bound to the IQ peptide of $Ca_V2.1$ channels revealed that CaM is able to exist in both parallel and antiparallel conformations [43, 98]. The ability of the C-lobe of CaM to bind in an antiparallel conformation upstream of the IQ domain in $Ca_V2.1$, as opposed to a parallel conformation for L-type channels, led to the proposition that this difference in binding conformation underlies the difference in CDF or CDI [98]. Structural analysis of the Ca^{2+}/CaM-$Ca_V2.1$ complex along with alanine scanning mutagenesis and chimeric channel experiments suggested that differences in CaM C-lobe mediated effects could be due to differences in multiple interactions with binding sites within and upstream of the IQ domain. Chimeric channel experiments pointed to an important role for the EF-hand domain, preIQ-domain, and the IQ domain for CDF to occur, as substitution of any of the abovementioned regions with the corresponding region from $Ca_V1.3$ resulted in a decrease in facilitation [43]. Coincidentally, a recent chimeric channel experiment involving $Ca_V2.1$ and $Ca_V2.2$ also demonstrated the importance of these three domains [42]. Furthermore, alternative splicing of exon 37 of $Ca_V2.1$ results in mutually exclusive splice variants, of which one preserves CDF while the other splice variant does not support CDF [40].

These studies, while mainly centered on the IQ-domain as being critical to Ca^{2+}/CaM regulation of VGCC, also demonstrate the importance of other regions in the vicinity of the IQ-domain, which will be important to consider when investigating the effects of Ca^{2+}/CaM-mediated regulation.

5.2.7.4 $Ca_V2.1$ Channelopathies

Human mutations in $Ca_V2.1$ α_1-subunit result in several neurological disorders in an autosomal-dominant inheritance fashion, such as episodic ataxia type 2 (EA2), familial hemiplegic migraine type 1 (FHM1), and spinocerebellar ataxia type 6 (SCA6) [49, 113]. It must be noted that with advances in RNA sequencing, changes in alternative splicing of $Ca_V2.1$ have been implicated in psychiatric disorders such as autism spectrum disorder (ASD) [114, 115].

5.2.8 Episodic Ataxia Type 2 (EA2)

EA2 is a neurological disorder that commonly affects patients in their childhood and is characterized by paroxysmal attacks of ataxia, vertigo, nausea, and may be accompanied by migraine. Symptoms may last from minutes to days. Emotional and physical stress as well as caffeine and alcohol may trigger attacks. In between attacks, EA2 patients may exhibit a progressive cerebellar syndrome with nystagmus as well as cerebellar atrophy in the vermis [116, 117]. Most of the EA2-related *CACNA1A* mutations reported disrupting the open reading frame resulting in a truncated channel. The EA2 mutations that do not affect the open reading frame are mainly located in the pore-forming loops of the channel [117, 118].

5.2.9 Familial Hemiplegic Migraine Type 1 (FHM1)

Clinically, patients with FHM1 exhibit hemiplegic migraines with an aura that might be accompanied by other symptoms such as nausea and ataxia and may last for hours to weeks. Some FHM1 patients exhibit permanent cerebellar symptoms comprising progressive cerebellar ataxia that might be accompanied by nystagmus [119, 120]. Mutations in *CACNA1A* have been identified as the underlying genetic cause of FHM1 [113]. The majority of the FHM1 mutations in *CACNA1A* affect conserved amino acid residues in critical functional regions of $Ca_V2.1$ such as the pore-lining loop as well as the S4 voltage sensors (see [121, 122] for reviews and references). The functional changes brought about by the FHM1 mutations have been probed in heterologous expression systems and reveal that the mutations bring about a gain-of-function where there is increased Ca^{2+} influx as a result of increased channel open probability and a decrease in voltage activation of the channel [123–125]. A similar result was also observed in knock-in mouse models of R192Q and S218L mice, where there was a larger $Ca_V.2.1$ current density in cerebellar granule cells despite no change in membrane expression of functional $Ca_V2.1$ channels [126, 127].

Interestingly, alternative splicing of *CACNA1A* at exon 47 affects the impact of FHM1 mutations on the channel kinetics. Using three FHM1 mutations, R192Q, S218L, and K1336E, expressed in a $Ca_V2.1$ (+47) and $Ca_V2.1$ (Δ47) background, it was shown that the FHM1 mutations result in different biophysical changes in channel properties depending on which splice variant the mutations are expressed in. Depending on the proportion of splice variants expressed in various neuronal subtypes, the same FHM1 mutations could result in varying degrees of symptoms [128].

5.2.10 Spinocerebellar Ataxia Type 6 (SCA6)

SCA6 is a polyglutamine disease caused by expansion of the trinucleotide (CAG) repeat in exon 47 of the *CACNA1A* gene, with longer expansions resulting in earlier onset of the disease [129]. SCA6 patients have a late onset of mild but slowly progressive cerebellar ataxia affecting the limbs and gait as well as slight vibratory and proprioceptive sensory loss and may be accompanied by nystagmus and dysarthria. Post-mortem examination of the cerebellum revealed a severe loss of Purkinje cells particularly in the vermis, moderate loss of granule cells, dentate nucleus neurons as well as neuronal loss in the inferior olive [49, 130]. In situ hybridization with probes for detecting exons 46–47 of human *CACNA1A* mRNA as well as RT-PCR showed that Purkinje cells had the highest expression of $Ca_V2.1$ protein with an expanded polyglutamine tract and partially explains the selective neurodegeneration observed in the cerebellum of SCA6 patients [131]. Unlike most polyglutamine diseases, SCA6 rarely results in intranuclear inclusions but instead, cytoplasmic aggregates are more commonly observed and were found to be sufficient to result in cell death [132, 133]. It was initially thought that the polyglutamine expansion of $Ca_V2.1$ would affect channel function and subsequently result in cell death [53–55]. However, as discussed earlier, the knockin mouse models of SCA6 showed no differences in $Ca_V2.1$ channel kinetics in cerebellar Purkinje cells [50–52].

Apart from its channel function, another interesting role for $Ca_V2.1$ was discovered. Similar to the C-terminal fragment of L-type channels translocating to the nucleus and functioning as a transcription factor [134], a 75 kDA C-terminal fragment of $Ca_V2.1$ (α1ACT) was found in the nucleus of Purkinje cells and was found to be toxic when the fragment contained the SCA6 polyglutamine expansion [135]. *CACNA1A* was later found to be bicistronic, with an internal ribosomal entry site (IRES) encoding for α1ACT that also served as a transcription factor for a number of genes found to be important for proper Purkinje cell and neural development. When α1ACT contained the SCA6 polyglutamine expansion, its normal gene expression regulation was abolished and resulted in increased cell death [136].

5.2.11 Psychiatric Disorders

Genome-wide association studies (GWAS) have uncovered a strong link between SNPs in *CACNA1C* and psychiatric disorders such as schizophrenia (SCZ), bipolar disorder, and major depressive disorder, which is not entirely unexpected given its high expression in affected brain regions such as the hippocampus [114, 137–141]. $Ca_V2.1$ was however not shown to be linked to any psychiatric disorders in the GWAS studies although some FHM1 and EA2 patients also presented with SCZ symptoms, difficulties in learning, and attention deficit disorder [142]. Interestingly, analysis of transcript isoforms from the PsychENCODE study showed changes in alternative splicing of *CACNA1A* in ASD [114].

5.3 Conclusion

In this chapter, we presented the varied roles in the body that VGCCs are involved in, from the heart to skeletal muscles and the brain. Given its importance, many mechanisms exist to modulate channel function, preventing excitotoxicity due to Ca^{2+} overload. Changes in splice isoforms have also been shown to alter channel kinetics and function and may be a way in which the cell modulates Ca^{2+} entry. Much is still unknown about the combinatorial effects of the splice variants as well as the splice factors responsible for the different splice variant expressions. More work will definitely be required if we wish to modulate VGCCs for therapeutic treatments.

References

1. Jeftinija DM et al (2007) The Ca(V) 1.2 Ca(2+) channel is expressed in sarcolemma of type I and IIa myofibers of adult skeletal muscle. Muscle Nerve 36:482–490. https://doi.org/10.1002/mus.20842

2. Takahashi M, Seagar MJ, Jones JF, Reber BF, Catterall WA (1987) Subunit structure of dihydropyridine-sensitive calcium channels from skeletal muscle. Proc Natl Acad Sci U S A 84: 5478–5482. https://doi.org/10.1073/pnas.84.15.5478
3. Gez LS, Hagalili Y, Shainberg A, Atlas D (2012) Voltage-driven Ca(2+) binding at the L-type Ca(2+) channel triggers cardiac excitation-contraction coupling prior to Ca(2+) influx. Biochemistry 51:9658–9666. https://doi.org/10.1021/bi301124a
4. Buraei Z, Yang J (2013) Structure and function of the β subunit of voltage-gated Ca^{2+} channels. Biochim Biophys Acta (BBA) – Biomembr 1828:1530–1540. https://doi.org/10.1016/j.bbamem.2012.08.028
5. Hidalgo P, Gonzalez-Gutierrez G, Garcia-Olivares J, Neely A (2006) The alpha1-beta-subunit interaction that modulates calcium channel activity is reversible and requires a competent alpha-interaction domain. J Biol Chem 281:24104–24110. https://doi.org/10.1074/jbc.M605930200
6. Hu Z et al (2018) Regulation of blood pressure by targeting CaV1.2-Galectin-1 protein interaction. Circulation 138:1431–1445. https://doi.org/10.1161/circulationaha.117.031231
7. Mangoni ME et al (2003) Functional role of L-type Cav1.3 Ca2+ channels in cardiac pacemaker activity. Proc Natl Acad Sci U S A 100:5543–5548. https://doi.org/10.1073/pnas.0935295100
8. Barrett CF, Tsien RW (2008) The Timothy syndrome mutation differentially affects voltage- and calcium-dependent inactivation of CaV1.2 L-type calcium channels. Proc Natl Acad Sci U S A 105:2157–2162. https://doi.org/10.1073/pnas.0710501105
9. Splawski I et al (2005) Severe arrhythmia disorder caused by cardiac L-type calcium channel mutations. Proc Natl Acad Sci U S A 102:8089–8096.; discussion 8086-8088. https://doi.org/10.1073/pnas.0502506102
10. Antzelevitch C et al (2007) Loss-of-function mutations in the cardiac calcium channel underlie a new clinical entity characterized by ST-segment elevation, short QT intervals, and sudden cardiac death. Circulation 115:442–449. https://doi.org/10.1161/circulationaha.106.668392
11. Simms BA, Zamponi GW (2012) The Brugada syndrome mutation A39V does not affect surface expression of neuronal rat Cav1.2 channels. Mol Brain 5:9. https://doi.org/10.1186/1756-6606-5-9
12. Baig SM et al (2011) Loss of Ca(v)1.3 (CACNA1D) function in a human channelopathy with bradycardia and congenital deafness. Nat Neurosci 14:77–84. https://doi.org/10.1038/nn.2694
13. Loh KWZ, Liang MC, Soong TW, Hu Z (2020) Regulation of cardiovascular calcium channel activity by post-translational modifications or interacting proteins. Pflügers Arch: Eur J Physiol. https://doi.org/10.1007/s00424-020-02398-x
14. Patriarchi T et al (2016) Phosphorylation of Cav1.2 on S1928 uncouples the L-type Ca^{2+} channel from the β2 adrenergic receptor. EMBO J 35:1330–1345. https://doi.org/10.15252/embj.201593409
15. Lerner M et al (2007) The RBCC gene RFP2 (Leu5) encodes a novel transmembrane E3 ubiquitin ligase involved in ERAD. Mol Biol Cell 18:1670–1682. https://doi.org/10.1091/mbc.e06-03-0248
16. Mori Y et al (1991) Primary structure and functional expression from complementary DNA of a brain calcium channel. Nature 350:398–402. https://doi.org/10.1038/350398a0
17. Dubel SJ et al (1992) Molecular cloning of the alpha-1 subunit of an omega-conotoxin-sensitive calcium channel. Proc Natl Acad Sci U S A 89:5058–5062. https://doi.org/10.1073/pnas.89.11.5058
18. Williams ME et al (1992) Structure and functional expression of alpha 1, alpha 2, and beta subunits of a novel human neuronal calcium channel subtype. Neuron 8:71–84. https://doi.org/10.1016/0896-6273(92)90109-q
19. Llinas R, Sugimori M, Lin JW, Cherksey B (1989) Blocking and isolation of a calcium channel from neurons in mammals and cephalopods utilizing a toxin fraction (FTX) from funnel-web spider poison. Proc Natl Acad Sci U S A 86:1689–1693. https://doi.org/10.1073/pnas.86.5.1689
20. Llinas RR, Sugimori M, Cherksey B (1989) Voltage-dependent calcium conductances in mammalian neurons. The P channel. Ann N Y Acad Sci 560: 103–111. https://doi.org/10.1111/j.1749-6632.1989.tb24084.x
21. Randall A, Tsien RW (1995) Pharmacological dissection of multiple types of Ca2+ channel currents in rat cerebellar granule neurons. J Neurosci 15:2995–3012
22. Wheeler DB, Randall A, Tsien RW (1994) Roles of N-type and Q-type Ca2+ channels in supporting hippocampal synaptic transmission. Science 264:107–111. https://doi.org/10.1126/science.7832825
23. Hillman D et al (1991) Localization of P-type calcium channels in the central nervous system. Proc Natl Acad Sci U S A 88:7076–7080. https://doi.org/10.1073/pnas.88.16.7076
24. Mintz IM et al (1992) P-type calcium channels blocked by the spider toxin omega-Aga-IVA. Nature 355:827–829. https://doi.org/10.1038/355827a0
25. Usowicz MM, Sugimori M, Cherksey B, Llinas R (1992) P-type calcium channels in the somata and dendrites of adult cerebellar Purkinje cells. Neuron 9:1185–1199. https://doi.org/10.1016/0896-6273(92)90076-p
26. Starr TV, Prystay W, Snutch TP (1991) Primary structure of a calcium channel that is highly expressed in the rat cerebellum. Proc Natl Acad Sci U S A 88:5621–5625. https://doi.org/10.1073/pnas.88.13.5621
27. Stea A et al (1994) Localization and functional properties of a rat brain alpha 1A calcium channel reflect similarities to neuronal Q- and P-type

channels. Proc Natl Acad Sci U S A 91:10576–10580. https://doi.org/10.1073/pnas.91.22.10576
28. Snutch TP, Reiner PB (1992) Ca2+ channels: diversity of form and function. Curr Opin Neurobiol 2: 247–253. https://doi.org/10.1016/0959-4388(92)90111-w
29. Tsien RW, Ellinor PT, Horne WA (1991) Molecular diversity of voltage-dependent Ca^{2+} channels. Trends Pharmacol Sci 12:349–354. https://doi.org/10.1016/0165-6147(91)90595-j
30. Mermelstein PG et al (1999) Properties of Q-type calcium channels in neostriatal and cortical neurons are correlated with beta subunit expression. J Neurosci 19:7268–7277
31. Bourinet E et al (1999) Splicing of alpha 1A subunit gene generates phenotypic variants of P- and Q-type calcium channels. Nat Neurosci 2:407–415. https://doi.org/10.1038/8070
32. Kanumilli S et al (2006) Alternative splicing generates a smaller assortment of CaV2.1 transcripts in cerebellar Purkinje cells than in the cerebellum. Physiol Genomics 24:86–96. https://doi.org/10.1152/physiolgenomics.00149.2005
33. Soong TW et al (2002) Systematic identification of splice variants in human P/Q-type channel alpha1 (2.1) subunits: implications for current density and Ca2+-dependent inactivation. J Neurosci 22:10142–10152
34. Hans M et al (1999) Structural elements in domain IV that influence biophysical and pharmacological properties of human alpha1A-containing high-voltage-activated calcium channels. Biophys J 76: 1384–1400. https://doi.org/10.1016/S0006-3495(99)77300-5
35. Harvey RJ, Napper RM (1988) Quantitative study of granule and Purkinje cells in the cerebellar cortex of the rat. J Comp Neurol 274:151–157. https://doi.org/10.1002/cne.902740202
36. Wetts R, Herrup K (1983) Direct correlation between Purkinje and granule cell number in the cerebella of lurcher chimeras and wild-type mice. Brain Res 312: 41–47. https://doi.org/10.1016/0165-3806(83)90119-0
37. Kretsinger RH (1976) Calcium-binding proteins. Ann Rev Biochem 45:239–266. https://doi.org/10.1146/annurev.bi.45.070176.001323
38. Babitch J (1990) Channel hands. Nature 346:321–322. https://doi.org/10.1038/346321b0
39. Kretsinger RH (2013) In: William J. Lennarz & M. Daniel Lane (Eds.), Encyclopedia of biological chemistry, 2nd ed., Academic Press, pp. 316–321.
40. Chaudhuri D et al (2004) Alternative splicing as a molecular switch for Ca^{2+}/calmodulin-dependent facilitation of P/Q-type Ca^{2+} channels. J Neurosci 24:6334–6342. https://doi.org/10.1523/JNEUROSCI.1712-04.2004
41. Chaudhuri D, Issa JB, Yue DT (2007) Elementary mechanisms producing facilitation of Cav2.1 (P/Q-type) channels. J Gen Physiol 129:385–401. https://doi.org/10.1085/jgp.200709749
42. Thomas JR, Hagen J (2018) Soh, D. & Lee, A. Molecular moieties masking Ca(2+)-dependent facilitation of voltage-gated Cav2.2 Ca(2+) channels. J Gen Physiol 150:83–94. https://doi.org/10.1085/jgp.201711841
43. Mori MX, Vander Kooi CW, Leahy DJ, Yue DT (2008) Crystal structure of the CaV2 IQ domain in complex with Ca2+/calmodulin: high-resolution mechanistic implications for channel regulation by Ca2+. Structure 16:607–620. https://doi.org/10.1016/j.str.2008.01.011
44. Chaudhuri D, Alseikhan BA, Chang SY, Soong TW, Yue DT (2005) Developmental activation of calmodulin-dependent facilitation of cerebellar P-type Ca2+ current. J Neurosci 25:8282–8294. https://doi.org/10.1523/JNEUROSCI.2253-05.2005
45. Chang SY et al (2007) Age and gender-dependent alternative splicing of P/Q-type calcium channel EF-hand. Neuroscience 145:1026–1036. https://doi.org/10.1016/j.neuroscience.2006.12.054
46. Kano M, Watanabe T, Uesaka N, Watanabe M (2018) Multiple phases of climbing fiber synapse elimination in the developing cerebellum. Cerebellum 17:722–734. https://doi.org/10.1007/s12311-018-0964-z
47. Thalhammer A et al (2017) Alternative Splicing of P/Q-Type Ca(2+) Channels Shapes Presynaptic Plasticity. Cell Rep 20:333–343. https://doi.org/10.1016/j.celrep.2017.06.055
48. Hirano M et al (2017) C-terminal splice variants of P/Q-type Ca(2+) channel CaV2.1 alpha1 subunits are differentially regulated by Rab3-interacting molecule proteins. J Biol Chem 292:9365–9381. https://doi.org/10.1074/jbc.M117.778829
49. Zhuchenko O et al (1997) Autosomal dominant cerebellar ataxia (SCA6) associated with small polyglutamine expansions in the alpha 1A-voltage-dependent calcium channel. Nat Genet 15:62–69. https://doi.org/10.1038/ng0197-62
50. Watase K et al (2008) Spinocerebellar ataxia type 6 knockin mice develop a progressive neuronal dysfunction with age-dependent accumulation of mutant CaV2.1 channels. Proc Natl Acad Sci U S A 105: 11987–11992. https://doi.org/10.1073/pnas.0804350105
51. Aikawa T et al (2017) Alternative splicing in the C-terminal tail of Cav2.1 is essential for preventing a neurological disease in mice. Hum Mol Genet 26: 3094–3104. https://doi.org/10.1093/hmg/ddx193
52. Saegusa H et al (2007) Properties of human Cav2.1 channel with a spinocerebellar ataxia type 6 mutation expressed in Purkinje cells. Mol Cell Neurosci 34: 261–270. https://doi.org/10.1016/j.mcn.2006.11.006
53. Matsuyama Z et al (1999) Direct alteration of the P/Q-type Ca2+ channel property by polyglutamine expansion in spinocerebellar ataxia 6. J Neurosci 19:RC14

54. Piedras-Renteria ES et al (2001) Increased expression of alpha 1A Ca2+ channel currents arising from expanded trinucleotide repeats in spinocerebellar ataxia type 6. J Neurosci 21:9185–9193
55. Toru S et al (2000) Spinocerebellar ataxia type 6 mutation alters P-type calcium channel function. J Biol Chem 275:10893–10898. https://doi.org/10.1074/jbc.275.15.10893
56. Lacerda AE et al (1991) Normalization of current kinetics by interaction between the alpha 1 and beta subunits of the skeletal muscle dihydropyridine-sensitive Ca2+ channel. Nature 352:527–530. https://doi.org/10.1038/352527a0
57. Varadi G, Lory P, Schultz D, Varadi M, Schwartz A (1991) Acceleration of activation and inactivation by the beta subunit of the skeletal muscle calcium channel. Nature 352:159–162. https://doi.org/10.1038/352159a0
58. Sokolov S, Weiss RG, Timin EN, Hering S (2000) Modulation of slow inactivation in class A Ca2+ channels by beta-subunits. J Physiol 527(Pt 3): 445–454. https://doi.org/10.1111/j.1469-7793.2000.t01-1-00445.x
59. Shi C, Soldatov NM (2002) Molecular determinants of voltage-dependent slow inactivation of the Ca2+ channel. J Biol Chem 277:6813–6821. https://doi.org/10.1074/jbc.M110524200
60. Isom LL, De Jongh KS, Catterall WA (1994) Auxiliary subunits of voltage-gated ion channels. Neuron 12:1183–1194. https://doi.org/10.1016/0896-6273(94)90436-7
61. Eckert R, Chad JE (1984) Inactivation of Ca channels. Prog Biophys Mol Biol 44:215–267. https://doi.org/10.1016/0079-6107(84)90009-9
62. Zhang JF, Ellinor PT, Aldrich RW, Tsien RW (1994) Molecular determinants of voltage-dependent inactivation in calcium channels. Nature 372:97–100. https://doi.org/10.1038/372097a0
63. Stotz SC, Hamid J, Spaetgens RL, Jarvis SE, Zamponi GW (2000) Fast inactivation of voltage-dependent calcium channels. A hinged-lid mechanism? J Biol Chem 275:24575–24582. https://doi.org/10.1074/jbc.M000399200
64. Kraus RL, Sinnegger MJ, Glossmann H, Hering S, Striessnig J (1998) Familial hemiplegic migraine mutations change alpha1A Ca2+ channel kinetics. J Biol Chem 273:5586–5590. https://doi.org/10.1074/jbc.273.10.5586
65. Kraus RL et al (2000) Three new familial hemiplegic migraine mutants affect P/Q-type Ca(2+) channel kinetics. J Biol Chem 275:9239–9243. https://doi.org/10.1074/jbc.275.13.9239
66. Hering S et al (1996) Transfer of high sensitivity for benzothiazepines from L-type to class A (BI) calcium channels. J Biol Chem 271:24471–24475. https://doi.org/10.1074/jbc.271.40.24471
67. Hering S, Berjukow S, Aczel S, Timin EN (1998) Ca2+ channel block and inactivation: common molecular determinants. Trends Pharmacol Sci 19: 439–443. https://doi.org/10.1016/s0165-6147(98)01258-9
68. Stotz SC, Zamponi GW (2001) Identification of inactivation determinants in the domain IIS6 region of high voltage-activated calcium channels. J Biol Chem 276:33001–33010. https://doi.org/10.1074/jbc.M104387200
69. Splawski I et al (2004) Ca(V)1.2 calcium channel dysfunction causes a multisystem disorder including arrhythmia and autism. Cell 119:19–31. https://doi.org/10.1016/j.cell.2004.09.011
70. Berrou L, Bernatchez G, Parent L (2001) Molecular determinants of inactivation within the I-II linker of alpha1E (CaV2.3) calcium channels. Biophys J 80: 215–228. https://doi.org/10.1016/S0006-3495(01)76008-0
71. Herlitze S, Hockerman GH, Scheuer T, Catterall WA (1997) Molecular determinants of inactivation and G protein modulation in the intracellular loop connecting domains I and II of the calcium channel alpha1A subunit. Proc Natl Acad Sci U S A 94:1512–1516. https://doi.org/10.1073/pnas.94.4.1512
72. Stotz SC, Jarvis SE, Zamponi GW (2004) Functional roles of cytoplasmic loops and pore lining transmembrane helices in the voltage-dependent inactivation of HVA calcium channels. J Physiol 554:263–273. https://doi.org/10.1113/jphysiol.2003.047068
73. Beyl S et al (2009) Different pathways for activation and deactivation in CaV1.2: a minimal gating model. J Gen Physiol 134:231–241.; S231–232. https://doi.org/10.1085/jgp.200910272
74. Hering S et al (2018) Calcium channel gating. Pflugers Arch 470:1291–1309. https://doi.org/10.1007/s00424-018-2163-7
75. Wu J et al (2016) Structure of the voltage-gated calcium channel Ca(v)1.1 at 3.6 A resolution. Nature 537:191–196. https://doi.org/10.1038/nature19321
76. Wu J et al (2015) Structure of the voltage-gated calcium channel Cav1.1 complex. Science 350: aad2395. https://doi.org/10.1126/science.aad2395
77. Brehm P, Eckert R (1978) Calcium entry leads to inactivation of calcium channel in paramecium. Science 202:1203–1206. https://doi.org/10.1126/science.103199
78. Brehm P, Dunlap K, Eckert R (1978) Calcium-dependent repolarization in paramecium. J Physiol 274:639–654. https://doi.org/10.1113/jphysiol.1978.sp012171
79. Eckert R, Tillotson DL (1981) Calcium-mediated inactivation of the calcium conductance in caesium-loaded giant neurones of Aplysia californica. J Physiol 314:265–280. https://doi.org/10.1113/jphysiol.1981.sp013706
80. Imredy JP, Yue DT (1994) Mechanism of Ca(2+)-sensitive inactivation of L-type Ca2+ channels. Neuron 12:1301–1318. https://doi.org/10.1016/0896-6273(94)90446-4
81. DeMaria CD, Soong TW, Alseikhan BA, Alvania RS, Yue DT (2001) Calmodulin bifurcates the local

Ca2+ signal that modulates P/Q-type Ca2+ channels. Nature 411:484–489. https://doi.org/10.1038/35078091
82. Kass RS, Sanguinetti MC (1984) Inactivation of calcium channel current in the calf cardiac Purkinje fiber. Evidence for voltage- and calcium-mediated mechanisms. J Gen Physiol 84:705–726. https://doi.org/10.1085/jgp.84.5.705
83. Mentrard D, Vassort G, Fischmeister R (1984) Calcium-mediated inactivation of the calcium conductance in cesium-loaded frog heart cells. J Gen Physiol 83:105–131. https://doi.org/10.1085/jgp.83.1.105
84. Lee KS, Marban E, Tsien RW (1985) Inactivation of calcium channels in mammalian heart cells: joint dependence on membrane potential and intracellular calcium. J Physiol 364:395–411. https://doi.org/10.1113/jphysiol.1985.sp015752
85. de Leon M et al (1995) Essential Ca(2+)-binding motif for Ca(2+)-sensitive inactivation of L-type Ca2+ channels. Science 270:1502–1506. https://doi.org/10.1126/science.270.5241.1502
86. Zhou J et al (1997) Feedback inhibition of Ca2+ channels by Ca2+ depends on a short sequence of the C terminus that does not include the Ca2+ -binding function of a motif with similarity to Ca2+ -binding domains. Proc Natl Acad Sci U S A 94:2301–2305. https://doi.org/10.1073/pnas.94.6.2301
87. Bernatchez G, Talwar D, Parent L (1998) Mutations in the EF-hand motif impair the inactivation of barium currents of the cardiac alpha1C channel. Biophys J 75:1727–1739. https://doi.org/10.1016/S0006-3495(98)77614-3
88. Zuhlke RD, Reuter H (1998) Ca2+-sensitive inactivation of L-type Ca2+ channels depends on multiple cytoplasmic amino acid sequences of the alpha1C subunit. Proc Natl Acad Sci U S A 95:3287–3294. https://doi.org/10.1073/pnas.95.6.3287
89. Peterson BZ, DeMaria CD, Adelman JP, Yue DT (1999) Calmodulin is the Ca2+ sensor for Ca2+ -dependent inactivation of L-type calcium channels. Neuron 22:549–558. https://doi.org/10.1016/s0896-6273(00)80709-6
90. Erickson MG, Alseikhan BA, Peterson BZ, Yue DT (2001) Preassociation of calmodulin with voltage-gated Ca(2+) channels revealed by FRET in single living cells. Neuron 31:973–985. https://doi.org/10.1016/s0896-6273(01)00438-x
91. Erickson MG, Liang H, Mori MX, Yue DT (2003) FRET two-hybrid mapping reveals function and location of L-type Ca2+ channel CaM preassociation. Neuron 39:97–107. https://doi.org/10.1016/s0896-6273(03)00395-7
92. Pitt GS et al (2001) Molecular basis of calmodulin tethering and Ca2+-dependent inactivation of L-type Ca2+ channels. J Biol Chem 276:30794–30802. https://doi.org/10.1074/jbc.M104959200
93. Lee A et al (1999) Ca2+/calmodulin binds to and modulates P/Q-type calcium channels. Nature 399: 155–159. https://doi.org/10.1038/20194
94. Lee A, Scheuer T, Catterall WA (2000) Ca2+/calmodulin-dependent facilitation and inactivation of P/Q-type Ca2+ channels. J Neurosci 20:6830–6838
95. Liang H et al (2003) Unified mechanisms of Ca2+ regulation across the Ca2+ channel family. Neuron 39:951–960. https://doi.org/10.1016/s0896-6273(03)00560-9
96. Yue DT, Backx PH, Imredy JP (1990) Calcium-sensitive inactivation in the gating of single calcium channels. Science 250:1735–1738. https://doi.org/10.1126/science.2176745
97. Lee A, Zhou H, Scheuer T, Catterall WA (2003) Molecular determinants of Ca(2+)/calmodulin-dependent regulation of Ca(v)2.1 channels. Proc Natl Acad Sci U S A 100:16059–16064. https://doi.org/10.1073/pnas.2237000100
98. Kim EY et al (2008) Structures of CaV2 Ca2+/CaM-IQ domain complexes reveal binding modes that underlie calcium-dependent inactivation and facilitation. Structure 16:1455–1467. https://doi.org/10.1016/j.str.2008.07.010
99. Dunlap K (2007) Calcium channels are models of self-control. J Gen Physiol 129:379–383. https://doi.org/10.1085/jgp.200709786
100. Dick IE et al (2008) A modular switch for spatial Ca2+ selectivity in the calmodulin regulation of CaV channels. Nature 451:830–834. https://doi.org/10.1038/nature06529
101. Zuhlke RD, Pitt GS, Deisseroth K, Tsien RW, Reuter H (1999) Calmodulin supports both inactivation and facilitation of L-type calcium channels. Nature 399: 159–162. https://doi.org/10.1038/20200
102. Dzhura I, Wu Y, Colbran RJ, Balser JR, Anderson ME (2000) Calmodulin kinase determines calcium-dependent facilitation of L-type calcium channels. Nat Cell Biol 2:173–177. https://doi.org/10.1038/35004052
103. Hudmon A et al (2005) CaMKII tethers to L-type Ca2+ channels, establishing a local and dedicated integrator of Ca2+ signals for facilitation. J Cell Biol 171:537–547. https://doi.org/10.1083/jcb.200505155
104. Lee TS et al (2006) Calmodulin kinase II is involved in voltage-dependent facilitation of the L-type Cav1.2 calcium channel: Identification of the phosphorylation sites. J Biol Chem 281:25560–25567. https://doi.org/10.1074/jbc.M508661200
105. Abiria SA, Colbran RJ (2010) CaMKII associates with CaV1.2 L-type calcium channels via selected beta subunits to enhance regulatory phosphorylation. J Neurochem 112:150–161. https://doi.org/10.1111/j.1471-4159.2009.06436.x
106. Koval OM et al (2010) CaV1.2 beta-subunit coordinates CaMKII-triggered cardiomyocyte death and afterdepolarizations. Proc Natl Acad Sci U S A

107:4996–5000. https://doi.org/10.1073/pnas.0913760107
107. Grueter CE, Abiria SA, Wu Y, Anderson ME, Colbran RJ (2008) Differential regulated interactions of calcium/calmodulin-dependent protein kinase II with isoforms of voltage-gated calcium channel beta subunits. Biochemistry 47:1760–1767. https://doi.org/10.1021/bi701755q
108. Walikonis RS et al (2001) Densin-180 forms a ternary complex with the (alpha)-subunit of Ca2+/calmodulin-dependent protein kinase II and (alpha)-actinin. J Neurosci 21:423–433
109. Strack S, Robison AJ, Bass MA, Colbran RJ (2000) Association of calcium/calmodulin-dependent kinase II with developmentally regulated splice variants of the postsynaptic density protein densin-180. J Biol Chem 275:25061–25064. https://doi.org/10.1074/jbc.C000319200
110. Sahu G, Asmara H, Zhang FX, Zamponi GW, Turner RW (2017) Activity-dependent facilitation of CaV1.3 calcium channels promotes KCa3.1 activation in hippocampal neurons. J Neurosci 37:11255–11270. https://doi.org/10.1523/JNEUROSCI.0967-17.2017
111. Forsythe ID, Tsujimoto T, Barnes-Davies M, Cuttle MF, Takahashi T (1998) Inactivation of presynaptic calcium current contributes to synaptic depression at a fast central synapse. Neuron 20:797–807. https://doi.org/10.1016/s0896-6273(00)81017-x
112. Borst JG, Sakmann B (1998) Facilitation of presynaptic calcium currents in the rat brainstem. J Physiol 513(Pt 1):149–155. https://doi.org/10.1111/j.1469-7793.1998.149by.x
113. Ophoff RA et al (1996) Familial hemiplegic migraine and episodic ataxia type-2 are caused by mutations in the Ca2+ channel gene CACNL1A4. Cell 87:543–552. https://doi.org/10.1016/s0092-8674(00)81373-2
114. Gandal MJ et al (2018) Transcriptome-wide isoform-level dysregulation in ASD, schizophrenia, and bipolar disorder. Science:362. https://doi.org/10.1126/science.aat8127
115. Andrade A et al (2019) Genetic associations between voltage-gated calcium channels and psychiatric disorders. Int J Mol Sci:20. https://doi.org/10.3390/ijms20143537
116. Yue Q, Jen JC, Nelson SF, Baloh RW (1997) Progressive ataxia due to a missense mutation in a calcium-channel gene. Am J Hum Genet 61:1078–1087. https://doi.org/10.1086/301613
117. Jen J, Kim GW, Baloh RW (2004) Clinical spectrum of episodic ataxia type 2. Neurology 62:17–22. https://doi.org/10.1212/01.wnl.0000101675.61074.50
118. Jen JC et al (2007) Primary episodic ataxias: diagnosis, pathogenesis and treatment. Brain 130:2484–2493. https://doi.org/10.1093/brain/awm126
119. Ducros A et al (2001) The clinical spectrum of familial hemiplegic migraine associated with mutations in a neuronal calcium channel. N Engl J Med 345:17–24. https://doi.org/10.1056/NEJM200107053450103
120. Haan J et al (2005) Migraine genetics: An update. Curr Pain Headache Rep 9:213–220. https://doi.org/10.1007/s11916-005-0065-9
121. Pietrobon D (2010) CaV2.1 channelopathies. Pflugers Arch 460:375–393. https://doi.org/10.1007/s00424-010-0802-8
122. Russell MB, Ducros A (2011) Sporadic and familial hemiplegic migraine: pathophysiological mechanisms, clinical characteristics, diagnosis, and management. Lancet Neurol 10:457–470. https://doi.org/10.1016/S1474-4422(11)70048-5
123. Tottene A et al (2002) Familial hemiplegic migraine mutations increase Ca(2+) influx through single human CaV2.1 channels and decrease maximal CaV2.1 current density in neurons. Proc Natl Acad Sci U S A 99:13284–13289. https://doi.org/10.1073/pnas.192242399
124. Hans M et al (1999) Functional consequences of mutations in the human alpha1A calcium channel subunit linked to familial hemiplegic migraine. J Neurosci 19:1610–1619
125. Tottene A et al (2005) Specific kinetic alterations of human CaV2.1 calcium channels produced by mutation S218L causing familial hemiplegic migraine and delayed cerebral edema and coma after minor head trauma. J Biol Chem 280:17678–17686. https://doi.org/10.1074/jbc.M501110200
126. van den Maagdenberg AM et al (2010) High cortical spreading depression susceptibility and migraine-associated symptoms in Ca(v)2.1 S218L mice. Ann Neurol 67:85–98. https://doi.org/10.1002/ana.21815
127. van den Maagdenberg AM et al (2004) A Cacna1a knockin migraine mouse model with increased susceptibility to cortical spreading depression. Neuron 41:701–710. https://doi.org/10.1016/s0896-6273(04)00085-6
128. Adams PJ et al (2009) Ca(V)2.1 P/Q-type calcium channel alternative splicing affects the functional impact of familial hemiplegic migraine mutations: implications for calcium channelopathies. Channels (Austin) 3:110–121. https://doi.org/10.4161/chan.3.2.7932
129. Ishikawa K et al (1997) Japanese families with autosomal dominant pure cerebellar ataxia map to chromosome 19p13.1-p13.2 and are strongly associated with mild CAG expansions in the spinocerebellar ataxia type 6 gene in chromosome 19p13.1. Am J Hum Genet 61:336–346. https://doi.org/10.1086/514867
130. Yang Q et al (2000) Morphological Purkinje cell changes in spinocerebellar ataxia type 6. Acta Neuropathol 100:371–376. https://doi.org/10.1007/s004010000201
131. Ishikawa K et al (1999) Abundant expression and cytoplasmic aggregations of [alpha]1A voltage-dependent calcium channel protein associated with neurodegeneration in spinocerebellar ataxia type 6. Hum Mol Genet 8:1185–1193. https://doi.org/10.1093/hmg/8.7.1185

132. Ishiguro T et al (2010) The carboxy-terminal fragment of alpha(1A) calcium channel preferentially aggregates in the cytoplasm of human spinocerebellar ataxia type 6 Purkinje cells. Acta Neuropathol 119: 447–464. https://doi.org/10.1007/s00401-009-0630-0
133. Takahashi M et al (2013) Cytoplasmic location of alpha1A voltage-gated calcium channel C-terminal fragment (Cav2.1-CTF) aggregate is sufficient to cause cell death. PLoS One 8:e50121. https://doi.org/10.1371/journal.pone.0050121
134. Gomez-Ospina N, Tsuruta F, Barreto-Chang O (2006) Hu, L. & Dolmetsch, R. The C terminus of the L-type voltage-gated calcium channel Ca(V)1.2 encodes a transcription factor. Cell 127:591–606. https://doi.org/10.1016/j.cell.2006.10.017
135. Kordasiewicz HB, Thompson RM, Clark HB, Gomez CM (2006) C-termini of P/Q-type Ca2+ channel alpha1A subunits translocate to nuclei and promote polyglutamine-mediated toxicity. Hum Mol Genet 15:1587–1599. https://doi.org/10.1093/hmg/ddl080
136. Du X et al (2013) Second cistron in CACNA1A gene encodes a transcription factor mediating cerebellar development and SCA6. Cell 154:118–133. https://doi.org/10.1016/j.cell.2013.05.059
137. Hamshere ML et al (2013) Genome-wide significant associations in schizophrenia to ITIH3/4, CACNA1C and SDCCAG8, and extensive replication of associations reported by the Schizophrenia PGC. Mol Psychiatry 18:708–712. https://doi.org/10.1038/mp.2012.67
138. Moskvina V et al (2009) Gene-wide analyses of genome-wide association data sets: evidence for multiple common risk alleles for schizophrenia and bipolar disorder and for overlap in genetic risk. Mol Psychiatry 14:252–260. https://doi.org/10.1038/mp.2008.133
139. Ripke S et al (2013) Genome-wide association analysis identifies 13 new risk loci for schizophrenia. Nat Genet 45:1150–1159. https://doi.org/10.1038/ng.2742
140. Cross-Disorder Group of the Psychiatric Genomics, C (2013) Identification of risk loci with shared effects on five major psychiatric disorders: a genome-wide analysis. Lancet 381:1371–1379. https://doi.org/10.1016/S0140-6736(12)62129-1
141. Green EK et al (2010) The bipolar disorder risk allele at CACNA1C also confers risk of recurrent major depression and of schizophrenia. Mol Psychiatry 15: 1016–1022. https://doi.org/10.1038/mp.2009.49
142. Indelicato E et al (2019) The neuropsychiatric phenotype in CACNA1A mutations: a retrospective single center study and review of the literature. Eur J Neurol 26:66–e67. https://doi.org/10.1111/ene.13765

Structure–Function of TMEM16 Ion Channels and Lipid Scramblases

6

Son C. Le and Huanghe Yang

Abstract

The TMEM16 protein family comprises two novel classes of structurally conserved but functionally distinct membrane transporters that function as Ca^{2+}-dependent Cl^- channels (CaCCs) or dual functional Ca^{2+}-dependent ion channels and phospholipid scramblases. Extensive functional and structural studies have advanced our understanding of TMEM16 molecular mechanisms and physiological functions. TMEM16A and TMEM16B CaCCs control transepithelial fluid transport, smooth muscle contraction, and neuronal excitability, whereas TMEM16 phospholipid scramblases mediate the flip-flop of phospholipids across the membrane to allow phosphatidylserine externalization, which is essential in a plethora of important processes such as blood coagulation, bone development, and viral and cell fusion. In this chapter, we summarize the major methods in studying TMEM16 ion channels and scramblases and then focus on the current mechanistic understanding of TMEM16 Ca^{2+}- and voltage-dependent channel gating as well as their ion and phospholipid permeation.

Keywords

TMEM16 · Anoctamin · CaCC · Scramblase · CaPLSase · Gating · Permeation

6.1 Introduction

Cell membranes are a bilayer structure made up of amphipathic phospholipid molecules. The fatty acid acyl chains of phospholipids form a hydrophobic core that creates a huge energy barrier for the transport of ions, glucose, and amino acids as well as spontaneous flip-flopping of phospholipids. To maintain cellular homeostasis, communication, and survival, numerous membrane proteins have evolved to overcome this energy barrier to facilitate membrane transport [1] (Fig. 6.1a). TMEM16 transmembrane proteins are a recently discovered family of membrane transport proteins that passively permeate ions, phospholipids, or both (Fig. 6.1b). Over the past decade, tremendous advances have been made to understand these mysterious TMEM16 proteins and their roles in human health and diseases. In this chapter, we briefly discuss the concise history of the molecular identification of TMEM16 proteins and then primarily focus on the current mechanistic understanding of how TMEM16 proteins work in response to Ca^{2+} and voltage to catalyze the permeation of two structurally distinct substrates: ions and phospholipids. This chapter by no means can cover all of the elegant

S. C. Le · H. Yang (✉)
Department of Biochemistry, Duke University School of Medicine, Durham, NC, USA
e-mail: huanghe.yang@duke.edu

L. Zhou (ed.), *Ion Channels in Biophysics and Physiology*, Advances in Experimental Medicine and Biology 1349, https://doi.org/10.1007/978-981-16-4254-8_6

works in the field and the physiological aspects of TMEM16 biology. We hope that our summary can give the readers an overview of the TMEM16 proteins and an up-to-date summary of the current understanding of TMEM16 molecular mechanisms. Interested readers can refer to several excellent reviews for more detailed insights into TMEM16 proteins and their biology [3–19].

6.2 Molecular Identifications of TMEM16 Proteins

6.2.1 TMEM16A and TMEM16B Form the Canonical Ca^{2+}-Activated Chloride Channels

TMEM16A/DOG1 with the unknown function was first found to be highly expressed in gastrointestinal stromal tumor (GIST) in 2004 [20]. However, systematic investigations of the TMEM16 family of proteins were not reported until 2008, when three groups independently identified TMEM16A as well as its closely related member TMEM16B as the long sought after Ca^{2+}-activated chloride channels (CaCCs) [21–23].

The Oh group searched for membrane proteins having multiple TMs with unknown functions and came across the TMEM16 family. They showed that co-expressing of TMEM16A with G_q-coupled receptors (G_qPCRs) gave rise to robust CaCC conductance following the application of receptor agonists. TMEM16A's biophysical properties strikingly resemble those of the canonical CaCCs [24] exemplified by the outward-rectifying current at low Ca^{2+} and linear current–voltage relationship at high Ca^{2+} (Fig. 6.1c) and especially with an anion selectivity sequence of $NO_3^- > I^- > Br^- > Cl^- > F^-$ [23]. In addition, TMEM16A was found to be highly expressed in the epithelial cells of pulmonary bronchioles, pancreatic acinar cells, epithelia of the renal tubules of the kidney, the outer nuclear layer of the retina, small-diameter sensory neurons of dorsal root ganglia (DRG), acinar cells of submandibular glands, and Leydig cells of testes. These patterns of tissue expression of TMEM16A are consistent with the reported expression of endogenous CaCCs [24].

The Galietta group discovered TMEM16A based on the previous observations that prolonged stimulation of human bronchial epithelial cells with interleukin-4 or -13 (IL-4 or IL-13) upregulated the expression of CaCCs [25, 26]. Using gene expression microarrays, the authors identified a list of genes that are upregulated following IL-4 and IL-13 stimulation. Among these candidates, only siRNA knockdown of *TMEM16A* gene significantly attenuated CaCC activity as measured via iodide influx, short-circuit current, and patch-clamp recordings of several cell lines known to highly express CaCCs such as bronchial CFBE41o, pancreatic CFPAC-1 as well as primary human bronchial epithelial cultures [21].

The Jan group took advantage of the Axolotl oocytes, which do not express endogenous CaCCs for an expression cloning strategy [22]. Size-fractionated mRNAs extracted from Xenopus oocytes, which highly express CaCCs, were injected into Axolotl oocytes for measurement of Ca^{2+}-activated Cl^- currents. After cycles of functional screening, the *TMEM16A* gene was identified to confer robust CaCC currents in Axolotl oocytes. Heterologous expression of TMEM16A as well as TMEM16B in HEK293 cells further confirmed that these two proteins are the *bona fide* CaCCs. The identification of TMEM16B as a CaCC was subsequently confirmed by other groups [27–29].

6.2.2 TMEM16F Encodes a Dual Functional CaPLSase and Nonselective Ion Channel

It was surprising and puzzling that TMEM16F does not function as a CaCC despite the fact that it shares about 45% of sequence identity with TMEM16A/B CaCCs. Through a series of elegant experiments, the Nagata group made a surprising discovery that TMEM16F (ANO6) plays an indispensable role in mediating Ca^{2+}-activated phospholipid scrambling to catalyze the flip-flop of phospholipids across the

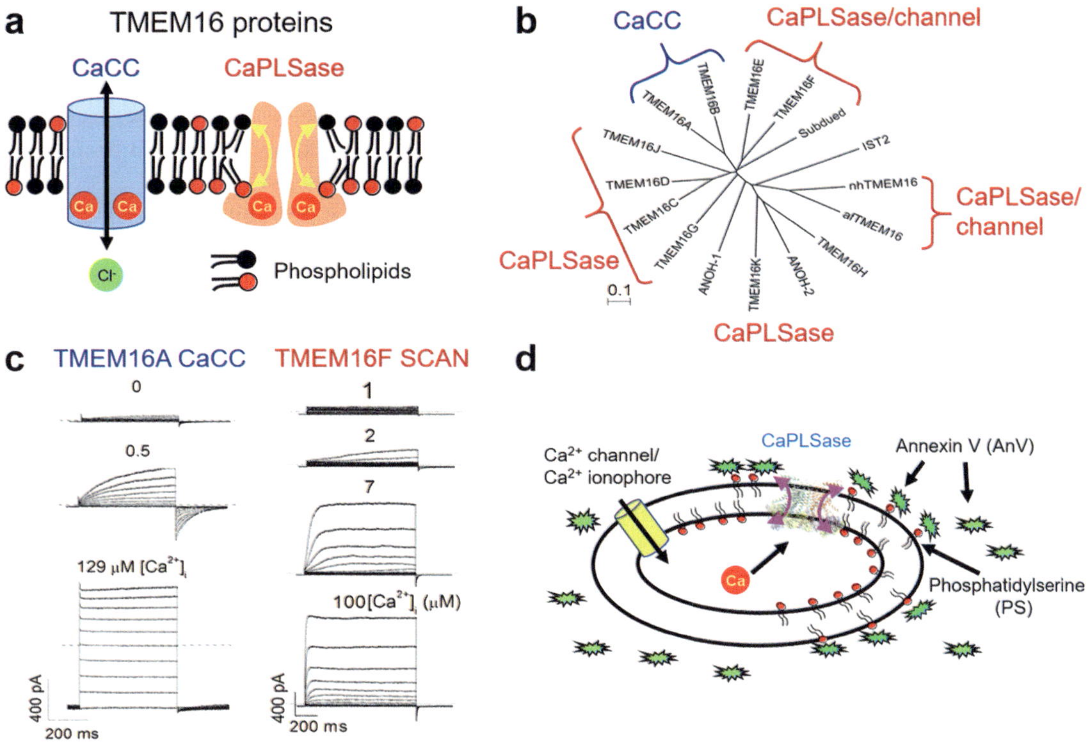

Fig. 6.1 TMEM16 protein family. (**a**) TMEM16 proteins can function as Ca^{2+}-activated chloride channels (CaCCs) or Ca^{2+}-dependent phospholipid scramblases (CaPLSases). (**b**) TMEM16 protein family with diverse functionality. Note that the yeast IST2 and *C. elegans* ANOH-1 and -2 have not been assigned as CaCCs or CaPLSases. (**c**) TMEM16A CaCC and TMEM16F small-conductance, Ca^{2+}-activated nonselective channel (SCAN) at different Ca^{2+} concentrations [2]. (**d**) A fluorescence assay to monitor CaPLSase activity

membrane [30]. The authors utilized the mouse Ba/F3 cell line and carried multiple rounds of annexin V (AnV)-based fluorescence-activated cell sorting (FACS) to identify a subpopulation of cells displaying enhanced phosphatidylserine (PS) externalization upon Ca^{2+} ionophore treatment (Fig. 6.1d). A gain-of-function mutation of TMEM16F, D409G, located in the first intracellular loop between TM2 and TM3 was identified to give rise to this enhanced PS exposure by this subpopulation. Interestingly, loss-of-function mutations of TMEM16F are responsible for Scott syndrome, a rare inherited bleeding disorder caused by a defect in Ca^{2+}-activated phospholipid scramblase (CaPLSase)-mediated PS exposure in platelets both in humans [30–32] and in dogs [33]. The Jan group subsequently demonstrated that TMEM16F plays a key role in blood coagulation and the TMEM16F-null mice recapitulate the human Scott syndrome phenotype of prolonged bleeding [2]. Remarkably, the TMEM16F-null mice resist strong thrombotic challenges induced by $FeCl_3$, suggesting that TMEM16F could serve as a novel anticoagulant target to prevent thrombotic disorders such as stroke, heart attack, and venous thromboembolism. These early findings suggested that TMEM16F may serve as a CaPLSase or function as a key element for the CaPLSase.

One of the reasons why TMEM16F was not immediately considered as a *bona fide* CaPLSase was that it also has an ion channel function (Fig. 6.1c). Different studies from several groups suggested that TMEM16F may function as a small conductive nonselective cation channel (SCAN) [2, 3], an outward-rectifying chloride channel [34], volume-gated anion channel [35], a CaCC [36–40], or a CaCC of delayed activation [41]. Despite the controversial aspect of TMEM16F's ion selectivity (see discussion in a later section), the proposal that TMEM16 proteins can support scrambling function was bolstered by the elegant structural and functional studies from the Dutzler and Accardi groups in which they were able to purify the TMEM16 fungal homologs, nhTMEM16 and afTMEM16, and demonstrated that they not only mediate Ca^{2+}-dependent phospholipid scrambling [42, 43] but also conduct ions [43, 44]. Furthermore, the Hartzell and Galietta groups showed that the mammalian TMEM16F exhibits both ion channel and lipid scrambling activities in mammalian cells [45, 46]. Now it is generally accepted that TMEM16F is indeed a dual functional ion channel and lipid scramblase [10]. Although the biophysical properties of TMEM16C/D/G/J as lipid scramblases have not been fully characterized [47], TMEM16E and TMEM16K were recently shown to display both scrambling and ion channel activities [48–51].

6.3 Structure and Function of TMEM16 Proteins

6.3.1 Biophysical Properties of TMEM16 Ion Channels

Consistent with studies on endogenous CaCCs [52–54], channel gating of TMEM16A and TMEM16B involves the synergistic action of Ca^{2+} and membrane depolarization (Fig. 6.1c). Both channels display higher apparent Ca^{2+} sensitivity at more depolarizing voltages and higher concentrations of Ca^{2+} shift the conductance-voltage (*G–V*) relationship curves toward more negative voltages. TMEM16A is highly sensitive to Ca^{2+} with an estimated Ca^{2+} EC_{50} in the low micromolar range from 0.4 to 1 μM at positive membrane potentials or from 0.7 to 6 μM at negative membrane potentials [23, 42, 55–62]. Despite sharing ~82% sequence identity with TMEM16A, TMEM16B displays a relatively lower Ca^{2+} sensitivity with an estimated Ca^{2+} EC_{50} ranging from 1.2 to 3.3 μM at positive membrane potentials and from 1.8 to 4.9 μM at negative potentials [3, 27, 28, 63, 64], all of which are good agreement with studies on endogenous CaCC encoded by TMEM16B from olfactory sensory neurons [54, 65, 66].

Under low open probability (i.e., <1 μM Ca^{2+} for TMEM16A or 1–2 μM for TMEM16B), TMEM16A and TMEM16B currents show pronounced voltage-dependent outward rectification in addition to time-dependent activation and deactivation kinetics, with TMEM16B displaying faster activation and deactivation kinetics [21, 23, 27, 28]. The time dependence and voltage dependence quickly disappear under saturating Ca^{2+} (i.e., >1 μM for TMEM16A or >10 μM for TMEM16B), and the currents follow Ohm's law. Interestingly, similar to endogenous CaCCs [24], permeant anions also strongly impact the channel's gating and kinetics. Larger and more permeable anions such as SCN^-, NO_3^-, or I^- pronouncedly enhance the channel's open probability and accelerate channel activation kinetics [61, 67, 68]. This phenomenon could be explained by their higher occupancy within the pore which stabilizes the open state to allosterically modulate Ca^{2+}-dependent gating [24].

In contrast to the CaCCs TMEM16A and TMEM16B, the small conductance and nonselective ion channel TMEM16F appears to be much less sensitive to Ca^{2+} with reported Ca^{2+} EC_{50} values ranging from 3.4 up to 105 μM [2, 38, 41, 45, 69–72]. These large variations could be partly due to the pronounced desensitization of TMEM16F as well as the differences in experimental approaches including voltage-clamp protocols used and/or recording configurations. Nevertheless, similar to TMEM16A/B CaCCs, TMEM16F also exhibits voltage-dependent Ca^{2+} activation with depolarizing potentials enhancing its apparent Ca^{2+} sensitivity [2]. Most importantly, while TMEM16A/B CaCCs become constitutively open under saturating micromolar Ca^{2+}, TMEM16F current remains strongly outward rectifying even at high Ca^{2+} and always requires membrane depolarization for activation (Fig. 6.1c).

Relatively small single-channel conductance values of mammalian TMEM16 channels hamper detailed single-channel analysis using patch clamp. TMEM16A' single-channel conductance reportedly ranges from 2.6 pS [58], 3.5 pS [3] to 8.3 pS [23], whereas that of TMEM16B ranges from 1.2 to 3.5 pS [3, 27, 28]. TMEM16F has the smallest single-channel conductance of ~0.5 pS based on noise analysis [2]. Interestingly, the fungal afTMEM16 was reported to have ~300 pS of single-channel conductance based on reconstituted planar bilayer measurement [43]. Although nhTMEM16 was reported as a nonselective ion channel recently [44], despite the initial report suggesting that it functions as a sole CaPLSase without channel activity [42], its single-channel conductance remains unknown. Therefore, it is not clear whether the marked differences in single-channel conductance between the mammalian TMEM16 channels and the fungal afTMEM16 reflect the evolutionary divergence of different organisms or are solely due to the difference in the measurement methods.

6.3.2 Fluorescence Methods Enable Biophysical Characterization of TMEM16 CaPLSases

Due to the difficulties in monitoring phospholipid flip-flop at high resolution and high sensitivity, systematic functional characterization of TMEM16 CaPLSases is technically challenging. Flow cytometry is commonly used to monitor scramblase-mediated PS exposure from a large population of cells. In this method, Ca^{2+} ionophore-induced Ca^{2+} elevation activates CaPLSases, which translocate PS to the cell surface (Fig. 6.1d). The cell surface-exposed PS then recruits fluorescently tagged AnV, a PS-specific binding protein that serves as a readout of CaPLSase activity. Alternatively, fluorescently labeled phospholipids can be visualized via their internalization by CaPLSases. After bovine serum albumin (BSA) extraction of the uninternalized fluorescent phospholipids, the BSA-resistant fluorescence from the internalized phospholipids is measured by flow cytometry and used as a readout of CaPLSase activity. Nevertheless, the flow cytometry method lacks single-cell resolution and temporal resolution in monitoring PS exposure, which limits its application in studying the molecular mechanisms of TMEM16 CaPLSases.

The Hartzell group was first to demonstrate the feasibility of applying live-cell microscopy to monitor CaPLSase activities at single-cell resolution using microscopy [46]. In this method, the CaPLSase TMEM16F-mediated PS exposure was monitored by PS-binding probes LactoglobulinC2 fused to the Clover fluorescent protein (or LactC2) or AlexaFluor-conjugated Annexin V following activation by the Ca^{2+} ionophore A23187. This approach requires incubation of A23187 for 5 min in Ca^{2+}-free solution followed by removal of A23187 and addition of 5 mM Ca^{2+} to mobilize Ca^{2+} entry via store-operated channels. Following activation, TMEM16F mediates PS externalization, and the binding of LactC2 or AnV to surface-exposed PS gives rise to the gradual increase in fluorescence signal at the cell membrane. Phospholipid scrambling of TMEM16F can be detected within several minutes following its activation. By combining this fluorescence-based assay with patch-clamp recordings in which Ca^{2+} was included in the pipette solution, the Hartzell group was also able to spontaneously measure CaPLSase and ion channel activities of TMEM16F for the first time. Also, similar to previous observations [38], even at high 200 μM Ca^{2+}, TMEM16F's current displayed a pronounced (nearly 10 min) delay before the development of both scrambling and channel activity.

Nevertheless, endogenous TMEM16F is robustly expressed in HEK293 and various other commonly used cell lines [37]. Therefore, endogenous CaPLSase activity can significantly contaminate the measurements of heterologously expressed TMEM16 CaPLSases. This presents a serious concern for reliably assessing loss-of-function CaPLSase mutations or mutations claimed to convert TMEM16A CaCC to a CaPLSase. To circumvent this issue, we generated a TMEM16F-null HEK293T cell line and optimized a microscopy assay by utilizing the fluorescently conjugated AnV to study TMEM16F's Ca^{2+}-dependent scrambling at a single-cell resolution [70]. Furthermore, instead of relying on Ca^{2+} entry via store-operated channels as previously done by Yu et al., we optimized the concentration of the Ca^{2+} ionophore ionomycin that can trigger sufficient Ca^{2+} entry as well as Ca^{2+} release from internal stores for TMEM16F activation. This assay enabled us to reliably interrogate the structure–function relationship of both the mammalian TMEM16F and the Drosophila TMEM16 homolog Subdued without the confounding effects from endogenous TMEM16F [73].

The Jan lab developed a new approach in studying TMEM16F scrambling activity by taking advantage of the fact that cells lacking TMEM16F displayed a defect in microvesicle release [74–76]. By chemically inducing mouse embryonic fibroblasts (MEF) or HEK293 cells using a combination of paraformaldehyde (PFA), dithiothreitol (DTT), and Ca^{2+}, robust generation of giant plasma membrane vesicles (GPMVs) can be observed [77]. Importantly, this generation of GPMVs begins after the initiation of PS exposure and requires both TMEM16F-dependent phospholipid scrambling and TMEM16F-mediated Ca^{2+} entry. By stably expressing TMEM16F, this approach allows the characterization of either gain-of-function or loss-of-function TMEM16F mutants. The time-dependent TMEM16F-mediated Ca^{2+} entry can also be monitored using a Ca^{2+} indicator, and the onset of Ca^{2+} influx usually takes place between 10 and 30 min. While this novel method allows measurements of both TMEM16F-dependent GPMV generation and Ca^{2+} entry, it is apparent that these cells may undergo apoptosis during the very long (up to 60 min) time of monitoring. As we showed previously, Ca^{2+} ionophore stimulation of overexpressed TMEM16F in HEK293 cells could lead cell death within 20–30 min after treatment [70]. Also, as this approach only monitors the generation of GPMVs, direct monitoring of PS exposure cannot be examined.

The Accardi and Dutzler groups first succeeded in purifying the fungal homologs afTMEM16 and nhTMEM16 and reconstituting them in liposomes composed of phospholipids conjugated with the fluorophore 7-nitro-2,1,3-benzoxadiazol (or NBD) [78]. Dithionite is added to quench the NBD-labeled phospholipids at the outer leaflet. In the absence of a scramblase, dithionite will reduce the fluorescence signal by half; when an active scramblase is present, additional quenching will be observed as inner leaflet NBD-labeled lipids are being transported to the outer leaflet for quenching by dithionite. This approach bypasses the challenges in overexpressing fungal TMEM16 scramblases in mammalian cells for scrambling assay. However, similar to TMEM16 ion channels, given the importance of membrane lipids such as PIP_2 on TMEM16 functions [57, 79–84], removal of these proteins from their native lipid environment could potentially lead to alterations of their functionality. Furthermore, it was often observed that while Ca^{2+} enhances lipid scrambling of reconstituted fungal scramblases, they can spontaneously mediate lipid scrambling in the absence of Ca^{2+} [85]. Whether this is due to an intrinsic property of these scramblases or the artificial environment remains to be established. Finally, while fungal TMEM16 scramblases have been extensively studied in reconstituted systems, the mammalian TMEM16 scramblases, namely TMEM16F, have yet to be successfully studied using this system. This could be due to the potential instability of the purified mammalian counterparts and/or their complex regulatory properties.

The Nagata group pioneered a single-molecule approach in studying TMEM16F-dependent phospholipid scrambling [86]. TMEM16F protein is purified from Ba/F3 cells stably expressing a high level of mouse TMEM16F. The purified TMEM16F is integrated into a microarray containing membrane bilayers with asymmetrically distributed fluorescently labeled phospholipids. Single-molecule scramblase assay is carried out in which TMEM16F-mediated translocation of phospholipids is activated by infusion of 100 μM Ca^{2+}. This assay revealed for the first time the remarkable efficiency of TMEM16F scramblase, which transports lipids at an estimated rate of ~45,000 lipids/s. This experiment also confirmed the "channel-like" biophysical property of TMEM16F and that it non-specifically transports lipids down their concentration gradients. Despite some requirements such as purification of TMEM16F with high homogeneity and stability as well as experimental setup and optimization of the microarray of the lipid bilayer, this novel single-molecule approach provides valuable information that cannot be otherwise attained from live-cell imaging or liposome-reconstituted assays.

6.3.3 Overall Architecture of TMEM16 Proteins

The ground-breaking X-ray structure of the fungal scramblase nhTMEM16 by the Dutzler group provided unprecedented details into the structural understanding of TMEM16 [42] (Fig. 6.2a, b). Subsequent structural studies on the fungal afTMEM16, mouse TMEM16A, mouse TMEM16F, and human TMEM16K revealed that these proteins adopt a highly similar overall architecture [42, 69, 87–91]. Consistent with previous biochemical and biophysical characterizations [60, 62, 92–94], TMEM16 proteins are assembled as homodimers in which the transmembrane domain consists of 10 TM segments flanked by a large N-terminal cytosolic domain (NCD) and a short C-terminal extension of TM10 (Fig. 6.2a). In TMEM16A and TMEM16F structures, there is a large extracellular domain formed by the extracellular loops of TM1-2, 3-4, 5-6, 7-8, and 9-10 and is stabilized by four disulfide bonds whose disruptions led to dysfunctional channels [62]. Interestingly, this large extracellular domain is not observed in the fungal nhTMEM16 and afTMEM16 as well as the human endoplasmic reticulum (ER) TMEM16K scramblases [48], which may reflect the different phospholipid environments in fungi or within the ER. Also, except for nhTMEM16 whose "domain-swapped" organization of the N- and C-termini endows intensive dimer interactions, inter-subunit interaction in TMEM16A, F, and K and afTMEM16 is mediated mostly by the extracellular halves of TM10. As a result, two large hydrophobic cavities, or dimer cavities, are present at the central axis of all TMEM16 proteins. As will be discussed later, these dimer cavities could provide phospholipid-binding sites and play critical roles in the lipid-dependent modulation of TMEM16 proteins.

Each monomer functions as an independent ion- and/or lipid-conducting unit [58, 95] whose activation is controlled by two highly conserved Ca^{2+} binding sites located within TMs 6–8 (Fig. 6.2a, b). The asymmetric hourglass-shaped substrate permeation pathway is formed by numerous hydrophilic as well as nonpolar residues from TMs 3–7 in which a central constriction site is formed just above the Ca^{2+} binding sites. A large body of studies have supported this hydrophilic cavity as the non-selective permeation pathway for ions and lipids in both TMEM16 scramblases and ion channels [42, 48, 70, 85, 88, 90, 91, 96, 97]. The hydrophilic grooves of both fungal afTMEM16 and nhTMEM16 and the human TMEM16K have been observed in an "open" conformation, which could represent a lipid-conductive state. Nevertheless, all current Ca^{2+}-bound structures of TMEM16A and TMEM16F adopt a closed permeation pathway in which the pores are too narrow to allow ion passage. These diverse functional states will be discussed in detail in the later section.

6.3.4 Ca^{2+}-Dependent Activation Mechanism

Prior to structural determination, mutagenesis studies have already provided valuable clues into the Ca^{2+}-dependent activation of TMEM16A. Yu et al. identified E698 and E701 (TM7) and Tien et al. identified three additional residues, E650 (TM6) and E730 and D734 (TM8), as the potential Ca^{2+} binding residues in TMEM16A [60, 62] (Fig. 6.2c, numbering based on TMEM16A(a) isoform lacking the EAVK

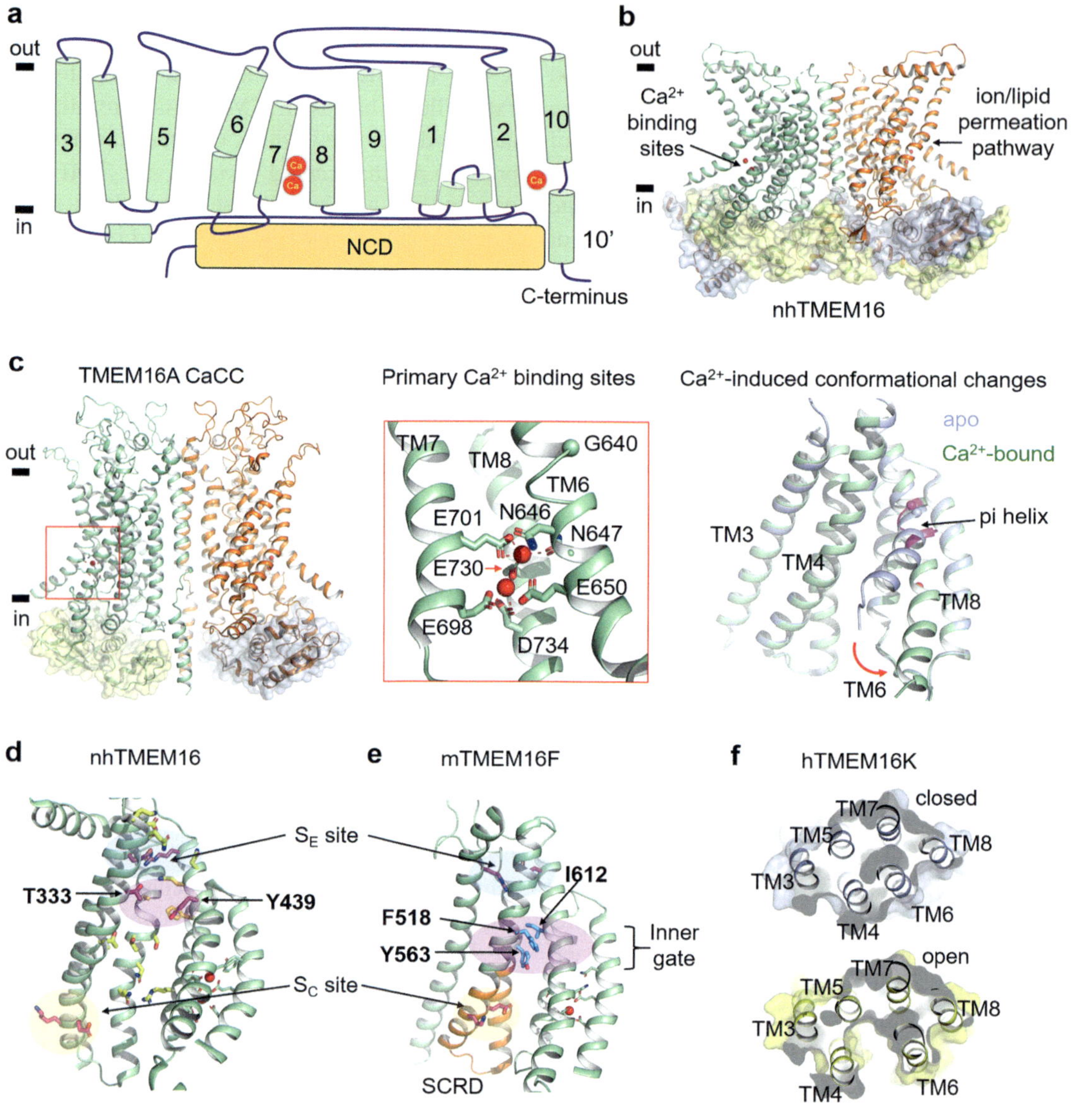

Fig. 6.2 Structural basis of TMEM16 proteins. (**a**) Topology of TMEM16 proteins. (**b**) Structure of the Ca^{2+}-bound fungal nhTMEM16 (PDB 4WIS). (**c**) Structure of the Ca^{2+}-bound mouse TMEM16A showing its overall architecture (left), Ca^{2+} coordinating residues (middle), and conformational change of TM6 upon Ca^{2+} binding. (**d**, **e**) Residues that are implicated in controlling phospholipid scrambling in the fungal nhTMEM16 (PDB 4WIS) and mouse TMEM16F (PDB 6QP6). (**f**) The lipid permeation pathway of the human TMEM16K as viewed from the ER lumen in its closed state (PDB 6R6X) and open state (PDB 5OC9)

segment in the first intracellular loop). Subsequent structural studies not only validated these electrophysiological findings but also revealed three additional asparagine residues (N646 and N647 of TM6 and N726 of TM8) as important for Ca^{2+} binding [42, 90, 96]. Within each TMEM16 monomer, these acidic and asparagine residues together form two highly conserved Ca^{2+} sites, S1 and S2. While S1 is located toward the intracellular side and is coordinated by the carboxylate groups of E650, E698, and D734, S2 is coordinated by E701, D730, and the asparagine N647 (also partially by N646 and N726 [90]) (Fig. 6.2c). In an excellent agreement with previous predictions [24], the Ca^{2+} binding sites of TMEM16 are located within the membrane field. Another important feature of TMEM16 proteins is that their Ca^{2+} binding sites are located within the immediate vicinity permeation pathway, which not only confers their efficient Ca^{2+}-dependent activation but also explains the synergistic coupling between permeant anions and Ca^{2+} ligands.

The ion conduction pathway of TMEM16A is partially enclosed by the interaction at the extracellular portions of TM4 and TM6 (Fig. 6.2c). The peripheral TM6 has been implicated in Ca^{2+}-dependent channel activation of TMEM16A [90, 96]. In the absence of Ca^{2+}, TM6 is entirely helical and adopts a kinked confirmation at the highly conserved G640, which causes the C-terminal segment of TM6 to swing away from TM7 and TM8. The kinked conformation of TM6 exposes both Ca^{2+} sites accessible to the cytosolic environment for Ca^{2+} binding from the intracellular side. The unliganded Ca^{2+} binding sites also make the intracellular vestibule of TMEM16A highly electronegative, which acts to impede Cl^- entry from the intracellular side [91, 98]. It was suggested that binding of Ca^{2+} ions to 4 highly acidic residues from TM7 and TM8 takes place first, and this then allows TM6 to form stabilizing interactions with the bound Ca^{2+} ions via N647 and E650 residues. The interactions between N647 and E650 with S2 Ca^{2+} and S1 Ca^{2+}, respectively, cause a slight rotation of TM6 and result in the formation of a π-helix at the G640 position that is stabilized by the interaction between the carbonyl of Q642 with S2 Ca^{2+} ion (Fig. 6.2c). Ca^{2+}-dependent rearrangement of TM6 triggers partial widening of the central constriction site, although the captured structure is still in a nonconductive state. Consistent with the importance of TM6 as a gating element, disrupting its G640 hinge via alanine and especially proline substitutions not only enhances the channel's apparent Ca^{2+} sensitivity but also confers basal channel activity in the absence of Ca^{2+} [68, 90]. Also, Q645A and I637A mutations on TM6 strongly enhance the channel's Ca^{2+} sensitivity and allow channel activation via membrane depolarization in the complete absence of intracellular Ca^{2+} [68, 96, 98]; by contrast, the P654A mutation, located toward the cytosolic end of TM6, markedly reduces the channel's Ca^{2+} sensitivity.

In the dual-functional TMEM16F, a similar yet different conformational transition of the gating TM6 was also observed. In the absence of Ca^{2+}, the cytosolic end of TM6 appeared mobile and moved away from TM4 in a direction opposite to that seen in TMEM16A's TM6. The binding of two Ca^{2+} ions neutralizes the negatively charged and polar Ca^{2+} binding acidic residues, including N620, N621, and E624 of TM6, E667 and E670 of TM7, and E699 and D703 of TM8, all of which strikingly resemble those in TMEM16A. This binding allows TM6 to approach TM7 and TM8 via a rigid body swinging movement around the highly conserved G615, which is equivalent to TMEM16A'a G640. However, because TMEM16F's TM6 lacks the insertion of a residue near its G615 hinge, Ca^{2+} binding did not result in the partial unwinding and π-helix formation of TM6. A similar transition from bent to straight conformations of TM6 was also observed in the structures of TMEM16F with zero or 1 Ca^{2+} bound, respectively [69]. While the fungal afTMEM16 and nhTMEM16 scramblases lack the conserved glycine hinge in mammalian TMEM16, their TM6 also undergoes a similar swinging movement around the equivalent region upon Ca^{2+} binding [88, 89]. Together, these studies underscore the functional importance of TM6 in both TMEM16 ion channel and lipid scramblase gating.

6.3.5 Voltage-Dependent Activation of TMEM16 Ion Channels

Compared to Ca^{2+} activation, the mechanism of voltage-dependent activation of TMEM16 channels is still under debate. This is mainly because TMEM16 proteins do not possess canonical voltage sensors as seen in the 6-TM cation channels, and it is challenging to separate the Ca^{2+}- and voltage-dependent activation. In contrast to the Ca^{2+}- and voltage-activated BK K^+ channel, voltage alone cannot activate TMEM16 channels in the absence of Ca^{2+}. As early studies predicted and recent structural and functional studies demonstrated [42, 52, 53, 90, 99], multiple Ca^{2+} binding sites are physically located in the membrane field electrical field. It is thus plausible that voltage-dependent Ca^{2+} binding is likely responsible for the voltage dependence of TMEM16 channels sites [52, 53, 99]. Nevertheless, a number of gain-of-function (GOF) mutations in the ion permeation pore such as Q645A or G640P challenged this mechanism as they can be activated solely by membrane depolarization without Ca^{2+} binding [68, 90]. To further understand the paradoxical voltage dependence of TMEM16A, Lam and Dutzler proposed that the apparent voltage dependence is derived from the highly negatively charged nature of the Ca^{2+} binding residues in TMs 6-8 which serve as an electrostatic gate in the ligand-free state [98]. The authors provided compelling electrophysiological and Poisson-Boltzmann calculations to demonstrate that binding of Ca^{2+} affects anion conduction by altering the electrostatics at the intracellular opening of the narrow neck via long-range Coulombic interactions. This mechanism nicely explains the strongly outward rectifying currents observed in Q645A and G640P GOF mutant CaCCs, in which impediment of Cl^- entry to the intracellular vestibule may be caused by the highly negatively charged residues of the Ca^{2+} binding sites whereas membrane depolarization favors Cl^- entry from the extracellular side. However, strong voltage dependence is also a hallmark of a number of TMEM16 dual-functional scramblase channels, some of which are more permeable to cations [2, 72, 73]. It is therefore unclear whether the same voltage-dependent mechanism proposed by Lam and Dutzler can also apply to the other TMEM16 channels that are not Cl^- selective.

By carefully characterizing two GOF mutations Q645A and I637A, which confer voltage-dependent activation without the requirement of Ca^{2+} binding, the Jan group proposed that TMEM16A could adopt multiple different open states depending on the Ca^{2+} occupancy at its binding sites [68]. The authors suggested that binding of a single Ca^{2+} triggers a partial opening of the steric gate to allow Cl^- to enter the channel pore, which then can undergo a fast voltage-dependent conformational change in TM6 to eventually allow Cl^- conduction to give rise to the outward rectifying current activation. More permeable ions such as I^- or SCN^- have a longer dwell time within the pore and thus could augment this voltage-dependent conformational change to enhance channel activation. The binding of the second Ca^{2+} allows TM6 to adopt a fully activated conformation and result in a linear Cl^- conductance. Nevertheless, the underlying voltage-sensing residue(s) remain unknown, although the authors suggested that K641 or even the acidic Ca^{2+} binding residues may play a role.

Interestingly, a mutagenesis study by Xiao et al. showed that neutralization of the 4 consecutive glutamates (444-EEEE-447) in the cytosolic TM2-3 loop, while not affecting Ca^{2+} dependent gating, significantly reduced the voltage-dependent activation of TMEM16A by right shifting the G–V curve [61]. On the other hand, deletion of the adjacent "c" splicing segment (448-EAVK-451) reduced both the Ca^{2+} sensitivity and the voltage-dependent channel activation in addition to enhancing the dissociation of Ca^{2+}. These results are consistent with a previous study from the Galietta group who showed that skipping the EAVK segment reduced the voltage sensitivity of human TMEM16A [56]. Interestingly, E367 residue and equivalent acidic stretch 386-EEEEE-390 in the first intracellular loop of TMEM16B were also found to be important for voltage-dependent activation [63]. These findings highlight the importance of the TM2-3 loop in regulating TMEM16A's voltage-dependent activation. However, how an intracellular loop affects voltage dependence is still unclear.

Taken together, while these structural and functional studies have provided important insights into the voltage-dependent activation mechanism of TMEM16A CaCC, future mechanistic studies are required for a comprehensive understanding of the voltage-dependent gating mechanism of both TMEM16 CaCCs and TMEM16 dual-functional channels and CaPLSases.

6.3.6 Ion Selectivity of the TMEM16A/B CaCCs

The anion selectivity of TMEM16A/B CaCCs partly arises from the presence of numerous basic residues located at both extracellular and intracellular vestibules of the channel that create an attractive environment for anions [91]. The central constricted region is formed by both polar and nonpolar residues. During permeation, these polar residues could compensate for the loss of anion-coordinating water molecules, whereas hydrophobic residues increase the energetic penalty for smaller anions and thus favor larger anions [90]. This mechanism explains the lyotropic permeability sequence of $SCN^- > NO_3^- > I^- > Br^- > Cl^- > F^-$ seen in both TMEM16A and TMEM16B [3, 22, 23, 27, 28, 59, 61, 64, 100], which strikingly resembles endogenous CaCCs [24, 67, 101, 102]. Also, large anions can shed their hydration shells faster than Cl^-, allowing them to enter the pore quickly [24]. However, these large ions display poor conductance, which reflects how fast the ions dissociate from the pore and traverse the channel. In other words, large anions enter quickly but get lodged within the pore. It should be noted that anionic conductance in CaCCs displays a bell-shaped relationship with their hydration energies in which anions with the lowest or highest hydration energies have the lowest conductance; this is simply because anions with the lowest permeabilities cannot easily enter the pore due to their large hydration energies, and anions with the highest permeabilities enter the pore quickly but interact strongly with the pore and are thus poorly conducted. Finally, hydrophobic anions such as SCN^- or $C(CN)_3^-$, which are highly permeable to CaCCs, exert pore blockage effects on CaCCs, consistent with the presence of hydrophobic residues within the channels' permeation pathway [90, 96, 102].

The polar and non-polar pore-lining residues that form the central constriction neck in TMEM16A have been shown to be important for ion selectivity in TMEM16A [96]. Alanine mutations including N542A, D550A, N587A, V595A, Q705A, and F712A further enhanced the channel's permeability toward large anions such as I^- and SCN^- by increasing P_I/P_{Cl} and P_{SCN}/P_{Cl} ratios, whereas S635A reduced both. Notably, in addition to their roles in discriminating anions [68, 96], mutations of these pore-lining residues were also found to modulate the channel's Ca^{2+}-dependent activation, perhaps via their allosteric effects on the channel activation gate [68, 70, 96]. Dang et al. showed that while N542A, I546A, Y589A, I592A, and F708A significantly enhanced, V595A and L639A reduced the channel's apparent Ca^{2+} sensitivity [96]. Remarkably, I546A and I637A mutations, located in TM4 and TM6, respectively, not only enhance the Ca^{2+} sensitivity but also allow the channel to conduct basal Cl^- currents in the complete absence of Ca^{2+} [68, 90]. We showed that the L543K mutation in TM4 of TMEM16A, which provides prominent steric hindrance to the central constriction gate, not only endows voltage-dependent activation but remarkably also turns the TMEM16A mutant channel into a constitutively active lipid scramblase [70]. Together with previous observations from endogenous CaCCs [24], these studies suggest that in TMEM16A/B CaCCs, anion occupancy within the channel's amphipathic central constricted region strongly influences Ca^{2+}-dependent channel gating.

6.3.7 Ion Selectivity of the Dual-Functional TMEM16F

The nature of ion selectivity of TMEM16 scramblases, most notably TMEM16F, remains highly controversial. We previously showed that TMEM16F is actually more permeable toward cations such as Ca^{2+} and Na^+ than anions with a P_{Na}/P_{Cl} of ~6.8 [2]. However, other studies reported that TMEM16F is rather a poorly selective channel with a P_{Na}/P_{Cl} of ~1.3 [71] or ~1.4 [46] or even more selective towards Cl^- with a P_{Na}/P_{Cl} of 0.5 [45] or 0.3 [41]. The discrepancies in ion selectivity values of TMEM16F among different labs may reflect the different methodologies as well as the assumption of using "impermeant" ions such as Cs^+, aspartate, or $NMDG^+$. In fact, TMEM16F was shown to display significant permeability toward $NMDG^+$ [2, 46] and aspartate with a P_{Asp}/P_{Cl} = 0.5 [41]. The fact that TMEM16F conducts Cs^+, $NMDG^+$, or aspartate implies a possibility that the presumably "impermeant" ions such as Cs^+, $NMDG^+$, and MES^- used in these studies could contribute significantly to the measured currents and therefore confound the permeability measurements. This promiscuous selectivity of TMEM16F may be explained by the "lipidic pore" model proposed by the Hartzell group who suggested that the observed current in TMEM16F could be a result of a leaky product during lipid permeation through the large hydrophilic cavity [103]. Adding to this complexity is the recent finding that TMEM16F's ion permeation undergoes a dynamic change in its ion selectivity [72]. Ye et al. showed that this change is not due to different channel activation states but is likely caused by an alteration in the electrostatic field of the permeation pathway. Given these observations, it is tempting to speculate that TMEM16F pore is a highly dynamic structure that can undergo significant widening or opening to accommodate large molecules such as phospholipids and different ions. Future studies are needed to further understand the dynamic nature of ion permeation through the TMEM16 dual-functional channels and CaPLSases.

6.3.8 Phospholipid Permeation Through TMEM16 CaPLSases

How phospholipids and ions permeate through TMEM16 CaPLSases has been extensively studied using various structural, functional, and computational methods. Different models have been proposed to describe the dual permeation of phospholipids and ions. In this section, we summarize the major findings that support these models.

6.3.8.1 Classical "Credit Card" Model for Phospholipid Permeation

The X-ray structure of the fungal scramblase nhTMEM16 by the Dutzler lab [42] beautifully demonstrated that TMEM16 CaPLSases may also utilize the "credit card" mechanism (Fig. 6.2b), a prevailing model for transporter-mediated phospholipid flip-flop [104], to conduct phospholipid permeation. According to this "credit card" mechanism, the phospholipid headgroup slides through a hydrophilic protein groove facing the hydrophobic membrane core with the phospholipid acyl tails remaining within the membrane core (Fig. 6.1a). Indeed, each nhTMEM16 monomer has a hydrophilic furrow or groove that consists of both polar and charged residues and faces the lipid environment with TM4 and TM6 at the protein–lipid interface [42]. Coarse-grained molecular dynamics simulations of nhTMEM16 in phosphatidylcholine from the Samson group first showed that lipid headgroups could indeed occupy this hydrophilic groove [105].

To understand how lipid permeation may take place in nhTMEM16 and how lipids interact with the hydrophilic groove, Bethel and Grabe performed atomistic molecular dynamics simulations and continuum membrane-bending calculations using the nhTMEM16 structure (PDB 4WIS) as a template [106]. They discovered two lipid-interacting sites flanking the hydrophilic groove that could serve as "stepping stones" for lipid permeation: an extracellular S_E site of E313 (on TM3) and R432 (on TM6) and a cytosolic S_C site of E352 and K353 (both at the intracellular end of TM4) (Fig. 6.2d). While residues constituting the S_E site appear to be switched in positions in mammalian TMEM16, they are highly conserved with the exceptions of TMEM16H and TMEM16K. Consistent with their roles in lipid scrambling, Feng et al. showed that the equivalent S_E site R478A mutation delayed the onset of GPMV generation, whereas Gyobu et al. showed that both R478A and E604C mutations significantly impaired PS exposure in TMEM16F [69, 107]. In TMEM16A CaCC, S_E

sites residues R511 and E619 are equivalent to E313 and K432, respectively, and R511 was previously shown to be important for TMEM16A's ion selectivity [100] while E619 plays an important role in proton sensing [55]. The S_C site, which resides in the scrambling domain previously proposed by the Hartzell group [46], serves to nucleate headgroup-dipole stacking interactions and to promote lipid penetration into the groove (Fig. 6.2e). Using a chimeric approach, the Hartzell group narrowed down to a stretch of 35 amino acids from N525 to Q559 of TMEM16F, which is equivalent to TMEM16A D550 to K584, that function as a "scrambling domain" (SCRD) in TMEM16F [46] (Fig. 6.2e). Remarkably, the introduction of this segment to TMEM16A endows its ability to scramble lipids while retaining its Cl^- selective ion conduction. Gyobu et al. further showed that substitution of the equivalent SCRD of the ER TMEM16E scramblase also confers scrambling activity to TMEM16A [50]. However, we later found that a single mutation TMEM16A L543K, which is outside of this SCRD, is sufficient to convert this CaCC into a constitutively active scramblase [70]. This raises the question of whether the scrambling activity in TMEM16 scramblases indeed requires this 35 amino acid stretch and how the residues within the long stretch control lipid scrambling.

6.3.8.2 Membrane Bending/Distortion Is a Common Feature in TMEM16 Scramblases

Molecular dynamics and structural studies all revealed an interesting phenomenon in which TMEM16 scramblases pronouncedly deform the membrane, albeit at different regions. Using fast and continuum membrane-bending calculations, Bethel and Grabe first showed that nhTMEM16 induces large-scale deformation of the membrane, which results in thinning of the bilayer across the hydrophilic groove of ~36%, thus significantly shortening the travel distance for lipids. Similar membrane distortions were also observed in several simulations studies by other groups [85, 97]. Remarkably, this phenomenon was subsequently confirmed by structural studies of nhTMEM16 and afTMEM16 either in a lipid-like nanodisc environment or detergent [88, 89]. Importantly, this bending of the membrane by TMEM16 scramblases does not depend on Ca^{2+} binding as it was observed in all nanodisc-reconstituted TMEM16 scramblases. Falzone et al., on the other hand, observed pronounced membrane bending at the dimer cavity, which is likely caused by the fact that at this dimer interface, TM3 and TM5 of one monomer are longer than TM1 and TM2 of the other monomer. Interestingly, in the open active afTMEM16 scramblase, the electron density of the nanodisc at the subunit cavity appears weaker, which is consistent with a thinner and/or disordered membrane that facilitates lipid permeation. In fact, in the ceramide-inhibited afTMEM16 structure, although the Ca^{2+}-induced conformation is similar to that of the active scramblase, the density of nanodisc at the subunit cavity is stronger, presumably explaining its inhibitory effect by modulating the lipid environment and not Ca^{2+}-dependent conformational transition. In the structures of the mouse TMEM16F, Alvadia et al. and Feng et al. did not observe any obvious membrane distortion in the nanodisc-reconstituted mTMEM16F. However, Feng et al. did observe that PIP_2 supplementation allows TM6 to adopt a kink conformation that in turn causes membrane thinning at TM3 and TM4 and that this membrane distortion is essential to lipid scrambling in TMEM16F [69]. Taken together, these atomistic and structural analyses suggest a possibility that TMEM16 CaPLSases have evolved a special arrangement at their protein–lipid interface to facilitate phospholipid permeation by membrane thinning and deformation.

6.3.8.3 "Lipidic Pore" Dual Permeation Model Derived from the "Credit-Card" Model

Experimental studies and MD simulations from the Hartzell and Tajkhorshid groups suggested that at least in the nhTMEM16 scramblase, both ions and lipids traverse the subunit cavity or hydrophilic groove [97] (Fig. 6.3a). The authors found three different sites for lipid interaction:

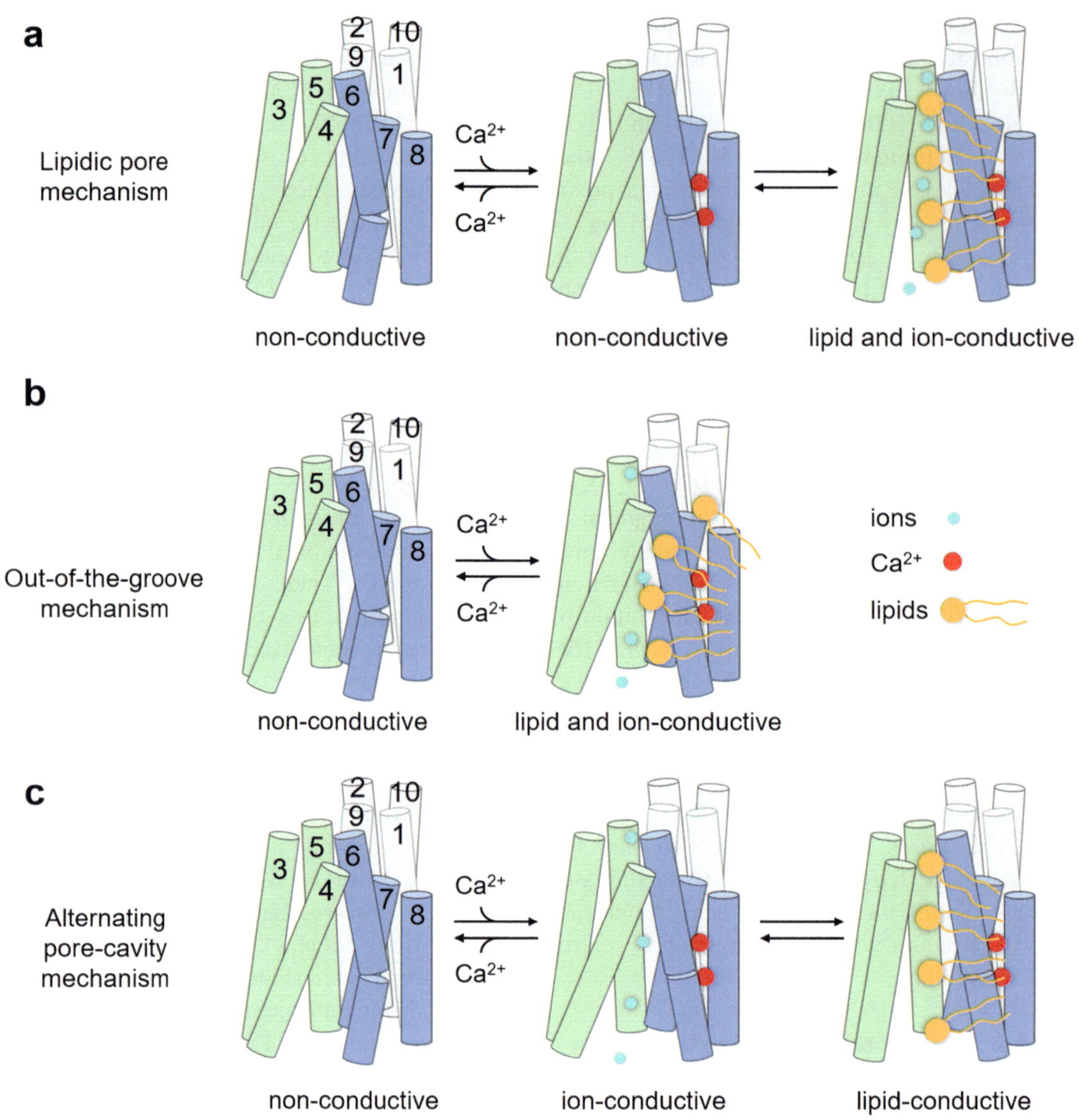

Fig. 6.3 Mechanistic models of phospholipid and ion permeation by dual-functional TMEM16 CaPLSases/channels. The proposed models for lipidic pore (**a**), out-of-the-groove (**b**), and alternating pore-cavity (**c**) mechanisms are shown

S_{int} of Q374 and N378 of TM5, R505 of TM7; S_{cen} (about one third into the membrane from the intracellular side) of N378, T381, S382 (TM5), and T340 (TM4); and S_{ext} (near the extracellular entrance) of E313, N317 (TM3), K325, Q326, T333 (TM3), R432, N435, and Y439 (TM6). R432 of TM6 is part of the S_E site identified in Bethel and Grabe. Notably, two full events of PS lipid permeation from the inner leaflet to the outer leaflet were observed during which the PS headgroup interacts mostly with the oxygen or nitrogen atoms of R505 of the S_{int} site. Furthermore, T333 and Y439 of the S_{ext} appear to form a steric gate to control the lipid permeation pathway. In the absence of Ca^{2+}, TM4 transitions closer toward TM6 to establish an interaction between T333 and Y439, thereby forming a constriction site to close the lipid pathway. This observation was supported by the greatly reduced scrambling activity by the T333V mutation, likely due to the increased hydrophobicity. However, in a reconstituted system, T333W showed no obvious reduction in scrambling [85]. Interestingly, the single TMEM16A mutations V543S, V543T, and K588N, when expressed in HEK293 cells that robustly express endogenous TMEM16F [37, 73], convert the ion channel into a scramblase [97]. These observations are interesting but are not readily reconcilable as V543 is not part of the previously proposed scrambling domain.

By performing Trp mutagenesis on residues lining the hydrophilic groove of nhTMEM16, Lee et al. found three distinct regions that are important for lipid permeation: (1) the lower constriction site of L302, T430, T381, and S382 near the midpoint of the membrane, (2) the extracellular entrance comprising E313 and R432, and (3) A395, Q436, Y439, and F440 residues between the lower constriction site and the extracellular entrance [85]. Their MD simulations studies further showed that a triad of charged residues, E313, E318, and R432, are critical for lipid permeation gating in nhTMEM16. These residues not only interact with the permeating lipids but also undergo dynamic rearrangement during the gating process. In fact, the interaction of phospholipids with R432 of TM6 promotes sequential disengagement from E313 and E318 of TM3 during permeation. Due to this disengagement, TM6 rotates away from TM3 and allows Y439 of TM6 to move away from T333 of TM4, thereby widening the groove to enable lipid permeation. Furthermore, Y439 is not only important for gating but is also critical for lipid coordination. Thus, Lee et al. proposed that lipid translocation requires a widening of the hydrophilic groove, a process that likely cannot be achieved by channel-only TMEM16 proteins such as TMEM16A and TMEM16B. Interestingly, Trp mutations that result in reduced scrambling also affect the ion conduction of nhTMEM16, which is consistent with the hypothesis that ions and lipids share a common permeation pathway formed by the hydrophilic groove.

6.3.8.4 "Ions-in-the-Pore and Lipids-Out-of-the-Groove" Dual Permeation Model

The Accardi group reconstituted purified afTMEM16 and nhTMEM16 scramblases into liposomes and investigated whether they can transport phospholipids whose headgroups are derivatized with polyethylene glycol (PEG) moieties of various sizes [43]. Remarkably, both fungal scramblases can scramble lipids conjugated with PEG with sizes ranging from 2000 to 5000 Da, which are equivalent to roughly 8–40 Å in diameter and are thus much larger than the hydrophilic groove. Notably, these large lipids are transported at rates that are comparable to those of normal-sized lipids. This led to the proposal that these large lipids surf on the surface of TMEM16 scramblases without having to permeate through the hydrophilic cavity, or via an "out-of-the-groove" mechanism (Fig. 6.3b), resembling the proposed phospholipid permeation mechanism for the Ca^{2+}-independent scramblase opsin [108]. Using the lipid pathway R432W mutant, Malvezzi et al. showed this mutation more strongly impaired scrambling of the smaller lipids than PEG-conjugated lipids, which is consistent with PEG-conjugated lipids permeating outside the groove whereas permeation of small lipids occurs within the groove. As TMEM16 scramblases also mediate ion transport, believed to occur through the hydrophilic groove [69, 87, 90, 91, 96], translocation of large lipids via this out-of-the-groove model should therefore impose minimal effects on their ion permeation. Indeed, afTMEM16-mediated scrambling of PEG-conjugated lipids does not affect ion permeation or ion selectivity to $NMDG^+$ of afTMEM16, further supporting this out-of-the-groove model. This also suggests that ion permeation in TMEM16 proteins likely takes place via a proteinaceous pore (ions-in-the-pore) and that permeation of large lipids does not induce dilation of the pathway.

The recent structural studies suggested that the "ions-in-the-pore and lipids-out-of-the-groove" model may also apply to mammalian TMEM16 CaPLSases. Alvadia et al. reported that TM4 is in contact with TM6 to enclose the permeation pathway in both unliganded and Ca^{2+}-bound TMEM16F structures [87]. The authors thus speculated that in mammalian TMEM16 scramblases, at least in the case of TMEM16F, lipid permeation could take place via the "out-of-the-groove" mechanism. The Jan and Cheng groups also made similar observations for TMEM16F structures in either digitonin or nanodisc conditions [69]. Feng et al. observed that while PIP_2 supplementation promoted widening of the intracellular portion of the pore, the putative lipid permeation pathway still adopted an enclosed conformation analogous to that observed in TMEM16A [90, 91, 96] and TMEM16F [87]. Owing to the lack of an open hydrophilic groove, Feng et al. proposed a different "out-of-the-groove" model that is dependent upon PIP_2 binding, which causes a kink in TM6 as well as membrane distortion and thinning. The authors suggested that TMEM16F harbors distinct permeation pathways for ions and lipids, and that membrane distortion plays a key role in lipid scrambling and that lipid scrambling may take place even in the absence of an open hydrophilic groove.

6.3.8.5 "Alternating Pore-Cavity" Dual Permeation Model

To reconcile the difference between the "lipidic pore" model and the "out-of-the-groove" model, an alternative model of the "alternating pore-cavity" model was proposed [87] (Fig. 6.3c). In this model, ions and lipids are proposed to traverse by distinct conformational states of TMEM16F. Lipid permeation requires a widened hydrophilic groove as that seen in the structures of nhTMEM16 [42, 89], afTMEM16 [88], or human TMEM16K [48], whereas ion conduction is mediated by an enclosed pathway, which is likely captured in the Ca^{2+}-bound TMEM16F [69, 87] and resembles that of TMEM16A structures [90, 96]. Although the "alternating pore-cavity mechanism" helps reconcile the discrepancy between the "lipidic core" model and the "ions-in-the-pore and lipids-out-of-the-groove" model, it is unclear how CaPLSases can precisely control the fast switching between the ion permeation state and the phospholipid permeation state with high efficiency and high fidelity. Additional functional and computational studies are needed to validate this model.

6.3.9 TMEM16 CaPLSase Gating

Scramblases and ion channels are passive membrane transporters that permeate their substrates at high speed. This requires that their permeation must be tightly controlled or gated between closed and open conformations. Numerous recent structural and functional studies have provided valuable insights into TMEM16 lipid permeation gating.

Owing to the lack of a mammalian TMEM16 scramblase structure at the time, we utilized the structures of nhTMEM16 (4WIS) and TMEM16A (5OYB and 5OYG) as templates to generate homology models of TMEM16F scramblase in an open, intermediate, or closed state [70]. We then performed atomistic MD simulations of these TMEM16F models to observe lipid permeation as well as the potential conformational transitions between the open and the closed states. We noticed that three hydrophobic residues, F518 (TM4), Y563 (TM5), and I612 (TM6), together form a hydrophobic constriction gate that impedes lipid permeation. Interestingly, whereas the side chains of these residues are in proximity in the closed state, they separate to widen the hydrophilic groove in the Ca^{2+}-bound open state as if they serve as an inner gate at the cytosolic mouth of the permeation pathway. Indeed, by applying our optimized microscopy-based lipid scrambling assay, we found that removing steric hindrance of the inner gate via alanine substitutions of F518, Y563, and I612 strongly promote lipid scrambling in TMEM16F following Ca^{2+} ionophore treatment. Consistent with this observation, introducing large hydrophobic side chains (i.e. F518L or Y563W mutations) strongly reduces their scrambling rates. By contrast, introducing polar or charged mutations to these three hydrophobic residues make TMEM16F constitutively open. In the case of F518K or Y563K, the gain-of-function effects are highly potent such that they result in constitutive scrambling activities even when the Ca^{2+} binding sites are disrupted. These results led us to propose a "clamp shell" gating model for TMEM16 CaPLSases. We suggested that the interface between TM4 and TM6 can open and close like a clam shell to control the accessibility of phospholipids to the interior of the hydrophilic groove. F518 in TM4 and I612 in TM6 likely serve as gate-keepers for the opening of this interface, whereas Y563 in TM5 likely serves as a cap that stabilizes the inner gate and obstructs phospholipid permeation in the closed state. Ca^{2+} binding-induced conformational changes lead to the separation of the TM4–TM6 interface as well as the inner gate residues, subsequently exposing the interior of the hydrophilic groove to the surrounding phospholipids such that their phospholipid headgroups can enter and translocate via the credit-card mechanism. On the one hand, introducing smaller, polar, or charged residues to these critical locations removes the steric hindrance, resulting in enhanced phospholipid permeation. On the other hand, the inner activation gates of F518K and Y563K mutants are severely disrupted that phospholipids can freely go through the constitutively open gate in the absence of Ca^{2+} binding.

Interestingly, we found that channel opening by two TMEM16A mutations L543K and I637K, which are equivalent to TMEM16F F518K and I612K, respectively, can be elicited solely via membrane depolarization without Ca^{2+}. Strikingly, L543K also conferred constitutively active lipid scrambling activity to the CaCC in the absence of Ca^{2+}. These experiments support the notion that TMEM16 CaCCs and CaPLSases may utilize a similar set of hydrophobic residues at their inner activation gate to control ion and phospholipid permeation. Instead of opening like a "clam shell" in TMEM16 CaPLSases to separate TM4 and TM6, the helices maintain in contact in TMEM16 CaCCs to exclude phospholipid headgroups to penetrate and permeate. Ca^{2+} binding only dilates the proteinaceous pore to allow Cl^- to go through. When a positive charge is introduced to L543 in TM4, the interaction between TM4 and TM6 is likely weakened such that the mutant TMEM16A can open like a clam shell to allow spontaneous permeation of phospholipids. Our findings thus suggest that TMEM16 CaPLSases and CaCCs may share similar Ca^{2+}-dependent gating mechanisms and overall similar design in their inner activation gates. Their distinct substrate selectivity and permeation may be partially derived from the differences on how widely the putative inner activation gate can open.

Bushell et al. applied X-ray crystallography and cryo-EM to capture the human TMEM16K in multiple different states, including closed, intermediate, and open states [48]. Most interestingly, their X-ray structure of the Ca^{2+}-bound TMEM16K revealed an open hydrophilic groove, which is thus far the only open groove confirmation of a mammalian TMEM16 scramblase (Fig. 6.2f). The transition from an open to a closed groove conformation is associated with several prominent structural rearrangements, including a ~10-degree rotation of the N-terminal cytosolic domain (NCD), separation of the interdimeric interaction between the C-terminal segments of TM10 (TM10' or αTM10) that results in a ~30% reduction in the dimer interface, and finally the pivoting movements of TM3 and TM4 around TM5, all of which culminate in the closure of the hydrophilic groove at the ER luminal side. In this closed groove conformation, an intensive network of residues from TM4-7, including Y366, A367, L416, S415, T435, L436, and Y507, form a hydrophobic gate and likely occlude lipid permeation.

By determining the cryo-EM structures of the afTMEM16 scramblase, Falzone et al. observed global conformational changes upon Ca^{2+} binding [88]. In the absence of Ca^{2+}, the bending of both TM4 and TM6 results in a complete enclosure of the subunit cavity that is accompanied by the upward movement of TM3 and the dilation of the Ca^{2+} binding sites. Notably, Ca^{2+} binding results in the straightening of both TM4 and

TM6, allowing TM6 to disengage from TM4 to open the lipid translocation pathway. The pronounced bending of TM4 is likely mediated by P324 and P333 residues (equivalent to nhTMEM16 P332 and P341), which was also observed in the nhTMEM16 scramblase [89]. Falzone et al. speculated that afTMEM16 could also adopt an "intermediate" state in which TM4 is bent while TM6 is straight, and that this intermediate state is only ion conductive.

To understand the structural transition during Ca^{2+}-dependent gating of the nhTMEM16 scramblase, the Paulino and Dutzler groups determined the Ca^{2+}-bound and Ca^{2+}-free cryo-EM structures of nhTMEM16 in detergent DDM and lipid-like nanodisc (2N2) [89]. These structures revealed the remarkable conformational changes that nhTMEM16 may undergo upon Ca^{2+} binding. In the absence of Ca^{2+}, TM4 undergoes a pronounced movement toward TM6 to enclose the subunit cavity from the membrane's hydrophobic environment, resembling the apo structures of TMEM16A. In the presence of Ca^{2+}, nhTMEM16 can adopt three major conformations at its subunit cavity, tentatively assigned as "Ca^{2+}-bound closed," "intermediate," and "Ca^{2+}-bound open." First, in the "Ca^{2+}-bound closed" state, nhTMEM16 structure highly resembles that of the Ca^{2+}-free closed state, and the subunit cavity is shielded from the membrane via tight interactions between TM4 and TM6. Second, in its "intermediate" conformation, while the subunit cavity is still enclosed from the membrane, the cavity has widened relative to that of the closed state. Third, in the "Ca^{2+}-bound open" state, the subunit cavity of nhTMEM16 is exposed to the lipid environment reminiscent of the Ca^{2+}-bound open structure in detergent. This widening of the subunit cavity is likely mediated by the pronounced orientation of both TM3 and TM4 toward TM10 the adjacent subunit. In transitioning from the closed to the open state, TM4 undergoes the largest helical rearrangements around the potential pivot points consisting of P332 and P431 as well as G339. In fact, mutations of P431 and G439 both significantly reduced nhTMEM16's lipid scrambling both in the absence and in the presence of Ca^{2+}. This open state is believed to represent the lipid scrambling-competent state of nhTMEM16. On the other hand, observation of the intermediate state, which likely does not promote scrambling, led Kalienkova et al. to propose an equilibrium model for the Ca^{2+}-bound and Ca^{2+}-free states in which the partially enclosed "intermediate" state represents the ion-conductive state whereas the full open cavity represents the lipid-conductive state. In agreement with these observations, Khelashvilli et al. showed that nhTMEM16 can transition from an open to an intermediate state which only allows ion conduction based on atomistic simulations and structural determination [109]. The middle region of TM4 moves toward TM6 to partially enclose the hydrophilic groove, thus impeding lipid permeation. They also revealed that the hydrophobic interaction between TM3 and TM4, mediated by L302 of TM3 and I343 and L347 of TM4, helps stabilize the open membrane-exposed lipid pathway. Consistent with this observation, disrupting this hydrophobic interaction via the L302A mutation which, despite strongly impairing lipid permeation, retains its ion channel activity. The cryo-EM structure of L302A mutant nhTMEM16 further confirmed that the scramblase indeed adopts an intermediate state that strikingly resembles their MD simulated structure and is similar to the intermediate state of the WT nhTMEM16 reported by Kalienkova et al. Interestingly, in the nhTMEM16 L302A structure, the interaction between E313 and R432 (the S_E site) is disrupted, resulting in an open extracellular gate as well as a continuous pore that is wide enough to accommodate water, Na^+ or K^+ ions.

6.4 Future Prospective

Since the identification of TMEM16A as a CaCC in 2008, the collective efforts of the TMEM16 field have greatly advanced our understanding of these intriguing membrane transporters and their contributions to human health and diseases. Given the wealth of structural information available and various established functional assays, we are at the exciting stage to comprehensively understand the molecular mechanisms of different TMEM16 proteins, including (1) the precise biophysical properties of the intracellular TMEM16 proteins; (2) the molecular basis of TMEM16 voltage dependence and its cooperativity with Ca^{2+} in activating TMEM16s; (3) the allosteric gating mechanisms that couple different parts of the proteins; (4) the relationship between ion and phospholipid permeation through the dual-functional TMEM16 proteins and their underlying structural basis; (5) the regulatory mechanisms of TMEM16 proteins by various factors and cell signaling pathways; (6) the identification of TMEM16-specific pharmacological reagents and their working mechanisms to influence TMEM16 functions. At the cellular level, we need to gain a better understanding of (1) the cellular functions of the intracellular TMEM16 CaPLSases; (2) how TMEM16 proteins are activated and regulated under physiological and pathological conditions; and (3) the cellular functions mediated by TMEM16 channels and CaPLSases in various cell types. There is also an urgent need to decipher the physiological roles of TMEM16 proteins using animal models as well as the pathophysiology of human TMEM16 mutations. We anticipate that answers to these questions will significantly promote our understanding of this important family of membrane transport proteins and facilitate the future design of TMEM16-specific therapies to treat numerous diseases such as heart attack, stroke, asthma, epilepsy, ataxia, muscular dystrophy, viral infection, pregnancy complications, and cancer.

Acknowledgments This work was supported by grants National Institutes of Health NIH-DP2-GM126898 (to H. Y.) and the American Heart Association Pre-Doctoral Fellowship 19PRE34380456 (to S.C.L.).

References

1. Almen MS, Nordstrom KJ, Fredriksson R, Schioth HB (2009) Mapping the human membrane proteome: a majority of the human membrane proteins can be classified according to function and evolutionary origin. BMC Biol 7:50
2. Yang H, Kim A, David T, Palmer D, Jin T, Tien J, Huang F, Cheng T, Coughlin SR, Jan YN et al (2012) TMEM16F forms a Ca^{2+}-activated cation channel required for lipid scrambling in platelets during blood coagulation. Cell 151:111–122
3. Adomaviciene A, Smith KJ, Garnett H, Tammaro P (2013) Putative pore-loops of TMEM16/anoctamin channels affect channel density in cell membranes. J Physiol 591:3487–3505
4. Berg J, Yang H, Jan LY (2012) Ca2+-activated Cl-channels at a glance. J Cell Sci 125:1367–1371
5. Bevers EM, Williamson PL (2016) Getting to the outer leaflet: physiology of phosphatidylserine exposure at the plasma membrane. Physiol Rev 96:605–645
6. Brunner JD, Schenck S, Dutzler R (2016) Structural basis for phospholipid scrambling in the TMEM16 family. Curr Opin Struct Biol 39:61–70
7. Crottes D, Jan LY (2019) The multifaceted role of TMEM16A in cancer. Cell Calcium 82:102050
8. Duran C, Hartzell HC (2011) Physiological roles and diseases of Tmem16/Anoctamin proteins: are they all chloride channels? Acta Pharmacol Sin 32:685–692
9. Duran C, Thompson CH, Xiao Q, Hartzell HC (2010) Chloride channels: often enigmatic, rarely predictable. Annu Rev Physiol 72:95–121
10. Falzone ME, Malvezzi M, Lee BC, Accardi A (2018) Known structures and unknown mechanisms of TMEM16 scramblases and channels. J Gen Physiol 150:933–947
11. Hartzell HC, Yu K, Xiao Q, Chien LT, Qu Z (2009) Anoctamin/TMEM16 family members are Ca2+-activated Cl- channels. J Physiol 587:2127–2139
12. Huang F, Wong X, Jan LY (2012) International Union of Basic and Clinical Pharmacology. LXXXV: calcium-activated chloride channels. Pharmacol Rev 64:1–15
13. Kunzelmann K, Nilius B, Owsianik G, Schreiber R, Ousingsawat J, Sirianant L, Wanitchakool P, Bevers EM, Heemskerk JW (2014) Molecular functions of anoctamin 6 (TMEM16F): a chloride channel, cation channel, or phospholipid scramblase? Pflugers Arch 466:407–414
14. Nagata S, Suzuki J, Segawa K, Fujii T (2016) Exposure of phosphatidylserine on the cell surface. Cell Death Differ 23:952–961
15. Oh U, Jung J (2016) Cellular functions of TMEM16/anoctamin. Pflugers Arch 468:443–453
16. Pedemonte N, Galietta LJ (2014) Structure and function of TMEM16 proteins (anoctamins). Physiol Rev 94:419–459
17. Picollo A, Malvezzi M, Accardi A (2015) TMEM16 proteins: unknown structure and confusing functions. J Mol Biol 427:94–105
18. Whitlock JM, Hartzell HC (2017) Anoctamins/TMEM16 proteins: chloride channels flirting with lipids and extracellular vesicles. Annu Rev Physiol 79:119–143

19. Yang H, Jan LY (2016) Chapter 7: TMEM16 Membrane proteins in health and disease. In: Pitt GS (ed) Ion channels in health and disease. Boston, Academic Press, pp 165–197
20. West RB, Corless CL, Chen X, Rubin BP, Subramanian S, Montgomery K, Zhu S, Ball CA, Nielsen TO, Patel R et al (2004) The novel marker, DOG1, is expressed ubiquitously in gastrointestinal stromal tumors irrespective of KIT or PDGFRA mutation status. Am J Pathol 165:107–113
21. Caputo A, Caci E, Ferrera L, Pedemonte N, Barsanti C, Sondo E, Pfeffer U, Ravazzolo R, Zegarra-Moran O, Galietta LJ (2008) TMEM16A, a membrane protein associated with calcium-dependent chloride channel activity. Science 322:590–594
22. Schroeder BC, Cheng T, Jan YN, Jan LY (2008) Expression cloning of TMEM16A as a calcium-activated chloride channel subunit. Cell 134:1019–1029
23. Yang YD, Cho H, Koo JY, Tak MH, Cho Y, Shim WS, Park SP, Lee J, Lee B, Kim BM et al (2008) TMEM16A confers receptor-activated calcium-dependent chloride conductance. Nature 455:1210–1215
24. Hartzell C, Putzier I, Arreola J (2005) Calcium-activated chloride channels. Annu Rev Physiol 67:719–758
25. Danahay H, Atherton H, Jones G, Bridges RJ, Poll CT (2002) Interleukin-13 induces a hypersecretory ion transport phenotype in human bronchial epithelial cells. Am J Physiol Lung Cell Mol Physiol 282: L226–L236
26. Galietta LJ, Pagesy P, Folli C, Caci E, Romio L, Costes B, Nicolis E, Cabrini G, Goossens M, Ravazzolo R et al (2002) IL-4 is a potent modulator of ion transport in the human bronchial epithelium in vitro. J Immunol 168:839–845
27. Pifferi S, Dibattista M, Menini A (2009) TMEM16B induces chloride currents activated by calcium in mammalian cells. Pflugers Arch 458:1023–1038
28. Stephan AB, Shum EY, Hirsh S, Cygnar KD, Reisert J, Zhao H (2009) ANO2 is the cilial calcium-activated chloride channel that may mediate olfactory amplification. Proc Natl Acad Sci U S A 106:11776–11781
29. Stohr H, Heisig JB, Benz PM, Schoberl S, Milenkovic VM, Strauss O, Aartsen WM, Wijnholds J, Weber BH, Schulz HL (2009) TMEM16B, a novel protein with calcium-dependent chloride channel activity, associates with a presynaptic protein complex in photoreceptor terminals. J Neurosci 29:6809–6818
30. Suzuki J, Umeda M, Sims PJ, Nagata S (2010) Calcium-dependent phospholipid scrambling by TMEM16F. Nature 468:834–838
31. Castoldi E, Collins PW, Williamson PL, Bevers EM (2011) Compound heterozygosity for 2 novel TMEM16F mutations in a patient with Scott syndrome. Blood 117:4399–4400
32. Satta N, Toti F, Fressinaud E, Meyer D, Freyssinet JM (1997) Scott syndrome: an inherited defect of the procoagulant activity of platelets. Platelets 8:117–124
33. Brooks MB, Catalfamo JL, MacNguyen R, Tim D, Fancher S, McCardle JA (2015) A TMEM16F point mutation causes an absence of canine platelet TMEM16F and ineffective activation and death-induced phospholipid scrambling. J Thromb Haemost 13:2240–2252
34. Martins JR, Faria D, Kongsuphol P, Reisch B, Schreiber R, Kunzelmann K (2011) Anoctamin 6 is an essential component of the outwardly rectifying chloride channel. Proc Natl Acad Sci U S A 108:18168–18172
35. Almaca J, Tian Y, Aldehni F, Ousingsawat J, Kongsuphol P, Rock JR, Harfe BD, Schreiber R, Kunzelmann K (2009) TMEM16 proteins produce volume-regulated chloride currents that are reduced in mice lacking TMEM16A. J Biol Chem 284:28571–28578
36. Juul CA, Grubb S, Poulsen KA, Kyed T, Hashem N, Lambert IH, Larsen EH, Hoffmann EK (2014) Anoctamin 6 differs from VRAC and VSOAC but is involved in apoptosis and supports volume regulation in the presence of Ca2+. Pflugers Arch 466:1899–1910
37. Schreiber R, Uliyakina I, Kongsuphol P, Warth R, Mirza M, Martins JR, Kunzelmann K (2010) Expression and function of epithelial anoctamins. J Biol Chem 285:7838–7845
38. Shimizu T, Iehara T, Sato K, Fujii T, Sakai H, Okada Y (2013) TMEM16F is a component of a Ca2+-activated Cl- channel but not a volume-sensitive outwardly rectifying Cl- channel. Am J Physiol Cell Physiol 304:C748–C759
39. Szteyn K, Schmid E, Nurbaeva MK, Yang W, Munzer P, Kunzelmann K, Lang F, Shumilina E (2012) Expression and functional significance of the Ca(2+)-activated Cl(−) channel ANO6 in dendritic cells. Cell Physiol Biochem 30:1319–1332
40. Tian Y, Schreiber R, Kunzelmann K (2012) Anoctamins are a family of Ca2+-activated Cl-channels. J Cell Sci 125:4991–4998
41. Grubb S, Poulsen KA, Juul CA, Kyed T, Klausen TK, Larsen EH, Hoffmann EK (2013) TMEM16F (Anoctamin 6), an anion channel of delayed Ca(2+) activation. J Gen Physiol 141:585–600
42. Brunner JD, Lim NK, Schenck S, Duerst A, Dutzler R (2014) X-ray structure of a calcium-activated TMEM16 lipid scramblase. Nature 516:207–212
43. Malvezzi M, Chalat M, Janjusevic R, Picollo A, Terashima H, Menon AK, Accardi A (2013) Ca2+-dependent phospholipid scrambling by a reconstituted TMEM16 ion channel. Nat Commun 4:2367

44. Lee BC, Menon AK, Accardi A (2016) The nhTMEM16 scramblase is also a nonselective ion channel. Biophys J 111:1919–1924
45. Scudieri P, Caci E, Venturini A, Sondo E, Pianigiani G, Marchetti C, Ravazzolo R, Pagani F, Galietta LJ (2015) Ion channel and lipid scramblase activity associated with expression of TMEM16F/ ANO6 isoforms. J Physiol 593:3829–3848
46. Yu K, Whitlock JM, Lee K, Ortlund EA, Cui YY, Hartzell HC (2015) Identification of a lipid scrambling domain in ANO6/TMEM16F. Elife 4:e06901
47. Suzuki J, Denning DP, Imanishi E, Horvitz HR, Nagata S (2013) Xk-related protein 8 and CED-8 promote phosphatidylserine exposure in apoptotic cells. Science 341:403–406
48. Bushell SR, Pike ACW, Falzone ME, Rorsman NJG, Ta CM, Corey RA, Newport TD, Christianson JC, Scofano LF, Shintre CA et al (2019) The structural basis of lipid scrambling and inactivation in the endoplasmic reticulum scramblase TMEM16K. Nat Commun 10:3956
49. Di Zanni E, Gradogna A, Scholz-Starke J, Boccaccio A (2018) Gain of function of TMEM16E/ANO5 scrambling activity caused by a mutation associated with gnathodiaphyseal dysplasia. Cell Mol Life Sci 75:1657–1670
50. Gyobu S, Miyata H, Ikawa M, Yamazaki D, Takeshima H, Suzuki J, Nagata S (2016) A role of TMEM16E carrying a scrambling domain in sperm motility. Mol Cell Biol 36:645–659
51. Whitlock JM, Yu K, Cui YY, Hartzell HC (2018) Anoctamin 5/TMEM16E facilitates muscle precursor cell fusion. J Gen Physiol 150:1498–1509
52. Arreola J, Melvin JE, Begenisich T (1996) Activation of calcium-dependent chloride channels in rat parotid acinar cells. J Gen Physiol 108:35–47
53. Kuruma A, Hartzell HC (2000) Bimodal control of a Ca(2+)-activated Cl(−) channel by different Ca(2+) signals. J Gen Physiol 115:59–80
54. Reisert J, Bauer PJ, Yau KW, Frings S (2003) The Ca-activated Cl channel and its control in rat olfactory receptor neurons. J Gen Physiol 122:349–363
55. Cruz-Rangel S, De Jesus-Perez JJ, Arechiga-Figueroa IA, Rodriguez-Menchaca AA, Perez-Cornejo P, Hartzell HC, Arreola J (2017) Extracellular protons enable activation of the calcium-dependent chloride channel TMEM16A. J Physiol 595:1515–1531
56. Ferrera L, Caputo A, Ubby I, Bussani E, Zegarra-Moran O, Ravazzolo R, Pagani F, Galietta LJ (2009) Regulation of TMEM16A chloride channel properties by alternative splicing. J Biol Chem 284:33360–33368
57. Le SC, Jia Z, Chen J, Yang H (2019a) Molecular basis of PIP2-dependent regulation of the Ca(2+)-activated chloride channel TMEM16A. Nat Commun 10:3769
58. Lim NK, Lam AK, Dutzler R (2016) Independent activation of ion conduction pores in the double-barreled calcium-activated chloride channel TMEM16A. J Gen Physiol 148:375–392
59. Ni YL, Kuan AS, Chen TY (2014) Activation and inhibition of TMEM16A calcium-activated chloride channels. PLoS One 9:e86734
60. Tien J, Peters CJ, Wong XM, Cheng T, Jan YN, Jan LY, Yang H (2014) A comprehensive search for calcium binding sites critical for TMEM16A calcium-activated chloride channel activity. Elife 3: e02772
61. Xiao Q, Yu K, Perez-Cornejo P, Cui Y, Arreola J, Hartzell HC (2011) Voltage- and calcium-dependent gating of TMEM16A/Ano1 chloride channels are physically coupled by the first intracellular loop. Proc Natl Acad Sci U S A 108:8891–8896
62. Yu K, Duran C, Qu Z, Cui YY, Hartzell HC (2012) Explaining calcium-dependent gating of anoctamin-1 chloride channels requires a revised topology. Circ Res 110:990–999
63. Cenedese V, Betto G, Celsi F, Cherian OL, Pifferi S, Menini A (2012) The voltage dependence of the TMEM16B/anoctamin2 calcium-activated chloride channel is modified by mutations in the first putative intracellular loop. J Gen Physiol 139:285–294
64. Pifferi S (2017) Permeation mechanisms in the TMEM16B calcium-activated chloride channels. PLoS One 12:e0169572
65. Kaneko H, Putzier I, Frings S, Kaupp UB, Gensch T (2004) Chloride accumulation in mammalian olfactory sensory neurons. J Neurosci 24:7931–7938
66. Reisert J, Lai J, Yau KW, Bradley J (2005) Mechanism of the excitatory Cl- response in mouse olfactory receptor neurons. Neuron 45:553–561
67. Perez-Cornejo P, De Santiago JA, Arreola J (2004) Permeant anions control gating of calcium-dependent chloride channels. J Membr Biol 198:125–133
68. Peters CJ, Gilchrist JM, Tien J, Bethel NP, Qi L, Chen T, Wang L, Jan YN, Grabe M, Jan LY (2018) The sixth transmembrane segment is a major gating component of the TMEM16A calcium-activated chloride channel. Neuron 97(1063–1077):e1064
69. Feng S, Dang S, Han TW, Ye W, Jin P, Cheng T, Li J, Jan YN, Jan LY, Cheng Y (2019) Cryo-EM studies of TMEM16F calcium-activated ion channel suggest features important for lipid scrambling. Cell Rep 28(567–579):e564
70. Le T, Jia Z, Le SC, Zhang Y, Chen J, Yang H (2019b) An inner activation gate controls TMEM16F phospholipid scrambling. Nat Commun 10:1846
71. Nguyen DM, Chen LS, Yu WP, Chen TY (2019) Comparison of ion transport determinants between a TMEM16 chloride channel and phospholipid scramblase. J Gen Physiol 151:518–531
72. Ye W, Han TW, He M, Jan YN, Jan LY (2019) Dynamic change of electrostatic field in TMEM16F permeation pathway shifts its ion selectivity. Elife 8
73. Le T, Le SC, Yang H (2019c) *Drosophila* subdued is a moonlighting transmembrane protein

16 (TMEM16) that transports ions and phospholipids. J Biol Chem 294(12):4529–4537
74. Ehlen HW, Chinenkova M, Moser M, Munter HM, Krause Y, Gross S, Brachvogel B, Wuelling M, Kornak U, Vortkamp A (2013) Inactivation of anoctamin-6/Tmem16f, a regulator of phosphatidylserine scrambling in osteoblasts, leads to decreased mineral deposition in skeletal tissues. J Bone Miner Res 28:246–259
75. Fujii T, Sakata A, Nishimura S, Eto K, Nagata S (2015) TMEM16F is required for phosphatidylserine exposure and microparticle release in activated mouse platelets. Proc Natl Acad Sci U S A 112:12800–12805
76. Headland SE, Jones HR, Norling LV, Kim A, Souza PR, Corsiero E, Gil CD, Nerviani A, Dell'Accio F, Pitzalis C et al (2015) Neutrophil-derived microvesicles enter cartilage and protect the joint in inflammatory arthritis. Sci Transl Med 7:315ra190
77. Han TW, Ye W, Bethel NP, Zubia M, Kim A, Li KH, Burlingame AL, Grabe M, Jan YN, Jan LY (2019) Chemically induced vesiculation as a platform for studying TMEM16F activity. Proc Natl Acad Sci U S A 116:1309–1318
78. Falzone ME, Accardi A (2020) Reconstitution of proteoliposomes for phospholipid scrambling and nonselective channel assays. Methods Mol Biol 2127:207–225
79. De Jesus-Perez JJ, Cruz-Rangel S, Espino-Saldana AE, Martinez-Torres A, Qu Z, Hartzell HC, Corral-Fernandez NE, Perez-Cornejo P, Arreola J (2018) Phosphatidylinositol 4,5-bisphosphate, cholesterol, and fatty acids modulate the calcium-activated chloride channel TMEM16A (ANO1). Biochim Biophys Acta Mol Cell Biol Lipids 1863:299–312
80. Ko W, Jung SR, Kim KW, Yeon JH, Park CG, Nam JH, Hille B, Suh BC (2020) Allosteric modulation of alternatively spliced Ca(2+)-activated Cl(−) channels TMEM16A by PI(4,5)P2 and CaMKII. Proc Natl Acad Sci U S A 17(48):30787–30798
81. Pritchard HA, Leblanc N, Albert AP, Greenwood IA (2014) Inhibitory role of phosphatidylinositol 4,5-bisphosphate on TMEM16A-encoded calcium-activated chloride channels in rat pulmonary artery. Br J Pharmacol 171:4311–4321
82. Ta CM, Acheson KE, Rorsman NJG, Jongkind RC, Tammaro P (2017) Contrasting effects of phosphatidylinositol 4,5-bisphosphate on cloned TMEM16A and TMEM16B channels. Br J Pharmacol 174:2984–2999
83. Tembo M, Wozniak KL, Bainbridge RE, Carlson AE (2019) Phosphatidylinositol 4,5-bisphosphate (PIP2) and Ca(2+) are both required to open the Cl(−) channel TMEM16A. J Biol Chem 294:12556–12564
84. Yu K, Jiang T, Cui Y, Tajkhorshid E, Hartzell HC (2019) A network of phosphatidylinositol 4,5-bisphosphate binding sites regulates gating of the Ca(2+)-activated Cl(−) channel ANO1 (TMEM16A). Proc Natl Acad Sci U S A 116:19952–19962
85. Lee BC, Khelashvili G, Falzone M, Menon AK, Weinstein H, Accardi A (2018) Gating mechanism of the extracellular entry to the lipid pathway in a TMEM16 scramblase. Nat Commun 9:3251
86. Watanabe R, Sakuragi T, Noji H, Nagata S (2018) Single-molecule analysis of phospholipid scrambling by TMEM16F. Proc Natl Acad Sci U S A 115:3066–3071
87. Alvadia C, Lim NK, Clerico Mosina V, Oostergetel GT, Dutzler R, Paulino C (2019) Cryo-EM structures and functional characterization of the murine lipid scramblase TMEM16F. Elife 8:e44365. https://doi.org/10.7554/eLife.44365
88. Falzone ME, Rheinberger J, Lee BC, Peyear T, Sasset L, Raczkowski AM, Eng ET, Di Lorenzo A, Andersen OS, Nimigean CM et al (2019) Structural basis of Ca(2+)-dependent activation and lipid transport by a TMEM16 scramblase. Elife 8:e43229. https://doi.org/10.7554/eLife.43229
89. Kalienkova V, Clerico Mosina V, Bryner L, Oostergetel GT, Dutzler R, Paulino C (2019) Stepwise activation mechanism of the scramblase nhTMEM16 revealed by cryo-EM. Elife 8:e44364
90. Paulino C, Kalienkova V, Lam AKM, Neldner Y, Dutzler R (2017a) Activation mechanism of the calcium-activated chloride channel TMEM16A revealed by cryo-EM. Nature 552:421–425
91. Paulino C, Neldner Y, Lam AK, Kalienkova V, Brunner JD, Schenck S, Dutzler R (2017b) Structural basis for anion conduction in the calcium-activated chloride channel TMEM16A. Elife 6:e26232
92. Fallah G, Romer T, Detro-Dassen S, Braam U, Markwardt F, Schmalzing G (2011) TMEM16A(a)/anoctamin-1 shares a homodimeric architecture with CLC chloride channels. Mol Cell Proteomics 10 (M110):004697
93. Sheridan JT, Worthington EN, Yu K, Gabriel SE, Hartzell HC, Tarran R (2011) Characterization of the oligomeric structure of the Ca(2+)-activated Cl-channel Ano1/TMEM16A. J Biol Chem 286:1381–1388
94. Tien J, Lee HY, Minor DL Jr, Jan YN, Jan LY (2013) Identification of a dimerization domain in the TMEM16A calcium-activated chloride channel (CaCC). Proc Natl Acad Sci U S A 110:6352–6357
95. Jeng G, Aggarwal M, Yu WP, Chen TY (2016) Independent activation of distinct pores in dimeric TMEM16A channels. J Gen Physiol 148:393–404
96. Dang S, Feng S, Tien J, Peters CJ, Bulkley D, Lolicato M, Zhao J, Zuberbuhler K, Ye W, Qi L et al (2017a) Cryo-EM structures of the TMEM16A calcium-activated chloride channel. Nature 552:426–429
97. Jiang T, Yu K, Hartzell HC, Tajkhorshid E (2017) Lipids and ions traverse the membrane by the same physical pathway in the nhTMEM16 scramblase. Elife 6:e28671

98. Lam AK, Dutzler R (2018) Calcium-dependent electrostatic control of anion access to the pore of the calcium-activated chloride channel TMEM16A. Elife 7:e39122. https://doi.org/10.7554/eLife.39122
99. Nilius B, Prenen J, Voets T, Van den Bremt K, Eggermont J, Droogmans G (1997) Kinetic and pharmacological properties of the calcium-activated chloride-current in macrovascular endothelial cells. Cell Calcium 22:53–63
100. Peters CJ, Yu H, Tien J, Jan YN, Li M, Jan LY (2015) Four basic residues critical for the ion selectivity and pore blocker sensitivity of TMEM16A calcium-activated chloride channels. Proc Natl Acad Sci U S A 112:3547–3552
101. Evans MG, Marty A (1986) Calcium-dependent chloride currents in isolated cells from rat lacrimal glands. J Physiol 378:437–460
102. Qu Z, Hartzell HC (2000) Anion permeation in Ca (2+)-activated Cl(-) channels. J Gen Physiol 116:825–844
103. Whitlock JM, Hartzell HC (2016) A Pore Idea: the ion conduction pathway of TMEM16/ANO proteins is composed partly of lipid. Pflugers Arch 468:455–473
104. Pomorski T, Menon AK (2006) Lipid flippases and their biological functions. Cell Mol Life Sci 63:2908–2921
105. Stansfeld PJ, Goose JE, Caffrey M, Carpenter EP, Parker JL, Newstead S, Sansom MS (2015) MemProtMD: automated insertion of membrane protein structures into explicit lipid membranes. Structure 23:1350–1361
106. Bethel NP, Grabe M (2016) Atomistic insight into lipid translocation by a TMEM16 scramblase. Proc Natl Acad Sci U S A 113:14049–14054
107. Gyobu S, Ishihara K, Suzuki J, Segawa K, Nagata S (2017) Characterization of the scrambling domain of the TMEM16 family. Proc Natl Acad Sci U S A 114:6274–6279
108. Menon I, Huber T, Sanyal S, Banerjee S, Barre P, Canis S, Warren JD, Hwa J, Sakmar TP, Menon AK (2011) Opsin is a phospholipid flippase. Curr Biol 21:149–153
109. Khelashvili G, Falzone ME, Cheng X, Lee BC, Accardi A, Weinstein H (2019) Dynamic modulation of the lipid translocation groove generates a conductive ion channel in Ca(2+)-bound nhTMEM16. Nat Commun 10:4972

Distribution and Assembly of TRP Ion Channels

7

Wei Cheng and Jie Zheng

Abstract

In the last several decades, a large family of ion channels have been identified and studied intensively as cellular sensors for diverse physical and/or chemical stimuli. Named transient receptor potential (TRP) channels, they play critical roles in various aspects of cellular physiology. A large number of human hereditary diseases are found to be linked to TRP channel mutations, and their dysregulations lead to acute or chronical health problems. As TRP channels are named and categorized mostly based on sequence homology rather than functional similarities, they exhibit substantial functional diversity. Rapid advances in TRP channel study have been made in recent years and reported in a vast body of literature; a summary of the latest advancements becomes necessary. This chapter offers an overview of current understandings of TRP channel distribution and subunit assembly.

W. Cheng
Dalian Medical University, Institute of Cancer Stem Cell, Dalian, Liaoning, China

J. Zheng (✉)
Department of Physiology and Membrane Biology, University of California Davis School of Medicine, Davis, CA, USA
e-mail: jzheng@ucdavis.edu

Keywords

Channelopathy · Trafficking · Protein complex · Mutation · Homomultimer · Heteromultimer

7.1 Introduction

The transient receptor potential (TRP) channels are cellular sensors for a wide spectrum of physical and chemical stimuli, and hence serve many critical physiological functions [1, 2]. As thus, they are attractive potential drug targets. The TRP channel superfamily is made of extremely diverse members. It contains six subfamilies and 28 members of cation channels in mammalian genomes—TRPC, TRPM, TRPV, TRPA, TRPML, and TRPP. While all TRP channels are cation permeable, they display different ion selectivity—from highly Ca^{2+} and Mg^{2+} selective TRPV5 and TRPV6 to monovalent cation selective TRPM4 and TRPM5, to many nonselective TRP channels. The majority of TRP channels are widely distributed in both neuronal and non-neuronal tissues and organs. They contribute to a wide range of pathological functions. TRP channels exhibit diverse activation mechanisms ranging from ligand binding to physical stimuli (such as temperature, pressure, etc.). TRP ion channels respond to heat, mechanical force, taste, light, osmolarity, pheromones, sound waves, and various noxious stimuli. Thus, these

L. Zhou (ed.), *Ion Channels in Biophysics and Physiology*, Advances in Experimental Medicine and Biology 1349, https://doi.org/10.1007/978-981-16-4254-8_7

ion channels play important roles in thermoregulation, chemo- and mechano-sensation, hyperalgesia, taste, pain, vision, hearing, and homeostasis.

TRP channels are tetrameric protein complexes, made of either identical or different subunits. A TRP channel subunit typically contains six transmembrane segments (S1–S6), with both C and N termini located in the intracellular compartment. The loop between S5 and S6 forms the pore region for determining selective cation permeation. Many TRP channel subunits contain an ankyrin repeat domain in the N terminus and a TRP domain in the C terminus. The long N and C termini also contain other conserved regulatory domains that may play an important role for trafficking or protein-protein interaction critical for subunits assemblies (Fig. 7.1), such as coiled-coil domain in TRPC, TRPM, and TRPP channels. Moreover, TRPC and TRPV channel subunits contain a PDZ domain in their C terminus. TRPC, TRPV, and TRPM channel subunits encompass one or multiple calmodulin binding domains. Four TRP channel subunits are necessary to form a functional channel protein complex, however, TRP channel subunits can assemble into homomeric channels or heteromeric channels either within the same subfamily or between different subfamilies.

In this chapter, mammalian TRP channels distribution in various tissues and organs as well as assembly of TRP channel subunits will be summarized. Physiological and pathophysiological implications related to TRP channels distribution and assembly will be highlighted.

7.2 Distribution of TRP Channels and Its Implications for Health

7.2.1 Cellular Distribution of TRP Channels

TRP channels can function in either the plasma membrane or intracellular organelle membrane. Most TRP channels predominantly function on plasma membranes for regulating ion homeostasis and producing electrical or chemical signaling. TRPML channels, for example, are intracellular ion channels. Studies showed that TRPML1 channels are almost exclusively localized in the membrane of late endosomes and lysosomes, where they may participate in regulating membrane fission/fusion, signal transduction, and ionic homeostasis in mammalian cells. Similarly, TRPML2 channels express in lysosomes, late endosomes, and recycling endosomes. TRPML3 channels mostly reside in membranes of organelles of cytoplasmic vesicles. Specifically, TRPML3 expression has been detected in an isolated endolysosomal membrane in a heterologous expression system from inward rectifying currents [3]. TRPP2 has been detected in different subcellular compartments, such as the endoplasmic reticulum (ER) and the plasma membrane. When deleting a retention motif in the carboxy-terminal (amino acids 787–820), TRPP2 can traffic from ER to the plasma membrane where they exhibit channel activities [4, 5]. Co-assembly of TRPP2 with polycystin-1 can form a functional receptor channel complex, which also promotes TRPP2 trafficking to the plasma membrane [6].

The abundance of TRP channel proteins in the plasma and intracellular membranes are subject to physiological and pathological regulations. Dynamic changes in the distribution of plasma and intracellular TRP channels, as well as the abundance of TRP channel proteins in these locations have been gaining increasing attention. For example, stimuli that activate TRPA1, as well as TRPM8 can induce these ion channels to traffic from the intracellular compartment to the plasma membrane [7, 8]. The presence of TRPV4 in the plasma membrane is regulated by a ubiquitous protein of OS-9, which interacts with TRPV4 monomers at the N terminus (amino acids 438–468) within the endoplasmic reticulum, thus reducing the abundance of TRPV4 channels on the membrane [9]. Studies also revealed that several proteins, for example, Rab11a, as well as KIF13B (kinesin-3 family member 13B) can modulate TRPV5, TRPV6, and TRPV1 ion

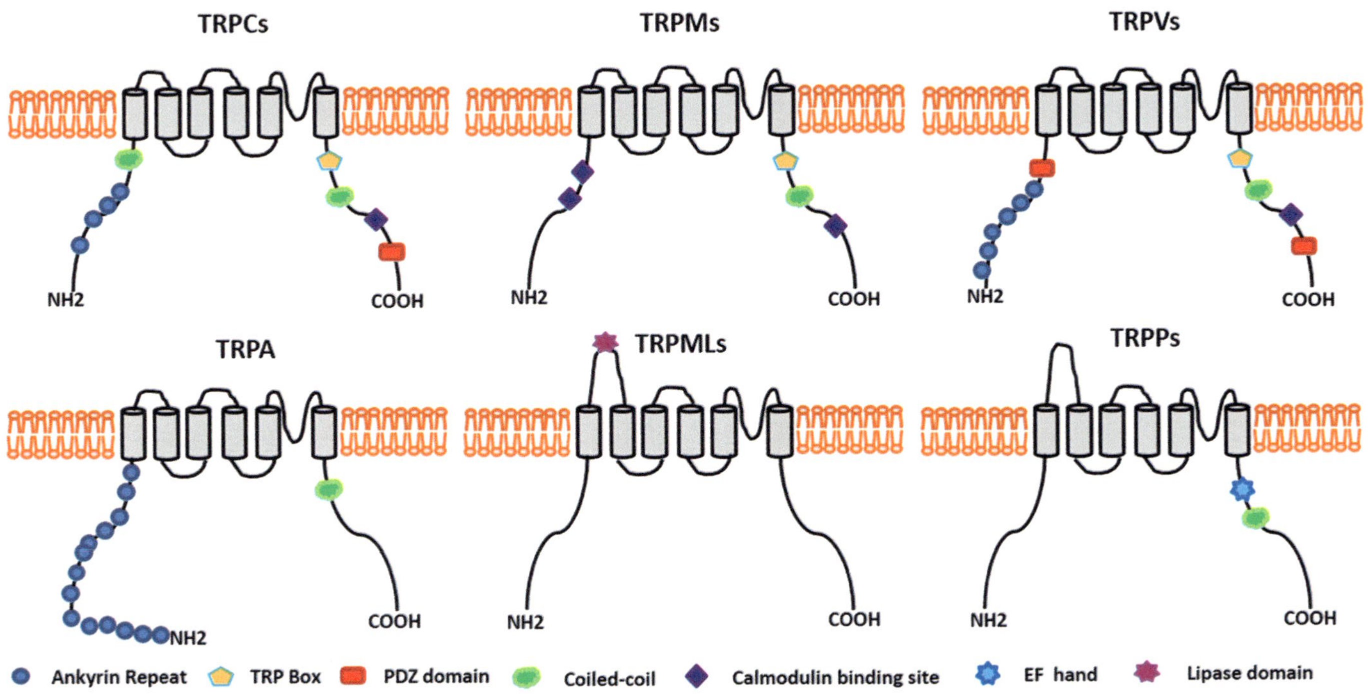

Fig. 7.1 Specific Regulatory Domains in TRP channels
TRP channel subunits contain six transmembrane segments with N and C termini both localized intracellularly. The two termini contain a number of regulatory domains that can be involved in regulating trafficking and assembly

channel abundance on the plasma membrane via interacting with these TRP channels [10, 11]. Furthermore, modulation of the TRP channel subunits at their cytoplasmic N or C termini can regulate their presence on the plasma membrane. For example, glycosylation of N651 in the pore-forming loop between transmembrane segments of 5 and 6 may enhance TRPV4 traffic to the plasma membrane and phosphorylation of Y199 in the N terminus of TRPV1 could enhance channel expression on the surface of the membrane [12, 13]. Meanwhile, studies show that infection by respiratory syncytial virus or measles virus up-regulates TRPV1 and TRPA1 expression on airway cells [14]. The presence of TRPV1/TRPA1 was upregulated by the application of bleomycin, which induced pulmonary fibrosis in the chronic cough in guinea pigs [15]. Similarly, three types of TRP ion channels (TRPA1, TRPV4, and TRPV3) were upregulated in burn scars with postburn pruritus [16].

In addition to dynamic changes in the total amount of TRP channel proteins, the cellular distribution of TRP channels can be regulated by physical or chemical stimuli, such as under certain pathological conditions. There is sometimes an intracellular pool of TRP channels that can be rapidly recruited to the plasma membrane. For example, the majority of TRPM1 proteins are present in the ER of retinal on-bipolar cells, from which it can potentially be transported to the dendritic tips as needed for light responses [17]. A large pool of TRPV1 proteins is located in the intracellular compartment under unstimulated circumstances. Activation of TRPV1 either by agonists or by inflammatory stimuli can trigger rapid translocation of functional TRPV1 to the plasma membrane, leading to an increase in the channel response to recurring, or prolonged stimuli [18]. Furthermore, oxidative stress promotes the redistribution of TRPM2 to the plasma membrane in hepatocytes [19]. Therefore, the study of TRP channel distribution can give important clues on the physiological state of host cells.

7.2.2 TRP Channels Distribution in Healthy Tissues and Organs

7.2.2.1 Distribution of TRPCs in Mammals

TRPC (canonical) ion channels subfamily comprises seven members: TRPC1, TRPC2, TRPC3, TRPC4, TRPC5, TRPC6, and TRPC7. TRPC ion channels are widely distributed in the brain, DRG neurons, heart, and kidney where they participate in the function of the nervous system, digestive system, cardiovascular system, as well as urinary system.

TRPC1

TRPC1 was the first mammalian TRP channel gene to be cloned. Studies revealed that TRPC1 proteins express in the mammalian brain including hippocampal neurons [20, 21]. Several independent studies characterized TRPC1 distribution in the mammalian brain including human fetal and adult brain [22], rat brain [23, 24], mouse cortical astrocytes [21], as well as mouse embryonic brain [25]. Moreover, TRPC1 expresses in human osteoblasts as well as mouse pre-osteoclasts, where they function in regulating osteoclastogenesis and bone mass [26, 27]. TRPC1 mRNA is widely expressed in mouse trigeminal ganglion (TG), brain, lung, kidney, human CNS, and peripheral tissues, including brain, heart, bone, lung, liver, kidney, stomach, intestine, spleen, adipose, pancreas, prostate, placenta, cartilage, muscle, pituitary gland, and esophagus [28–31]. TRPC1 protein has been detected in mouse skeletal muscle in which it may be involved in regulating cell differentiation and osteoclast formation via modulating calcium homeostasis [32, 33]. Human TRPC1 has been identified in myometrial tissue, primary cultured myometrial smooth muscle cells, and sperm [34–36]. TRPC1 seems to be present in pancreatic-β cells [37]. A recent study found that TRPC1 proteins are highly expressed in all adipocyte depots including brown adipose tissue (BAT) and that *Trpc1*-deficient mice are prone to gain weight and reduce metabolic control [38].

TRPC2

TRPC2 is a pseudogene in humans. TRPC2 is present in the rodent vomeronasal organ (VNO), mouse testis, brain, heart, lung, kidney, liver, bovine spleen, liver, testis, and rat thyroid cells [30, 39–42]. TRPC2 expression on the plasma membrane has been found in murine erythroid cells [43].

TRPC3

TRPC3 is expressed in the mammalian brain [21, 44, 45]. Human TRPC3 mRNA has been detected in esophagus, pituitary gland, sperm, and myometrium [28, 31, 34, 35]. Mouse TRPC3 gene has been detected in TG, DRG, brain, heart, lung, kidney, testis, and skeletal muscle [30]. In addition, TRPC3, as well as heteromeric TRPC3/6 have been detected in immune cells (including human parotid gland ductal cells or T-lymphocytes) [46–48], in which these ion channels are thought to participate in Ca^{2+} signaling.

TRPC4

Human TRPC4 has been detected in the brain, heart, lung, kidney, stomach, intestine, adipose, pancreas, prostate, placenta, bone, pituitary gland, adrenal gland, sperm, esophagus, myometrial tissue, and primary myometrial smooth muscle cells [28, 31, 34–36, 49]. Mouse TRPC4 has been found widely distribution in the brain, heart, kidney and skeletal muscle [30, 33, 50]. Also, TRPC4 has been detected in murine jejunum, stomach, and colon [51], as well as smooth muscle of canine colon, antrum, jejunum, renal artery, and pulmonary artery [51]. Moreover, TRPC4 has been observed in renal epithelial cells [52], preglomerular resistance vessels [53], bladder and urothelium [54], where the channel may involve in regulating mechanotransduction in the urinary system. In addition, TRPC4 has been observed in coronary artery endothelial cells of the cardiovascular system, in which it may regulate endothelial permeability and agonist-dependent vasorelaxation [55, 56].

TRPC5

TRPC5 is a cold transducer in peripheral nervous system [57]. Human TRPC5 presents in the brain, heart, liver, muscle, adipose, pancreas, cartilage, bone, pituitary gland, kidney, and esophagus [28, 31]. Mouse TRPC5 mRNA distributes in DRG, brain, lung, kidney, liver, testis, and uterus [30, 58]. TRPC5 has been reported to exhibit much lower expression in the murine small intestine [59]. A recent study reported that TRPC5 proteins are distributed in primary mouse intestinal mesenteric vascular endothelial cells in which they regulate angiogenesis and promote recovery from ischemic injury in mice [60].

TRPC6

TRPC6 is expressed in the mouse brain, lung, kidney, testis, heart, stomach, intestine, colon, jejunum, and liver [30, 51, 61, 62]. Human TRPC6 has been detected in the brain, heart, lung, liver, kidney, muscle, stomach, spleen, intestine, adipose, pancreas, prostate, placenta, ovary, bone, pituitary gland, sperm, and esophagus [28, 31, 34, 63]. TRPC6 distributions in smooth muscles of stomach, colon, myometrium and esophagus have been identified by several studies [31, 36, 51, 62]. TRPC6 has also been identified in immune cells of the lung like alveolar macrophages [64]. Interestingly, *Trpc6* $^{-/-}$ mice challenged by ovalbumin via bronchoalveolar lavage exhibit reduced allergic responses [65].

TRPC7

TRPC7 expresses in the mammalian brain [66]. Apparent expressions of TRPC7 in intestine, lung, kidney, prostate, cartilage, pituitary gland, testis, and, at a low level, in the heart, liver, and stomach have been detected [28, 30, 66]. TRPC7 also expresses in rat and human myometrium [35, 67]. In addition, TRPC7 mRNA has been detected in human coronary artery endothelial cells [55] (Table 7.1).

Table 7.1 Distribution of TRPCs in mammalian tissues and organs

TRPCs	Tissues and organs	References
TRPC1	TG, Brain, hippocampal neurons, bone, heart, skeletal muscle, liver, kidney, lung, stomach, intestine, spleen, pancreas, prostate, placenta, cartilage, pituitary gland, esophagus, myometrium, sperm, brown adipose tissue, pancreatic-β cells, adipocytes, osteoblasts, and cortical astrocytes	[20–38]
TRPC2	Bovine spleen, testis, liver, rodent vomeronasal organ, murine testis, brain, heart, lung, kidney, liver, erythroblasts, and thyroid cells	[30, 39–43]
TRPC3	TG, DRG, Brain, heart, kidney, pituitary gland, lung, esophagus, sperm, myometrium, testis, skeletal muscle, parotid gland ductal cells, and T-lymphocytes	[21, 28, 30, 31, 34, 35, 44–48]
TRPC4	Brain, heart, kidney, lung, adipose, bone, pituitary gland, adrenal gland, esophagus, smooth muscle, skeletal muscle, stomach, intestine, antrum, pancreas, jejunum, colon, prostate, sperm, myometrium, placenta, bladder, urothelium, preglomerular resistance vessels, renal artery, pulmonary artery, renal epithelial cells, and coronary artery endothelium cells	[28, 30, 31, 33–36, 49–56]
TRPC5	DRG, Brain, heart, kidney, small intestine, testis, uterus, lung, liver, muscle, adipose, pancreas, cartilage, bone, pituitary gland, esophagus, and intestinal mesenteric vascular endothelia cells	[28, 30, 31, 58–60]
TRPC6	Brain, heart, kidney, lung, smooth muscle, stomach, colon, esophagus, spleen, lung, placenta, ovary, adipose, testis, intestine, jejunum, liver, pancreas, bone, pituitary gland, sperm, myometrium, and alveolar macrophages	[28, 30, 31, 34, 36, 39, 51, 61, 62, 64, 65]
TRPC7	Brain, intestine, lung, kidney, stomach, myometrium, prostate, testis, heart, Cartilage, pituitary gland, and coronary artery endothelial cells	[28, 30, 35, 55, 66, 67]

7.2.2.2 Distribution of TRPMs in Mammals

TRPM (melastatin) channel subfamily includes eight members: TRPM1, TRPM2, TRPM3, TRPM4, TRPM5, TRPM6, TRPM7, and TRPM8. TRPMs are ubiquitously expressed in mammalian tissues and organs spreading over the endocrine system, respiratory system, urinary system, cardiovascular system, reproductive system, immune system, digestive system, skeletal system, and nervous system.

TRPM1

TRPM1 is the founding member of TRPM subfamily, and it is known to be a tumor suppressor gene. There are not many studies of TRPM1 distribution. TRPM1 has been reported to be positively distributed in rat brain, human brain, heart, and macrophages [68, 69], mouse eyes, as well as murine and human melanocyte lineage cell lines. Loss of TRPM1 expression in melanoma metastases may be an indicator of melanoma aggressiveness [70, 71].

TRPM2

TRPM2 is thermo-sensitive TRP ion channel, which can be directly evoked by warm temperature [72]. A recent study reveals that TRPM2 is involved in maintaining behavioral flexibility with the potentiation of susceptibility to stress [73]. TRPM2 is widely expressed in mammalian brain, bone marrow, placenta, adipose, spleen, intestine, stomach, skeletal muscle, kidney, lung, pituitary gland, heart, prostate, endometrium, and pancreas [69, 74, 75]. TRPM2 has been found in the mucosa and muscle layer of the rat stomach, jejunum, ileum and colon, where it may be involved in the pathogenesis of gastrointestinal diseases [76]. A study also reports that TRPM2 expresses in salivary gland and is involved in salivary gland fluid secretion [77]. TRPM2 is detected in pancreatic β cells [37] in which it is involved in regulating insulin release by heat stimulation, as well as glucose-application [72, 78]. And H_2O_2-mediated cell death in pancreatic β-cells via TRPM2 ion channel [79]. Furthermore, TRPM2 has also been found in immune cells (including neutrophils, megakaryocytes, monocytes, macrophages, B lymphoblast cells, T lymphocytes, dendritic cells, and mast cells) [80–87] in which it may act as a mediator for inflammation via stimulus-induced Ca^{2+} influx.

TRPM3

TRPM3 is a nociceptor channel for detecting noxious heat [88]. TRPM3 expresses in mouse TG, brain, kidney, testis and human brain, pituitary gland, kidney, and adipose [30, 69]. TRPM3 also distributes in human ovary and pancreas [89], rat spermatogenic cells [90], and rat prostatic tissues [91]. A recent study reveals that TRPM3 expresses in human pancreatic beta cells [37].

TRPM4

TRPM4 has been proposed to serve as a temperature-sensitive TRP channel, like TRPM5 [92]. TRPM4 is detected in mouse brain, heart, lung, testis, skeletal muscle, kidney, uterus, spleen, bladder [30, 54, 93], rat prostate, sperm [90, 91] and human intestine, prostate, kidney, stomach, adipose, bone, skeletal muscle, lung, heart, pituitary gland, placenta, pancreas, brain, liver, spleen, and macrophages [37, 69]. Furthermore, TRPM4 has been observed in T cells and it may be relevant to immune response for interleukin-2 production [94]. TRPM4 also presents in mast cells and participates in cell migration by regulating Ca^{2+}-dependent actin cytoskeleton rearrangements [95, 96]. In addition, TRPM4 has been detected in exocrine β-cells from the pancreas [37]. A recent study using live cell imaging and behavioral tests in KO mice identified that both TRPM4 and TRPM5 are distributed in peripheral taste receptor cells in which they are required for transduction of taste-evoked signaling by bitter, sweet, or umami stimuli [97]. TRPM4 expressed in keratinocytes and modulated cell proliferation and differentiation in affected skin tissues [98].

TRPM5

TRPM5 is known to participate in the sensation of smell, taste, and temperature. Studies have identified that TRPM5 is expressed in taste buds where it is involved in taste sensation for sweet, bitter, and umami, but not for sour or salty tastes [99]. TRPM5 is detected in mouse brain, bladder [54, 100], human intestine, pancreas, prostate, kidney, stomach, and pituitary gland [69]. TRPM5 expresses in mouse pancreatic islets where it is involved in the regulation of glucose-stimulated insulin secretion [101, 102]. Furthermore, TRPM5 has been found to distribute in sparse chemosensory cells located throughout the digestive tract, taste buds, vomeronasal organ, olfactory epithelium, and may even involve in chemosensation [103, 104].

TRPM6

TRPM6 is detected in mouse lung and bladder [30, 54], rat prostate [91], human intestine, brain, kidney, pituitary gland, adipose, and prostate [69]. TRPM6 transcripts have been detected in mouse duodenum, jejunum and renal tubule cell, and testis [105, 106]. Human TRPM6 mRNA is also present in leukocytes [106].

TRPM7

Human TRPM7 has been detected in mouse TG, brain, heart, lung, kidney, testis, liver, spleen, epididymis, ovary, bladder, rat prostate, sperm [30, 54, 90, 91, 93], human heart, adipose, pituitary gland, bone, spleen, prostate, intestine, skeletal muscle, kidney, brain, lung, macrophages, pancreas, placenta, stomach, liver, bone marrow, and cartilage [37, 69]. A recent study reveals that TRPM7 regulates Mg^{2+} ions to promote the osteoinduction of human osteoblast via the PI3K pathway [107]. Moreover, TRPM7 has been found in rat odontoblasts and may be involved in maintaining Mg^{2+} homeostasis with colocalization of Mg^{2+} transporter CNNM4 [108].

TRPM8

TRPM8 is a cold-sensitive channel that was originally cloned from prostate tissue [109, 110]. TRPM8 has been found in mouse TG, DRG, testis, human DRG confined to small/medium diameter neurons, peripheral nerves in large caliber fibres, prostate, and liver [30, 69, 110, 111]. TRPM8 can be a biomarker for prostate cancer because upregulation of TRPM8

Table 7.2 Distribution of TRPMs in tissues and organs in mammalians

TRPMs	Tissues and organs	References
TRPM1	Brain, heart, eye, macrophages, melanocytes	[68–71]
TRPM2	Brain, stomach, intestine, salivary gland, bone marrow, spleen, lung, kidney, placenta, pancreas, adipose, skeletal muscle, pituitary gland, heart, prostate, jejunum, ileum, colon, macrophages, pancreatic β cells, neutrophils, megakaryocytes, monocytes, B lymphoblast cells, T lymphocytes, dendritic cells, and mast cells	[37, 69, 72, 74–87]
TRPM3	TG, brain, kidney, ovary, testis, prostate, pancreas, adipose, pituitary gland, spermatogenic cells, and pancreatic β cells	[30, 37, 69, 88–91]
TRPM4	Brain, lung, kidney, heart, testis, skeletal muscle, uterus, bladder, prostate, sperm, intestine, stomach, bone, pituitary gland, placenta, liver, spleen, pancreas, tongue, white and brown adipose tissues, macrophages, T cells, mast cells, pancreatic β cells, taste receptor cells, and keratinocytes	[30, 37, 54, 69, 90–98]
TRPM5	Brain, kidney, bladder, prostate, pituitary gland, pancreas, digestive tract, stomach, intestine, colon, tongue, taste buds, vomeronasal organ, and olfactory epithelium	[54, 69, 99–104]
TRPM6	Brain, lung, kidney, intestine, bladder, prostate, pituitary gland, testis, adipose, duodenum, jejunum, renal tubule cells, and leukocytes	[30, 54, 69, 91, 105, 106]
TRPM7	TG, brain, bone, lung, kidney, bone marrow, spleen, testis, heart, liver, epididymis, ovary, bladder, prostate, sperm, adipose, pituitary gland, intestine, skeletal muscle, pancreas, placenta, stomach, cartilage, osteoblasts, odontoblasts, and macrophages	[30, 37, 54, 69, 90, 91, 93, 107, 108]
TRPM8	TG, DRG, liver, prostate, testis, sperm, bladder urothelium, urogenital tract, lung, and brown adipose tissues	[30, 69, 109–115]

transcript has been originally identified in cancerous testis and prostate tissues [109]. In addition, TRPM8 has also been detected in human sperm [112]. TRPM8 has been characterized in the bladder urothelium and male urogenital tract [113]. In human lung epithelium and bronchial epithelial cells, TRPM8 has been found to localize in the endoplasmic reticulum [114]. A recent study reported that TRPM8 expresses in brown adipose tissues in which TRPM8 may be involved in the regulation of clock gene and clock-controlled genes expressions and thermogenesis in brown adipose tissues [115] (Table 7.2).

7.2.2.3 Distribution of TRPVs in Mammals

TRPV (vanilloid) channels subfamily comprises six members: TRPV1, TRPV2, TRPV3, TRPV4, TRPV5, and TRPV6. TRPV1, TRPV2, TRPV3, and TRPV4 together with TRPM2, TRPM3, TRPM4, TRPM5, TRPC5, TRPM8, and TRPA1 are characterized as thermosensors (Fig. 7.2). To

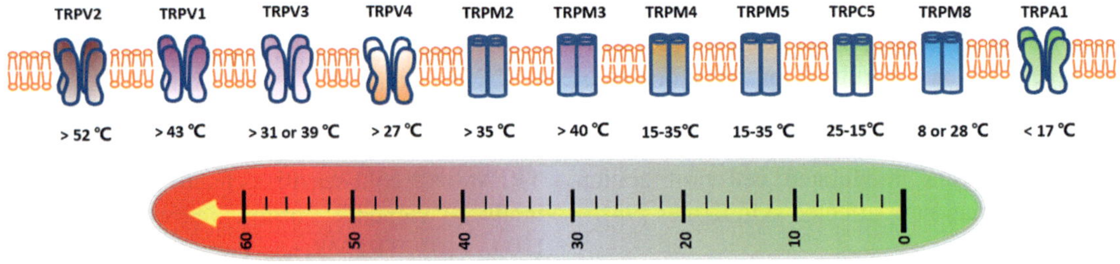

Fig. 7.2 Thermo TRPs with different temperature thresholds
Thermo TRPs with TRPV1, TRPV2, TRPV3, TRPV4, TRPM2, TRPM3, TRPM4, and TRPM5 yield an increasing current with the temperature warming up, while TRPC5, TRPM8, and TRPA1 yield an increasing current with temperature cooling down

date, accumulated information indicates that TRPV channels are widely expressed in the nervous system, skeletal system, digestive system, immune system, reproductive system, cardiovascular system, urinary system, respiratory system, and endocrine system.

TRPV1

TRPV1 was originally cloned from trigeminal ganglia (TG) and dorsal root ganglia (DRG) sensory neurons where it exhibits high expressions [116–118]. TRPV1 also expresses in the hypothalamus where it is thought to participate in thermoregulation of body core temperature [119–121]. However, whether TRPV1 expresses in the other parts of the brain is debated, with some reports of broadly distributed TRPV1 proteins and mRNA and associated roles in higher brain functions [122–130], whereas other reports found restricted TRPV1 expression in brain to mostly hypothalamus [131, 132]. Further, TRPV1 has been observed in mouse olfactory epithelium [133], guinea pig trachea epithelium [134], and mouse laryngeal epithelium [135]. In arteriolar smooth muscle, TRPV1 is expressed in thermoregulatory tissues, including the cremaster muscle, trachea, ear, tongue, skin, and dura where it may help to mediate vessel constriction and control blood flow [131, 136]. TRPV1 expression has been reported in the bladder urothelium, where the channel regulates bladder contractions via ATP releasing in response to stretch stimuli [137]. A recent study reveals that TRPV1 expresses in brown adipose tissues in which TRPV1 participates in the modulation of clock gene oscillations in response to the light-dark cycle [138].

TRPV2

TRPV2 channel can be activated by noxious heat and various ligands. It is highly expressed in a subset of medium to large diameter sensory neurons [139]. TRPV2 expresses in rat brain, the spinal cord, the dorsal horn, lung, spleen, and intestine [139]. The transcripts of TRPV2 have been found in mouse TG, DRG, brain, heart, kidney, lung, liver, testis [30], respiratory, and olfactory epithelium [133]. TRPV2 has been detected in osteoclasts of bone, where it be involved in osteoclastogenesis [140]. TRPV2 is abundantly expressed in cells of the immune system. Studies indicated that TRPV2 was strongly expressed in macrophages and Kupffer cells [141]. Furthermore, TRPV2 has been observed in mast cells [142, 143], neutrophils [144], as well as both T and B lymphocytes [145, 146]. TRPV2 is also observed in hematopoietic stem cells with CD34+/CD45+/CD133+/CD73+ [147]. Moreover, TRPV2 is widely expressed in various types of endocrine cells, such as the pituitary, neuroendocrine cells with chromogranin-positive in the stomach, duodenum, and intestine [148]. Meanwhile, TRPV2 expresses in insulin-producing β-cells in pancreas [149]. In addition, TRPV2 has been found to express in murine arterial smooth muscle cells [150], rat pulmonary venous smooth muscle cells [151], human pulmonary artery endothelial cells [152], mouse bladder epithelial cells [54], and cardiomyocytes [153], and may act as a stretch sensor in these tissues.

TRPV3

TRPV3 is also a thermo-sensitive Ca^{2+} permeable nonselective cation channel, which is reminiscent of other TRPV channels. It has been observed in rat skin and keratinocytes where it acts as a warm/heat receptor [154]. TRPV3 is detected in mouse olfactory epithelium [133]. In the respiratory epithelium, TRPV3 has been observed in both the epithelial layer and the lamina propria [133]. Furthermore, TRPV3 distributes in DRG, nodose ganglia, super cervical ganglia, stomach, small intestine, palate, tongue and nose in mice, and transcripts in human and monkey tissues including DRG, TG, brain, skin, testis, stomach, trachea, small intestine, placenta, tongue, and superior cervical ganglia [155–157]. In the epithelium of mouse distal colon, TRPV3 mRNA has also been detected [158]. A recent study reports that TRPV3 is functionally expressed in human epidermal keratinocytes and plays a role in the cutaneous inflammatory process [159].

TRPV4

Both physical (thermo, mechanical, and osmotic) and chemical (endogenous, synthetic, and plant-derived) stimuli can modulate TRPV4 ion channel. TRPV4 expresses in mouse TG, brain, heart, lung, kidney, liver, testis, skeletal muscle [30], olfactory epithelium [133], as well as intestine [160]. TRPV4 has been detected in mouse bladder epithelial cells where it may be involved in sensing mechanical stimuli [54]. Furthermore, TRPV4 has been detected in mandibular condylar chondrocytes and mandibular condylar cartilage tissues of rats and may contribute to mastication-associated pain at the temporomandibular joint [161, 162]. Moreover, TRPV4 also distributes in osteoclasts, osteoblasts, and chondrocytes and regulates cell differentiation [163, 164] or affects bone mass [165]. Despite studies that identified that TRPV4 mutations can lead to osteoarticular pathology, the mechanism underlying TRPV4 modulation remains unclear. Furthermore, TRPV4 has been detected in mouse kidneys, heart, testis, liver [166], and rat bile duct [167] in which it may be involved in the regulation of cellular volume. TRPV4 is also expressed in pancreatic islets and insulin-secreting β cells of the pancreas, where it may sense stimuli from the cellular milieu of physical or chemical resources in type II diabetes [168]. Moreover, TRPV4 has been detected in mammary glands [169], endolymphatic sac [170], mast cells [171]. In addition, TRPV4 presents in bronchi [172], larynx, and trachea [135], where it may be involved in receptor-operated calcium entry [173].

TRPV5

TRPV5 exhibits high selectivity to calcium ions. It expresses in mouse kidneys and testis [30]. Furthermore, TRPV5 has been detected in the ruffled border membrane of murine osteoclasts and it is essential for proper osteoclastic bone resorption [174]. Study shows that TRPV5 expressed abundantly in duodenum, jejunum, kidney, and placenta of rabbit [175]. In addition, TRPV5 presents in lymphocytes, Jurkat leukemia T cells [176], and leukemia K562 cells [177].

TRPV6

TRPV6 is the closest relative of TRPV5 with high selectivity to calcium ions. TRPV6 distributes in mouse brain, heart, kidney, testis, bladder, and lung [30, 54]. TRPV6 also expresses in lymphocytes, Jurkat leukemia T cells [176], and leukemia K562 cells [177]. Wide expression of TRPV6 has been observed in bone cells but it is not crucial for bone mineralization in mice [178]. The murine and human TRPV6 transcripts seem to be present in pancreas, placenta, salivary gland, prostate, liver, kidney, testis, and gastrointestinal tract (including esophagus, stomach, duodenum, jejunum, ileum, and colon) and involved in mediating intestinal calcium uptake [179, 180] (Table 7.3).

7.2.2.4 Distribution of TRPA1 in Mammals

TRPA (ankyrin) subfamily only comprises one member, namely TRPA1. TRPA1 is recognized as a nociceptor and mechanosensor. It is expressed in the nervous, respiratory, digestive, immune, and reproductive systems. Studies reveal extensively distribution of TRPA1 in mammalian sensory ganglia, trigeminal sensory afferents, and spinal dorsal horn, even in astrocytes [181, 182]. It is expressed in a subset of TRPV1-positive nociceptive DRG neurons and transduce cold stimuli [183]. TRPA1 has also been found to be present in nonneuronal tissues including mouse lung, human bladder, prostate, tongue, pharynx, epiglottis, vagina, clitoral tissue, and skin (epidermal keratinocytes, melanocytes, and fibroblasts), rat pancreatic β-cells [184–189]. Furthermore, a study has identified TRPA1 expression in primary extrinsic afferent nerves innervating the esophagus, stomach, intestine, and colon [190–192]. A recent study reveals that TRPA1 is expressed endogenously in mature spermatozoa of multiple species and may benefit reproductive function in mammalians [193].

7.2.2.5 Distribution of TRPMLs in Mammals

The three members of the TRPML (mucolipin) subfamily are widely expressed in the nervous,

Table 7.3 Distribution of TRPVs in mammalian tissues and organs

TRPVs	Tissues and organs	References
TRPV1	TG, DRG, brain, adipose, trachea, ear, tongue, skin, arteriolar smooth muscle, cremaster muscle, dura, olfactory epithelia, trachea epithelia, larynx epithelia, and bladder urothelium	[116–118, 131, 133–138]
TRPV2	TG, DRG, brain, spinal cord, the dorsal horn, heart, kidney, lung, liver, testis, olfactory epithelium, bone, spleen, stomach, duodenum, intestine, pancreas, bladder, pituitary gland, arterial smooth muscle, venous smooth muscle, olfactory epithelium, respiratory epithelium, macrophages, Kupffer cells, mast cells, neutrophils, T lymphocytes, B lymphocytes, hematopoietic stem cells, pituitary cells, and neuroendocrine cells	[30, 54, 133, 139–153]
TRPV3	TG, DRG, brain, skin, nodose ganglia, super cervical ganglia, stomach, small intestine, palate, tongue, nose, testis, trachea, placenta, colon, olfactory epithelium, keratinocytes, palate epithelium, and respiratory epithelium	[133, 154–159]
TRPV4	TG, brain, heart, lung, kidney, liver, testis, skeletal muscle, intestine, bile duct, bladder, bone, cartilage, pancreas, mammary gland, bronchi, larynx, trachea, endolymphatic sac, olfactory epithelium, bladder epithelium, chondrocytes, osteoclasts, osteoblasts, pancreatic β cells, and mast cells	[30, 54, 133, 135, 160–173]
TRPV5	Kidney, testis, bone, duodenum, jejunum, placenta, osteoclasts, lymphocytes, Jurkat leukemia T cells, and leukemia K562 cells	[30, 174–177]
TRPV6	Brain, heart, kidney, testis, bladder, lung, bone, pancreas, placenta, prostate, liver, salivary gland, esophagus, stomach, duodenum, jejunum, ileum, colon, lymphocytes, Jurkat leukemia T cells, and leukemia K562 cells	[30, 54, 176–180]

digestive, immune, reproductive, cardiovascular, urinary, respiratory systems. Specifically, TRPML1, TRPML2, and TRPML3 are all found expressed in mouse brain, lung, kidney, spleen, and liver [194–196]. In addition, TRPML1 transcripts also have been detected in mouse skeletal muscle and testis [195]. TRPML2 also has been found in the stomach, colon, small intestine, thymus, and pancreas [196]. Meanwhile, low-level TRPML3 ia present in the colon, stomach, testis, eye, and cerebellum of mice [194]. TRPML2 mRNA has been detected in B cells, T cells, mastocytoma, myeloma cell lines, and primary splenocytes [197]. A recent study reports that TRPML1 and TRPML3 are expressed in human pancreatic β cells [37].

7.2.2.6 Distribution of TRPPs in Mammals

The TRPP (polycystic) subfamily includes TRPP2 (PKD2), TRPP3 (PKD2L1), and TRPP5 (PKD2L2). Briefly, TRPP channels are present in the reproductive, urinary, respiratory systems. Specifically, both TRPP2 and TRPP3 transcripts have been found in the brain, heart, testis, ovary, kidney, and lung [198–201]. In contrast, TRPP5 transcripts appear to be mostly restricted to the testis [199]. However, the roles of TRPPs in these tissues and organs remain to be elucidated. A recent study reports that TRPP2 was found in human pancreatic β cells [37].

7.2.3 TRP Channels in Abnormal Tissues and Organs

7.2.3.1 TRP Channels Distribution in Cancers

Abnormal expressions of many TRP channels in cancer have been noticed since their discoveries and described in detail in previous reviews [202–207]. Very recently, TRPA1 has been found enriched in breast and lung cancer tissues and spheroids by analysis of The Cancer Genome Atlas (TCGA) datasets. And TRPA1 is critical for the survival of inner cells of tumor spheroids formed by either breast cancer cells of HCC1569 or lung cancer cells of H1792 that exhibit reactive oxygen species (ROS) accumulation. Moreover, TRPA1 promotes resistance to ROS-producing chemotherapies [208]. TRPA1 is also upregulated in human nasopharyngeal carcinoma which was associated with advanced primary tumor and progression in the clinical stage [209]. A study indicates that TRPM7 is expressed in the human leukemia cell line and it is involved in regulating the proliferation and differentiation of K562 cells via spontaneous Ca^{2+} entry [210]. TRPV4 is found upregulated in human hepatocellular carcinoma tumor tissues and inhibition of channel activity exhibits antitumor effects [211]. Similarly, TRPV3 is overexpressed in non-small cell lung cancer and correlates with lung cancer progression [212].

7.2.3.2 TRP Channels in Other Diseases

Studies indicate that TRP channels also participate in many pathological processes. TRPV4 and TRPA1 have been detected in pancreatic nerve fibers and DRG neurons innervating the pancreas, which may contribute to inflammatory pain in mice [213]. Although TRPC1 has not been detected in the immune system yet, *Trpc1* deletion reduces T helper type 2 (Th2) cells in response to allergen (methacholine) challenge in vivo. In vitro, *Trpc1* knockout splenocytes show reduced proliferation and receptor-induced IL-2 production in T cells. These results indicate that TRPC1 may be involved in pro-inflammation and could be a therapeutic target in asthma and immune diseases [214]. Moreover, TRPA1 and TRPV1 are present on the plasma membrane of CD4+ T cells in colonic biopsies from inflammatory bowel disease patients; in particular, TRPA1 plays a protective role in T-cell-mediated colitis by inhibition of TRPV1 channel activity [215]. Furthermore, TRPV4 is present in metabolically active tissues including adipose tissue, liver, gastrointestinal tract, brain (hypothalamus), pancreas, and skeletal muscle, which suggests a potential role in metabolic disorders including obesity [216]. Three related irritant receptors of TRPV1, TRPA1, and TRPM8 are functionally expressed in the mouse oral mucosa and participate in inducing neurogenic inflammation [217]. Available evidence suggests that TRPC3, TRPC6, TRPV1, TRPV3, TRPV4, TRPA1, TRPM6, and TRPM7 may play central roles in the progression and/or prevention of fibroproliferative disorders in vital visceral organs such as lung, heart, liver, kidney, and bowel, as well as brain, blood vessels, and skin, and may contribute to both acute and chronic inflammatory processes [218]. Furthermore, a study in rheumatoid arthritis synovial fibroblasts reveals the distribution of TRPV1, TRPV2, TRPV4, TRPA1, and TRPM8, which may be involved in the modulation of inflammation [219].

TRPA1 has been found to be upregulated in human oral lichen planus [220] and maybe involved in lichen planus pathomechanism. Activation of TRPA1 by non-histaminergic pruritogens elicits thermal and mechanical hyperalgesia and may be implicated in pruritus [221]. A recent study demonstrated that functional expression of TRPV1 affects multiple pathways in pruritus, and TRPV1 is involved in maintaining skin barrier function [222]. Moreover, aberrant TRPM2 function has been implicated in several neurological disorders including ischemia/stroke, Alzheimer's disease, neuropathic pain, Parkinson's disease, and bipolar disorder [223].

7.3 Assembly of TRP Channels

Ion channel subunit assembly generally produces a novel functional channel complex with either intermediate conductance/gating properties or very different pharmacological properties from their homomeric ones. And the formation of heteromeric channel complexes via channel subunits assembly can enhance the novel channel maturation, as well as the expression on the plasma membrane, such as TRPP trimer combine with one subunit of PKD1 family to form a mature channel complex and traffic to the membrane surface. Meanwhile, the assembly of channel subunits and trafficking of novel channels to the plasma membrane increases the recycle process inside the cell. As we know, there are 28 TRP genes/subunits, with co-assembly, the number of TRP channels is much larger than 28. Thus, the widespread heteromultimerization not only extends TRP channel heterogeneity in tissues and organs in mammalians but also widen their physiological functions in modulating cationic homeostasis and somatosensations.

In addition, TRP subunits are associated with auxiliary proteins such as CaM (calmodulin), PIRT (Phosphoinositide-interacting protein), and so on. But this topic will not be discussed in this chapter. Interested readers may refer to Refs. [224–227].

7.3.1 Intra-Subunit Interactions Affecting TRP Channel Assembly and Trafficking

TRP channel assembly have been governed by some molecular mechanisms, such as Ankyrin repeat domain, a coiled-coil domain, TRP domain, C-terminus, N-terminus, even pore region, transmembrane, as well as PDZ domain are all candidates for the molecular determinants of TRP channel subunits association. Readers are referred to reviews for more detailed information [228, 229]. TRPC1 and TRPC3 assemble to form heteromeric channels, the ankyrin repeats (AR) region of TRPC3 could mediate the heteromeric TRPC1/TRPC3 formation [230]. TRPV4 can form heteromeric channels with TRPC1 in vascular endothelial cells. And Ca^{2+} store depletion enhances the trafficking of TRPV4/TRPC1 channels into the plasma membrane [231]. A TRPP3 C-terminal coiled-coil domain forms a trimer in solution and crystal and has a crucial role in the assembly and surface expression of the TRPP3/PKD1L3 complex [232]. Recent study found that extracellular loops between the first and second transmembrane segments of TRPP2 and TRPP3 associate with the extracellular loops between the sixth and seventh transmembrane segments of polycystin-1 and PKD1L3, respectively. The associations between these loops are essential for heteromeric channel complexes assembly and trafficking [233].

7.3.2 Assembly of TRP Channels Within and Between Subfamilies

7.3.2.1 Assembly Within TRP Subfamilies

Assembly of TRP ion channel subunits has been studied extensively. Among the TRPC subfamily, the formation of heteromeric complexes of TRPC1/TRPC4 and TRPC1/TRPC5 have been identified in mammalian cells [234]. Heteromeric TRPC1/TRPC4 channel displays dynamic gating property depending on TRPC1 isoform subtypes and receptor stimulation system [235]. Similarly, heteromerization of TRPC1 with TRPC3 has been identified in skeletal muscle, and the heteromeric TRPC1/TRPC3 plays a role in regulating the resting cytosolic Ca^{2+} levels [230]. Similar associations have been observed between TRPC1 and TRPC3 via an N-termini domain interaction in salivary gland cells lines [46]. Furthermore, co-assembly of TRPC1 and TRPC5 in hippocampal neurons and HEK293 cells produces a novel nonselective cation channel with voltage dependence [24]. Assembly of TRPC3 and TRPC4 seems to form a channel with a distinct pore structure [236]. Moreover, TRPC1/TRPC3/TRPC7 can interact to form a store-operated channel complex [237].

Wide-spread interactions between TRPV1, TRPV2, TRPV3, and TRPV4 ion channel subunits have been identified by using fluorescence resonance energy transfer (FRET) as well as single-channel recording [238]. Heteromeric thermosensitive TRPV channel channels exhibit intermediate conductance and gating properties compared to homomeric channels. Moreover, colocalization of TRPV1 and TRPV2, TRPV1, and TRPV3, TRPV5, and TRPV6 can produce heteromeric channel complexes [239–241]. Recent study indicates that TRPV4 could interact with TRPV1 to participate in itch signaling in some sensory neurons [242].

Assembly of TRPM subunits remains lacking. Only heteromeric TRPM6/TRPM7 channels with intermediate conductance and gating properties have been identified [243]. Recent study indicates heteromeric TRPM6/TRPM7 channels with altered pharmacology and sensitivity to intracellular Mg-ATP. Furthermore, the sensitivity of heteromeric channels to intracellular Mg-ATP concentrations has been modulated by TRPM6 kinase domain [244].

TRPMLs can interact to form heteromultimers with intermediate conductance and kinetic properties [245]. Moreover, the presence of either

TRPML1 or TRPML2 specifically modulates TRPML3 trafficking from the endoplasmic reticulum to lysosomes [246].

7.3.2.2 Assembly Between TRP Subfamilies

Studies reveal extensive assembly between mammalian TRP ion channel subfamilies. Heteromeric TRPC1/TRPP2 channel complexes with a stoichiometry of 2:2 exhibit a new receptor-operated channel property [247]. Heteromeric TRPC1/TRPP2 and TRPC4/TRPP2 channel complexes mediate angiotensin II-induced Ca^{2+} responses in mesangial cells [248]. Heteromeric TRPC3/TRPP2 or TRPC7/TRPP2 protein with TRPP2 mutant can regulate cell growth by potentiation of receptor-activated Ca^{2+} influx in autosomal dominant polycystic kidney disease[249]. Interaction between TRPC1 and TRPP3, TRPC1 and TRPP5, TRPC5 and TRPP2, TRPC5 and TRPP3, TRPC5 and TRPP5 are also identified to form heteromic channel complexes [229]. Study reveals that TRPV4 and TRPP2 formed heteromeric channel complex with a 2: 2 stoichiometry [250]. Moreover, TRPV4 can form heteromeric channels with TRPC1 in vascular endothelial cells [231]. Functional interaction of TRPV6 with TRPC1 negatively regulates Ca^{2+} influx in HEK293 cells [251]. Heteromeric TRPV5/TRPML3 channel complexes with novel conductance are detected under conditions that did not activate either TRPML3 or TRPV5 [252]. Meanwhile, TRPV4 can form a heteromeric channel with TRPC6 in the pulmonary artery smooth muscle cell [253]. Recent study reveals that TRPP2 and TRPM3 may form channel complexes in renal primary cilia [254].

Novel combinations of triad TRP ion channels have been observed in native tissues and heterologous expression systems. Heteromeric TRPC1/TRPC6/TRPV4 channel complex may mediate mechanical hyperalgesia and primary afferent nociceptor sensitization [255]. Similarly, TRPC1, TRPV4, and TRPP2 have been reported to form a flow-sensitive heteromeric channel in primary cultured rat mesenteric artery endothelial cells, as well as HEK293 cells [256]. Recent study indicates that acute noxious heat sensing in mice depends on the combination of TRPM3, TRPV1, and TRPA1 ion channels. The robust somatosensory heat responsiveness at the cellular and behavioral levels is observed only if at least one of these TRP channels is functional [257].

7.3.3 Assembly Between TRPP Channels and Receptor-like Polycystin-1 Family Proteins

TRPP ion channels and polycystin-1 family proteins are all related to autosomal dominant polycystic kidney disease (ADPKD). Three members of TRPP2 (encoding gene *PKD2*), TRPP3 (encoding gene *PKD2L1*), and TRPP5 (encoding gene *PKD2L2*) belong to the TRPP subfamily. While polycystin-1 family proteins comprise five members including polycystin-1 (encoding gene *PKD1*), PKD1L1 (encoding gene *PKD1L1*), PKD1L2 (encoding gene *PKD1L1*), PKD1L3 (encoding gene *PKD1L3*), and PKDREJ (encoding gene *PKDREJ*). TRPP subunits can assemble into functional homomeric ion channels. Meanwhile, TRPP proteins co-assemble with Polycystin-1 family proteins to form receptor-channel complexes.

Assembly of TRPP2 with polycystin-1 has been identified [258]. Then studies reveal that TRPP2 physically interacts with PKD1L1 in the cilium of mice embryonic node and may involve in sensing nodal flow [259]. The interaction between TRPP3 and polycystin-1 is essential for TRPP3 trafficking and channel formation [201]. Meanwhile, studies show that the association of TRPP2 or TRPP3 with PKD1L3 or PKDREJ result in TRPP2/PKD1L3, TRPP2/PKDREJ, TRPP3/PKD1L3, TRPP3/PKDREJ channels complexes [233, 260]. Heteromeric channel complex of TRPP3/PKD1L3 functions as an acid-sensor, moreover, point mutations in the putative pore region of both proteins can alter the ion selectivity of this TRPP3/PKD1L3 channel [232]. Meanwhile, heteromeric TRPP3/

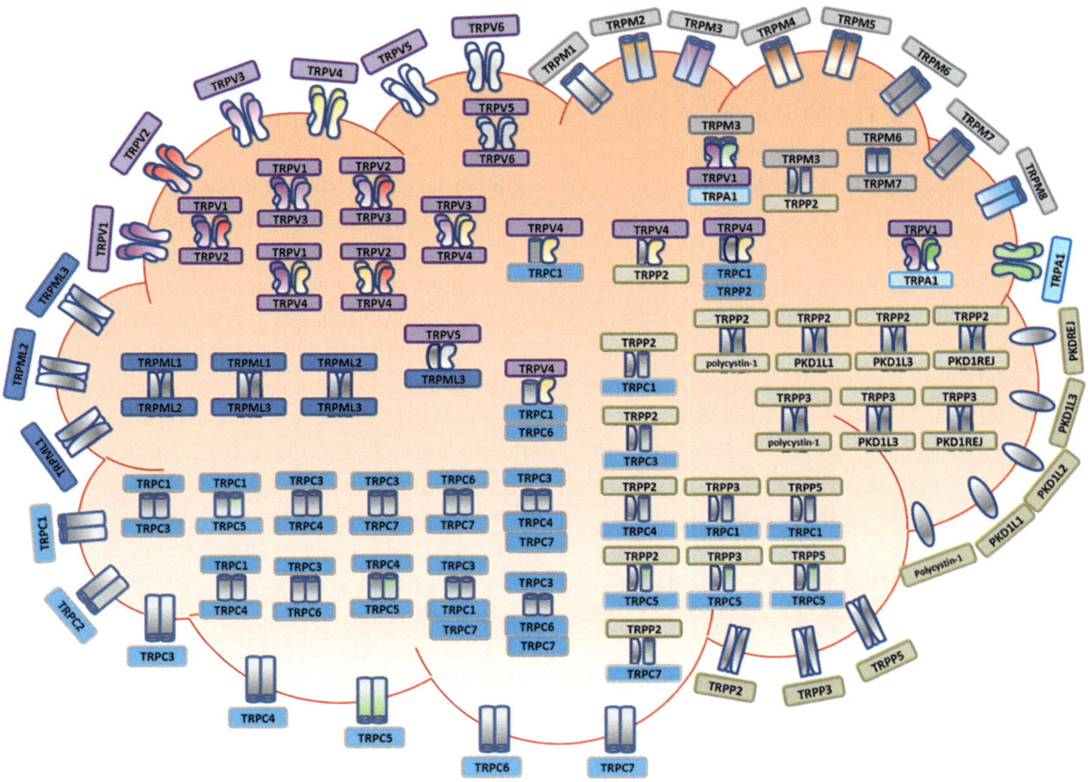

Fig. 7.3 Heteromerization of mammalian TRP ion channel subunits
Assembly of TRPs within the same subfamily and among different subfamilies, all thermo TRPs are marked with different hues indicated their sensing of temperature warming up or cooling down.

PKD1L3 channel complex in mice and humans regulates cilia through modulation of ciliary calcium concentration [261] (Fig. 7.3).

7.3.4 Specificity of TRP Channel Subunits Co-Assembly

Co-assembly of ion channel subunits yields a variety of diverse channel complexes. Heteromerization among mammalian TRP subunits produces novel channel types with functional properties distinct from their homomeric counterparts. Based on the current documents, we found that TRPC1, TRPV4, and TRPP2 are the most active coordinators in TRP channel heteromultimerization. They participate in either pair group or triad combination to form novel channel complexes. Interestingly, the three ion channel subunits combine to form a flow-sensitive heteromeric TRPC1/TRPV4/TRPP2 channel complex in cultured rat mesenteric artery endothelial cells as well as HEK293 cells. Moreover, heteromeric TRPC1/TRPV4/TRPP2 channels function as mediating the flow-induced Ca^{2+} increase in native vascular endothelial cells [256].

Meanwhile, we notice that thermo TRPs in TRP channel subunit assembly exhibit some sort of inclination to their thermosensitive compatibility. As the noxious cold-sensitive TRPA1 channel subunit may functionally interact with the noxious heat-sensitive TRPV1 channel subunit, despite their substantial sequence dissimilarity. It is found that co-expression of TRPA1 and TRPV1 contributes to TRPA1-mediated responses in trigeminal sensory neurons [262]. Furthermore, TRPA1 with TRPV1 functionally co-express in primary adult mouse ventricular cardiomyocytes throughout the endocardium, myocardium, and epicardium, and the crosstalk between TRPA1 and TRPV1 may be important in mediating cellular signaling events in cardiac muscle [263]. In addition, the novel combination of triad TRP ion channels of TRPV1/TRPA1/TRPM3 is responsible for acute noxious heat sensing in mice. Moreover, the robust somatosensory heat responsiveness at the cellular and behavioral levels is observed only if at least one of these TRP channels is functional [257].

7.4 Discussion and Outlook

7.4.1 Relevance of TRP Channel Distribution to Their Function

TRP ion channels are widely distributed in the tissues and organs of mammalians. Especially, extensive distributions of TRPCs, TRPMs, and TRPVs appear prominently in mammalians. These three TRP subfamilies expressions disperse to almost all the major organ systems. Thus, their functions should be diverse as their distributions. However, many studies are limited to TRP channels expression profiles but lacking their roles in the specific location.

In the nervous system, TRPCs, TRPMs, TRPVs, TRPA1, and TRPMLs are all present. These distributions are better explanations for TRPs as cell sensors for classical sensory transduction. Moreover, studies also identify that TRPs not only act as polymodal cell sensors, which are involved in cellular sensing from somatosensation, hearing, taste, and olfaction [264], they are also involved in many other physiological and pathological processes from metabolism, learning, homeostasis, pheromone modulation, vasorelaxation, respiratory rhythm regulation, inflammation, and even carcinogenesis. Despite all this so far, our knowledge of TRPs function regarding their distribution in tissues and organs remain limited. Further investigations regarding the functional relevance of TRPs need to be dug deeply and broadly.

7.4.2 Deciphering TRP Channel Assembly for a Better Understanding of Their Distribution and Functions

The intracellular distributions of TRP channels may be dynamically regulated by cytosolic changes. Studies indicate that assembly of TRP channel subunits can promote heteromeric channel complex relocation from an intracellular pool to the plasma membrane, such as polycystin-1 located in the plasma membrane, when heteromeric channel complexes form, polycystin-2 is recruited from the intracellular compartment to the plasma membrane [6]. Furthermore, the association between TRPC1 and TRPV4 enhance the insertion of the TRPC1/TRPV4 channel complex to the plasma membrane [231].

TRP ion channel is a superfamily with 28 members in mammalians. Presumed that each two or three can produce a novel combination of channel complex, then the channel family will be extensively expanded. Despite many studies focused on TRP channel assembly, the whole scope regarding TRP assembly remains limited. To date, most published studies focused on the static TRP subunits assembly. Further study regarding the assembly and distribution of TRPs, dynamic influences in living cell systems have to be considered. Furthermore, a practical approach and advanced technique to monitor dynamic TRP ion channel subunits assembly should be explored.

7.4.3 Summary

In this chapter, we review studies in the distribution and assembly of TRP ion channels based on recent references. Assembly of TRP ion channel subunits clearly expand channel subtypes and redistribute/translocate TRP channel complexes. The redistribution of TRPs includes alterations not only in TRP abundance and location in tissues and organs but also include TRP translocation inside the cells. These alterations of TRP distributions highlight the dynamic influences by cellular stimuli (extra or intra) or pathological processes. Meanwhile, the scope of TRP subunits assembly needs further investigation. With new technical approaches application, many more aspects will be deciphered regarding TRP ion channels distribution and assembly.

References

1. Clapham DE (2003) TRP channels as cellular sensors. Nature 426(6966):517–524
2. Zheng J (2013) Molecular mechanism of TRP channels. Compr Physiol 3(1):221–242
3. Grimm C, Jors S, Saldanha SA, Obukhov AG, Pan B, Oshima K, Cuajungco MP, Chase P, Hodder P, Heller S (2010) Small molecule activators of TRPML3. Chem Biol 17(2):135–148
4. Chen XZ, Segal Y, Basora N, Guo L, Peng JB, Babakhanlou H, Vassilev PM, Brown EM, Hediger MA, Zhou J (2001) Transport function of the naturally occurring pathogenic polycystin-2 mutant, R742X. Biochem Biophys Res Commun 282(5): 1251–1256
5. Cai Y, Maeda Y, Cedzich A, Torres VE, Wu G, Hayashi T, Mochizuki T, Park JH, Witzgall R, Somlo S (1999) Identification and characterization of polycystin-2, the PKD2 gene product. J Biol Chem 274(40):28557–28565
6. Hanaoka K, Qian F, Boletta A, Bhunia AK, Piontek K, Tsiokas L, Sukhatme VP, Guggino WB, Germino GG (2000) Co-assembly of polycystin-1 and -2 produces unique cation-permeable currents. Nature 408(6815):990–994
7. Schmidt M, Dubin AE, Petrus MJ, Earley TJ, Patapoutian A (2009) Nociceptive signals induce trafficking of TRPA1 to the plasma membrane. Neuron 64(4):498–509
8. Erler I, Al-Ansary DM, Wissenbach U, Wagner TF, Flockerzi V, Niemeyer BA (2006) Trafficking and assembly of the cold-sensitive TRPM8 channel. J Biol Chem 281(50):38396–38404
9. Wang Y, Fu X, Gaiser S, Kottgen M, Kramer-Zucker A, Walz G, Wegierski T (2007) OS-9 regulates the transit and polyubiquitination of TRPV4 in the endoplasmic reticulum. J Biol Chem 282(50):36561–36570
10. van de Graaf SF, Chang Q, Mensenkamp AR, Hoenderop JG, Bindels RJ (2006) Direct interaction with Rab11a targets the epithelial Ca2+ channels TRPV5 and TRPV6 to the plasma membrane. Mol Cell Biol 26(1):303–312
11. Xing BM, Yang YR, Du JX, Chen HJ, Qi C, Huang ZH, Zhang Y, Wang Y (2012) Cyclin-dependent kinase 5 controls TRPV1 membrane trafficking and the heat sensitivity of nociceptors through KIF13B. J Neurosci 32(42):14709–14721
12. Zhang X, Huang J, McNaughton PA (2005) NGF rapidly increases membrane expression of TRPV1 heat-gated ion channels. EMBO J 24(24):4211–4223
13. Xu H, Fu Y, Tian W, Cohen DM (2006) Glycosylation of the osmoresponsive transient receptor potential channel TRPV4 on Asn-651 influences membrane trafficking. Am J Physiol Renal Physiol 290(5):F1103–F1109
14. Omar S, Clarke R, Abdullah H, Brady C, Corry J, Winter H, Touzelet O, Power UF, Lundy F, McGarvey LP, Cosby SL (2017) Respiratory virus infection up-regulates TRPV1, TRPA1 and ASICS3 receptors on airway cells. PLoS One 12(2):e0171681
15. Guo Y, Ying S, Zhao X, Liu J, Wang Y (2019) Increased expression of lung TRPV1/TRPA1 in a cough model of bleomycin-induced pulmonary fibrosis in Guinea pigs. BMC Pulm Med 19(1):27
16. Yang YS, Cho SI, Choi MG, Choi YH, Kwak IS, Park CW, Kim HO (2015) Increased expression of three types of transient receptor potential channels (TRPA1, TRPV4 and TRPV3) in burn scars with post-burn pruritus. Acta Derm Venereol 95(1):20–24
17. Agosto MA, Anastassov IA, Robichaux MA, Wensel TG (2018) A large endoplasmic reticulum-resident pool of TRPM1 in retinal ON-bipolar cells. eNeuro 5(3)
18. Vetter I, Cheng W, Peiris M, Wyse BD, Roberts-Thomson SJ, Zheng J, Monteith GR, Cabot PJ (2008) Rapid, opioid-sensitive mechanisms involved in transient receptor potential vanilloid 1 sensitization. J Biol Chem 283(28):19540–19550
19. Kheradpezhouh E, Zhou FH, Barritt GJ, Rychkov GY (2018) Oxidative stress promotes redistribution of TRPM2 channels to the plasma membrane in hepatocytes. Biochem Biophys Res Commun 503(3):1891–1896
20. Broker-Lai J, Kollewe A, Schindeldecker B, Pohle J, Nguyen Chi V, Mathar I, Guzman R, Schwarz Y, Lai A, Weissgerber P, Schwegler H, Dietrich A, Both M, Sprengel R, Draguhn A, Kohr G, Fakler B, Flockerzi V, Bruns D, Freichel M (2017) Heteromeric channels formed by TRPC1, TRPC4 and TRPC5 define hippocampal synaptic transmission and working memory. EMBO J 36(18): 2770–2789
21. Belkacemi T, Niermann A, Hofmann L, Wissenbach U, Birnbaumer L, Leidinger P, Backes C, Meese E, Keller A, Bai X, Scheller A, Kirchhoff F, Philipp SE, Weissgerber P, Flockerzi V, Beck A (2017) TRPC1- and TRPC3-dependent Ca(2+) signaling in mouse cortical astrocytes affects injury-evoked astrogliosis in vivo. Glia 65(9):1535–1549
22. Wes PD, Chevesich J, Jeromin A, Rosenberg C, Stetten G, Montell C (1995) TRPC1, a human homolog of a Drosophila store-operated channel. Proc Natl Acad Sci U S A 92(21):9652–9656

23. Sinkins WG, Goel M, Estacion M, Schilling WP (2004) Association of immunophilins with mammalian TRPC channels. J Biol Chem 279(33): 34521–34529
24. Strubing C, Krapivinsky G, Krapivinsky L, Clapham DE (2001) TRPC1 and TRPC5 form a novel cation channel in mammalian brain. Neuron 29(3):645–655
25. Ma R, Rundle D, Jacks J, Koch M, Downs T, Tsiokas L (2003) Inhibitor of myogenic family, a novel suppressor of store-operated currents through an interaction with TRPC1. J Biol Chem 278(52): 52763–52772
26. Ong EC, Nesin V, Long CL, Bai CX, Guz JL, Ivanov IP, Abramowitz J, Birnbaumer L, Humphrey MB, Tsiokas L (2013) A TRPC1 protein-dependent pathway regulates osteoclast formation and function. J Biol Chem 288(31):22219–22232
27. Abed E, Labelle D, Martineau C, Loghin A, Moreau R (2009) Expression of transient receptor potential (TRP) channels in human and murine osteoblast-like cells. Mol Membr Biol 26(3):146–158
28. Riccio A, Medhurst AD, Mattei C, Kelsell RE, Calver AR, Randall AD, Benham CD, Pangalos MN (2002) mRNA distribution analysis of human TRPC family in CNS and peripheral tissues. Brain Res Mol Brain Res 109(1–2):95–104
29. Brereton HM, Chen J, Rychkov G, Harland ML, Barritt GJ (2001) Maitotoxin activates an endogenous non-selective cation channel and is an effective initiator of the activation of the heterologously expressed hTRPC-1 (transient receptor potential) non-selective cation channel in H4-IIE liver cells. Biochim Biophys Acta 1540(2):107–126
30. Jang Y, Lee Y, Kim SM, Yang YD, Jung J, Oh U (2012) Quantitative analysis of TRP channel genes in mouse organs. Arch Pharm Res 35(10):1823–1830
31. Wang J, Laurier LG, Sims SM, Preiksaitis HG (2003) Enhanced capacitative calcium entry and TRPC channel gene expression in human LES smooth muscle. Am J Physiol Gastrointest Liver Physiol 284(6): G1074–G1083
32. Berbey C, Weiss N, Legrand C, Allard B (2009) Transient receptor potential canonical type 1 (TRPC1) operates as a sarcoplasmic reticulum calcium leak channel in skeletal muscle. J Biol Chem 284(52):36387–36394
33. Sabourin J, Lamiche C, Vandebrouck A, Magaud C, Rivet J, Cognard C, Bourmeyster N, Constantin B (2009) Regulation of TRPC1 and TRPC4 cation channels requires an alpha1-syntrophin-dependent complex in skeletal mouse myotubes. J Biol Chem 284(52):36248–36261
34. Castellano LE, Trevino CL, Rodriguez D, Serrano CJ, Pacheco J, Tsutsumi V, Felix R, Darszon A (2003) Transient receptor potential (TRPC) channels in human sperm: expression, cellular localization and involvement in the regulation of flagellar motility. FEBS Lett 541(1–3):69–74
35. Dalrymple A, Slater DM, Beech D, Poston L, Tribe RM (2002) Molecular identification and localization of Trp homologues, putative calcium channels, in pregnant human uterus. Mol Hum Reprod 8(10): 946–951
36. Ku CY, Babich L, Word RA, Zhong M, Ulloa A, Monga M, Sanborn BM (2006) Expression of transient receptor channel proteins in human fundal myometrium in pregnancy. J Soc Gynecol Investig 13(3):217–225
37. Marabita F, Islam MS (2017) Expression of transient receptor potential channels in the purified human pancreatic beta-cells. Pancreas 46(1):97–101
38. Wolfrum C, Kiehlmann E, Pelczar P (2018) TRPC1 regulates brown adipose tissue activity in a PPARgamma-dependent manner. Am J Physiol Endocrinol Metab 315(5):E825–E832
39. Hofmann T, Schaefer M, Schultz G, Gudermann T (2000) Cloning, expression and subcellular localization of two novel splice variants of mouse transient receptor potential channel 2. Biochem J 351(Pt 1): 115–122
40. Sukumaran P, Lof C, Kemppainen K, Kankaanpaa P, Pulli I, Nasman J, Viitanen T, Tornquist K (2012) Canonical transient receptor potential channel 2 (TRPC2) as a major regulator of calcium homeostasis in rat thyroid FRTL-5 cells: importance of protein kinase C delta (PKCdelta) and stromal interaction molecule 2 (STIM2). J Biol Chem 287(53): 44345–44360
41. Wissenbach U, Schroth G, Philipp S, Flockerzi V (1998) Structure and mRNA expression of a bovine trp homologue related to mammalian trp2 transcripts. FEBS Lett 429(1):61–66
42. Liman ER, Corey DP, Dulac C (1999) TRP2: a candidate transduction channel for mammalian pheromone sensory signaling. Proc Natl Acad Sci U S A 96(10):5791–5796
43. Chu X, Cheung JY, Barber DL, Birnbaumer L, Rothblum LI, Conrad K, Abrasonis V, Chan YM, Stahl R, Carey DJ, Miller BA (2002) Erythropoietin modulates calcium influx through TRPC2. J Biol Chem 277(37):34375–34382
44. Roedding AS, Gao AF, Wu AM, Li PP, Kish SJ, Warsh JJ (2009) TRPC3 protein is expressed across the lifespan in human prefrontal cortex and cerebellum. Brain Res 1260:1–6
45. Fusco FR, Martorana A, Giampa C, De March Z, Vacca F, Tozzi A, Longone P, Piccirilli S, Paolucci S, Sancesario G, Mercuri NB, Bernardi G (2004) Cellular localization of TRPC3 channel in rat brain: preferential distribution to oligodendrocytes. Neurosci Lett 365(2):137–142
46. Liu X, Bandyopadhyay BC, Singh BB, Groschner K, Ambudkar IS (2005) Molecular analysis of a store-operated and 2-acetyl-sn-glycerol-sensitive non-selective cation channel. Heteromeric assembly of TRPC1-TRPC3. J Biol Chem 280(22): 21600–21606

47. Philipp S, Strauss B, Hirnet D, Wissenbach U, Mery L, Flockerzi V, Hoth M (2003) TRPC3 mediates T-cell receptor-dependent calcium entry in human T-lymphocytes. J Biol Chem 278(29): 26629–26638
48. Carrillo C, Hichami A, Andreoletti P, Cherkaoui-Malki M, del Mar Cavia M, Abdoul-Azize S, Alonso-Torre SR, Khan NA (2012) Diacylglycerol-containing oleic acid induces increases in [Ca(2+)](i) via TRPC3/6 channels in human T-cells. Biochim Biophys Acta 1821(4):618–626
49. McKay RR, Szymeczek-Seay CL, Lievremont JP, Bird GS, Zitt C, Jungling E, Luckhoff A, Putney JW Jr (2000) Cloning and expression of the human transient receptor potential 4 (TRP4) gene: localization and functional expression of human TRP4 and TRP3. Biochem J 351(Pt 3):735–746
50. Zechel S, Werner S, von Bohlen Und Halbach O (2007) Distribution of TRPC4 in developing and adult murine brain. Cell Tissue Res 328(3):651–656
51. Walker RL, Hume JR, Horowitz B (2001) Differential expression and alternative splicing of TRP channel genes in smooth muscles. Am J Physiol Cell Physiol 280(5):C1184–C1192
52. Veliceasa D, Ivanovic M, Hoepfner FT, Thumbikat P, Volpert OV, Smith ND (2007) Transient potential receptor channel 4 controls thrombospondin-1 secretion and angiogenesis in renal cell carcinoma. FEBS J 274(24):6365–6377
53. Facemire CS, Mohler PJ, Arendshorst WJ (2004) Expression and relative abundance of short transient receptor potential channels in the rat renal microcirculation. Am J Physiol Renal Physiol 286(3):F546–F551
54. Yu W, Hill WG, Apodaca G, Zeidel ML (2011) Expression and distribution of transient receptor potential (TRP) channels in bladder epithelium. Am J Physiol Renal Physiol 300(1):F49–F59
55. Yip H, Chan WY, Leung PC, Kwan HY, Liu C, Huang Y, Michel V, Yew DT, Yao X (2004) Expression of TRPC homologs in endothelial cells and smooth muscle layers of human arteries. Histochem Cell Biol 122(6):553–561
56. Freichel M, Suh SH, Pfeifer A, Schweig U, Trost C, Weissgerber P, Biel M, Philipp S, Freise D, Droogmans G, Hofmann F, Flockerzi V, Nilius B (2001) Lack of an endothelial store-operated Ca2+ current impairs agonist-dependent vasorelaxation in TRP4-/- mice. Nat Cell Biol 3(2):121–127
57. Zimmermann K, Lennerz JK, Hein A, Link AS, Kaczmarek JS, Delling M, Uysal S, Pfeifer JD, Riccio A, Clapham DE (2011) Transient receptor potential cation channel, subfamily C, member 5 (TRPC5) is a cold-transducer in the peripheral nervous system. Proc Natl Acad Sci U S A 108(44): 18114–18119
58. Okada T, Shimizu S, Wakamori M, Maeda A, Kurosaki T, Takada N, Imoto K, Mori Y (1998) Molecular cloning and functional characterization of a novel receptor-activated TRP Ca2+ channel from mouse brain. J Biol Chem 273(17):10279–10287
59. Lee JH, Kim SY, Kwon YK, Kim BJ, So I (2013) Characteristics of the cholecystokinin-induced depolarization of pacemaking activity in cultured interstitial cells of Cajal from murine small intestine. Cell Physiol Biochem 31(4-5):542–554
60. Zhu Y, Gao M, Zhou T, Xie M, Mao A, Feng L, Yao X, Wong WT, Ma X (2019) The TRPC5 channel regulates angiogenesis and promotes recovery from ischemic injury in mice. J Biol Chem 294(1):28–37
61. Xu P, Xu J, Li Z, Yang Z (2012) Expression of TRPC6 in renal cortex and hippocampus of mouse during postnatal development. PLoS One 7(6): e38503
62. Lee YM, Kim BJ, Kim HJ, Yang DK, Zhu MH, Lee KP, So I, Kim KW (2003) TRPC5 as a candidate for the nonselective cation channel activated by muscarinic stimulation in murine stomach. Am J Physiol Gastrointest Liver Physiol 284(4):G604–G616
63. Hofmann T, Schaefer M, Schultz G, Gudermann T (2000) Transient receptor potential channels as molecular substrates of receptor-mediated cation entry. J Mol Med (Berl) 78(1):14–25
64. Finney-Hayward TK, Popa MO, Bahra P, Li S, Poll CT, Gosling M, Nicholson AG, Russell RE, Kon OM, Jarai G, Westwick J, Barnes PJ, Donnelly LE (2010) Expression of transient receptor potential C6 channels in human lung macrophages. Am J Respir Cell Mol Biol 43(3):296–304
65. Sel S, Rost BR, Yildirim AO, Sel B, Kalwa H, Fehrenbach H, Renz H, Gudermann T, Dietrich A (2008) Loss of classical transient receptor potential 6 channel reduces allergic airway response. Clin Exp Allergy 38(9):1548–1558
66. Nagamine K, Kudoh J, Minoshima S, Kawasaki K, Asakawa S, Ito F, Shimizu N (1998) Molecular cloning of a novel putative Ca2+ channel protein (TRPC7) highly expressed in brain. Genomics 54(1):124–131
67. Babich LG, Ku CY, Young HW, Huang H, Blackburn MR, Sanborn BM (2004) Expression of capacitative calcium TrpC proteins in rat myometrium during pregnancy. Biol Reprod 70(4): 919–924
68. Rodella L, Rezzani R, Bianchi R (1996) Perivascular localization of TRPM1- and TRPM2-positive cells in the rat brain. Boll Soc Ital Biol Sper 72(7-8):223–226
69. Fonfria E, Murdock PR, Cusdin FS, Benham CD, Kelsell RE, McNulty S (2006) Tissue distribution profiles of the human TRPM cation channel family. J Recept Signal Transduct Res 26(3):159–178
70. Duncan LM, Deeds J, Hunter J, Shao J, Holmgren LM, Woolf EA, Tepper RI, Shyjan AW (1998) Down-regulation of the novel gene melastatin correlates with potential for melanoma metastasis. Cancer Res 58(7):1515–1520
71. Miller AJ, Du J, Rowan S, Hershey CL, Widlund HR, Fisher DE (2004) Transcriptional regulation of the

melanoma prognostic marker melastatin (TRPM1) by MITF in melanocytes and melanoma. Cancer Res 64(2):509–516
72. Togashi K, Hara Y, Tominaga T, Higashi T, Konishi Y, Mori Y, Tominaga M (2006) TRPM2 activation by cyclic ADP-ribose at body temperature is involved in insulin secretion. EMBO J 25(9): 1804–1815
73. Andoh C, Nishitani N, Hashimoto E, Nagai Y, Takao K, Miyakawa T, Nakagawa T, Mori Y, Nagayasu K, Shirakawa H, Kaneko S (2019) TRPM2 confers susceptibility to social stress but is essential for behavioral flexibility. Brain Res 1704: 68–77
74. Olah ME, Jackson MF, Li H, Perez Y, Sun HS, Kiyonaka S, Mori Y, Tymianski M, MacDonald JF (2009) Ca2+-dependent induction of TRPM2 currents in hippocampal neurons. J Physiol 587 (Pt 5):965–979
75. Hiroi H, Momoeda M, Watanabe T, Ito M, Ikeda K, Tsutsumi R, Hosokawa Y, Koizumi M, Zenri F, Muramatsu M, Taketani Y, Inoue S (2013) Expression and regulation of transient receptor potential cation channel, subfamily M, member 2 (TRPM2) in human endometrium. Mol Cell Endocrinol 365(2):146–152
76. Matsumoto K, Takagi K, Kato A, Ishibashi T, Mori Y, Tashima K, Mitsumoto A, Kato S, Horie S (2016) Role of transient receptor potential melastatin 2 (TRPM2) channels in visceral nociception and hypersensitivity. Exp Neurol 285(Pt A):41–50
77. Liu X, Cotrim A, Teos L, Zheng C, Swaim W, Mitchell J, Mori Y, Ambudkar I (2013) Loss of TRPM2 function protects against irradiation-induced salivary gland dysfunction. Nat Commun 4:1515
78. Uchida K, Dezaki K, Damdindorj B, Inada H, Shiuchi T, Mori Y, Yada T, Minokoshi Y, Tominaga M (2011) Lack of TRPM2 impaired insulin secretion and glucose metabolisms in mice. Diabetes 60(1): 119–126
79. Ishii M, Shimizu S, Hara Y, Hagiwara T, Miyazaki A, Mori Y, Kiuchi Y (2006) Intracellular-produced hydroxyl radical mediates H2O2-induced Ca2+ influx and cell death in rat beta-cell line RIN-5F. Cell Calcium 39(6):487–494
80. Carter RN, Tolhurst G, Walmsley G, Vizuete-Forster M, Miller N, Mahaut-Smith MP (2006) Molecular and electrophysiological characterization of transient receptor potential ion channels in the primary murine megakaryocyte. J Physiol 576(Pt 1): 151–162
81. Kashio M, Sokabe T, Shintaku K, Uematsu T, Fukuta N, Kobayashi N, Mori Y, Tominaga M (2012) Redox signal-mediated sensitization of transient receptor potential melastatin 2 (TRPM2) to temperature affects macrophage functions. Proc Natl Acad Sci U S A 109(17):6745–6750
82. Lange I, Penner R, Fleig A, Beck A (2008) Synergistic regulation of endogenous TRPM2 channels by adenine dinucleotides in primary human neutrophils. Cell Calcium 44(6):604–615
83. Magnone M, Bauer I, Poggi A, Mannino E, Sturla L, Brini M, Zocchi E, De Flora A, Nencioni A, Bruzzone S (2012) NAD+ levels control Ca2+ store replenishment and mitogen-induced increase of cytosolic Ca2+ by Cyclic ADP-ribose-dependent TRPM2 channel gating in human T lymphocytes. J Biol Chem 287(25):21067–21081
84. Oda S, Uchida K, Wang X, Lee J, Shimada Y, Tominaga M, Kadowaki M (2013) TRPM2 contributes to antigen-stimulated Ca(2)(+) influx in mucosal mast cells. Pflugers Arch 465(7):1023–1030
85. Roedding AS, Gao AF, Au-Yeung W, Scarcelli T, Li PP, Warsh JJ (2012) Effect of oxidative stress on TRPM2 and TRPC3 channels in B lymphoblast cells in bipolar disorder. Bipolar Disord 14(2): 151–161
86. Sumoza-Toledo A, Lange I, Cortado H, Bhagat H, Mori Y, Fleig A, Penner R, Partida-Sanchez S (2011) Dendritic cell maturation and chemotaxis is regulated by TRPM2-mediated lysosomal Ca2+ release. FASEB J 25(10):3529–3542
87. Yamamoto S, Shimizu S, Kiyonaka S, Takahashi N, Wajima T, Hara Y, Negoro T, Hiroi T, Kiuchi Y, Okada T, Kaneko S, Lange I, Fleig A, Penner R, Nishi M, Takeshima H, Mori Y (2008) TRPM2-mediated Ca2+influx induces chemokine production in monocytes that aggravates inflammatory neutrophil infiltration. Nat Med 14(7):738–747
88. Vriens J, Owsianik G, Hofmann T, Philipp SE, Stab J, Chen X, Benoit M, Xue F, Janssens A, Kerselaers S, Oberwinkler J, Vennekens R, Gudermann T, Nilius B, Voets T (2011) TRPM3 is a nociceptor channel involved in the detection of noxious heat. Neuron 70(3):482–494
89. Grimm C, Kraft R, Sauerbruch S, Schultz G, Harteneck C (2003) Molecular and functional characterization of the melastatin-related cation channel TRPM3. J Biol Chem 278(24):21493–21501
90. Li SL, Wang XH, Wang HP, Yang ZH, Gao WC, Pu XY (2008) Expression of TRPM and TRPV channel family mRNA in rat spermatogenic cells. Nan Fang Yi Ke Da Xue Xue Bao 28(12):2150–2153
91. Wang HP, Pu XY, Wang XH (2007) Distribution profiles of transient receptor potential melastatin-related and vanilloid-related channels in prostatic tissue in rat. Asian J Androl 9(5):634–640
92. Talavera K, Yasumatsu K, Voets T, Droogmans G, Shigemura N, Ninomiya Y, Margolskee RF, Nilius B (2005) Heat activation of TRPM5 underlies thermal sensitivity of sweet taste. Nature 438(7070): 1022–1025
93. Kunert-Keil C, Bisping F, Kruger J, Brinkmeier H (2006) Tissue-specific expression of TRP channel genes in the mouse and its variation in three different mouse strains. BMC Genomics 7:159
94. Launay P, Cheng H, Srivatsan S, Penner R, Fleig A, Kinet JP (2004) TRPM4 regulates calcium

oscillations after T cell activation. Science 306(5700):1374–1377
95. Vennekens R, Olausson J, Meissner M, Bloch W, Mathar I, Philipp SE, Schmitz F, Weissgerber P, Nilius B, Flockerzi V, Freichel M (2007) Increased IgE-dependent mast cell activation and anaphylactic responses in mice lacking the calcium-activated nonselective cation channel TRPM4. Nat Immunol 8(3): 312–320
96. Shimizu T, Owsianik G, Freichel M, Flockerzi V, Nilius B, Vennekens R (2009) TRPM4 regulates migration of mast cells in mice. Cell Calcium 45(3): 226–232
97. Dutta Banik, D., L.E. Martin, M. Freichel, A.M. Torregrossa, K.F. Medler, TRPM4 and TRPM5 are both required for normal signaling in taste receptor cells. Proc Natl Acad Sci U S A, 2018. 115(4): E772-E781.
98. Wang H, Xu Z, Lee BH, Vu S, Hu L, Lee M, Bu D, Cao X, Hwang S, Yang Y, Zheng J, Lin Z (2019) Gain-of-function mutations in TRPM4 activation gate cause progressive symmetric erythrokeratodermia. J Invest Dermatol 139(5):1089–1097
99. Zhang Y, Hoon MA, Chandrashekar J, Mueller KL, Cook B, Wu D, Zuker CS, Ryba NJ (2003) Coding of sweet, bitter, and umami tastes: different receptor cells sharing similar signaling pathways. Cell 112(3):293–301
100. Lei YT, Thuault SJ, Launay P, Margolskee RF, Kandel ER, Siegelbaum SA (2014) Differential contribution of TRPM4 and TRPM5 nonselective cation channels to the slow afterdepolarization in mouse prefrontal cortex neurons. Front Cell Neurosci 8:267
101. Colsoul B, Jacobs G, Philippaert K, Owsianik G, Segal A, Nilius B, Voets T, Schuit F, Vennekens R (2014) Insulin downregulates the expression of the Ca2+-activated nonselective cation channel TRPM5 in pancreatic islets from leptin-deficient mouse models. Pflugers Arch 466(3):611–621
102. Colsoul B, Schraenen A, Lemaire K, Quintens R, Van Lommel L, Segal A, Owsianik G, Talavera K, Voets T, Margolskee RF, Kokrashvili Z, Gilon P, Nilius B, Schuit FC, Vennekens R (2010) Loss of high-frequency glucose-induced Ca2+ oscillations in pancreatic islets correlates with impaired glucose tolerance in Trpm5−/− mice. Proc Natl Acad Sci U S A 107(11):5208–5213
103. Bezencon C, Furholz A, Raymond F, Mansourian R, Metairon S, Le Coutre J, Damak S (2008) Murine intestinal cells expressing Trpm5 are mostly brush cells and express markers of neuronal and inflammatory cells. J Comp Neurol 509(5):514–525
104. Kaske S, Krasteva G, Konig P, Kummer W, Hofmann T, Gudermann T, Chubanov V (2007) TRPM5, a taste-signaling transient receptor potential ion-channel, is a ubiquitous signaling component in chemosensory cells. BMC Neurosci 8:49
105. Schlingmann KP, Weber S, Peters M, Niemann Nejsum L, Vitzthum H, Klingel K, Kratz M, Haddad E, Ristoff E, Dinour D, Syrrou M, Nielsen S, Sassen M, Waldegger S, Seyberth HW, Konrad M (2002) Hypomagnesemia with secondary hypocalcemia is caused by mutations in TRPM6, a new member of the TRPM gene family. Nat Genet 31(2):166–170
106. Walder RY, Landau D, Meyer P, Shalev H, Tsolia M, Borochowitz Z, Boettger MB, Beck GE, Englehardt RK, Carmi R, Sheffield VC (2002) Mutation of TRPM6 causes familial hypomagnesemia with secondary hypocalcemia. Nat Genet 31(2):171–174
107. Zhang X, Zu H, Zhao D, Yang K, Tian S, Yu X, Lu F, Liu B, Yu X, Wang B, Wang W, Huang S, Wang Y, Wang Z, Zhang Z (2017) Ion channel functional protein kinase TRPM7 regulates Mg ions to promote the osteoinduction of human osteoblast via PI3K pathway: in vitro simulation of the bone-repairing effect of Mg-based alloy implant. Acta Biomater 63: 369–382
108. Won J, Kim JH, Oh SB (2018) Molecular expression of Mg(2+) regulator TRPM7 and CNNM4 in rat odontoblasts. Arch Oral Biol 96:182–188
109. Tsavaler L, Shapero MH, Morkowski S, Laus R (2001) Trp-p8, a novel prostate-specific gene, is up-regulated in prostate cancer and other malignancies and shares high homology with transient receptor potential calcium channel proteins. Cancer Res 61(9):3760–3769
110. Peier AM, Moqrich A, Hergarden AC, Reeve AJ, Andersson DA, Story GM, Earley TJ, Dragoni I, McIntyre P, Bevan S, Patapoutian A (2002) A TRP channel that senses cold stimuli and menthol. Cell 108(5):705–715
111. Facer P, Casula MA, Smith GD, Benham CD, Chessell IP, Bountra C, Sinisi M, Birch R, Anand P (2007) Differential expression of the capsaicin receptor TRPV1 and related novel receptors TRPV3, TRPV4 and TRPM8 in normal human tissues and changes in traumatic and diabetic neuropathy. BMC Neurol 7:11
112. Gibbs GM, Orta G, Reddy T, Koppers AJ, Martinez-Lopez P, de la Vega-Beltran JL, Lo JC, Veldhuis N, Jamsai D, McIntyre P, Darszon A, O'Bryan MK (2011) Cysteine-rich secretory protein 4 is an inhibitor of transient receptor potential M8 with a role in establishing sperm function. Proc Natl Acad Sci U S A 108(17):7034–7039
113. Stein RJ, Santos S, Nagatomi J, Hayashi Y, Minnery BS, Xavier M, Patel AS, Nelson JB, Futrell WJ, Yoshimura N, Chancellor MB, De Miguel F (2004) Cool (TRPM8) and hot (TRPV1) receptors in the bladder and male genital tract. J Urol 172(3): 1175–1178
114. Sabnis AS, Shadid M, Yost GS, Reilly CA (2008) Human lung epithelial cells express a functional cold-sensing TRPM8 variant. Am J Respir Cell Mol Biol 39(4):466–474
115. Moraes MN, de Assis LVM, Henriques FDS, Batista ML Jr, Guler AD, Castrucci AML (2017) Cold-

sensing TRPM8 channel participates in circadian control of the brown adipose tissue. Biochim Biophys Acta 1864(12):2415–2427
116. Caterina MJ, Schumacher MA, Tominaga M, Rosen TA, Levine JD, Julius D (1997) The capsaicin receptor: a heat-activated ion channel in the pain pathway. Nature 389(6653):816–824
117. Toth A, Boczan J, Kedei N, Lizanecz E, Bagi Z, Papp Z, Edes I, Csiba L, Blumberg PM (2005) Expression and distribution of vanilloid receptor 1 (TRPV1) in the adult rat brain. Brain Res Mol Brain Res 135(1-2):162–168
118. Helliwell RJ, McLatchie LM, Clarke M, Winter J, Bevan S, McIntyre P (1998) Capsaicin sensitivity is associated with the expression of the vanilloid (capsaicin) receptor (VR1) mRNA in adult rat sensory ganglia. Neurosci Lett 250(3):177–180
119. Iida T, Shimizu I, Nealen ML, Campbell A, Caterina M (2005) Attenuated fever response in mice lacking TRPV1. Neurosci Lett 378(1):28–33
120. Roberts JC, Davis JB, Benham CD (2004) [3H] Resiniferatoxin autoradiography in the CNS of wild-type and TRPV1 null mice defines TRPV1 (VR-1) protein distribution. Brain Res 995(2): 176–183
121. Molinas AJR, Desmoulins LD, Hamling BV, Butcher SM, Anwar IJ, Miyata K, Enix CL, Dugas CM, Satou R, Derbenev AV, Zsombok A (2019) Interaction between TRPV1-expressing neurons in the hypothalamus. J Neurophysiol 121(1):140–151
122. Jeong JH, Lee DK, Liu SM, Chua SC Jr, Schwartz GJ, Jo YH (2018) Activation of temperature-sensitive TRPV1-like receptors in ARC POMC neurons reduces food intake. PLoS Biol 16(4):e2004399
123. Tian YH, Lee SY, Kim HC, Jang CG (2010) Repeated methamphetamine treatment increases expression of TRPV1 mRNA in the frontal cortex but not in the striatum or hippocampus of mice. Neurosci Lett 472(1):61–64
124. Gavva NR, Bannon AW, Surapaneni S, Hovland DN Jr, Lehto SG, Gore A, Juan T, Deng H, Han B, Klionsky L, Kuang R, Le A, Tamir R, Wang J, Youngblood B, Zhu D, Norman MH, Magal E, Treanor JJ, Louis JC (2007) The vanilloid receptor TRPV1 is tonically activated in vivo and involved in body temperature regulation. J Neurosci 27(13): 3366–3374
125. Li HB, Mao RR, Zhang JC, Yang Y, Cao J, Xu L (2008) Antistress effect of TRPV1 channel on synaptic plasticity and spatial memory. Biol Psychiatry 64(4):286–292
126. Marsch R, Foeller E, Rammes G, Bunck M, Kossl M, Holsboer F, Zieglgansberger W, Landgraf R, Lutz B, Wotjak CT (2007) Reduced anxiety, conditioned fear, and hippocampal long-term potentiation in transient receptor potential vanilloid type 1 receptor-deficient mice. J Neurosci 27(4):832–839
127. Almeida V, Levin R, Peres FF, Suiama MA, Vendramini AM, Santos CM, Silva ND, Zuardi AW, Hallak JEC, Crippa JA, Abilio VC (2019) Role of the endocannabinoid and endovanilloid systems in an animal model of schizophrenia-related emotional processing/cognitive deficit. Neuropharmacology 155:44–53
128. Back FP, Carobrez AP (2018) Periaqueductal gray glutamatergic, cannabinoid and vanilloid receptor interplay in defensive behavior and aversive memory formation. Neuropharmacology 135:399–411
129. Balaban H, Naziroglu M, Demirci K, Ovey IS (2017) The Protective role of selenium on scopolamine-induced memory impairment, oxidative stress, and apoptosis in aged rats: the involvement of TRPM2 and TRPV1 channels. Mol Neurobiol 54(4): 2852–2868
130. Bannazadeh M, Fatehi F, Fatemi I, Roohbakhsh A, Allahtavakoli M, Nasiri M, Azin M, Shamsizadeh A (2017) The role of transient receptor potential vanilloid type 1 in unimodal and multimodal object recognition task in rats. Pharmacol Rep 69(3): 526–531
131. Cavanaugh DJ, Chesler AT, Jackson AC, Sigal YM, Yamanaka H, Grant R, O'Donnell D, Nicoll RA, Shah NM, Julius D, Basbaum AI (2011) Trpv1 reporter mice reveal highly restricted brain distribution and functional expression in arteriolar smooth muscle cells. J Neurosci 31(13):5067–5077
132. Benninger F, Freund TF, Hajos N (2008) Control of excitatory synaptic transmission by capsaicin is unaltered in TRPV1 vanilloid receptor knockout mice. Neurochem Int 52(1–2):89–94
133. Ahmed MK, Takumida M, Ishibashi T, Hamamoto T, Hirakawa K (2009) Expression of transient receptor potential vanilloid (TRPV) families 1, 2, 3 and 4 in the mouse olfactory epithelium. Rhinology 47(3): 242–247
134. Watanabe N, Horie S, Michael GJ, Spina D, Page CP, Priestley JV (2005) Immunohistochemical localization of vanilloid receptor subtype 1 (TRPV1) in the guinea pig respiratory system. Pulm Pharmacol Ther 18(3):187–197
135. Hamamoto T, Takumida M, Hirakawa K, Takeno S, Tatsukawa T (2008) Localization of transient receptor potential channel vanilloid subfamilies in the mouse larynx. Acta Otolaryngol 128(6):685–693
136. Kark T, Bagi Z, Lizanecz E, Pasztor ET, Erdei N, Czikora A, Papp Z, Edes I, Porszasz R, Toth A (2008) Tissue-specific regulation of microvascular diameter: opposite functional roles of neuronal and smooth muscle located vanilloid receptor-1. Mol Pharmacol 73(5):1405–1412
137. Birder LA, Nakamura Y, Kiss S, Nealen ML, Barrick S, Kanai AJ, Wang E, Ruiz G, De Groat WC, Apodaca G, Watkins S, Caterina MJ (2002) Altered urinary bladder function in mice lacking the vanilloid receptor TRPV1. Nat Neurosci 5(9): 856–860
138. Moraes MN, Mezzalira N, de Assis LV, Menaker M, Guler A, Castrucci AM (2017) TRPV1 participates in

the activation of clock molecular machinery in the brown adipose tissue in response to light-dark cycle. Biochim Biophys Acta 1864(2):324–335

139. Caterina MJ, Rosen TA, Tominaga M, Brake AJ, Julius D (1999) A capsaicin-receptor homologue with a high threshold for noxious heat. Nature 398(6726):436–441
140. Kajiya H, Okamoto F, Nemoto T, Kimachi K, Toh-Goto K, Nakayana S, Okabe K (2010) RANKL-induced TRPV2 expression regulates osteoclastogenesis via calcium oscillations. Cell Calcium 48(5):260–269
141. Link TM, Park U, Vonakis BM, Raben DM, Soloski MJ, Caterina MJ (2010) TRPV2 has a pivotal role in macrophage particle binding and phagocytosis. Nat Immunol 11(3):232–239
142. Stokes AJ, Shimoda LM, Koblan-Huberson M, Adra CN, Turner H (2004) A TRPV2-PKA signaling module for transduction of physical stimuli in mast cells. J Exp Med 200(2):137–147
143. Zhang D, Spielmann A, Wang L, Ding G, Huang F, Gu Q, Schwarz W (2012) Mast-cell degranulation induced by physical stimuli involves the activation of transient-receptor-potential channel TRPV2. Physiol Res 61(1):113–124
144. Heiner I, Eisfeld J, Luckhoff A (2003) Role and regulation of TRP channels in neutrophil granulocytes. Cell Calcium 33(5-6):533–540
145. Saunders CI, Kunde DA, Crawford A, Geraghty DP (2007) Expression of transient receptor potential vanilloid 1 (TRPV1) and 2 (TRPV2) in human peripheral blood. Mol Immunol 44(6):1429–1435
146. Santoni G, Farfariello V, Liberati S, Morelli MB, Nabissi M, Santoni M, Amantini C (2013) The role of transient receptor potential vanilloid type-2 ion channels in innate and adaptive immune responses. Front Immunol 4:34
147. Park KS, Pang B, Park SJ, Lee YG, Bae JY, Park S, Kim I, Kim SJ (2011) Identification and functional characterization of ion channels in CD34(+) hematopoietic stem cells from human peripheral blood. Mol Cells 32(2):181–188
148. Kowase T, Nakazato Y, Yoko OH, Morikawa A, Kojima I (2002) Immunohistochemical localization of growth factor-regulated channel (GRC) in human tissues. Endocr J 49(3):349–355
149. Hisanaga E, Nagasawa M, Ueki K, Kulkarni RN, Mori M, Kojima I (2009) Regulation of calcium-permeable TRPV2 channel by insulin in pancreatic beta-cells. Diabetes 58(1):174–184
150. Muraki K, Iwata Y, Katanosaka Y, Ito T, Ohya S, Shigekawa M, Imaizumi Y (2003) TRPV2 is a component of osmotically sensitive cation channels in murine aortic myocytes. Circ Res 93(9):829–838
151. Peng G, Lu W, Li X, Chen Y, Zhong N, Ran P, Wang J (2010) Expression of store-operated Ca2+ entry and transient receptor potential canonical and vanilloid-related proteins in rat distal pulmonary venous smooth muscle. Am J Physiol Lung Cell Mol Physiol 299(5):L621–L630
152. Fantozzi I, Zhang S, Platoshyn O, Remillard CV, Cowling RT, Yuan JX (2003) Hypoxia increases AP-1 binding activity by enhancing capacitative Ca2+ entry in human pulmonary artery endothelial cells. Am J Physiol Lung Cell Mol Physiol 285(6): L1233–L1245
153. Iwata Y, Katanosaka Y, Arai Y, Komamura K, Miyatake K, Shigekawa M (2003) A novel mechanism of myocyte degeneration involving the Ca2+-permeable growth factor-regulated channel. J Cell Biol 161(5):957–967
154. Peier AM, Reeve AJ, Andersson DA, Moqrich A, Earley TJ, Hergarden AC, Story GM, Colley S, Hogenesch JB, McIntyre P, Bevan S, Patapoutian A (2002) A heat-sensitive TRP channel expressed in keratinocytes. Science 296(5575):2046–2049
155. Xu H, Delling M, Jun JC, Clapham DE (2006) Oregano, thyme and clove-derived flavors and skin sensitizers activate specific TRP channels. Nat Neurosci 9(5):628–635
156. Xu H, Ramsey IS, Kotecha SA, Moran MM, Chong JA, Lawson D, Ge P, Lilly J, Silos-Santiago I, Xie Y, DiStefano PS, Curtis R, Clapham DE (2002) TRPV3 is a calcium-permeable temperature-sensitive cation channel. Nature 418(6894):181–186
157. Zhang L, Jones S, Brody K, Costa M, Brookes SJ (2004) Thermosensitive transient receptor potential channels in vagal afferent neurons of the mouse. Am J Physiol Gastrointest Liver Physiol 286(6): G983–G991
158. Ueda T, Yamada T, Ugawa S, Ishida Y, Shimada S (2009) TRPV3, a thermosensitive channel is expressed in mouse distal colon epithelium. Biochem Biophys Res Commun 383(1):130–134
159. Szollosi AG, Vasas N, Angyal A, Kistamas K, Nanasi PP, Mihaly J, Beke G, Herczeg-Lisztes E, Szegedi A, Kawada N, Yanagida T, Mori T, Kemeny L, Biro T (2018) Activation of TRPV3 regulates inflammatory actions of human epidermal keratinocytes. J Invest Dermatol 138(2):365–374
160. D'Aldebert E, Cenac N, Rousset P, Martin L, Rolland C, Chapman K, Selves J, Alric L, Vinel JP, Vergnolle N (2011) Transient receptor potential vanilloid 4 activated inflammatory signals by intestinal epithelial cells and colitis in mice. Gastroenterology 140(1):275–285
161. Chen Y, Williams SH, McNulty AL, Hong JH, Lee SH, Rothfusz NE, Parekh PK, Moore C, Gereau RWT, Taylor AB, Wang F, Guilak F, Liedtke W (2013) Temporomandibular joint pain: a critical role for Trpv4 in the trigeminal ganglion. Pain 154(8): 1295–1304
162. Hu F, Zhu W, Wang L (2013) MicroRNA-203 up-regulates nitric oxide expression in temporomandibular joint chondrocytes via targeting TRPV4. Arch Oral Biol 58(2):192–199

163. Muramatsu S, Wakabayashi M, Ohno T, Amano K, Ooishi R, Sugahara T, Shiojiri S, Tashiro K, Suzuki Y, Nishimura R, Kuhara S, Sugano S, Yoneda T, Matsuda A (2007) Functional gene screening system identified TRPV4 as a regulator of chondrogenic differentiation. J Biol Chem 282(44): 32158–32167
164. Masuyama R, Vriens J, Voets T, Karashima Y, Owsianik G, Vennekens R, Lieben L, Torrekens S, Moermans K, Vanden Bosch A, Bouillon R, Nilius B, Carmeliet G (2008) TRPV4-mediated calcium influx regulates terminal differentiation of osteoclasts. Cell Metab 8(3):257–265
165. Masuyama R, Mizuno A, Komori H, Kajiya H, Uekawa A, Kitaura H, Okabe K, Ohyama K, Komori T (2012) Calcium/calmodulin-signaling supports TRPV4 activation in osteoclasts and regulates bone mass. J Bone Miner Res 27(8):1708–1721
166. Strotmann R, Harteneck C, Nunnenmacher K, Schultz G, Plant TD (2000) OTRPC4, a nonselective cation channel that confers sensitivity to extracellular osmolarity. Nat Cell Biol 2(10):695–702
167. Gradilone SA, Masyuk AI, Splinter PL, Banales JM, Huang BQ, Tietz PS, Masyuk TV, Larusso NF (2007) Cholangiocyte cilia express TRPV4 and detect changes in luminal tonicity inducing bicarbonate secretion. Proc Natl Acad Sci U S A 104(48): 19138–19143
168. Casas S, Novials A, Reimann F, Gomis R, Gribble FM (2008) Calcium elevation in mouse pancreatic beta cells evoked by extracellular human islet amyloid polypeptide involves activation of the mechanosensitive ion channel TRPV4. Diabetologia 51(12):2252–2262
169. Jung C, Fandos C, Lorenzo IM, Plata C, Fernandes J, Gene GG, Vazquez E, Valverde MA (2009) The progesterone receptor regulates the expression of TRPV4 channel. Pflugers Arch 459(1):105–113
170. Kumagami H, Terakado M, Sainoo Y, Baba A, Fujiyama D, Fukuda T, Takasaki K, Takahashi H (2009) Expression of the osmotically responsive cationic channel TRPV4 in the endolymphatic sac. Audiol Neurootol 14(3):190–197
171. Kim KS, Shin DH, Nam JH, Park KS, Zhang YH, Kim WK, Kim SJ (2010) Functional expression of TRPV4 cation channels in Human Mast Cell Line (HMC-1). Korean J Physiol Pharmacol 14(6): 419–425
172. Li J, Kanju P, Patterson M, Chew WL, Cho SH, Gilmour I, Oliver T, Yasuda R, Ghio A, Simon SA, Liedtke W (2011) TRPV4-mediated calcium influx into human bronchial epithelia upon exposure to diesel exhaust particles. Environ Health Perspect 119(6): 784–793
173. Lorenzo IM, Liedtke W, Sanderson MJ, Valverde MA (2008) TRPV4 channel participates in receptor-operated calcium entry and ciliary beat frequency regulation in mouse airway epithelial cells. Proc Natl Acad Sci U S A 105(34):12611–12616
174. van der Eerden BC, Hoenderop JG, de Vries TJ, Schoenmaker T, Buurman CJ, Uitterlinden AG, Pols HA, Bindels RJ, van Leeuwen JP (2005) The epithelial Ca2+ channel TRPV5 is essential for proper osteoclastic bone resorption. Proc Natl Acad Sci U S A 102(48):17507–17512
175. Hoenderop JG, van der Kemp AW, Hartog A, van de Graaf SF, van Os CH, Willems PH, Bindels RJ (1999) Molecular identification of the apical Ca2+ channel in 1, 25-dihydroxyvitamin D3-responsive epithelia. J Biol Chem 274(13):8375–8378
176. Vassilieva IO, Tomilin VN, Marakhova II, Shatrova AN, Negulyaev YA, Semenova SB (2013) Expression of transient receptor potential vanilloid channels TRPV5 and TRPV6 in human blood lymphocytes and Jurkat leukemia T cells. J Membr Biol 246(2): 131–140
177. Semenova SB, Vassilieva IO, Fomina AF, Runov AL, Negulyaev YA (2009) Endogenous expression of TRPV5 and TRPV6 calcium channels in human leukemia K562 cells. Am J Physiol Cell Physiol 296(5):C1098–C1104
178. van der Eerden BC, Weissgerber P, Fratzl-Zelman N, Olausson J, Hoenderop JG, Schreuders-Koedam M, Eijken M, Roschger P, de Vries TJ, Chiba H, Klaushofer K, Flockerzi V, Bindels RJ, Freichel M, van Leeuwen JP (2012) The transient receptor potential channel TRPV6 is dynamically expressed in bone cells but is not crucial for bone mineralization in mice. J Cell Physiol 227(5):1951–1959
179. Peng JB, Chen XZ, Berger UV, Weremowicz S, Morton CC, Vassilev PM, Brown EM, Hediger MA (2000) Human calcium transport protein CaT1. Biochem Biophys Res Commun 278(2):326–332
180. Zhuang L, Peng JB, Tou L, Takanaga H, Adam RM, Hediger MA, Freeman MR (2002) Calcium-selective ion channel, CaT1, is apically localized in gastrointestinal tract epithelia and is aberrantly expressed in human malignancies. Lab Invest 82(12):1755–1764
181. Kim YS, Son JY, Kim TH, Paik SK, Dai Y, Noguchi K, Ahn DK, Bae YC (2010) Expression of transient receptor potential ankyrin 1 (TRPA1) in the rat trigeminal sensory afferents and spinal dorsal horn. J Comp Neurol 518(5):687–698
182. Lee SM, Cho YS, Kim TH, Jin MU, Ahn DK, Noguchi K, Bae YC (2012) An ultrastructural evidence for the expression of transient receptor potential ankyrin 1 (TRPA1) in astrocytes in the rat trigeminal caudal nucleus. J Chem Neuroanat 45(1-2):45–49
183. Story GM, Peier AM, Reeve AJ, Eid SR, Mosbacher J, Hricik TR, Earley TJ, Hergarden AC, Andersson DA, Hwang SW, McIntyre P, Jegla T, Bevan S, Patapoutian A (2003) ANKTM1, a TRP-like channel expressed in nociceptive neurons, is activated by cold temperatures. Cell 112(6): 819–829
184. Nassenstein C, Kwong K, Taylor-Clark T, Kollarik M, Macglashan DM, Braun A, Undem BJ

(2008) Expression and function of the ion channel TRPA1 in vagal afferent nerves innervating mouse lungs. J Physiol 586(6):1595–1604
185. Atoyan R, Shander D, Botchkareva NV (2009) Non-neuronal expression of transient receptor potential type A1 (TRPA1) in human skin. J Invest Dermatol 129(9):2312–2315
186. Cao DS, Zhong L, Hsieh TH, Abooj M, Bishnoi M, Hughes L, Premkumar LS (2012) Expression of transient receptor potential ankyrin 1 (TRPA1) and its role in insulin release from rat pancreatic beta cells. PLoS One 7(5):e38005
187. Uckert S, Sonnenberg JE, Albrecht K, Kuczyk MA, Hedlund P (2015) Expression and distribution of the transient receptor potential cationic channel ankyrin 1 (TRPA1) in the human vagina. Int J Impot Res 27(1):16–19
188. Alvarez-Berdugo D, Rofes L, Farre R, Casamitjana JF, Enrique A, Chamizo J, Padron A, Navarro X, Clave P (2016) Localization and expression of TRPV1 and TRPA1 in the human oropharynx and larynx. Neurogastroenterol Motil 28(1):91–100
189. Uckert S, Albrecht K, Bannowsky A, Sohn M, Kuczyk MA, Hedlund P (2017) Expression and distribution of the transient receptor potential cationic channel A1 (TRPA1) in the human clitoris-comparison to male penile erectile tissue. Int J Impot Res 29(5):179–183
190. Zhao H, Sprunger LK, Simasko SM (2010) Expression of transient receptor potential channels and two-pore potassium channels in subtypes of vagal afferent neurons in rat. Am J Physiol Gastrointest Liver Physiol 298(2):G212–G221
191. Yu S, Gao G, Peterson BZ, Ouyang A (2009) TRPA1 in mast cell activation-induced long-lasting mechanical hypersensitivity of vagal afferent C-fibers in guinea pig esophagus. Am J Physiol Gastrointest Liver Physiol 297(1):G34–G42
192. Brierley SM, Hughes PA, Page AJ, Kwan KY, Martin CM, O'Donnell TA, Cooper NJ, Harrington AM, Adam B, Liebregts T, Holtmann G, Corey DP, Rychkov GY, Blackshaw LA (2009) The ion channel TRPA1 is required for normal mechanosensation and is modulated by algesic stimuli. Gastroenterology 137(6):2084–2095. e3
193. Saha S, Sucharita S, Majhi RK, Tiwari A, Ghosh A, Pradhan SK, Patra BK, Dash RR, Nayak RN, Giri SC, Routray P, Kumar A, Kumar GP, Goswami C (2019) TRPA1 is selected as a semi-conserved channel during vertebrate evolution due to its involvement in spermatogenesis. Biochem Biophys Res Commun
194. Cuajungco MP, Samie MA (2008) The varitint-waddler mouse phenotypes and the TRPML3 ion channel mutation: cause and consequence. Pflugers Arch 457(2):463–473
195. Falardeau JL, Kennedy JC, Acierno JS Jr, Sun M, Stahl S, Goldin E, Slaugenhaupt SA (2002) Cloning and characterization of the mouse Mcoln1 gene reveals an alternatively spliced transcript not seen in humans. BMC Genomics 3:3
196. Samie MA, Grimm C, Evans JA, Curcio-Morelli C, Heller S, Slaugenhaupt SA, Cuajungco MP (2009) The tissue-specific expression of TRPML2 (MCOLN-2) gene is influenced by the presence of TRPML1. Pflugers Arch 459(1):79–91
197. Lindvall JM, Blomberg KE, Wennborg A, Smith CI (2005) Differential expression and molecular characterisation of Lmo7, Myo1e, Sash1, and Mcoln2 genes in Btk-defective B-cells. Cell Immunol 235(1):46–55
198. Mochizuki T, Wu G, Hayashi T, Xenophontos SL, Veldhuisen B, Saris JJ, Reynolds DM, Cai Y, Gabow PA, Pierides A, Kimberling WJ, Breuning MH, Deltas CC, Peters DJ, Somlo S (1996) PKD2, a gene for polycystic kidney disease that encodes an integral membrane protein. Science 272(5266): 1339–1342
199. Veldhuisen B, Spruit L, Dauwerse HG, Breuning MH, Peters DJ (1999) Genes homologous to the autosomal dominant polycystic kidney disease genes (PKD1 and PKD2). Eur J Hum Genet 7(8): 860–872
200. Nomura H, Turco AE, Pei Y, Kalaydjieva L, Schiavello T, Weremowicz S, Ji W, Morton CC, Meisler M, Reeders ST, Zhou J (1998) Identification of PKDL, a novel polycystic kidney disease 2-like gene whose murine homologue is deleted in mice with kidney and retinal defects. J Biol Chem 273(40):25967–25973
201. Murakami M, Ohba T, Xu F, Shida S, Satoh E, Ono K, Miyoshi I, Watanabe H, Ito H, Iijima T (2005) Genomic organization and functional analysis of murine PKD2L1. J Biol Chem 280(7):5626–5635
202. Bodding M (2007) TRP proteins and cancer. Cell Signal 19(3):617–624
203. Prevarskaya N, Zhang L, Barritt G (2007) TRP channels in cancer. BBA-Mol Basis Dis 1772(8): 937–946
204. Lehen'kyi V, Prevarskaya N (2011) Oncogenic TRP channels. Transient Recept Potential Channels 704: 929–945
205. Gautier M, Dhennin-Duthille I, Ay AS, Rybarczyk P, Korichneva I, Ouadid-Ahidouch H (2014) New insights into pharmacological tools to TR(i)P cancer up. Br J Pharmacol 171(10):2582–2592
206. Arcangeli A, Becchetti A (2015) Novel perspectives in cancer therapy: targeting ion channels. Drug Resist Updat 21-22:11–19
207. Shapovalov G, Ritaine A, Skryma R, Prevarskaya N (2016) Role of TRP ion channels in cancer and tumorigenesis. Semin Immunopathol 38(3):357–369
208. Takahashi N, Chen HY, Harris IS, Stover DG, Selfors LM, Bronson RT, Deraedt T, Cichowski K, Welm AL, Mori Y, Mills GB, Brugge JS (2018) Cancer cells co-opt the neuronal redox-sensing channel TRPA1 to promote oxidative-stress tolerance. Cancer Cell 33(6):985–1003. e7

209. Wu YT, Yen SL, Li CF, Chan TC, Chen TJ, Lee SW, He HL, Chang IW, Hsing CH, Shiue YL (2016) Overexpression of transient receptor protein cation channel Subfamily A Member 1, confers an independent prognostic indicator in nasopharyngeal carcinoma. J Cancer 7(10):1181–1188
210. Takahashi K, Umebayashi C, Numata T, Honda A, Ichikawa J, Hu Y, Yamaura K, Inoue R (2018) TRPM7-mediated spontaneous Ca(2+) entry regulates the proliferation and differentiation of human leukemia cell line K562. Physiol Rep 6(14): e13796
211. Fang Y, Liu G, Xie C, Qian K, Lei X, Liu Q, Liu G, Cao Z, Fu J, Du H, Liu S, Huang S, Hu J, Xu X (2018) Pharmacological inhibition of TRPV4 channel suppresses malignant biological behavior of hepatocellular carcinoma via modulation of ERK signaling pathway. Biomed Pharmacother 101:910–919
212. Li X, Zhang Q, Fan K, Li B, Li H, Qi H, Guo J, Cao Y, Sun H (2016) Overexpression of TRPV3 correlates with tumor progression in non-small cell lung cancer. Int J Mol Sci 17(4):437
213. Ceppa E, Cattaruzza F, Lyo V, Amadesi S, Pelayo JC, Poole DP, Vaksman N, Liedtke W, Cohen DM, Grady EF, Bunnett NW, Kirkwood KS (2010) Transient receptor potential ion channels V4 and A1 contribute to pancreatitis pain in mice. Am J Physiol Gastrointest Liver Physiol 299(3):G556–G571
214. Yildirim E, Carey MA, Card JW, Dietrich A, Flake GP, Zhang Y, Bradbury JA, Rebolloso Y, Germolec DR, Morgan DL, Zeldin DC, Birnbaumer L (2012) Severely blunted allergen-induced pulmonary Th2 cell response and lung hyperresponsiveness in type 1 transient receptor potential channel-deficient mice. Am J Physiol Lung Cell Mol Physiol 303(6):L539–L549
215. Bertin S, Aoki-Nonaka Y, Lee J, de Jong PR, Kim P, Han T, Yu T, To K, Takahashi N, Boland BS, Chang JT, Ho SB, Herdman S, Corr M, Franco A, Sharma S, Dong H, Akopian AN, Raz E (2017) The TRPA1 ion channel is expressed in CD4+ T cells and restrains T-cell-mediated colitis through inhibition of TRPV1. Gut 66(9):1584–1596
216. Bishnoi M, Khare P, Brown L, Panchal SK (2018) Transient receptor potential (TRP) channels: a metabolic TR(i)P to obesity prevention and therapy. Obes Rev 19(9):1269–1292
217. Kichko TI, Neuhuber W, Kobal G, Reeh PW (2018) The roles of TRPV1, TRPA1 and TRPM8 channels in chemical and thermal sensitivity of the mouse oral mucosa. Eur J Neurosci 47(3):201–210
218. Inoue R, Kurahara LH, Hiraishi K (2018) TRP channels in cardiac and intestinal fibrosis. Semin Cell Dev Biol
219. Lowin T, Bleck J, Schneider M, Pongratz G (2018) Selective killing of proinflammatory synovial fibroblasts via activation of transient receptor potential ankyrin (TRPA1). Biochem Pharmacol 154:293–302
220. Kun J, Perkecz A, Knie L, Setalo G Jr, Tornoczki T, Pinter E, Ban A (2017) TRPA1 receptor is upregulated in human oral lichen planus. Oral Dis 23(2):189–198
221. Nozadze I, Tsiklauri N, Gurtskaia G, Tsagareli M (2018) The role of transient receptor potential (Trpa1) channel in pruritus. Georgian Med News (280–281):134–137
222. Bonchak JG, Swerlick RA (2018) Emerging therapies for atopic dermatitis: TRPV1 antagonists. J Am Acad Dermatol 78(3 Suppl 1):S63–S66
223. Belrose JC, Jackson MF (2018) TRPM2: a candidate therapeutic target for treating neurological diseases. Acta Pharmacol Sin 39(5):722–732
224. Nilius B, Owsianik G, Voets T (2008) Transient receptor potential channels meet phosphoinositides. EMBO J 27(21):2809–2816
225. Saul S, Stanisz H, Backes CS, Schwarz EC, Hoth M (2014) How ORAI and TRP channels interfere with each other: interaction models and examples from the immune system and the skin. Eur J Pharmacol 739: 49–59
226. Shin YC, Shin SY, Chun JN, Cho HS, Lim JM, Kim HG, So I, Kwon D, Jeon JH (2012) TRIP database 2.0: a manually curated information hub for accessing TRP channel interaction network. PLoS One 7(10): e47165
227. Zhu MX (2005) Multiple roles of calmodulin and other Ca(2+)-binding proteins in the functional regulation of TRP channels. Pflugers Arch 451(1): 105–115
228. Lepage PK, Boulay G (2007) Molecular determinants of TRP channel assembly. Biochem Soc Trans 35 (Pt 1):81–83
229. Cheng W, Sun C, Zheng J (2010) Heteromerization of TRP channel subunits: extending functional diversity. Protein Cell 1(9):802–810
230. Woo JS, Lee KJ, Huang M, Cho CH, Lee EH (2014) Heteromeric TRPC3 with TRPC1 formed via its ankyrin repeats regulates the resting cytosolic Ca2+ levels in skeletal muscle. Biochem Biophys Res Commun 446(2):454–459
231. Ma X, Cheng KT, Wong CO, O'Neil RG, Birnbaumer L, Ambudkar IS, Yao X (2011) Heteromeric TRPV4-C1 channels contribute to store-operated Ca(2+) entry in vascular endothelial cells. Cell Calcium 50(6):502–509
232. Yu Y, Ulbrich MH, Li MH, Dobbins S, Zhang WK, Tong L, Isacoff EY, Yang J (2012) Molecular mechanism of the assembly of an acid-sensing receptor ion channel complex. Nat Commun 3:1252
233. Salehi-Najafabadi Z, Li B, Valentino V, Ng C, Martin H, Yu Y, Wang Z, Kashyap P, Yu Y (2017) Extracellular loops are essential for the assembly and function of polycystin receptor-ion channel complexes. J Biol Chem 292(10):4210–4221

234. Riccio A, Li Y, Tsvetkov E, Gapon S, Yao GL, Smith KS, Engin E, Rudolph U, Bolshakov VY, Clapham DE (2014) Decreased anxiety-like behavior and Galphaq/11-dependent responses in the amygdala of mice lacking TRPC4 channels. J Neurosci 34(10): 3653–3667
235. Kim J, Kwak M, Jeon JP, Myeong J, Wie J, Hong C, Kim SY, Jeon JH, Kim HJ, So I (2014) Isoform- and receptor-specific channel property of canonical transient receptor potential (TRPC)1/4 channels. Pflugers Arch 466(3):491–504
236. Poteser M, Graziani A, Rosker C, Eder P, Derler I, Kahr H, Zhu MX, Romanin C, Groschner K (2006) TRPC3 and TRPC4 associate to form a redox-sensitive cation channel. Evidence for expression of native TRPC3-TRPC4 heteromeric channels in endothelial cells. J Biol Chem 281(19):13588–13595
237. Zagranichnaya TK, Wu X, Villereal ML (2005) Endogenous TRPC1, TRPC3, and TRPC7 proteins combine to form native store-operated channels in HEK-293 cells. J Biol Chem 280(33):29559–29569
238. Cheng W, Yang F, Takanishi CL, Zheng J (2007) Thermosensitive TRPV channel subunits coassemble into heteromeric channels with intermediate conductance and gating properties. J Gen Physiol 129(3): 191–207
239. Smith GD, Gunthorpe MJ, Kelsell RE, Hayes PD, Reilly P, Facer P, Wright JE, Jerman JC, Walhin JP, Ooi L, Egerton J, Charles KJ, Smart D, Randall AD, Anand P, Davis JB (2002) TRPV3 is a temperature-sensitive vanilloid receptor-like protein. Nature 418(6894):186–190
240. Rutter AR, Ma QP, Leveridge M, Bonnert TP (2005) Heteromerization and colocalization of TrpV1 and TrpV2 in mammalian cell lines and rat dorsal root ganglia. Neuroreport 16(16):1735–1739
241. Hellwig N, Albrecht N, Harteneck C, Schultz G, Schaefer M (2005) Homo- and heteromeric assembly of TRPV channel subunits. J Cell Sci 118(Pt 5): 917–928
242. Kim S, Barry DM, Liu XY, Yin S, Munanairi A, Meng QT, Cheng W, Mo P, Wan L, Liu SB, Ratnayake K, Zhao ZQ, Gautam N, Zheng J, Karunarathne WK, Chen ZF (2016) Facilitation of TRPV4 by TRPV1 is required for itch transmission in some sensory neuron populations. Sci Signal 9(437): ra71
243. Jiang LH (2007) Subunit interaction in channel assembly and functional regulation of transient receptor potential melastatin (TRPM) channels. Biochem Soc Trans 35(Pt 1):86–88
244. Zhang Z, Yu H, Huang J, Faouzi M, Schmitz C, Penner R, Fleig A (2014) The TRPM6 kinase domain determines the Mg.ATP sensitivity of TRPM7/M6 heteromeric ion channels. J Biol Chem 289(8): 5217–5227
245. Curcio-Morelli, C., P. Zhang, B. Venugopal, F.A. Charles, M.F. Browning, H.F. Cantiello, and S.A. Slaugenhaupt, Functional multimerization of mucolipin channel proteins. J Cell Physiol, 2010. 222(2): 328-335.
246. Venkatachalam K, Hofmann T, Montell C (2006) Lysosomal localization of TRPML3 depends on TRPML2 and the mucolipidosis-associated protein TRPML1. J Biol Chem 281(25):17517–17527
247. Kobori T, Smith GD, Sandford R, Edwardson JM (2009) The transient receptor potential channels TRPP2 and TRPC1 form a heterotetramer with a 2: 2 stoichiometry and an alternating subunit arrangement. J Biol Chem 284(51):35507–35513
248. Du J, Ding M, Sours-Brothers S, Graham S, Ma R (2008) Mediation of angiotensin II-induced Ca2+ signaling by polycystin 2 in glomerular mesangial cells. Am J Physiol Renal Physiol 294(4):F909–F918
249. Miyagi K, Kiyonaka S, Yamada K, Miki T, Mori E, Kato K, Numata T, Sawaguchi Y, Numaga T, Kimura T, Kanai Y, Kawano M, Wakamori M, Nomura H, Koni I, Yamagishi M, Mori Y (2009) A pathogenic C terminus-truncated polycystin-2 mutant enhances receptor-activated Ca2+ entry via association with TRPC3 and TRPC7. J Biol Chem 284(49): 34400–34412
250. Stewart AP, Smith GD, Sandford RN, Edwardson JM (2010) Atomic force microscopy reveals the alternating subunit arrangement of the TRPP2-TRPV4 heterotetramer. Biophys J 99(3):790–797
251. Schindl R, Fritsch R, Jardin I, Frischauf I, Kahr H, Muik M, Riedl MC, Groschner K, Romanin C (2012) Canonical transient receptor potential (TRPC) 1 acts as a negative regulator for vanilloid TRPV6-mediated Ca2+ influx. J Biol Chem 287(42):35612–35620
252. Guo Z, Grimm C, Becker L, Ricci AJ, Heller S (2013) A novel ion channel formed by interaction of TRPML3 with TRPV5. PLoS One 8(2):e58174
253. Goldenberg NM, Wang L, Ranke H, Liedtke W, Tabuchi A, Kuebler WM (2015) TRPV4 is required for hypoxic pulmonary vasoconstriction. Anesthesiology 122(6):1338–1348
254. Kleene SJ, Siroky BJ, Landero-Figueroa JA, Dixon BP, Pachciarz NW, Lu L, Kleene NK (2019) The TRPP2-dependent channel of renal primary cilia also requires TRPM3. PLoS One 14(3):e0214053
255. Alessandri-Haber N, Dina OA, Chen X, Levine JD (2009) TRPC1 and TRPC6 channels cooperate with TRPV4 to mediate mechanical hyperalgesia and nociceptor sensitization. J Neurosci 29(19): 6217–6228
256. Du J, Ma X, Shen B, Huang Y, Birnbaumer L, Yao X (2014) TRPV4, TRPC1, and TRPP2 assemble to form a flow-sensitive heteromeric channel. FASEB J 28(11):4677–4685
257. Vandewauw I, De Clercq K, Mulier M, Held K, Pinto S, Van Ranst N, Segal A, Voet T, Vennekens R, Zimmermann K, Vriens J, Voets T (2018) A TRP channel trio mediates acute noxious heat sensing. Nature 555(7698):662–666
258. Qian F, Germino FJ, Cai Y, Zhang X, Somlo S, Germino GG (1997) PKD1 interacts with PKD2

through a probable coiled-coil domain. Nat Genet 16(2):179–183
259. Field S, Riley KL, Grimes DT, Hilton H, Simon M, Powles-Glover N, Siggers P, Bogani D, Greenfield A, Norris DP (2011) Pkd1l1 establishes left-right asymmetry and physically interacts with Pkd2. Development 138(6):1131–1142
260. Sutton KA, Jungnickel MK, Ward CJ, Harris PC, Florman HM (2006) Functional characterization of PKDREJ, a male germ cell-restricted polycystin. J Cell Physiol 209(2):493–500
261. Delling M, DeCaen PG, Doerner JF, Febvay S, Clapham DE (2013) Primary cilia are specialized calcium signalling organelles. Nature 504(7479): 311–314
262. Salas MM, Hargreaves KM, Akopian AN (2009) TRPA1-mediated responses in trigeminal sensory neurons: interaction between TRPA1 and TRPV1. Eur J Neurosci 29(8):1568–1578
263. Andrei SR, Sinharoy P, Bratz IN, Damron DS (2016) TRPA1 is functionally co-expressed with TRPV1 in cardiac muscle: co-localization at z-discs, costameres and intercalated discs. Channels (Austin) 10(5): 395–409
264. Voets T, Talavera K, Owsianik G, Nilius B (2005) Sensing with TRP channels. Nat Chem Biol 1(2): 85–92

Regulation of Ion Channel Function by Gas Molecules

8

Nikhil Shah and Lei Zhou

Abstract

Our understanding of the gaseous signaling molecules that play important roles in diverse physiological processes keeps expanding. These gas molecules, also called gasotransmitters, include NO, H_2S, $^1O_2$, CO, and CO_2 and are generated within the cell through enzymatic pathways and photochemical reactions. These molecules are chemically unstable and directly react with amino acids such as cysteine, histidine, and so on. Compared to well-characterized reactive oxygen species (ROS), including H_2O_2, ONOO−, O_2^-, and OH·, the gasotransmitters are in general less polar and show higher solubility in hydrophobic environments like the lipid membrane. Correspondingly, accumulating evidence has begun to unveil the broad impacts of these gaseous molecules on the function of membrane proteins, including ion channels. This review summarizes the major physicochemical characteristics of representative gasotransmitters and their regulation of ion channel functions.

Keywords

Ion channel · Gasotransmitters · Reactive oxygen species · Nitric oxide · Hydrogen sulfide · Singlet oxygen

8.1 Ion Channels in General

Ion channels are membrane proteins that, upon opening, produce ion-conducting pores through the lipid membrane. The opening and closing of ion channels are under the control of factors including membrane potential, ligands, mechanical force, temperature, and so on. The movement of ions across the membrane through the pore is driven by an electrochemical gradient. Ion channels function as the molecular basis for most electrical activities in excitable cells such as neurons in the brain and muscle cells in the heart. In non-excitable cells, ion channels are also broadly expressed and fulfill important physiological functions such as regulating Ca^{2+} homeostasis in immune and cancer cells.

The life journey of an ion channel protein starts from protein translation from mRNA, followed by protein folding and assembly in the endoplasmic reticulum, transportation through secretory pathways, and final targeting to specific subcellular regions. During these stages, ion channel proteins can be modified by different chemical mechanisms that directly impact their biogenesis and function. Most posttranslational

N. Shah
Shenzhen Bay Laboratory, Institute of Molecular Physiology, Shenzhen, Guangdong Province, China
e-mail: shahnn4@mymail.vcu.edu

L. Zhou (✉)
Virginia Commonwealth University School of Medicine, Richmond, VA, USA
e-mail: zhoulei@szbl.ac.cn

L. Zhou (ed.), *Ion Channels in Biophysics and Physiology*, Advances in Experimental Medicine and Biology 1349, https://doi.org/10.1007/978-981-16-4254-8_8

modifications (PDM) are mediated by specific protein enzymes that target certain types of amino acids. Other than well-studied protein phosphorylation, ion channel proteins can be modified through the processes of acetylation, glycosylation, ubiquitylation, palmitoylation, and so on.

Under physiological conditions, ion channels can also be modified by chemically reactive compounds, such as reactive oxygen species (ROS). In contrast to protein enzyme-mediated PDM, the modifications made by small reactive compounds are in general less specific and quite often target certain types of amino acids without any reference to the primary sequence or 3D structure. This review focuses on the modification of ion channels by reactive gas molecules, including NO, H_2S, $^1O_2$, and CO, which are also called gasotransmitters, a term that was first introduced in 2002 [1].

8.2 Chemical Physics of Gasotransmitters in General

For gas signaling molecules to reach ion channels, especially the transmembrane domain that anchors in the lipid membrane, gas molecules need to diffuse from the production site, traverse the cytoplasm, and then embed in the membrane. Although the presence of gaseous molecules is likely transient and the reported physiological concentrations vary dramatically, permeation of gas molecules can be quantified by the following equation:

$$P_{\mathrm{m}} = \frac{D_{\mathrm{m}} \cdot K_{\mathrm{m}}}{d} = \frac{D_{\mathrm{m}} \cdot \frac{S_{\mathrm{m}}}{S_{\mathrm{w}}}}{d} = \frac{D_{\mathrm{m}} \cdot e^{(-\Delta G_{\mathrm{w}\to m}/RT)}}{d},$$

where P_{m} is the permeability coefficient (m/s), D_{m} is the diffusion coefficient in the membrane (m^2/s), K_{m} is the partition coefficient and equals to the ratio of its solubility in the membrane (S_{m}) to its solubility in the aqueous phase (S_{w}), and d is the thickness of the membrane (m). This relationship was first established by Meyer and Overton more than 100 years ago and has since been validated for thousands of compounds including most gas molecules [2]. Protons, CO_2, and a few compounds relying on facilitated transport are among a limited number of known exceptions for the Meyer-Overton rule.

In general, it had been believed that diffusion across cell membranes composes the rate-limiting step for gas molecules. However, recent theoretical and experimental studies suggest that diffusion within the cytoplasm, affected by both trajectory and tortuosity, may present significant challenges to gas molecules, including O_2 [3]. Moreover, the diffusion coefficients of small gas molecules are known to be comparable, including the same molecule under different physical states [4, 5]. For example, the D_{m} value for O_2 was determined to be 1.6×10^{-5} cm^2 s^{-1}, a value that is very close to that in water (2×10^{-5} cm^2 s^{-1}) and regardless of the membrane composition and physical states [5]. Therefore, a commonly recognized notion is that the lipid membrane does not act as a diffusion barrier for most gasotransmitters.

On the other hand, the partition coefficient seems to play a major role in affecting the permeability of gas molecules. Indeed, the correlation between permeability and solubility for gas molecules to pass through cell membranes has been well established for more than 100 years [2]. K_{m} equals the ratio of solubilities of the membrane phase to the water phase and is directly correlated with the free energy difference of transferring between two phases ($G_{\mathrm{w}\to\mathrm{m}}$). Most gasotransmitters are non-polar, such as $^1O_2$, or weakly polar, such as H_2S. Thus, they have weak interactions with water molecules and show increased solubility in the lipid phase. The P_{m} of O_2 was determined to be ~130 cm/s, about twice the permeability coefficient of passing a 4 nm slab of water [4].

Compared to other gas signaling molecules, H_2S shows some unique physiochemical features. H_2S is classified as a nonpolar molecule because of the small difference in electronegativity between hydrogen (electronegativity number or EN = 2.20) and sulfur (EN = 2.55), in contrast to that of the water molecule (O_{EN} = 3.44). The solubility of H_2S in water is about 80 mM at 37 °C [6]. H_2S is a weak acid with the value of pKa_1 ($H_2S \rightarrow HS^- + H^+$) about 6.76, which indicates that only about a quarter of the total ensemble exists as H_2S with the majority stays in the form of hydrosulfide ion (HS^-). The gas form of H_2S is lipophilic, with its solubility in lipid membrane about twice the value of that in water, and can easily diffuse across the cell membrane without any aid of a facilitator ($P_M > 0.5$ cm/s) [7]. However, HS^- cannot freely traverse the cell membrane, and the understanding of its physiological role has been limited. An HS^--permeable ion channel was identified from the formate/nitrite transport family in the bacterium *Clostridium difficile* [8]. The eukaryotic homolog of this bacteria HS^- channel remains to be identified.

8.3 Ion Channel Modification Via NO-Mediated S-Nitrosylation

Nitric oxide (NO) was discovered to be the molecule responsible for endothelial smooth muscle relaxation over three decades ago [9]. It was initially thought that, like other similar gaseous molecules, NO was a by-product of cellular metabolism. This idea was overturned upon the discovery of specific nitric oxide synthases (NOS) that generate NO by metabolizing L-arginine. Three types of NOS catalyze the chemical production of NO: neuronal NOS (nNOS, encoded by *NOS1*, activated by Ca^{2+}/CaM), inducible NOS (iNOS, encoded by *NOS2,* activated by inflammatory stimulation; Ca^{2+} independent), and endothelial NOS (eNOS, encoded by *NOS3*, activated by Ca^{2+}/CaM). NO synthesis via NOS and its subsequent activity on downstream cellular targets is often described in the literature as canonical or noncanonical.

The canonical pathway by which NO acts is through binding soluble guanylate cyclase (sGC), which converts GTP to the second messenger cGMP. The production and amplification of cGMP results in the activation of protein kinases and thus desired cellular responses based on cell types. In the noncanonical pathways, NO can act as a signaling factor via posttranslational modification of target proteins. S-nitrosylation has emerged as a critical posttranslational modification mechanism that is ubiquitously expressed and therefore involved in many physiological systems and processes [10]. S-nitrosylation is described as the addition of NO to thiol moieties of cysteine residues within a protein, which forms S-nitrosothiols (SNO) (Fig. 8.1) [11]. Importantly, not all cysteines in a target protein undergo S-nitrosylation. Rather, S-nitrosylation favors cysteine residues confined within an SNO-motif, which is a tertiary or quaternary consensus structure of amino acids punctuated by acidic and basic residues [10, 12, 13].

Over the last two decades, mounting evidence has revealed that posttranslational modification of proteins by NO-mediated S-nitrosylation has regulatory mechanisms on a protein's conformational state and its interaction with other proteins. Because NO is known to have pro-oxidant and antioxidant properties, the S-nitrosylation of proteins is also dependent on the redox status of cellular processes and physiological states. Indeed, dysregulation of S-nitrosylation is associated with a broad array of disease states related to ischemia–reperfusion, neurogenesis, and synaptic transmission [14]. The following section will briefly introduce various ion channels, describe their structural and functional modulation by NO-mediated S-nitrosylation, and note any physiological and pathophysiological implications. The reader is encouraged to reference Table 8.1 as needed.

8.3.1 NMDA Receptors

The N-methyl-D-aspartate receptors (NMDARs) are neurotransmitter receptors expressed in nervous tissue and are embedded within the postsynaptic membranes of neurons. NMDARs are involved in signal transduction through long-term potentiation of action potentials as they receive signals across synapses from previous neurons. These channels are known to have important roles in neurotransmission, neuroplasticity, and memory. Upon binding of glutamate and glycine to NMDAR, the channel activates and permits the flow of non-selective cations, such as Na^+ and Ca^{2+}, resulting in depolarization of neuronal membranes. When Ca^{2+} traverses through NMDAR channel pores, it

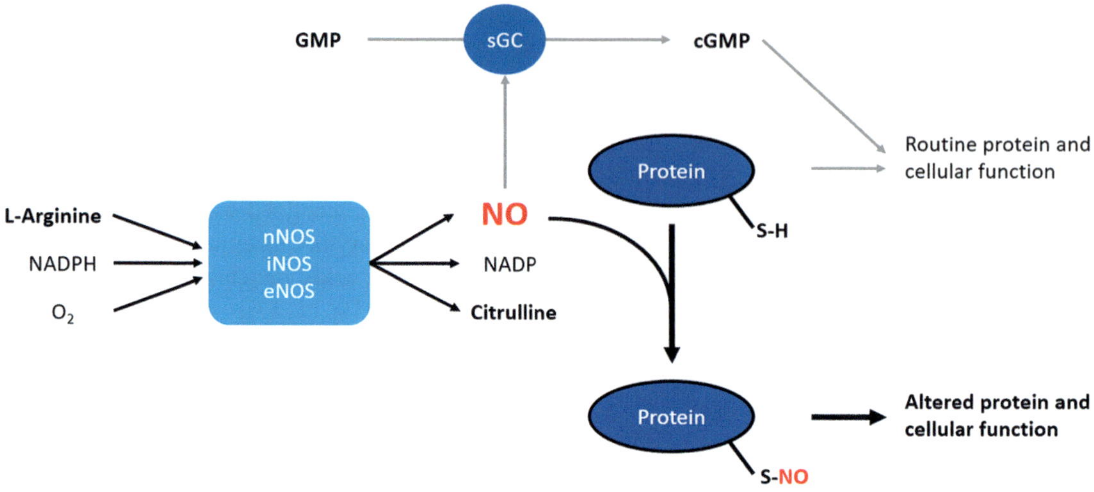

Fig. 8.1 Nitric Oxide synthesis and S-nitrosylation of proteins: Nitric oxide (NO) is synthesized in different cell types via three major isoforms of nitric oxide synthase: neuronal NOS (nNOS; encoded by NOS1), inducible NOS (iNOS; encoded by NOS2), and endothelial NOS (eNOS; encoded by NOS3). Conversion of L-arginine into citrulline by NOS using NADPH and oxygen as cofactors results in the production of NO. Aside from the canonical activation of soluble guanylyl cyclase, NO can also directly alter protein function via non-enzymatic reactions with thiol groups (−SH) on cysteine residues, a process referred to as S-nitrosylation. The newly formed covalent bond of NO to a thiol group forms an S-nitrosothiol (SNO) protein. The resulting SNO protein alters the normal function of the protein and, in some cases, modulates cellular function (e.g., inhibition or activation of different ion channels)

directly stimulates the activity of nNOS [28], which itself is directly tethered to NMDAR via the scaffolding protein PSD95 on the postsynaptic side of neurons [29]. This physical juxtaposition of NMDAR and nNOS ensures a higher likelihood of NMDAR modification by S-nitrosylation [15]. Through the application of an endogenous NO donor, it has been demonstrated that NMDAR function is inhibited via S-nitrosylation [30–32]. These observations were supported with site-directed mutagenesis of C399 in the NR2A subunit of NMDAR, which abolished the downregulation of NMDAR by S-nitrosylation [33, 34].

The mammalian brain typically exists in a state of relative hypoxia, with the concentration of O_2 in the range of 10–20 mmHg. This effectively produces a mild reducing environment that favors free thiol rather than the formation of disulfide bonds [16]. As such, pathological states of hypoxia such as those seen in stroke patients favor the formation of free thiol forms, thus allowing more cysteine residues to undergo S-nitrosylation. In fact, hypoxia-induced downregulation of NMDAR activity through S-nitrosylation of cysteine residues in the ligand-binding domain of NR1 (C744, C798) and NR2 subunits (C87, C320, C399) has been demonstrated, suggesting that ligand-binding domain restriction prevents excessive Ca^{2+} influx and is therefore neuroprotective [16]. Thus, S-nitrosylation of NMDARs may serve as a basis for therapeutic development for treatments in patients with neurodegenerative diseases [35].

8.3.2 Voltage-Gated Na^+ Channels

Sodium channels are integral membrane ion channels expressed in excitable cells found in muscle and nervous tissue. The conduction of sodium through the cellular membranes is necessary for the rising phase of an action potential and thus transmission of depolarization throughout individual cells and cellular conduction systems. Throughout the nervous system, Na^+ channel

Table 8.1 A summative table that highlights notable investigations where site-directed mutagenesis of critical cysteine residues resulted in alteration of channel activity that was different from the effect of NO-mediated S-nitrosylation

Study	Ion Channel	Preparation	Effect
Choi et al. [15]	NMDAR	Oocytes expressing mutant cRNA NR1/NR2A (C399A)	Abolishment of inhibition of NMDA-evoked currents
Takahashi et al. [16]	NMDAR	Oocytes expressing mutant cRNA of NR1/NR2A (C399A), or NR1(C744A, C798A)/NR2A	C399A mutant showed loss of NO-mediated inhibition under ambient and hypoxic conditions; C744A, C798A showed only loss of hypoxic mediated enhancement of NO inhibition
Chen et al. [17]	NaV1.5	HEK-293 cells expressing mutant $Na_V1.5$/ΔKPQ-$Na_V1.5$ (C489A, C1135A, together or separate)	C489A or C1135A displayed a 40% reduction in I_{NaL}; C489A/C1135A displayed a 75% reduction in I_{NaL}
Broillet [18]	CNG2α	HEK-293 cells expressing mutant CNG2α (C460S, C484S, C520S, and C552S)	C460S resulted in complete loss of sensitivity to NO; no alteration of NO sensitivity was seen with C484S, C520S, and C552S
Yoshida et al. [19]	TRPC5	HEK cells and chicken DT40 B lymphocytes expressing mutant TRPC5 (C553S and C558S	SNAP treatment resulted in increased Ca^{2+} influx by TRPC5. S-nitrosylation of TRPC5 was drastically suppressed by C553S and less extensively by C558S
Lee et al. [20]	TRPV4	HEK-293 cells expressing mutant TRPV4 (C853A)	Enhancement of channel activity, presumably due to loss of protein-protein interaction with calmodulin
Miyamoto et al. [21]	TRPA1	HEK-293T cells expressing mutant TRPA1 (C665S)	Reduced current conductance
Asada et al. [22]	KCNQ1	HEK-293 cells expressing mutant KNCQ1 (C445A, D446N, together or separate) in the presence or absence of calmodulin	Loss of current augmentation from C445A and D446N (together or separate) and in the absence of calmodulin
Gómez et al. [23]	Kir2.1	CHO cells expressing mutant Kir2.1 (C76L)	NO donation failed to increase inward and outward $I_{Kir2.1}$
Kawano et al. [24]	K_{ATP}	COS7 cells expressing mutant SUR1 subunits (C717S)	Decreased SNAP mediated current enhancement
Zhou et al. [25]	$Ca_V2.2$	HEK-293 cells expressing mutant $Ca_V2.2$: *III pore loop:* (C805A, C930A, C1045A); *C-terminal:* (C1835A, C2145A); *Cavβ3:* (C346A)	Loss of SNAP mediated inhibition: notably, C805A, C930A, C1835A, and C2145A diminished SNAP effect; C1045A and C346A reduced the inhibitory effect of SNAP by ~60%, and ~50% respectively
Lee et al. [26]	$Ca_V3.2$	HEK-293 cells expressing mutant $Ca_V3.2$ (C123A, C128A, C133A, and C939A)	Complete loss of GSNO mediated inhibition
Sun et al. [27]	RyR1	HEK-293 cells expressing mutant RyR1 (C3635A)	Loss of responsiveness to NO at low and high concentrations (see text for details)

Of note, each site-directed mutation (e.g., CXXX → CXXXA) was studied independently unless noted
Abbreviations: 2-ME (2-mercaptoethanol); CHO (Chinese hamster ovary); GSNO (S-nitrosoglutathione); HEK (human embryonic kidney cells); NO (nitric oxide); SNAP (S-Nitroso-N-Acetyl-D,L-Penicillamine)

activity can be modulated independent of sGC/cGMP pathways by NO-mediated S-nitrosylation. For example, potentiation of persistent Na^+ current ($I_{Na,P}$) and thus depolarization of the membrane potential has been shown by endogenous NO-mediated S-nitrosylation [36–39].

Paradoxically, Na^+ currents can also be inhibited by S-nitrosylation. For example, in small C-type dorsal root ganglion neurons known to be involved in nociception [40], three types of voltage-dependent sodium currents exist: rapidly inactivating tetrodotoxin-sensitive (TTX-S), slowly-inactivating tetrodotoxin-resistant (TTX-R), and persistent TTX-R [41, 42]. These Na^+ currents were all blocked by exogenous NO donors [43]. Similarly, in baroreceptors, TTX-S and TTX-R currents carried by sodium channels are inhibited by S-nitrosylation from either exogenous application of NO or endogenous NO activity [44, 45]. Although the mechanism of regulation that produces opposing effects observed in these studies is not entirely clear, it likely reflects the heterogeneous expression of different subtypes of sodium channels.

In cardiac myocytes, it was found that S-nitrosylation of cysteine residues C489 and C1135 in the sodium channel NaV1.5 (also known as SCN5A) increases the late sodium current ($I_{Na,L}$) [17]. Site-directed mutagenesis of these cysteine residues (C489A or C1135A) resulted in a 40% reduction in $I_{Na,L}$, whereas double mutagenesis resulted in a 75% reduction in $I_{Na,L}$. NaV1.5 is part of a macromolecular complex that also contains syntrophins and dystrophin [46]. The α1-Syntrophin (SNTA1) protein acts as an anchor for nNOS and links nNOS to its own inhibitor, the plasma membrane Ca-ATPase subtype 4b (PMCA4b) [47, 48]. Site-directed mutagenesis (A390V in SNTA1) and subsequent co-transfection of this macromolecular complex (SCN5A-SNTA1-nNOS) into HEK293 has provided insight into a mutation that causes long QT syndrome [49, 50]. This mutation disrupts the binding between nNOS and its inhibitor PMCA4b, thereby releasing the inhibition of nNOS. The resulting excessive S-nitrosylation of SCN5A increased late sodium currents. Indeed, prolongation of cardiac action potentials is an electrophysiological signature of Long QT syndrome [49]. Caveolin-3, a member of the caveolae family of membrane lipid rafts, was also found to associate with the SCN5A-SNTA1-nNOS macromolecular complex and inhibit nNOS activity [51]. Like mutant SNTA1, mutant Caveolin-3 (F97C) augmented NO production and thus increased late sodium currents through direct S-nitrosylation of SCN5A. Therefore, downregulation of nNOS activity in cardiac tissues may serve a cardioprotective role during pathophysiologic states of prolonged cardiac action potentials [49].

8.3.3 Acid Sensing Ion Channels (ASICs)

ASICs are voltage-insensitive sodium channels that are mostly expressed in the central and peripheral nervous systems. Their primary function is to provoke the desired response of neurons under conditions of increased extracellular protons [52–54]. ASICs are activated during conditions of low extracellular pH, and they are thought to invoke neuronal apoptosis and progression of acidic disease states such as acid-mediated nociception and brain ischemia-related pathology [55, 56]. ASICs have a cysteine-rich extracellular loop, containing 14 highly conserved cysteine residues [57, 58]. Therefore, it was postulated that ASICS may be targets for S-nitrosylation. It was demonstrated in DRG neurons and CHO cells containing recombinant ASIC channels that ASIC currents could be potentiated with the application of donor NO, independent of the cGMP/PKG pathway [59].

Other than sodium, ASICs are also permeable to Ca^{2+}, especially the ASICa subunit. The ASICa subunit, together with voltage-gated Ca^{2+} channels or NMDARs, facilitates a significant increase in intracellular calcium that results in neuronal cell death [60, 61]. Because ASIC1a channels uniquely function in response to tissue acidosis, it is thought that the direct modification and potentiation of ASICs by NO contributes to pathological states such as those seen in brain ischemia [55, 56]. Indeed, ASICs are the subject of pharmacotherapies related to these types of disease states [60, 61]

8.3.4 Cyclic-Nucleotide-Gated Ion Channels

Hyperpolarization-activated cyclic-nucleotide-gated (HCN) channels are a subgroup of cyclic nucleotide-regulated cation channels within the superfamily of voltage-gated K^+ channels [62–65]. The hyperpolarization-activated I_h (also known as I_f for funny current) carried by HCN channels has been characterized in both cardiac and neuronal tissues [66–69]. HCN channels have important physiological functions such as cardiac pacemaking, pain sensation, learning, and memory. The most significant physiological attributions of HCN channels are the generation and regulation of the heartbeat [70, 71] and control of intrinsic, rhythmic oscillations in neuronal circuitry [72–74].

HCN channels are also modulated by NO-mediated S-nitrosylation in addition to the canonical cGMP/PKG pathway [75, 76]. In magnocellular neurons of the supraoptic nucleus, S-nitrosylation of HCN channels caused a significant reduction in I_h currents. [76]. In the same study, the reducing agent ascorbate was used to remove endogenous NO linked to cysteine residues, which resulted in an increased magnitude of I_h currents. The physiological implications of this study suggest that NO directly modulates HCN channel activity in magnocellular neurons, whose electrical activity determines the rate of vasopressin and oxytocin secretion and thus serum osmolality [76, 77]. However, in hypoglossal motor neurons, application of an NO donor (DEA) increased I_h currents as well as the voltage-insensitive HCN component [76]. These findings illustrate opposing yet critical physiological roles for NO-mediated S-nitrosylation of HCN channels.

The closely related family of cyclin-nucleotide-gated Ca^{2+}-ion channels (CNGcs) play critical roles in the signal transduction of retinal photoreceptors and olfactory neurons [78]. These channels have also been shown to undergo direct modulation by NO-mediated S-nitrosylation [18]. Site-directed mutagenesis of a single cysteine residue (C460S) in the C-linker region completely suppressed the ion channel's response to the application of NO. Similar to the modulation of NMDAR through S-nitrosylation of cysteine residues in the ligand-binding domain of the NR1 subunit [16], S-nitrosylation of C460 in the C-linker region of CNGcs may confer conformational changes that interfere with the nearby cyclic-nucleotide-binding domain, thereby altering the function of the channel [18].

8.3.5 Transient Receptor Potential Channels

Transient receptor potential (TRP) channels are a family of nonselective cationic channels ubiquitously expressed in numerous cell types. They are activated in response to a wide range of physical and chemical stimuli including pain, temperature, and taste. TRPC channels are Ca^{2+}-permeable and can be activated by signaling pathways involving phospholipase C, diacylglycerol, mechanical forces, and Ca^{2+} depletion from internal stores or directly by intracellular Ca^{2+} [79–81]. It has been demonstrated that NO-mediated S-nitrosylation of select cysteine residues in TRPC5 channels results in Ca^{2+} influx [19]. In this study, TRPC5 cysteine residues were subjected to mutagenesis or deletion until it was found that C553S and C558S had noticeable suppression of receptor activity during S-nitrosylation assays.

Unlike other members of the TRP family, TRPV channels are selectively permeable to calcium ions. For the TRPV4 channel, NO donor studies have shown that S-nitrosylation occurs on C853 in the intracellular C-terminus of the channel [20]. C853 is flanked by lysine and aspartic acid residues, supporting the SNO-motif enhancement of S-nitrosylation. Mutagenesis experiments (C853A) confirmed that S-nitrosylation occurs at the C853 residue and modulates channel activity through negative feedback by limiting the interaction with calmodulin. This suggests that S-nitrosylation protects against apoptosis by preventing intracellular responses to excessive Ca^{2+} influx [20]. In the channel pore of

TRPV1, C616 and C621 are thought to undergo S-nitrosylation, although it is not required for channel activation [21]. Therefore, how S-nitrosylation of TRPV1 modulates channel activity remains to be clarified [19, 21, 82, 83]. Still, it has recently been demonstrated in rat gastric tissue that S-nitrosylation of TRPV1 channels may have anti-inflammatory properties by enhancing the electrical activity of mesenteric afferent vagal nerve [84].

TRPA1 channels are known to be involved in various forms of sensory detection such as in noxious stimuli, cold perception, and inner ear function [85, 86]. Located throughout the cytoplasmic N-terminal ankyrin repeats of the TRPA1 channel, three cysteine residues C621, C641, and C665 were found to be susceptible to S-nitrosylation [87, 88]. SNAP donor application showed that site-directed mutagenesis of these cysteine residues had variable modulation of TRPA1 activity [88]. Of the three, C665S significantly reduced the sensitivity to two chemically distinct NO donors, SNAP and NOR3 [21].

8.3.6 Voltage-Gated K^+ Channels

Voltage-gated potassium channels play critical roles in the setting of the resting membrane potential and the reconstitution of depolarized cells to the resting state. In cardiac tissues, several types of potassium currents are required for the maintenance of normal cell excitability: activating and inactivating transient outward current (I_{to}), ultra-rapid (I_{Kur}) current, the rapid (I_{Kr}) and slow (I_{Ks}) currents of the delayed rectifier, and the inward rectifier (I_{K1}) current [89, 90]. Modulation of these potassium currents by direct S-nitrosylation has profound impacts on cardiac physiology through which the cellular redox state affects the maintenance, duration, and integrity of cellular action potentials [91].

It has been demonstrated that S-nitrosylation augments the I_{Ks} current carried by potassium channels and shortens the duration of action potentials [92]. Site-directed mutagenesis of a cysteine residue (C445A) and a negatively charged residue (D446) within the α-subunit of the C-terminus of an I_{Ks} channel (KCNQ1), separately or together, largely abolished the effect of S-nitrosylation by NO and provided more support regarding the significance of an SNO motif [10, 12, 13]. In a more recent study, the overexpression of CAPON, an inducer of NOS1 activity, increased I_{Ks} current; however, the involvement of S-nitrosylation was not determined [93].

For the Kir2.1 channel, S-nitrosylation of C76 increased the macroscopic I_{K1} currents by augmenting the channel's open probability [91]. Importantly, the effect of NO was found to be dependent on the redox state of the cell as current enhancement from NO was ablated by the addition of the thiol-specific reducing agent DTT [23].

In contrast to the augmentation of I_{Ks} and I_{K1} currents, NO via direct S-nitrosylation does not appear to alter the I_{Kur} and I_{to} currents mediated by Kv1.5 and Kv4.3 channels, respectively [94]. Based on molecular modeling, it was proposed that S-nitrosylation of two cysteine residues (C331 and C346) located in the voltage sensor region of the Kv1.5 channel inhibited channel activity due to the formation of hydrogen bonds between neighboring residues [94]. However, site-directed mutagenesis (C331A and C346A) in Kv1.5 channels did not result in a significant degree of inhibition by NO [95]. In Kv4.3 channels, although S-nitrosylation was confirmed through the application of an NO donor, the magnitude of inhibition of I_{to} currents by NO donors was not found to be significantly different in the presence or absence of DTT [23]. I_{Kr} currents carried by the human *Ether-à-go-go*-Related Gene (hERG) channel were inhibited by an sGC/cGMP independent fashion when HERG channels were expressed in frog oocytes, but it was not assessed whether this was due to direct S-nitrosylation [96].

Moreover, while S-nitrosylation appears to enhance I_{Ks} and I_{K1} currents, the data are less clear regarding the pathway by which NO results in inhibition of I_{Kur}, I_{to}, and I_{Kr} and currents. Given that each of these currents plays a critical role in maintaining action potential integrity in electrical cardiac cells, it is not surprising that dysregulation of NO is implicated in a variety of cardiac arrhythmias and pathologic disease states [93, 97].

8.3.7 ATP-Sensitive Potassium Channels

ATP-sensitive potassium (K_{ATP}) channels are found in many different tissues throughout the body, such as smooth muscle [98], heart [99, 100], and nervous tissue [101]. These inwardly rectifying potassium ion conductors are perhaps most well known for their role in pancreatic β-cells [102] in which the state of glucose metabolism and subsequent ATP production is coupled with K_{ATP} channel activity: ATP directly binds to the channel and inhibits channel activity [103, 104]. K_{ATP} channels are hetero-octomers consisting of four regulatory subunits responsive to sulfonylurea (SUR1, SUR2A, and SUR2B) and four ATP-sensitive pore-forming subunits (Kir 6.1 and Kir 6.2) [105]. In large DRG neurons, S-nitrosylation of native K_{ATP} channels activated the channel by decreasing the sensitivity to ATP [24]. Site-directed mutagenesis (C717S) in the nucleotide-binding domain of SUR1 significantly reduced channel activation upon application of SNAP but did not completely abolish channel activity, implying the involvement of other cysteine sites. Notably, these findings are in contrast with previous studies that attributed the potentiation effect to the NO-cGMP-PKG pathway [106].

8.3.8 Large-Conductance Ca^{2+}-Activated K^+ Channels

BK (big potassium) channels are large-conductance calcium-activated potassium channels that are activated either by cell membrane voltage or increased intracellular calcium. They are ubiquitous channels but have been characterized in major physiological processes such as vascular tonicity and neuronal excitation [107, 108]. The primary function of the BK channel is to repolarize the cell membranes via potassium efflux in response to membrane depolarization or increased intracellular calcium concentrations. It has been well established that BK channel activity is enhanced by both sGC/cGMP pathways and NO-mediated S-nitrosylation [36]. Studies have revealed that BK channel activity is enhanced directly by S-nitrosylation in smooth muscle cells [109], chromaffin cells [110], and in brain tissue [110–112]. Notably, in these studies, patch-clamp recordings of excised BK channels without cGMP confirmed the enhancement of channel activity with the exogenous application of NO. The physiological significance of NO-mediated S-nitrosylation of BK channels has been implicated as a mechanism for smooth muscle vasodilation [109], catecholamine secretion [110], and neuropeptide release [110–112]. As of this publication, the localization of exact sites of S-nitrosylation on BK channels remains unknown. However, it has been shown that a cysteine-rich motif located within the C-terminus of BK channels confers sensitivity to hypoxic conditions [113]. Still, the relationship between BK channel enhancement and NO-mediated S-nitrosylation is encouraging for therapeutic developments. In fact, it has been demonstrated that uncoupling of NO and BK channels leads to hippocampal damage in patients with sleep apnea [114].

8.3.9 Voltage-Gated Ca^{2+} Channels

Intracellular calcium ions are critical second messengers in excitable and non-excitable cells. Voltage-gated calcium channels are heteromeric membrane protein complexes composed of the primary α_1 subunit and auxiliary β and $\alpha_2\delta$ subunits [115, 116].

S-nitrosylation of the L-type Ca^{2+} channel (LTCC) was first observed over 20 years ago when it was shown that the canonical cGMP pathway inhibited LTCCs while S-nitrosylation from an NO donor stimulated channel activity [117]. However, other studies where LTCCs were expressed in heterologous systems suggest that S-nitrosylation decreases channel activity

[118–120]. For example, it has been observed in mouse heart cells that, under normal physiological conditions, the α_1 subunit of LTCC is constitutively S-nitrosylated by eNOS [120]. In this study, the application of exogenous NO donor increased the degree of S-nitrosylation and reduced the damage associated with ischemia–reperfusion. This suggests that S-nitrosylation of the LTCC serves as a cardioprotective mechanism during ischemic conditions, which is consistent with other observations [118–123]. The specific residues and location of S-nitrosylation remain unknown, although strong evidence indicates that the likely location is within the C-terminal tail of LTCC, which, interestingly, is located adjacent to sites of phosphorylation. This suggests that the modulation of the channel may also be required for appropriate regulation during β-adrenergic stimulation [120, 124].

For N-type CaV2.2 channels, S-nitrosylation of five cysteine residues (C805, C930, C1045 in the II-III loop and C1835, C2145 in the C-terminus) has an inhibitory effect on channel function by positively shifting voltage-dependent activation [25]. In this study, it was suggested that these cysteine residues function as NO sensors, thereby providing a negative feedback mechanism for increasing intracellular calcium concentrations in neurons. When Cav2.2 was expressed with the Cavβ subunit with site-directed mutagenesis of C346, S-nitrosylation also caused channel inhibition [25]. As N-type calcium channels are highly expressed in nervous tissue and dorsal root ganglion neurons, they are thought to be necessary for the transmission of nociception [125]. Indeed, negative regulation of N-type CaV channel by NO-mediated S-nitrosylation can be potential targets for pharmacotherapy [126–128].

8.3.10 Ryanodine Receptors

Three ryanodine receptors (RyRs) are broadly expressed in various physiological systems: RyR1 in skeletal muscle, RyR2 in the heart, and RyR3 in brain tissues [129]. RyRs are situated in sarcoplasmic and endoplasmic reticular membranes and permit the efflux of calcium from intracellular stores into the cytosol, a critical step in excitation–contraction coupling in heart and skeletal muscles. In skeletal tissue, activation of RyR1 occurs through a physical link with the dihydropyridine receptor (a voltage-gated L-type calcium channel), whereas in the heart, RyR2 is activated by calcium-induced calcium release [130, 131]. In healthy cardiac and skeletal tissue, nNOS is colocalized with the RyRs in subcellular locations and can be co-immunoprecipitated [132, 133]. Thus, for RyR1 and RyR2, nNOS provides NO for thiol modification via S-nitrosylation and the subsequent modification of channel activity [134–136]. NO-mediated S-nitrosylation mostly enhances RyR activity by increasing the open probability of the channels and thus Ca^{2+} release from intracellular stores [137, 138]. However, channel inhibition has been demonstrated with a supraphysiological application of NO-donation [27, 139, 140].

Four cysteines (C315, C811, C906, and C3635) are sites of S-nitrosylation in RyR1 [141]. It has been suggested that this channel has a redox-sensor function, as S-nitrosylation favors select cysteine residues depending on the redox state of the cell [141]. The C3635 residue of RyR1 is located within a hydrophobic pocket and is readily available for S-nitrosylation under hypoxic conditions: when oxygen levels are high, C315, C811, and C906 are oxidized, thus preventing S-nitrosylation of the channel [27, 142]. Moreover, NO-mediated S-nitrosylation of C3635 on the RyR1 channel displaces Ca^{2+}-Calmodulin, thereby activating the channel [142]. These findings are in concordance with disease states related to the dysregulation of intracellular calcium. For example, in mouse models of malignant hyperthermia, mutant RyR1 channels (Y522S) leak calcium, which promotes a positive feed-forward effect of increased oxidative stress, S-nitrosylation of the channel, and augmented calcium leakage, thereby increasing the risk of sudden heat death [143]. In genetically altered mice with a C3636A substitution in the RyR1 channel (corresponding to C3635 in the RyR1 channel in humans), inhibition of S-nitrosylation at this critical cysteine residue was found to have neuroprotective effects in mouse hippocampus after an induced seizure, suggesting therapeutic utility in RyR1 inhibition [144].

In cardiac myocytes, the RyR2 receptor is responsible for mediating sarcoplasmic calcium release upon calcium influx via activated L-type calcium channels (calcium-induced calcium release). The C3635 residue in RyR1 is a conserved cysteine among all three RyR isoforms; in the case of RyR2, the corresponding cysteine residue is C3602 [145, 146]. In contrast to RyR1 channel activation by NO via S-nitrosylation of C3635, NO-mediated S-nitrosylation of C3602 in RyR2 does not alter channel activity; however, nitrosylation by S-nitrosoglutathione (GSNO), another biological source of NO, enhances the open probability of the channel [145, 146]. Although the specific sites of S-nitrosylation in RyR2 channels have yet to be elucidated, it has been shown in nNOS knockout mice that hyponitrosylation of the RyR2 channel leads to arrhythmogenesis causing sudden cardiac death in the mice [135]. In support of this, nNOS has been shown to translocate to the sarcolemma during heart failure, indicating that RyR channel modulation mediated by S-nitrosylation is critical for muscle cell calcium homeostasis in cardiac and skeletal tissues [132].

8.4 Ion Channel Modification Via H_2S and S-Sulfhydration

The human body produces small amounts of H_2S, which primarily functions as a signaling molecule in the vascular and nervous systems [147]. Enzymatic production of H_2S is facilitated by an array of enzymes [148]. Cystathionine β-synthase (CBS) is the primary enzyme responsible for H_2S generation in the central nervous system while cystathionine γ-lyase (CSE) predominates in peripheral tissues, including the cardiovascular smooth muscles and pancreas. 3-Mercaptopyruvate sulfurtransferase (MPST) activity is ubiquitous and leads to the production of H_2S, but it is primarily active with CBS in the central nervous system.

The interest in H_2S and its physiological significance has increased over the last two decades as it is now recognized as the third gaseous signaling molecule (gasotransmitter), along with NO and CO [1]. Modification of proteins by H_2S can be thought of as comparable to S-nitrosylation by NO: The side chain of cysteine residues on target proteins can undergo S-sulfhydration to form a persulfide (–SSH) [149]. The chemical groups of thiols (R-SH) and hydropersulfide (R-SSH) are of drastically different chemical properties. As such, hydropersulfides are more nucleophilic and therefore more reactive, and the pKa of an R-SSH is at least 1–2 units lower than the corresponding R-SH [150]. Therefore, under physiological conditions, hydropersulfides are more acidic and thus serve as more effective hydrogen donors than thiols [151, 152].

The physiological relevance of H_2S signaling in the cardiovascular system was discovered decades ago, and today, it is well established that H_2S signaling results in blood vessel dilation, thereby serving as a cardioprotective mechanism during conditions of ischemia [153]. However, several studies have delegated other critical physiological regulatory processes for H_2S such as cellular redox status, programmed cell death, and inflammatory responses [154]. Interestingly, H_2S signaling in the nervous system appears to confer neurotoxic effects during hypoxic conditions, which is in stark contrast to its protective role in hypoxic cardiovascular tissue. For example, H_2S appears to augment seizure-like events in rat epileptic seizure models [155]. Accordingly, removal of H_2S through the inhibition of CBS was found to attenuate neuronal ischemia in ischemic stroke rat models [156]. The nature of how H_2S signaling has drastically opposing effects between these two physiological systems has recently been the subject of investigation. This section will provide a summary of the structural and functional modulation of ion channels known to undergo S-sulfhydration by H_2S at specific cysteine residues. Physiological implications are briefly discussed as well (Fig. 8.2).

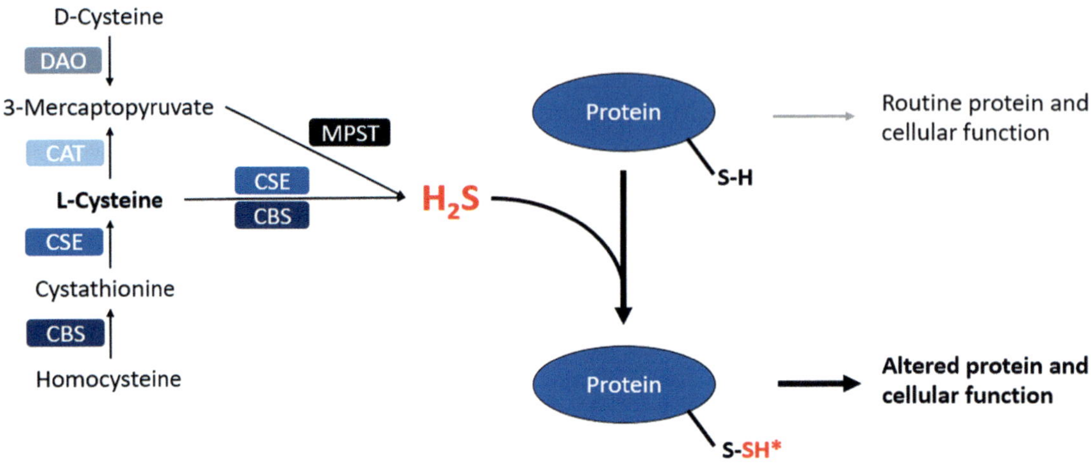

Fig. 8.2 Pathways of enzymatic formation of endogenous H_2S. Abbreviations: CBS, cystathionine β-synthase; CSE, cystathionine γ-lyase; CAT, cysteine aminotransferase; DAO, D-amino acid oxidase; MPST, 3-mercaptopyruvate sulfurtransferase. *H_2S cannot directly react with thiols on proteins. The reader is encouraged to see the excellent review article by Zhang et al. detailing the proposed mechanisms for the formation of sulfhydrated proteins [152]

8.4.1 ATP-Sensitive Potassium Channels

ATP-sensitive potassium (K_{ATP}) channels have been characterized extensively regarding their modulation by H_2S. Activation of the K_{ATP} channel by H_2S leads to cellular hyperpolarization, which, in vascular smooth muscle, has the downstream effect of decreasing voltage-gated Ca^{2+} influx, thus facilitating vasodilation [157]. Earlier studies revealed that H_2S increases the open probability of native K_{ATP} channels in rat insulin-secreting cells [158], although the mechanism remained elusive. In subsequent studies, expression of heterologous K_{IR} subunits lacking their sulfonylurea receptors (SUR) in HEK 293 cells abolished the enhancement of channel function by H_2S, which suggested that the K_{ATP} channel-associated SURs may contain critical cysteine residues targeted for S-sulfhydration [153]. Indeed, site-directed mutagenesis of select cysteine residues (C6S and C26S) located in the extracellular N-terminal of the SUR1 subunit abrogated heterologously expressed K_{ATP} channel modulation by H_2S [153]. While Jiang et al. and colleagues propose that modification of C6 and C26 by H_2S alters K_{ATP} channel configuration to promote opening of the channel pore, it was not specified whether this was due to direct S-sulfhydration of each cysteine residue or disruption of a disulfide bond between the two cysteines. A subsequent breakthrough study established the relationship between H_2S-mediated S-sulfhydration of K_{ATP} channels regarding the regulation of channel opening and downstream cellular processes [149]. With site-directed mutagenesis of a single cysteine residue (C43S) in the K_{IR} 6.1 subunit, and with blood vessels of mutant mice lacking CSE, Mustafa et al. and colleagues found that S-sulfhydration is required at C43 for membrane hyperpolarization and that H_2S is an endothelial-derived hyperpolarization factor required for vasorelaxation, respectively [149]. It is thought that C43 is situated in the ATP binding region, which itself is adjacent to the PIP2 binding region of the channel [149, 159]. Moreover, these findings strongly suggest that an inaccessible ATP-binding domain due to S-sulfhydration facilitates enhanced channel activity by making the adjacent binding site more favorable for PIP2, a known activator of K_{IR} 6.1 [149].

As mentioned in the section introduction, H_2S has opposing effects in different physiological systems. In the cardiovascular system, H_2S causes hyperpolarization of heart muscle cells, which serves as cardioprotection during conditions of ischemia. It has been established for some time now that H_2S modulation of K_{ATP} channels is the pathway by which cardioprotection is conferred in mouse models of ischemia–reperfusion injuries [160]. However, in the nervous system, H_2S enhances neuronal depolarization, which has been implicated in the progression of ischemic insults to cerebral tissue during pathological states, such as seizure and cerebrovascular infarction [155, 156]. Until recently, the involvement of specific ion channels in nervous tissue that were susceptible to H_2S modulation had not been elucidated. A recent study has now demonstrated that other members of the K_{IR} family, $K_{IR}2$ and $K_{IR}3$, are inhibited by H_2S, in stark contrast to the closely related K_{ATP} channel [161]. In the $K_{IR}3$ channel, the cytoplasmic cysteine residue C65 is the corresponding cysteine to the C43 residue in the $K_{IR}6.1$ channel. Site-directed mutagenesis of C65 and another cytoplasmic cysteine (C321) in the $K_{IR}3$ channel abolished the inhibitory effect of the H_2S donor sodium hydrosulfide (NaSH) [161]. Indeed, this observation is striking when considering the homology between K_{IR} family members. Although other explanations cannot be ruled out, computational models of the $K_{IR}3$ channel suggest that sulfhydration of these two cysteine residues also modulates the effect of PIP2 on channel activation [161]. These molecular models suggest that PIP2 binding affinity is decreased upon sulfhydration of $K_{IR}2$ and $K_{IR}3$ channels due to a resultant increased distance between PIP2 and a critical lysine residue required for PIP2 channel activation, completely opposite of that which was seen for K_{ATP} channels [149, 161]. It remains to be elucidated as to how the downstream effect of sulfhydration favors either cellular hyperpolarization or depolarization, given the ubiquitous expression of the K_{IR} family of ion channels in both cardiac and nervous tissue.

8.4.2 Transient Receptor Potential Channels

TRP channels are nonselective ion channels that are ubiquitously expressed in various mammalian cell types and primarily function as mediators of various environmental stimuli. Over the last decade, different members of the TRP channel family have been found to undergo posttranslational modification by H_2S. It has been demonstrated in astrocytes that excitation of neurons results in the release of H_2S, which augments the Ca^{2+} influx by TRPA1 channels thereby facilitating propagation of the Ca^{2+} signal to surrounding cells [162]. Interestingly, polysulfides (a mix of compounds with variable numbers of sulfurs [e.g., H_2S_n]) appear to activate TRPA1 channels much more potently than H2S, although the physiological function and tissue distribution of polysulfides have not been extensively investigated [163, 164]. In HEK 293 cells expressing a double cysteine mutant (C422S and C622S) mouse TRPA1 channel, Ca^{2+} influx was diminished, suggesting that these two N-terminal cysteines are the target sites for sulfhydration required for H_2S modulation of TRPA1 activity [165]. Notably, the effect of H_2S on TRPA1 was enhanced at lower pH (6.8). As TRPA1 channels are known to be involved in propagating nociceptive signals, the results of this study provide a basis for pharmacological development targeting pain relief associated with H_2S-mediated TRPA1 activation.

CBS deficiency is an autosomal recessive, multisystem disease characterized by hyperhomocysteinemia and homocystinuria, lens dislocation, osteoporosis, intellectual deficits, and significantly increased risk for thromboembolism [166]. It was discovered that bone marrow mesenchymal stem cells (BMMSCs) biosynthesize H_2S via CBS, which not only appears to be critical for the self-renewal and osteogenic differentiation of BMMSCs but also augments Ca^{2+} influx via TRP channels [167]. In this study, Liu et al. and colleagues used siRNA knockdown on three candidate TRP channels (TRPV3, TRPV6, and TRPM4) containing cysteine residues, which decreased Ca^{2+} influx by more than 50% with the application of NaSH. Mass spectrometry was used to identify sites of sulfhydration and subsequent site-directed mutagenesis of two cysteines (C172 and C329) in TRPV6 (together, and C329 alone) decreased Ca^{2+} influx during NaSH application. Mass spectrometry also identified sulfhydration at C131 of TRPV3 and C168 of TRPM4, although mutagenesis studies were not conducted on these channels. Moreover, H_2S modulation of TRP channels via sulfhydration maintains normal channel activity in BMMSCs that appear to be required for downstream signaling pathways involved in osteogenesis [167]. Expectedly, H_2S may have therapeutic implications in patients with erosive bone disorders [168].

8.4.3 L-Type Calcium Channels

In cardiac myocytes, L-type calcium channels (LTCC) play a key role in excitation–contraction coupling by permitting extracellular Ca^{2+} influx. H_2S has been recognized to confer cardioprotective effects against oxidative stress, apoptosis, and necrosis [154]. In rat heart tissue, application of NaSH resulted in an inhibition of LTCC function and was found to induce negative inotropy in cardiomyocytes [145, 146]. In a later study, dithiothreitol (DTT; a reducing agent that breaks disulfide bridges between cysteine residues into sulfhydryl groups) was found to reverse the inhibition of NaSH on LTCCs while diamide (DM; an oxidizing agent that creates sulfide bridges between cysteine groups) prevented alterations of NaSH on LTCC calcium currents [169]. The significance of this is clear considering that the inotropic effects of H_2S are mediated by the oxidative state of sulfhydryl groups within LTCCs. To date, only one study has established that LTCC modulation is likely mediated through direct S-sulfhydration of cysteine residues, although no specific cysteine residues have been identified [170].

In the nervous system, although the physiological significance between H_2S and LTCCs is less clear, the consensus is that H_2S enhances Ca^{2+} currents carried by neuronal LTCCs, contrasting with the effects noted above in myocardial tissues. For example, in human neuroblastoma and cerebellar granule cultures, the application of NaSH increased calcium influx and thus subsequent calcium release from internal stores [171, 172]. Interestingly, in cerebellar granular cells, it was observed that excessive intracellular Ca^{2+} influx resulted in glutamate-induced excitotoxicity. This observation is confounding considering that endogenous H_2S biosynthesis is thought to confer neuroprotection during oxidative stress [173].

In gastric tissue, H_2S modulation on LTCCs maybe even more complicated. For example, in colonic tissue from rats, NaSH inhibited LTCC activity and halted spontaneous muscle contractions in the circular and longitudinal muscles [174]. However, in the gastric fundus of mice, physiological concentrations of H_2S increased gastric tonicity and contraction [175]. Moreover, it remains unclear how and if H_2S mediates opposing effects on LTCCs expressed in different physiological systems through direct sulfhydration of cysteine residues.

8.5 Singlet Oxygen

Most life forms on the earth require molecular oxygen (O_2). Electrons of O_2 have three different configurations: the triplet ground state ($^3\Sigma$) and the first ($^1\Delta$) and the second ($^1\Sigma$) singlet excited states (Fig. 8.3a) [176, 178]. Singlet oxygen ($^1O_2$) is a volatile ROS and directly oxidizes many different types of molecules, including nucleic acids, protein, and lipids. Chemically, $^1O_2$ is different from other well-studied ROS, including hydrogen peroxide (H_2O_2), superoxide, hydroxyl radical, and nitric oxide [179]. In the presence of photosensitizer, oxygen, and light, $^1O_2$ can be generated through photodynamic processes [180]. In cells that are naturally exposed to sunlight, such as those in the eye and skin, $^1O_2$ can be produced through photodynamic processes as compounds including flavins and NADH/ NADPH can function as photosensitizers [181]. Alternatively, certain metabolic processes

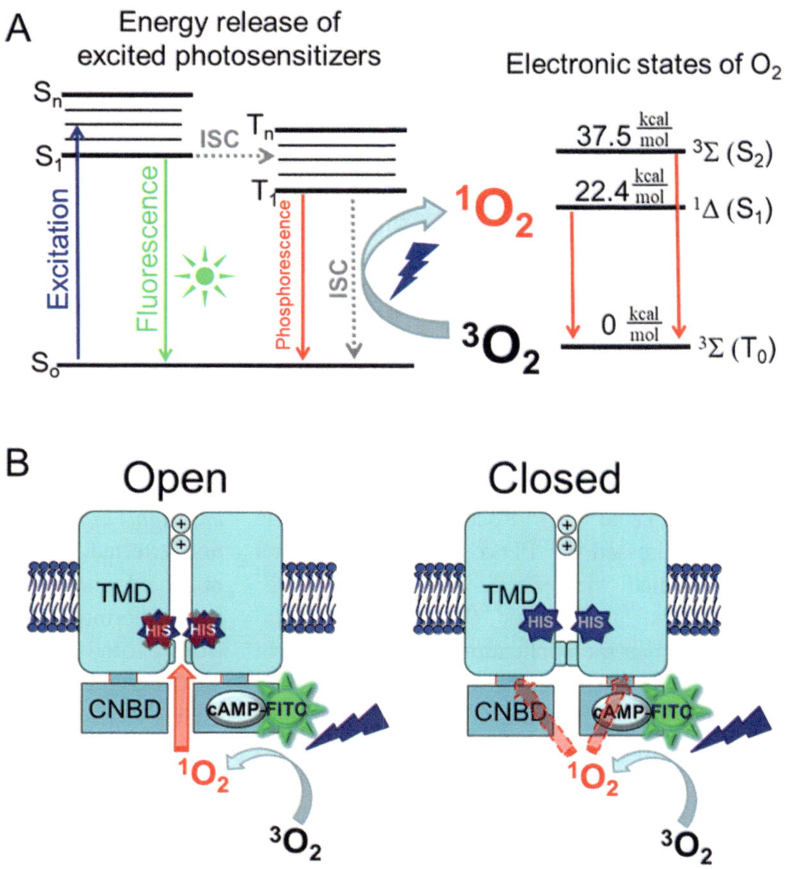

Fig. 8.3 Photodynamic production of $^1O_2$ and a working model for studying $^1O_2$ modification of ion channels. (**a**) The energy release of excited photosensitizers can be through the release of photon (fluorescence) or intersystem crossing (ISC) from excited singlet (S) to triplet (T) spin multiplicity states [176]. (**b**) A working model based on excised membrane patches is developed for studying $^1O_2$-mediated photodynamic modification of ion channels. Adapted from [177]

involving enzymes and other ROS can also produce $^1O_2$ [178, 182–184]. Enzymatic generation of $^1O_2$ has been confirmed in stimulated neutrophils, macrophages, and plant cells [185, 186]. In liver cells, $^1O_2$ can be generated in the absence of light through peroxidase catalyzed oxidation of triplet carbonyls [182]. Notably, excessive $^1O_2$ causes detrimental effects to the cell and has been linked to diseases including aging and cancer [187, 188]. Low levels of $^1O_2$ may function as a signaling factor through modifying the function of molecules within their vicinity [185, 189]. The role of $^1O_2$ as a signaling factor remains to be clarified.

Compared to other ROS, $^1O_2$ is unique due to its short lifetime in the microseconds and short working distance in nanometers, which has been utilized to specifically knockout the function of target molecules or cells without much collateral damage. Commonly used fluorescent molecules (such as FITC) and fluorescence proteins (such as KillerRed) are effective photosensitizers and have been used for this purpose. In chromophore-assisted light inactivation (CALI), fluorescent-labeled antibodies recognize target proteins, and subsequent light exposure generates toxic $^1O_2$ in the vicinity [190–192]. This approach has been extended to clinical practice, such as the FDA-approved procedure known as photodynamic therapy [193]. Photosensitizers are first administered to the patient, followed by light exposure to the target tissue. It is believed that $^1O_2$ damages target cells, but a combination of apoptosis, necrosis, and acutely triggered local immune responses contributes to the final therapeutic effects. Photodynamic therapy has been used in the treatment of cancers in the esophagus, lung, and skin as well as other diseases related to the skin and eye.

In addition to HCN channels, many other channels and transporters appear to be sensitive to photodynamic modification, potentially with $^1O_2$ as the major player [194–198] (*see Stief [197] and the references within for a comprehensive list of ion-channels that have been studied with $^1O_2$ generated by chloramine-T*®). Previous studies on $^1O_2$-mediated photodynamic modification (PDM) mainly used exogenous photosensitizers such as Rose Bengal. For ion channels, PDM leads to the blockage and suppression of ionic currents (NaV, CaV, KaV, and Gramacidin), incomplete or slowed inactivation (NaV), shifts in *I–V* curves (NaV), and potentiation of ionic currents in NMAD and GABAa receptors (Table 8.2). Recently, two separate studies addressed the PDM of TRPA1 channel but attributed the effects to different ROS: hydroxyl radicals [204] or $^1O_2$ [194, 207]. Notably, other than the application of protoporphyrin as the photosensitizer, no direct evidence was provided for the involvement of $^1O_2$. Furthermore, most of the studies on PDM of ion channels were carried out at the whole-cell or tissue level and required prolonged light exposure time in minutes. Indeed, complicated intracellular signaling pathways, likely with the involvement of many other ROS, present a great challenge. It has been difficult to specifically tease out the effect of $^1O_2$, given the complex signaling pathways within the cell. Thus, an in-depth investigation of $^1O_2$-mediated PDM of ion channel function using a well-defined working model is warranted.

In our previous study of dynamic interaction between ligand and full-length, functional channels on the membrane, we applied the patch-clamp fluorometry (PCF) technique and used cAMP molecules tagged with different fluorophores to track the binding of cAMP to HCN channels [208–211]. Surprisingly, we found that HCN channels are very sensitive to PDM. With FITC-cAMP applied to the intracellular side of the HCN2 channel, we discovered that the application of laser pulses slowed down the phase of channel deactivation and increased the voltage-insensitive, instantaneous (I_{inst}) component [177, 212]. I_{inst} refers to the component showing an immediate response to voltage step, without the time dependency typically observed with voltage-dependent gating. These two alterations in channel functions were found to be related to H434, which is located near the intracellular end of S6. The alanine replacement of H434 (H434A) abolished the delay in channel activation delay as well as the generation of I_{inst} after the photodynamic process. To further clarify the involvement of $^1O_2$, we fused miniSOG, a genetically encoded singlet oxygen generating protein to the C-terminal end of the CNBD domain on mHCN2. MiniSOG was originated from the light, oxygen, voltage (LOV) domain of a blue-light photoreceptor from *Arabidopsis thaliana* [213]. We discovered that light irradiation exerted strikingly similar effects to those observed with mHCN2 and FITC-cAMP.

Furthermore, we extended the investigation from photodynamic modification of HCN channel from the mouse HCN2 channel to the sea urchin HCN channel, spHCN [208, 209]. In the absence of cAMP, spHCN channels open only briefly in response to a hyperpolarizing voltage step and then quickly inactivate. This voltage-dependent channel inactivation is due to reclosure of the activation gate and is caused by loose coupling between the voltage sensor and the gate in the spHCN channel [214, 215]. Surprisingly, we discovered that photodynamic modification abolished the voltage-dependent inactivation of spHCN channel and increased the voltage-insensitive I_{inst} component. To examine the involvement of $^1O_2$, we exposed the intracellular end of spHCN channel to a chemical mixture of hypochlorite and hydrogen peroxide, a chemical reaction known for generating $^1O_2$ in the absence of light and photosensitizer. Indeed, we observed changes in the spHCN channel that were comparable to the effects of photodynamic modification. Moreover, we tested $^1O_2$ quenchers (NaN_3, Trolox-C) and reagents that generate other ROS. These lines of evidence supported the conclusion that $^1O_2$ is the major player in the photodynamic modification of mHCN2 and spHCN channels [208, 209].

Remarkably, we found that in both mHCN2 and spHCN channels, the highly conserved

Table 8.2 A review on the photodynamic modification of ion channels and cell surface receptors

Protein	Preparation	Photosensitizer	Excitation light	Major effects
NaV Oxford et al. [199]	Squid giant axon	Eosin Y, Rose Bengal	4–6 s	Irreversible block; incomplete inactivation
CaV, Kv, NaV Tarr and Valenzeno [200]	Single frog atrial myocyte	Rose Bengal (0.5 μM)	6.5 mW/cm^2; 525 nm; 2–4 s	Current suppression; slowed NaV inactivation; NaV I-V shift
CCK1 GPCR Cui and Kanno [201]	Rat pancreatic acinar cell	aluminum phthalocyanine	1 min (55,000 lux)	Permanent activation
CaV, Kv, NaV Valenzeno and Tarr [202]	Mouse pituitary, GH3, cells	Rose Bengal	52–300 s; 6.5 mW/cm^2	Block of CaV, Kdelay, Kca; increase in I_{leak}
Kv4.2 Sack et al. [203]	CHO-K1 cells	Porphyrin	100 s; 395–440 nm; 8–12 mW 10× obj.	Current ablation
TRPA1 Hill and Schaefer [204]	HEK293	Acridine orange	490 nm; 60 s	Channel activation; hydroxyl radicals?
TRPA1; TRPV1 Babes et al. [194]	DRG or HEK293t	PpIX; ALA	390 nm; 85 s	Channel activation
	Artificial bilayer	PpIX (1 microM)	405 nm laser; 0.45 mW/mm^2; > 60 s.	Single-channel activation
NMDA receptor Eisenman et al. [205]	HEK293	NBD or FITC	480 nm; 156 mW/mm^2; 1–5 min	Current potentiation
GABAa receptor Eisenman et al. [195]	Hippocampal neuron	NBD	30–60 s; 480 nm; 37 mW; 40× objective	Current potentiation
Gramacidin Rokitskaya et al. [206]	Lipid bilayer	Rose Bengal	400 mJ/cm^2; 2 ms	Photo-inactivation of current

histidine residue (H434 in mHCN2 or H462 in spHCN) located near the intracellular end of S6 appears to be critical for the effects of photodynamic modification (Fig. 8.3b). Alanine replacement of that histidine residue in both mHCN2 and spHCN channels abolished most of the effects elicited by light pulses. Notably, the decrease in I_h current amplitude became more significant in mHCN2/H434A mutant channel. For the spHCN/H462A channel, light pulses applied before the voltage step (while most channels are in the closed state) resulted in a slight increase in current amplitude but had almost no effect on the voltage-dependent channel inactivation [208, 209]. These discrepancies between alanine replacement mutant and WT channels could be attributed to modifications made to other residues by $^1O_2$.

Finally, in both mHCN2 and spHCN channels, we observed strong state dependency of PDM. For mHCN2 channels, PDM of the open state slows down channel deactivation and increases the component of I_{inst}, whereas the major effect of PDM of closed channels is the decrease in I_h [177]. For spHCN channels, PDM of both closed and inactivated channels abolishes the voltage-dependent channel inactivation. Moreover, PDM of closed spHCN channels results in more pronounced increases in I_h but only moderate increases in I_{inst} and steady-state current [208, 209]. These state-dependent responses to PDM by HCN channels indicate that $^1O_2$-mediated modification of protein molecules is sensitive to protein conformation changes, which, although subtle, may alter the diffusion trajectory of $^1O_2$ and therefore the accessibility of critical residues such as the histidine residue near the activation gate in HCN channels.

8.6 Carbon Monoxide (CO) as a Gasotransmitter and Crosstalk Among Different Regulatory Pathways

Within the cell, heme oxygenase (HO) degrades heme and produces biliverdin, Fe^{2+}, and CO. A growing number of ion channels have been discovered to be regulated/modulated by CO. CO has been reported to inhibit the function of T-type calcium channels [216] and increase the activity of BK channels [217]. On the other hand, it becomes commonly recognized that the same type of ion channels can be modulated by all major gasotransmitters. For BK channels, application of NaSH (an H_2S donor) inhibited the activity of BKCaα channel subunits transfected into HEK 293 cells, producing a rightward shift in the channels activation curve [218], an opposite effect observed after application of CO [219]. Furthermore, the application of KCN (CO donor) eliminated the impacts on channel activity by CO but had no effect on the inhibitory effect of H_2S, suggesting these regulatory gases do not compete with each other and act on different structural elements within the channel.

Notably, CO can stimulate the production of NO and ROS, and thus a complex crosstalk among these gaseous channel regulators exists. The modulatory functions of NO and H_2S have some overlap, which is supported by the discovery of thionitrous acid and nitrosothiol (HSNO) signaling factors that contribute to diverse intracellular signaling pathways and may directly connect H_2S and NO pathways [220–222]. Another possibility of synergistic effects of H_2S and NO is through the production of polysulfides (H_2S_n), which activates TRPA1 channels [223]. All of these processes at the molecular level probably explain the observation at the physiological level such as the synergistic vascular relaxation functions by H_2S and NO.

Finally, gaseous signaling pathways are intimately connected to well-characterized factors that regulate channel function, such as Ca^{2+} and PIP2. PIP2 is mainly distributed on the inner leaflet of the cell membrane and exerts essential regulatory effects on almost every type of ion channel and transporter [224, 225]. The regulatory effects of PIP2 and the chemical modifications made by gaseous signaling molecules can be synergistic or anti-synergistic. For example, under conditions of strong channel-PIP2 interaction, the regulatory effect by H_2S is weakened, and vice versa, suggesting a common structural element targeted by both regulatory pathways [161]. Alternatively, PIP2 directly inhibits the activity of NOS and contributes to the regulation of NO level [226].

In summary, this review provides a limited peak into the modification and regulation of ion channels by gaseous molecules. This is an exciting research field, and its pace of advancing is fast. Other than traditional electrophysiology and pharmacology methods, contemporary approaches of theoretical and experimental system biology, especially quantitative proteomics, are being introduced into the study of gaseous molecules. It is expected that new insights will be gained from two frontiers in the very near future: the chemical nature of modifications at the atomic level and the integrated interplay of signaling pathways at the cellular and tissue levels.

References

1. Wang R (2002) Two's company, three's a crowd: can H2S be the third endogenous gaseous transmitter? FASEB J 16(13):1792–1798
2. Missner A, Pohl P (2009) 110 of the Meyer-Overton-rule: predicting membrane permeability of gases and other small compounds. ChemPhysChem 10(9–10): 1405–1414
3. Richardson SL, Hulikova A, Proven M, Hipkiss R, Akanni M, Roy NBA, Swietach P (2020) Single-cell O_2 exchange imaging shows that cytoplasmic diffusion is a dominant barrier to efficient gas transport in red blood cells. Proc Natl Acad Sci U S A 117(18): 10067–10078
4. Moller MN, Cuevasanta E, Orrico F, Lopez AC, Thomson L, Denicola A (2019) Diffusion and transport of reactive species across cell membranes. Adv Exp Med Biol 1127:3–19
5. Moller MN, Li Q, Chinnaraj M, Cheung HC, Lancaster JR Jr, Denicola A (2016) Solubility and diffusion of oxygen in phospholipid membranes. Biochim Biophys Acta 1858(11):2923–2930
6. Li Q, Lancaster JR Jr (2013) Chemical foundations of hydrogen sulfide biology. Nitric Oxide 35:21–34

7. Mathai JC, Missner A, Kugler P, Saparov SM, Zeidel ML, Lee JK, Pohl P (2009) No facilitator required for membrane transport of hydrogen sulfide. Proc Natl Acad Sci U S A 106(39):16633–16638
8. Czyzewski BK, Wang DN (2012) Identification and characterization of a bacterial hydrosulphide ion channel. Nature 483(7390):494–497
9. Ignarro LJ, Buga GM, Wood KS, Byrns RE, Chaudhuri G (1987) Endothelium-derived relaxing factor produced and released from artery and vein is nitric oxide. Proc Natl Acad Sci U S A 84(24): 9265–9269
10. Hess DT, Matsumoto A, Kim S-O, Marshall HE, Stamler JS (2005) Protein S-nitrosylation: purview and parameters. Nat Rev Mol Cell Biol 6(2):150–166
11. Stamler JS, Simon DI, Osborne JA, Mullins ME, Jaraki O, Michel T, Singel DJ, Loscalzo J (1992) S-nitrosylation of proteins with nitric oxide: synthesis and characterization of biologically active compounds. Proc Natl Acad Sci U S A 89(1): 444–448
12. Hess DT, Matsumoto A, Nudelman R, Stamler JS (2001) S-nitrosylation: spectrum and specificity. Nat Cell Biol 3(2):E46–E49
13. Stamler JS, Toone EJ, Lipton SA, Sucher NJ (1997) (S)NO Signals: translocation, regulation, and a consensus motif. Neuron 18(5):691–696
14. Mnatsakanyan R, Markoutsa S, Walbrunn K, Roos A, Verhelst SHL, Zahedi RP (2019) Proteome-wide detection of S-nitrosylation targets and motifs using bioorthogonal cleavable-linker-based enrichment and switch technique. Nat Commun 10(1):2195
15. Choi YB, Tenneti L, Le DA, Ortiz J, Bai G, Chen HS, Lipton SA (2000) Molecular basis of NMDA receptor-coupled ion channel modulation by S-nitrosylation. Nat Neurosci 3(1):15–21
16. Takahashi H, Shin Y, Cho S-J, Zago WM, Nakamura T, Gu Z, Ma Y, Furukawa H, Liddington R, Zhang D, Tong G, Chen H-SV, Lipton SA (2007) A novel thiol oxygen sensor: hypoxia enhances S-nitrosylation—mediated inhibition of NMDA receptor activity. Neuron 53(1):53–64
17. Chen X, Cheng J, Kyle John W, Wiedmeyer Brandi A, Adsit Graham S, Valdivia Carmen R, Alyson KF, Jonathan CM (2017) Abstract 16578: two cysteine S-nitrosylation sites on the cardiac sodium channel are responsible for increased late sodium current in long QT syndromes type 9 & type 12 but not type 3. Circulation 136(Suppl. 1): A16578–A16578
18. Broillet MC (2000) A single intracellular cysteine residue is responsible for the activation of the olfactory cyclic nucleotide-gated channel by NO. J Biol Chem 275(20):15135–15141
19. Yoshida T, Inoue R, Morii T, Takahashi N, Yamamoto S, Hara Y, Tominaga M, Shimizu S, Sato Y, Mori Y (2006) Nitric oxide activates TRP channels by cysteine S-nitrosylation. Nat Chem Biol 2(11):596–607
20. Lee EJ, Shin SH, Hyun S, Chun J, Kang SS (2011) Mutation of a putative S-nitrosylation site of TRPV4 protein facilitates the channel activates. Anim Cells Systems 15(2):95–106
21. Miyamoto T, Dubin AE, Petrus MJ, Patapoutian A (2009) TRPV1 and TRPA1 mediate peripheral nitric oxide-induced nociception in mice. PLoS One 4(10): e7596
22. Asada K, Kurokawa J, Furukawa T (2009) Redox- and Calmodulin-dependent S-Nitrosylation of the KCNQ1 Channel. J Biol Chem 284(9):6014–6020
23. Gómez R, Núñez L, Vaquero M, Amorós I, Barana A, de Prada T, Macaya C, Maroto L, Rodríguez E, Caballero R, López-Farré A, Tamargo J, Delpón E (2008) Nitric oxide inhibits Kv4.3 and human cardiac transient outward potassium current (Ito1). Cardiovasc Res 80(3):375–384
24. Kawano T, Zoga V, Kimura M, Liang M-Y, Wu H-E, Gemes G, McCallum JB, Kwok W-M, Hogan QH, Sarantopoulos CD (2009) Nitric oxide activates ATP-sensitive potassium channels in mammalian sensory neurons: action by direct S-nitrosylation. Mol Pain 5:12
25. Zhou M-H, Bavencoffe A, Pan H-L (2015) Molecular basis of regulating high voltage-activated calcium channels by S-nitrosylation. J Biol Chem 290(51): 30616–30623
26. Lee J, Nelson MT, Rose KE, Todorovic SM (2013) Redox mechanism of S-nitrosothiol modulation of neuronal CaV3.2 T-type calcium channels. Mol Neurobiol 48(2):274–280
27. Sun J, Xin C, Eu JP, Stamler JS, Meissner G (2001) Cysteine-3635 is responsible for skeletal muscle ryanodine receptor modulation by NO. Proc Natl Acad Sci U S A 98(20):11158–11162
28. Bredt DS, Glatt CE, Hwang PM, Fotuhi M, Dawson TM, Snyder SH (1991) Nitric oxide synthase protein and mRNA are discretely localized in neuronal populations of the mammalian CNS together with NADPH diaphorase. Neuron 7:615–624
29. Brenman JE, Chao DS, Gee SH, McGee AW, Craven SE, Santillano DR, Wu Z, Huang F, Xia H, Peters MF, Froehner SC, Bredt DS (1996) Interaction of nitric oxide synthase with the postsynaptic density protein PSD-95 and α1-syntrophin mediated by PDZ domains. Cell 84(5):757–767
30. Lei SZ, Pan ZH, Aggarwal SK, Chen HS, Hartman J, Sucher NJ, Lipton SA (1992) Effect of nitric oxide production on the redox modulatory site of the NMDA receptor-channel complex. Neuron 8(6): 1087–1099
31. Lipton SA, Choi YB, Pan ZH, Lei SZ, Chen HS, Sucher NJ, Loscalzo J, Singel DJ, Stamler JS (1993) A redox-based mechanism for the neuroprotective and neurodestructive effects of nitric oxide and related nitroso-compounds. Nature 364(6438):626–632
32. Lipton SA, Rosenberg PA (1994) Excitatory amino acids as a final common pathway for neurologic disorders. N Engl J Med 330(9):613–622

33. Kendrick KM, Guevara-Guzman R, de la Riva C, Christensen J, Ostergaard K, Emson PC (1996) NMDA and kainate-evoked release of nitric oxide and classical transmitters in the rat striatum: in vivo evidence that nitric oxide may play a neuroprotective role. Eur J Neurosci 8(12):2619–2634
34. Manzoni O, Bockaert J (1993) Nitric oxide synthase activity endogenously modulates NMDA receptors. J Neurochem 61(1):368–370
35. Nakamura T, Lipton SA (2016) Protein S-nitrosylation as a therapeutic target for neurodegenerative diseases. Trends Pharmacol Sci 37(1):73–84
36. Ahern GP, Klyachko VA, Jackson MB (2002) cGMP and S-nitrosylation: two routes for modulation of neuronal excitability by NO. Trends Neurosci 25(10):510–517
37. Artinian L, Zhong L, Yang H, Rehder V (2012) Nitric oxide as intracellular modulator: internal production of NO increases neuronal excitability via modulation of several ionic conductances. Eur J Neurosci 36(10): 3333–3343
38. Hammarström AKM, Gage PW (1998) Inhibition of oxidative metabolism increases persistent sodium current in rat CA1 hippocampal neurons. J Physiol 510(3):735–741
39. Hammarström AKM, Gage PW (1999) Nitric oxide increases persistent sodium current in rat hippocampal neurons. J Physiol 520(Pt 2):451–461
40. Dubin AE, Patapoutian A (2010) Nociceptors: the sensors of the pain pathway. J Clin Invest 120(11): 3760–3772
41. Benn SC, Costigan M, Tate S, Fitzgerald M, Woolf CJ (2001) Developmental expression of the TTX-resistant voltage-gated sodium channels Nav1.8 (SNS) and Nav1.9 (SNS2) in primary sensory neurons. J Neurosci 21(16):6077–6085
42. Pinto V, Derkach VA, Safronov BV (2008) Role of TTX-sensitive and TTX-resistant sodium channels in Aδ- and C-fiber conduction and synaptic transmission. J Neurophysiol 99(2):617–628
43. Renganathan M, Cummins TR, Waxman SG (2002) Nitric oxide blocks fast, slow, and persistent Na+ channels in C-type DRG neurons by S-nitrosylation. J Neurophysiol 87:761–775
44. Li Z, Chapleau MW, Bates JN, Bielefeldt K, Lee H-C, Abboud FM (1998) Nitric oxide as an autocrine regulator of sodium currents in baroreceptor neurons. Neuron 20(5):1039–1049
45. Matsuda T, Bates James N, Lewis Stephen J, Abboud Francois M, Chapleau Mark W (1995) Modulation of baroreceptor activity by nitric oxide and S-nitrosocysteine. Circ Res 76(3):426–433
46. Gavillet B, Rougier J-S, Domenighetti AA, Behar R, Boixel C, Ruchat P, Lehr H-A, Pedrazzini T, Abriel H (2006) Cardiac sodium channel Nav1.5 is regulated by a multiprotein complex composed of syntrophins and dystrophin. Circ Res 99(4):407–414
47. Oceandy D, Cartwright EJ, Emerson M, Prehar S, Baudoin FM, Zi M, Alatwi N, Venetucci L, Schuh K, Williams JC, Armesilla AL, Neyses L (2007) Neuronal nitric oxide synthase signaling in the heart is regulated by the sarcolemmal calcium pump 4b. Circulation 115(4):483–492
48. Williams JC, Armesilla AL, Mohamed TMA, Hagarty CL, McIntyre FH, Schomburg S, Zaki AO, Oceandy D, Cartwright EJ, Buch MH, Emerson M, Neyses L (2006) The sarcolemmal calcium pump, alpha-1 syntrophin, and neuronal nitric-oxide synthase are parts of a macromolecular protein complex. J Biol Chem 281(33):23341–23348
49. Song W, Shou W (2012) Cardiac sodium channel Nav1.5 mutations and cardiac arrhythmia. Pediatr Cardiol 33(6):943–949
50. Ueda K, Valdivia C, Medeiros-Domingo A, Tester DJ, Vatta M, Farrugia G, Ackerman MJ, Makielski JC (2008) Syntrophin mutation associated with long QT syndrome through activation of the nNOS–SCN5A macromolecular complex. Proc Natl Acad Sci U S A 105(27):9355–9360
51. Cheng J, Valdivia CR, Vaidyanathan R, Balijepalli RC, Ackerman MJ, Makielski JC (2013) Caveolin-3 suppresses late sodium current by inhibiting nNOS-dependent S-nitrosylation of SCN5A. J Mol Cell Cardiol 61:102–110
52. Gründer S, Pusch M (2015) Biophysical properties of acid-sensing ion channels (ASICs). Neuropharmacology 94:9–18
53. Waldmann R, Champigny G, Bassilana F, Heurteaux C, Lazdunski M (1997) A proton-gated cation channel involved in acid-sensing. Nature 386(6621):173–177
54. Yermolaieva O, Leonard AS, Schnizler MK, Abboud FM, Welsh MJ (2004) Extracellular acidosis increases neuronal cell calcium by activating acid-sensing ion channel 1a. Proc Natl Acad Sci U S A 101(17):6752–6757
55. Jetti SK, Swain SM, Majumder S, Chatterjee S, Poornima V, Bera AK (2010) Evaluation of the role of nitric oxide in acid sensing ion channel mediated cell death. Nitric Oxide 22(3):213–219
56. Wang JQ, Chu X-P, Guo M-L, Jin D-Z, Xue B, Berry TJ, Fibuch EE, Mao L-M (2012) Modulation of ionotropic glutamate receptors and acid-sensing ion channels by nitric oxide. Front Physiol 3
57. Kellenberger S, Schild L (2002) Epithelial sodium channel/degenerin family of ion channels: a variety of functions for a shared structure. Physiol Rev 82(3): 735–767
58. Waldmann R, Lazdunski M (1998) H(+)-gated cation channels: neuronal acid sensors in the NaC/DEG family of ion channels. Curr Opin Neurobiol 8(3): 418–424
59. Cadiou H, Studer M, Jones NG, Smith ESJ, Ballard A, McMahon SB, McNaughton PA (2007) Modulation of acid-sensing ion channel activity by nitric oxide. J Neurosci 27(48):13251–13260
60. Wang Y-Z, Xu T-L (2011) Acidosis, acid-sensing ion channels, and neuronal cell death. Mol Neurobiol 44(3):350–358

61. Xiong Z-G, Pignataro G, Li M, Chang S-y, Simon RP (2008) Acid-Sensing Ion Channels (ASICs) as pharmacological targets for neurodegenerative diseases. Curr Opin Pharmacol 8(1):25–32
62. Gauss R, Seifert R, Kaupp UB (1998) Molecular identification of a hyperpolarization-activated channel in sea urchin sperm. Nature 393(6685):583–587
63. Jan LY, Jan YN (1990) A superfamily of ion channels. Nature 345(6277):672
64. Ludwig A, Zong X, Jeglitsch M, Hofmann F, Biel M (1998) A family of hyperpolarization-activated mammalian cation channels. Nature 393(6685):587–591
65. Santoro B, Liu DT, Yao H, Bartsch D, Kandel ER, Siegelbaum SA, Tibbs GR (1998) Identification of a gene encoding a hyperpolarization-activated pacemaker channel of brain. Cell 93(5):717–729
66. Biel M, Wahl-Schott C, Michalakis S, Zong X (2009) Hyperpolarization-activated cation channels: from genes to function. Physiol Rev 89(3):847–885
67. Noma A, Irisawa H (1976) Membrane currents in the rabbit sinoatrial node cell as studied by the double microelectrode method. Pflugers Arch 364(1):45–52
68. Robinson RB, Siegelbaum SA (2003) Hyperpolarization-activated cation currents: from molecules to physiological function. Annu Rev Physiol 65:453–480
69. Stevens DR, Seifert R, Bufe B, Muller F, Kremmer E, Gauss R, Meyerhof W, Kaupp UB, Lindemann B (2001) Hyperpolarization-activated channels HCN1 and HCN4 mediate responses to sour stimuli. Nature 413(6856):631–635
70. Brown HF, Difrancesco D, Noble SJ (1979) How does adrenaline accelerate the heart? Nature 280(5719):235–236
71. DiFrancesco D, Ducouret P, Robinson RB (1989) Muscarinic modulation of cardiac rate at low acetylcholine concentrations. Science (New York, N.Y.) 243(4891):669–671
72. Dossi RC, Nuñez A, Steriade M (1992) Electrophysiology of a slow (0.5-4 Hz) intrinsic oscillation of cat thalamocortical neurones in vivo. J Physiol 447:215–234
73. Leresche N, Lightowler S, Soltesz I, Jassik-Gerschenfeld D, Crunelli V (1991) Low-frequency oscillatory activities intrinsic to rat and cat thalamocortical cells. J Physiol 441:155–174
74. McCormick DA, Pape HC (1990) Properties of a hyperpolarization-activated cation current and its role in rhythmic oscillation in thalamic relay neurones. J Physiol 431:291–318
75. Pires da Silva M, de Almeida Moraes DJ, Mecawi ADS, Rodrigues JA, Varanda WA (2016) Nitric oxide modulates HCN channels in magnocellular neurons of the supraoptic nucleus of rats by an S-nitrosylation-dependent mechanism. J Neurosci 36(44):11320–11330
76. Wenker IC, Benoit JP, Chen X, Liu H, Horner RL, Mulkey DK (2012) Nitric oxide activates hypoglossal motoneurons by cGMP-dependent inhibition of TASK channels and cGMP-independent activation of HCN channels. J Neurophysiol 107(5):1489–1499
77. Brown CH, Bains JS, Ludwig M, Stern JE (2013) Physiological regulation of magnocellular neurosecretory cell activity: integration of intrinsic, local and afferent mechanisms. J Neuroendocrinol 25(8): 678–710
78. Kaupp UB, Seifert R (2002) Cyclic nucleotide-gated ion channels. Physiol Rev 82(3):769–824
79. Ambudkar IS, Ong HL (2007) Organization and function of TRPC channelosomes. Pflugers Arch: Eur J Physiol 455(2):187–200
80. Cheng KT, Ong HL, Liu X, Ambudkar IS (2011) Contribution of TRPC1 and Orai1 to Ca(2+) entry activated by store depletion. Adv Exp Med Biol 704: 435–449
81. Salido GM, Sage SO, Rosado JA (2009) TRPC channels and store-operated Ca(2+) entry. Biochim Biophys Acta 1793(2):223–230
82. Salazar H, Llorente I, Jara-Oseguera A, García-Villegas R, Munari M, Gordon SE, Islas LD, Rosenbaum T (2008) A single N-terminal cysteine in TRPV1 determines activation by pungent compounds from onion and garlic. Nat Neurosci 11(3):255–261
83. Xu S-Z, Sukumar P, Zeng F, Li J, Jairaman A, English A, Naylor J, Ciurtin C, Majeed Y, Milligan CJ, Bahnasi YM, Al-Shawaf E, Porter KE, Jiang L-H, Emery P, Sivaprasadarao A, Beech DJ (2008) TRPC channel activation by extracellular thioredoxin. Nature 451(7174):69–72
84. Han T, Tang Y, Li J, Xue B, Gong L, Li J, Yu X, Liu C (2017) Nitric oxide donor protects against acetic acid-induced gastric ulcer in rats via S-nitrosylation of TRPV1 on vagus nerve. Sci Rep 7(1):2063
85. Moparthi L, Survery S, Kreir M, Simonsen C, Kjellbom P, Högestätt ED, Johanson U, Zygmunt PM (2014) Human TRPA1 is intrinsically cold- and chemosensitive with and without its N-terminal ankyrin repeat domain. Proc Natl Acad Sci U S A 111(47):16901–16906
86. Paulsen CE, Armache J-P, Gao Y, Cheng Y, Julius D (2015) Structure of the TRPA1 ion channel suggests regulatory mechanisms. Nature 520(7548):511–517
87. Gaudet R (2008) A primer on ankyrin repeat function in TRP channels and beyond. Mol Biosyst 4(5): 372–379
88. Takahashi N, Mizuno Y, Kozai D, Yamamoto S, Kiyonaka S, Shibata T, Uchida K, Mori Y (2008) Molecular characterization of TRPA1 channel activation by cysteine-reactive inflammatory mediators. Channels 2(4):287–298
89. Jeevaratnam K, Chadda KR, Huang CLH, Camm AJ (2018) Cardiac potassium channels: physiological insights for targeted therapy. J Cardiovasc Pharmacol Ther 23(2):119–129
90. Yost SC (1999) Potassium channels basic aspects, functional roles, and medical significance. Anesthesiology 90(4):1186–1203

91. Gómez R, Caballero R, Barana A, Amorós I, Calvo E, López JA, Klein H, Vaquero M, Osuna L, Atienza F, Almendral J, Pinto A, Tamargo J, Delpón E (2009) Nitric oxide increases cardiac IK1 by nitrosylation of cysteine 76 of Kir2.1 channels. Circ Res 105(4):383–392
92. Bai C-X, Takahashi K, Masumiya H, Sawanobori T, Furukawa T (2004) Nitric oxide-dependent modulation of the delayed rectifier K+ current and the L-type Ca^{2+} current by ginsenoside Re, an ingredient of Panax ginseng, in guinea-pig cardiomyocytes. Br J Pharmacol 142(3):567–575
93. Chang K-C, Barth AS, Sasano T, Kizana E, Kashiwakura Y, Zhang Y, Foster DB, Marbán E (2008) CAPON modulates cardiac repolarization via neuronal nitric oxide synthase signaling in the heart. Proc Natl Acad Sci U S A 105(11):4477–4482
94. Núñez L, Vaquero M, Gómez R, Caballero R, Mateos-Cáceres P, Macaya C, Iriepa I, Gálvez E, López-Farré A, Tamargo J, Delpón E (2006) Nitric oxide blocks hKv1.5 channels by S-nitrosylation and by a cyclic GMP-dependent mechanism. Cardiovasc Res 72(1):80–89
95. Al-Owais MM, Hettiarachchi NT, Boyle JP, Scragg JL, Elies J, Dallas ML, Lippiat JD, Steele DS, Peers C (2017) Multiple mechanisms mediating carbon monoxide inhibition of the voltage-gated K^+ channel Kv1.5. Cell Death Dis 8(11):e3163
96. Taglialatela M, Pannaccione A, Iossa S, Castaldo P, Annunziato L (1999) Modulation of the K(+) channels encoded by the human ether-a-gogo-related gene-1 (hERG1) by nitric oxide. Mol Pharmacol 56(6):1298–1308
97. Aarnoudse A-JLHJ, Newton-Cheh C, de Bakker PIW, Straus SMJM, Kors JA, Hofman A, Uitterlinden AG, Witteman JCM, Stricker BHC (2007) Common NOS1AP variants are associated with a prolonged QTc interval in the Rotterdam Study. Circulation 116(1):10–16
98. Standen NB, Quayle JM, Davies NW, Brayden JE, Huang Y, Nelson MT (1989) Hyperpolarizing vasodilators activate ATP-sensitive K+ channels in arterial smooth muscle. Science (New York, N.Y.) 245(4914):177–180
99. Noma A (1983) ATP-regulated K+ channels in cardiac muscle. Nature 305(5930):147–148
100. Trube G, Hescheler J (1984) Inward-rectifying channels in isolated patches of the heart cell membrane: ATP-dependence and comparison with cell-attached patches. Pflugers Arch: Eur J Physiol 401(2):178–184
101. Bernardi H, Fosset M, Lazdunski M (1988) Characterization, purification, and affinity labeling of the brain [3H]glibenclamide-binding protein, a putative neuronal ATP-regulated K+ channel. Proc Natl Acad Sci U S A 85(24):9816–9820
102. Cook DL, Hales CN (1984) Intracellular ATP directly blocks K+ channels in pancreatic B-cells. Nature 311(5983):271–273
103. Akrouh A, Halcomb SE, Nichols CG, Sala-Rabanal M (2009) Molecular biology of KATP channels and implications for health and disease. IUBMB Life 61(10):971–978
104. McTaggart JS, Clark RH, Ashcroft FM (2010) The role of the KATP channel in glucose homeostasis in health and disease: more than meets the islet. J Physiol 588(Pt 17):3201–3209
105. Inagaki N, Gonoi T, Clement JP, Namba N, Inazawa J, Gonzalez G, Aguilar-Bryan L, Seino S, Bryan J (1995) Reconstitution of IKATP: an inward rectifier subunit plus the sulfonylurea receptor. Science 270(5239):1166–1170
106. Han J, Kim N, Joo H, Kim E, Earm YE (2002) ATP-sensitive K(+) channel activation by nitric oxide and protein kinase G in rabbit ventricular myocytes. American Journal of Physiology. Heart Circ Physiol 283(4):H1545–H1554
107. Hill MA, Yang Y, Ella SR, Davis MJ, Braun AP (2010) Large conductance, Ca2+-activated K+ channels (BKCa) and arteriolar myogenic signaling. FEBS Lett 584(10):2033–2042
108. Jackson WF (2017) Potassium channels in regulation of vascular smooth muscle contraction and growth. Adv Pharmacol (San Diego, CA) 78:89–144
109. Bolotina VM, Najibi S, Palacino JJ, Pagano PJ, Cohen RA (1994) Nitric oxide directly activates calcium-dependent potassium channels in vascular smooth muscle. Nature 368(6474):850–853
110. Chen CH, Houchi H, Ohnaka M, Sakamoto S, Niwa Y, Nakaya Y (1998) Nitric oxide activates Ca2+-activated K+ channels in cultured bovine adrenal chromaffin cells. Neurosci Lett 248(2):127–129
111. Ahern GP, Hsu S-F, Jackson MB (1999) Direct actions of nitric oxide on rat neurohypophysial K+ channels. J Physiol 520(Pt 1):165–176
112. Shin JH, Chung S, Park EJ, Uhm D-Y, Suh CK (1997) Nitric oxide directly activates calcium-activated potassium channels from rat brain reconstituted into planar lipid bilayer. FEBS Lett 415(3):299–302
113. McCartney CE, McClafferty H, Huibant JM, Rowan EG, Shipston MJ, Rowe IC (2005) A cysteine-rich motif confers hypoxia sensitivity to mammalian large conductance voltage- and Ca-activated K (BK) channel alpha-subunits. Proc Natl Acad Sci U S A 102(49):17870–17876
114. Tjong YW, Li M, Hung MW, Wang K, Fung ML (2008) Nitric oxide deficit in chronic intermittent hypoxia impairs large conductance calcium-activated potassium channel activity in rat hippocampal neurons. Free Radic Biol Med 44(4):547–557
115. Buraei Z, Yang J (2010) The ß subunit of voltage-gated Ca2+ channels. Physiol Rev 90(4):1461–1506
116. Catterall WA (2000) Structure and regulation of voltage-gated Ca2+ channels. Annu Rev Cell Dev Biol 16:521–555
117. Campbell DL, Stamler JS, Strauss HC (1996) Redox modulation of L-type calcium channels in ferret

ventricular myocytes. Dual mechanism regulation by nitric oxide and S-nitrosothiols. J Gen Physiol 108(4):277–293
118. Hu H, Chiamvimonvat N, Yamagishi T, Marban E (1997) Direct inhibition of expressed cardiac L-type Ca2+ channels by S-nitrosothiol nitric oxide donors. Circ Res 81(5):742–752
119. Poteser M, Romanin C, Schreibmayer W, Mayer B, Groschner K (2001) S-Nitrosation controls gating and conductance of the α1 subunit of class C L-type Ca2+ channels. J Biol Chem 276(18):14797–14803
120. Sun J, Picht E, Ginsburg Kenneth S, Bers Donald M, Steenbergen C, Murphy E (2006) Hypercontractile female hearts exhibit increased S-nitrosylation of the L-type Ca2+ channel α1 subunit and reduced ischemia/reperfusion injury. Circ Res 98(3):403–411
121. Gonzalez DR, Treuer A, Sun Q-A, Stamler JS, Hare JM (2009) S-nitrosylation of cardiac ion channels. J Cardiovasc Pharmacol 54(3):188–195
122. Sun J, Morgan M, Shen R-F, Steenbergen C, Murphy E (2007) Preconditioning results in S-Nitrosylation of proteins involved in regulation of mitochondrial energetics and calcium transport. Circ Res 101(11): 1155–1163
123. Tong G, Aponte AM, Kohr MJ, Steenbergen C, Murphy E, Sun J (2014) Postconditioning leads to an increase in protein S-nitrosylation. Am J Physiol-Heart Circ Physiol 306(6):H825–H832
124. Yue Z-J, Xu P-T, Jiao B, Chang H, Song Z, Xie M-J, Yu Z-B (2015) Nitric oxide protects L-type calcium channel of cardiomyocyte during long-term isoproterenol stimulation in tail-suspended rats. BioMed Research International
125. Catterall WA, Few AP (2008) Calcium channel regulation and presynaptic plasticity. Neuron 59(6): 882–901
126. McGivern JG (2006) Targeting N-type and T-type calcium channels for the treatment of pain. Drug Discov Today 11(5–6):245–253
127. Park JF, Luo ZD (2010) Calcium channel functions in pain processing. Channels 4(6):510–517
128. Snutch TP (2005) Targeting chronic and neuropathic pain: the N-type calcium channel comes of age. NeuroRx 2(4):662–670
129. Lanner JT (2012) Ryanodine receptor physiology and its role in disease. Adv Exp Med Biol 740:217–234
130. Fabiato A (1983) Calcium-induced release of calcium from the cardiac sarcoplasmic reticulum. Am J Physiol 245(1):C1–C14
131. Melzer W, Herrmann-Frank A, Lüttgau HC (1995) The role of Ca2+ ions in excitation-contraction coupling of skeletal muscle fibres. Biochim Biophys Acta 1241(1):59–116
132. Bendall JK, Damy T, Ratajczak P, Loyer X, Monceau V, Marty I, Milliez P, Robidel E, Marotte F, Samuel J-L, Heymes C (2004) Role of myocardial neuronal nitric oxide synthase-derived nitric oxide in beta-adrenergic hyporesponsiveness after myocardial infarction-induced heart failure in rat. Circulation 110:2368–2375
133. Salanova M, Schiffl G, Rittweger J, Felsenberg D, Blottner D (2008) Ryanodine receptor type-1 (RyR1) expression and protein S-nitrosylation pattern in human soleus myofibres following bed rest and exercise countermeasure. Histochem Cell Biol 130:105–118
134. Eu JP, Sun J, Xu L, Stamler JS, Meissner G (2000) The skeletal muscle calcium release channel: coupled O2 sensor and NO signaling functions. Cell 102(4): 499–509
135. Gonzalez DR, Beigi F, Treuer AV, Hare JM (2007) Deficient ryanodine receptor S-nitrosylation increases sarcoplasmic reticulum calcium leak and arrhythmogenesis in cardiomyocytes. Proc Natl Acad Sci U S A 104(51):20612–20617
136. Xu L, Eu JP, Meissner G, Stamler JS (1998) Activation of the cardiac calcium release channel (ryanodine receptor) by poly-S-nitrosylation. Science (New York, N.Y.) 279(5348):234–237
137. Kakizawa S, Yamazawa T, Chen Y, Ito A, Murayama T, Oyamada H, Kurebayashi N, Sato O, Watanabe M, Mori N, Oguchi K, Sakurai T, Takeshima H, Saito N, Iino M (2012) Nitric oxide-induced calcium release via ryanodine receptors regulates neuronal function. EMBO J 31(2):417–428
138. Stoyanovsky D, Murphy T, Anno PR, Kim Y-M, Salama G (1997) Nitric oxide activates skeletal and cardiac ryanodine receptors. Cell Calcium 21(1): 19–29
139. Hart JD, Dulhunty AF (2000) Nitric oxide activates or inhibits skeletal muscle ryanodine receptors depending on its concentration, membrane potential and ligand binding. J Membr Biol 173(3):227–236
140. Mészáros LG, Minarovic I, Zahradnikova A (1996) Inhibition of the skeletal muscle ryanodine receptor calcium release channel by nitric oxide. FEBS Lett 380(1–2):49–52
141. Aracena-Parks P, Goonasekera SA, Gilman CP, Dirksen RT, Hidalgo C, Hamilton SL (2006) Identification of cysteines involved in S-nitrosylation, S-glutathionylation, and oxidation to disulfides in ryanodine receptor type 1. J Biol Chem 281(52): 40354–40368
142. Aracena P, Tang W, Hamilton SL, Hidalgo C (2005) Effects of S-glutathionylation and S-nitrosylation on calmodulin binding to triads and FKBP12 binding to type 1 calcium release channels. Antioxid Redox Signal 7(7-8):870–881
143. Durham WJ, Aracena-Parks P, Long C, Rossi AE, Goonasekera SA, Boncompagni S, Galvan DL, Gilman CP, Baker MR, Shirokova N, Protasi F, Dirksen R, Hamilton SL (2008) RyR1 S-nitrosylation underlies environmental heat stroke and sudden death in Y522S RyR1 knockin mice. Cell 133(1):53–65
144. Mikami Y, Kanemaru K, Okubo Y, Nakaune T, Suzuki J, Shibata K, Sugiyama H, Koyama R, Murayama T, Ito A, Yamazawa T, Ikegaya Y, Sakurai T, Saito N, Kakizawa S, Iino M (2016) Nitric oxide-induced activation of the type 1 ryanodine

receptor is critical for epileptic seizure-induced neuronal cell death. EBioMedicine 11:253–261

145. Sun J, Yamaguchi N, Xu L, Eu JP, Stamler JS, Meissner G (2008a) Regulation of the cardiac muscle ryanodine receptor by O2 tension and S-nitrosoglutathione. Biochemistry 47(52): 13985–13990
146. Sun Y-G, Cao Y-X, Wang W-W, Ma S-F, Yao T, Zhu Y-C (2008b) Hydrogen sulphide is an inhibitor of L-type calcium channels and mechanical contraction in rat cardiomyocytes. Cardiovasc Res 79(4): 632–641
147. Bos EM, van Goor H, Joles JA, Whiteman M, Leuvenink HGD (2015) Hydrogen sulfide: physiological properties and therapeutic potential in ischaemia. Br J Pharmacol 172(6):1479–1493
148. Kimura H (2014) Production and physiological effects of hydrogen sulfide. Antioxid Redox Signal 20(5):783–793
149. Mustafa AK, Gadalla MM, Sen N, Kim S, Mu W, Gazi SK, Barrow RK, Yang G, Wang R, Snyder SH (2009) H2S signals through protein S-sulfhydration. Sci Signal 2(96):ra72
150. Fukuto JM, Ignarro LJ, Nagy P, Wink DA, Kevil CG, Feelisch M, Cortese-Krott MM, Bianco CL, Kumagai Y, Hobbs AJ, Lin J, Ida T, Akaike T (2018) Biological hydropersulfides and related polysulfides - a new concept and perspective in redox biology. FEBS Lett 592(12):2140–2152
151. Paul BD, Snyder SH (2012) H(2)S signalling through protein sulfhydration and beyond. Nat Rev Mol Cell Biol 13(8):499–507
152. Zhang D, Du J, Tang C, Huang Y, Jin H (2017) H2S-Induced sulfhydration: biological function and detection methodology. Front Pharmacol 8:608
153. Jiang HL, Wu HC, Li ZL, Geng B, Tang CS (2005) Changes of the new gaseous transmitter H2S in patients with coronary heart disease. Di Yi Jun Yi Da Xue Xue Bao 25(8):951–954
154. Salloum FN (2015) Hydrogen sulfide and cardioprotection—mechanistic insights and clinical translatability. Pharmacol Ther 152:11–17
155. Luo Y, Wu PF, Zhou J, Xiao W, He JG, Guan XL, Zhang JT, Hu ZL, Wang F, Chen JG (2014) Aggravation of seizure-like events by hydrogen sulfide: involvement of multiple targets that control neuronal excitability. CNS Neurosci Ther 20(5):411–419
156. McCune CD, Chan SJ, Beio ML, Shen W, Chung WJ, Szczesniak LM, Chai C, Koh SQ, Wong PT, Berkowitz DB (2016) Zipped synthesis by cross-metathesis provides a cystathionine beta-synthase inhibitor that attenuates cellular H2S levels and reduces neuronal infarction in a rat ischemic stroke model. ACS Cent Sci 2(4):242–252
157. Brayden JE (2002) Functional roles of KATP channels in vascular smooth muscle. Clin Exp Pharmacol Physiol 29(4):312–316
158. Yang W, Yang G, Jia X, Wu L, Wang R (2005) Activation of KATP channels by H2S in rat insulin-secreting cells and the underlying mechanisms. J Physiol 569(Pt 2):519–531
159. Hilgemann DW, Feng S, Nasuhoglu C (2001) The complex and intriguing lives of PIP2 with ion channels and transporters. Sci STKE 2001:re19
160. Johansen D, Ytrehus K, Baxter GF (2006) Exogenous hydrogen sulfide (H2S) protects against regional myocardial ischemia-reperfusion injury--evidence for a role of K ATP channels. Basic Res Cardiol 101(1):53–60
161. Ha J, Xu Y, Kawano T, Hendon T, Baki L, Garai S, Papapetropoulos A, Thakur GA, Plant LD, Logothetis DE (2018) Hydrogen sulfide inhibits Kir2 and Kir3 channels by decreasing sensitivity to the phospholipid phosphatidylinositol 4,5-bisphosphate (PIP2). J Biol Chem 293(10): 3546–3561
162. Nagai Y, Tsugane M, Oka J, Kimura H (2004) Hydrogen sulfide induces calcium waves in astrocytes. FASEB J 18(3):557–559
163. Greiner R, Pálinkás Z, Bäsell K, Becher D, Antelmann H, Nagy P, Dick TP (2013) Polysulfides link H2S to protein thiol oxidation. Antioxid Redox Signal 19(15):1749–1765
164. Kimura H (2013) Physiological role of hydrogen sulfide and polysulfide in the central nervous system. Neurochem Int 63(5):492–497
165. Ogawa H, Takahashi K, Miura S, Imagawa T, Saito S, Tominaga M, Ohta T (2012) H2S functions as a nociceptive messenger through transient receptor potential ankyrin 1 (TRPA1) activation. Neuroscience 218:335–343
166. Sacharow SJ, Picker JD, Levy HL (1993). Homocystinuria caused by cystathionine beta-synthase deficiency. In: Adam MP, Ardinger HH, Pagon RA, et al. (Eds.), GeneReviews((R)). Seattle, WA
167. Liu Y, Yang R, Liu X, Zhou Y, Qu C, Kikuiri T, Wang S, Zandi E, Du J, Ambudkar IS, Shi S (2014) Hydrogen sulfide maintains mesenchymal stem cell function and bone homeostasis via regulation of Ca2+ channel sulfhydration. Cell Stem Cell 15(1): 66–78
168. Gambari L, Lisignoli G, Cattini L, Manferdini C, Facchini A, Grassi F (2014) Sodium hydrosulfide inhibits the differentiation of osteoclast progenitor cells via NRF2-dependent mechanism. Pharmacol Res 87:99–112
169. Zhang R, Sun Y, Tsai H, Tang C, Jin H, Du J (2012) Hydrogen sulfide inhibits L-type calcium currents depending upon the protein sulfhydryl state in rat cardiomyocytes. PLoS One 7(5)
170. Dai L, Qian Y, Zhou J, Zhu C, Jin L, Li S (2019) Hydrogen sulfide inhibited L-type calcium channels (CaV1.2) via up-regulation of the channel sulfhydration in vascular smooth muscle cells. Eur J Pharmacol 858:172455
171. García-Bereguiaín MA, Samhan-Arias AK, Martín-Romero FJ, Gutiérrez-Merino C (2008) Hydrogen

sulfide raises cytosolic calcium in neurons through activation of L-type Ca2+ channels. Antioxid Redox Signal 10(1):31–42
172. Yong QC, Choo CH, Tan BH, Low C-M, Bian J-S (2010) Effect of hydrogen sulfide on intracellular calcium homeostasis in neuronal cells. Neurochem Int 56(3):508–515
173. Kimura Y, Kimura H (2004) Hydrogen sulfide protects neurons from oxidative stress. FASEB J 18(10):1165–1167
174. Quan X, Luo H, Liu Y, Xia H, Chen W, Tang Q (2015) Hydrogen sulfide regulates the colonic motility by inhibiting both L-type calcium channels and BKCa channels in smooth muscle cells of rat colon. PLoS One 10:e0121331
175. Meng X-M, Huang X, Zhang C-M, Liu D-H, Lu H-L, Kim Y, Xu W-X (2015) Hydrogen sulfide-induced enhancement of gastric fundus smooth muscle tone is mediated by voltage-dependent potassium and calcium channels in mice. World J Gastroenterol 21: 4840–4851
176. Ogilby PR (2010) Singlet oxygen: there is indeed something new under the sun. Chem Soc Rev 39(8): 3181–3209
177. Gao W, Su Z, Liu Q, Zhou L (2014) State-dependent and site-directed photodynamic transformation of HCN2 channel by singlet oxygen. J Gen Physiol 143(5):633–644
178. Kanofsky JR, Hoogland H, Wever R, Weiss SJ (1988) Singlet oxygen production by human eosinophils. J Biol Chem 263(20):9692–9696
179. Schweitzer C, Schmidt R (2003) Physical mechanisms of generation and deactivation of singlet oxygen. Chem Rev 103(5):1685–1757
180. DeRosa MC, Crutchley RJ (2002) Photosensitized singlet oxygen and its applications. Coord Chem Rev 233–234:351–371
181. da Silva EFF, Pimenta FM, Pedersen BW, Blaikie FH, Bosio GN, Breitenbach T, Westberg M, Bregnhøj M, Etzerodt M, Arnaut LG, Ogilby PR (2016) Intracellular singlet oxygen photosensitizers: on the road to solving the problems of sensitizer degradation, bleaching and relocalization. Integr Biol: Quant Biosci Nano to Macro 8(2): 177–193
182. Mano CM, Prado FM, Massari J, Ronsein GE, Martinez GR, Miyamoto S, Cadet J, Sies H, Medeiros MHG, Bechara EJH, Di Mascio P (2014) Excited singlet molecular $O_2(^1\Delta g)$ is generated enzymatically from excited carbonyls in the dark. Sci Rep 4:5938
183. Onyango AN (2016) Endogenous generation of singlet oxygen and ozone in human and animal tissues: mechanisms, biological significance, and influence of dietary components. Oxid Med Cell Longev 2016: 2398573
184. Prasad A, Ferretti U, Sedlářová M, Pospíšil P (2016) Singlet oxygen production in *Chlamydomonas reinhardtii* under heat stress. Sci Rep 6:20094
185. Klotz L-O, Kröncke K-D, Sies H (2003) Singlet oxygen-induced signaling effects in mammalian cells. Photochem Photobiol Sci 2(2):88–94
186. Steinbeck MJ, Khan AU, Karnovsky MJ (1992) Intracellular singlet oxygen generation by phagocytosing neutrophils in response to particles coated with a chemical trap. J Biol Chem 267: 13425–13433
187. Baier J, Maisch T, Maier M, Engel E, Landthaler M, Bäumler W (2006) Singlet oxygen generation by UVA light exposure of endogenous photosensitizers. Biophys J 91(4):1452–1459
188. Bäumler W, Regensburger J, Knak A, Felgenträger A, Maisch T (2012) UVA and endogenous photosensitizers--the detection of singlet oxygen by its luminescence. Photochem Photobiol Sci 11(1):107–117
189. Kochevar IE (2004) Singlet oxygen signaling: from intimate to global. Science's STKE: Signal Transduction knowl Environ 2004(221):pe7
190. Liao JC, Roider J, Jay DG (1994) Chromophore-assisted laser inactivation of proteins is mediated by the photogeneration of free radicals. Proc Natl Acad Sci U S A 91(7):2659–2663
191. Tour O, Meijer RM, Zacharias DA, Adams SR, Tsien RY (2003) Genetically targeted chromophore-assisted light inactivation. Nat Biotechnol 21(12): 1505–1508
192. Wojtovich AP, Wei AY, Sherman TA, Foster TH, Nehrke K (2016) Chromophore-assisted light inactivation of mitochondrial electron transport chain complex II in caenorhabditis elegans. Sci Rep 6:29695
193. Agostinis P, Berg K, Cengel KA, Foster TH, Girotti AW, Gollnick SO, Hahn SM, Hamblin MR, Juzeniene A, Kessel D, Korbelik M, Moan J, Mroz P, Nowis D, Piette J, Wilson BC, Golab J (2011) Photodynamic therapy of cancer: an update. CA Cancer J Clin 61(4):250–281
194. Babes A, Sauer SK, Moparthi L, Kichko TI, Neacsu C, Namer B, Filipovic M, Zygmunt PM, Reeh PW, Fischer MJ (2016) Photosensitization in porphyrias and photodynamic therapy involves TRPA1 and TRPV1. J Neurosci 36(19):5264–5278
195. Eisenman LN, Shu H-J, Akk G, Wang C, Manion BD, Kress GJ, Evers AS, Steinbach JH, Covey DF, Zorumski CF, Mennerick S (2007) Anticonvulsant and anesthetic effects of a fluorescent neurosteroid analog activated by visible light. Nat Neurosci 10(4): 523–530
196. Jiang HN, Li Y, Cui ZJ (2017) Photodynamic physiology-photonanomanipulations in cellular physiology with protein photosensitizers. Front Physiol 8:191
197. Stief TW (2003) The physiology and pharmacology of singlet oxygen. Med Hypotheses 60(4):567–572
198. Valenzeno DP, Tarr M (1991) Membrane photomodification of cardiac myocytes: potassium and leakage currents. Photochem Photobiol 53(2): 195–201

199. Oxford GS, Pooler JP, Narahashi T (1977) Internal and external application of photodynamic sensitizers on squid giant axons. J Membr Biol 36(2-3):159–173
200. Tarr M, Valenzeno DP (1991) Modification of cardiac ionic currents by photosensitizer-generated reactive oxygen. J Mol Cell Cardiol 23(5):639–649
201. Cui ZJ, Kanno T (1997) Photodynamic triggering of calcium oscillation in the isolated rat pancreatic acini. J Physiol 504(Pt 1):47–55
202. Valenzeno DP, Tarr M (1998) GH3 cells, ionic currents and cell killing: photomodification sensitized by Rose Bengal. Photochem Photobiol 68(4): 519–526
203. Sack JT, Stephanopoulos N, Austin DC, Francis MB, Trimmer JS (2013) Antibody-guided photoablation of voltage-gated potassium currents. J Gen Physiol 142(3):315–324
204. Hill K, Schaefer M (2009) Ultraviolet light and photosensitising agents activate TRPA1 via generation of oxidative stress. Cell Calcium 45(2):155–164
205. Eisenman LN, Shu HJ, Wang C, Aizenman E, Covey DF, Zorumski CF, Mennerick S (2009) NMDA potentiation by visible light in the presence of a fluorescent neurosteroid analogue. J Physiol 587 (Pt 12):2937–2947
206. Rokitskaya TI, Firsov AM, Kotova EA, Antonenko YN (2015) Photodynamic inactivation of gramicidin channels in bilayer lipid membranes: protective efficacy of singlet oxygen quenchers depends on photosensitizer location. Biochemistry (Mosc) 80(6): 745–751
207. Rossi LM, Silva PR, Vono LL, Fernandes AU, Tada DB, Baptista MS (2008) Protoporphyrin IX nanoparticle carrier: preparation, optical properties, and singlet oxygen generation. Langmuir 24(21): 12534–12538
208. Idikuda V, Gao W, Grant K, Su Z, Liu Q, Zhou L (2018a) Singlet oxygen modification abolishes voltage-dependent inactivation of the sea urchin spHCN channel. J Gen Physiol 150(9):1273–1286
209. Idikuda V, Gao W, Su Z, Liu Q, Zhou L (2018b) cAMP binds to closed, inactivated, and open sea urchin HCN channels in a state-dependent manner. J Gen Physiol
210. Wu S, Gao W, Xie C, Xu X, Vorvis C, Marni F, Hackett AR, Liu Q, Zhou L (2012) Inner activation gate in S6 contributes to the state-dependent binding of cAMP in full-length HCN2 channel. J Gen Physiol 140(1):29–39
211. Wu S, Vysotskaya ZV, Xu X, Xie C, Liu Q, Zhou L (2011) State-dependent cAMP binding to functioning HCN channels studied by patch-clamp fluorometry. Biophys J 100(5):1226–1232
212. Proenza C, Angoli D, Agranovich E, Macri V, Accili EA (2002) Pacemaker channels produce an instantaneous current. J Biol Chem 277:5101–5109
213. Shu X, Lev-Ram V, Deerinck TJ, Qi Y, Ramko EB, Davidson MW, Jin Y, Ellisman MH, Tsien RY (2011) A genetically encoded tag for correlated light and electron microscopy of intact cells, tissues, and organisms. PLoS Biol 9(4):e1001041
214. Ryu S, Yellen G (2012) Charge movement in gating-locked HCN channels reveals weak coupling of voltage sensors and gate. J Gen Physiol 140(5):469–479
215. Shin KS, Maertens C, Proenza C, Rothberg BS, Yellen G (2004) Inactivation in HCN channels results from reclosure of the activation gate: desensitization to voltage. Neuron 41(5):737–744
216. Boycott HE, Dallas ML, Elies J, Pettinger L, Boyle JP, Scragg JL, Gamper N, Peers C (2013) Carbon monoxide inhibition of Cav3.2 T-type Ca2+ channels reveals tonic modulation by thioredoxin. FASEB J 27(8):3395–3407
217. Wilkinson WJ, Kemp PJ (2011) Carbon monoxide: an emerging regulator of ion channels. J Physiol-London 589(13):3055–3062
218. Telezhkin V, Brazier SP, Cayzac S, Müller CT, Riccardi D, Kemp PJ (2009) Hydrogen sulfide inhibits human BKCa channels. In: Gonzalez C, Nurse CA, Peers C et al (eds) Arterial chemoreceptors: arterial chemoreceptors. Dordrecht, pp 65–72
219. Williams SE, Brazier SP, Baban N, Telezhkin V, Müller CT, Riccardi D, Kemp PJ (2008) A structural motif in the C-terminal tail of slo1 confers carbon monoxide sensitivity to human BK Ca channels. Pflugers Arch: Eur J Physiol 456(3):561–572
220. Filipovic MR, Miljkovic J, Nauser T, Royzen M, Klos K, Shubina T, Koppenol WH, Lippard SJ, Ivanovic-Burmazovic I (2012) Chemical characterization of the smallest S-nitrosothiol, HSNO; cellular cross-talk of H2S and S-nitrosothiols. J Am Chem Soc 134(29):12016–12027
221. Ivanovic-Burmazovic I, Filipovic MR (2019) Saying NO to H2S: a story of HNO, HSNO, and SSNO(). Inorg Chem 58(7):4039–4051
222. Paul BD, Snyder SH (2018) Gasotransmitter hydrogen sulfide signaling in neuronal health and disease. Biochem Pharmacol 149:101–109
223. Miyamoto R, Koike S, Takano Y, Shibuya N, Kimura Y, Hanaoka K, Urano Y, Ogasawara Y, Kimura H (2017) Polysulfides (H2Sn) produced from the interaction of hydrogen sulfide (H2S) and nitric oxide (NO) activate TRPA1 channels. Sci Rep 7:45995
224. Rodriguez-Menchaca AA, Adney SK, Zhou L, Logothetis DE (2012) Dual regulation of voltage-sensitive ion channels by PIP(2). Front Pharmacol 3:170
225. Suh BC, Hille B (2008) PIP2 is a necessary cofactor for ion channel function: how and why? Annu Rev Biophys 37:175–195
226. Yamamoto Y, Katsumata O, Furuyama S, Sugiya H (2004) Ca2+, calmodulin and phospholipids regulate nitricoxide synthase activity in the rabbit submandibular gland. J Comp Physiol B 174(8):593–599

DEG/ENaC Ion Channels in the Function of the Nervous System: From Worm to Man

9

Laura Bianchi

Abstract

DEG/ENaC channels are voltage-independent Na^+/Ca^{2+} channels that are conserved across species and are expressed in many different cell types and tissues, where they contribute to a wide array of physiological functions from transepithelial Na^+ transport, to sensory perception, and learning and memory. In this chapter, we focus on the members of this family that are expressed in the nervous system, grouping them based on their function. Structurally, DEG/ENaC channels are trimers formed by either identical or homologous subunits, each one protruding from the plasma membrane like a clenched hand. Crystallographic studies on chicken ASIC1a in the closed, inactivated, and open states revealed important details about the gating and permeation properties of these channels, and overall they show that the extracellular domain of the channel undergoes large conformational changes during gating. The vast majority of the channel's extracellular domain is conserved across different members and species; however, key changes including the insertion of extra loops near the finger and palm domains most likely confers gating specificity. Indeed, DEG/ENaC channels are gated by a wide range of stimuli, including mechanical forces, protons, and peptides, owing to the wide array of physiological functions they serve. Interestingly, DEG/ENaC channels are not only expressed in neurons but also in glia. Work in *C. elegans* is now beginning to shed new light on the role of glial DEG/ENaC in the function of the nervous system and suggests that they may be implicated in controlling ionic concentrations in the extracellular microenvironment. Finally, DEG/ENaC channels can become toxic and cause neuronal death when they are hyperactivated by genetic mutations or prolonged acidosis causing them to contribute to neuronal demise in stroke and ischemia. Taken together, molecular, structural, and behavioral work on DEG/ENaC channels expressed in the nervous system of different species highlights the crucial role of these channels in neuronal function. These data place DEG/ENaC channels in an excellent position for being considered as drug targets for the treatment of several neurological conditions and disorders from pain to epilepsy and ischemia.

Keywords

DEG/ENaC · ASIC · MEC-4 · Stomatin · Paraoxonase · Ischemia · Touch · *C. elegans* · Drosophila

L. Bianchi (✉)
Department Physiology and Biophysics, University of Miami, Miller School of Medicine, Miami, FL, USA
e-mail: lbianchi@med.miami.edu

L. Zhou (ed.), *Ion Channels in Biophysics and Physiology*, Advances in Experimental Medicine and Biology 1349, https://doi.org/10.1007/978-981-16-4254-8_9

9.1 Introduction

DEG/ENaC channels are metazoan-specific Na^+/Ca^{2+} channels that operate in a wide range of physiological functions from Na^+ reabsorption, to sensory perception, and learning and memory. In this chapter, we will focus on DEG/ENaCs expressed in the nervous system of *C. elegans*, Drosophila, and mammals. We will review what we have learned so far about the structure/function of these channels and the molecular mechanisms in which they are involved, both in physiological and in pathological conditions, grouping channels across species based on their function.

The first two DEG/ENaC family members that were cloned were the *C. elegans* degenerin *mec-4* (the DEG part of the name) and the alpha subunit of the rat epithelial Na^+ channel (ENaC) [1, 2]. The Chalfie lab was studying mutationally induced neurodegeneration in the nematode *C. elegans*. *mec-4* was identified by positional cloning as the gene that caused degeneration of six gentle touch neurons when mutated from Alanine to Valine at position 713 (*mec-4(d)*) [2]. Two years later, Cecilia Canessa cloned a homolog of *mec-4* from epithelial cells of the rat colon. The channel subunit, which was named alpha ENaC, produced voltage-independent Na^+ currents blocked by the antidiuretic amiloride when expressed in *Xenopus* oocytes [1]. Alpha ENaC homologs Acid Sensing Ion channels (ASICs) were discovered by sequence homology in 1996 [3] and in 1997 were identified as proton gated cation channels [4]. Two years prior, in 1995, a DEG/ENaC channel directly gated by FMRFamide peptides was cloned from the snail *Helix aspersa*. This channel increases the excitability of snail neurons when exposed to FMRFamide peptides [5] (Fig. 9.1a, b).

Over the years, many other DEG/ENaC subunits have been identified both in mammals and in invertebrates. Interestingly, it was determined that while the mammalian family of DEG/ENaCs includes only eight members, four of which are expressed in the nervous system, invertebrate families encode many more DEG/ENaCs (the *C. elegans* and Drosophila genomes encode as many as 30 and 31 genes homologous to DEG/ENaC channels respectively) [8, 9]. A possible explanation for this number could be that invertebrate DEG/ENaCs are more specialized with each subunit, performing only a specific part of the physiological role handled by their human counterpart [10].

9.2 DEG/ENaC Channels Structure

Hydrophobicity sequence and biochemical analysis show that DEG/ENaC channel subunits, which are 550–950 long polypeptides, span the membrane twice, leaving N- and C-termini in the intracellular space [11–13]. The sequence also features a series of highly conserved cysteines (10–14) in the extracellular domain. Structure/function and genetic studies initially suggested that the channel was formed by three to nine DEG/ENaC subunits [14–21]. The nature of the channel arrangement was finally revealed in 2007 by the resolution of the crystal structure of chicken ASIC1a which is one of the neuronally expressed family members [6].

The crystal of a truncated chicken ASIC1a, missing the intracellular termini, was solved at 1.9 Å resolution and revealed that the channel is formed by three subunits arranged around a central pore (Fig. 9.2a). More than 65% of each subunit is located extracellularly with a structure resembling a clenched hand protruding from the plasma membrane. The clenched hand is formed by 7 alpha helices and 12 beta-sheets, with one of the beta-sheets forming almost the entire palm domain (Fig. 9.2b, c). The highly conserved cysteines are located in the thumb domain of the channel and form 7 cysteine bridges that confer rigidity to this part of the protein. Stimuli that gate DEG/ENaCs, such as H^+, interact with channel residues near the thumb domain and cause the channel to open by a conformational change of this domain, which is then transmitted down to the channel wrist, thanks to the stiffness of the thumb domain [24] (Fig. 9.3a). In the wrist domain of ASIC1a, two aromatic residues,

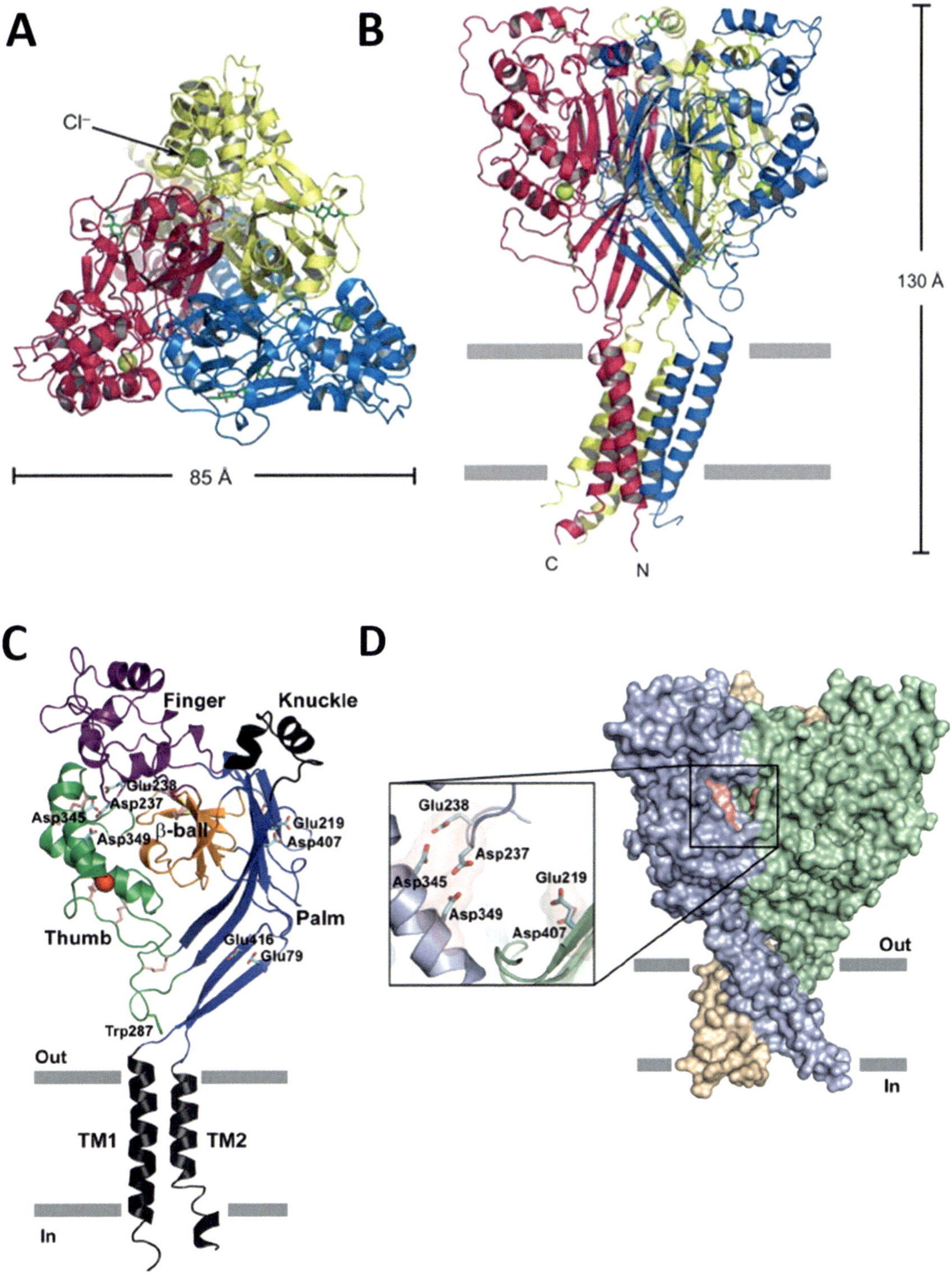

Fig. 9.1 Crystal structure of chicken ASIC1a. (**a** and **b**) Top and side view of truncated chicken ASIC1a shown in ribbon representation based on crystals solved at 1.9 A resolution. Each of the three subunits is labeled with a different color. In green, the Cl^- ions that were found bound to each subunit. N-linked carbohydrates are shown in green and in stick representation (figure adapted and printed with permission from [6]). (**c**) Ribbon representation of one chicken ASIC1a subunit. Each domain of the "clenched hand" structure is labeled and depicted in a

Tyr72, located right after the first transmembrane domain, and Trp288, located in a loop between the palm and the thumb domains, interact like a ball and a socket to allow channel gating [22]. Indeed, mutations that disrupt this interaction shift the proton sensitivity of the channel toward more acidic pHs. Conversely, swapping these two residues (Trp72/Tyr288) preserves channel function, confirming that these two amino acids need to interact to gate ASIC1a.

The resolution of the crystal structure of chicken ASIC1a revealed other important features of this family of channels: (1) in its inactivated state (ASIC1a initial crystals were formed at pH 5, a pH in which this channel is inactivated), the second transmembrane domain is tilted. Subsequent structure/function studies using MTSEA reagents suggested that the opening of the pore might derive from the straightening of this tilt [40]. (2) At the interface between the finger and the thumb domains, there is a pocket containing acidic residues (Glu219, Asp237, Glu238, Asp345, Asp349, and Asp407 in chicken ASIC1a, Fig. 9.2c, d). This acidic pocket was proposed to bind H^+. Site-directed mutagenesis indeed showed that neutralization of the acidic residues in this pocket causes a shift in proton sensitivity [6]. Later studies showed that extracellular calcium, which modulates channel gating, also binds to this acidic pocket [41, 42]. (3) At the base of the thumb domain, the crystal structure reveals a bound chloride ion [6, 41]. Disrupting Cl^- binding at this site alters pH-dependent gating, speeds desensitization, and attenuates channel tachyphylaxis [43], a phenomenon by which repeated stimulations of the channel leads to smaller and smaller currents, although enough time between stimulations should allow for recovery from desensitization.

Further crystallographic studies of ASIC1a in a complex with the snake toxin MitTx, which locks the channel in an open configuration, revealed that the helix that constitutes the second transmembrane domain is discontinuous at the selectivity filter defined by the 'GAS' belt (the GAS motif is (Gly/Ser)-X-Ser) and that the selectivity mechanism is based on discrimination of the hydrated ion size [44]. More recent crystallographic analysis of chicken ASIC1a in resting as well as desensitized states, at high and low pH, shows that divalent cations bind the channel, both at the finger domain and in the channel vestibule [41]. While binding at the finger domain is state-independent, the binding in the vestibule is detected only in the resting state (at high pH). Similarly, Cl^- ions that were seen bound to the channel in its open and inactivated states [6, 44] cannot be detected in the channel resting state (at high pH). Presumably, this is due to the disruption of the Cl^- binding site generated by the rearrangement of thumb helices α4 and α5 in the channel resting conformation. Overall, the comparison of the structure of ASIC1a in crystals obtained under different conditions favoring the open, closed, or desensitized states, respectively, highlights the extent of the conformational changes the channel undergoes during gating.

The pore contains three Na^+ binding sites that allow Na^+ ions to permeate in a single file, where not all the sites are occupied at once. Interestingly, while the residues that coordinate Na^+ binding in chicken ASIC1a are glycines (G432, G436, G439, and G443), these residues are

Fig. 9.1 (continued) different color. The two transmembrane domains are in black and labeled TM1 and TM2. The amino acids that constitute the acidic pocket (E219, E238, D237, D345, D349, and D407) are in stick representation. The bound Cl^- ion is shown here in red (This research was originally published in the Journal of Biological Chemistry [7]). (**d**) Solvent accessible surface representation of the chicken ASIC1a trimer. Each subunit is depicted in a different color, the acidic pocket is shown in pink. The inset shows a close-up view of the acidic pocket. Within the acidic pocket, there are three pairs of carboxyl-carboxylate interactions between the side chains of aspartate or glutamate residues (This research was originally published in the Journal of Biological Chemistry [7])

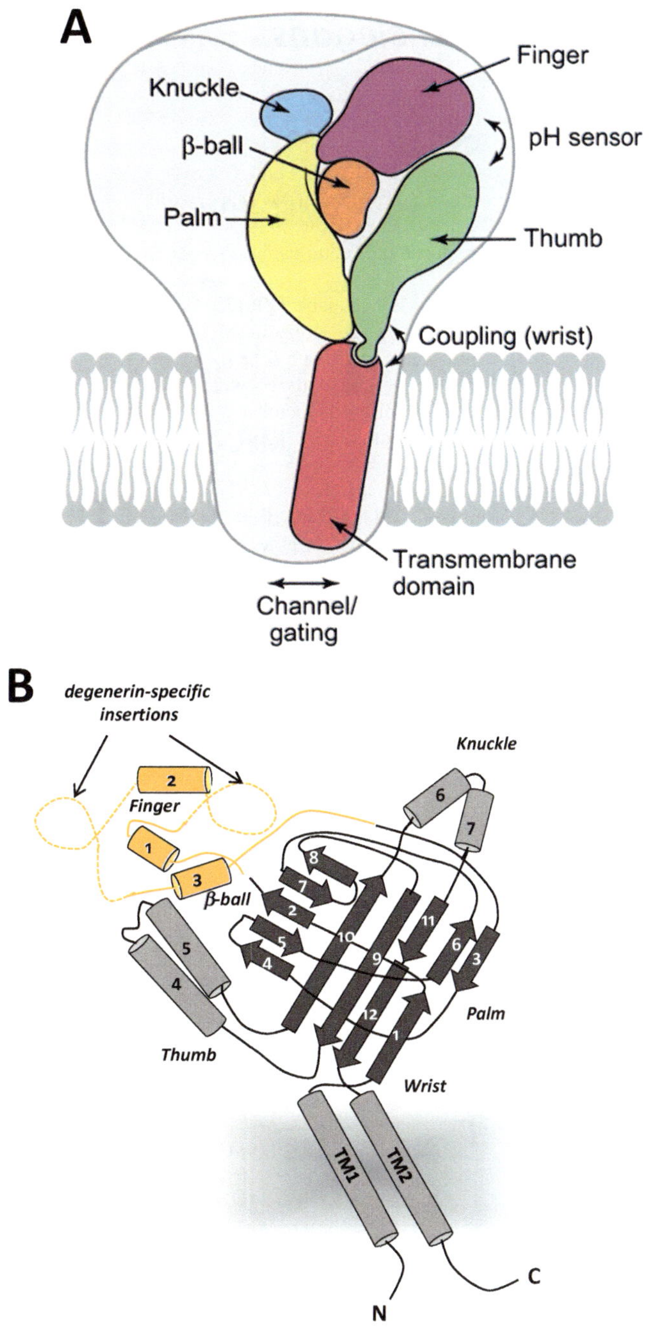

Fig. 9.2 ASIC1a gating and degenerin insertion. (**a**) Schematic representation of chicken ASIC1a depicting the channel domains in different colors. Protons bind in the acidic pocket at the interface between the thumb and the finger domains causing a conformational change transmitted down to the wrist domain via the rigid thumb domain. At the wrist, a ball-socket type of interaction between W288, in a loop protruding from the palm domain, and Y72, at the end of the first transmembrane domain leads to channel gating (figure adapted from [6]) [22]. (**b**) Schematic representation of the structure of *C. elegans* MEC-4 which was inferred by threading MEC-4 sequence onto the structure of chicken ASIC1a. Cylinders and arrows represent α-helixes and β-sheets, respectively, which are labeled 1 through 7 and 1 through 12. In yellow, the finger domain with the degenerin-specific domains shown as dotted lines (figure adapted from [23])

serines in ENaCs [44]. This point is interesting because ENaCs are 10 times more selective to Na^+ than ASICs, with ASIC1a even found to permeate Ca^{2+} and H^+ ions [45], suggesting that the carboxyl groups of serine residues allow for more strict selectivity.

Whether all DEG/ENaC channels have this structure is not yet known; however, a good degree of sequence homology across species (>20% identity) suggests that this is most likely the case. Threading of even distantly related *C. elegans* MEC-4 sequence over the sequence

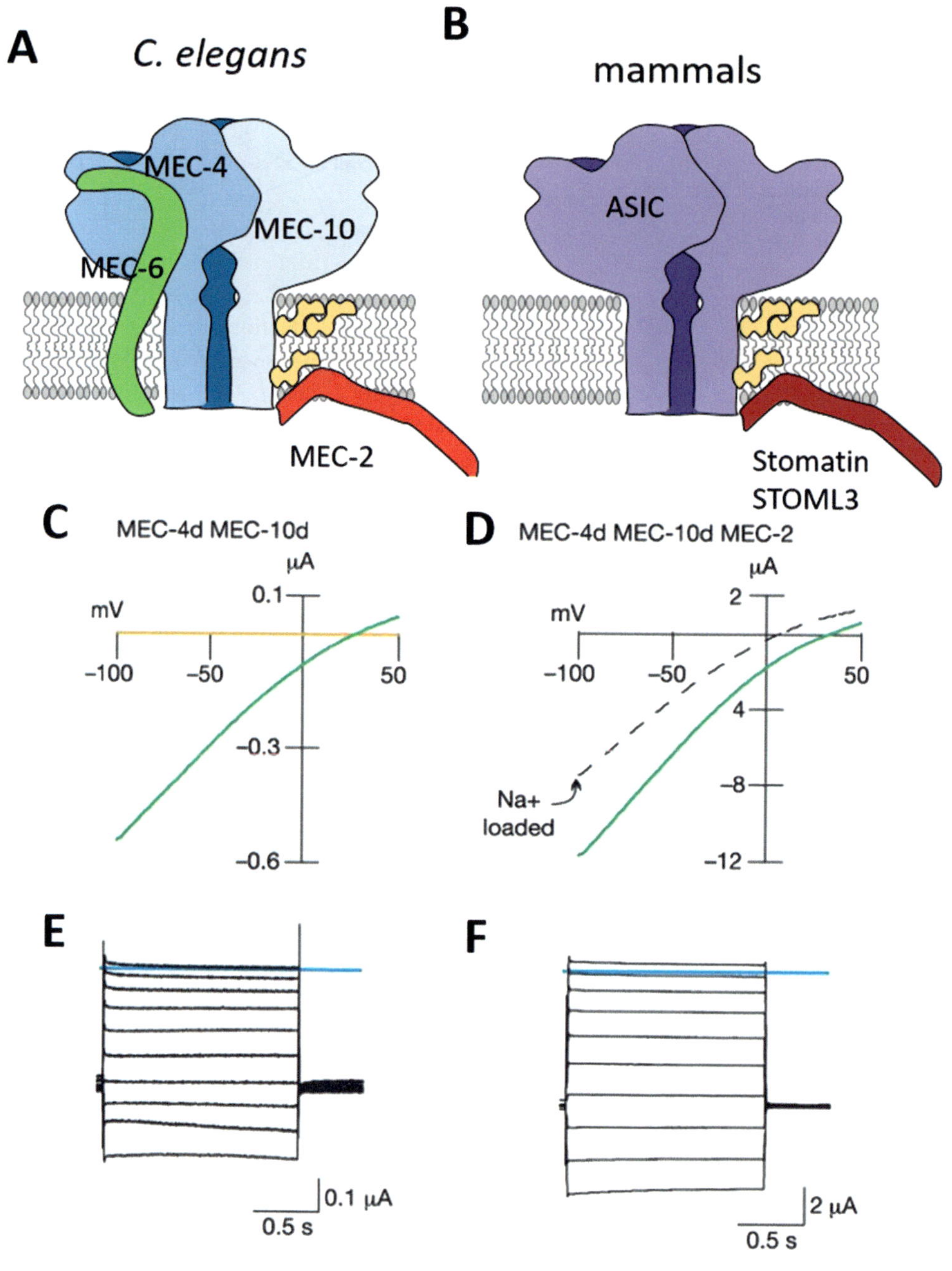

Fig. 9.3 The DEG/ENaC channel complex and regulation by stomatin-like proteins. (**a**) Genetic, biochemical, and functional studies in heterologous expression systems and *C. elegans*, revealed that MEC-4 co-assembles with MEC-10 to form a heteromeric channel [25–30]. MEC-4 also interacts with stomatin-like protein MEC-2 and paraoxonase-like protein MEC-6 [28, 31–34]. While MEC-2 appears to mediate enrichment of cholesterol around the channel by directly binding this lipid, MEC-6 functions as a chaperone, facilitating channel trafficking to the cell surface [35, 36]. Via these two different mechanisms, MEC-2 and MEC-6 enhance MEC-4 current amplitude. (**b**) In mammals, ASIC channels are either homomeric or heteromeric and, at least some of them, are regulated in their function by stomatin and stomatin-like protein STOML3 [37–39]. **C-F.** MEC-4 currents are enhanced by co-expression with MEC-2. Here the hyperactive mutants MEC-4(d) and MEC-10(d) were expressed in *Xenopus* oocytes (figure adapted and reprinted with permission from [28]). (**c**) Current-voltage relationships of an oocyte expressing MEC-4(d) + MEC-10(d) (green line) and of one expressing MEC-10(d) alone (yellow

of chicken ASIC1a reveals, indeed, that there is strong structural alignment between the two polypeptides [46]. There is one segment though, near the finger domain, which is unique to MEC-4 and does not have a corresponding protein sequence in ASIC1a (Fig. 9.3b). This segment, which was named "degenerin insertion" and which does not have an alpha or beta-sheet secondary structure, is present also in UNC-8, another *C. elegans* degenerin [46]. Studies from our laboratory, using chimeric MEC-4/UNC-8 proteins, revealed that the degenerin insertion is important both for gating regulation by divalent cations and for interaction with chaperon protein MEC-6 [23]. Likewise, at least nine members of the Drosophila DEG/ENaC family have a disordered loop protruding from the palm domain that does not align with the ASIC1a sequence. Interestingly, this loop is present in at least three PPK proteins implicated in mechanosensation, PPK, RPK, and PPK26, suggesting that it may help tether the channel to extracellular matrix proteins needed for mechanical gating [9].

9.3 Modulation of DEG/ENaCs by Homologous and Accessory Subunits

DEG/ENaC channels consist of either homomeric or heteromeric trimers, and at least some DEG/ENaC channels are regulated in their function by auxiliary subunits [25, 28, 29, 32–34, 37, 38, 47–56] (Fig. 9.4a, b). Homomeric and heteromeric channels were shown to have different properties [28, 47, 51, 54]. For example, mammalian heteromeric ASIC1a/ASIC2a desensitizes more rapidly than the homomeric channels formed by each subunit alone [48]. ASIC2b does not form channels by itself; however, it modulates the properties of other ASICs by forming heteromeric channels [51]. Similarly, *C. elegans* MEC-10 subunit does not function when expressed alone in *Xenopus* oocytes, but it can associate with MEC-4 to modulate channel properties, including sensitivity to amiloride [26, 28]. Consistent with this finding, the hyperactivated *mec-10(d)* mutation, which corresponds to the A713V mutation in *mec-4* (*mec-4(d)*), induces less extensive degeneration when compared to *mec-4(d)* [29].

DEG/ENaC subunits may also interact with other types of proteins, including stomatin and paraoxonase-like proteins [28, 32–34, 37, 38, 49, 50, 52, 53]. Stomatin is a membrane-bound protein that attaches to the plasma membrane and the cytoskeleton, acting as an inhibitor of cation transport in red blood cells [57]. In mammals, stomatin co-localizes with ASICs, reduces ASIC3 current in sensory neurons, and accelerates desensitization of ASIC2 and heteromeric ASICs [53]. Studies in stomatin knock-out mice revealed that stomatin is needed for the sensitivity of D-hair receptors [58]. These data suggest that stomatin might regulate D-hair receptors mechanosensitivity via interaction with ASICs.

In contrast, the *C. elegans* stomatin-like protein MEC-2 positively regulates homomeric MEC-4(d) and heteromeric MEC-4(d)/MEC-10(d) channels, increasing the number of functional channels at the cells surface in *Xenopus* oocytes [28, 59] (Fig. 9.4c, d). A study investigating MEC-2 mutants, that cause touch insensitivity, shows that touch sensitivity requires cholesterol binding by MEC-2. Thus, this suggests a role for

Fig. 9.3 (continued) line). (**d**) Current-voltage relationships of oocytes expressing MEC-4(d) + MEC-10(d) +MEC-2 cultured in a solution with and without amiloride (green line and dashed lines respectively). Oocytes expressing large MEC-4(d) currents display Na^+ overload which results in a shift of the reversal potential of the current towards less positive values. (**e**) Examples of currents recorded from an oocyte expressing MEC-4(d) + MEC-10(d). Currents were elicited by voltage steps from −100 to +35 mV, in 15 mV increments. The blue line represents the zero current level. (**f**) Same as in E for an oocyte expressing MEC-4(d) + MEC-10(d) +MEC-2. Note the difference in the scale bars, which indicates that MEC-2 increases current amplitude ~20 times

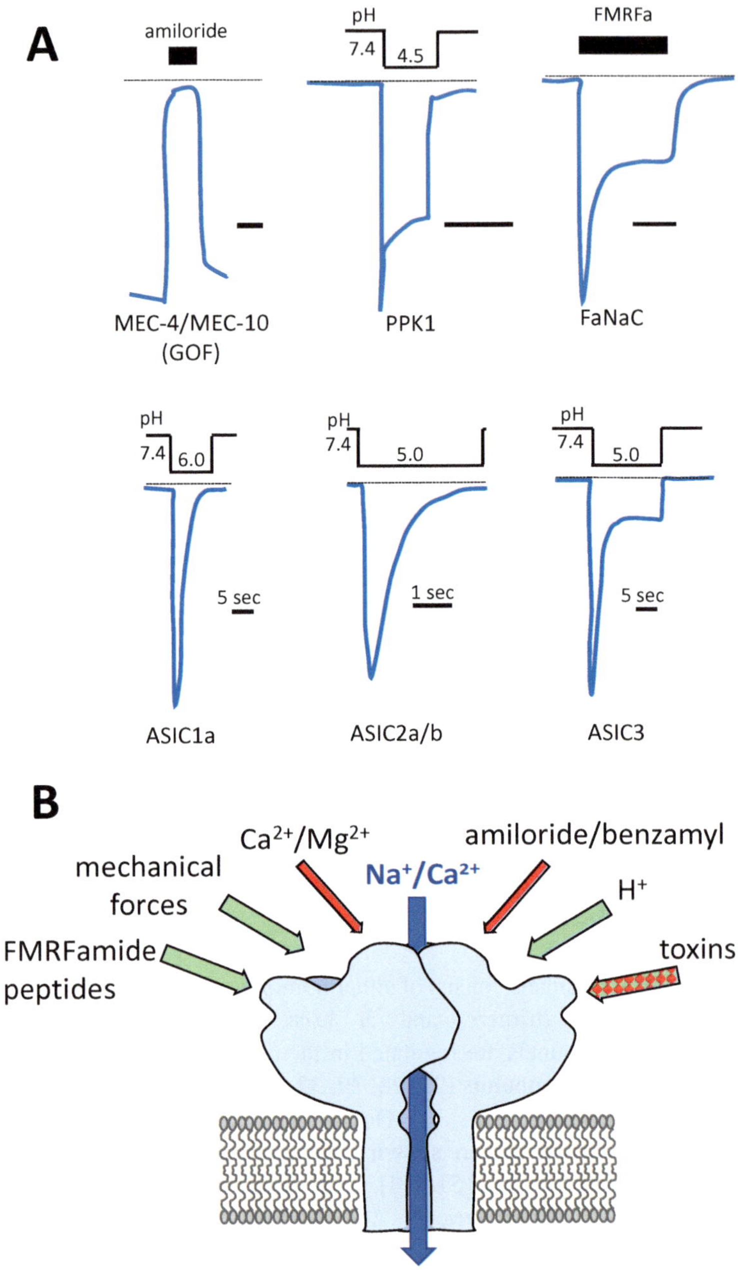

Fig. 9.4 Neuronal DEG/ENaC currents and DEG/ENaC channels' activators and blockers. (**a**) Schematic representation of representative DEG/ENaC currents activated either by acidic pH or FMRFa peptides (*Helix aspersa*). Currents produced by expression of gain of function (GOF) mutants MEC-4(d) + MEC-10(d) are also shown. In this case, since the currents are constitutively active, the effect of the blocker amiloride is shown. (**b**) A schematic representation of a DEG/ENaC channel is shown here with some of the stimuli that open the channel (green arrows) and some of the blockers/ inhibitors (red arrows)

MEC-2 in recruiting or maintaining cholesterol in the vicinity of the MEC-4 channel complex in vivo [36]. Another study shows MEC-2 co-localizes with MEC-4 in vivo and that MEC-2 co-immunoprecipitates with MEC-4 in oocytes. The interaction between MEC-2 and MEC-4 occurs through the MEC-2 stomatin-like region and MEC-4 intracellular N-terminus. Mutations that disrupt this interaction also disrupt the punctate localization of the channel along with the neuronal processes of touch neurons as well as the increase in current amplitude [60]. Interestingly, genetic screens for mutations that affect anesthetic sensitivity in *C. elegans* revealed that a second stomatin, UNC-1, interacts with DEG/ENaC channel UNC-8. Subsequent biochemical studies revealed that UNC-1 and UNC-8 physically interact in membrane lipid rafts, which are enriched in cholesterol. Within the rafts, another stomatin named UNC-24 is found. UNC-24 also affects anesthetic sensitivity and regulates UNC-1 distribution [61].

Another study in mammals found the interaction of stomatin-like protein STOML3 with both stomatin and ASIC channels (ASIC1b, ASIC2a and b, and ASIC3) within a highly mobile vesicle pool in DRG neurons and CHO cells [37]. The authors proposed that this vesicle pool might represent a putative transductome. Indeed, studies in stomatin, ASICs, and STOML3 single and double knock-out mice, revealed that stomatin–ASIC interactions have profound consequences on the function of mechanoreceptors and nociceptors. In particular, the lower mechanosensitivity of Aδ-mechanonociceptors in the ASIC3 knock-out mouse was exacerbated by the knock-out of STOML3, and mechanosensitivity of A-δ-mechanonociceptors was reduced in ASIC2 −/−;stomatin −/− double mutant mice but not in knock-out mice for the individual genes. Also, C-MH fibers fired less to very intense mechanical stimuli in ASC3 knock-out mice, and this defect was even more pronounced in mice that were knocked out for both ASIC3 and stomatin [38]. The importance of stomatin-like proteins in touch sensation in mammals has been recently underscored by the finding that STOML3 inhibitors can reverse mechanical hypersensitivity in mice [39].

The paraoxonase-like protein MEC-6 exerts an effect similar to the one exerted by MEC-2, on MEC-4(d) and MEC-4(d)/ MEC-10(d) channels in *Xenopus* oocytes: It increases the number of functional channels at the cell surface [32, 59]. Initially, it was suggested that MEC-6 might colocalize with the MEC-4/ MEC-10 channel at the plasma membrane [32]. More recent work from the Chalfie lab, however, shows that MEC-6 and another paraoxonase-like protein, POLM-1, act as ER chaperones for MEC-4 trafficking to the cell surface and they are not per se channel accessory subunits [35]. Work from our laboratory showed that DEG/ENaC channel subunits are regulated by MEC-6 via the channel finger domain [23]. Given that the majority of the MEC-6 protein is extracellular, these data suggest that a physical interaction may occur between the finger domain of the channel and the extracellular portion of MEC-6. Since the finger domain is the most divergent region across the DEG/ENaC family, these results also suggest that the finger domain influences channel trafficking and function uniquely, depending upon the channel subunit. Importantly, loss-of-function mutations in both *mec-2* and *mec-6* suppress *mec-4(d)* and *mec-10(d)*-induced neuronal death in vivo, indicating that these interactions are physiologically relevant [62, 63]. Furthermore, both *mec-2* and *mec-6* are required for normal channel function as assayed by Ca^{2+} imaging and electrophysiological methods in vivo [64, 65]. In mammals, the proteins with the closest similarity to worm MEC-6 are paraoxonases (PON proteins). The human genome contains three genes encoding PON proteins, which have been shown to interact with lipids and to prevent oxidation of low-density lipoproteins [66, 67]. A manuscript published in 2017 provides evidence that ENaC channels, normally expressed in epithelial tissues, interact with and precipitate with PON-2. This interaction between PON-2 and ENaCs leads to negative regulation of the expression of these channels at the cell surface both in HEK293 cells and in *Xenopus* oocytes [68]. There are no data yet though on whether PON proteins interact with or regulates the trafficking of neuronally expressed DEG/ENaC channels in mammals.

9.4 Neuronal DEG/ENaC Channels in Mechanosensation

C. elegans MEC-4 and MEC-10. In the 1980s, a genetic screen in *C. elegans* designed to discover genes involved in gentle touch sensation identified the DEG/ENaC channel subunits *mec-4* and *mec-10* (mec for mechanosensory abnormal) [2, 29, 69]. *mec-4* and *mec-10* encode proteins that are 48% identical and that are co-expressed in gentle touch neurons, with *mec-10* being additionally expressed in other worm mechanosensory neurons, that detect harsh touch (FLPs and PVDs) [29]. Loss-of-function alleles of *mec-4* and *mec-10* produce touch insensitivity in animals with normal structure and normal development of touch neurons [69]. Interestingly, more recent work has shown that *mec-10* is required closer to the cell body of touch neurons for normal mechanosensitivity [27]. The requirement of DEG/ENaC subunits *mec-4* and *mec-10* for gentle touch sensation in *C. elegans* led to the hypothesis that the proteins encoded by these two genes may form a channel responsive to mechanical forces.

One of the mutant alleles identified in this screen induces swelling and death of touch neurons [2]. This mutant allele, named *mec-4(d)*, encodes the substitution at position 713 of an Alanine for a Valine. More recent crystallographic studies show that A713 is located in the pore of MEC-4 and most likely allows tight packing of the transmembrane domains during channel desensitization. When this Alanine is substituted with a bulkier amino acid like Valine, or Threonine, tight packing of the transmembrane domains during desensitization is hindered resulting in a constitutively open channel [6, 44]. This mutant that gave the name to this family of channels in *C. elegans* (degenerins) was indeed postulated to give rise to hyperactive channels [2, 70]). Later work in which MEC-4 (d) was reconstituted in *Xenopus* oocytes confirmed this prediction [26, 31] (Fig. 9.4b). MEC-4(d), expressed in oocytes, gives rise to amiloride-sensitive currents that resemble currents produced by the expression of ENaC [1, 31]. Interestingly, co-expression with MEC-10 dampens current amplitude and changes amiloride sensitivity, suggesting that MEC-10 functions more as a modulatory subunit. In addition to being hyperactive, MEC-4(d) is also Ca^{2+} permeable. This Ca^{2+} permeability may cause neurodegeneration by conducting toxic levels of Ca^{2+} into the cell [26]. However, both MEC-4(d) and wild-type MEC-4 are not mechanosensitive in *Xenopus* oocytes, suggesting that other proteins/factors present in vivo, but not in *Xenopus* oocytes, may be needed to confer mechanosensitivity to the channel.

The first hint that MEC-4 might have been part of a mechanosensory channel came from studies employing the genetically encoded calcium sensor called “chameleon” [65]. Experiments employing chameleon expressed in touch neurons revealed indeed that the MEC channel is specifically required for responses to gentle touch and not for the general function of touch neurons. However, while experiments using chameleon positioned the MEC channel very close to the actual mechanosensor, the definitive demonstration that the MEC channel is indeed activated by mechanical forces came from elegant electrophysiological experiments on in situ touch neurons [71]. O' Hagan and colleagues indeed showed that a mechanically gated Na^+ current normally present in wild-type *C. elegans* touch neurons was absent in null *mec-4*, *mec-2*, and *mec-6* mutant animals and had altered selectivity properties in animals expressing pore variants of *mec-4* (Fig. 9.5a–c).

Interestingly, the mechanoreceptor current recorded in touch neurons in situ and carried by the MEC-4 channel displays features that are similar to the current evoked in the Pacinian corpuscles upon the application of mechanical forces. First, it is sensitive to pressure (force per unit area) rather than force per se, and second, it is activated both by the application and by the release of pressure, but not while constant pressure is applied. These results suggested at the time, that DEG/ENaC channels might encode mechanosensitive channels in mammalian touch receptors as well.

DEG-1. Much of the work conducted on nose touch behavior in *C. elegans* has focused on the involvement of the ASH neuron, as laser ablation studies revealed that ASH accounts for the highest fraction of nose touch response [73]. The Goodman lab recently identified the DEG/ENaC DEG-1 as the major mechanotransduction channel expressed in ASH neurons [74]. Using in vivo patch clamp, they showed that a significant decrease in mechanoreceptor currents was observed in *deg-1* mutants or when currents were recorded in Na^+-free or amiloride-containing solutions. Conversely, mutations in UNC-8, another DEG/ENaC

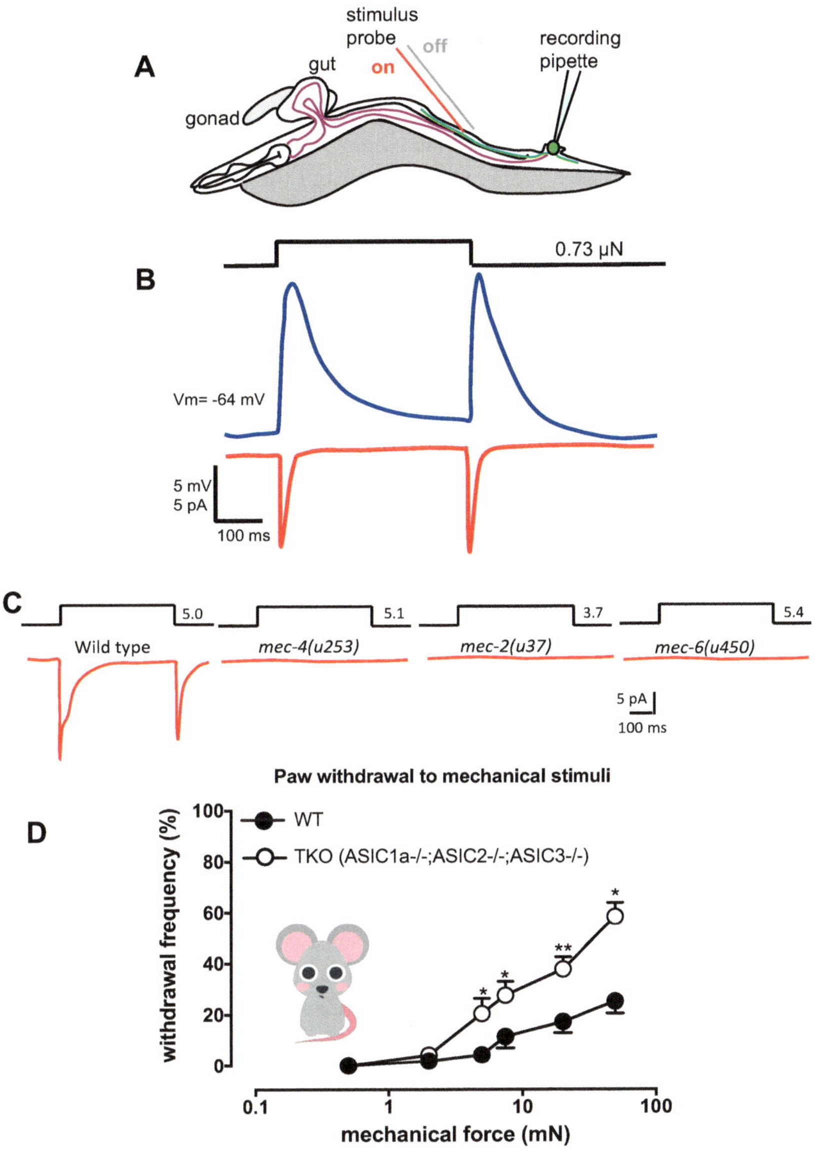

Fig. 9.5 DEG/ENaC channels in touch sensation. (**a**) Drawing of the set up used to record currents from *C. elegans* gentle touch neurons. An adult *C. elegans* is glued onto an agarose pad on a glass coverslip. A hole is then punctured through the cuticle to release some of the internal pressure causing the release to the outside of part of the gonad and intestine. Another hole is then punctured to release the cell body of a touch neuron labeled with GFP, from which patch-clamp recordings of currents and voltages may be then obtained. (**b**) Schematic representation of representative voltage (in red) and current (in blue) recordings from a *C. elegans* gentle touch neuron stimulated with a force of 73 nN. (**c**) Force generated currents in in situ gentle touch neurons of the indicated *C. elegans* strains, including wild type, *mec-4*, *mec-2*, and *mec-6* null strains (*mec-4(u253)*, *mec-2(u37)*, and *mec-6 (u450)*) (B and C are adapted from [71]). (**d**) Paw withdrawal frequencies upon application of forces of different intensities in wild type and ASIC triple knock-out mice (ASIC1a−/−; ASIC2a−/−; ASIC3−/−). The data show that in the ASIC channels triple knock-out mice, touch sensation is enhanced, suggesting a role of ASIC channels in touch sensation but most likely not as direct mechanotransducers (adapted and reprinted with permission from [72]

expressed in ASH, have no effect on mechanosensitive currents or nose touch responses in vivo.

UNC-8. Analysis of the locomotion phenotype of *C. elegans unc-8* null mutants suggested that this channel might be mechanosensitive as well. Degenerin *unc-8* is expressed in interneurons, motoneurons, and some sensory neurons [75, 76]. *unc-8* loss-of-function mutants inscribe shallower tracks on the bacterial lawn in which they travel when compared to the wild type. This phenotype suggested that UNC-8 may mediate proprioception by functioning as a stretch receptor. This idea is supported by the anatomical features of motoneurons. Indeed, the distal part of the neuronal process of motoneurons does not appear to have synapses. Thus, this suggested that this distal part might be stretch sensitive via the activity of UNC-8 channels and signal the extent and timing of body bends.

Drosophila DEG/ENaC channels in mechanosensation. *ppk1* is expressed in the peripheral class IV multidendritic (md) abdominal sensory neurons that form an extensive dendritic network just below the epidermis of the fruit fly. These sensory neurons have been proposed to be polymodal, involved in both proprioception and nociception [77]. Upon stimulation with low pH, *ppk1* channels become activated in class IV multidendritic (md) neurons, increasing action potentials firing. It is not known whether *ppk1* is directly activated by mechanical forces, but analysis of the *ppk1* mutant flies suggests that this channel might be. Indeed, *ppk1* mutants display loss of wandering, a behavior that involves mechanosensation and that is critical for feeding at the larval stage [78].

In 2014, another group reported the requirement of another DEG/ENaC channel subunit in class IV multidendritic (md) for mechanical but not thermal nociception [79]. *ppk26* is specifically expressed in these multimodal sensory neurons, and its knock-out significantly reduces rolling behavior upon application of 47 mN mechanical forces, an intensity of stimulation that is considered to be in the nociception range. Importantly, *ppk26* mutant Drosophila larvae display normal thermal nociception, indicating that the sensory deficit caused by *ppk26* knock-out is limited to the detection of mechanical forces. Using genetic approaches, the authors went on to test whether *ppk26* and *ppk1* are part of the same pathway or possibly the same channel complex. They found that *ppk26/ppk1* double mutant larvae display touch deficits similar to the ones caused by the single knock-out, suggesting that indeed PPK26 and PPK1 might work in the same pathway.

Interestingly, class II and class III neurons in Drosophila, that function as gentle touch receptors, require the function of another DEG/ENaC channel for behavioral responses to touch. Tsubouchi and colleagues used cell-specific RNAi to knock-down ion channel genes in class II and class III neurons and test the effect of the knock-downs on gentle touch response in Drosophila larvae. They found that knock-down of DEG/ENaC channel RPK significantly decreased responses to touch [80].

Mouse ASIC1, ASIC2, and ASIC3. Given the degree of sequence conservation between *C. elegans* degenerins MEC-4 and MEC-10, and mammalian ASIC channels, it was hypothesized that ASICs might be involved in mechanosensation. This idea was supported by the finding that almost all ASICs are expressed in sensory neurons [81–86]. Moreover, immunochemical studies showed that at least ASIC2 and ASIC3 are localized in specialized cutaneous mechanosensors, including the Meissner's corpuscles, the Merkle cells/neurite complexes, and the lanceolate nerve endings that wrap around the hair shaft [84, 87].

To test this hypothesis, knock-out mice models were generated for ASIC1, ASIC2, and ASIC3. ASIC1 knock-out mice did not display any defect in mechanosensory responses [88]. However, ASIC2 and ASIC3 knock-out mice did display touch defects, at least to a certain degree. In nerve–skin preparations of the ASIC2 knock-out mouse, two types of low threshold mechanosensory fibers, the rapidly-adapting or RA, and the slowly-adapting or SA, responded abnormally to mechanical stimuli. While normally RA and SA fibers respond to larger displacements by steeply increasing their firing rate, RA and SA fibers from ASIC2 knock-out mouse showed a reduction in this response. The function of other nerve fibers known to mediate other types of sensations (pain, harsher touch) remained unaffected in ASIC2 knock-out mice [84]. Because RA and SA mechanosensors are thought to play a role in gentle touch sensation, these results initially suggested that ASIC2 might be the mammalian orthologue of MEC-4. However, a closer look at the results revealed that the defects are subtle, when compared to the complete loss of touch sensitivity in *mec-4* mutant *C. elegans*.

Similarly, in nerve/skin preparations of ASIC3 knock-out mice, AM fibers (mechanociceptors) cannot increase action potentials frequency following large displacements. Interestingly, though, the response of RA fibers is twice as large as in the wild type. This result suggested that ASIC3 may function as a negative regulator in RA fibers. However, when ASIC3 knock-out mice were tested in behavioral assays, no defects were found in touch responses, and only a mild defect was observed in response to mechanical stimulation following acid injection into the skeletal muscle (mechanical hyperalgesia) [85]. Taken together, results on ASIC2 and ASIC3 knock-out mice revealed mild defects at best, in sharp contrast with the touch insensitive phenotype of *mec-4* mutants *in C. elegans*. These results warranted exploration of the potential of extensive redundancy.

To test for this possibility, Kang and colleagues generated a triple mutant mouse in which ASIC1, ASC2, and ASIC3 were knocked out [72]. First, they confirmed that in the DRG neurons of these mice, acid-evoked currents were completely absent and then systematically tested mechanosensory responses using both electrophysiology and behavioral assays. Surprisingly, there were no defects in touch responses but rather heightened sensitivity to mechanical forces. In nerve/skin preparations, they found that mechanical stimulation generated enhanced activity in A-mechanonociceptors, and in behavioral assays, they reported increased paw withdrawal frequencies when animals were mechanically stimulated with von Frey filaments (Fig. 9.5d). These results support that ASICs somewhat influence cutaneous mechanosensitivity, yet they are not as essential for touch sensation as they are in invertebrates. Taken together, these data point perhaps toward a divergence of the function of mammalian ASICs when compared to *C. elegans* degenerins and Drosophila pickpocket genes, most likely with other ion channel families taking over the role of mechanosensors in mammalian skin. Interestingly, though, in sharp contrast with their small effect in the skin, ASIC1, ASIC2, and ASC3 play a major role in visceral mechanosensitivity as measured both by electrophysiology and analysis of digestive function in vivo [88].

9.5 Neuronal DEG/ENaC Channels in Other Sensory Modalities

ASICs and pain. ASIC channels are gated by extracellular H^+ ions (Fig. 9.1a). They are activated by increasing the concentration of H^+ with a Kd that is specific for each ASIC subunit. However, the acid gating of ASIC channels is still not fully understood. The extracellular pH at which the majority of ASICs become activated is outside the physiological range; so, it remains uncertain whether acidic pH is the physiological stimulus that gates these channels. However, in pathological conditions, such as inflammation and ischemia, extracellular pH can decrease significantly and even reach the levels required for activation of ASICs. Thus, given that ASICs are expressed in DRG neurons, these channels have been proposed to mediate pain in conditions in which there is tissue acidosis. Indeed, in several models of pain, treatment with the DEG/ENaC channel blocker amiloride reduces pain and inhibits activation by acidic pH of the C-type fibers in rats [89–91]. Furthermore, treatment with more specific ASIC blockers, such as the venom from the black mamba and the psalmotoxin 1 (PcTx1) from tarantula, has a strong analgesic effect [92–94]. On the other hand, the toxin MitTx isolated from the Texas coral snake potentiates ASIC1 and ASIC2a currents and produces intense pain by activating

ASIC1 channels on capsaicin-sensitive nerve fibers [44, 95]. Interestingly, in *C. elegans*, avoidance of acidic environments requires the activity of DEG/ENaC channels DEG-1 and ACD-1, expressed in sensory neurons and associated amphid sheath glia cells, respectively, suggesting the conserved role of this family's members in nociception across species [96].

Drosophila ppk channels in salt detection and courtship. ppk11 and *ppk19* are expressed in peripheral taste sensory organs, both in the larva and in the adult fly. Liu and colleagues disrupted the function of both *ppk11* and *ppk19* and found that larvae could no longer discriminate low concentrations of Na^+ or K^+. They also found that the electrophysiologic responses to low salt concentrations were attenuated in these mutant larvae. Furthermore, in both larvae and adults, disrupting the function of *ppk11* and *ppk19* leads to changes in behavioral responses to high salt concentrations [97].

ppks play a role in pheromone detection as well [98–102]. Lin and colleagues showed that *ppk25* is expressed in a single sexually dimorphic gustatory neuron needed for courtship behavior and that its knockdown significantly impairs male response to females [98]. Later the same group reported expression of a homolog of *ppk25*, called *nope*, in gustatory neurons of the labellum wings, and legs, including all gustatory neurons in which *ppk25* functions [99]. They further showed that gustatory-specific knockdown of *nope* impairs male courtship of females. Interestingly, though, *nope* and *ppk25* have nonredundant functions, which suggest that they may be part of the same channel complex. Indeed, *nope* and *ppk25* form specific complexes when co-expressed in cultured cells. Recent work from the same group showed that *ppk25* is required for pheromone detection because it mediates Ca^{2+}-dependent signal amplification of olfactory receptors. *ppk25* mediates this amplification via intracellular calmodulin-binding motif [103].

More direct evidence that PPK channels are activated by pheromones derived from a study in which *ppk25*, *ppk29*, and *ppk23* were ectopically expressed [104]. "M" cells inhibit male–male courtship by being activated by male but not by female aphrodisiac pheromones. Using calcium imaging, Liu and colleagues showed that ectopic expression of these PPK channel proteins leads to calcium responses to the female aphrodisiac pheromone 7,11-heptacosadiene (7,11-HD) in M cells. Taken together, these data highlight that PPK proteins may function as ligand-gated channels.

9.6 DEG/ENaC Channels in Neurotoxicity and Axonal Degeneration

Members of the DEG/ENaC family such as mammalian ASICs, and *C. elegans* homologs MEC-4, MEC-10, DEG-1, UNC-8, and UNC-105 may become hyperactivated as a result of extracellular acidification or genetic dominant gain-of-function mutations designated "d" [2, 3, 45, 62, 75, 105, 106]. This hyperactivation induces swelling and death of neurons that express these channels [107] (Fig. 9.6a). The pharmacological block or genetic mutations that render the channel inactive protect neurons from death (Fig. 9.6a–d). For example, neuronal survival is enhanced by treatment with DEG/ENaC channel blockers amiloride or psalmotoxin1, a more specific ASIC1a channel blocker, in a mouse model of ischemia. Furthermore, even more, robust neuronal survival is observed in ASIC1a knock-out mice under the same ischemic conditions [26, 45] (Fig. 9.6e, f). Similarly, in *C. elegans*, point mutations that disrupt ion flux through the channel pore prevent *mec-4(d)*-induced neuronal death, supporting the idea that increased cation influx is essential for neuronal death, induced by hyperactivation of these channels [109, 110]. Thus, ASICs are the proton sensors that may provide a voltage and glutamate-independent pathway for calcium entry to contribute to neurotoxicity. It is also noteworthy that the time window effectiveness of the treatment using NMDA antagonists is about 1 hour following ischemic insult, while treatment using ASIC1a antagonists is effective even at 5 hours after ischemic insult and persists for 7 days after

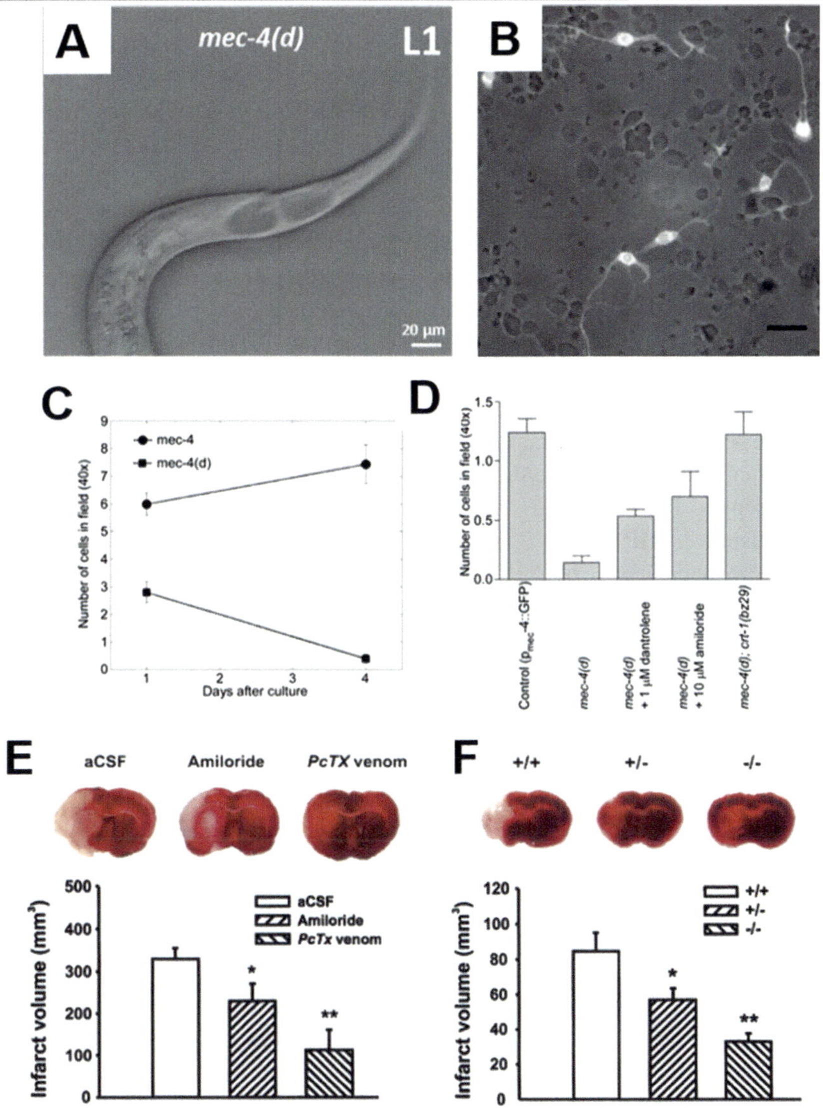

Fig. 9.6 DEG/ENaC channels in neuronal death. (**a**) Photograph of a *mec-4(d)* mutant *C. elegans* at the L1 larval stage showing vacuolated necrotic touch neurons in the tail. *mec-4(d)* starts to be expressed early in development so by the time the animal reaches the first larval stage, most of the gentle touch neurons are already necrotic. (**b**) Chameleon expressing *C. elegans* touch neurons cultured in vitro [65]. Pictures taken in bright field and using a GFP filter were superimposed to show that ~ 0.5% of the cells express GFP, as expected from the ratio of touch neurons/total cells in the *C. elegans* embryos. (**c**) The number of chameleons expressing touch neurons/field was quantified for cultures obtained from wild-type animals and *mec-4(d)* mutant animals, demonstrating the lower number and rapid loss over time of touch neurons cultured from *mec-4(d)* mutants. (**d**) Loss of mec-4(d) neurons can be partially rescued by the cultivation of the cells in media containing dantrolene, an antagonist of the ryanodine receptor which blocks the release of calcium from the endoplasmic reticulum, or amiloride a DEG/ENaC channel blocker. Rescue from cell death can be also seen in mutants for the endoplasmic reticulum (ER) calcium-binding protein calreticulin, underscoring that release of calcium from the ER is needed for *mec-4(d)*-induced neuronal death in *C. elegans* [70]. B-D are adapted from [108]. (**e** and **f**) Neuroprotection in a model of ischemia in the mouse by ASIC1a blockage and knock-out. (**e**) 2,3,5-triphenyltetrazolium hydrochloride (TTC) staining of brain slices obtained from mice subjected to transient focal ischemia. The three slices represent examples obtained from mice that were injected intracerebroventricularly 30 min before and after the induction of ischemia with artificial cerebral spinal fluid (aCSF), aCSF + amiloride, or aCSF + PcTX. The bar graph below shows the quantification of the infarct volume as determined by TTC staining (areas in which TTC is not enzymatically reduced to red by dehydrogenases present in living tissue). (**f**) A similar reduction in infarct volume is also observed in ASIC1a

treatment [111]. Thus, the lack of effectiveness of NMDA receptor antagonists in stroke patients may be due to the inability to intervene during this short time window, and blocking ASICs may be therapeutically more relevant than NMDA receptor blockade alone.

However, a study published in 2015 challenged the idea that ion flux through ASIC1a is required for neuronal death in ischemia [112]. Wang and colleagues indeed reported that acidosis can trigger neuronal death via ASIC1a independently from its cation conducting function. Indeed, these authors showed that acidosis induced by ischemia recruits the serine/threonine kinase death receptor RIP1 to the ASIC1a C-terminus, causing RIP1 to become phosphorylated. Phosphorylation of RIP1 then leads to neuronal death which remains unaffected by pretreatment with mild acidic conditions that fully inactivate ASIC1a channels and by substitution of Na^+, in the extracellular solution, with the impermeant ion N-methyl-D-glucamine (NMDG). The deletion of the ASIC1a gene significantly decreases RIP1 phosphorylation and brain damage, suggesting ASIC1a-mediated RIP1 activation has an important role in ischemic neuronal injury.

Intriguingly, another study reported that ASIC1a is also localized in the mitochondria of mouse cortical neurons, where it physically interacts with the adenine nucleotide translocase, a protein that resides in the inner mitochondrial membrane. Mitochondria extracted from ASIC1a knock-out mice are more efficient at retaining and taking up calcium in the mitochondria. Moreover, when cortical neurons of ASIC1a knock-out mice are challenged with hydrogen peroxide (H2O2), they show reduced levels of cytochrome c (Cyt C) release and reduced inner mitochondrial membrane depolarization when compared to wild-type mice. This phenotype correlates with reduced H_2O_2-induced neuronal death in ASIC1a knock-out when compared to wild-type mice [113]. Taken together, these results show that ASIC1a may influence neuronal survival in disease conditions by functioning both at the plasma membrane and in mitochondria.

The swelling and death induced by these hyperactive channels are diagnostic morphological features of necrosis and depend on intracellular calcium overload and activation of calcium-dependent cathepsins and calpains [70, 107, 114, 115]. Cytosolic calcium overload is thought to be caused by both entry of calcium from the extracellular compartment, through calcium-permeable DEG/ENaC channels, and by the calcium-induced release of calcium from the endoplasmic reticulum [26, 45, 70, 106]. Indeed, hyperactive MEC-4(d) which causes swelling and death of gentle touch neurons in *C. elegans* is calcium-permeable [26]. Similarly, mammalian neuronal-expressed ASIC1a, hyperactivated in ischemic conditions by prolonged acidosis, is also calcium-permeable [45, 106, 116]. However, we recently challenged this assumption by reporting that UNC-8, another *C. elegans* DEG/ENaC channel, hyperactivated by genetic mutations, is toxic both in vivo and in *Xenopus* oocytes, without being calcium-permeable [117, 118]. Thus, our data suggest that the calcium permeability of DEG/ENaC channels is not essential for neuronal death.

ASIC1a is also involved in axonal degeneration in a mouse model of multiple sclerosis. Friese and colleagues induced experimental autoimmune encephalomyelitis (EAE) in wild-type and ASIC1a knock-out mice and found that the severity of the disease was reduced in animals that were lacking ASIC1a [119]. Importantly, the authors find that the spinal cord tissue of EAE mice has a pH of 6.5, while the same tissue has a pH of 7.2 in healthy mice. Given that ASIC1a channels have a $pH_{0.5}$ of 6.5, these data clearly show that tissue acidosis in EAE mice reaches

Fig. 9.6 (continued) knock-out mice and a smaller reduction is seen in ASIC1a heterozygotes. E and F are adapted and reprinted with permission from [45]

levels that are sufficient for robust activation of ASIC1a. The effects of ASIC1a knock-out are recapitulated both in vivo and in vitro by treatment with amiloride, suggesting that DEG/ENaC channel blockers may be considered for the treatment of multiple sclerosis [119–121].

9.7 *C. elegans* DEG/ENaC Channel UNC-8 is Involved in Synaptic Remodeling

The Miller lab, in collaboration with our lab and the Richmond lab, recently showed that the function of DEG/ENaC UNC-8 is required in *C. elegans* motoneurons for synaptic remodeling [122]. The Miller lab used a well-defined paradigm of synaptic remodeling in *C. elegans* to identify genes that control the removal of synapses during development. In *C. elegans* DD-type GABAergic motoneurons initially form synapses onto the ventral muscles. Later in development, these synapses are removed and new ones are established onto the dorsal muscles. In another set of motoneurons, the VD neurons, synaptic remodeling is blocked by the transcription factor UNC-55, homologous to mammalian COUP-TFII transcription factors. So, in VD motoneurons, synapses are retained ventrally.

To identify genes that are transcriptionally regulated by UNC-55, the Miller lab used the mRNA poly-A tail tagging strategy to collect and compare transcripts from GABAergic motor neurons extracted from wild type and *unc-55* mutants. One of the transcripts found, in wild type and not in *unc-55* mutants, was *unc-8*. Using synaptic markers, they showed that DD ventral synapses, which are normally removed during development in wild-type animals, persist in *unc-8* mutant worms (Fig. 9.7a, b). Conversely, in *unc-55* mutants, VD synapses ectopically remodel but are retained in the ventral muscles of *unc-55:unc-8* double mutant *C. elegans*.

We further showed that these effects depend on UNC-8 channel activity. Indeed, the synaptic remodeling phenotype seen in *unc-8* mutants is mimicked by cultivating worms in the presence of DEG/ENaC channel blockers amiloride and benzamil. Using genetic and imaging approaches, we established that synaptic remodeling mediated by UNC-8 channels requires the function of N/P/Q type calcium channel UNC-2 of the calcium-sensitive phosphatase calcineurin TAX-6 and of the apoptotic protein CED-4. Thus, a possible model is that depolarization of GABAergic neurons during neuronal activity leads to the activation of the voltage-gated calcium channel UNC-2, which allows calcium entry. Intracellular calcium then activates the calcium/calmodulin-dependent phosphatase calcineurin TAX-6 which, in turn, activates UNC-8. Activation of UNC-8 leads to further depolarization and activation of more calcium channels, which cause further elevation in intracellular calcium. This positive feedback loop of calcium elevation eventually leads to the removal of presynaptic components via the function of the apoptotic protein CED-4 (Fig. 9.7c). The model is still under investigation and other molecular components of this pathway are likely going to be identified, but this work supports that DEG/ENaC channels are involved in long-term cellular processes in the nervous system such as synaptic remodeling.

9.8 DEG/ENaC Channels in Synaptic Transmission

Members of the ASIC family have also been implicated in synaptic activity by functioning either presynaptically or postsynaptically. In mice, ASIC1a is expressed in various regions of the central nervous system including the hippocampus, the amygdala, and the cerebellum. Especially in the hippocampus, ASIC1a is highly enriched at the synapses. Not surprisingly then, knock-out of ASIC1a impairs long-term potentiation (LTP) in the hippocampus, indicating that the activity of this channel is critical for synaptic plasticity [123]. The activity of ASIC1a is also needed in the amygdala, one of the key brain structures involved in fear conditioning. Here, disruption of ASIC1a leads to attenuation of learned fear behavior and overexpression of ASIC1a enhances fear conditioning [124]. The proposed mechanism is that protons released by

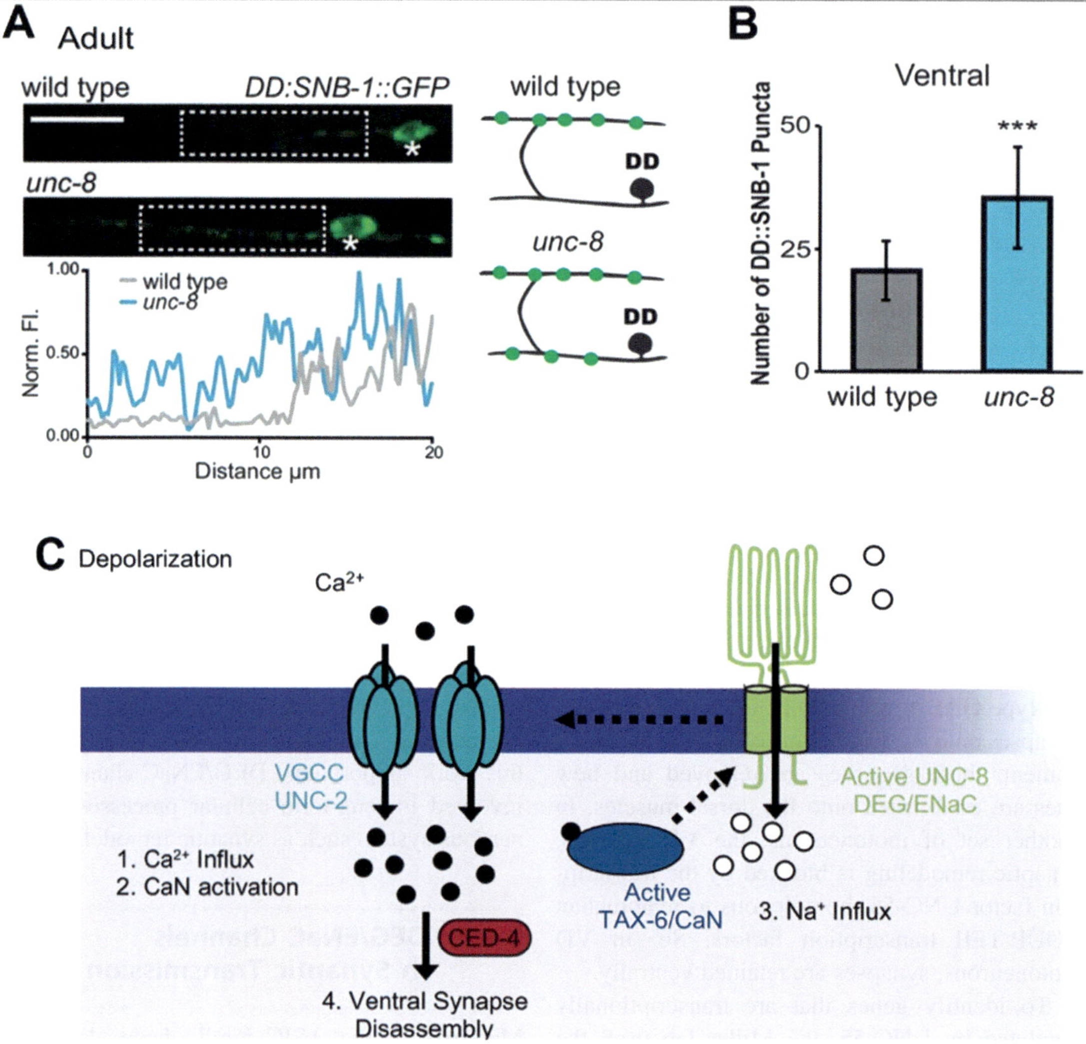

Fig. 9.7 *C. elegans* DEG/ENaC channel UNC-8 is needed for synaptic remodeling. (**a**) Micrographs of ventral DD synapses labeled by expression of GFP-tagged synaptobrevin (*pflp-13::SNB-1::GFP*) in wild-type and *unc-8* mutant *C. elegans*. The asterisk denotes the cell body of the DD5 neuron. Scale bar is 10 μm. The inset shows the pixel intensity of the region indicated by dashed line boxes. Data show that in *unc-8* mutants, a larger number of synapses are retained ventrally in remodeling DD neurons (see graphic representation on the right), indicating that DEG/ENaC channel UNC-8 is needed for the removal of DD synapses from the ventral muscle (**b**). (**c**) Model of *unc-8*-dependent mechanism for removal of synapses. In active GABAergic neurons, depolarization of the plasma membrane leads to the activation of voltage-gated calcium channels UNC-2. UNC-2 channels conduct calcium (black circles) into the neuron leading to activation of the calcium/calmodulin-dependent phosphatase, calcineurin (CaN/TAX-6). In turn, TAX-6 may activate UNC-8 which conducts Na^+ ions (white circles) into the cell leading to further depolarization and further activation of UNC-2 calcium channels. This positive feedback loop is predicted to lead to even further elevation of intracellular calcium. Future work will clarify the exact molecular mechanism involving UNC-8 which leads to the removal of synapses, however, data so far show that the apoptotic protein CED-4 is required for the process of synaptic removal. Adapted from [122]

the presynaptic terminal, via fusion of the acidic synaptic vesicles, bind ASIC1a in the postsynaptic membrane leading to activation of the channel and therefore membrane depolarization. The depolarization of the postsynaptic membrane then leads to the release of the Mg^{2+} block from

NMDA glutamate receptors, whose activity is required in learning and memory.

Interestingly a similar role of ASICs in the function of the postsynaptic neurons has been shown in Drosophila. *ppk29* contributes to spontaneous neurotransmitter release at the fly neuromuscular junction by functioning in the postsynaptic cell. The effect is on the function and transcription of postsynaptic glutamate receptors but not on the size and structure of the synapses. These effects of *ppk29* on glutamate receptors are translated into animal behavior; indeed, larvae that lack *ppk29* show defects in rollover behavior [125].

Previously, fly *ppk11* and *ppk16* were found to regulate synaptic plasticity presynaptically rather than postsynaptically, by altering glutamate release at the neuromuscular junction [126, 127]. Similarly, Voglis and colleagues found that *C. elegans* ASIC-1 is localized at the presynaptic terminals of dopaminergic neurons where it functions to enhance dopamine release. In doing so, ASIC-1 amplifies dopaminergic signaling needed for associative learning [128, 129]. Taken together, studies in worms and mammals show that members of the ASIC subfamily localize at synapses, presynaptically or postsynaptically, where they function to enhance synaptic function.

Obviously, depending on the neuronal type, activation of ASICs at synapses could have opposing effects on the activity of the nervous system. This is exactly what Ziemann and colleagues observed in a mouse model of seizures. They injected mice with glutamate agonist kainate to induce tonic–clonic seizures and found that knock-out of ASIC1a increased the severity of seizures, while overexpression of ASIC1a had the opposite effect. Since seizures have been known to be associated with acidosis, the authors tested the hypothesis that ASIC1a may facilitate seizure termination by being activated by acidosis. Indeed, they found that CO_2 inhalation, which is known to cause acidosis in the nervous system and to inhibit seizures, required the activity of ASIC1a to interrupt seizures. Finally, the authors hypothesized that these results could be explained if ASIC1a were expressed in inhibitory neurons. Using patch-clamp, they indeed demonstrated that inhibitory neurons, acutely isolated from the mouse hippocampus, are endowed with a large proton-activated cationic current, absent in interneurons isolated from ASIC1a knock-out mice. Finally, the authors showed that this acid-evoked current plays a key role in acid-induced neuronal excitability by demonstrating that acid-induced excitability of interneurons is greatly reduced in ASIC1a knock-out mice [130]. Taken together, studies from worm to mammals underscore the key role of neuronally expressed DEG/ENaC channels in regulating the excitability and synaptic function of different neuronal populations.

9.9 Expression and Function of DEG/ENaC Channels in Glia

DEG/ENaC channels have also been shown to be expressed in glial cells. Using immunocytochemical techniques, Golestaneh and colleagues showed expression of α-ENaC in cultured Muller glial cells isolated from rat retina [131]. Similarly, using immunohistochemical approaches and whole-cell patch-clamp, Brockway and colleagues provided evidence for the expression of α-ENaC subunits and amiloride-sensitive currents in rabbit Muller glial cells [132]. A functional role of DEG/ENaCs in Muller cells was also suggested. The authors showed that the electroretinograms recorded from anesthetized rabbits are changed in the presence of amiloride, suggesting that DEG/ENaC channels in Muller cells play an important role in retinal function [132]. Other groups showed the expression of ASIC channels in oligodendrocyte-lineage cells and gliomas and recorded amiloride-sensitive sodium conductance in high-grade glioblastoma cells [133–137]. In particular, Tian and colleagues analyzed tissues of patients with glioblastomas and showed that higher expression of ASIC1a and ASIC2 correlated with improved survival, suggesting that activation of ASICs by acidosis, which is known to be present in malignant tumors such as glioblastomas, maintains the tumor sensitive to the acidic environment perhaps limiting the growth or malignancy of the tumor [137]. In addition, Hitomi and colleagues showed the localization of β-ENaC in the specialized Schwann cells associated with the periodontal Ruffini endings of rat incisor [138]. Finally, Calavia and colleagues demonstrated that ASIC2 is expressed in the inner lamellae, which are of Schwan cell origin, of human cutaneous Pacinian corpuscles [139, 140]. Taken together these studies show the expression of DEG/ENaC channels in glial cells. However, no clear role for these channels in glial cells had been proposed.

Using the model organism *C. elegans*, our lab recently shed some light on the role of glial DEG/ENaC channels in the function of the nervous system. We published that DEG/ENaC channel ACD-1 is expressed in the amphid sheath glia in *C. elegans*. These glial cells are located in the head of the worm and extend processes to the tip of the nose, where they wrap around or envelop the dendritic ends of 12 pairs of sensory neurons. Knock-out of *acd-1* exacerbates sensory deficits caused by mutations in *deg-1* and *tax-2*, two neuronally expressed channels belonging to the DEG/ENaC and cycling nucleotide-gated channel families, respectively [96, 141]. More specifically, sensory deficits are found in response to the tastants lysine and Na^+, to the odor isoamyl alcohol, and to acidic pH, a noxious stimulus for *C. elegans*. Using in vivo calcium imaging, we showed that *acd-1 tax-2* double mutant animals fail to undergo changes in intracellular calcium in AWC sensory neurons when exposed to the odor isoamyl alcohol at low concentrations [141]. These results indicate that ACD-1 in glia supports sensory neurons activity. Indeed, we found that we could bypass these sensory deficits by expressing the calcium-permeable capsaicin-gated channel TRPV1 in AWC sensory neurons of *acd-1 tax-2* double mutants and exposing animals to capsaicin, thereby artificially increasing the levels of intracellular calcium. This work on ACD-1 revealed that glial DEG/ENaC channels play a key role in supporting the activity of associated neurons.

But how is this regulation taking place? Our analysis of the other two glial DEG/ENaCs in *C. elegans* suggested a mechanism. DELM-1 and DELM-2 share homology with ACD-1 and are expressed in the OLQ and IL socket glial cells (Fig. 9.8a, b). These glial cells wrap around the sensory dendrites of OLQ and IL nose touch-sensing neurons. Knock-out of either *delm-1* and *delm-2* causes nose touch insensitivity (Fig. 9.8c–f). Analysis of the sensory endings, using electron microscopy, revealed that these have normal morphology in *delm* mutants. These results clearly indicated that the glial DEG/ENaC channels are involved in a mechanism that regulates neuronal output rather than one controlling the morphology or development of neurons, just as suggested by our work on ACD-1. To gain insight into the underlying mechanism, we characterized DELM-1 channels in *Xenopus* oocytes using electrophysiological techniques. We found that DELM-1 expresses a constitutively active Na^+ current resembling current produced by the expression of epithelial ENaCs (Fig. 9.8g, h). In our work with ACD-1, we had obtained similar results when we expressed ACD-1 in *Xenopus* oocytes. These results suggested that DEG/ENaC channels might function in glia of *C. elegans* like ENaCs function in epithelial cells, by allowing K^+ excretion. Indeed, epithelial ENaC channels, by conducting Na^+ inside the cell, create a favorable driving force for K^+ excretion via inward rectifier K^+ channels. This function of epithelial ENaCs is underscored by human disorders caused by mutations in ENaCs, which are characterized by hypokalemia or hyperkalemia (Liddle syndrome and Pseudohypoaldosteronism type I respectively). To test this idea, we overexpressed worm IRK-2 K^+ channel in glia of *delm-1* mutants. We found that in these transgenic *C. elegans* nose touch sensitivity was restored supporting our hypothesis that glial DELM channels sustain K^+ excretion [142]. But, into which compartment is K^+ being excreted? Work on the cat Pacinian corpuscle suggests that K^+ might be secreted by glia into the microenvironment between glia and neurons.

Indeed, work on the Pacinian corpuscles of the cat showed that the concentration of K^+ in the microenvironment between the nerve terminal and the internal lamellae of Schwan cells origin is higher than in the blood. Ilyinsky and

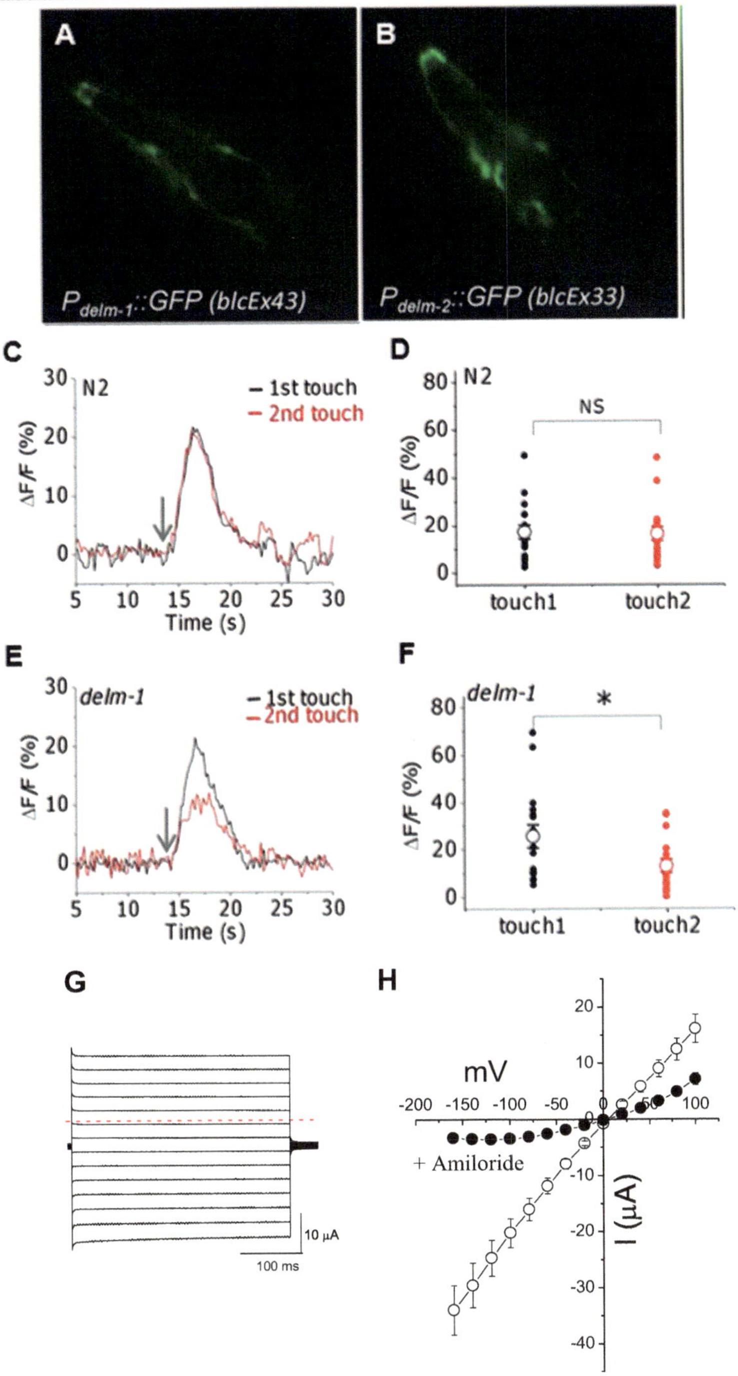

Fig. 9.8 Glial DEG/ENaC channels in nervous system function and behavior. (**a** and **b**) *C. elegans* DEG/ENaC channels *delm-1* and *delm-2* are co-expressed in OLQ and IL glial socket cells as determined by GFP expression driven by *delm-1* and *delm-2* promoters. In parenthesis the extrachromosomal DNA array designations. (**c–f**) The activity of OLQ mechanosensory neurons, which are ensheathed by the OLQ glial socket cells, is impaired by knock-out of *delm-1*. Calcium changes were recorded in OLQ neurons of wild type and *delm-1* mutant using the genetically encoded calcium sensor GCamP 3.0 upon stimulation of the animal's nose with a glass probe (8um indentation). While wild-type (N2) OLQ neurons respond robustly to two consecutive touches (1.5 min apart), *delm-1* mutants show an attenuated calcium response upon the second touch stimulation. Representative calcium

colleagues isolated Pacinian corpuscles from adult cats and demonstrated, using integrative ultramicroflame photometry, that the concentration of K^+ in the fluid surrounding the nerve ending is higher than in the blood and surrounding tissues (6.19 mM versus 2.78 mM). They also showed that such high K^+ concentration results in the enhancement of receptor mechanosensitivity, due to the lowered threshold for the generation of action potentials [143]. Moreover, their electrophysiological experiments showed that the increase in mechanosensitivity was observed only with a moderate increase in K^+ concentration; high concentrations of K^+ (above 12 mM) were found to decrease the sensitivity of the receptors. Hence, there is a range of K^+ concentrations that is optimal for the function of the Pacinian corpuscles. While this study demonstrated the important role of extracellular K^+ in Pacinian corpuscle function, no molecular mechanism explaining the higher K^+ concentration in the fluid surrounding the nerve ending was suggested. Thus, our work may have uncovered the molecular mechanism that leads to this phenomenon. Future studies will focus on experiments designed to test this model, including establishing whether DEG/ENaC channels are localized at the apical membrane of *C. elegans* glia, which would support the K^+ excretion model, and measuring K^+ concentration in the microenvironment between glia and neurons. Given that DEG/ENaC channels are expressed in different types of glia including Muller cells of the retina and astrocytes, we propose that this mechanism of regulation of neuronal output by these channels might be operative in other areas of the nervous system.

Acknowledgments I thank all the trainees and colleagues who have contributed to the work which was conducted in my laboratory and is cited in this book chapter. I also thank Nicole Encalada for critical reading of the manuscript. Work in my laboratory has been supported by the National Institute of Health (NS105616, NS106951, NS081259, NS070969, and NS049511) and the American Cancer Society (RGS-09-043-01-DDC).

References

1. Canessa CM, Horisberger JD, Rossier BC (1993) Epithelial sodium channel related to proteins involved in neurodegeneration. Nature 361:467–470
2. Driscoll M, Chalfie M (1991) The mec-4 gene is a member of a family of Caenorhabditis elegans genes that can mutate to induce neuronal degeneration. Nature 349:588–593
3. Waldmann R, Champigny G, Voilley N, Lauritzen I, Lazdunski M (1996) The mammalian degenerin Mdeg, an amiloride-sensitive cation channel activated by mutations causing neurodegeneration in Caenorhabditis elegans. J Biol Chem 271:10433–10436
4. Waldmann R, Champigny G, Bassilana F, Heurteaux C, Lazdunski M (1997) A proton-gated cation channel involved in acid-sensing. Nature 386:173–177
5. Lingueglia E, Champigny G, Lazdunski M, Barbry P (1995) Cloning of the amiloride-sensitive FMRFamide peptide-gated sodium channel. Nature 378:730–733
6. Jasti J, Furukawa H, Gonzales EB, Gouaux E (2007) Structure of acid-sensing ion channel 1 at 1.9 A resolution and low pH. Nature 449:316–323
7. Krauson AJ, Rued AC, Carattino MD (2013) Independent contribution of extracellular proton binding sites to ASIC1a activation. J Biol Chem 288:34375–34383
8. Goodman MB, Schwarz EM (2003) Transducing touch in Caenorhabditis elegans. Annu Rev Physiol 65:429–452
9. Zelle KM, Lu B, Pyfrom SC, Ben-Shahar Y (2013) The genetic architecture of degenerin/epithelial sodium channels in Drosophila. G3 (Bethesda) 3:441–450

Fig. 9.8 (continued) transients are shown in C and E and quantifications are shown in D and F. (**g–i**) Characterization of DELM-1 currents in *Xenopus* oocytes. (**g**) Example of currents recorded in oocytes expressing DELM-1. Currents were stimulated by voltage steps from −160 to + 100 mV in 20 mV increments. The dotted line represents the zero current level. (**h**) Current–voltage relationships of DELM-1 currents recorded in oocytes perfused with saline (open symbols) and saline containing the DEG/ENaC channel blocker amiloride (filled symbols) (500 μM). Adapted from [142]

10. Ben-Shahar Y (2011) Sensory functions for degenerin/epithelial sodium channels (DEG/ENaC). Adv Genet 76:1–26
11. Canessa CM, Merillat AM, Rossier BC (1994) Membrane topology of the epithelial sodium channel in intact cells. Am J Physiol 267:C1682–C1690
12. Renard S, Lingueglia E, Voilley N, Lazdunski M, Barbry P (1994) Biochemical analysis of the membrane topology of the amiloride-sensitive Na+ channel. J Biol Chem 269:12981–12986
13. Snyder PM, McDonald FJ, Stokes JB, Welsh MJ (1994) Membrane topology of the amiloride-sensitive epithelial sodium channel. J Biol Chem 269:24379–24383
14. Anantharam A, Palmer LG (2007) Determination of epithelial Na+ channel subunit stoichiometry from single-channel conductances. J Gen Physiol 130:55–70
15. Berdiev BK, Karlson KH, Jovov B, Ripoll PJ, Morris R, Loffing-Cueni D, Halpin P, Stanton BA, Kleyman TR, Ismailov II (1998) Subunit stoichiometry of a core conduction element in a cloned epithelial amiloride-sensitive Na+ channel. Biophys J 75:2292–2301
16. Coscoy S, Lingueglia E, Lazdunski M, Barbry P (1998) The Phe-Met-Arg-Phe-amide-activated sodium channel is a tetramer. J Biol Chem 273:8317–8322
17. Dijkink L, Hartog A, Van OS, van Os CH, Bindels RJ (2002) The epithelial sodium channel (ENaC) is intracellularly located as a tetramer. Pflugers Arch 444:549–555
18. Firsov D, Gautschi I, Merillat AM, Rossier BC, Schild L (1998) The heterotetrameric architecture of the epithelial sodium channel (ENaC). EMBO J 17:344–352
19. Kosari F, Sheng S, Li J, Mak DO, Foskett JK, Kleyman TR (1998) Subunit stoichiometry of the epithelial sodium channel. J Biol Chem 273:13469–13474
20. Snyder PM, Cheng C, Prince LS, Rogers JC, Welsh MJ (1998) Electrophysiological and biochemical evidence that DEG/ENaC cation channels are composed of nine subunits. J Biol Chem 273:681–684
21. Staruschenko A, Medina JL, Patel P, Shapiro MS, Booth RE, Stockand JD (2004) Fluorescence resonance energy transfer analysis of subunit stoichiometry of the epithelial Na+ channel. J Biol Chem 279:27729–27734
22. Li T, Yang Y, Canessa CM (2009) Interaction of the aromatics Tyr-72/Trp-288 in the interface of the extracellular and transmembrane domains is essential for proton gating of acid-sensing ion channels. J Biol Chem 284:4689–4694
23. Matthewman C, Johnson CK, Miller DM 3rd, Bianchi L (2018) Functional features of the "finger" domain of the DEG/ENaC channels MEC-4 and UNC-8. Am J Physiol Cell Physiol 315:C155–C163
24. Yoder N, Yoshioka C, Gouaux E (2018) Gating mechanisms of acid-sensing ion channels. Nature 555:397–401
25. Arnadottir J, O'hagan R, Chen Y, Goodman MB, Chalfie M (2011) The DEG/ENaC protein MEC-10 regulates the transduction channel complex in Caenorhabditis elegans touch receptor neurons. J Neurosci 31:12695–12704
26. Bianchi L, Gerstbrein B, Frokjaer-Jensen C, Royal DC, Mukherjee G, Royal MA, Xue J, Schafer WR, Driscoll M (2004) The neurotoxic MEC-4(d) DEG/ENaC sodium channel conducts calcium: implications for necrosis initiation. Nat Neurosci 7:1337–1344
27. Chatzigeorgiou M, Grundy L, Kindt KS, Lee WH, Driscoll M, Schafer WR (2010) Spatial asymmetry in the mechanosensory phenotypes of the C. elegans DEG/ENaC gene mec-10. J Neurophysiol 104:3334–3344
28. Goodman MB, Ernstrom GG, Chelur DS, O'hagan R, Yao CA, Chalfie M (2002) MEC-2 regulates C. elegans DEG/ENaC channels needed for mechanosensation. Nature 415:1039–1042
29. Huang M, Chalfie M (1994) Gene interactions affecting mechanosensory transduction in *Caenorhabditis elegans*. Nature 367:467–470
30. Zhang W, Bianchi L, Lee WH, Wang Y, Israel S, Driscoll M (2008) Intersubunit interactions between mutant DEG/ENaCs induce synthetic neurotoxicity. Cell Death Differ 15:1794–1803
31. Brown AL, Fernandez-Illescas SM, Liao Z, Goodman MB (2007) Gain-of-function mutations in the MEC-4 DEG/ENaC sensory mechanotransduction channel alter gating and drug blockade. J Gen Physiol 129:161–173
32. Chelur DS, Ernstrom GG, Goodman MB, Yao CA, Chen L, Hagan RO, Chalfie M (2002) The mechanosensory protein MEC-6 is a subunit of the *C. elegans* touch-cell degenerin channel. Nature 420:669–673
33. Chen Y, Bharill S, Isacoff EY, Chalfie M (2015) Subunit composition of a DEG/ENaC mechanosensory channel of Caenorhabditis elegans. Proc Natl Acad Sci U S A 112:11690–11695
34. Huang M, Gu G, Ferguson EL, Chalfie M (1995) A stomatin-like protein necessary for mechanosensation in *C. elegans*. Nature 378:292–295
35. Chen Y, Bharill S, Altun Z, O'hagan R, Coblitz B, Isacoff EY, Chalfie M (2016) Caenorhabditis elegans paraoxonase-like proteins control the functional expression of DEG/ENaC mechanosensory proteins. Mol Biol Cell 27:1272–1285
36. Huber TB, Schermer B, Muller RU, Hohne M, Bartram M, Calixto A, Hagmann H, Reinhardt C, Koos F, Kunzelmann K, Shirokova E, Krautwurst D, Harteneck C, Simons M, Pavenstadt H, Kerjaschki D, Thiele C, Walz G, Chalfie M, Benzing T (2006) Podocin and MEC-2

bind cholesterol to regulate the activity of associated ion channels. Proc Natl Acad Sci U S A 103:17079–17086
37. Lapatsina L, Jira JA, Smith ES, Poole K, Kozlenkov A, Bilbao D, Lewin GR, Heppenstall PA (2012b) Regulation of ASIC channels by a stomatin/STOML3 complex located in a mobile vesicle pool in sensory neurons. Open Biol 2:120096
38. Wetzel C, Hu J, Riethmacher D, Benckendorff A, Harder L, Eilers A, Moshourab R, Kozlenkov A, Labuz D, Caspani O, Erdmann B, Machelska H, Heppenstall PA, Lewin GR (2007) A stomatin-domain protein essential for touch sensation in the mouse. Nature 445:206–209
39. Wetzel C, Pifferi S, Picci C, Gok C, Hoffmann D, Bali KK, Lampe A, Lapatsina L, Fleischer R, Smith ES, Begay V, Moroni M, Estebanez L, Kuhnemund J, Walcher J, Specker E, Neuenschwander M, Von Kries JP, Haucke V, Kuner R, Poulet JF, Schmoranzer J, Poole K, Lewin GR (2017) Small-molecule inhibition of STOML3 oligomerization reverses pathological mechanical hypersensitivity. Nat Neurosci 20:209–218
40. Li T, Yang Y, Canessa CM (2011) Outlines of the pore in open and closed conformations describe the gating mechanism of ASIC1. Nat Commun 2:399
41. Yoder N, Gouaux E (2018) Divalent cation and chloride ion sites of chicken acid sensing ion channel 1a elucidated by x-ray crystallography. PloS One 13: e0202134
42. Zhang P, Sigworth FJ, Canessa CM (2006) Gating of acid-sensitive ion channel-1: release of Ca2+ block vs. allosteric mechanism. J Gen Physiol 127:109–117
43. Kusama N, Harding AM, Benson CJ (2010) Extracellular chloride modulates the desensitization kinetics of acid-sensing ion channel 1a (ASIC1a). J Biol Chem 285:17425–17431
44. Baconguis I, Bohlen CJ, Goehring A, Julius D, Gouaux E (2014) X-ray structure of acid-sensing ion channel 1-snake toxin complex reveals open state of a Na(+)-selective channel. Cell 156:717–729
45. Xiong ZG, Zhu XM, Chu XP, Minami M, Hey J, Wei WL, Macdonald JF, Wemmie JA, Price MP, Welsh MJ, Simon RP (2004) Neuroprotection in ischemia: blocking calcium-permeable acid-sensing ion channels. Cell 118:687–698
46. Gessmann R, Kourtis N, Petratos K, Tavernarakis N (2010) Molecular modeling of mechanosensory ion channel structural and functional features. PLoS One 5:e12814
47. Askwith CC, Wemmie JA, Price MP, Rokhlina T, Welsh MJ (2004) Acid-sensing ion channel 2 (ASIC2) modulates ASIC1 H+-activated currents in hippocampal neurons. J Biol Chem 279:18296–18305
48. Baron A, Waldmann R, Lazdunski M (2002) ASIC-like, proton-activated currents in rat hippocampal neurons. J Physiol 539:485–494
49. Brand J, Smith ES, Schwefel D, Lapatsina L, Poole K, Omerbasic D, Kozlenkov A, Behlke J, Lewin GR, Daumke O (2012) A stomatin dimer modulates the activity of acid-sensing ion channels. EMBO J 31:3635–3646
50. Lapatsina L, Brand J, Poole K, Daumke O, Lewin GR (2012a) Stomatin-domain proteins. Eur J Cell Biol 91:240–245
51. Lingueglia E, DE Weille JR, Bassilana F, Heurteaux C, Sakai H, Waldmann R, Lazdunski M (1997) A modulatory subunit of acid sensing ion channels in brain and dorsal root ganglion cells. J Biol Chem 272:29778–29783
52. Moshourab RA, Wetzel C, Martinez-Salgado C, Lewin GR (2013) Stomatin-domain protein interactions with acid-sensing ion channels modulate nociceptor mechanosensitivity. J Physiol 591:5555–5574
53. Price MP, Thompson RJ, Eshcol JO, Wemmie JA, Benson CJ (2004) Stomatin modulates gating of acid-sensing ion channels. J Biol Chem 279:53886–53891
54. Sherwood TW, Lee KG, Gormley MG, Askwith CC (2011) Heteromeric acid-sensing ion channels (ASICs) composed of ASIC2b and ASIC1a display novel channel properties and contribute to acidosis-induced neuronal death. J Neurosci 31:9723–9734
55. Vukicevic M, Kellenberger S (2004) Modulatory effects of acid-sensing ion channels on action potential generation in hippocampal neurons. Am J Physiol Cell Physiol 287:C682–C690
56. Wu LJ, Duan B, Mei YD, Gao J, Chen JG, Zhuo M, Xu L, Wu M, Xu TL (2004) Characterization of acid-sensing ion channels in dorsal horn neurons of rat spinal cord. J Biol Chem 279:43716–43724
57. Stewart GW, Hepworth-Jones BE, Keen JN, Dash BC, Argent AC, Casimir CM (1992) Isolation of cDNA coding for an ubiquitous membrane protein deficient in high Na+, low K+ stomatocytic erythrocytes. Blood 79:1593–1601
58. Martinez-Salgado C, Benckendorff AG, Chiang LY, Wang R, Milenkovic N, Wetzel C, Hu J, Stucky CL, Parra MG, Mohandas N, Lewin GR (2007) Stomatin and sensory neuron mechanotransduction. J Neurophysiol 98:3802–3808
59. Brown AL, Liao Z, Goodman MB (2008) MEC-2 and MEC-6 in the Caenorhabditis elegans sensory mechanotransduction complex: auxiliary subunits that enable channel activity. J Gen Physiol 131:605–616
60. Zhang S, Arnadottir J, Keller C, Caldwell GA, Yao CA, Chalfie M (2004) MEC-2 is recruited to the putative mechanosensory complex in *C. elegans* touch receptor neurons through its stomatin-like domain. Curr Biol 14:1888–1896
61. Sedensky MM, Siefker JM, Koh JY, Miller DM 3rd, Morgan PG (2004) A stomatin and a degenerin interact in lipid rafts of the nervous system of *Caenorhabditis elegans*. Am J Physiol Cell Physiol 287:C468–C474

62. Chalfie M, Wolinsky E (1990) The identification and suppression of inherited neurodegeneration in Caenorhabditis elegans. Nature 345:410–416
63. Harbinder S, Tavernarakis N, Herndon LA, Kinnell M, Xu SQ, Fire A, Driscoll M (1997) Genetically targeted cell disruption in *Caenorhabditis elegans*. Proc Natl Acad Sci U S A 94:13128–13133
64. O'hagan R, Chalfie M (2006) Mechanosensation in Caenorhabditis elegans. Int Rev Neurobiol 69:169–203
65. Suzuki H, Kerr R, Bianchi L, Frokjaer-Jensen C, Slone D, Xue J, Gerstbrein B, Driscoll M, Schafer WR (2003) In vivo imaging of *C. elegans* mechanosensory neurons demonstrates a specific role for the MEC-4 channel in the process of gentle touch sensation. Neuron 39:1005–1017
66. Mackness MI, Mackness B, Durrington PN, Fogelman AM, Berliner J, Lusis AJ, Navab M, Shih D, Fonarow GC (1998) Paraoxonase and coronary heart disease. Curr Opin Lipidol 9:319–324
67. Reddy ST, Wadleigh DJ, Grijalva V, Ng C, Hama S, Gangopadhyay A, Shih DM, Lusis AJ, Navab M, Fogelman AM (2001) Human paraoxonase-3 is an HDL-associated enzyme with biological activity similar to paraoxonase-1 protein but is not regulated by oxidized lipids. Arterioscler Thromb Vasc Biol 21:542–547
68. Shi S, Buck TM, Kinlough CL, Marciszyn AL, Hughey RP, Chalfie M, Brodsky JL, Kleyman TR (2017) Regulation of the epithelial Na+ channel by paraoxonase-2. J Biol Chem 292:15927–15938
69. Chalfie M, Horvitz HR, Sulston JE (1981) Mutations that lead to reiterations in the cell lineages of *C. elegans*. Cell 24:59–69
70. Xu K, Tavernarakis N, Driscoll M (2001) Necrotic cell death in *C. elegans* requires the function of calreticulin and regulators of Ca(2+) release from the endoplasmic reticulum. Neuron 31:957–971
71. O'hagan R, Chalfie M, Goodman MB (2005) The MEC-4 DEG/ENaC channel of Caenorhabditis elegans touch receptor neurons transduces mechanical signals. Nat Neurosci 8:43–50
72. Kang S, Jang JH, Price MP, Gautam M, Benson CJ, Gong H, Welsh MJ, Brennan TJ (2012) Simultaneous disruption of mouse ASIC1a, ASIC2 and ASIC3 genes enhances cutaneous mechanosensitivity. PLoS One 7:e35225
73. Kaplan JM, Horvitz HR (1993) A dual mechanosensory and chemosensory neuron in Caenorhabditis elegans. Proc Natl Acad Sci U S A 90:2227–2231
74. Geffeney SL, Cueva JG, Glauser DA, Doll JC, Lee TH, Montoya M, Karania S, Garakani AM, Pruitt BL, Goodman MB (2011) DEG/ENaC but not TRP channels are the major mechanoelectrical transduction channels in a C. elegans nociceptor. Neuron 71:845–857
75. Shreffler W, Magardino T, Shekdar K, Wolinsky E (1995) The unc-8 and sup-40 genes regulate ion channel function in Caenorhabditis elegans motorneurons. Genetics 139:1261–1272
76. Tavernarakis N, Driscoll M (1997) Molecular modeling of mechanotransduction in the nematode *Caenorhabditis elegans*. Annu Rev Physiol 59:659–689
77. Adams CM, Anderson MG, Motto DG, Price MP, Johnson WA, Welsh MJ (1998) Ripped pocket and pickpocket, novel Drosophila DEG/ENaC subunits expressed in early development and in mechanosensory neurons. J Cell Biol 140:143–152
78. Ainsley JA, Pettus JM, Bosenko D, Gerstein CE, Zinkevich N, Anderson MG, ADAMS CM, Welsh MJ, Johnson WA (2003) Enhanced locomotion caused by loss of the Drosophila DEG/ENaC protein Pickpocket1. Curr Biol 13:1557–1563
79. Guo Y, Wang Y, Wang Q, Wang Z (2014) The role of PPK26 in Drosophila larval mechanical nociception. Cell Rep 9:1183–1190
80. Tsubouchi A, Caldwell JC, Tracey WD (2012) Dendritic filopodia, Ripped Pocket, NOMPC, and NMDARs contribute to the sense of touch in Drosophila larvae. Curr Biol 22:2124–2134
81. Alvarez de la Rosa D, Zhang P, Shao D, White F, Canessa CM (2002) Functional implications of the localization and activity of acid-sensitive channels in rat peripheral nervous system. Proc Natl Acad Sci U S A 99:2326–2331
82. Dusenkova S, Ru F, Surdenikova L, Nassenstein C, Hatok J, Dusenka R, Banovcin P Jr, Kliment J, Tatar M, Kollarik M (2014) The expression profile of acid-sensing ion channel (ASIC) subunits ASIC1a, ASIC1b, ASIC2a, ASIC2b, and ASIC3 in the esophageal vagal afferent nerve subtypes. Am J Physiol Gastrointest Liver Physiol 307:G922–G930
83. Garcia-Anoveros J, Samad TA, Zuvela-Jelaska L, Woolf CJ, Corey DP (2001) Transport and localization of the DeG/ENaC ion channel BNaC1alpha to peripheral mechanosensory terminals of dorsal root ganglia neurons. J Neurosci 21:2678–2686
84. Price MP, Lewin GR, Mcilwrath SL, Cheng C, Xie J, Heppenstall PA, Stucky CL, Mannsfeldt AG, Brennan TJ, Drummond HA, Qiao J, Benson CJ, Tarr DE, Hrstka RF, Yang B, Williamson RA, Welsh MJ (2000) The mammalian sodium channel BNC1 is required for normal touch sensation. Nature 407:1007–1011
85. Price MP, Mcilwrath SL, Xie J, Cheng C, Qiao J, Tarr DE, Sluka KA, Brennan TJ, Lewin GR, Welsh MJ (2001) The DRASIC cation channel contributes to the detection of cutaneous touch and acid stimuli in mice. Neuron 32:1071–1083
86. Xie J, Price MP, Wemmie JA, Askwith CC, Welsh MJ (2003) ASIC3 and ASIC1 mediate FMRFamide-related peptide enhancement of H+-gated currents in cultured dorsal root ganglion neurons. J Neurophysiol 89:2459–2465
87. Cabo R, Galvez MA, San Jose I, Laura R, Lopez-Muniz A, Garcia-Suarez O, Cobo T, Insausti R, Vega

JA (2012) Immunohistochemical localization of acid-sensing ion channel 2 (ASIC2) in cutaneous Meissner and Pacinian corpuscles of *Macaca fascicularis*. Neurosci Lett 516:197–201
88. Page AJ, Brierley SM, Martin CM, Martinez-Salgado C, Wemmie JA, Brennan TJ, Symonds E, Omari T, Lewin GR, Welsh MJ, Blackshaw LA (2004) The ion channel ASIC1 contributes to visceral but not cutaneous mechanoreceptor function. Gastroenterology 127:1739–1747
89. Jones NG, Slater R, Cadiou H, McNaughton P, McMahon SB (2004) Acid-induced pain and its modulation in humans. J Neurosci 24:10974–10979
90. Rocha-Gonzalez HI, Herrejon-Abreu EB, Lopez-Santillan FJ, Garcia-Lopez BE, Murbartian J, Granados-Soto V (2009) Acid increases inflammatory pain in rats: effect of local peripheral ASICs inhibitors. Eur J Pharmacol 603:56–61
91. Ugawa S, Ueda T, Ishida Y, Nishigaki M, Shibata Y, Shimada S (2002) Amiloride-blockable acid-sensing ion channels are leading acid sensors expressed in human nociceptors. J Clin Invest 110:1185–1190
92. Diochot S, Baron A, Salinas M, Douguet D, Scarzello S, Dabert-Gay AS, Debayle D, Friend V, Alloui A, Lazdunski M, Lingueglia E (2012) Black mamba venom peptides target acid-sensing ion channels to abolish pain. Nature 490:552–555
93. Mazzuca M, Heurteaux C, Alloui A, Diochot S, Baron A, Voilley N, Blondeau N, Escoubas P, Gelot A, Cupo A, Zimmer A, Zimmer AM, Eschalier A, Lazdunski M (2007) A tarantula peptide against pain via ASIC1a channels and opioid mechanisms. Nat Neurosci 10:943–945
94. Verkest C, Piquet E, Diochot S, Dauvois M, Lanteri-Minet M, Lingueglia E, Baron A (2018) Effects of systemic inhibitors of acid-sensing ion channels 1 (ASIC1) against acute and chronic mechanical allodynia in a rodent model of migraine. Br J Pharmacol 175:4154–4166
95. Bohlen CJ, Chesler AT, Sharif-Naeini R, Medzihradszky KF, Zhou S, King D, Sanchez EE, Burlingame AL, Basbaum AI, Julius D (2011) A heteromeric Texas coral snake toxin targets acid-sensing ion channels to produce pain. Nature 479:410–414
96. Wang Y, Apicella A Jr, Lee SK, Ezcurra M, Slone RD, Goldmit M, Schafer WR, Shaham S, Driscoll M, Bianchi L (2008) A glial DEG/ENaC channel functions with neuronal channel DEG-1 to mediate specific sensory functions in *C. elegans*. EMBO J 27:2388–2399
97. Liu L, Leonard AS, Motto DG, Feller MA, Price MP, Johnson WA, Welsh MJ (2003) Contribution of Drosophila DEG/ENaC genes to salt taste. Neuron 39:133–146
98. Lin H, Mann KJ, Starostina E, Kinser RD, Pikielny CW (2005) A Drosophila Deg/ENaC channel subunit is required for male response to female pheromones. Proc Natl Acad Sci U S A 102:12831–12836
99. Liu T, Starostina E, Vijayan V, Pikielny CW (2012) Two Drosophila DEG/ENaC channel subunits have distinct functions in gustatory neurons that activate male courtship. J Neurosci 32:11879–11889
100. Pikielny CW (2012) Sexy DEG/ENaC channels involved in gustatory detection of fruit fly pheromones. Sci Signal 5:e48
101. Starostina E, Liu T, Vijayan V, Zheng Z, Siwicki KK, Pikielny CW (2012) A Drosophila DEG/ENaC subunit functions specifically in gustatory neurons required for male courtship behavior. J Neurosci 32:4665–4674
102. Vijayan V, Thistle R, Liu T, Starostina E, Pikielny CW (2014) Drosophila pheromone-sensing neurons expressing the ppk25 ion channel subunit stimulate male courtship and female receptivity. PLoS Genet 10:e1004238
103. Ng R, Salem SS, Wu ST, Wu M, Lin HH, Shepherd AK, Joiner WJ, Wang JW, Su CY (2019) Amplification of Drosophila olfactory responses by a DEG/ENaC channel. Neuron
104. Liu T, Wang Y, Tian Y, Zhang J, Zhao J, Guo A (2018) The receptor channel formed by ppk25, ppk29 and ppk23 can sense the Drosophila female pheromone 7,11-heptacosadiene. Genes Brain Behav: e12529
105. Garcia-Anoveros J, Garcia JA, Liu JD, Corey DP (1998) The nematode degenerin UNC-105 forms ion channels that are activated by degeneration- or hypercontraction-causing mutations. Neuron 20:1231–1241
106. Yermolaieva O, Leonard AS, Schnizler MK, Abboud FM, Welsh MJ (2004) Extracellular acidosis increases neuronal cell calcium by activating acid-sensing ion channel 1a. Proc Natl Acad Sci U S A 101:6752–6757
107. Hall DH, Gu G, Garcia-Anoveros J, Gong L, Chalfie M, Driscoll M (1997) Neuropathology of degenerative cell death in *Caenorhabditis elegans*. J Neurosci 17:1033–1045
108. Bianchi L, Driscoll M (2006) Heterologous expression of C. elegans ion channels in Xenopus oocytes. WormBook:1–16
109. Hong K, Driscoll M (1994) A transmembrane domain of the putative channel subunit MEC-4 influences mechanotransduction and neurodegeneration in C. elegans. Nature 367:470–473
110. Hong K, Mano I, Driscoll M (2000) In vivo structure-function analyses of *Caenorhabditis elegans* MEC-4, a candidate mechanosensory ion channel subunit. J Neurosci 20:2575–2588
111. Pignataro G, Simon RP, Xiong ZG (2007) Prolonged activation of ASIC1a and the time window for neuroprotection in cerebral ischaemia. Brain 130:151–158
112. Wang YZ, Wang JJ, Huang Y, Liu F, Zeng WZ, Li Y, Xiong ZG, Zhu MX, Xu TL (2015) Tissue acidosis

induces neuronal necroptosis via ASIC1a channel independent of its ionic conduction. Elife 4
113. Wang YZ, Zeng WZ, Xiao X, Huang Y, Song XL, Yu Z, Tang D, Dong XP, Zhu MX, Xu TL (2013b) Intracellular ASIC1a regulates mitochondrial permeability transition-dependent neuronal death. Cell Death Differ 20:1359–1369
114. Artal-Sanz M, Samara C, Syntichaki P, Tavernarakis N (2006) Lysosomal biogenesis and function is critical for necrotic cell death in Caenorhabditis elegans. J Cell Biol 173:231–239
115. Syntichaki P, Xu K, Driscoll M, Tavernarakis N (2002) Specific aspartyl and calpain proteases are required for neurodegeneration in *C. elegans*. Nature 419:939–944
116. Mari Y, Katnik C, Cuevas J (2010) ASIC1a channels are activated by endogenous protons during ischemia and contribute to synergistic potentiation of intracellular Ca(2+) overload during ischemia and acidosis. Cell Calcium 48:70–82
117. Matthewman C, Miller-Fleming TW, Miller DMR, Bianchi L (2016) Ca2+ permeability and Na+ conductance in cellular toxicity caused by hyperactive DEG/ENaC channels. Am J Physiol Cell Physiol 311:C920–C930
118. Wang Y, Matthewman C, Han L, Miller T, Miller DM 3rd, Bianchi L (2013a) Neurotoxic unc-8 mutants encode constitutively active DEG/ENaC channels that are blocked by divalent cations. J Gen Physiol 142:157–169
119. Friese MA, Craner MJ, Etzensperger R, Vergo S, Wemmie JA, Welsh MJ, Vincent A, Fugger L (2007) Acid-sensing ion channel-1 contributes to axonal degeneration in autoimmune inflammation of the central nervous system. Nat Med 13:1483–1489
120. Vergo S, Craner MJ, Etzensperger R, Attfield K, Friese MA, Newcombe J, Esiri M, Fugger L (2011) Acid-sensing ion channel 1 is involved in both axonal injury and demyelination in multiple sclerosis and its animal model. Brain 134:571–584
121. Wang IC, Chung CY, Liao F, Chen CC, Lee CH (2017) Peripheral sensory neuron injury contributes to neuropathic pain in experimental autoimmune encephalomyelitis. Sci Rep 7:42304
122. Miller-Fleming TW, Petersen SC, Manning L, Matthewman C, Gornet M, Beers A, Hori S, Mitani S, Bianchi L, Richmond J, Miller DM (2016) The DEG/ENaC cation channel protein UNC-8 drives activity-dependent synapse removal in remodeling GABAergic neurons. Elife 5
123. Wemmie JA, Chen J, Askwith CC, Hruska-Hageman AM, Price MP, Nolan BC, Yoder PG, Lamani E, Hoshi T, Freeman JH, Welsh MJ (2002) The acid-activated ion channel ASIC contributes to synaptic plasticity learning and memory. Neuron 34(3):463–477. https://doi.org/10.1016/S0896-6273(02)00661-X
124. Wemmie JA, Coryell MW, Askwith CC, Lamani E, Leonard AS, Sigmund CD, Welsh MJ (2004) Overexpression of acid-sensing ion channel 1a in transgenic mice increases acquired fear-related behavior. Proc Natl Acad Sci 101(10):3621–3626. https://doi.org/10.1073/pnas.0308753101
125. Hill A, Zheng X, Li X, McKinney R, Dickman D, Ben-Shahar Y (2017) The drosophila postsynaptic DEG/ENaC Channel ppk29 contributes to excitatory neurotransmission. J Neurosci 37:3171–3180
126. Orr BO, Gorczyca D, Younger MA, Jan LY, Jan YN, Davis GW (2017) Composition and control of a DEG/ENaC channel during presynaptic homeostatic plasticity. Cell Rep 20:1855–1866
127. Younger MA, Muller M, Tong A, Pym EC, Davis GW (2013) A presynaptic ENaC channel drives homeostatic plasticity. Neuron 79:1183–1196
128. Formisano R, Mersha MD, Caplan J, Singh A, Rankin CH, Tavernarakis N, Dhillon HS (2020) Synaptic vesicle fusion is modulated through feedback inhibition by dopamine auto-receptors. Synapse 74: e22131
129. Voglis G, Tavernarakis N (2008) A synaptic DEG/ENaC ion channel mediates learning in *C. elegans* by facilitating dopamine signalling. EMBO J 27:3288–3299
130. Ziemann AE, Schnizler MK, Albert GW, Severson MA, Howard MA 3rd, Welsh MJ, Wemmie JA (2008) Seizure termination by acidosis depends on ASIC1a. Nat Neurosci 11:816–822
131. Golestaneh N, DE Kozak Y, Klein C, Mirshahi M (2001) Epithelial sodium channel and the mineralocorticoid receptor in cultured rat Muller glial cells. Glia 33:160–168
132. Brockway LM, Zhou ZH, Bubien JK, Jovov B, Benos DJ, Keyser KT (2002) Rabbit retinal neurons and glia express a variety of ENaC/DEG subunits. Am J Physiol Cell Physiol 283:C126–C134
133. Berdiev BK, Xia J, Mclean LA, Markert JM, Gillespie GY, Mapstone TB, Naren AP, Jovov B, Bubien JK, Ji HL, Fuller CM, Kirk KL, Benos DJ (2003) Acid-sensing ion channels in malignant gliomas. J Biol Chem 278:15023–15034
134. Bubien JK, Ji HL, Gillespie GY, Fuller CM, Markert JM, Mapstone TB, Benos DJ (2004) Cation selectivity and inhibition of malignant glioma Na+ channels by Psalmotoxin 1. Am J Physiol Cell Physiol 287: C1282–C1291
135. Kapoor N, Bartoszewski R, Qadri YJ, Bebok Z, Bubien JK, Fuller CM, Benos DJ (2009) Knockdown of ASIC1 and epithelial sodium channel subunits inhibits glioblastoma whole cell current and cell migration. J Biol Chem 284:24526–24541
136. Ross SB, Fuller CM, Bubien JK, Benos DJ (2007) Amiloride-sensitive Na+ channels contribute to regulatory volume increases in human glioma cells. Am J Physiol Cell Physiol 293:C1181–C1185
137. Tian Y, Bresenitz P, Reska A, El Moussaoui L, Beier CP, Grunder S (2017) Glioblastoma cancer stem cell lines express functional acid sensing ion channels ASIC1a and ASIC3. Sci Rep 7:13674

138. Hitomi Y, Suzuki A, Kawano Y, Nozawa-Inoue K, Inoue M, Maeda T (2009) Immunohistochemical detection of ENaCbeta in the terminal Schwann cells associated with the periodontal Ruffini endings of the rat incisor. Biomed Res 30:113–119
139. Calavia MG, Montano JA, Garcia-Suarez O, Feito J, Guervos MA, Germana A, Del Valle M, Perez-Pinera P, Cobo J, Vega JA (2010) Differential localization of Acid-sensing ion channels 1 and 2 in human cutaneus pacinian corpuscles. Cell Mol Neurobiol 30:841–848
140. Montano JA, Calavia MG, Garcia-Suarez O, Suarez-Quintanilla JA, Galvez A, Perez-Pinera P, Cobo J, Vega JA (2009) The expression of ENa(+)C and ASIC2 proteins in Pacinian corpuscles is differently regulated by TrkB and its ligands BDNF and NT-4. Neurosci Lett 463:114–118
141. Wang Y, D'urso G, Bianchi L (2012) Knockout of glial channel ACD-1 exacerbates sensory deficits in a *C. elegans* mutant by regulating calcium levels of sensory neurons. J Neurophysiol 107:148–158
142. Han L, Wang Y, Sangaletti R, D'urso G, Lu Y, Shaham S, Bianchi L (2013) Two novel DEG/ENaC channel subunits expressed in glia are needed for nose-touch sensitivity in *Caenorhabditis elegans*. J Neurosci 33:936–949
143. Ilyinsky OB, Akoev GN, Krasnikova TL, Elman SI (1976) K and Na ion content in the Pacinian corpuscle fluid and its role in the activity of receptors. Pflugers Arch 361:279–285

Part II

Physiological Function

Glial Chloride Channels in the Function of the Nervous System Across Species

10

Jesus Fernandez-Abascal, Bianca Graziano, Nicole Encalada, and Laura Bianchi

Abstract

In the nervous system, the concentration of Cl^- in neurons that express GABA receptors plays a key role in establishing whether these neurons are excitatory, mostly during early development, or inhibitory. Thus, much attention has been dedicated to understanding how neurons regulate their intracellular Cl^- concentration. However, regulation of the extracellular Cl^- concentration by other cells of the nervous system, including glia and microglia, is as important because it ultimately affects the Cl^- equilibrium potential across the neuronal plasma membrane. Moreover, Cl^- ions are transported in and out of the cell, via either passive or active transporter systems, as counter ions for K^+ whose concentration in the extracellular environment of the nervous system is tightly regulated because it directly affects neuronal excitability. In this book chapter, we report on the Cl^- channel types expressed in the various types of glial cells focusing on the role they play in the function of the nervous system in health and disease. Furthermore, we describe the types of stimuli that these channels are activated by, the other solutes that they may transport, and the involvement of these channels in processes such as pH regulation and Regulatory Volume Decrease (RVD). The picture that emerges is one of the glial cells expressing a variety of Cl^- channels, encoded by members of different gene families, involved both in short- and long-term regulation of the nervous system function. Finally, we report data on invertebrate model organisms, such as *C. elegans* and Drosophila, that are revealing important and previously unsuspected functions of some of these channels in the context of living and behaving animals.

Keywords

Glial chloride channels · Channelopathies · Neuron · glia interaction · Nervous system development · ClC-2 · LRRC8 · SWELL1 · VRAC · Bestrophins · Maxi chloride channels · Pannexins

10.1 ClC-2

10.1.1 Structure and Function

Encoded by the *CLCN2* gene, ClC-2 is a plasma membrane voltage-gated chloride (Cl^-) channel that is expressed in most mammalian tissues [1]. In brain cells, ClC-2 protein and currents are present in neurons and glial cells such as

J. Fernandez-Abascal · B. Graziano · N. Encalada · L. Bianchi (✉)
Department Physiology and Biophysics, University of Miami, Miller School of Medicine, Miami, FL, USA
e-mail: jxf952@med.miami.edu; bxg561@miami.edu; nxe206@med.miami.edu; lbianchi@med.miami.edu

L. Zhou (ed.), *Ion Channels in Biophysics and Physiology*, Advances in Experimental Medicine and Biology 1349, https://doi.org/10.1007/978-981-16-4254-8_10

astrocytes and oligodendrocytes [2–5]. This channel is a member of the ClC family of voltage-gated Cl^- channels, characterized by a double-barreled structure with two independent pores, one in each subunit, and two intracellular conserved cystathionine-β-synthase (CBS) domains in the C-terminus, which are involved in gating regulation (Fig. 10.1a, b) [9, 10]. ClC-2 is an inward rectifying channel that remains closed at positive potentials and is activated by hyperpolarizing voltages (Fig. 10.1c) [11]. However, channels formed by one ClC-2 and one ClC-1 (or ClC-0) subunit are also partially open at positive potentials (Fig. 10.1c, d) [12, 13].

ClC-2 gating is also regulated by intracellular and extracellular ions. Intracellular Cl^- regulates ClC-2 gating by pore occupancy, promoting conformational changes to the gate that allow for the opening of the channel [14]. Conversely, extracellular protons (H^+) open the channel at low concentrations but block it at high concentrations [15, 16]. Concerning the role of Cl^- and H^+ on ClC-2 channel gating, Sanchez-Rodriguez and colleagues proposed that while intracellular Cl^- is responsible for the opening of ClC-2, extracellular H^+ stabilizes the open state of the channel (Sanchez-Rodriguez et al. [17]. Cell swelling is another mechanism by which ClC-2 is activated; yet, according to a recent review, this channel does not seem to be a major contributor to cell volume regulation [18, 19].

10.1.2 ClC-2 in the Vertebrate Brain

As mentioned earlier, ClC-2 is expressed both in neurons and in glia. In neurons, ClC-2 has been proposed to regulate neuronal excitability likely via regulation of Cl^- concentration inside and outside the cell, which consequently affects the neuronal membrane potential. Rinke and colleagues demonstrated that in mouse CA1 pyramidal cells, ClC-2 participates in Cl^- efflux [11]. One year later, Ratte and Prescott [20] reported opposite results for rat CA1 pyramidal cells and stated that ClC-2 participates in Cl^- influx, resulting in reduced neuronal excitability. This conflicting finding might be the result of differences in experimental conditions since in the first study neurons were loaded with high Cl^- concentrations, while in the second study physiological solutions were used.

Despite the proposed role of neuronal ClC-2 in regulating excitability, a link between ClC-2 and excitability disorders such as epilepsy has not been firmly established. ClC-2 was proposed to participate in controlling GABA neurons' excitability; however, the screening of epileptic patients for mutations in *CLCN2* provided inconclusive results. This suggests that mutations in *CLCN2* might increase susceptibility to epilepsy in individuals with other underlying conditions or with mutations in other genes [21]. *CLCN2* loss-of-function mutations have also been linked to cerebellar ataxia and minor cognitive defects that could be attributed to the ClC-2 role in either glia or neurons [22].

In glia, the function of ClC-2 is strongly supported by a correlation between mutations in the *CLCN2* gene and either phenotype in animal models or pathology in humans. In humans, mutations in *CLCN2* have been linked to a type of leukodystrophy called megalencephalic leukoencephalopathy with subcortical cysts (MLC) [23–25]. In this disease, the brain is enlarged showing signs of edema and the white matter is characterized by atrophy, vacuoles, and cysts that worsen with time. In these patients, mutations in ClC-2 cause mislocalization or dysfunction of the channel, both of which have been proposed to result in white matter pathology [6]. In an analytical study of patients with this disease, Depienne and colleagues observed that myelin vacuolization is present in both the brain and the spinal cord [23]. This neurological phenotype supports the role of ClC-2 in the function and survival of glia, in particular of oligodendrocytes throughout the central nervous system, but not in the peripheral nervous system. Similarly, in a mouse model in which the corresponding rodent gene has been knocked out, there is widespread vacuolation that progresses with age. Interestingly, as seen in humans, vacuolation is limited to the white matter and is not seen in the gray matter, again underscoring ClC-2 function in glia [6].

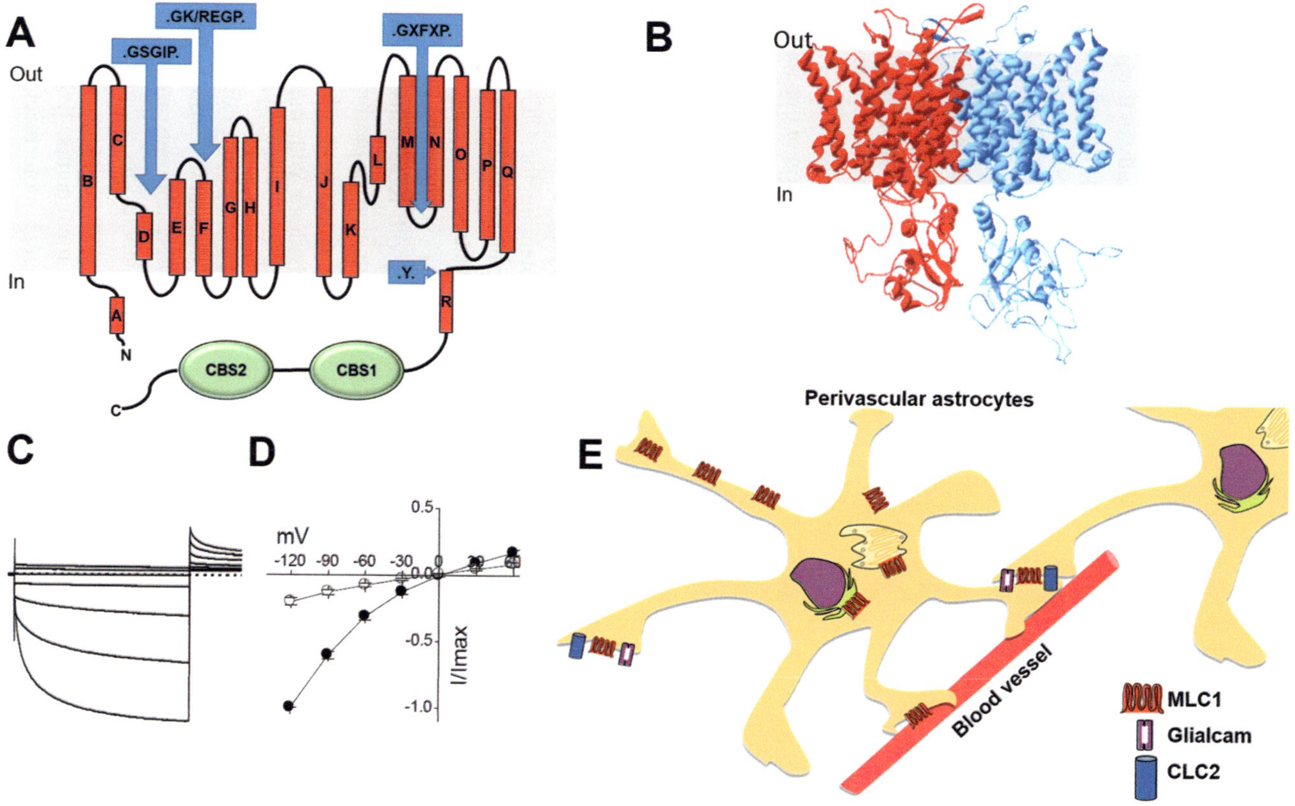

Fig. 10.1 ClC-2 structure and function. (**a**) Membrane topology of a ClC channel. The 18 α-helices are labeled by letters (A through R). The amino acid sequences that contribute to the Cl^- selectivity are designated by the blue arrows. The two CBS motifs are shown in green. (**b**) Ribbon rendition of rat ClC-2 channel. The two subunits of the dimer are represented in red and blue, respectively. In both A and B, the light gray shaded area represents the plasma membrane. (**c**) Currents elicited by voltage steps from −160 mV to +60 in 20 mV increments from a holding potential of −30 mV in a oocyte expressing rat ClC-2 perfused with a solution in which the main anion was Cl^-. (**d**) Average current/voltage relationship from oocytes injected with rat ClC-2 and perfused with a Cl^- solution (filled circles) and with the Cl^- solution containing 2 mM $CdCl_2$, a ClC-2 channel blocker (empty circles) ($n = 8$ for both, Sangaletti R., Johnson C.K., and Bianchi L., unpublished observations). (**e**) Perivascular astrocytes showcasing MLC1, Glialcam, and ClC2. All three proteins co-localize to the endfeet of perivascular astrocytes contacting blood vessels and astrocyte–astrocyte contacts [6, 7]. The localization and function of ClC-2 channels are controlled by the interaction with both GlialCAM and MLC1 [8]

Remarkably, megalencephalic leukoencephalopathy with subcortical cysts can be caused by mutations in other two genes: *MLC1* [8], encoding a protein predicted to have eight transmembrane domains and *GLIALCAM* [7, 26], which encodes the adhesion molecule GlialCAM of the immunoglobulin superfamily, highly expressed in glial cells. Based, on the similarity of the pathological manifestations and expression pattern of these three proteins, ClC-2, GLIALCAM, and MLC1 have been proposed to interact and regulate each other (Fig. 10.1e). Evidence that supports this idea are: (1) all three proteins colocalize to the endfeet contacting blood vessels and at astrocyte–astrocyte contacts [6, 7], (2) GlialCAM directs ClC-2 and MLC1 to cell–cell contacts of heterologously transfected cells, and (3) ClC-2 localization and function are controlled by the interaction with both GlialCAM and MLC1. Indeed, knock-out of either GlialCAM or MLC1 in mice results in impaired localization of ClC-2 in astrocytes and oligodendrocytes and in altered inward current rectification. In particular, in GlialCAM$^{-/-}$ or MLC1$^{-/-}$, ClC-2 is localized in the soma instead of the cellular processes [8]. Similarly in zebrafish, two ClC-2 orthologs, clc-2a and clc-2b, are expressed in astrocyte-like cells and interact with the GlialCAM paralog, glialcama, suggesting that targeting and stabilization of ClC-2 in the glial plasma membrane by GlialCAM-like proteins might be an evolutionary conserved mechanism [27].

Based on the mice knock-out phenotype and the neuroanatomical and neurological features observed in subjects with mutations in *CLCN2* and ClC-2 regulatory genes GlialCAM or MLC1, a function for ClC-2 has been suggested. It has been proposed that ClC-2 might function as a pathway for the release and reuptake of Cl^- from the cell that follows K^+ movement during high neuronal activity [4]. Indeed, the movement of ions across the plasma membrane during action potential discharge is normally followed by osmotically driven shifts in water. Thus, anything that interferes with the compensatory movement of Cl^- and water is expected to cause brain edema and leukodystrophy, both of which are observed in *CLCN2* knock-out mice and in patients with mutations in this gene [23].

This function of ClC-2 might be particularly important in brain regions with GABAergic synapses, where regulation of Cl^- concentration plays a key role in maintaining efficient GABAergic transmission. Indeed, Sik and colleagues, using immunostaining and electron microscopy, found that ClC-2 is expressed in CA1 pyramidal neurons, especially in the plasma membrane of dendrites closely associated with synaptic active zones [28]. ClC-2 was also present in the end feet of astrocytes ensheathing capillaries and blood vessels, and in the neuropil of the stratum pyramidale, in close proximity to GABAergic neurons. In addition, the level of expression of ClC-2 was polarized in astrocytes and it was layer-specific. These findings lead the authors to propose a role for glial ClC-2 in the reuptake of Cl^- and redistribution of this ion over the different brain areas, suggesting that ClC-2 actively participates in K^+ siphoning, the phenomenon by which K^+ is removed from the extracellular environment and dumped into the blood stream during high neuronal activity.

Interestingly, ClC-2 currents are smaller in in situ astrocytes of P19 versus P60 mice and lower in situ hybridization staining for ClC-2 is also observed in neonatal versus adult hippocampus [29, 30]. These observations correlate with the GABA switch during development, suggesting that ClC-2 participates with KCC transporters in regulating Cl^- concentration in the brain. They also correlate with the progressive white matter pathology seen in individuals with mutations in *CLCN2* gene and in ClC-2 knockout mice models.

Makara and colleagues also found downregulation of ClC-2 currents in reactive astrocytes around a stab lesion, suggesting, in this case, that dysregulation of the ClC-2 currents might contribute to neuronal demise [29]. In contrast, Zhao and colleagues more recently found that the injection of Cl^- channel blocker DIDS protects against white matter damage in a model of chronic cerebral ischemia–hypoxia in the rat via reduction of ClC-2 protein levels [31]. Similarly, in a diabetes rat model, high glucose concentrations were found to increase the ClC-2 activity and to promote white matter damage, which is ameliorated by the administration of DIDS [32]. These contrasting results might derive in the last two studies from the use of DIDS, a nonspecific Cl^- channel blocker.

The function of ClC-2 in controlling ionic homeostasis in the extracellular environment was suggested by Bosl and colleagues in other two organs, the eye and the testis [33]. Studies in these organs were prompted by the fact that ClC-2 knockout mice are blind and sterile. Careful analysis of the retina and testes extracted from ClC-2 knockout mice demonstrated that the cause of blindness and sterility was massive degeneration of the photoreceptors and the spermatocytes, respectively. Interestingly though, ClC-2 is not expressed in these cells. The investigators found that ClC-2 is expressed in supporting epithelial cells that are responsible in these two organs for creating an isolated microenvironment where photoreceptors and spermatocytes develop. These observations lead Bosl and colleagues to speculate that ClC-2 might be required to regulate ionic homeostasis in this microenvironment between support cells and principal cells [33]. This idea is in line with what is observed in the nervous system where ClC-2 is expressed in glia that are tasked with controlling ionic homeostasis in the microenvironment surrounding neurons.

Changes in ClC-2 expression levels near GABAergic synapses during the lifespan of the mouse clearly underscores that the function of ClC-2 might be more critical during certain life stages [11]. This is also evident when studying the pathology of knock-out models of ClC-2 regulatory genes MLC1 and GlialCAM. For example, using histological and electron microscopy approaches, Dubey and colleagues showed that vacuolization in the white matter in *Mlc1*-null mice begins at 3 months of age, suggesting that ClC-2 function is more critical in adults than in juvenile mice [34]. In a *Glialcam*-null mouse model, the vacuolization begins earlier (3 weeks of age), but this could be due to the fact that GlialCAM may have other functions that are independent of MLC1 and ClC-2 [35]. Importantly, both in patients and in megalencephalic leukoencephalopathy mice models, myelination appears normal [34, 35]. These findings support that oligodendrocytes lacking ClC-2 can still efficiently insulate axons during development.

10.1.3 Insights into the Function of Glial ClC Channels from Studies in Invertebrates

A study in the fruit fly Drosophila melanogaster suggests a role for ClC-2 type channels in early nervous system development that might explain the cerebellar ataxia and cognitive deficits found in some patients with null mutations in *CLCN2* [36]. Indeed, Plazaola and colleagues generated loss-of-function mutants for ClC-2 fly homolog ClC-a and observed that these flies have a smaller brain and defective photoreceptor guidance [37]. Using confocal microscopy and genetic approaches, they found that ClC-a is required in the niche of glial cells for the development of neuroepithelial cells and neuroblast as well as for the maturation of neurons outside the niche (Fig. 10.2a, b). The authors also found that ClC-a is important for the formation of a specific tissue called the medulla glia which is responsible for photoreceptor guidance. Plazaola-Sasieta and colleagues proposed two possible explanations for the phenotypes observed in the ClC-a knockout fly: (1) impaired secretion of glial factors, including tropic factors and guidance cues, due to imbalance in Cl^- homeostasis and (2) impaired proliferative capacity of stem cells due to alterations in pH regulation. ClC-2 type channels have been proposed to function in mediating Cl^- movement across the membrane in exchange for bicarbonate, the major pH buffering system in our body [39].

Interestingly, our lab showed that CLH-1, a *Caenorhabditis elegans* Cl^- channel that shares 37% identity with ClC-2, is expressed in glia and is permeable to bicarbonate (HCO_3^-) [38]. Using the pH sensor phluorin, we showed that *clh-1* knockout animals have impaired HCO_3^--dependent pH buffering in amphid sheath glia, which

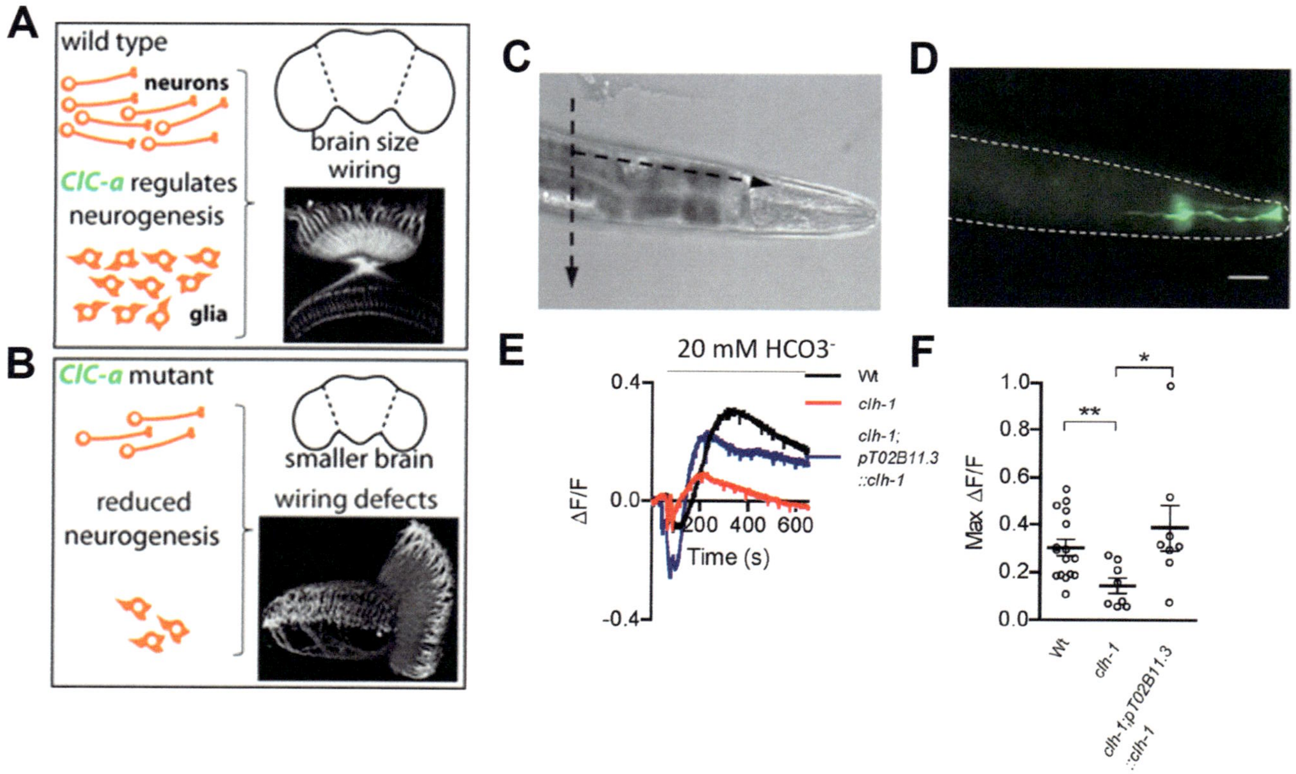

Fig. 10.2 ClC-2 type channels in Drosophila and *C. elegans*. (**a and b**) In Drosophila, ClC-a is expressed in the niche in cortex glia, which is associated with neurogenic tissues. Analysis of ClC-a mutant flies revealed that ClC-a controls both the wiring of the nervous system and the size of the fly brain. Indeed, in flies that are mutant for ClC-a, the brain is smaller and there are widespread wiring defects. Plazaola and colleagues proposed that ionic homeostasis mediated by glial ClC-a may nonautonomously affect neurogenesis and the assembly of neural circuits [37]. The photographs show representative confocal sections of adult optic lobes of wild type and ClC-a mutant photoreceptor arrays stained with anti-Chaoptin. Apparent is the axonal trajectory defect in the optic lobe of the ClC-a mutant fly. (**c**) Bright-field image of *C. elegans* glued on an agarose pad on a glass slide. The dashed arrows indicate the direction of the cuts made with a glass micropipette prior to pHlourin pH imaging experiments to expose the glia to the perfusing solution. (**d**) Fluorescent image of the same animal as in (**c**). The fluorescence is pHlourin expressed in an amphid sheath glial cell. Scale bars for C and D are 50 μm. (**e**) pHlourin-mediated pH imaging of amphid sheath glial cells in *C. elegans* perfused with an $HCO3^-$ buffer following baseline imaging perfusing with an $HCO3^-$ free and Cl^- free solution. Note the reduced alkalinization in the *clh-1* mutant animal (red line), which is restored in the rescue animal (*clh-1;pT02B11.3:clh-1*, blue line), indicating that CLH-1 is important for $HCO3^-$ permeation into *C. elegans* glia. (**f**) Average alkalinization expressed as DF/F for wild type, *clh-1* mutants, and *clh-1* rescue *C. elegans*, n was 16, 8, and 8 respectively. (**c–f**) were adapted from Grant et al. [38]

can be restored by the rescue of CLH-1 in these cells (Fig. 10.2c–f). By electrophysiological analysis, we also showed that CLH-1 is an inward rectifier channel activated by extracellular acidification. We thus proposed that CLH-1 might be activated by extracellular acidification and contribute to pH buffering of the extracellular environment via HCO_3^- permeation. Our study suggests that pH buffering mediated by ClC-2 type channels could be mediated via direct permeation of HCO_3^- and/or via regulation of bicarbonate transporters by exchange of Cl^- for HCO_3^-, as mentioned earlier.

A more recent study by Park et al. [40] reports a role for CLH-1 in the regulation of neuronal response to salt stimuli. Using Cl^- and Ca^{2+} sensors, Park and colleagues showed in vivo that the response of the ASER neuron to Na^+ is regulated by CLH-1. Using cell-specific promoters, they demonstrated that the role of CLH-1 in salt sensing is mediated by the expression of the channel in this sensory neuron and not in amphid sheath glia. This study suggests that the function of CLH-1 in neurons and glia might be distinct and may be responsible for regulating different aspects of the function of the nervous system.

10.2 Acid and Swelling-Activated Cl^- Channels (LRRC8 or SWELL1)

During neuronal activity, there are rapid changes in pH and cell volume, the last ones caused by the movement of K^+ and Cl^- ions, which need to be precisely corrected to ensure the functioning of the nervous system. To regulate cell volume and pH, astrocytes and microglia (among other cell types) are thought to use volume-sensitive anion channels (VRAC) [41], also known as volume-sensitive organic anion channels (VSOAC) and volume-activated chloride channels (VACC).

More specifically, VRAC is activated by cell swelling and plays a key role in regulatory volume decrease (RVD). RVD is a process by which all cells activate membrane transporters and channels to reduce their volume following events that cause cell swelling (Fig. 10.3a). VRAC, by mediating the efflux of Cl^- and organic osmolytes such as glutamate, taurine, and possibly ATP, coupled to efflux of K^+ via K^+ channels, bring cell volume down to control levels [44, 45]. Importantly, as we discuss below, in the nervous system, the release of glutamate, taurine, and ATP by VRAC, is thought to participate in gliotransmission.

Across cell types, VRAC has been implicated in processes such as cell cycle progression and migration, which are characterized by an increase in cell volume, and in apoptosis, which is on the contrary characterized by volume decrease [46]. In this case, VRAC is activated under isovolumetric conditions by mitochondrial-mediated apoptosis inducers (possibly src-like tyrosine kinase p56LcK) that are triggered by reactive oxygen species (ROS) production [47, 48].

The first evidence of Cl^- channel-mediated volume regulation in glial cells was reported by Pasantes-Morales and colleagues, who showed that exposure to hypoosmotic solutions caused the release of taurine from cultured rat astrocytes, which was blocked by Cl^- channels inhibitors DIDS, dipyramidole, and niflumic acid [49, 50]. A few years later, Bakhramov and colleagues described a weak outwardly rectifying Cl^- current in cultured astrocytoma activated by exposure to hypotonic solutions and suggested that this current was mediated by VRAC [51]. The current was activated slowly with a time course of 1–2 min, had a half-maximum inactivation of +50 mV, and was blocked by Cl^- channel blockers DIDS, SITS, and NPPB. Following these initial reports, many other investigators described Cl^- currents resembling VRAC in glial cells. For example, in a study using rat cortical astrocytes, Parkerson and colleagues proposed that VRAC was the major contributor to RVD in these cells [52]. This was further confirmed 2 years later by Abdullaev and colleagues who showed that both the release of excitatory amino acids (EAA) and Cl^- currents were specifically inhibited by VRAC blockers but not by other Cl^- channels/transporters blockers [53].

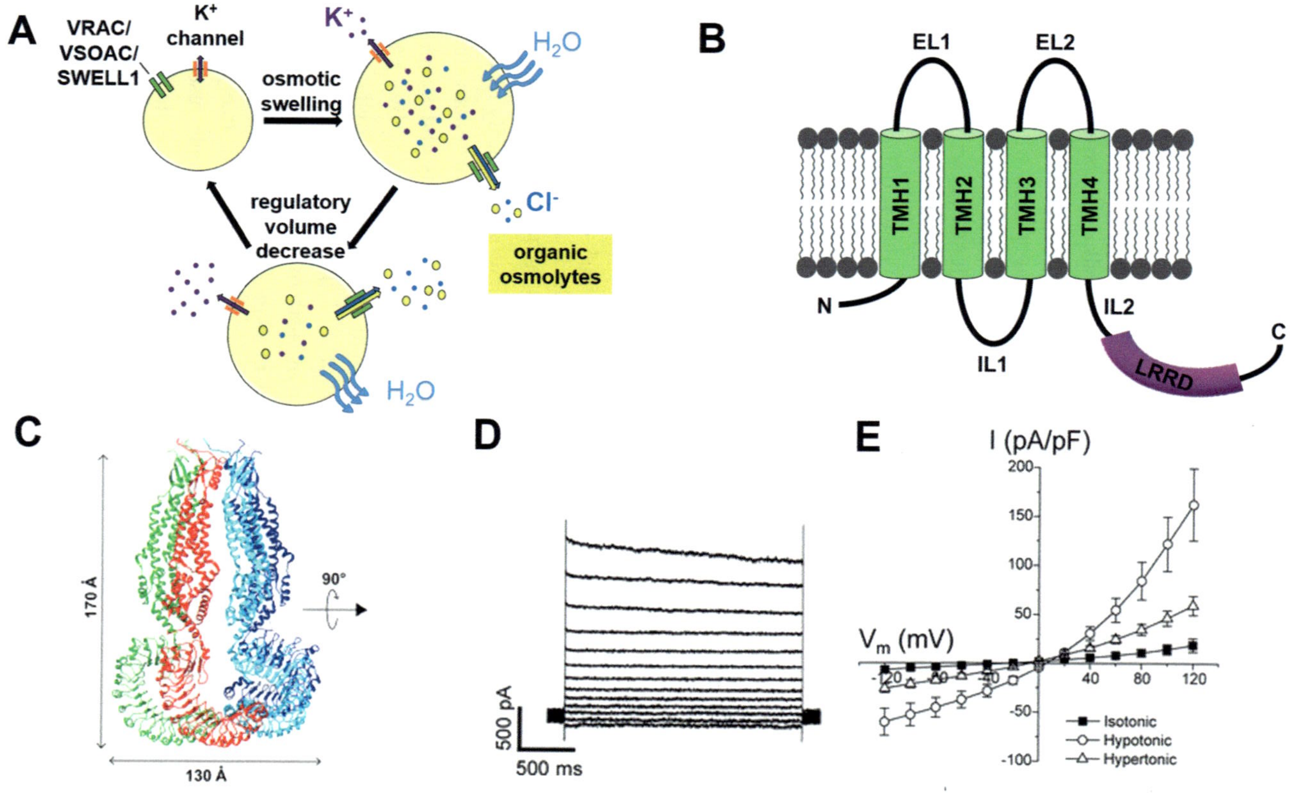

Fig. 10.3 LRRC8 structure and role in the regulation of volume decrease. (**a**) When cells sense hypotonic conditions they tend to swell causing VRAC/VSOAC/SWELL1 channels to open, allowing for an efflux of organic osmolytes and Cl^-. When the most substrate being transported is Cl^- then VRAC/VSOAC/SWELL1 causes efflux of K^+ via the K^+ channels to maintain electroneutrality. Water efflux across the plasma membrane is induced by the efflux of osmotically active substances. Water efflux is via the lipid bilayer or can be mediated by aquaporins. This mechanism allows the cellular volume to return to its original control level. (**b**) Schematic representation of the topology of an LRRC8 channel subunit. Transmembrane domains are labeled TMH1-4 (Transmembrane Helix 1–4), the extracellular loops are labeled EL1 and EL2 (Extracellular Loop 1 and 2), and the intracellular loops are labeled IL1 and IL2 (Intracellular Loop 1 and 2). The leucine-rich repeat domain (LRRD) is shown in purple. (**c**) Ribbon rendition of LRRC8A hexameric structure. Two subunits in the back are not shown for clarity (from [42]). Each subunit is shown in a different color. (**d**) Example of swelling activated currents in HCT166 cells coexpressing isoforms 8A and 8C of LRRC8. Currents were activated by voltage steps from −120 to +120 mV in 20 mV increments. (**e**) Current–voltage relationships of SWELL1 currents recorded in isotonic, hypertonic, and hypotonic solutions. (**c**, **d**) panels are adapted and reprinted with permission from Yamada and Strange [43]

How does VRAC become activated by volume increase? A few mechanisms have been suggested. For example, using immunoblotting, confocal microscopy, and patch-clamp experiments, Ando-Akatsuka and colleagues proposed that under isovolumetric conditions VRAC is inhibited by interaction with ATP-binding cassette transporter (ABCF2). When the cell is challenged by a hypotonic solution, its volume increases which cause the interaction between ATP-binding cassette transporter (ABCF2) and the channel to be disrupted. The disruption of this molecular interaction is due to the association of ABCF2 with α-actinin-4 (ACTN4) [54]. In another study, using patch-clamp and imaging techniques, Murana and colleagues proposed that activation of VRAC in mouse hippocampal microglia is caused by membrane stretch and is subsequently amplified by the raise of intracellular Ca^{2+} which is in turn mediated by activation P2Y purinergic receptors [44]. Importantly, the investigators proposed that P2Y purinergic receptors are themselves activated by ATP released by VRAC, suggesting that VRAC channels can be potentiated via a positive feedback mechanism. Interestingly, in another study VRAC amplification by ATP was shown to be mediated by two different Ca^{2+}-sensitive signaling cascades involving both PKC and CaMK II [55]. Finally, in a recent study, Konig and colleagues used Förster resonance energy transfer (FRET) to monitor the opening of VRAC channels and determined that changes in the intracellular ionic strength are not needed for channel activation. Rather, channel activation is dependent on the DAG-Protein Kinase D pathway [56]. To conclude, there are a few different hypotheses on the mechanism underlying VRAC activation, suggesting either that the mechanism is cell specific or that it is not fully understood. The recent cloning of the gene underlying VRAC and the resolution of the protein structure should help in solving this debate.

In 2014, the long-standing controversy on the molecular identity of VRAC came to an end when two groups independently reported that the Leucine-rich repeat containing 8A (LRRC8A), renamed by one of the two groups SWELL1, encoded VRAC [57, 58]. Using a clever method developed by Galietta and colleagues [59], which exploits the quenching properties of I^- on YFP fluorescence, both groups performed large siRNA screens and demonstrated that siRNA of LRRC8A in HEK cells resulted in knock-down of VRAC. LRRC8A is one of five related genes encoding homologous subunits that span the plasma membrane four times leaving N and C termini in the cytosol (Fig. 10.3b, c), a structure that resembles connexins, pannexins, and innexins with which LRRC8A share some degree of homology [18]. The name of the family derives from the intracellular LRR domain (LRRD) which contains up to 15–16 leucine repeats (Fig. 10.3b). Interestingly, while LRRC8A expresses on its own in heterologous expression systems, it still needs one of the other subunits for full expression of the current (Fig. 10.3d, e) [43, 58]. Moreover, functional expression of all the other subunits (LRRC8B-E) requires LRRC8A [58]. These results suggest that LRRC8B-E might function more as regulatory subunits.

The function of the LRRC8B-E subunits as regulators of channel properties is reflected in the fact that channel permeability is altered when LRRC8B-E subunits are removed from the channel complexes. For example, Schober and colleagues, using RNAi to knockdown individual subunits of the channel, studied the release of [^{3}H]taurine (inhibitory) and D-[^{14}C]aspartate (excitatory). They showed that while LRRC8A is an essential channel subunit, LRRC8B does not seem to participate in channel selectivity, and LRRC8D is a major contributing factor in conferring selectivity to [^{3}H]taurine [45]. In the same study, the investigators also showed that the combined knockdown of LRRC8C and LRRC8E inhibited the release of D-[^{14}C]aspartate but not [^{3}H]taurine. Based upon these results, the authors concluded that different regions of the nervous system, depending on the expression patterns of the LRRC8 subunits, might release different sets of gliotransmitters, including excitatory glutamate/aspartate and inhibitory taurine/glycine. The authors also propose that the regional specific release of these gliotransmitters might be altered in pathological conditions such as trauma, stroke, hyponatremia, and epilepsy all of which are characterized by cell swelling.

The excitatory/inhibitory amino acids that are released by glial-expressed VRAC are thought to mediate regulation of neuronal function/survival, especially during pathological conditions such as ischemia and trauma in which there are changes in cell volume. For example, Liu and colleagues showed in cultured rat astrocytes that cell swelling and ischemic stimuli cause the release of glutamate, which is at least in part inhibited by the VRAC inhibitor phloretin [60]. VRAC activation and release of glutamate can occur also under isovolumetric conditions. Indeed, Liu and colleagues showed in rat co-cultures that activation by bradykinin of astrocytic VRAC leads to an increase of intracellular Ca^{2+} in neurons which is mediated by NMDA receptors. These data indicate that VRAC mediates the release of glutamate from astrocytes which in turn mediates glia/neuron functional interaction [61].

Although VRAC has been shown to release excitatory glutamate/aspartate and inhibitory taurine/glycine, its role in gliotransmission is still controversial. Nevertheless, new insights into the modulation of neuronal function by glutamate released from astrocytes via VRAC have been published in a recent study by Yang and colleagues [62]. These authors generated a mouse model in which the VRAC subunit LRRC8A (SWELL1) had been knocked out specifically in astrocytes. Using a combination of electrophysiology and the "sniffer patch" technique to monitor both neuronal activity and release of glutamate, they showed that, upon osmotic challenges, glutamatergic synaptic transmission and synaptic plasticity in the hippocampus were impaired in these knock-out mice (Fig. 10.4). Furthermore, they showed that there is tonic glutamate release from astrocytes and this is mostly mediated by VRAC. Reduced tonic release of glutamate in the knock-out mice results in fewer action potentials in neurons, suggesting that it serves a function in regulating neuronal activity. Importantly, these cellular phenotypes are reflected in behavioral phenotypes. Indeed, LRRC8A knock-out mice showed impaired cognitive abilities including deficits in learning and memory, while locomotion and anxiety levels remained unaffected. These new data support a physiological role of astrocytic VRAC channels in modulating neuronal excitability by maintaining a tonic concentration of glutamate in the extracellular environment, a role that had been already suggested by previous studies using VRAC blockers [63, 64].

The release of glutamate by astrocytes mediated by VRAC can become toxic in pathological conditions. For example, during ischemia, hypoxic conditions induce swelling and consequently the excessive release of glutamate by astrocytes that in turn causes persistent elevation of intracellular Ca^{2+} concentration in neurons, ultimately causing cell death [65]. VRAC blockers such as tamoxifen and DCPIB were shown to reduce glutamate-induced neuronal death, suggesting that this channel mediates, at least in part, the persistent release of glutamate under these conditions [66]. Importantly, the above-mentioned LRRC8A (SWELL1) astrocytic knock-out mouse model confirms that VRAC is indeed responsible for at least some of the neuronal damage in stroke. Yang and colleagues used temporal middle cerebral artery occlusion to induce stroke and showed that the infarct volume was smaller in LRRC8A knock-out mice when compared to control, which was also reflected in a better neurological score in these mice [62].

Cell swelling and glutamate release occur also during epileptic seizures. However, the role of VRAC in releasing glutamate in epilepsy is currently controversial. In rat models of epilepsy, Tian and colleagues using in vivo two-photon imaging, electrophysiology, and HPLC showed that seizures can be induced or amplified by glutamate released from astrocytes [67]. Interestingly, under these conditions, glutamate is co-released with taurine suggesting that it might be via VRAC. However, more recently, Woo and colleagues using channel blockers and siRNA showed that glutamate release from astrocytes is via the glutamate permeable two pore domains K^+ channel TREK-1, following activation of G-protein coupled receptors (fast release), and via Cl^- channel Bestrophin-1 (Best1) (slow

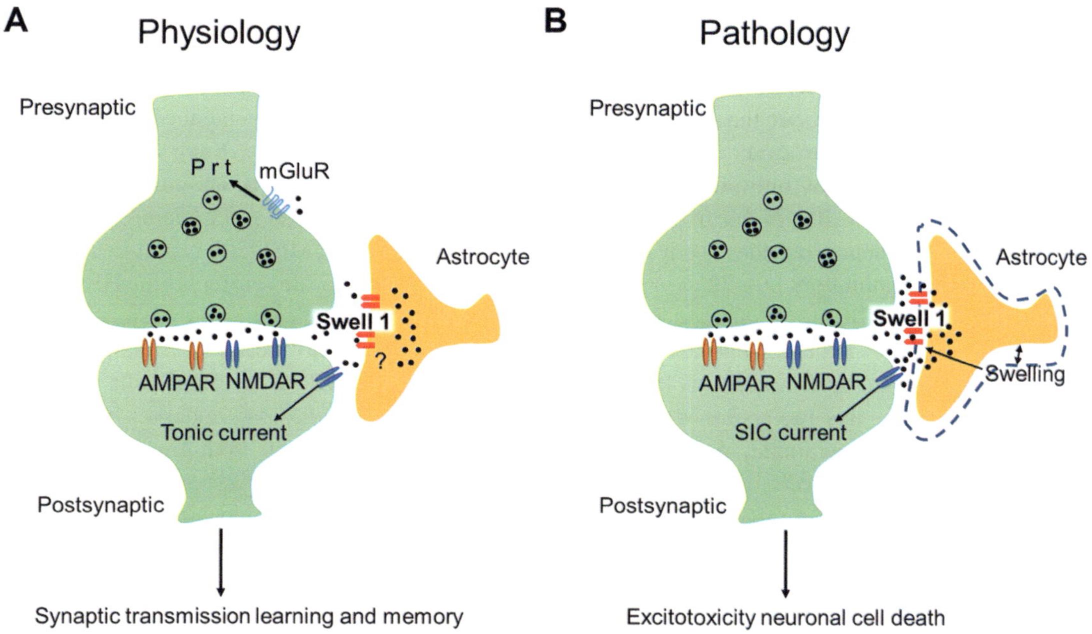

Fig. 10.4 Role of SWELL1 in learning and memory and in excitotoxicity. Astrocytes contribute to excitotoxicity and regulate synaptic transmission via glutamate release. The exact mechanisms by which astrocytic glutamate is released are not fully understood. VRAC/VSOAC/SWELL1 has been proposed to mediate non-vesicular glutamate release from astrocytes. Indeed, reduced ambient glutamate levels were observed in astrocyte-specific Swell1 knockout mice. These mutant mice exhibited impairment of learning and memory that was dependent on the hippocampus (**a**). Swell1 knockout mice were also protected from brain damage following an ischemic stroke due to reduced glutamate release from astrocytes (**b**) [62]

release) [68]. Taken together these reports highlight that glutamate (and other transmitters) might be released via VRAC but also via other types of glial ion channels, all of which might contribute under different conditions to physiological and aberrant gliotransmission in the healthy and diseased brain.

While model organisms have been instrumental in deepening our understanding of the role of glial ClC Cl^- channels in the function of the nervous system, they have not been helpful for understanding VRAC. The reason is that there are no clear homologs of LRRC8 in *C. elegans* or *Drosophila melanogaster*, where the function of VRAC may be carried out by bestrophins. In the future, the analysis or other cell-specific LRRC8 knock-out mice should help elucidate the function of VRAC in other types of glial cells such as oligodendrocytes, Schwann cells, and microglia both under physiological and pathological conditions.

10.3 Acid-Sensitive Outwardly Rectifying (ASOR) Anion Channels

Strong acidification at pHs lower than 5.5 evokes outwardly rectifying chloride currents with electrophysiological features similar to those observed in VRAC and ClC channels. Several studies have shown currents with these properties in different cell types, including glia [69]. For example, using whole-cell electrophysiology, Lambert and Oberwinkler reported an outwardly rectifying current activated at low extracellular pH in HEK cells. They showed that this current is voltage- and extracellular Cl^--dependent, needs the binding of three or four protons to be activated, and it is blocked by DIDS and niflumic acid. In the same study, these investigators recorded currents with similar electrophysiological and pharmacological properties in primary mice hippocampal astrocytes [70]. These pH-sensitive currents not only resembled ClC type channels and VARC in terms of voltage dependence and kinetics, but they were also sensitive to the same inhibitors [70–72].

For this reason, ClC-3, ClC-7, and LRRC8A were postulated to underlie this Cl^- current activated by strong acidification in astrocytes [73, 74]. However, this molecular identity was never confirmed. For example, Auzanneau and colleagues used whole-cell patch-clamp and RT-PCR to show that the outwardly rectifying acid-activated chloride currents evoked in rat Sertoli cells were not mediated by the ClC channels [71]. In another study, the involvement of LRRC8A in mediating ASOR currents was also discarded by pharmacological and small-interfering RNA (siRNA) approaches. Indeed, classical VRAC inhibitors, as well as silencing of the LRRC8 isoforms using siRNA had no effect on the outwardly rectifying acid-evoked currents recorded in HeLa cells [75]. Similar observations were also reported in microglia BV-2 cell cultures by Kittl and colleagues, who demonstrated by patch-clamp and fluorometric techniques that both ASOR and VRAC currents are activated in these cells under acidic conditions but with a different half maximal activation. So pH lower than 5 activates ASOR but inactivates VRAC. Moreover, the two Cl^- currents can be distinguished using pharmacological agents with only DIDS being equally effective in blocking both currents [76].

In two simultaneous studies recently published, the molecular basis of the ASOR channels was finally elucidated [77, 78]. Yang and colleagues used fluorescence to measure acid-induced Cl^- currents in HEK cells. The investigators then, knocked down by siRNA 2725 human proteins, predicted to have at least two transmembrane domains (Fig. 10.5a) and measured cellular fluorescence indicative of the amplitude of the Cl^- current. Using this method, the investigators identified the uncharacterized transmembrane protein 206 (TMEM206) as a possible candidate for the ASOR channel, as its knockdown caused reduction of acid-induced Cl^- currents. To test the involvement of TMEM206 in mediating ASOR currents, Yang and colleagues used a combination of CRISPR-cas9 to inactivate the gene and whole-cell patch clamp. They confirmed that indeed TMEM206 is required for acid activation of chloride currents in HEK cells [78]. Ullrich and colleagues used a similar fluorescent approach to perform a genome-wide screening with siRNA in HeLa cells and also concluded that TMEM206 underlies the ASOR current [77]. Importantly, both groups report that TMEM206 is highly expressed in CNS, especially in the cortex and hippocampus, and suggest a possible role of TMEM206 channels in acidotoxicity and ischemic brain injury [77–79]. No studies have been conducted yet to establish whether TMEM206 is expressed in neurons, glia, or both. However, previously published work suggests that TMEM206 might be expressed at least in astrocytes and microglia [70, 76].

Interestingly, both groups also report that TMEM206 is present in the genome of several other vertebrates including mouse, rat, chicken, frog, as well as zebrafish and conserves general functional features including activation by extracellular acidification and outward rectification (Fig. 10.5b). However, pH sensitivity and selectivity of TMEM206 and homologs appear to be more species specific. In other species such as the polychaete worm Capitella teleta and the sponge Amphimedon queenslandica, there are genes that have some degree of homology suggesting that TMEM206 may have appeared early in evolution. However, no clear homolog can be identified in *C. elegans* or Drosophila.

10.4 Maxi Chloride Channels

The maxi Cl^- channel (MAC) is a channel of high unitary current expressed in most human tissues, including the nervous system, where it has been observed in astrocytes and Schwann cells [80–83]. In addition to having a large

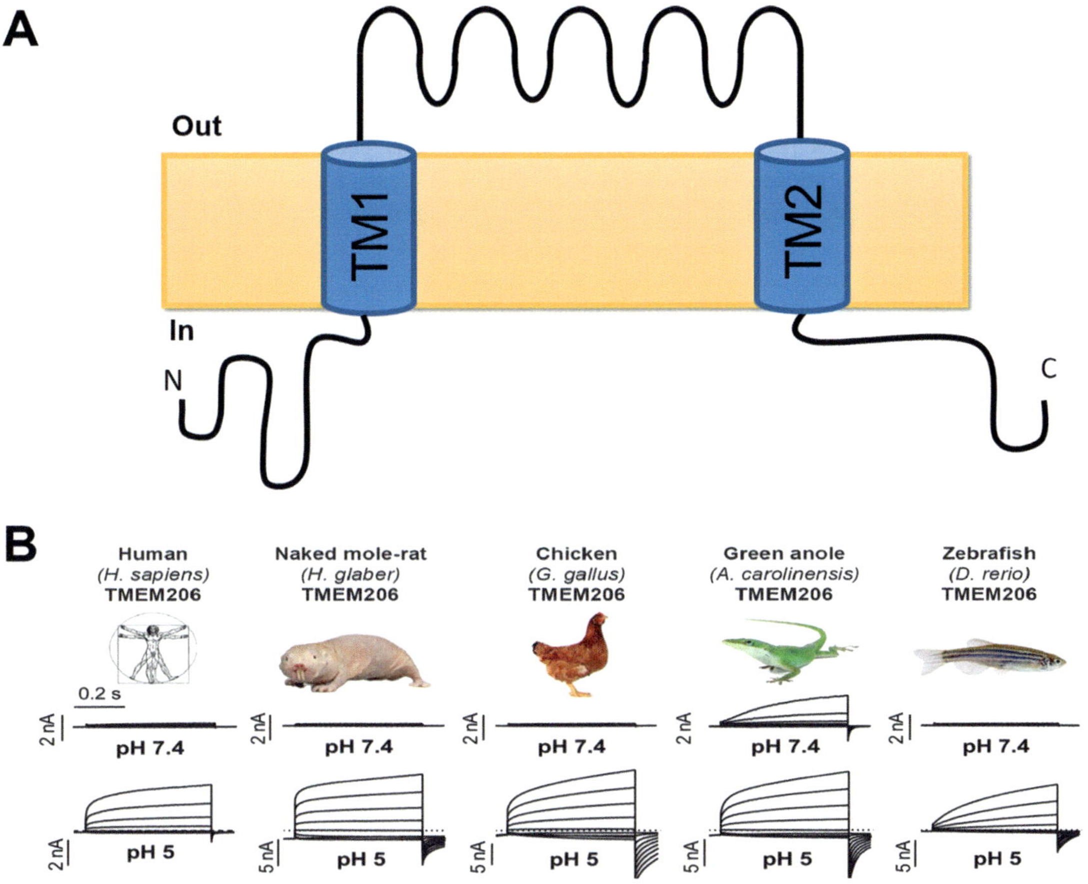

Fig. 10.5 TMEM206 encodes an acid activated Cl^- channel across species. (**a**) Schematic topology of TMEM206. Human TMEM206 is 411 aa long, it is predicted to have two transmembrane domains (TM1 and TM2) and intracellular N and C termini. (**b**) Human TMEM206 and the corresponding homologs from the naked mole rat, chicken, green anole, and zebrafish expressed in *TMEM206*$^{-/-}$ HEK cells generate Cl^- currents activated by perfusion with a solution at pH 5 [77]

conductance, MAC displays an ohmic current–voltage relationship, a strong selectivity for anions over cations ($PCl^-/PNa^+ > 8$), a sensitivity to Gd^{3+}, which blocks the channel, and voltage-dependent inactivation [81, 84]. The MAC channel is also inhibited by intracellular ATP via phosphorylation of tyrosine residues [85, 86]. Thus, the excision of a patch containing the MAC channel in a solution devoid of ATP induces its strongest activation [81, 86, 87]. In 2017 [88], Sabirov and colleagues using a combination of siRNA and electrophysiology discovered that the MAC channel is encoded by the *SLCO2A1* gene, a prostaglandin transporter (PGT) [88]. How might a prostaglandin transporter function also as a large conductance Cl^- channel? Sabirov and colleagues suggested that SLCO2A1 functions in two modes: as a PGT in the resting state, most likely with phosphorylated tyrosine residues, and as a large conductance Cl^- channel in the activated state, presumably following dephosphorylation of tyrosine residues [88].

In glial cells, the first report of high unitary anion conductance was by Gray and colleagues, who used patch clamp to study endogenous currents of rat cultured Schwann cells. The investigators found a Cl^- channel with a conductance of ~450 pS, a linear current–voltage relationship, and inactivation kinetics at small positive and negative voltages [80]. Later, other groups reported the presence of currents with similar electrophysiological properties in cultured mouse and rat astrocytes [81–83]. The channel observed in these preparations is not only permeable to chloride and other anions, with a permeability sequence of $I^- > Br^- > Cl^- > F^-$, but also to organic molecules such as glutamate and ATP [60, 89]. Indeed, Liu and colleagues demonstrated that in mouse astrocytes the release of glutamate under ischemic and hypoxia conditions is mainly carried out through MAC channels. Using a fluorometric glutamate assay, they observed that the MAC blocker Gd^{+3} significantly decreased the glutamate release, while the VRAC blocker phloretin was less efficient in blocking glutamate release [60]. In another study, Zhao and colleagues used a luciferin–luciferase assay to study the ATP release upon incubation with glutamate in cultured astrocytes and observed that Gd^{+3} reduced the amount of ATP release but other VRAC or P2X7 blockers did not [89]. These findings highlight the potential contribution of MAC channels to ischemic injury, suggesting that these channels could be considered as novel potential targets for prevention of neuronal death in ischemia.

Although MAC channels seem to become activated primarily in the presence of toxic stimuli, such as ischemia, hypoxia, and under salt stress [2], they have been suggested to function also in physiological conditions. For example, Quasthoff and colleagues, using patch-clamp in isolated rat spinal roots, observed anion channel currents with characteristics similar to those reported for MAC and suggested a possible role of this type of channels in Schwann cells in balancing K^+ concentrations in the extracellular space of the rat spinal cord [90]. However, in this case, the role of other chloride channels cannot be ruled out, since there is no evidence of *SLCO2A1* expression in Schwann cells. Indeed, Lu and colleagues, using northern blot analysis, found that human PGT was expressed at very low levels in the adult brain. Its expression in the fetal brain was much higher though [91].

Mutations in *SLCO2A1* have been linked to some cases of pachydermoperiostosis, a rare disorder that is characterized by clubbing of the fingers, excessive sweating (hyperhidrosis), and thickening of the skin of the face [92]. In a mouse model of PGT deletion, generated by using gene targeting approaches, pups die one day after birth probably because PGT is critical for maintaining prostaglandin E2 (PGE2) concentrations during development and is needed for the closure of the ductus arteriosus [93]. Taken together, these results suggest the MAC channel might not play a key role in the function of the nervous system. However, more studies in which the MAC channels are knocked out or knocked down specifically in neurons and glia are needed. One possibility is that the MAC channels might be involved in the regulation of ion concentration and other substances such as glutamate and ATP, therefore affecting more subtly the function of the nervous system.

The *C. elegans* gene F21G4.1 shares 28.8% of identity with *SLCO2A1* and is predicted to have sodium-independent organic anion transport activity. F21G4.1 is expressed in both glial amphid sheath and socket cells and was identified in a screen for genes that are upregulated during memory training in a CREB-dependent manner [38, 94]. Knock-down of F21G4.1 blocks long-term memory formation suggesting that glial socket cells via maxi Cl^- channels may play a role in long term memory [94]. These results are the first hint that the MAC channel in fact has a role in the physiological function of the nervous system in vivo and must be active under normal conditions. *C. elegans* represents an attractive model to further study the function of the MAC channel in glia in an in vivo context, exploiting the many advantages that this model organism offers. Another attractive genetic model is Drosophila whose genome also encodes a *SLCO2A1* homolog called Organic anion transporting polypeptide 30B (Oatp30B) of which nothing is known at this time.

10.5 Pannexins as Cl^- Channels

10.5.1 Structure and Function

Pannexins constitute another family of membrane channels expressed in glia that are permeable to Cl^- ions, at least under certain conditions. In mammals, there are three genes encoding pannexins: Panx1, 2, and 3, with Panx1 being the most studied both in vivo and in vitro. Pannexin subunits have four transmembrane domains, two short extracellular loops, and intracellular N- and C-termini. Recent structural studies have shown that Panx1 channels are heptamers [95–97], yet, previous studies have suggested that functional pannexins are formed by six subunits [98, 99]. These contrasting results suggest that techniques used in the past, often employing cross-linking agents, are prone to artifacts.

Despite their similarity with invertebrate gap junction proteins innexins, pannexins are nonjunctional membrane channels. The evidence that supports this conclusion is: first, pannexins are expressed in solitary cells such as erythrocytes; second, pannexins can be localized in nonjunctional membranes of polarized cells; third, pannexin channels are glycosylated, a post-translational modification that prevents close physical interaction which would be needed for gap-junction formation; and fourth, gap junctions formed by pannexins would physically overlap with those formed by connexins [100, 101].

The single-channel conductance and permeability properties of pannexins vary depending on the activation stimulus. Thus, pannexins have a small conductance and are permeable to Cl^- when activated by voltage (~50 pS [102, 103]) but have a large conductance (~500 pS) and are permeable to other ions as well as large endogenous molecules such as ATP and dyes when they are activated by low oxygen tension [104], mechanical forces [105], and high extracellular K^+ concentrations [106] (Fig. 10.6). A recent study has suggested that these two different pore sizes and permeability properties are the result of different channel conformations that can be reproduced by sequentially eliminating the C-terminus from each Panx1 subunit in a channel complex [108]. The same study demonstrated that a stepwise transition of the pore size can also occur via caspase 3-mediated cleavage of the C-terminus and via activation of the $\alpha 1$ adrenoceptors. Recent cryo-EM studies confirm that the Panx1 channel has a large pore with a diameter of up to 30 A but with a major constriction toward the extracellular side that might account for the small conductance state [95, 96]. The small conductance Cl^--permeable state of Panx1 is activated only by strong depolarizations at membrane potentials above +20 mV [102, 103]. This requirement for activation of this state of the channel limits its physiological role. One possibility is that voltage-dependent Cl^- permeable Panx1 becomes activated during sustained depolarizations such as trains of action potential in neurons [100] or during prolonged pathological depolarizations that characterize cell death.

The permeability of Pannexins to ATP, which is associated with its higher conductance state, is relatively more understood. Indeed, this is thought to constitute the major ATP release mechanism in cells like erythrocytes that lack vesicular release [104, 109] and to contribute to ATP release in other cell types, such as epithelial cells, astrocytes, neurons, and macrophages [105, 110, 111]. Panx1 permeability to ATP is thought to be the basis of calcium waves occurring in astrocytes in vivo [112, 113].

10.5.2 Pannexins in the Nervous System of Vertebrates

Two of the three pannexins isoforms, Panx1 and Panx2, are found in the nervous system and in particular the brain ([103, 114]); however, their expression patterns do not fully overlap spatially or developmentally. Panx1 is co-expressed with Panx2 in the hippocampus, cortex, cerebellum, and olfactory bulb as shown by Bruzzone and colleagues via northern blot and in situ hybridization experiments [103]. However, the same study showed that Panx1 is the only

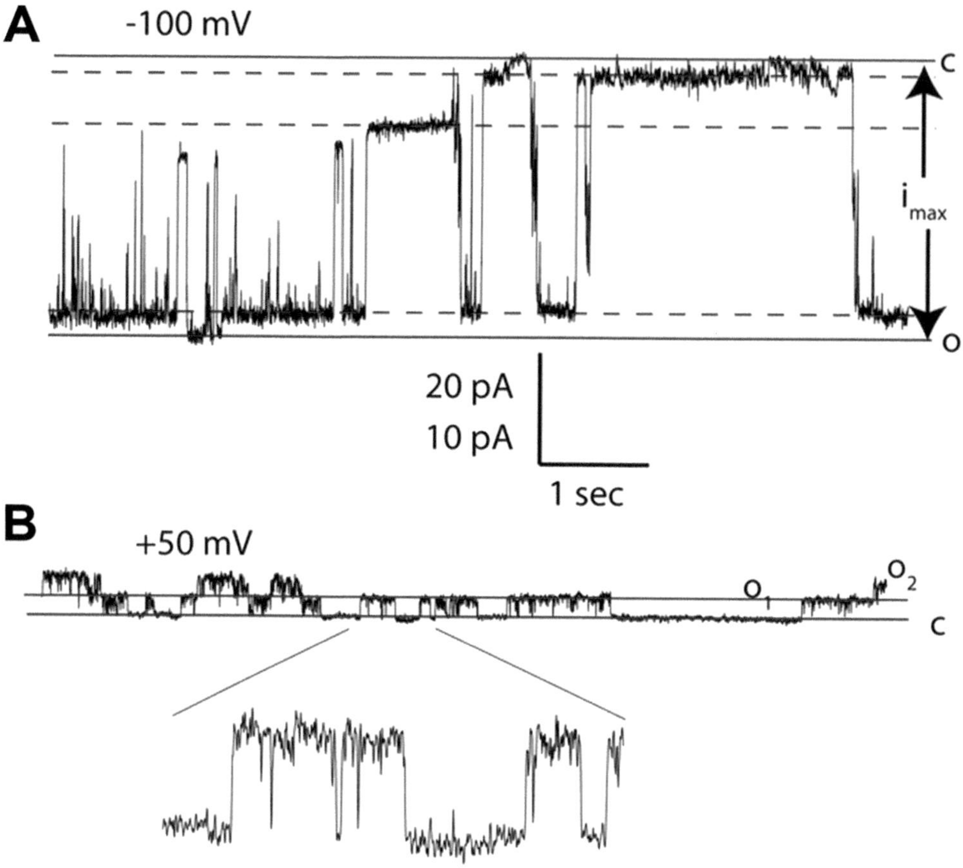

Fig. 10.6 Pax1 encodes a Cl^- channel when activated by voltage. (**a**) Panx1 opens as a large conductance channel (500 pS) when exposed to high extracellular K^+. The membrane patch was in the inside-out configuration and was clamped at −100 mV. The dashed and solid lines indicate subconductance and fully open and closed states, respectively. (**b**) Panx1 exhibits a low conductance state when activated by voltage. A membrane patch in the outside-out configuration was exposed to low K^+ and clamped at +50 mV. Two small conductance channels, indicated by the red lines and by O1 and O2, are activated under these conditions. One of the channels is also shown on a 5× scale. The scale shown in panel A applies also to panel B (10 pA). Modified from Wang et al. [107].

pannexin expressed in the white matter. These data suggest that there is a possible role of Panx1, independent from Panx2, in glial cells. In another study, Vogt and colleagues, also using in situ hybridization, reported that the temporal expression patterns of Panx1 and Panx2 do not overlap in the rat brain. In particular, Panx1 is highly expressed during the embryonic and postnatal stages, while Panx2 becomes more expressed in adult brains [115]. Taken together, these findings suggest that while Panx1 and Panx2 may function in the same cells, at least in some brain regions, they also function independently and may serve overall different functions. These data also suggest that Panx1 may serve different functions in the gray (neurons) versus the white (glia) matter [116–119].

For example, Scemes and colleagues demonstrated in mice that cell-specific deletion of Panx1 in astrocytes or neurons had opposite effects on seizures induced with kainic acid [119]. In particular, mice lacking Panx1 in neurons had lower seizure score, while mice lacking Panx1 in astrocytes had worse scores, suggesting that astrocytic Panx1 is protective. The authors also showed that the amount of ATP released, upon stimulation with 10 mM KCl, was lower in mice lacking Panx1 in astrocytes. Further, using immunocytochemistry, they showed that the expression of extracellular adenosine kinase (ADK), an enzyme that regulates extracellular levels of adenosine, was higher in glial Panx1 KO mice than in wild-type mice under conditions that induce a seizure. Furthermore, inhibiting ADK improved seizure outcomes. Taken together, these results suggest that Panx1 in astrocytes regulates the ATP/adenosine balance in the microenvironment between neurons and glia resulting in effects on aberrant neuronal function in seizure. Under these conditions, Panx1 might become activated in neurons and astrocytes by either extracellular K^+ or membrane depolarization. Thus, in this case, Cl^- permeability of glial and neuronal Panx1 might have a functional significance.

Activation of Panx1 in astrocytes and neurons by high extracellular K^+ has also been linked to activation of the inflammasome, a multiprotein intracellular complex that detects pathogens and other stressors and that leads to the activation of pro-inflammatory cytokines [106]. Silverman and colleagues showed that astrocytes and neurons have an increase in caspase-1 activation and IL-1β release upon incubation with high concentrations of K^+. Using the Panx1 blocker probenicid, the investigators established that these changes were dependent on Panx1. This data underscores the role of Panx1 in inflammasome activation which normally occurs in ischemic conditions and stroke [106, 120–122]. Under these conditions, though, the low conductance Cl^- permeable Panx1 channel state is likely not present. Rather the Panx1 channel is in the higher conducting state since it displays permeability to ATP and interleukins.

While Panx1 is the most abundant pannexin in the white matter of the brain, Panx2 is expressed in astrocytes at least under pathological conditions [123]. Using immunohistochemistry, Zappala and colleagues showed that Panx2 is normally expressed in neurons of the rat hippocampus but starts to become expressed in astrocytes after ischemia. Interestingly, at the same time, the expression of Panx2 in neurons decreases. The authors suggest that astrocytic Panx2 might be involved in the release of signaling molecules in a function similar to that of Panx1. Based on what is known for Panx1, the large conductance channel is likely involved under these conditions.

10.5.3 Invertebrate Innexins

Pannexins' homologs in invertebrates are called innexins. In fact, pannexins were discovered in vertebrates following the discovery of innexins, when homologs of innexins were searched in other species [124]. The vast majority of the studies conducted so far in invertebrates, in particular in Drosophila and *C. elegans*, support that innexins function as junctional channels allowing for the passage of ions and small molecules from cell to cell [125]. However, some studies suggest that innexins may function as hemichannels.

In the leech, two types of innexins are expressed in glial cells. Both *Hm-inx2* and *Hm-inx3* are expressed in neuropil glial cells and in the packet and connective macroglia [126]. A study employing the innexin/pannexin blocker carbenoxelon and *Hm-inx2* RNAi showed that *Hm-inx2* functions as hemichannels and releases ATP in the nervous system of the leech following nerve crush injury. ATP released by *Hm-inx2* in turn activates microglial cells and promotes their migration towards the injury site. This study, though, does not address whether *Hm-inx2* functions also as a small Cl^- conductance in the nervous system of the leech [127].

In *C. elegans*, we have previously published that innexins function as mechanosensitive large-conductance hemichannels (up to 2 nS) on the plasma membrane of touch neurons cultured in vitro and in situ in living worms [128]. Using a model of chemically induced ischemia and innexin blockers probenecid and brilliant blue G (BBG), we also showed that these innexin hemichannels promote cell death most likely functioning in their large conductance mode. However, in our electrophysiological recordings, we also found that innexins function as small conductance channels (63 pS) that can be induced to transition into large conductance state by the application of mechanical forces. Similar to the small Cl^- conductance of pannexins, the small innexin conductance we observed in *C. elegans* touch neurons is voltage-dependent and is activated by depolarizations above +20 mV. No experiments were specifically performed to address the role of this small conductance state in the physiology of touch neurons. Among the many innexins in this model organism, only INX-5 seems to be expressed almost exclusively in glial cells [129, 130]. In the future, it would be important to address the role of this glial innexin, especially its smaller Cl^- permeability state, using the wealth of genetic and molecular tools available in *C. elegans*.

10.6 Bestrophins

10.6.1 Structure and Function

Bestrophins are subunits of a family of Ca^{2+}-activated C^- channels involved in many physiological processes. In humans, bestrophins include four homologous proteins (Best1-4) encoded by the Vitelliform Macular Dystrophy (VMD) genes [131]. Interestingly, both the Drosophila and *C. elegans* genomes encode 25 genes homolog to bestrophins, perhaps suggesting that functional specificity in these invertebrates is achieved via expression of different bestrophin isoforms [131].

Human bestrophins contain between 473 and 668 amino acids and are predicted to have four α-helical transmembrane domains, with N and C termini protruding intracellularly (Fig. 10.7a). While the N-terminal region of bestrophins is highly conserved across species, the C- terminal tail is characterized by low sequence homology. Recent crystallographic studies of chicken Best1 and prokaryotic Klebsiella Pneumoniae Best revealed that this channel is barrel-shaped and is composed of five subunits positioned around a central pore. The pore is wider at the extracellular and intracellular entrances and constricts halfway through the membrane [134, 135]. A fundamental component of the bestrophin channel is the Ca^{2+} clasp, which corresponds to the Ca^{2+} binding site. The Ca^{2+} clasp is formed by a cluster of acidic residues spanning the fourth transmembrane domain and the helix-turn-helix motif in the first transmembrane domain of two adjacent subunits (in chicken Best1: Glu 300, Asp 301, Asp 302, Asp 303, and Asp 304).

Electrophysiological studies in heterologous expression systems have elucidated the basic functional properties of bestrophins. Sun and colleagues found that bestrophin channels are permeable to Cl^- ions and other anionic species by performing whole-cell recordings on human embryonic kidney cells (HEK 293) transfected with human (hBest1 and hBEST2), drosophila (dmBest1) and *C. elegans* (ceBest1) bestrophin cDNA [136]. With a pipette solution containing 148 mM CsCl, all forms of bestrophin channels displayed a reversal potential close to 0 independent of extracellular Na^+ concentration, which suggested that the current carried by the bestrophins is not generated by the movement of cations. Moreover, the fact that the reversal potential became more positive after the substitution of CsCl with gluconate was highly indicative of Cl^- conductance. The study by Sun and colleagues also showed that bestrophins' gating mechanism depends on intracellular Ca^{2+} and it is only slightly sensitive to voltage changes (Fig. 10.7b, c). To test for Ca^{2+} dependence, the authors used a photolyzable caged calcium compound called NPEGTA. Whole-cell recordings conducted during the application of voltage pulses in the presence or absence of a flash of light showed a ~6-fold increase in the current amplitude after the flash of light delivery, while the current did not change if a Ca^{2+} chelator was administered. Similar results were obtained later including using chicken Best1 reconstituted in lipid bilayers [134]. The study in lipid bilayers supports that bestrophins' Ca^{2+} dependent gating is an intrinsic property of the channel and it is not conferred by accessory subunits or associated proteins [134].

A second mechanism of regulation of Best channels is cell volume. Fischmeister and Hartzell showed that hBest1 and mBest2 from

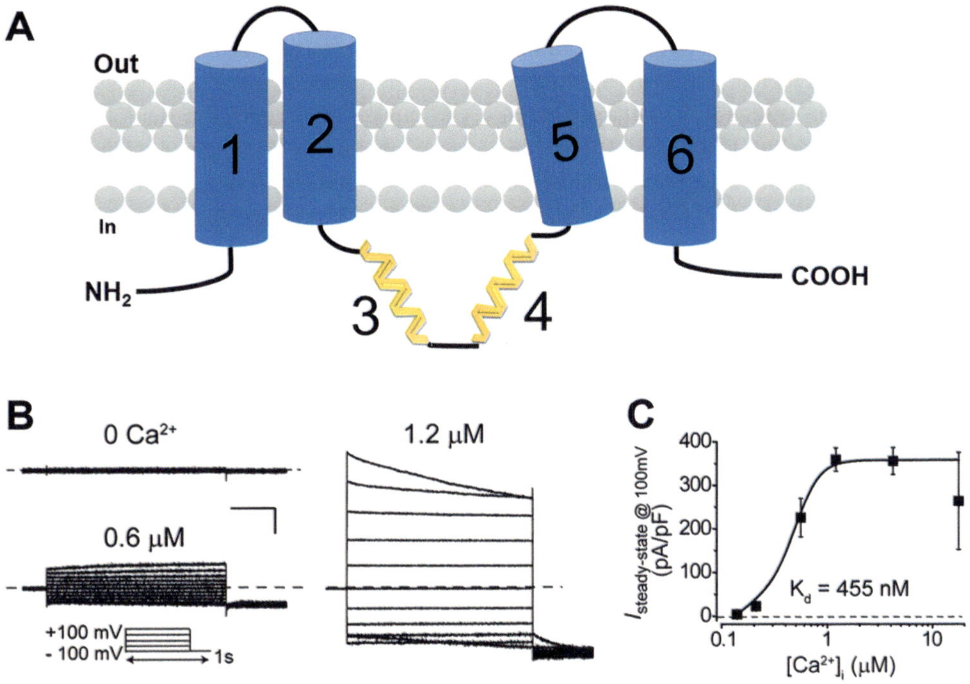

Fig. 10.7 Bestrophins structure and function. (**a**) Schematic representation of the topology of a Bestrophin channel based on studies conducted on Best1 [132]. There are six hydrophobic domains, however, domains 3 and 4 are expected to be intracellular. (**b**) Best1 currents were recorded in retinal pigment epithelial cells differentiated from induced pluripotent stem cells of a wild-type donor using intracellular solutions containing 0, 0.6 μM, and 1.2 μM Ca^{2+}. The voltage-clamp protocol is shown in the insert. The scale bar is 1 nA and 150 ms. (**c**) Ca^{2+} dose–response curve for Best1 currents similar to the ones shown in panel B. The number of cells tested was 5 or 6 for each data point. The dotted line represents the zero current level. Modified and reprinted with permission from Li et al. [133]

different cell lines are sensitive to changes in osmolarity. For example, they showed that a 20% increase in extracellular osmolarity leads to 70–80% reduction in bestrophin current [137]. On the contrary, hypo-osmolarity caused an increase in current amplitude, although the effect in this case was not as strong. However, the sensitivity of bestrophins to cell volume and their involvement in the regulation of cell volume including RVD have been controversial. Indeed, peritoneal cells from BEST1/BEST2 null mice were shown to have normal volume-activated anion currents (VRAC) [138]. A more recent study though has shown that BEST1 in fact underlies the VRAC current in in mouse sperm and human retinal pigmental epithelium (RPE) cells and that best1 null mice have severe male infertility [139]. Furthermore, RPE cells derived from patients expressing mutant forms of BEST1 (BEST1-A243V; BEST1-Q238R) show reduced VRAC and RVD function [139]. Taken together, these reports seem to support regulation of bestrophin currents by cell volume and in turn involvement of these Cl^- channels in RVD, at least in some tissues.

10.6.2 Bestrophins in the Mammalian Nervous System

The expression pattern of bestrophins has been determined using PCR, immunochemical, and electrophysiological methods. One of the early studies by Marquardt and colleagues showed that Best1 was highly expressed in the retinal pigmented epithelium (RPE) in humans

[140]. Similarly, Petrukhin and colleagues, also using RT-PCR and in situ hybridization, found that Best1 was highly expressed in RPE but also detected Best1 in the brain, testis, and spinal cord [141]. The expression of Best1 in RPE is preserved also in cell lines derived from this tissue as shown by Marmorstein and colleagues [142]. Outside the retina, Best1 was shown, by a combination of RT-PCR, Western Blot analysis and immunohistochemistry, to be expressed in the mouse trachea, human airway epithelial cells, mouse colon, mouse kidney, and in a mouse kidney epithelial cell line [143].

Electrophysiological experiments on mouse dorsal root ganglia (DRG) demonstrated the presence of a Ca^{2+}-dependent Cl^{-} current with properties resembling bestrophins [144]. Interestingly, the current was present only in medium-diameter (30–40 μm) neurons in control mice but became prominent also in large-diameter (40–50 μm) neurons in mice that had undergone transection of the sciatic nerve, suggesting ectopic upregulation of Best1 gene expression in these pathological conditions. Following these findings, RT-PCR and in situ hybridization studies showed that, although DRG neurons express three types of calcium-activated Cl^{-} channels (besthrophins, maxi-Cl channels, and TMEM16), Best1 is the only one that becomes upregulated after nerve injury [145]. Indeed, using electrophysiology, Best1 siRNA, and expression of Best1 mutants, Boudes and colleagues showed that Best1 underlies that Ca^{2+}-activated Cl^{-} current that becomes upregulated in large diameter neurons following nerve injury [146]. The authors speculate that Best1 might be involved in the regeneration of sensory neurons after injury. Pineda-Farias and colleagues also found upregulation of Best1 in DRG neurons of a neuropathic rat model [147].

Murine Best1 has also been found in the brain in both neurons and astrocytes [148]. Brain regions that showed particularly high expression were the olfactory bulb, the hippocampus, and the cerebellum. Park and colleagues showed the expression of Best1 in both neurons and astrocytes using single-cell RT-PCR and immunohistochemistry. Furthermore, using electrophysiology, the investigators characterized Best1 currents in the hippocampal astrocytes of CA1 stratum radiatum region. They showed that a Cl^{-} current with features resembling bestrophins was activated by an increase in intracellular Ca^{2+} following activation of G-protein coupled receptor PAR1. Moreover, they showed that this current was reduced after silencing mBest1 gene by shRNA. Another study employing RT-PCR and shRNA silencing provided additional evidence of mBest1 expression in cortical astrocytes [149].

Later studies using immunochemical approaches combined with electron microscopy revealed that Best1 is localized in the perisynaptic astrocytic microdomains at least in the hippocampal CA1 region of the mouse (Fig. 10.8a, b) [68, 151]. Given that Best1 is enriched in these regions and shows low levels of expression in the astrocytes' soma and processes, the authors speculate that Best1 may hold an important role in the regulation of glutamate concentration at the synapses, being Best1 a possible supply of glutamate release. In support of these conclusions are the results of a study conducted in a mouse model of Alzheimer's disease (APP/PS1) [152]. The authors of this study found that in this model Best1 is not localized in perisynaptic domains but rather it is found in abundance in the hippocampal astrocytes' cell body and processes. The authors suggest that the mislocalization of Best1 might be the result of astrocytes' reactivity, a feature of astrocytes associated with several neurological and psychiatric disorders including Alzheimer's, Parkinson's, schizophrenia, and addictive disorders [153, 154].

Interestingly, in the cerebellum, Best1 is not perisynaptic, but rather in the cell body and processes of another type of glia, the Bergman glia [150, 155, 156] (Fig. 10.8d, e). This difference in protein distribution between the hippocampal astrocytes and Bergman glia accounts for the shift in the reversal potential of the currents that Park and colleagues noticed in hippocampal astrocytes when performing whole-cell recordings (Fig. 10.8c, f) [150]. Indeed, in these cells, the reversal potential was 40 mV more negative than expected under those experimental conditions, which could be explained by space

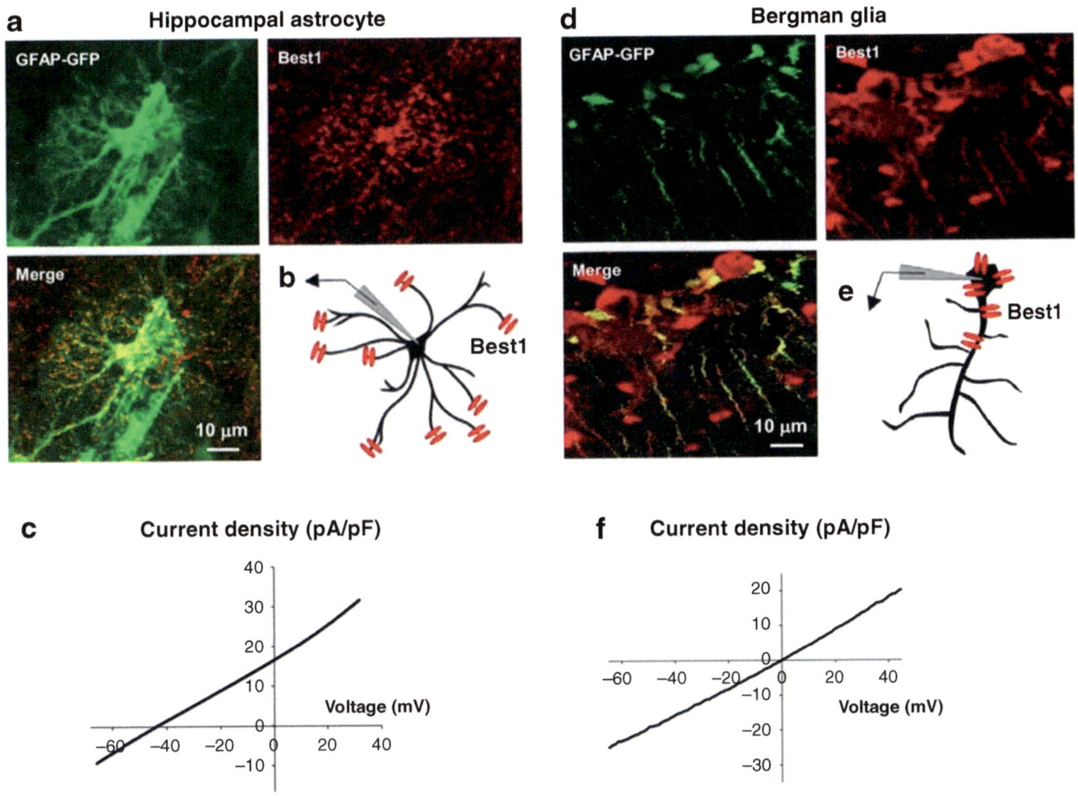

Fig. 10.8 Best1 localization in hippocampal astrocytes and in Bergman glia. (**a**) Immunohistochemical staining of GFAP-GFP (green) and of Best1(red), and a merge of the two images demonstrating exclusive expression of Best1 channels in microdomains of hippocampal astrocytes. (**b**) Schematic representation of the subcellular localization of Best1 (red) in hippocampal astrocytes. (**c**) Representative current–voltage relationship of NPPB-sensitive currents showing anion conductance in hippocampal astrocytes. (**d**) Immunohistochemical staining of GFAP-GFP (green) and of Best1(red), and a merge of the two images showing exclusive localization of Best1 in the soma of Bergmann glia. (**e**) Schematic representation of the subcellular localization of Best1 in Bergman glia. (**f**) Representative current–voltage relationship of NPPB-sensitive currents in Bergman glia. Note that the experiments shown in **c** and **f** were conducted in isometric Cl^- predicting a reversal potential of 0 mV. The more negative reversal potential observed in hippocampal astrocytes is due to the space clamp error caused by the localization of Best1 at the end of the cellular processes in this cell type. From Park et al. [150]

clamping issues that occur when the channels are not localized in the cell body. On the contrary, in Bergman cells, the reversal potential of the ionic currents was exactly as expected, supporting that in these cells Best1 is localized in the cell body. The investigators speculate that this striking difference in the localization of Best1 between the hippocampus and the cerebellum might highlight different functions. Indeed, Best1 in the cerebellum mediates tonic inhibition by releasing GABA, while in the hippocampus Best1 may release glutamate at the synapses [156].

As already hinted above, in addition to small anions, other anionic species permeate through bestrophin channels. For example, Qu and Hartzell using patch-clamp electrophysiology showed that all four human bestrophins and a mouse BEST2 [157] are highly permeable to $HCO3^-$ both at high and at physiological concentrations of this ion. These results suggest that bestrophins might participate to pH buffering in tissues. In addition to being permeable to larger anions, bestrophins are also permeable to large biologically active molecules such as glutamate and GABA, which is rather remarkable considering the size of these molecules. Park and colleagues showed in electrophysiological experiments by ion substitution method that bestrophins are permeable to glutamate, isethionate, bromine, iodine, and other small ions in mouse cultured astrocytes, with a relative permeability of glutamate to chloride ($P_{glutamate}$/P_{cl}) of 0.47 [148]. Similar permeability ratios of 0.67 and 0.53 were found for mBest1 expressed in HEK293 cells [68] and for endogenous bestrophins in CA1 hippocampal mouse astrocytes, respectively [150]. It is worth mentioning a study by Dickson and colleagues in which chicken Best1 was found to be impermeable to glutamate [134]. In this case, anionic permeability was estimated via the quenching of a fluorescent pH indicator, exploiting the fact that anionic entry into liposomes is accompanied by the entry of protons. The indirectness of the measurement may account for potential misinterpretations [158].

Among the studies investigating bestrophins' permeability to glutamate and GABA, several employed the sniffer patch-clamp technique [68, 152, 155]. This technique uses cells expressing a mutant form of GluR1 that does not get desensitized to detect glutamate released from astrocytes. This technique was developed by Lee and colleagues in a study in which they showed that astrocytic Best1, following activation of G-protein coupled receptors (GPCR), such as P2Y, bradykinin, and protease-activated receptor PAR1 (TFLLR), releases glutamate in concentrations that are sufficient for the activation of neuronal NMDRs at the synapses [159]. These results suggested that bestrophins may regulate NMDA-mediated synaptic transmission.

This hypothesis was indeed supported by Park and colleagues who showed that the release of glutamate via bestrophin channels, that are expressed in mouse hippocampal CA1 astrocytes, leads to an increase in the evoked excitatory post-synaptic potentials mediated by NMDA receptors [151]. The investigators further showed that Best1-mediated glutamate release increased the size of LTP and LTD and lowered the threshold of induction of long-term potentiation. Taken together, these results underscore that release of glutamate through Best1 expressed in astrocytes influence synaptic plasticity.

Best1 is permeable also to GABA. Lee and colleagues found a permeability ratio (P_{GABA}/P_{Cl}) of 0.27 for Best1 expressed in HEK293T cells and of 0.19 for native Best1 channels in mouse Bergmann glia [155]. Although these permeability values seem low, the investigators clarified that the current elicited by GABA permeation is carried by the less frequent anionic form of the molecule when GABA is mostly zwitterionic [158]. Interestingly, GABA release is substantial even at resting intracellular calcium concentration (100 nM), suggesting that bestrophins could leak GABA in basal conditions. To show that GABA released via Best1 can activate GABA receptors expressed on nearby cells, Lee and colleagues used the sniffer technique which employed the slowly deactivating receptor $GABA_C$ expressed in HEK cells. Finally, the investigators, using Best1 inhibitors and shRNA against Best1, showed that in native tissue GABA released by astrocytic Best1 produces an inhibitory current in cerebellar granule cells, a finding reported also by Yoon and colleagues [156].

Importantly, while under physiological conditions GABA release via glial Best1 occurs only in the cerebellum (Bergmann glia), under pathological conditions it is released also from astrocytes. For example, Jo and colleagues showed in a mouse model of Alzheimer disease (APP/PS1) that hippocampal astrocytes in the dentate gyrus release GABA upon stimulation of the PAR-1 receptors and that the astrocytes that are closest to the plaques have the highest GABA immunoreactivity [152]. This result suggests that astrocytes reactivity may be associated with GABA production and release. Indeed, Chun and colleagues showed that a stab wound injury inflicted to the CA1 hippocampal area induced both astrocytes reactivity and an increase in GABA production and release [160].

10.6.3 Bestrophins in Invertebrates

Bestrophins have not been studied in *C. elegans* and have only been investigated by a few in Drosophila. This leaves a lot of room for exploiting these powerful model organisms to advance our understanding of the role of this family of Cl^- channels in pathophysiology, especially as it pertains to the nervous system. Drosophila Best1 has been shown to be activated by both cell swelling and Ca^{2+}, with activation by cell swelling being independent of intracellular Ca^{2+} concentrations [161]. Furthermore, Stotz and Clapham showed that the first 64 N-terminal residues of Drosphila Best1 are important for volume sensitivity and can transfer this channel feature to volume insensitive Drosophila Best2 [162]. Chien and Hartzell using a point mutation in the pore region of Drosophila Best1 (F81C) finally demonstrated that the VRAC currents resulting from expression of Best1 in S2 insect cells are mediated by Best1 and are not the result of activation of another channel for which Best1 serves as an accessory subunit [161].

Acknowledgments I thank all the trainees and colleagues who have contributed to the work which was conducted in my laboratory and is cited in this book chapter. Work in my laboratory has been supported by the National Institute of Health (NS105616, NS106951, NS081259, NS070969, and NS049511) and the American Cancer Society (RGS-09-043-01-DDC).

References

1. Thiemann A, Grunder S, Pusch M, Jentsch TJ (1992) A chloride channel widely expressed in epithelial and non-epithelial cells. Nature 356:57–60
2. Elorza-Vidal X, Gaitan-Penas H, Estevez R (2019) Chloride channels in astrocytes: structure, roles in brain homeostasis and implications in disease. Int J Mol Sci 20
3. Hou X, Zhang R, Wang J, Li Y, Li F, Zhang Y, Zheng X, Shen Y, Wang Y, Zhou L (2018) CLC-2 is a positive modulator of oligodendrocyte precursor cell differentiation and myelination. Mol Med Rep 17:4515–4523
4. Sirisi S, Elorza-Vidal X, Arnedo T, Armand-Ugon M, Callejo G, Capdevila-Nortes X, Lopez-Hernandez T, Schulte U, Barrallo-Gimeno A, Nunes V, Gasull X, Estevez R (2017) Depolarization causes the formation of a ternary complex between GlialCAM, MLC1 and ClC-2 in astrocytes: implications in megalencephalic leukoencephalopathy. Hum Mol Genet 26:2436–2450
5. Walz W (2002) Chloride/anion channels in glial cell membranes. Glia 40:1–10
6. Blanz J, Schweizer M, Auberson M, Maier H, Muenscher A, Hubner CA, Jentsch TJ (2007) Leukoencephalopathy upon disruption of the chloride channel ClC-2. J Neurosci 27:6581–6589
7. Lopez-Hernandez T, Sirisi S, Capdevila-Nortes X, Montolio M, Fernandez-Duenas V, Scheper GC, Van Der Knaap MS, Casquero P, Ciruela F, Ferrer I, Nunes V, Estevez R (2011) Molecular mechanisms of MLC1 and GliALCAM mutations in megalencephalic leukoencephalopathy with subcortical cysts. Hum Mol Genet 20:3266–3277
8. Hoegg-Beiler MB, Sirisi S, Orozco IJ, Ferrer I, Hohensee S, Auberson M, Godde K, Vilches C, De Heredia ML, Nunes V, Estevez R, Jentsch TJ (2014) Disrupting MLC1 and GlialCAM and ClC-2 interactions in leukodystrophy entails glial chloride channel dysfunction. Nat Commun 5:3475
9. Dutzler R (2007) A structural perspective on ClC channel and transporter function. FEBS Lett 581:2839–2844
10. Ramjeesingh M, Li C, Huan LJ, Garami E, Wang Y, Bear CE (2000) Quaternary structure of the chloride channel ClC-2. Biochemistry 39:13838–13847
11. Rinke I, Artmann J, Stein V (2010) ClC-2 voltage-gated channels constitute part of the background conductance and assist chloride extrusion. J Neurosci 30:4776–4786
12. Stolting G, Fischer M, Fahlke C (2014) ClC-1 and ClC-2 form hetero-dimeric channels with novel protopore functions. Pflugers Arch 466:2191–2204
13. Weinreich F, Jentsch TJ (2001) Pores formed by single subunits in mixed dimers of different CLC chloride channels. J Biol Chem 276:2347–2353
14. De Jesus-Perez JJ, Castro-Chong A, Shieh RC, Hernandez-Carballo CY, De Santiago-Castillo JA, Arreola J (2016) Gating the glutamate gate of CLC-2 chloride channel by pore occupancy. J Gen Physiol 147:25–37
15. Arreola J, Begenisich T, Melvin JE (2002) Conformation-dependent regulation of inward rectifier chloride channel gating by extracellular protons. J Physiol 541:103–112
16. Niemeyer MI, Yusef YR, Cornejo I, Flores CA, Sepulveda FV, Cid LP (2004) Functional evaluation of human ClC-2 chloride channel mutations associated with idiopathic generalized epilepsies. Physiol Genomics 19:74–83
17. Sanchez-Rodriguez JE, De Santiago-Castillo JA, Contreras-Vite JA, Nieto-Delgado PG, Castro-Chong A, Arreola J (2012) Sequential interaction of chloride and proton ions with the fast gate steer the voltage-dependent gating in ClC-2 chloride channels. J Physiol 590:4239–4253

18. Okada Y, Okada T, Sato-Numata K, Islam MR, Ando-Akatsuka Y, Numata T, Kubo M, Shimizu T, Kurbannazarova RS, Marunaka Y, Sabirov RZ (2019) Cell volume-activated and volume-correlated anion channels in mammalian cells: their biophysical, molecular, and pharmacological properties. Pharmacol Rev 71:49–88
19. Roman RM, Smith RL, Feranchak AP, Clayton GH, Doctor RB, Fitz JG (2001) ClC-2 chloride channels contribute to HTC cell volume homeostasis. Am J Physiol Gastrointest Liver Physiol 280:G344–G353
20. Ratte S, Prescott SA (2011) ClC-2 channels regulate neuronal excitability, not intracellular chloride levels. J Neurosci 31:15838–15843
21. Galanopoulou AS (2010) Mutations affecting GABAergic signaling in seizures and epilepsy. Pflugers Arch 460:505–523
22. Guo Z, Lu T, Peng L, Cheng H, Peng F, Li J, Lu Z, Chen S, Qiu W (2019) CLCN2-related leukoencephalopathy: a case report and review of the literature. BMC Neurol 19:156
23. Depienne C, Bugiani M, Dupuits C, Galanaud D, Touitou V, Postma N, Van Berkel C, Polder E, Tollard E, Darios F, Brice A, De Die-Smulders CE, Vles JS, Vanderver A, Uziel G, Yalcinkaya C, Frints SG, Kalscheuer VM, Klooster J, Kamermans M, Abbink TE, Wolf NI, Sedel F, Van Der Knaap MS (2013) Brain white matter oedema due to ClC-2 chloride channel deficiency: an observational analytical study. Lancet Neurol 12:659–668
24. Estevez R, Elorza-Vidal X, Gaitan-Penas H, Perez-Rius C, Armand-Ugon M, Alonso-Gardon M, Xicoy-Espaulella E, Sirisi S, Arnedo T, Capdevila-Nortes X, Lopez-Hernandez T, Montolio M, Duarri A, Teijido O, Barrallo-Gimeno A, Palacin M, Nunes V (2018) Megalencephalic leukoencephalopathy with subcortical cysts: a personal biochemical retrospective. Eur J Med Genet 61:50–60
25. Gaitan-Penas H, Apaja PM, Arnedo T, Castellanos A, Elorza-Vidal X, Soto D, Gasull X, Lukacs GL, Estevez R (2017) Leukoencephalopathy-causing CLCN2 mutations are associated with impaired Cl(-) channel function and trafficking. J Physiol 595:6993–7008
26. Arnedo T, Aiello C, Jeworutzki E, Dentici ML, Uziel G, Simonati A, Pusch M, Bertini E, Estevez R (2014) Expanding the spectrum of megalencephalic leukoencephalopathy with subcortical cysts in two patients with GLIALCAM mutations. Neurogenetics 15:41–48
27. Perez-Rius C, Gaitan-Penas H, Estevez R, Barrallo-Gimeno A (2015) Identification and characterization of the zebrafish ClC-2 chloride channel orthologs. Pflugers Arch 467:1769–1781
28. Sik A, Smith RL, Freund TF (2000) Distribution of chloride channel-2-immunoreactive neuronal and astrocytic processes in the hippocampus. Neuroscience 101:51–65
29. Makara JK, Rappert A, Matthias K, Steinhauser C, Spat A, Kettenmann H (2003) Astrocytes from mouse brain slices express ClC-2-mediated Cl-currents regulated during development and after injury. Mol Cell Neurosci 23:521–530
30. Mladinic M, Becchetti A, Didelon F, Bradbury A, Cherubini E (1999) Low expression of the ClC-2 chloride channel during postnatal development: a mechanism for the paradoxical depolarizing action of GABA and glycine in the hippocampus. Proc Biol Sci 266:1207–1213
31. Zhao B, Quan H, Ma T, Tian Y, Cai Q, Li H (2015) 4,4′-Diisothiocyanostilbene-2,2′-disulfonic Acid (DIDS) ameliorates ischemia-hypoxia-induced white matter damage in neonatal rats through inhibition of the voltage-gated chloride channel ClC-2. Int J Mol Sci 16:10457–10469
32. He F, Peng Y, Yang Z, Ge Z, Tian Y, Ma T, Li H (2017) Activated ClC-2 Inhibits p-Akt to repress myelination in GDM newborn rats. Int J Biol Sci 13:179–188
33. Bosl MR, Stein V, Hubner C, Zdebik AA, Jordt SE, Mukhopadhyay AK, Davidoff MS, Holstein AF, Jentsch TJ (2001) Male germ cells and photoreceptors, both dependent on close cell-cell interactions, degenerate upon ClC-2 Cl(-) channel disruption. EMBO J 20:1289–1299
34. Dubey M, Bugiani M, Ridder MC, Postma NL, Brouwers E, Polder E, Jacobs JG, Baayen JC, Klooster J, Kamermans M, Aardse R, De Kock CP, Dekker MP, Van Weering JR, Heine VM, Abbink TE, Scheper GC, Boor I, Lodder JC, Mansvelder HD, Van Der Knaap MS (2015) Mice with megalencephalic leukoencephalopathy with cysts: a developmental angle. Ann Neurol 77:114–131
35. Bugiani M, Dubey M, Breur M, Postma NL, Dekker MP, Ter Braak T, Boschert U, Abbink TEM, Mansvelder HD, Min R, Van Weering JRT, Van Der Knaap MS (2017) Megalencephalic leukoencephalopathy with cysts: the Glialcam-null mouse model. Ann Clin Transl Neurol 4:450–465
36. Van Der Knaap MS, Depienne C, Sedel F, Abbink TEM (1993) CLCN2-Related leukoencephalopathy. In: Adam MP, Ardinger HH, Pagon RA, Wallace SE, Bean LJH, Stephens K, Amemiya A (eds) GeneReviews((R)), Seattle (WA)
37. Plazaola-Sasieta H, Zhu Q, Gaitan-Penas H, Rios M, Estevez R, Morey M (2019) Drosophila ClC-a is required in glia of the stem cell niche for proper neurogenesis and wiring of neural circuits. Glia 67:2374–2398
38. Grant J, Matthewman C, Bianchi L (2015) A novel mechanism of pH buffering in *C. elegans* Glia: bicarbonate transport via the voltage-gated ClC Cl- Channel CLH-1. J Neurosci 35:16377–16397
39. Sanchez-Rodriguez JE, De Santiago-Castillo JA, Arreola J (2010) Permeant anions contribute to voltage dependence of ClC-2 chloride channel by

interacting with the protopore gate. J Physiol 588:2545–2556
40. Park C, Sakurai Y, Sato H, Kanda S, Iino Y, Kunitomo H (2021) Roles of the ClC Chloride Channel CLH-1 in food-associated salt chemotaxis behavior of *C. elegans*. Elife 10:e255701. 2020.02.16.951368
41. Akita T, Okada Y (2014) Characteristics and roles of the volume-sensitive outwardly rectifying (VSOR) anion channel in the central nervous system. Neuroscience 275:211–231
42. Konig B, Stauber T (2019) Biophysics and structure-function relationships of LRRC8-formed volume-regulated anion channels. Biophys J 116:1185–1193
43. Yamada T, Strange K (2018) Intracellular and extracellular loops of LRRC8 are essential for volume-regulated anion channel function. J Gen Physiol 150:1003–1015
44. Murana E, Pagani F, Basilico B, Sundukova M, Batti L, Di Angelantonio S, Cortese B, Grimaldi A, Francioso A, Heppenstall P, Bregestovski P, Limatola C, Ragozzino D (2017) ATP release during cell swelling activates a Ca(2+)-dependent Cl(-) current by autocrine mechanism in mouse hippocampal microglia. Sci Rep 7:4184
45. Schober AL, Wilson CS, Mongin AA (2017) Molecular composition and heterogeneity of the LRRC8-containing swelling-activated osmolyte channels in primary rat astrocytes. J Physiol 595:6939–6951
46. Nunez R, Sancho-Martinez SM, Novoa JM, Lopez-Hernandez FJ (2010) Apoptotic volume decrease as a geometric determinant for cell dismantling into apoptotic bodies. Cell Death Differ 17:1665–1671
47. Hoffmann EK, Sorensen BH, Sauter DP, Lambert IH (2015) Role of volume-regulated and calcium-activated anion channels in cell volume homeostasis, cancer and drug resistance. Channels (Austin) 9:380–396
48. Shimizu T, Numata T, Okada Y (2004) A role of reactive oxygen species in apoptotic activation of volume-sensitive Cl(-) channel. Proc Natl Acad Sci U S A 101:6770–6773
49. Pasantes-Morales H, Moran J, Schousboe A (1990) Volume-sensitive release of taurine from cultured astrocytes: properties and mechanism. Glia 3:427–432
50. Sanchez-Olea R, Pena C, Moran J, Pasantes-Morales H (1993) Inhibition of volume regulation and efflux of osmoregulatory amino acids by blockers of Cl- transport in cultured astrocytes. Neurosci Lett 156:141–144
51. Bakhramov A, Fenech C, Bolton TB (1995) Chloride current activated by hypotonicity in cultured human astrocytoma cells. Exp Physiol 80:373–389
52. Parkerson KA, Sontheimer H (2004) Biophysical and pharmacological characterization of hypotonically activated chloride currents in cortical astrocytes. Glia 46:419–436
53. Abdullaev IF, Rudkouskaya A, Schools GP, Kimelberg HK, Mongin AA (2006) Pharmacological comparison of swelling-activated excitatory amino acid release and Cl-currents in cultured rat astrocytes. J Physiol 572:677–689
54. Ando-Akatsuka Y, Shimizu T, Numata T, Okada Y (2012) Involvements of the ABC protein ABCF2 and alpha-actinin-4 in regulation of cell volume and anion channels in human epithelial cells. J Cell Physiol 227:3498–3510
55. Mongin AA, Kimelberg HK (2005) ATP regulates anion channel-mediated organic osmolyte release from cultured rat astrocytes via multiple Ca2+-sensitive mechanisms. Am J Physiol Cell Physiol 288:C204–C213
56. Konig B, Hao Y, Schwartz S, Plested AJ, Stauber T (2019) A FRET sensor of C-terminal movement reveals VRAC activation by plasma membrane DAG signaling rather than ionic strength. Elife 8
57. Qiu Z, Dubin AE, Mathur J, Tu B, Reddy K, Miraglia LJ, Reinhardt J, Orth AP, Patapoutian A (2014) SWELL1, a plasma membrane protein, is an essential component of volume-regulated anion channel. Cell 157:447–458
58. Voss FK, Ullrich F, Munch J, Lazarow K, Lutter D, Mah N, Andrade-Navarro MA, Von Kries JP, Stauber T, Jentsch TJ (2014) Identification of LRRC8 heteromers as an essential component of the volume-regulated anion channel VRAC. Science 344:634–638
59. Galietta LJ, Haggie PM, Verkman AS (2001) Green fluorescent protein-based halide indicators with improved chloride and iodide affinities. FEBS Lett 499:220–224
60. Liu HT, Tashmukhamedov BA, Inoue H, Okada Y, Sabirov RZ (2006) Roles of two types of anion channels in glutamate release from mouse astrocytes under ischemic or osmotic stress. Glia 54:343–357
61. Liu HT, Akita T, Shimizu T, Sabirov RZ, Okada Y (2009) Bradykinin-induced astrocyte-neuron signalling: glutamate release is mediated by ROS-activated volume-sensitive outwardly rectifying anion channels. J Physiol 587:2197–2209
62. Yang J, Vitery MDC, Chen J, Osei-Owusu J, Chu J, Qiu Z (2019b) Glutamate-releasing SWELL1 channel in astrocytes modulates synaptic transmission and promotes brain damage in stroke. Neuron 102 (813–827):e6
63. Cavelier P, Attwell D (2005) Tonic release of glutamate by a DIDS-sensitive mechanism in rat hippocampal slices. J Physiol 564:397–410
64. Le Meur K, Galante M, Angulo MC, Audinat E (2007) Tonic activation of NMDA receptors by ambient glutamate of non-synaptic origin in the rat hippocampus. J Physiol 580:373–383
65. Kimelberg HK, Goderie SK, Higman S, Pang S, Waniewski RA (1990) Swelling-induced release of glutamate, aspartate, and taurine from astrocyte cultures. J Neurosci 10:1583–1591

66. Bowens NH, Dohare P, Kuo YH, Mongin AA (2013) DCPIB, the proposed selective blocker of volume-regulated anion channels, inhibits several glutamate transport pathways in glial cells. Mol Pharmacol 83:22–32
67. Tian GF, Azmi H, Takano T, Xu Q, Peng W, Lin J, Oberheim N, Lou N, Wang X, Zielke HR, Kang J, Nedergaard M (2005) An astrocytic basis of epilepsy. Nat Med 11:973–981
68. Woo DH, Han KS, Shim JW, Yoon BE, Kim E, Bae JY, Oh SJ, Hwang EM, Marmorstein AD, Bae YC, Park JY, Lee CJ (2012) TREK-1 and Best1 channels mediate fast and slow glutamate release in astrocytes upon GPCR activation. Cell 151:25–40
69. Capurro V, Gianotti A, Caci E, Ravazzolo R, Galietta LJ, Zegarra-Moran O (2015) Functional analysis of acid-activated Cl(-) channels: properties and mechanisms of regulation. Biochim Biophys Acta 1848:105–114
70. Lambert S, Oberwinkler J (2005) Characterization of a proton-activated, outwardly rectifying anion channel. J Physiol 567:191–213
71. Auzanneau C, Thoreau V, Kitzis A, Becq F (2003) A Novel voltage-dependent chloride current activated by extracellular acidic pH in cultured rat Sertoli cells. J Biol Chem 278:19230–19236
72. Fu ZJ, Li XZ, Wang QR, Shi L, Zhang LQ, Pan XL (2013) Extracellular acidic pH-activated, outward rectifying chloride currents can be regulated by reactive oxygen species in human THP-1 monocytes. Biochem Biophys Res Commun 432:701–706
73. Matsuda JJ, Filali MS, Collins MM, Volk KA, Lamb FS (2010) The ClC-3 Cl-/H+ antiporter becomes uncoupled at low extracellular pH. J Biol Chem 285:2569–2579
74. Wang HY, Shimizu T, Numata T, Okada Y (2007) Role of acid-sensitive outwardly rectifying anion channels in acidosis-induced cell death in human epithelial cells. Pflugers Arch 454:223–233
75. Sato-Numata K, Numata T, Inoue R, Okada Y (2016) Distinct pharmacological and molecular properties of the acid-sensitive outwardly rectifying (ASOR) anion channel from those of the volume-sensitive outwardly rectifying (VSOR) anion channel. Pflugers Arch 468:795–803
76. Kittl M, Helm K, Beyreis M, Mayr C, Gaisberger M, Winklmayr M, Ritter M, Jakab M (2019) Acid- and volume-sensitive chloride currents in microglial cells. Int J Mol Sci 20
77. Ullrich F, Blin S, Lazarow K, Daubitz T, Von Kries JP, Jentsch TJ (2019) Identification of TMEM206 proteins as pore of PAORAC/ASOR acid-sensitive chloride channels. Elife 8:e49187
78. Yang J, Chen J, Del Carmen Vitery M, Osei-Owusu J, Chu J, Yu H, Sun S, Qiu Z (2019a) PAC, an evolutionarily conserved membrane protein, is a proton-activated chloride channel. Science 364:395–399
79. Osei-Owusu J, Yang J, Del Carmen Vitery M, Tian M, Qiu Z (2020) PAC proton-activated chloride channel contributes to acid-induced cell death in primary rat cortical neurons. Channels (Austin) 14:53–58
80. Gray PT, Bevan S, Ritchie JM (1984) High conductance anion-selective channels in rat cultured Schwann cells. Proc R Soc Lond B Biol Sci 221:395–409
81. Jalonen T (1993) Single-channel characteristics of the large-conductance anion channel in rat cortical astrocytes in primary culture. Glia 9:227–237
82. Nowak L, Ascher P, Berwald-Netter Y (1987) Ionic channels in mouse astrocytes in culture. J Neurosci 7:101–109
83. Sonnhof U (1987) Single voltage-dependent K+ and Cl- channels in cultured rat astrocytes. Can J Physiol Pharmacol 65:1043–1050
84. Hazama A, Fan HT, Abdullaev I, Maeno E, Tanaka S, Ando-Akatsuka Y, Okada Y (2000) Swelling-activated, cystic fibrosis transmembrane conductance regulator-augmented ATP release and Cl- conductances in murine C127 cells. J Physiol 523(Pt 1):1–11
85. Islam MR, Uramoto H, Okada T, Sabirov RZ, Okada Y (2012) Maxi-anion channel and pannexin 1 hemichannel constitute separate pathways for swelling-induced ATP release in murine L929 fibrosarcoma cells. Am J Physiol Cell Physiol 303:C924–C935
86. Toychiev AH, Sabirov RZ, Takahashi N, Ando-Akatsuka Y, Liu H, Shintani T, Noda M, Okada Y (2009) Activation of maxi-anion channel by protein tyrosine dephosphorylation. Am J Physiol Cell Physiol 297:C990–C1000
87. Bosma MM (1989) Anion channels with multiple conductance levels in a mouse B lymphocyte cell line. J Physiol 410:67–90
88. Sabirov RZ, Merzlyak PG, Okada T, Islam MR, Uramoto H, Mori T, Makino Y, Matsuura H, Xie Y, Okada Y (2017) The organic anion transporter SLCO2A1 constitutes the core component of the Maxi-Cl channel. EMBO J 36:3309–3324
89. Zhao B, Gu L, Liu K, Zhang M, Liu H (2017) Maxi-anion channels play a key role in glutamate-induced ATP release from mouse astrocytes in primary culture. Neuroreport 28:380–385
90. Quasthoff S, Strupp M, Grafe P (1992) High conductance anion channel in Schwann cell vesicles from rat spinal roots. Glia 5:17–24
91. Lu R, Kanai N, Bao Y, Schuster VL (1996) Cloning, in vitro expression, and tissue distribution of a human prostaglandin transporter cDNA(hPGT). J Clin Invest 98:1142–1149
92. Diggle CP, Parry DA, Logan CV, Laissue P, Rivera C, Restrepo CM, Fonseca DJ, Morgan JE, Allanore Y, Fontenay M, Wipff J, Varret M, Gibault L, Dalantaeva N, Korbonits M, Zhou B,

Yuan G, Harifi G, Cefle K, Palanduz S, Akoglu H, Zwijnenburg PJ, Lichtenbelt KD, Aubry-Rozier B, Superti-Furga A, Dallapiccola B, Accadia M, Brancati F, Sheridan EG, Taylor GR, Carr IM, Johnson CA, Markham AF, Bonthron DT (2012) Prostaglandin transporter mutations cause pachydermoperiostosis with myelofibrosis. Hum Mutat 33:1175–1181
93. Chang HY, Locker J, Lu R, Schuster VL (2010) Failure of postnatal ductus arteriosus closure in prostaglandin transporter-deficient mice. Circulation 121:529–536
94. Lakhina V, Arey RN, Kaletsky R, Kauffman A, Stein G, Keyes W, Xu D, Murphy CT (2015) Genome-wide functional analysis of CREB/long-term memory-dependent transcription reveals distinct basal and memory gene expression programs. Neuron 85:330–345
95. Deng Z, He Z, Maksaev G, Bitter RM, Rau M, Fitzpatrick JAJ, Yuan P (2020) Cryo-EM structures of the ATP release channel pannexin 1. Nat Struct Mol Biol 27:373–381
96. Michalski K, Syrjanen JL, Henze E, Kumpf J, Furukawa H, Kawate T (2020) The Cryo-EM structure of pannexin 1 reveals unique motifs for ion selection and inhibition. Elife 9
97. Mou L, Ke M, Song M, Shan Y, Xiao Q, Liu Q, LI J, Sun K, Pu L, Guo L, Geng J, Wu J, Deng D (2020) Structural basis for gating mechanism of Pannexin 1 channel. Cell Res 30:452–454
98. Boassa D, Ambrosi C, Qiu F, Dahl G, Gaietta G, Sosinsky G (2007) Pannexin1 channels contain a glycosylation site that targets the hexamer to the plasma membrane. J Biol Chem 282:31733–31743
99. Boassa D, Qiu F, Dahl G, Sosinsky G (2008) Trafficking dynamics of glycosylated pannexin 1 proteins. Cell Commun Adhes 15:119–132
100. Dahl G (2018) The Pannexin1 membrane channel: distinct conformations and functions. FEBS Lett 592:3201–3209
101. Penuela S, Bhalla R, Gong XQ, Cowan KN, Celetti SJ, Cowan BJ, Bai D, Shao Q, Laird DW (2007) Pannexin 1 and pannexin 3 are glycoproteins that exhibit many distinct characteristics from the connexin family of gap junction proteins. J Cell Sci 120:3772–3783
102. Bao L, Locovei S, Dahl G (2004) Pannexin membrane channels are mechanosensitive conduits for ATP. FEBS Lett 572:65–68
103. Bruzzone R, Hormuzdi SG, Barbe MT, Herb A, Monyer H (2003) Pannexins, a family of gap junction proteins expressed in brain. Proc Natl Acad Sci U S A 100:13644–13649
104. Sridharan M, Adderley SP, Bowles EA, Egan TM, Stephenson AH, Ellsworth ML, Sprague RS (2010) Pannexin 1 is the conduit for low oxygen tension-induced ATP release from human erythrocytes. Am J Physiol Heart Circ Physiol 299:H1146–H1152
105. Ransford GA, Fregien N, Qiu F, Dahl G, Conner GE, Salathe M (2009) Pannexin 1 contributes to ATP release in airway epithelia. Am J Respir Cell Mol Biol 41:525–534
106. Silverman WR, De Rivero Vaccari JP, Locovei S, Qiu F, Carlsson SK, Scemes E, Keane RW, Dahl G (2009) The pannexin 1 channel activates the inflammasome in neurons and astrocytes. J Biol Chem 284:18143–18151
107. Wang J, Ambrosi C, Qiu F, Jackson DG, Sosinsky G, Dahl G (2014) The membrane protein Pannexin1 forms two open-channel conformations depending on the mode of activation. Sci Signal 7:ra69
108. Chiu YH, Jin X, Medina CB, Leonhardt SA, Kiessling V, Bennett BC, Shu S, Tamm LK, Yeager M, Ravichandran KS, Bayliss DA (2017) A quantized mechanism for activation of pannexin channels. Nat Commun 8:14324
109. Locovei S, Bao L, Dahl G (2006a) Pannexin 1 in erythrocytes: Function without a gap. Proc Natl Acad Sci U S A 103:7655–7659
110. Dahl G (2015) ATP release through pannexon channels. Philos Trans R Soc Lond B Biol Sci 370
111. Qu Y, Misaghi S, Newton K, Gilmour LL, Louie S, Cupp JE, Dubyak GR, Hackos D, Dixit VM (2011) Pannexin-1 is required for ATP release during apoptosis but not for inflammasome activation. J Immunol 186:6553–6561
112. Locovei S, Wang J, Dahl G (2006b) Activation of pannexin 1 channels by ATP through P2Y receptors and by cytoplasmic calcium. FEBS Lett 580:239–244
113. Suadicani SO, Iglesias R, Wang J, Dahl G, Spray DC, Scemes E (2012) ATP signaling is deficient in cultured Pannexin1-null mouse astrocytes. Glia 60:1106–1116
114. Baranova A, Ivanov D, Petrash N, Pestova A, Skoblov M, Kelmanson I, Shagin D, Nazarenko S, Geraymovych E, Litvin O, Tiunova A, Born TL, Usman N, Staroverov D, Lukyanov S, Panchin Y (2004) The mammalian pannexin family is homologous to the invertebrate innexin gap junction proteins. Genomics 83:706–716
115. Vogt A, Hormuzdi SG, Monyer H (2005) Pannexin1 and Pannexin2 expression in the developing and mature rat brain. Brain Res Mol Brain Res 141:113–120
116. Bennett MV, Garre JM, Orellana JA, Bukauskas FF, Nedergaard M, Saez JC (2012) Connexin and pannexin hemichannels in inflammatory responses of glia and neurons. Brain Res 1487:3–15
117. Hanstein R, Hanani M, Scemes E, Spray DC (2016) Glial pannexin1 contributes to tactile hypersensitivity in a mouse model of orofacial pain. Sci Rep 6:38266
118. Sanderson J, Dartt DA, Trinkaus-Randall V, Pintor J, Civan MM, Delamere NA, Fletcher EL, Salt TE, Grosche A, Mitchell CH (2014) Purines in the eye: recent evidence for the physiological and pathological role of purines in the RPE, retinal neurons,

astrocytes, Muller cells, lens, trabecular meshwork, cornea and lacrimal gland. Exp Eye Res 127:270–279
119. Scemes E, Velisek L, Veliskova J (2019) Astrocyte and neuronal pannexin1 contribute distinctly to seizures. ASN Neuro 11:1759091419833502
120. Adamson SE, Leitinger N (2014) The role of pannexin1 in the induction and resolution of inflammation. FEBS Lett 588:1416–1422
121. Crespo Yanguas S, Willebrords J, Johnstone SR, Maes M, Decrock E, De Bock M, Leybaert L, Cogliati B, Vinken M (2017) Pannexin1 as mediator of inflammation and cell death. Biochim Biophys Acta Mol Cell Res 1864:51–61
122. Pelegrin P, Surprenant A (2006) Pannexin-1 mediates large pore formation and interleukin-1beta release by the ATP-gated P2X7 receptor. Embo J 25:5071–5082
123. Zappala A, Li Volti G, Serapide MF, Pellitteri R, Falchi M, La Delia F, Cicirata V, Cicirata F (2007) Expression of pannexin2 protein in healthy and ischemized brain of adult rats. Neuroscience 148:653–667
124. Panchin Y, Kelmanson I, Matz M, Lukyanov K, Usman N, Lukyanov S (2000) A ubiquitous family of putative gap junction molecules. Curr Biol 10: R473–R474
125. Sanchez A, Castro C, Flores DL, Gutierrez E, Baldi P (2019) Gap junction channels of innexins and connexins: relations and computational perspectives. Int J Mol Sci 20
126. Dykes IM, Macagno ER (2006) Molecular characterization and embryonic expression of innexins in the leech Hirudo medicinalis. Dev Genes Evol 216:185–197
127. Samuels SE, Lipitz JB, Dahl G, Muller KJ (2010) Neuroglial ATP release through innexin channels controls microglial cell movement to a nerve injury. J Gen Physiol 136:425–442
128. Sangaletti R, Dahl G, Bianchi L (2014) Mechanosensitive unpaired innexin channels in C. elegans touch neurons. Am J Physiol Cell Physiol 307:C966–C977
129. Altun ZF, Chen B, Wang ZW, Hall DH (2009) High resolution map of Caenorhabditis elegans gap junction proteins. Dev Dyn 238:1936–1950
130. Hall DH (2017) Gap junctions in *C. elegans*: their roles in behavior and development. Dev Neurobiol 77:587–596
131. Hartzell HC, Qu Z, Yu K, Xiao Q, Chien LT (2008) Molecular physiology of bestrophins: multifunctional membrane proteins linked to best disease and other retinopathies. Physiol Rev 88:639–672
132. Milenkovic VM, Rivera A, Horling F, Weber BH (2007) Insertion and topology of normal and mutant bestrophin-1 in the endoplasmic reticulum membrane. J Biol Chem 282:1313–1321
133. Li Y, Zhang Y, Xu Y, Kittredge A, Ward N, Chen S, Tsang SH, Yang T (2017) Patient-specific mutations impair BESTROPHIN1's essential role in mediating Ca(2+)-dependent Cl(-) currents in human RPE. Elife 6
134. Kane Dickson V, Pedi L, Long SB (2014) Structure and insights into the function of a Ca(2+)-activated Cl(-) channel. Nature 516:213–218
135. Yang T, Liu Q, Kloss B, Bruni R, Kalathur RC, Guo Y, Kloppmann E, Rost B, Colecraft HM, Hendrickson WA (2014) Structure and selectivity in bestrophin ion channels. Science 346:355–359
136. Sun H, Tsunenari T, Yau KW, Nathans J (2002) The vitelliform macular dystrophy protein defines a new family of chloride channels. Proc Natl Acad Sci U S A 99:4008–4013
137. Fischmeister R, Hartzell HC (2005) Volume sensitivity of the bestrophin family of chloride channels. J Physiol 562:477–491
138. Chien LT, Hartzell HC (2008) Rescue of volume-regulated anion current by bestrophin mutants with altered charge selectivity. J Gen Physiol 132:537–546
139. Milenkovic A, Schmied D, Tanimoto N, Seeliger MW, Sparrow JR, Weber BHF (2019) The Y227N mutation affects bestrophin-1 protein stability and impairs sperm function in a mouse model of Best vitelliform macular dystrophy. Biol Open 8
140. Marquardt A, Stohr H, Passmore LA, Kramer F, Rivera A, Weber BH (1998) Mutations in a novel gene, VMD2, encoding a protein of unknown properties cause juvenile-onset vitelliform macular dystrophy (Best's disease). Hum Mol Genet 7:1517–1525
141. Petrukhin K, Koisti MJ, Bakall B, Li W, Xie G, Marknell T, Sandgren O, Forsman K, Holmgren G, Andreasson S, Vujic M, Bergen AA, Mcgarty-Dugan V, Figueroa D, Austin CP, Metzker ML, Caskey CT, Wadelius C (1998) Identification of the gene responsible for Best macular dystrophy. Nat Genet 19:241–247
142. Marmorstein AD, Marmorstein LY, Rayborn M, Wang X, Hollyfield JG, Petrukhin K (2000) Bestrophin, the product of the Best vitelliform macular dystrophy gene (VMD2), localizes to the basolateral plasma membrane of the retinal pigment epithelium. Proc Natl Acad Sci U S A 97:12758–12763
143. Barro Soria R, Spitzner M, Schreiber R, Kunzelmann K (2009) Bestrophin-1 enables Ca2+-activated Cl- conductance in epithelia. J Biol Chem 284:29405–29412
144. Andre S, Boukhaddaoui H, Campo B, Al-Jumaily M, Mayeux V, Greuet D, Valmier J, Scamps F (2003) Axotomy-induced expression of calcium-activated chloride current in subpopulations of mouse dorsal root ganglion neurons. J Neurophysiol 90:3764–3773
145. Al-Jumaily M, Kozlenkov A, Mechaly I, Fichard A, Matha V, Scamps F, Valmier J, Carroll P (2007) Expression of three distinct families of calcium-activated chloride channel genes in the mouse dorsal root ganglion. Neurosci Bull 23:293–299

146. Boudes M, Sar C, Menigoz A, Hilaire C, Pequignot MO, Kozlenkov A, Marmorstein A, Carroll P, Valmier J, Scamps F (2009) Best1 is a gene regulated by nerve injury and required for Ca2+-activated Cl- current expression in axotomized sensory neurons. J Neurosci 29:10063–10071
147. Pineda-Farias JB, Barragan-Iglesias P, Loeza-Alcocer E, Torres-Lopez JE, Rocha-Gonzalez HI, Perez-Severiano F, Delgado-Lezama R, Granados-Soto V (2015) Role of anoctamin-1 and bestrophin-1 in spinal nerve ligation-induced neuropathic pain in rats. Mol Pain 11:41
148. Park H, Oh SJ, Han KS, Woo DH, Park H, Mannaioni G, Traynelis SF, Lee CJ (2009) Bestrophin-1 encodes for the Ca2+-activated anion channel in hippocampal astrocytes. J Neurosci 29:13063–13073
149. Oh SJ, Han KS, Park H, Woo DH, Kim HY, Traynelis SF, Lee CJ (2012) Protease activated receptor 1-induced glutamate release in cultured astrocytes is mediated by Bestrophin-1 channel but not by vesicular exocytosis. Mol Brain 5:38
150. Park H, Han KS, Oh SJ, Jo S, Woo J, Yoon BE, Lee CJ (2013) High glutamate permeability and distal localization of Best1 channel in CA1 hippocampal astrocyte. Mol Brain 6:54
151. Park H, Han KS, Seo J, Lee J, Dravid SM, Woo J, Chun H, Cho S, Bae JY, An H, Koh W, Yoon BE, Berlinguer-Palmini R, Mannaioni G, Traynelis SF, Bae YC, Choi SY, Lee CJ (2015) Channel-mediated astrocytic glutamate modulates hippocampal synaptic plasticity by activating postsynaptic NMDA receptors. Mol Brain 8:7
152. Jo S, Yarishkin O, Hwang YJ, Chun YE, Park M, Woo DH, Bae JY, Kim T, Lee J, Chun H, Park HJ, Lee DY, Hong J, Kim HY, Oh SJ, Park SJ, Lee H, Yoon BE, Kim Y, Jeong Y, Shim I, Bae YC, Cho J, Kowall NW, Ryu H, Hwang E, Kim D, Lee CJ (2014) GABA from reactive astrocytes impairs memory in mouse models of Alzheimer's disease. Nat Med 20:886–896
153. Miguel-Hidalgo JJ (2009) The role of glial cells in drug abuse. Curr Drug Abuse Rev 2:76–82
154. Tarasov VV, Svistunov AA, Chubarev VN, Sologova SS, Mukhortova P, Levushkin D, Somasundaram SG, Kirkland CE, Bachurin SO, Aliev G (2019) Alterations of astrocytes in the context of schizophrenic dementia. Front Pharmacol 10:1612
155. Lee S, Yoon BE, Berglund K, Oh SJ, Park H, Shin HS, Augustine GJ, Lee CJ (2010) Channel-mediated tonic GABA release from glia. Science 330:790–796
156. Yoon BE, Jo S, Woo J, Lee JH, Kim T, Kim D, Lee CJ (2011) The amount of astrocytic GABA positively correlates with the degree of tonic inhibition in hippocampal CA1 and cerebellum. Mol Brain 4:42
157. Qu Z, Hartzell HC (2008) Bestrophin Cl- channels are highly permeable to HCO3. Am J Physiol Cell Physiol 294:C1371–C1377
158. Oh SJ, Lee CJ (2017) Distribution and function of the bestrophin-1 (Best1) channel in the brain. Exp Neurobiol 26:113–121
159. Lee CJ, Mannaioni G, Yuan H, Woo DH, Gingrich MB, Traynelis SF (2007) Astrocytic control of synaptic NMDA receptors. J Physiol 581:1057–1081
160. Chun H, An H, Lim J, Woo J, Lee J, Ryu H, Lee CJ (2018) Astrocytic proBDNF and Tonic GABA distinguish active versus reactive astrocytes in hippocampus. Exp Neurobiol 27:155–170
161. Chien LT, Hartzell HC (2007) Drosophila bestrophin-1 chloride current is dually regulated by calcium and cell volume. J Gen Physiol 130:513–524
162. Stotz SC, Clapham DE (2012) Anion-sensitive fluorophore identifies the Drosophila swell-activated chloride channel in a genome-wide RNA interference screen. PLoS One 7:e46865

Physiological and Pathological Relevance of Selective and Nonselective Ca^{2+} Channels in Skeletal and Cardiac Muscle

11

Jaime Balderas-Villalobos, Tyler W. E. Steele, and Jose M. Eltit

Abstract

Contraction of the striated muscle is fundamental for human existence. The action of voluntary skeletal muscle enables activities such as breathing, establishing body posture, and diverse body movements. Additionally, highly precise motion empowers communication, artistic expression, and other activities that define everyday human life. The involuntary contraction of striated muscle is the core function of the heart and is essential for blood flow. Several ion channels are important in the transduction of action potentials to cytosolic Ca^{2+} signals that enable muscle contraction; however, other ion channels are involved in the progression of muscle pathologies that can impair normal life or threaten it. This chapter describes types of selective and nonselective Ca^{2+} permeable ion channels expressed in the striated muscle, their participation in different aspects of muscle excitation and contraction, and their relevance to the progression of some pathological states.

Keywords

Calcium · Voltage-gated calcium channel · Ryanodine receptor · Store-operated Ca^{2+} entry · TRPC · Orai1 · STIM1 · Resting Ca^{2+} entry · Hypertrophy · Malignant hyperthermia · Dystrophy

11.1 Introduction

Calcium (Ca^{2+}) is a universal intracellular signaling messenger, where the cytosolic concentration of free Ca^{2+} ($[Ca^{2+}]_{cyt}$) is much lower than the concentration of other biologically relevant intracellular cations (i.e., Na^+, K^+, Mg^{2+}) [1–5]. In striated muscle, a reasonable estimation of $[Ca^{2+}]_{cyt}$ in inactive cells is approximately 100 nM [5]. By contrast, the extracellular Ca^{2+} concentration ($[Ca^{2+}]_{out}$) and the free Ca^{2+} concentration inside the sarcoplasmic reticulum ($[Ca^{2+}]_{SR}$), the main intracellular Ca^{2+} store, are in the millimolar (mM) range (~2 mM and ~0.4 mM, respectively) [6, 7]. To maintain $[Ca^{2+}]_{cyt}$ in such energetically unfavorable conditions, cells have an extremely low permeability to Ca^{2+} at rest, and any "leak" of Ca^{2+} through channels is exquisitely balanced by extruding mechanisms such as Ca^{2+} pumps and Ca^{2+} transporters that nullify any inward Ca^{2+} flux [8–10]. Conversely, the effects of the Ca^{2+} extruding machinery are overcome when Ca^{2+} channels are open during activity, which results

J. Balderas-Villalobos · T. W. E. Steele · J. M. Eltit (✉)
Department of Physiology and Biophysics, School of Medicine, Virginia Commonwealth University, Richmond, VA, USA
e-mail: Jaime.Villalobos@vcuhealth.org; twsteele2@vcu.edu; jose.eltit@vcuhealth.org

L. Zhou (ed.), *Ion Channels in Biophysics and Physiology*, Advances in Experimental Medicine and Biology 1349, https://doi.org/10.1007/978-981-16-4254-8_11

in a net augmentation in $[Ca^{2+}]_{cyt}$. Therefore, in the muscle, fast activation of Ca^{2+} channels causes a transient increase in $[Ca^{2+}]_{cyt}$, which activates the contractile machinery as part of the excitation–contraction (EC) coupling mechanism. Upon Ca^{2+} channel deactivation, when the channel conductance drops, Ca^{2+} extruding mechanisms in the plasma membrane and recapture mechanisms in the sarcoplasmic reticulum (SR) recover resting $[Ca^{2+}]_{cyt}$, and the muscle relaxes [11–13]. A schematic of Ca^{2+} fluxes present in muscle cells is shown in Fig. 11.1.

The first part of this chapter discusses seminal as well as recent findings that elucidate the key involvement of voltage-activated Ca^{2+} channels and intracellular Ca^{2+} channels in excitation–contraction (EC) coupling of the striated muscle. The second part discusses other Ca^{2+} channels expressed in the striated muscle that are not directly involved in EC coupling but are implicated in the progression of muscle pathologies.

11.2 L-Type Ca^{2+} Channels and Ryanodine Receptors form the Core Functional Unit of Excitation–Contraction Coupling

At rest or inactivity, electrical conductance in the plasma membrane is primarily driven by channels permeable to ions that have a negative equilibrium potential, that is, K^+ and Cl^-. Channels for Na^+ and Ca^{2+} ions, which have positive equilibrium potentials, remain mostly closed [7, 14]. The basal K^+ and Cl^- conductances result in a negative electrical potential across the plasma membrane. The resting membrane potential in mammalian adult skeletal muscle fibers, for example, is about −85 mV [5]. Excitable cells express voltage-gated ion channels, the main players in the progression of the action potential. The fleeting, quickly-propagated electrical depolarization of the plasma membrane during an action potential is the primary driver of muscle contraction in the striated muscle. This connected, bipartite process is called EC coupling. The current view of the action potential comes from classical work in the giant axon of the longfin inshore squid. In this model, the opening of tetrodotoxin-sensitive, Na^+ selective voltage-gated channels drives the fast depolarization phase of the action potential and is followed by a very fast Na^+ channel inactivation [15]. Meanwhile, slowly activating, K^+ selective voltage-gated channels contribute to the repolarizing phase of the action potential [7, 16, 17]. The squid giant axon action potential is fast and the full event lasts for only around a couple of milliseconds [7]. Adult skeletal muscle fibers display action potentials that kinetically resemble those in the squid giant axon. Action potentials in ventricular myocytes, however, are more complex. Myocytes contain voltage-gated Ca^{2+} channels and NCX, which can produce depolarizing currents, as well as several voltage-gated K^+ channels with different biophysical properties and repolarizing currents. These differences alter the kinetic shape of the action potential, and action potentials in the heart can last much longer than the canonical ones in the squid giant axon, persisting for up to ~200 ms [18].

The first step in EC coupling is electrical to chemical transduction, where an electrical signal in the membrane (an action potential) is transformed into an intracellular chemical signal (a transient increase in $[Ca^{2+}]_{cyt}$). Analogous to how a chemical signal is recognized by a membrane receptor, voltage-gated channels detect variations of the electrical potential across the membrane by conformationally changing as the electrical potential across the membrane fluctuates. Voltage-gated channels are composed of four repeats of six transmembrane segments (S1 to S6) and a pore loop between the segments S5 and S6. These repeats are coded differently in different channels: all four repeats in Ca^{2+} and Na^+ voltage-gated channels are encoded as a single gene product, whereas in K^+ voltage-gated channels, each of the four repeats is coded from the same gene [19]. The first four segments (S1–S4) of each repeat constitute the voltage-sensing domain (VSD), while the S5, a pore loop, and S6 of each repeat intermingle to form the unique

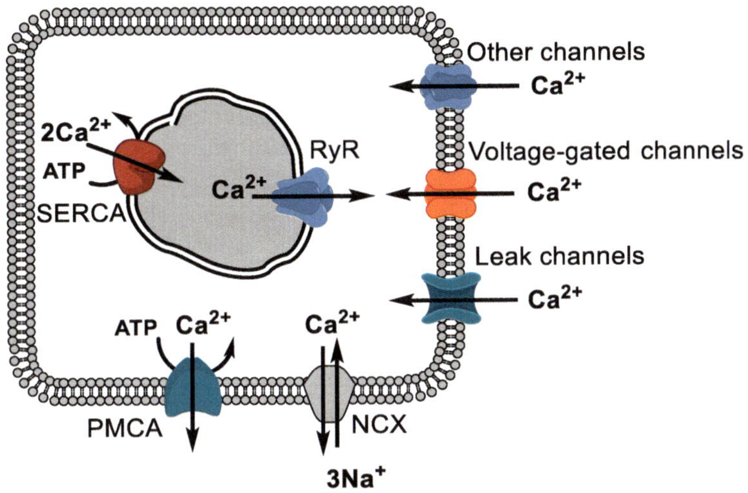

Fig. 11.1 Cartoon showing the main Ca^{2+} fluxes in muscle cells. Primary and secondary active transport mechanisms in the plasma membrane, plasma membrane Ca^{2+} ATPase (PMCA) and Na^+/Ca^{2+} exchanger (NCX), extrude Ca^{2+} from the cytosol against its electrochemical gradient and are coupled to ATP hydrolysis or the influx of Na^+ down its electrochemical gradient, respectively. PMCA and NCX are important for keeping the resting Ca^{2+} concentration in quiescent cells or reestablishing resting Ca^{2+} levels after channel activation. Ca^{2+} permeability in the plasma membrane is low at rest (leak channels), but ion channels in the plasma membrane can open in response to stimuli, including voltage-gated channels and other channels such as store-operated channels, stretch-activated channels, and channels activated by signaling molecules (all grouped in "other channels"). When Ca^{2+} permeable channels open, Ca^{2+} enters the cells down its electrochemical gradient. Sarco-endoplasmic reticulum Ca^{2+} ATPase (SERCA) is expressed in the membrane of the SR; it is a Ca^{2+} ATPase that concentrates Ca^{2+} in the SR lumen. Intracellular ion channels expressed in the SR membrane such as the ryanodine receptor (RyR) can open enabling Ca^{2+} efflux from the SR lumen down its chemical gradient

central structure that includes the selectivity filter and the ion gate. The fourth segment (S4) of the VSD is a transmembrane helix that has several basic, positively charged residues. The S4 is the primary structure subjected to electrical work upon membrane depolarization [20]. At rest, the inner face of the plasma membrane holds negative charges that are balanced by positive charges on its outer face, which establishes an electrical field across the membrane, akin to a capacitor. Inward currents during action potentials vary the charge stored in the membrane, thereby altering the electrical field across the plasma membrane. These changes in the membrane electrical field act on VSDs and produce work. The outward movement of the S4 promotes other conformational changes in the protein, ultimately opening the ion gate.

In the striated muscle, EC coupling proceeds by virtue of the voltage-gated Ca^{2+} channel. This membrane-bound protein senses membrane depolarizations during action potentials and is the first step in EC coupling. The voltage-gated Ca^{2+} channel is composed of four subunits, the $\alpha1$, β, $\alpha2\delta$, and γ [21]. The main subunit is the $\alpha1$ subunit, which contains the four VSDs and the central pore structure described above. The rest are auxiliary subunits that interact with the $\alpha1$ subunit. The most important auxiliary subunit is the β subunit, which improves the expression of the complex in the membrane and can modulate the biophysical properties of the channel [22]. Interestingly, skeletal and cardiac muscle express different isoforms of both the $\alpha1$ and the β subunits. These components confer important differences to the mechanism of EC coupling in the respective tissues. Over time, Ca^{2+} channels have been classified using differing criteria. First, voltage-gated Ca^{2+} channels activated by small depolarizations were identified (low voltage-activated). These displayed fast inactivation when the charge carrier was Ba^{2+} and were therefore classified as transient, "T-type" Ca^{2+}

channels. These were different than other voltage-gated Ca^{2+} channels that are activated by stronger depolarizations (high voltage-activated). High voltage-activated channels have higher unitary conductance and their Ba^{2+} currents inactivate more slowly. These channels were called "long-lasting" or "L-type" Ca^{2+} channels [17, 23, 24]. Additionally, L-type currents are pharmacologically inhibited by dihydropyridines, and the channel carrying this current came to be referred to as the dihydropyridine receptor or DHPR [21, 25, 26]. In the skeletal muscle, depolarizing voltage steps produce slowly activating Ca^{2+} currents that display very slow inactivation and have a fast deactivation upon repolarization [27, 28]. The gene CACNA1S codes the α1 subunit expressed in the skeletal muscle (α_{1S}), and the current nomenclature for this protein is $Ca_V1.1$ [29]. Cardiac ventricular myocytes also express L-type Ca^{2+} channels that are sensitive to dihydropyridines [30]. The L-type Ca^{2+} channel expressed in cardiac myocytes is coded by the CACNA1C gene, and the product of this gene is the cardiac α1 subunit (α_{1C}), which is currently known as $Ca_V1.2$ [29]. During an action potential, the strong depolarization of the plasma membrane drives the activation of the L-type Ca^{2+} channels. Interestingly, the strong but brief (~a few ms) skeletal action potential activates the VSDs of $Ca_V1.1$, but the onset of the current is relatively slow, and the Ca^{2+} influx through the channel is therefore negligible. In addition, quantitative experiments comparing dihydropyridine binding sites and current densities suggested that only ~5% of L-type channels are functional Ca^{2+} channels in the skeletal muscle [27, 31]. During the longer cardiac action potential (~200 ms), however, the faster $Ca_V1.2$ is an active contributor to the depolarizing current. In fact, the L-type Ca^{2+} current is in part responsible for the unusual extended shape of the cardiac action potential. Although the cardiac L-type Ca^{2+} current shows Ca^{2+}-induced inactivation, it accounts for 10–30% (species dependent) of the total cytosolic increase of Ca^{2+} during the cardiac activation cycle [32, 33].

Striated muscle cells have a unique architecture: the contractile machinery accounts for most of the cell volume and is surrounded by a highly organized membranous system (Fig. 11.2a). The plasma membrane (or sarcolemma) contains deep, periodic invaginations that extend the cell surface into the intracellular volume of the muscle cell [34, 35]. These orthogonal invaginations of the plasma membrane are called transversal tubules (t-tubules). Voltage-gated channels involved in the propagation of the action potential as well as the L-type Ca^{2+} channels are expressed in the t-tubules, and since each t-tubule is a projection of the plasma membrane, action potentials can propagate deep into the body of the cell [36]. In mammalian skeletal muscle, the t-tubules run around the boundary between the I and A bands of the contractile machinery [37], while in the cardiac muscle, they are located around the Z-disk [38]. The t-tubule is in intimate contact with the terminal cisterna of the SR (also called the junctional SR) [39]. In skeletal muscle, the t-tubule contacts two terminal cisternae, one at each side, in a structure called a triad (Fig. 11.2b) [40]. In the cardiac muscle, one t-tubule interacts with one terminal cisterna forming a structure called a dyad [41]. The cardiac muscle also contains peripheral couplings in which the membrane of the terminal cisterna can interact with the plasma membrane in the periphery of the cell [41]. The distance between the t-tubule and junctional SR membranes is approximately 14 nm. The junctional SR membrane of triads and dyads express a nonselective cationic ion channel called the ryanodine receptor (RyR) [42–46]. Three isoforms of RyR have been described: type 1 (RyR1), which is the most abundant isoform in the skeletal muscle, type 2 (RyR2), which is expressed in the cardiac muscle, and type 3 (RyR3), which is expressed in low quantities in some muscle fibers but is more abundant in other tissues including the brain [47–52]. At 2.2 MDa, RyR is one of the largest known ion channels. It is a homotetramer with fourfold symmetry and although it is not gated by voltage, its transmembrane spanning domain has an architecture that somewhat resembles that of voltage-gated ion channels [53]. RyR has a high-affinity Ca^{2+} binding site with several residues involved in the coordination of Ca^{2+} that when occupied

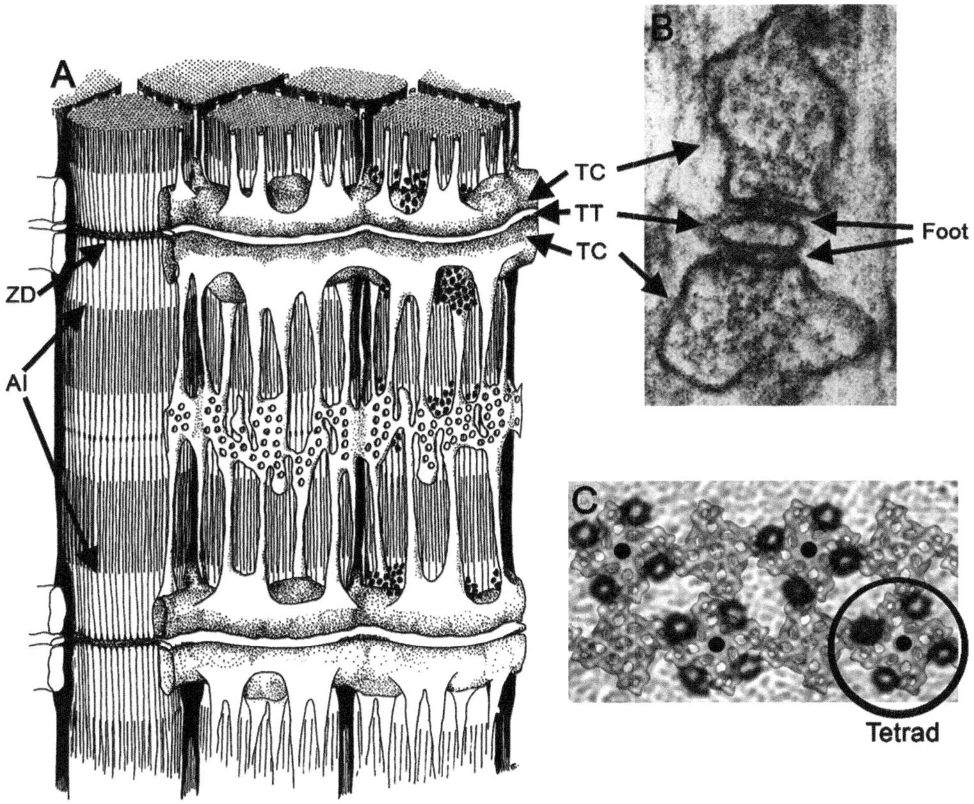

Fig. 11.2 Structural organization of the membranous system in skeletal muscle and the disposition of the $Ca_V1.1$ and RyR1 channels in the t-tubule and junctional SR membranes, respectively. (**a**) Drawing depicting the disposition of the membranous system with respect to the contractile machinery in amphibian muscle. Note that the t-tubule (TT) runs along the side of the Z-disk (ZD). Mammalian skeletal muscle has a similar disposition, but the t-tubule runs along between the A and I bands (AI) having two t-tubules per sarcomere (Drawing reproduced with publisher permission from Peachey, 1965 [35]). (**b**) Thin section of a triad visualized using electron microscopy. The t-tubule is clearly observed as surrounded by two terminal cisternae (TC) of the SR. Two densities between the junctional SR membrane and the t-tubule membrane are identifiable (two per junction, four in total). Each density is a "foot" that corresponds to the large cytoplasmic domain of the RyR1. It has been proposed that two rows of RyR1 run along each junctional SR (Image reproduced with publisher permission from Franzini-Armstrong, 1970 [55]). (**c**) Freeze fracture and rotary shadowing experiments were used to visualize the $Ca_V1.1$ particles along the longitudinal axis of the t-tubule. A group of four $Ca_V1.1$ particles is a tetrad; the image depicts four tetrads. In addition, a hypothetical array of RyR1s was overlayed considering the dimensions from CryoEM reconstructions. The best approximation for $Ca_V1.1$/RyR1 disposition in vivo is shown (Image reproduced with publisher permission from Paolini et al., 2004 [58])

promotes conformational transitions that open the channel [54]. The membrane-embedded domain of RyR extends to a massive cytosolic domain. Early imaging of thin sections of skeletal muscle by electron microscopy revealed periodic densities located in the space between the junctional SR and the t-tubule membranes. Those densities were called "feet" and each density corresponds to the large cytosolic domain of RyRs that are expressed there (Fig. 11.2b) [55].

Freeze fracture and rotary shadowing experiments have been used to visualize structures on the t-tubule membrane juxtaposed to the junctional SR in skeletal muscle. These experiments revealed that $Ca_V1.1$ channels are organized in groups of four in a square formation, with each corner corresponding to one $Ca_V1.1$ channel. This array is called a tetrad, and each complete tetrad faces the cytoplasmic domain of every RyR1, in an alternating fashion (Fig. 11.2c) [39]. The tetrad formation is contingent on RyR1 expression: RyR1-null skeletal muscle does not contain "feet" and is "dyspedic." Although $Ca_V1.1$ particles are present on the junctional t-tubule membrane in these cells, they are not expressed in the characteristic tetrad formation [56]. Interestingly, viral transduction of RyR1 cDNA into dyspedic skeletal muscle cells recovers not only RyR1 expression but also the characteristic $Ca_V1.1$ tetrad formation in t-tubules. The tetrad assembly is not recovered by the expression of other RyR isoforms [57]. The presence of tetrads strongly suggests that although $Ca_V1.1$ and RyR1 are expressed in different membranes, they form a complex in skeletal muscle [58].

There are important functional implications of the physical contact between $Ca_V1.1$ and RyR1. $Ca_V1.1$ acts as a sensor for action potentials running through the t-tubule and undergoes conformational changes in response that transduce this signal to physically associated RyR1s. Thus, "skeletal type" EC coupling is comprised of $Ca_V1.1$ working as a sensor and RyR1 working as a channel. The SR has a free Ca^{2+} concentration of approximately 0.4 mM and a stored component of Ca^{2+} that is in complex with calsequestrin [6]. When RyR1 opens, the large chemical gradient drives Ca^{2+} efflux from the SR into the cytosol to activate the nearby contractile machinery. Since the signal from $Ca_V1.1$ to RyR1 is transduced physically, the permeation of Ca^{2+} through $Ca_V1.1$ (L-type current) is not necessary for skeletal EC coupling. The experimental blockade of the L-type Ca^{2+} current by poisoning of the Ca^{2+} selectivity filter of the channel with Cd^{2+} does not block the depolarization-induced Ca^{2+} release from the SR and downstream contraction [59–61]. These poisoned channels still have functionally intact VSDs that can sense the depolarization and transduce a mechanical signal that opens RyR1. These findings are consistent with others mentioned above that suggest that the short-lived skeletal action potential and slow activation of the L-type Ca^{2+} current allows a narrow and insufficient window for the channel to open. Moreover, immature skeletal muscle expresses a splice variant of $Ca_V1.1$ that has a higher conductance than the adult form of $Ca_V1.1$, implying that the L-type Ca^{2+} current decreases as the muscle differentiates [62, 63]. Additionally, some fish species demonstrate perfectly efficient skeletal type EC coupling in their muscles despite having $Ca_V1.1$ isoforms that are completely impermeable to Ca^{2+} [64]. Finally, knock-in mice expressing $Ca_V1.1$ that is mutated in the pore region to mimic impermeant, piscine $Ca_V1.1$ isoforms, have normal muscle performance [65].

The L-type Ca^{2+} channel inhibitors, 1,4 dihydropyridines (e.g., nifedipine), work allosterically by binding to a hydrophobic site that is situated between two repeats towards the extracellular end of the S6 and the P-helices of the pore domain but away from the actual pore [66]. Nifedipine binding alters the normal function of the VSD of L-type channels as evidenced by a hindrance to the small intramembrane gating currents upon membrane depolarization that results from the movement of positive charges that line the S4 segment [31]. The gating currents that result from electrical stimulation of the muscle closely precede Ca^{2+} release from the SR [67]. The disruption in the voltage sensing activity of $Ca_V1.1$ by nifedipine also prevents the downstream effects of membrane depolarization in the skeletal muscle, such as the opening of RyR1 and its release of Ca^{2+} from the SR [31]. Additionally, the discovery of a naturally occurring mutation in mice that results in no expression of $Ca_V1.1$ was direct evidence that $Ca_V1.1$ acts as the voltage sensor for EC coupling in the skeletal muscle. These mice can live as heterozygotes, but mice that are homozygous for the so-called dysgenic (mdg) allele are smaller in size and die right after birth. Myoblasts isolated from homozygous mdg mice were cultured and differentiated into myotubes, in vitro. In contrast to myotubes obtained from wild-type mice, the homozygous mdg myotubes did not produce the L-type Ca^{2+} current, were not shown to release Ca^{2+}, and did not contract as a result of electrical stimulation

[68, 69]. Interestingly, the microinjection of Ca$_V$1.1 cDNA reestablished the L-type current, gating currents attributable to Ca$_V$1.1, and the contractile functionality upon depolarization in these mdg myotubes [28, 70]. Furthermore, this recovery was isoform-specific, as the expression of the cardiac Ca$_V$1.2 isoform did not recover the skeletal type EC coupling in mdg myotubes (non-dependent on external Ca^{2+} permeation), although the L-type Ca^{2+} current was recovered [71]. Similar to the mdg phenotype, β_{1a} (a skeletal isoform of the β subunit of the L-type Ca^{2+} channel) knockout mice and a zebra fish mutant lacking the expression of this protein has a stark reduction of Ca$_V$1.1 membrane expression and electrical stimulation fails to produce both Ca^{2+} release from the SR and muscle contraction [72, 73]. Treatment of these null-genotypes with β_{1a} cDNA completely restores Ca$_V$1.1 membrane expression as well as skeletal type EC coupling [74, 75]. The transfection of the β_{1a} knockout cells with the cardiac isoform β_{2a} reestablished the membrane expression of Ca$_V$1.1 and remarkably, it also marginally restored the skeletal type EC coupling. This suggests that not only is the β_{1a} important for the recovery of Ca$_V$1.1 membrane expression, but more importantly, it may also be structurally involved in the mechanical coupling between Ca$_V$1.1 and the RyR1 [74, 76, 77]. Similar experiments were performed in cultured RyR1-null (dyspedic) muscle cells, and as expected, these cells lacked Ca^{2+} transients when depolarized. However, the rescue of skeletal type EC coupling only occurred when cells were virally transduced or nuclearly injected with RyR1, but not RyR2 nor RyR3 cDNA, which demonstrates that this type of EC coupling is RyR isoform-specific [57, 78–81]. Furthermore, despite dyspedic myotubes showed normal expression of Ca$_V$1.1 as evidenced by normal levels of gating currents, the L-type Ca^{2+} current is strongly reduced. The recovery of RyR1 expression in these cells reestablishes both EC coupling and the L-type Ca^{2+} current, suggesting a retrograde signal from RyR1 to Ca$_V$1.1, which further supports the idea that there is a complex between RyR1/Ca$_V$1.1 [82].

The α subunit of the L-type Ca^{2+} channel, in addition to the intracellular N- and C- termini, has cytosolic loops that connect the four repeats. These include the I–II loop that connects repeats I and II, which is the binding site for the β subunit [83]. Ca$_V$1.1's II–III loop was identified early on as a structural feature required for skeletal type EC coupling [71]. In gain of function experiments, the N- and C- termini and all internal connecting loops of Ca$_V$1.1 were substituted in the Ca$_V$1.2 backbone one at a time. When these chimeric constructs were expressed in mdg myotubes, only the chimera that had the II–III loop skeletal isoform produced skeletal type EC coupling [71]. Additionally, there is an adaptor protein, Stac3 that interacts with the II–III loop of Ca$_V$1.1 and presumably supports the physical communication between Ca$_V$1.1 and RyR1 [84]. Stac3 is essential for EC coupling in the skeletal muscle, and knockout animals that lack Stac3 die right after birth as a result of skeletal muscle paralysis [85]. In humans, a missense mutation in the STAC3 gene disrupts normal EC-coupling and is causative of the congenital Native American myopathy [86]. Together the available data suggest that Ca$_V$1.1 is expressed in the t-tubule membrane and recruits the essential cytosolic proteins β_{1a} and Stack3 to the I–II and II–III loops, respectively, and this complex interacts with the large foot structures or cytoplasmic domains of RyR1s expressed on the junctional SR membrane. These proteins constitute the Ca^{2+} release unit, a quaternary complex responsible for sensing the depolarization of the t-tubule membrane and facilitating Ca^{2+} release from SR stores, thereby enabling the contraction of the skeletal muscle (Fig. 11.3).

The process of cardiac EC coupling includes a series of events that cause the cyclic increase and decrease in cytoplasmic Ca^{2+} concentration, resulting in contraction (systole), and relaxation (diastole) of the ventricular myocardium. EC coupling begins with the arrival of the action potential to the ventricular myocytes and its propagation into the cells through the system of t-tubular membranes [34]. This depolarization activates the Ca$_V$1.2 located in the t-tubules, which results in the fast influx of extracellular

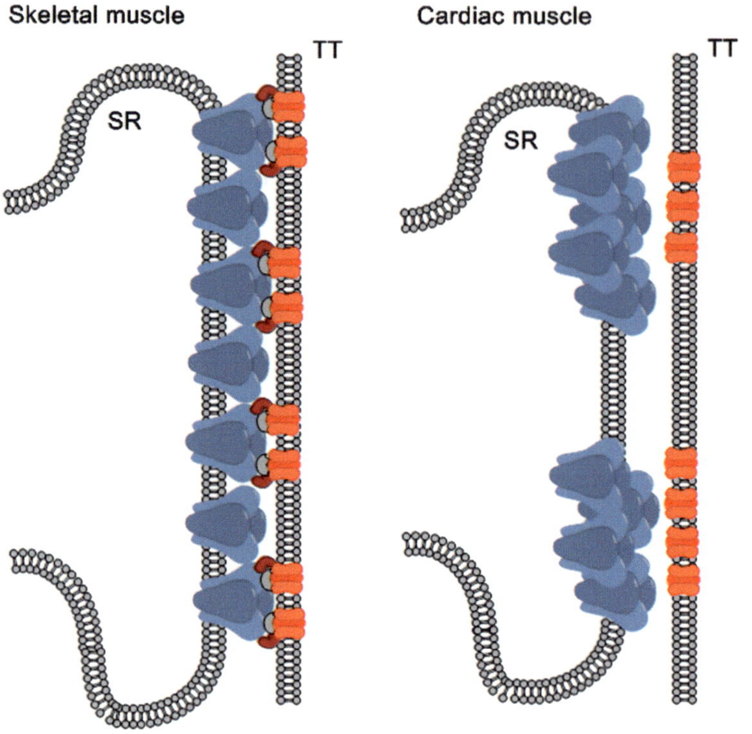

Fig. 11.3 Comparison between skeletal and cardiac EC coupling. The current model for skeletal type EC coupling hypothesizes physical communication between the L-type Ca^{2+} channel ($Ca_V1.1$, orange) and the RyR1 (blue) expressed in the t-tubule (TT) and sarcoplasmic reticulum (SR) membranes, respectively. The effective communication between $Ca_V1.1$ and the RyR1 requires two essential accessory proteins β_{1a} and Stac3. Other proteins are expressed in the junctional regions and may modulate skeletal EC coupling and some are relevant in skeletal muscle diseases (see [179]). In cardiac EC coupling, it is believed that $Ca_V1.2$ does not physically interact with the RyR2, but rather that clusters of these proteins are placed next to each other. Thus, a discrete group of $Ca_V1.2$ can activate a nearby cluster of RyR2 through a Ca^{2+}-induced Ca^{2+} release mechanism

Ca^{2+}. The RyR2 located in the membrane of the sarcoplasmic reticulum opens in response to cytosolic Ca^{2+}, which further increases cytosolic Ca^{2+} by the release of Ca^{2+} from stores in the SR. This process is called Ca^{2+}-induced Ca^{2+} release (CIRC). The increase in intracellular Ca^{2+} concentration amplified by CICR activates the contractile mechanisms and initiates ventricular contraction [87, 88].

Relaxation begins with the decrease of cytoplasmic Ca^{2+} concentration to diastolic levels, which occurs mainly through the following: (1) closure of $Ca_V1.2$ and RyR2 channels, (2) Ca^{2+} recapture through SERCA that is located in the longitudinal SR, and (3) Ca^{2+} transport outside of ventricular myocytes through NCX located in the sarcolemma (and the t-tubule) [33, 89]. SERCA-mediated Ca^{2+} recapture favors the accumulation of Ca^{2+} in the SR and determines, in part, how much Ca^{2+} will be available for the next contraction event. The intrinsic cyclical nature of heart function undermines the concept of using the resting $[Ca^{2+}]_{cyt}$ as a measure of Ca^{2+} homeostasis. In the heart, Ca^{2+} equilibrium has to be understood dynamically; the net influx of Ca^{2+} through $Ca_V1.2$ is exactly balanced by the efflux through NCX, and the Ca^{2+} released through RyR2 must be balanced with the Ca^{2+} recaptured by SERCA in each cardiac cycle, resulting in a null-Ca^{2+} gain or loss upon full integration of the cycle.

In cardiac ventricular myocytes, Ca^{2+} permeation through $Ca_V1.2$ is a *sine qua non* for muscle contraction [90]. In the rapid onset of the action potential mediated by the fast activation of voltage-gated Na^+ channels, the membrane depolarization activates $Ca_V1.2$. At the fast peak of the action potential, the amplitude of the Ca^{2+} current is diminished by its slower intrinsic Ca^{2+} current activation kinetic (compared to the Na^+ channel) and by the relatively low Ca^{2+} driving force resultant from the strong membrane depolarization [12]. Then, in the early partial repolarization phase of the action potential that results from fast inactivation of voltage-gated Na^+ channels and activation of the transient K^+ conductance (I_{to}), the Ca^{2+} driving force increases, which in turn potentiates the Ca^{2+} influx through $Ca_V1.2$. After activation, this L-type channel suffers from voltage and, more importantly, Ca^{2+}-dependent inactivation, which eventually stops the L-type inward Ca^{2+} current [12]. This works together with the repolarization mediated by slow-activated or delayed K^+ channels [91]. Thus, Ca^{2+} permeation through the L-type channels is not favorable during the fast-initial peak of the action potential, but it is important in the relatively long plateau of the cardiac action potential.

In the dyad, clusters of approximately 10–25 $Ca_V1.2$s expressed in the t-tubule are juxtaposed with clusters of approximately 100 RyR2s expressed in the junctional SR. Although it is not believed that $Ca_V1.2$ and RyR2 are in physical contact, both clusters are intimately located across from one other forming a functional unit called the "couplon" [92–94]. Such an arrangement ensures that during an action potential, when $Ca_V1.2$ channels are activated, Ca^{2+} influx sharply increases Ca^{2+} concentration in the nanodomains around the clusters of juxtaposed $Ca_V1.2$s and RyR2s [95]. This also ensures that Ca^{2+} binds and opens RyR2 in the cardiac muscle thereby increasing the Ca^{2+} permeability of the SR membrane, which causes further Ca^{2+} release from the SR [96]. The CICR mechanism effectively works as an amplifier of the initial L-type-mediated Ca^{2+} signal, and this overall Ca^{2+} elevation triggers the contraction of the nearby contractile machinery (Fig. 11.3) [33, 89]. In the cardiac muscle, RyR2 clusters can spontaneously activate, producing short-lived and localized Ca^{2+} release events called sparks [97]. Although still controversial, factors such as spacing between clusters and fast and localized depletion of Ca^{2+} in the SR have been hypothesized to terminate the spark without propagation to other couplons through CICR. This prevents the genesis of spontaneous and auto-regenerative slow arrhythmogenic Ca^{2+} waves in the myocyte [98]. Thus, in physiological conditions, the activation of the RyR is shackled to $Ca_V1.2$ opening, which in turn is strictly synchronized by the rhythmic action potential in every heartbeat.

11.3 Ca^{2+} Permeation Through Voltage-Insensitive Channels Is Altered in Striated Muscle Under Pathological States

While voltage-gated-Ca^{2+} channels in the sarcolemma (including the t-tubule) command striated muscle EC coupling, other Ca^{2+} permeation pathways that are not activated by changes in membrane potential are also present. One of these mechanisms is store-operated Ca^{2+} entry (SOCE). In the late 1980s, James Putney postulated that secondary to store depletion, Ca^{2+} permeation in the plasma membrane is part of the complex Ca^{2+} signals mediated by receptor-activated Gq/IP_3 signaling [99]. Using thapsigargin (an irreversible inhibitor of SERCA) he and his group showed that the depletion of intracellular Ca^{2+} stores is a sufficient signal to cause an increase in the permeability of the plasma membrane to extracellular Ca^{2+}, thereby indicating that SOCE is not directly activated by IP_3 [100]. A highly Ca^{2+} selective inwardly rectified current (I_{crac}) was discovered a few years later in mast cells. I_{crac} has a positive reversal potential and is activated by Ca^{2+} store depletion. Its discovery further supported the idea that SOCE should have a distinct channel with unique biophysical properties [101]. Several controversial hypotheses of the molecular mechanism underlying store depletion channel activation were suggested, but it was not until an

RNA interference-based screen in *Drosophila* was performed that the protein act as a Ca^{2+} sensor in the endoplasmic reticulum (ER) was identified [102]. This is stromal interaction molecule 1 (STIM1), a single transmembrane domain protein resident of the ER membrane and has an EF-hand Ca^{2+} binding domain in the luminal side that acts as the Ca^{2+} sensor responsible for SOCE activation [103]. The knockdown of STIM1 not only suppresses SOCE but also inhibits the activation of I_{crac} [102]. After the first description of STIM1 as the Ca^{2+} sensor in the ER, it became clear that it did not constitute the Ca^{2+} selective channel in the plasma membrane responsible for the I_{crac} current. A few months later, the plasma membrane channel responsible for the I_{crac} current was reported independently by three groups [104–106]. This channel is known as Orai1. The current view of the I_{crac} current activation holds that STIM1 is inactive at rest when its luminal sensing EF-hand domains are Ca^{2+}-bound, presenting a diffuse distribution within the ER membrane. The depletion of the luminal Ca^{2+} store results in STIM1 switching to a Ca^{2+}-free conformation, which promotes its migration and redistribution into sporadic and discrete agglomeration sites called puncta, which form on the ER near the plasma membrane [107]. Puncta formation is associated with physical interactions between STIM1 in the ER and Orai1 in junctional domains of the plasma membrane that activate the channel [108]. For some time it was thought that the Orai1 channel was a homotetramer [109, 110], but when the crystal structure of Orai1 was resolved it revealed a hexameric channel structure [111]. This crystallography-based assembly was further supported by later functional studies [112]. The nature of the quaternary STIM1-Orai1 interaction is still controversial but it is commonly accepted that STIM1 dimers interact with Orai1 hexamers in the activation of I_{crac} [113]. Other isoforms of Orai have been described (Orai2 and Orai3) and they also generate I_{crac} when overexpressed with STIM1 in expression systems, but much less is known about their function in vivo [114–116]. STIM2 has a decreased sensitivity to Ca^{2+} (Kd ~400 μM compared to 200 μM for STIM1). This difference suggests that STIM2 can sense small changes of the free luminal Ca^{2+} concentration in the ER, implying participation in the control of Ca^{2+} homeostasis at rest [117, 118]. However, this view is somewhat controversial since other STIM2 functions have been reported [119, 120].

The classical I_{crac} current is important for immune cell activation; indeed mutation of Orai1, resulting in defective I_{crac} and altered T cell activation patterns, is associated with severe combined immunodeficiency (SCID) syndrome [105, 115]. Interestingly, the study of SOCE in other cell types has revealed that in many cases the biophysical properties of the currents initiated by store depletion can differ from I_{crac}. For instance, the currents can be less selective for Ca^{2+} [121, 122]. These observations suggest that SOCE may be a more complex process and might include activation of nonselective cationic channels such as canonical transient receptor potential (TRPC) channels, which may also use STIM1 as the luminal Ca^{2+} sensor in the ER [123].

As described earlier in this chapter, Ca^{2+} entry through L-type Ca^{2+} channels in the skeletal muscle does not contribute to overall Ca^{2+} homeostasis, even during muscle activity [65]. We know that the skeletal muscle fibers have an enormous Ca^{2+} reserve in the SR that makes them less susceptible to the effects of low external Ca^{2+}. During sustained muscle activity, most of the Ca^{2+} released through RyR1 has to be recaptured by SERCA, but a 100% efficient cycle is unlikely, and some fraction of the tetanic Ca^{2+} has to be extruded out of the cell by active mechanisms. Similarly, a voltage-independent Ca^{2+} entry mechanism may play a role in fine-tuning the intracellular Ca^{2+} homeostasis in the long run, especially during extensive tetanic activity. Also, other Ca^{2+} pathways that are not strictly controlled by voltage may be involved in signaling and gene regulation. The study of voltage-independent Ca^{2+} entry in the skeletal muscle has been controversial and is not a mainstream research topic, but with the development of molecular and pharmacological tools and better detection technologies, several studies have better established the role of voltage-independent Ca^{2+}

entry in the physiology and pathology of the skeletal muscle. One of the first descriptions of voltage-insensitive Ca^{2+} leak channels in the plasma membrane of cultured skeletal muscle (myotubes) was developed by Richard Steinhardt in the early 1990s. This work described how these channels are involved in establishing the resting $[Ca^{2+}]_{cyt}$ in normal conditions while demonstrating that this leak channel contributed to chronically elevated resting $[Ca^{2+}]_{cyt}$ in dystrophic human myotubes [124]. Furthermore, these channels were involved in Ca^{2+} entry at rest (using Mn^{2+} entry as a Ca^{2+} surrogate) and were further activated by Ca^{2+} store depletion [125]. Although the identities of these leak channels in muscle are not known, they produce a unitary conductance of ~14 pS, which is too large to be a canonical I_{crac} (Orai1-mediated) conductance. The latter has a small characteristic unitary conductance of ~10 fS [126]. Interestingly, we now know that functional Orai1 channels are expressed in skeletal muscle. Both the expression of dominant-negative forms of Orai1 and the knockdown of STIM1 in myotubes abolished most of the Ca^{2+} entry activated by store depletion [127]. Ca^{2+} entry in skeletal muscle is not restricted to Orai1; several forms of TRPC channels are expressed in skeletal muscle that may contribute to Ca^{2+} permeation, especially in pathological states [128, 129]. As mentioned above, other experimental models have clearly shown that Ca^{2+} entry mediated by store depletion is consistently reduced when specific TRPC channels are knocked down or out, suggesting a complex interplay between several channels in regulating Ca^{2+} entry [123]. In adult skeletal muscle fibers subjected to a gradual decrease in Ca^{2+} store content by several cycles of high K^+ depolarization in zero extracellular Ca^{2+} and in the presence of a SERCA inhibitor, a clearly enhanced Ca^{2+} entry (SOCE) is observed. This entry was blocked by depolarization and by Ni^{2+}, which is similar to properties shown in Orai1 and some TRPC channels [130]. Quantitative studies performed by Launikonis and Ríos unexpectedly showed a fast (less than a second) activation of SOCE in skinned skeletal muscle fiber preparations as a result of SR-Ca^{2+} release induced by lowered cytosolic Mg^{2+} (an endogenous inhibitor of the RyR1) or application of caffeine (RyR1 agonist) [131]. Additionally, it was convincingly shown that full Ca^{2+} depletion is not required for SOCE activation in the skeletal muscle. Such fast SOCE activation suggests that the channel or channels responsible must be expressed in the tubular system close to the junctional SR where the SR Ca^{2+} sensor must be located (STIM1) [131]. Consistent with these findings, a long variant of STIM1 (STIM1L) was found to be expressed in skeletal muscle and is responsible for the fast mode of SOCE activation [132]. The permanent deletion of Orai1 in a murine knockout model resulted in compromised skeletal muscle performance and a reduction in maximal and sustained force. When a similar assessment was made using an inducible knockout model where Orai1 was deleted in adulthood, muscle performance was identical to controls, suggesting that the Orai1-mediated Ca^{2+} pathway is not important in the overall contractile performance of muscle [133]. Rather, it was shown that the permanent deletion of Orai1 altered the expression of other proteins responsible for the development of the normal muscular phenotype. This suggested that Orai1-mediated Ca^{2+} permeation might not be important for the Ca^{2+} cycling related to muscle contractile efficiency, but it may be important for other types of Ca^{2+} signaling mechanisms related to protein expression during development [133]. In fact, STIM1/Orai1 and SOCE are implicated in myoblast differentiation, a process important for muscle development and maintenance [134]. Furthermore, both "loss of function" and "gain of function" mutations in STIM1/Orai1 genes, that produce SOCE impairment or its constitutive activation, respectively, result in skeletal muscle weakness phenotypes in humans [135–137].

Ultrastructural studies of adult muscle fibers were conducted to identify structures where SOCE could be taking place. In this work, fibers of sedentary and exercise-conditioned mice were studied. Boncompagni and co-workers showed that exercised muscle fibers developed some resistance to fatigue in part because of an enhanced contribution from external Ca^{2+} entry. The experimental evidence shows that enhanced resistance to fatigue was inhibited when extracellular Ca^{2+} was removed, or SOCE was blocked with a high concentration of BTP-2 or 2-APB (nonselective blockers of SOCE). Interestingly, the ultrastructural organization of the triad changes upon exercise; the t-tubule appeared to extend and interact with additional SR membrane to enhance the area of interaction in de novo arrays of membranes that may facilitate STIM-Orai1 communication and Ca^{2+} entry [138].

Ca^{2+} entry is important for controlling Ca^{2+} homeostasis in skeletal muscle at rest. A preliminary study in myotubes clearly described a small permeability to Ca^{2+} at rest [125]. Later, through the use of nonselective pharmacological inhibitors of SOCE and a dominant-negative form of Orai1, it was shown that part of the resting Ca^{2+} entry in myotubes was through an Orai1 dependent pathway [128]. In these experimental conditions, TRPC channel participation was not excluded. These observations were also replicated in adult skeletal muscle fibers [139]. RyR1 is an important indirect regulator of the Ca^{2+} permeability in the plasmalemma and myotubes derived from RyR1-null mice show a decreased resting Ca^{2+} entry and a reduced setpoint for $[Ca^{2+}]_{cyt}$ at rest [140]. Moreover, a functional link between RyR1 and TRPC3 channel activity has been proposed [141]. Interestingly, the Ca^{2+} entry at rest is enhanced in knock-in mice models expressing RyR1 mutations linked to malignant hyperthermia (MH) syndrome, (RyR1 R163C and RyR1 G2435R) [142, 143]. Most MH mutations of RyR1 result in leaky RyR1 channels, and the high-resolution structure of one such RyR1 (rabbit R164C, equivalent to human R163C) was recently solved showing a conformational change that resembles a partial transition to the open state in the large cytoplasmic domain and a dilated but closed pore at rest, suggesting lower stability of the closed state [144]. Leaky RyR1s may decrease SR Ca^{2+} stores, thereby increasing the plasma membrane's permeability to Ca^{2+}. This results in a chronically elevated $[Ca^{2+}]_{cyt}$ at rest. This chronic elevation produces oxidative stress, resulting in a metabolic toll on the muscle fiber [145]. Interestingly, the experimental data suggest that the presence of leaky RyR1 may not only activate the STIM-Orai pathway but may also promote a compensatory effect in the overexpression of nonselective TRPCs, thereby further contributing to the misregulation of resting Ca^{2+} [142, 143, 146]. Accordingly, the cytosolic concentration of Na^+ ($[Na^+]_{cyt}$) is also elevated in MH skeletal muscle [142, 143, 147], which shifts the bioenergetics of NCX function toward its reversal, and contributes to the higher $[Ca^{2+}]_{cyt}$ (Fig. 11.4) [148, 149]. When MH susceptible individuals or experimental animals are exposed to volatile anesthetics (e.g., halothane), the skeletal muscle develops an aberrant and massive cytosolic Ca^{2+} elevation in the absence of depolarization, contributing to fatal rhabdomyolysis. Experiments performed in skinned skeletal muscle fibers from MH susceptible patients showed that halothane induces both Ca^{2+} release from internal stores and increases the Ca^{2+} permeability of the t-tubule, consistent with SOCE activation [150]. Interestingly, measurements performed in susceptible MH mice in vivo showed that ~70% of the persistent abnormal Ca^{2+} response upon halothane exposure can be inhibited using nonselective SOCE blockers (affecting both Orai1 and TRPCs) [142]. Moreover, a double transgenic mouse model bearing the RyR1-R163C MH mutation and a muscle specific dominant-negative form of TRPC6 further implicate TRPCs in the Ca^{2+} and Na^+ dysregulation observed in skeletal muscle at rest and during the MH crisis [146]. Although the expression of the dominant-negative form of TRPC6 did not abolish the lethal effect of halothane in these animals, it delayed the fatal outcome [146]. Taken together, these data imply that although voltage-independent Ca^{2+} entry mechanisms may not be a pharmacological target to replace the well-known "MH antidote" dantrolene to treat the acute MH crisis, Ca^{2+} entry may be a target to intervene in other chronic phenotypes associated to RyR1-MH mutations [151, 152]. Similar studies can be extended to central core disease, a muscle-debilitating disease also caused by mutations in RyR1.

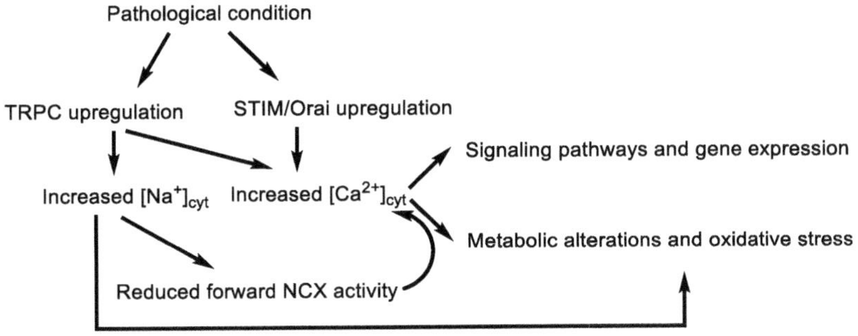

Fig. 11.4 Na^+ and Ca^{2+} dysregulation can contribute to maladaptive changes in striated muscle pathologies. In several pathological models of striated muscle, upregulation or gain on the function of some ion channels not activated by voltage (e.g., TRPC, STIM/Orai) can promote Ca^{2+} and Na^+ dysregulation. Alterations in the homeostasis of these ions may influence downstream signaling pathways, metabolic changes, and oxidative stress that can contribute to the final pathological phenotype

The permeation of Ca^{2+} through voltage-insensitive pathways has largely been studied in models of Duchene Muscular Dystrophy (MD). It has been well documented that Ca^{2+} homeostasis is altered in MD and is in part responsible for the loss of myofibers and the overall muscular deterioration in this genetic disease caused by the lack of expression of the scaffolding protein dystrophin [153]. Early studies showed that the resting $[Ca^{2+}]_{cyt}$ is augmented in adult muscle and in myotubes derived from humans bearing this disease and an MD mice model [124, 154]. Later work described elevated $[Ca^{2+}]_{cyt}$ and $[Na^+]_{cyt}$ in skeletal muscle at rest in vivo [155, 156]. Interestingly, TRPC3 overexpression in the skeletal muscle increased the abundance of muscle fibers presenting central nuclei and fibrosis, which are hallmarks of MD [129]. In addition, muscle fibers from an MD mouse model have increased SOCE, as demonstrated after a robust Ca^{2+} store depletion protocol. SOCE was strongly inhibited by knocking down TRPC6 in this MD model. Moreover, the decreased expression of TRPC6 ameliorates several biomarkers of the disease, indicating that TRPCs are important contributors to the Ca^{2+} and Na^+ dysregulation observed in MD [129]. Additionally, MD muscles overexpress NCX, and taken together with the increased intracellular Na^+ concentration and overall decreased Na^+ gradient, this condition would promote Ca^{2+} entry through the reverse mode of the NCX [149, 155]. When STIM1 was overexpressed in skeletal muscles of mice, it resulted in increased Ca^{2+} entry after depletion (SOCE) and increased $[Ca^{2+}]_{cyt}$ in muscle fibers at rest compared to controls. Interestingly, the dysregulation evoked by STIM1 overexpression was enough to mimic a dystrophic phenotype in the skeletal muscle. More importantly, the knockdown of Orai1 in two mouse models of MD strongly attenuated the pathological phenotype in the skeletal muscle, further supporting the idea that Ca^{2+} entry is a determinant factor responsible for MD pathogenesis (Fig. 11.4) [157].

The contribution of SOCE to the intrinsic Ca^{2+} cycling and overall Ca^{2+} homeostasis in cardiac muscle is a matter of debate. Early studies conducted by Takeshima and coworkers described the presence of robust Ca^{2+} entry induced by Ca^{2+} store depletion in cultured embryonic myocytes. They showed that the magnitude of SOCE sharply declines during maturation. Initially, it decreases by half in cultured neonatal myocytes and virtually disappears in adult myocytes [158]. Accordingly, STIM1 is abundant in cultured neonatal cardiomyocytes and is strongly downregulated in adult cardiomyocytes [159]. This suggests that under physiological conditions SOCE is not a prevalent Ca^{2+} homeostatic mechanism in adult ventricular myocytes. This situation completely changes in pathological states. STIM1 is upregulated in a pressure overload-induced cardiac hypertrophic model (transverse aortic constriction model), which results in a reemergence of SOCE in adult cardiomyocytes [159, 160]. Enhanced SOCE upon store depletion in ventricular adult myocytes was observed in a transgenic mouse model that overexpresses STIM1. Interestingly, these cells were hyperreactive to increases in extracellular Ca^{2+} concentration as evidenced by higher spontaneous activity, suggesting elevated Ca^{2+} permeability at rest. In addition, their hearts were hypertrophic and the animals suffered heart failure concomitant with increased sudden death [161]. Together, these findings indicate that STIM1 may be a potential molecular mediator in heart disease. Accordingly, when STIM1 was silenced in adult mice using adeno-associated virus gene delivery, it produced only mild changes in the heart. As expected, this knockdown of STIM1 completely blunted the adaptive hypertrophic response due to pressure overload in the transverse aortic constriction model. These STIM1 knockdown mice developed left ventricle dilation and had decreased systolic function and heart failure when subjected to pressure overload [162]. This suggested that the STIM1-mediated adaptive hypertrophic response has a beneficial effect. It was proposed that STIM1 produces this hypertrophy response in the pressure overload model by activating the mTOR/Akt/GSK3 signaling pathway [162]. A newer study showed that STIM1 was not upregulated in the pressure overload model, but instead, increases in Orai1, TRPC6, Orai3 and STIM2 mRNA, and protein levels were reported [163]. As expected, a transgenic model that expresses a dominant-negative form of Orai1 that suppresses the endogenous Orai1 current did not generate any cardiac phenotype. However, when these animals were subjected to transverse aortic constriction, the development of cardiac hypertrophy was unaffected and the mice were protected from the development of systolic dysfunction [163]. Together these studies suggest divergent actions of STIM1 and Orai1 in the cardiac pressure overload model, wherein STIM1 favors the development of cardiac hypertrophy and Orai1 is involved in reducing ventricular contractile performance.

The function of STIM1 in the heart is enigmatic. In one study, STIM1 was downregulated in adult mice using a tamoxifen-inducible knockdown system specific for cardiac myocytes. A week after the initiation of the tamoxifen regime, the animal survival rate declined ~50%, concomitant with an 80% decrease in STIM1 expression in the adult myocytes. Further evaluation of these animals suggested an increased risk for the development of arrhythmias and ventricular fibrillation [164]. Thus, although STIM1 is strongly downregulated in the adult mouse heart upon development, it appears that the remaining protein has a role. However, further experiments are necessary to understand its function. Noncanonical actions of STIM1 have been reported as it can modulate SR Ca^{2+} fluxes by sequestering phospholamban, thereby indirectly activating SERCA activity in the heart [165].

A well-established signaling mechanism involved in the progression of pathological hypertrophy is the calcineurin/nuclear factor of activated T cells (NFAT) axis [166]. Calcineurin is a Ca^{2+}-activated serine/threonine phosphatase that, upon activation, dephosphorylates NFAT. NFAT is a transcription factor that when dephosphorylated translocates to the nucleus to modulate the transcription of genes involved in the hypertrophic program in the heart. The calcineurin/NFAT pathway is not unique to cardiac cells. It was first identified as part of the activation mechanism in lymphocytes, in which Ca^{2+} entry results in the calcineurin-mediated NFAT activation and promotes the expression of genes involved in the immune response [167–170]. Even though the relevance of I_{crac} (and SOCE) as a Ca^{2+} source for activation of immune cells is well established [105], the identification of the Ca^{2+} source involved in the calcineurin/NFAT activation pathway in hypertrophic hearts has been more difficult. This is in part because prominent Ca^{2+} transients are constitutively present due to the rhythmic contractile heart activity in live organisms. A large body of evidence links TRPC channels to the progression of pathological cardiac hypertrophy induced either by G-protein-coupled receptor (GPCR) ligands (e.g., angiotensin II or endothelin 1) or by pressure overload. This likely occurs through activation of Ca^{2+} microdomains that in turn can activate signaling pathways involved in the hypertrophic response [171]. TRPC channels are divided into two subfamilies, TRPC1/4/5 and TRPC3/6/7. The activation mechanism of these proteins is controversial but is believed that Ca^{2+} store depletion, mechanical stretch, and oxidation can contribute to the activation of these channels [171]. These stimuli are critically involved in pathophysiological responses in several cardiac pathologies. In addition, TRPC3 and TRPC6 are activated by diacylglycerol [172, 173]. This is a direct Ca^{2+} permeation mechanism downstream of the Gq signaling that can be relevant in the hypertrophy induced by GPCR ligands. Consistent with this notion, TRPC1 upregulation has been reported in a model of cardiac hypertrophy [174]. Meanwhile, in a TRPC1 knockout mouse model, pressure overload no longer produced the hypertrophic response, thereby preventing fibrosis and promoting maintenance of normal contractile function [175]. Moreover, voltage-clamp experiments show an increase in a nonselective cationic current in isolated adult ventricular myocytes from pressure overload induced hypertrophic hearts. This current is strongly reduced in TRPC1 knockout cardiomyocytes. Experimental evidence also suggests, that neuroendocrine stimulation that promotes hypertrophy in the heart, as well as mechanical stretch can produce at least part of the hypertrophic response through a TRPC dependent mechanism [175]. The cardiac expression of TRPC6 is increased in mouse models of pressure overload, and interestingly, TRPC6's promotor has response elements for NFAT. In fact, in transgenic mice overexpressing calcineurin that also show a hypertrophic cardiac phenotype, TRPC6 is upregulated. Conversely, the overexpression of TRPC6 in murine hearts also promotes the activation of the calcineurin/NFAT pathway and pathological hypertrophy, suggesting that TRPC6 constitutes a positive regulator of hypertrophic markers during the progression toward pathological hypertrophy [176]. Similarly, in a mouse model in which TRPC3 was selectively overexpressed in the heart, SOCE was elevated in isolated adult myocytes, suggesting that TRPC3 can support Ca^{2+} entry induced by store depletion [177]. Additionally, the overexpression of cardiac TRPC3 resulted in increased lethality, which correlated with the level of TRPC3 expression and the overexpression of protein markers of cardiac stress. The animals that survived for 12 months showed a dilated cardiac hypertrophy phenotype [177]. Also, mice overexpressing TRPC3 have a higher basal activity of NFAT, which is more easily activated by hypertrophic ligands [177]. Interestingly, cardiac hypertrophy induced by ligands or pressure overload was much greater in TRPC3 overexpressing animals as compared to controls [177]. Concurrently, the expression of a dominant-negative form of TRPC3 decreases the hypertrophy induced by ligands and pressure overload [178]. Comparable effects were observed in mice expressing dominant-negative forms of TRPC6 or TRPC4 [178]. Together, these observations support a model in which several TRPC channels coordinate the progression of a pathological hypertrophic program in the heart in a calcineurin/NFAT-dependent mechanism. Furthermore, they suggest that TRPCs might act as mediators of hypertrophy at least by causing an increase in SOCE (Fig. 11.4).

Acknowledgments The writing of this chapter was funded in part by NIH grants R01 HL139874 and R01 AR067738, and Department of Veterans Affairs grant 1I01BX004861.

References

1. Berridge MJ (1997) Elementary and global aspects of calcium signalling. J Exp Biol 200(Pt 2):315–319
2. Aickin CC (1987) Investigation of factors affecting the intracellular sodium activity in the smooth muscle of guinea-pig ureter. J Physiol 385:483–505. https://doi.org/10.1113/jphysiol.1987.sp016503
3. Powell T, Terrar DA, Twist VW (1980) Electrical properties of individual cells isolated from adult rat ventricular myocardium. J Physiol 302:131–153. https://doi.org/10.1113/jphysiol.1980.sp013234
4. Lopez JR, Alamo L, Caputo C, DiPolo R, Vergara S (1983) Determination of ionic calcium in frog

skeletal muscle fibers. Biophys J 43(1):1–4. https://doi.org/10.1016/S0006-3495(83)84316-1
5. Lopez JR, Sanchez V, Lopez I, Ryan JF, Mendoza M, Sreter FA et al (1990) The effects of extracellular magnesium on myoplasmic [Ca^{2+}] in malignant hyperthermia susceptible swine. Anesthesiology 73 (1):109–117. https://doi.org/10.1097/00000542-199007000-00016
6. Sztretye M, Yi J, Figueroa L, Zhou J, Royer L, Rios E (2011) D4cpv-calsequestrin: a sensitive ratiometric biosensor accurately targeted to the calcium store of skeletal muscle. J Gen Physiol 138(2):211–229. https://doi.org/10.1085/jgp.201010591
7. Keynes RD, Aidley DJ (2001) Nerve and muscle, Studies in biology, 3rd edn. Cambridge University Press, Cambridge
8. Berridge MJ, Bootman MD, Roderick HL (2003) Calcium signalling: dynamics, homeostasis and remodelling. Nat Rev Mol Cell Biol 4(7):517–529. https://doi.org/10.1038/nrm1155
9. Rios E (2010) The cell boundary theorem: a simple law of the control of cytosolic calcium concentration. J Physiol Sci 60(1):81–84. https://doi.org/10.1007/s12576-009-0069-z
10. Friel DD, Tsien RW (1992) A caffeine- and ryanodine-sensitive Ca^{2+} store in bullfrog sympathetic neurones modulates effects of Ca^{2+} entry on [Ca^{2+}]i. J Physiol 450:217–246. https://doi.org/10.1113/jphysiol.1992.sp019125
11. Berridge MJ, Lipp P, Bootman MD (2000) The versatility and universality of calcium signalling. Nat Rev Mol Cell Biol 1(1):11–21. https://doi.org/10.1038/35036035
12. Bers DM (2008) Calcium cycling and signaling in cardiac myocytes. Annu Rev Physiol 70:23–49. https://doi.org/10.1146/annurev.physiol.70.113006.100455
13. Lamb GD (2000) Excitation-contraction coupling in skeletal muscle: comparisons with cardiac muscle. Clin Exp Pharmacol Physiol 27(3):216–224. https://doi.org/10.1046/j.1440-1681.2000.03224.x
14. Palade PT, Barchi RL (1977) Characteristics of the chloride conductance in muscle fibers of the rat diaphragm. J Gen Physiol 69(3):325–342. https://doi.org/10.1085/jgp.69.3.325
15. Armstrong CM, Bezanilla F, Rojas E (1973) Destruction of sodium conductance inactivation in squid axons perfused with pronase. J Gen Physiol 62 (4):375–391. https://doi.org/10.1085/jgp.62.4.375
16. Hodgkin AL, Huxley AF (1952) A quantitative description of membrane current and its application to conduction and excitation in nerve. J Physiol 117 (4):500–544. https://doi.org/10.1113/jphysiol.1952.sp004764
17. Hille B (2001) Ion channels of excitable membranes, 3rd edn. Sinauer, Sunderland, MA
18. Huang CL (2017) Murine electrophysiological models of cardiac arrhythmogenesis. Physiol Rev 97(1):283–409. https://doi.org/10.1152/physrev.00007.2016
19. Catterall WA (1995) Structure and function of voltage-gated ion channels. Annu Rev Biochem 64:493–531. https://doi.org/10.1146/annurev.bi.64.070195.002425
20. Bezanilla F (2008) How membrane proteins sense voltage. Nat Rev Mol Cell Biol 9(4):323–332. https://doi.org/10.1038/nrm2376
21. Curtis BM, Catterall WA (1984) Purification of the calcium antagonist receptor of the voltage-sensitive calcium channel from skeletal muscle transverse tubules. Biochemistry 23(10):2113–2118. https://doi.org/10.1021/bi00305a001
22. Singer D, Biel M, Lotan I, Flockerzi V, Hofmann F, Dascal N (1991) The roles of the subunits in the function of the calcium channel. Science 253 (5027):1553–1557. https://doi.org/10.1126/science.1716787
23. Nilius B, Hess P, Lansman JB, Tsien RW (1985) A novel type of cardiac calcium channel in ventricular cells. Nature 316(6027):443–446. https://doi.org/10.1038/316443a0
24. Bean BP (1985) Two kinds of calcium channels in canine atrial cells. Differences in kinetics, selectivity, and pharmacology. J Gen Physiol 86(1):1–30. https://doi.org/10.1085/jgp.86.1.1
25. Fosset M, Jaimovich E, Delpont E, Lazdunski M (1983) [3H]nitrendipine receptors in skeletal muscle. J Biol Chem 258(10):6086–6092
26. Gould RJ, Murphy KM, Snyder SH (1982) [3H] nitrendipine-labeled calcium channels discriminate inorganic calcium agonists and antagonists. Proc Natl Acad Sci U S A 79(11):3656–3660. https://doi.org/10.1073/pnas.79.11.3656
27. Schwartz LM, McCleskey EW, Almers W (1985) Dihydropyridine receptors in muscle are voltage-dependent but most are not functional calcium channels. Nature 314(6013):747–751. https://doi.org/10.1038/314747a0
28. Tanabe T, Beam KG, Powell JA, Numa S (1988) Restoration of excitation-contraction coupling and slow calcium current in dysgenic muscle by dihydropyridine receptor complementary DNA. Nature 336(6195):134–139. https://doi.org/10.1038/336134a0
29. Ertel EA, Campbell KP, Harpold MM, Hofmann F, Mori Y, Perez-Reyes E et al (2000) Nomenclature of voltage-gated calcium channels. Neuron 25 (3):533–535. https://doi.org/10.1016/s0896-6273(00)81057-0
30. Hess P, Lansman JB, Nilius B, Tsien RW (1986) Calcium channel types in cardiac myocytes: modulation by dihydropyridines and beta-adrenergic stimulation. J Cardiovasc Pharmacol 8(Suppl 9):S11–S21
31. Rios E, Brum G (1987) Involvement of dihydropyridine receptors in excitation-contraction coupling in skeletal muscle. Nature 325 (6106):717–720. https://doi.org/10.1038/325717a0

32. Bers DM (2000) Calcium fluxes involved in control of cardiac myocyte contraction. Circ Res 87 (4):275–281. https://doi.org/10.1161/01.res.87.4.275
33. Bers DM (2002) Cardiac excitation-contraction coupling. Nature 415(6868):198–205. https://doi.org/10.1038/415198a
34. Franzini-Armstrong C, Porter KR (1964) Sarcolemmal invaginations constituting the T system in fish muscle fibers. J Cell Biol 22:675–696. https://doi.org/10.1083/jcb.22.3.675
35. Peachey LD (1965) The sarcoplasmic reticulum and transverse tubules of the frog's sartorius. J Cell Biol 25(3 Suppl):209–231. https://doi.org/10.1083/jcb.25.3.209
36. Franzini-Armstrong C, Peachey LD (1981) Striated muscle-contractile and control mechanisms. J Cell Biol 91(3 Pt 2):166s–186s. https://doi.org/10.1083/jcb.91.3.166s
37. Franzini-Armstrong C (1991) Simultaneous maturation of transverse tubules and sarcoplasmic reticulum during muscle differentiation in the mouse. Dev Biol 146(2):353–363. https://doi.org/10.1016/0012-1606(91)90237-w
38. McNary TG, Spitzer KW, Holloway H, Bridge JH, Kohl P, Sachse FB (2012) Mechanical modulation of the transverse tubular system of ventricular cardiomyocytes. Prog Biophys Mol Biol 110 (2–3):218–225. https://doi.org/10.1016/j.pbiomolbio.2012.07.010
39. Block BA, Imagawa T, Campbell KP, Franzini-Armstrong C (1988) Structural evidence for direct interaction between the molecular components of the transverse tubule/sarcoplasmic reticulum junction in skeletal muscle. J Cell Biol 107(6 Pt 2):2587–2600. https://doi.org/10.1083/jcb.107.6.2587
40. Franzini-Armstrong C (1975) Membrane particles and transmission at the triad. Fed Proc 34 (5):1382–1389
41. Franzini-Armstrong C, Protasi F, Tijskens P (2005) The assembly of calcium release units in cardiac muscle. Ann N Y Acad Sci 1047:76–85. https://doi.org/10.1196/annals.1341.007
42. Pessah IN, Waterhouse AL, Casida JE (1985) The calcium-ryanodine receptor complex of skeletal and cardiac muscle. Biochem Biophys Res Commun 128 (1):449–456. https://doi.org/10.1016/0006-291x(85)91699-7
43. Pessah IN, Francini AO, Scales DJ, Waterhouse AL, Casida JE (1986) Calcium-ryanodine receptor complex. Solubilization and partial characterization from skeletal muscle junctional sarcoplasmic reticulum vesicles. J Biol Chem 261(19):8643–8648
44. Inui M, Saito A, Fleischer S (1987) Isolation of the ryanodine receptor from cardiac sarcoplasmic reticulum and identity with the feet structures. J Biol Chem 262(32):15637–15642
45. Inui M, Saito A, Fleischer S (1987) Purification of the ryanodine receptor and identity with feet structures of junctional terminal cisternae of sarcoplasmic reticulum from fast skeletal muscle. J Biol Chem 262 (4):1740–1747
46. Imagawa T, Smith JS, Coronado R, Campbell KP (1987) Purified ryanodine receptor from skeletal muscle sarcoplasmic reticulum is the Ca^{2+}-permeable pore of the calcium release channel. J Biol Chem 262 (34):16636–16643
47. Penner R, Neher E, Takeshima H, Nishimura S, Numa S (1989) Functional expression of the calcium release channel from skeletal muscle ryanodine receptor cDNA. FEBS Lett 259(1):217–221. https://doi.org/10.1016/0014-5793(89)81532-7
48. Imagawa T, Takasago T, Shigekawa M (1989) Cardiac ryanodine receptor is absent in type I slow skeletal muscle fibers: immunochemical and ryanodine binding studies. J Biochem 106(2):342–348. https://doi.org/10.1093/oxfordjournals.jbchem.a122855
49. Otsu K, Willard HF, Khanna VK, Zorzato F, Green NM, MacLennan DH (1990) Molecular cloning of cDNA encoding the Ca^{2+} release channel (ryanodine receptor) of rabbit cardiac muscle sarcoplasmic reticulum. J Biol Chem 265(23):13472–13483
50. Hakamata Y, Nakai J, Takeshima H, Imoto K (1992) Primary structure and distribution of a novel ryanodine receptor/calcium release channel from rabbit brain. FEBS Lett 312(2–3):229–235. https://doi.org/10.1016/0014-5793(92)80941-9
51. Sorrentino V, Giannini G, Malzac P, Mattei MG (1993) Localization of a novel ryanodine receptor gene (RYR3) to human chromosome 15q14-q15 by in situ hybridization. Genomics 18(1):163–165. https://doi.org/10.1006/geno.1993.1446
52. Sorrentino V, Volpe P (1993) Ryanodine receptors: how many, where and why? Trends Pharmacol Sci 14 (3):98–103. https://doi.org/10.1016/0165-6147(93)90072-r
53. Samso M, Feng W, Pessah IN, Allen PD (2009) Coordinated movement of cytoplasmic and transmembrane domains of RyR1 upon gating. PLoS Biol 7(4):e85. https://doi.org/10.1371/journal.pbio.1000085
54. des Georges A, Clarke OB, Zalk R, Yuan Q, Condon KJ, Grassucci RA et al (2016) Structural basis for gating and activation of RyR1. Cell 167(1):145–57.e17. https://doi.org/10.1016/j.cell.2016.08.075
55. Franzini-Armstrong C (1970) STUDIES OF THE TRIAD: I. Structure of the junction in frog twitch fibers. J Cell Biol 47(2):488–499. https://doi.org/10.1083/jcb.47.2.488
56. Takekura H, Nishi M, Noda T, Takeshima H, Franzini-Armstrong C (1995) Abnormal junctions between surface membrane and sarcoplasmic reticulum in skeletal muscle with a mutation targeted to the ryanodine receptor. Proc Natl Acad Sci U S A 92 (8):3381–3385. https://doi.org/10.1073/pnas.92.8.3381
57. Protasi F, Takekura H, Wang Y, Chen SR, Meissner G, Allen PD et al (2000) RYR1 and

RYR3 have different roles in the assembly of calcium release units of skeletal muscle. Biophys J 79 (5):2494–2508. https://doi.org/10.1016/S0006-3495(00)76491-5
58. Paolini C, Protasi F, Franzini-Armstrong C (2004) The relative position of RyR feet and DHPR tetrads in skeletal muscle. J Mol Biol 342(1):145–153. https://doi.org/10.1016/j.jmb.2004.07.035
59. Tanabe T, Mikami A, Numa S, Beam KG (1990) Cardiac-type excitation-contraction coupling in dysgenic skeletal muscle injected with cardiac dihydropyridine receptor cDNA. Nature 344 (6265):451–453. https://doi.org/10.1038/344451a0
60. Fu WM, Day SY, Lin-Shiau SY (1989) Studies on cadmium-induced myotonia in the mouse diaphragm. Naunyn Schmiedeberg's Arch Pharmacol 340 (2):191–195. https://doi.org/10.1007/BF00168968
61. Forshaw PJ (1977) The inhibitory effect of cadmium on neuromuscular transmission in the rat. Eur J Pharmacol 42(4):371–377. https://doi.org/10.1016/0014-2999(77)90171-6
62. Tuluc P, Molenda N, Schlick B, Obermair GJ, Flucher BE, Jurkat-Rott K (2009) A CaV1.1 Ca^{2+} channel splice variant with high conductance and voltage-sensitivity alters EC coupling in developing skeletal muscle. Biophys J 96(1):35–44. https://doi.org/10.1016/j.bpj.2008.09.027
63. Flucher BE, Tuluc P (2011) A new L-type calcium channel isoform required for normal patterning of the developing neuromuscular junction. Channels (Austin) 5(6):518–524. https://doi.org/10.4161/chan.5.6.17951
64. Schredelseker J, Shrivastav M, Dayal A, Grabner M (2010) Non-Ca^{2+}–conducting Ca^{2+} channels in fish skeletal muscle excitation-contraction coupling. Proc Natl Acad Sci U S A 107(12):5658–5663. https://doi.org/10.1073/pnas.0912153107
65. Dayal A, Schrotter K, Pan Y, Fohr K, Melzer W, Grabner M (2017) The Ca(2+) influx through the mammalian skeletal muscle dihydropyridine receptor is irrelevant for muscle performance. Nat Commun 8 (1):475. https://doi.org/10.1038/s41467-017-00629-x
66. Tang L, Gamal El-Din TM, Swanson TM, Pryde DC, Scheuer T, Zheng N et al (2016) Structural basis for inhibition of a voltage-gated Ca(2+) channel by Ca (2+) antagonist drugs. Nature 537(7618):117–121. https://doi.org/10.1038/nature19102
67. Kovacs L, Rios E, Schneider MF (1979) Calcium transients and intramembrane charge movement in skeletal muscle fibres. Nature 279(5712):391–396. https://doi.org/10.1038/279391a0
68. Powell JA, Fambrough DM (1973) Electrical properties of normal and dysgenic mouse skeletal muscle in culture. J Cell Physiol 82(1):21–38. https://doi.org/10.1002/jcp.1040820104
69. Beam KG, Knudson CM, Powell JA (1986) A lethal mutation in mice eliminates the slow calcium current in skeletal muscle cells. Nature 320(6058):168–170. https://doi.org/10.1038/320168a0
70. Adams BA, Tanabe T, Mikami A, Numa S, Beam KG (1990) Intramembrane charge movement restored in dysgenic skeletal muscle by injection of dihydropyridine receptor cDNAs. Nature 346 (6284):569–572. https://doi.org/10.1038/346569a0
71. Tanabe T, Beam KG, Adams BA, Niidome T, Numa S (1990) Regions of the skeletal muscle dihydropyridine receptor critical for excitation-contraction coupling. Nature 346(6284):567–569. https://doi.org/10.1038/346567a0
72. Gregg RG, Messing A, Strube C, Beurg M, Moss R, Behan M et al (1996) Absence of the beta subunit (cchb1) of the skeletal muscle dihydropyridine receptor alters expression of the alpha 1 subunit and eliminates excitation-contraction coupling. Proc Natl Acad Sci U S A 93(24):13961–13966. https://doi.org/10.1073/pnas.93.24.13961
73. Schredelseker J, Di Biase V, Obermair GJ, Felder ET, Flucher BE, Franzini-Armstrong C et al (2005) The beta 1a subunit is essential for the assembly of dihydropyridine-receptor arrays in skeletal muscle. Proc Natl Acad Sci U S A 102(47):17219–17224. https://doi.org/10.1073/pnas.0508710102
74. Schredelseker J, Dayal A, Schwerte T, Franzini-Armstrong C, Grabner M (2009) Proper restoration of excitation-contraction coupling in the dihydropyridine receptor beta1-null zebrafish relaxed is an exclusive function of the beta1a subunit. J Biol Chem 284(2):1242–1251. https://doi.org/10.1074/jbc.M807767200
75. Beurg M, Sukhareva M, Strube C, Powers PA, Gregg RG, Coronado R (1997) Recovery of Ca^{2+} current, charge movements, and Ca^{2+} transients in myotubes deficient in dihydropyridine receptor beta 1 subunit transfected with beta 1 cDNA. Biophys J 73 (2):807–818. https://doi.org/10.1016/S0006-3495(97)78113-X
76. Beurg M, Sukhareva M, Ahern CA, Conklin MW, Perez-Reyes E, Powers PA et al (1999) Differential regulation of skeletal muscle L-type Ca^{2+} current and excitation-contraction coupling by the dihydropyridine receptor beta subunit. Biophys J 76 (4):1744–1756. https://doi.org/10.1016/S0006-3495(99)77336-4
77. Eltit JM, Franzini-Armstrong C, Perez CF (2014) Amino acid residues 489–503 of dihydropyridine receptor (DHPR) beta1a subunit are critical for structural communication between the skeletal muscle DHPR complex and type 1 ryanodine receptor. J Biol Chem 289(52):36116–36124. https://doi.org/10.1074/jbc.M114.615526
78. Nakai J, Ogura T, Protasi F, Franzini-Armstrong C, Allen PD, Beam KG (1997) Functional nonequality of the cardiac and skeletal ryanodine receptors. Proc Natl Acad Sci U S A 94(3):1019–1022. https://doi.org/10.1073/pnas.94.3.1019

79. Nakai J, Sekiguchi N, Rando TA, Allen PD, Beam KG (1998) Two regions of the ryanodine receptor involved in coupling with L-type Ca^{2+} channels. J Biol Chem 273(22):13403–13406. https://doi.org/10.1074/jbc.273.22.13403
80. Fessenden JD, Wang Y, Moore RA, Chen SR, Allen PD, Pessah IN (2000) Divergent functional properties of ryanodine receptor types 1 and 3 expressed in a myogenic cell line. Biophys J 79(5):2509–2525. https://doi.org/10.1016/S0006-3495(00)76492-7
81. Sheridan DC, Takekura H, Franzini-Armstrong C, Beam KG, Allen PD, Perez CF (2006) Bidirectional signaling between calcium channels of skeletal muscle requires multiple direct and indirect interactions. Proc Natl Acad Sci U S A 103(52):19760–19765. https://doi.org/10.1073/pnas.0609473103
82. Nakai J, Dirksen RT, Nguyen HT, Pessah IN, Beam KG, Allen PD (1996) Enhanced dihydropyridine receptor channel activity in the presence of ryanodine receptor. Nature 380(6569):72–75. https://doi.org/10.1038/380072a0
83. Pragnell M, De Waard M, Mori Y, Tanabe T, Snutch TP, Campbell KP (1994) Calcium channel beta-subunit binds to a conserved motif in the I–II cytoplasmic linker of the alpha 1-subunit. Nature 368 (6466):67–70. https://doi.org/10.1038/368067a0
84. Polster A, Nelson BR, Papadopoulos S, Olson EN, Beam KG (2018) Stac proteins associate with the critical domain for excitation-contraction coupling in the II–III loop of CaV1.1. J Gen Physiol 150 (4):613–624. https://doi.org/10.1085/jgp.201711917
85. Nelson BR, Wu F, Liu Y, Anderson DM, McAnally J, Lin W et al (2013) Skeletal muscle-specific T-tubule protein STAC3 mediates voltage-induced Ca^{2+} release and contractility. Proc Natl Acad Sci U S A 110(29):11881–11886. https://doi.org/10.1073/pnas.1310571110
86. Horstick EJ, Linsley JW, Dowling JJ, Hauser MA, McDonald KK, Ashley-Koch A et al (2013) Stac3 is a component of the excitation-contraction coupling machinery and mutated in Native American myopathy. Nat Commun 4:1952. https://doi.org/10.1038/ncomms2952
87. Fabiato A, Fabiato F (1975) Contractions induced by a calcium-triggered release of calcium from the sarcoplasmic reticulum of single skinned cardiac cells. J Physiol 249(3):469–495. https://doi.org/10.1113/jphysiol.1975.sp011026
88. Fabiato A (1983) Calcium-induced release of calcium from the cardiac sarcoplasmic reticulum. Am J Phys 245(1):C1–C14. https://doi.org/10.1152/ajpcell.1983.245.1.C1
89. Eisner DA, Caldwell JL, Kistamas K, Trafford AW (2017) Calcium and excitation-contraction coupling in the heart. Circ Res 121(2):181–195. https://doi.org/10.1161/CIRCRESAHA.117.310230
90. Morgan JP, Wier WG, Hess P, Blinks JR (1983) Influence of Ca^{++}-channel blocking agents on calcium transients and tension development in isolated mammalian heart muscle. Circ Res 52(2 Pt 2):I47–I52
91. Grandi E, Sanguinetti MC, Bartos DC, Bers DM, Chen-Izu Y, Chiamvimonvat N et al (2017) Potassium channels in the heart: structure, function and regulation. J Physiol 595(7):2209–2228. https://doi.org/10.1113/JP272864
92. Franzini-Armstrong C, Protasi F, Ramesh V (1999) Shape, size, and distribution of Ca(2+) release units and couplons in skeletal and cardiac muscles. Biophys J 77(3):1528–1539. https://doi.org/10.1016/S0006-3495(99)77000-1
93. Soeller C, Crossman D, Gilbert R, Cannell MB (2007) Analysis of ryanodine receptor clusters in rat and human cardiac myocytes. Proc Natl Acad Sci U S A 104(38):14958–14963. https://doi.org/10.1073/pnas.0703016104
94. Scriven DR, Asghari P, Moore ED (2013) Microarchitecture of the dyad. Cardiovasc Res 98(2):169–176. https://doi.org/10.1093/cvr/cvt025
95. Asghari P, Scriven DR, Hoskins J, Fameli N, van Breemen C, Moore ED (2012) The structure and functioning of the couplon in the mammalian cardiomyocyte. Protoplasma 249(Suppl 1):S31–S38. https://doi.org/10.1007/s00709-011-0347-5
96. Cleemann L, Wang W, Morad M (1998) Two-dimensional confocal images of organization, density, and gating of focal Ca^{2+} release sites in rat cardiac myocytes. Proc Natl Acad Sci U S A 95 (18):10984–10989. https://doi.org/10.1073/pnas.95.18.10984
97. Cheng H, Lederer WJ, Cannell MB (1993) Calcium sparks: elementary events underlying excitation-contraction coupling in heart muscle. Science 262 (5134):740–744. https://doi.org/10.1126/science.8235594
98. Stern MD, Rios E, Maltsev VA (2013) Life and death of a cardiac calcium spark. J Gen Physiol 142 (3):257–274. https://doi.org/10.1085/jgp.201311034
99. Putney JW Jr (1986) A model for receptor-regulated calcium entry. Cell Calcium 7(1):1–12. https://doi.org/10.1016/0143-4160(86)90026-6
100. Takemura H, Hughes AR, Thastrup O, Putney JW Jr (1989) Activation of calcium entry by the tumor promoter thapsigargin in parotid acinar cells. Evidence that an intracellular calcium pool and not an inositol phosphate regulates calcium fluxes at the plasma membrane. J Biol Chem 264 (21):12266–12271
101. Hoth M, Penner R (1992) Depletion of intracellular calcium stores activates a calcium current in mast cells. Nature 355(6358):353–356. https://doi.org/10.1038/355353a0
102. Roos J, DiGregorio PJ, Yeromin AV, Ohlsen K, Lioudyno M, Zhang S et al (2005) STIM1, an essential and conserved component of store-operated Ca^{2+} channel function. J Cell Biol 169(3):435–445. https://doi.org/10.1083/jcb.200502019

103. Putney JW Jr (2005) Capacitative calcium entry: sensing the calcium stores. J Cell Biol 169 (3):381–382. https://doi.org/10.1083/jcb.200503161
104. Zhang SL, Yeromin AV, Zhang XH, Yu Y, Safrina O, Penna A et al (2006) Genome-wide RNAi screen of Ca(2+) influx identifies genes that regulate Ca(2+) release-activated Ca(2+) channel activity. Proc Natl Acad Sci U S A 103 (24):9357–9362. https://doi.org/10.1073/pnas.0603161103
105. Feske S, Gwack Y, Prakriya M, Srikanth S, Puppel SH, Tanasa B et al (2006) A mutation in Orai1 causes immune deficiency by abrogating CRAC channel function. Nature 441(7090):179–185. https://doi.org/10.1038/nature04702
106. Vig M, Peinelt C, Beck A, Koomoa DL, Rabah D, Koblan-Huberson M et al (2006) CRACM1 is a plasma membrane protein essential for store-operated Ca^{2+} entry. Science 312(5777):1220–1223. https://doi.org/10.1126/science.1127883
107. Luik RM, Wu MM, Buchanan J, Lewis RS (2006) The elementary unit of store-operated Ca^{2+} entry: local activation of CRAC channels by STIM1 at ER-plasma membrane junctions. J Cell Biol 174 (6):815–825. https://doi.org/10.1083/jcb.200604015
108. Cahalan MD (2009) STIMulating store-operated Ca (2+) entry. Nat Cell Biol 11(6):669–677. https://doi.org/10.1038/ncb0609-669
109. Penna A, Demuro A, Yeromin AV, Zhang SL, Safrina O, Parker I et al (2008) The CRAC channel consists of a tetramer formed by Stim-induced dimerization of Orai dimers. Nature 456(7218):116–120. https://doi.org/10.1038/nature07338
110. Mignen O, Thompson JL, Shuttleworth TJ (2008) Orai1 subunit stoichiometry of the mammalian CRAC channel pore. J Physiol 586(2):419–425. https://doi.org/10.1113/jphysiol.2007.147249
111. Hou X, Pedi L, Diver MM, Long SB (2012) Crystal structure of the calcium release-activated calcium channel Orai. Science 338(6112):1308–1313. https://doi.org/10.1126/science.1228757
112. Yen M, Lokteva LA, Lewis RS (2016) Functional analysis of Orai1 concatemers supports a hexameric stoichiometry for the CRAC channel. Biophys J 111 (9):1897–1907. https://doi.org/10.1016/j.bpj.2016.09.020
113. Yen M, Lewis RS (2019) Numbers count: how STIM and Orai stoichiometry affect store-operated calcium entry. Cell Calcium 79:35–43. https://doi.org/10.1016/j.ceca.2019.02.002
114. DeHaven WI, Smyth JT, Boyles RR, Putney JW Jr (2007) Calcium inhibition and calcium potentiation of Orai1, Orai2, and Orai3 calcium release-activated calcium channels. J Biol Chem 282 (24):17548–17556. https://doi.org/10.1074/jbc.M611374200
115. Gwack Y, Srikanth S, Feske S, Cruz-Guilloty F, Oh-hora M, Neems DS et al (2007) Biochemical and functional characterization of Orai proteins. J Biol Chem 282(22):16232–16243. https://doi.org/10.1074/jbc.M609630200
116. Lis A, Peinelt C, Beck A, Parvez S, Monteilh-Zoller-M, Fleig A et al (2007) CRACM1, CRACM2, and CRACM3 are store-operated Ca^{2+} channels with distinct functional properties. Curr Biol 17(9):794–800. https://doi.org/10.1016/j.cub.2007.03.065
117. Brandman O, Liou J, Park WS, Meyer T (2007) STIM2 is a feedback regulator that stabilizes basal cytosolic and endoplasmic reticulum Ca^{2+} levels. Cell 131(7):1327–1339. https://doi.org/10.1016/j.cell.2007.11.039
118. Bhardwaj R, Hediger MA, Demaurex N (2016) Redox modulation of STIM-ORAI signaling. Cell Calcium 60(2):142–152. https://doi.org/10.1016/j.ceca.2016.03.006
119. Shalygin A, Skopin A, Kalinina V, Zimina O, Glushankova L, Mozhayeva GN et al (2015) STIM1 and STIM2 proteins differently regulate endogenous store-operated channels in HEK293 cells. J Biol Chem 290(8):4717–4727. https://doi.org/10.1074/jbc.M114.601856
120. Soboloff J, Spassova MA, Hewavitharana T, He LP, Xu W, Johnstone LS et al (2006) STIM2 is an inhibitor of STIM1-mediated store-operated Ca^{2+} entry. Curr Biol 16(14):1465–1470. https://doi.org/10.1016/j.cub.2006.05.051
121. Wayman CP, McFadzean I, Gibson A, Tucker JF (1997) Cellular mechanisms underlying carbachol-induced oscillations of calcium-dependent membrane current in smooth muscle cells from mouse anococcygeus. Br J Pharmacol 121(7):1301–1308. https://doi.org/10.1038/sj.bjp.0701279
122. Lopez JJ, Albarran L, Gomez LJ, Smani T, Salido GM, Rosado JA (2016) Molecular modulators of store-operated calcium entry. Biochim Biophys Acta 1863(8):2037–2043. https://doi.org/10.1016/j.bbamcr.2016.04.024
123. Bodnar D, Chung WY, Yang D, Hong JH, Jha A, Muallem S (2017) STIM-TRP pathways and microdomain organization: Ca(2+) influx channels: the Orai-STIM1-TRPC complexes. Adv Exp Med Biol 993:139–157. https://doi.org/10.1007/978-3-319-57732-6_8
124. Fong PY, Turner PR, Denetclaw WF, Steinhardt RA (1990) Increased activity of calcium leak channels in myotubes of Duchenne human and mdx mouse origin. Science 250(4981):673–676. https://doi.org/10.1126/science.2173137
125. Hopf FW, Reddy P, Hong J, Steinhardt RA (1996) A capacitative calcium current in cultured skeletal muscle cells is mediated by the calcium-specific leak channel and inhibited by dihydropyridine compounds. J Biol Chem 271(37):22358–22367. https://doi.org/10.1074/jbc.271.37.22358
126. Prakriya M, Lewis RS (2003) CRAC channels: activation, permeation, and the search for a molecular identity. Cell Calcium 33(5–6):311–321. https://doi.org/10.1016/s0143-4160(03)00045-9

127. Lyfenko AD, Dirksen RT (2008) Differential dependence of store-operated and excitation-coupled Ca^{2+} entry in skeletal muscle on STIM1 and Orai1. J Physiol 586(20):4815–4824. https://doi.org/10.1113/jphysiol.2008.160481
128. Li H, Ding X, Lopez JR, Takeshima H, Ma J, Allen PD et al (2010) Impaired Orai1-mediated resting Ca^{2+} entry reduces the cytosolic [Ca2+] and sarcoplasmic reticulum Ca^{2+} loading in quiescent junctophilin 1 knock-out myotubes. J Biol Chem 285(50):39171–39179. https://doi.org/10.1074/jbc.M110.149690
129. Millay DP, Goonasekera SA, Sargent MA, Maillet M, Aronow BJ, Molkentin JD (2009) Calcium influx is sufficient to induce muscular dystrophy through a TRPC-dependent mechanism. Proc Natl Acad Sci U S A 106(45):19023–19028. https://doi.org/10.1073/pnas.0906591106
130. Kurebayashi N, Ogawa Y (2001) Depletion of Ca^{2+} in the sarcoplasmic reticulum stimulates Ca^{2+} entry into mouse skeletal muscle fibres. J Physiol 533 (Pt 1):185–199. https://doi.org/10.1111/j.1469-7793.2001.0185b.x
131. Launikonis BS, Rios E (2007) Store-operated Ca^{2+} entry during intracellular Ca^{2+} release in mammalian skeletal muscle. J Physiol 583(Pt 1):81–97. https://doi.org/10.1113/jphysiol.2007.135046
132. Darbellay B, Arnaudeau S, Bader CR, Konig S, Bernheim L (2011) STIM1L is a new actin-binding splice variant involved in fast repetitive Ca^{2+} release. J Cell Biol 194(2):335–346. https://doi.org/10.1083/jcb.201012157
133. Carrell EM, Coppola AR, McBride HJ, Dirksen RT (2016) Orai1 enhances muscle endurance by promoting fatigue-resistant type I fiber content but not through acute store-operated Ca^{2+} entry. FASEB J 30(12):4109–4119. https://doi.org/10.1096/fj.201600621R
134. Darbellay B, Arnaudeau S, Konig S, Jousset H, Bader C, Demaurex N et al (2009) STIM1- and Orai1-dependent store-operated calcium entry regulates human myoblast differentiation. J Biol Chem 284(8):5370–5380. https://doi.org/10.1074/jbc.M806726200
135. Bohm J, Chevessier F, Maues De Paula A, Koch C, Attarian S, Feger C et al (2013) Constitutive activation of the calcium sensor STIM1 causes tubular-aggregate myopathy. Am J Hum Genet 92 (2):271–278. https://doi.org/10.1016/j.ajhg.2012.12.007
136. Bohm J, Laporte J (2018) Gain-of-function mutations in STIM1 and ORAI1 causing tubular aggregate myopathy and Stormorken syndrome. Cell Calcium 76:1–9. https://doi.org/10.1016/j.ceca.2018.07.008
137. Feske S (2019) CRAC channels and disease - from human CRAC channelopathies and animal models to novel drugs. Cell Calcium 80:112–116. https://doi.org/10.1016/j.ceca.2019.03.004
138. Boncompagni S, Michelucci A, Pietrangelo L, Dirksen RT, Protasi F (2017) Exercise-dependent formation of new junctions that promote STIM1-Orai1 assembly in skeletal muscle. Sci Rep 7 (1):14286. https://doi.org/10.1038/s41598-017-14134-0
139. Meizoso-Huesca A, Launikonis BS (2021) The Orai1 inhibitor BTP2 has multiple effects on Ca^{2+} handling in skeletal muscle. J Gen Physiol 153(1): e202012747. https://doi.org/10.1085/jgp.202012747
140. Eltit JM, Yang T, Li H, Molinski TF, Pessah IN, Allen PD et al (2010) RyR1-mediated Ca^{2+} leak and Ca^{2+} entry determine resting intracellular Ca^{2+} in skeletal myotubes. J Biol Chem 285 (18):13781–13787. https://doi.org/10.1074/jbc.M110.107300
141. Kiselyov KI, Shin DM, Wang Y, Pessah IN, Allen PD, Muallem S (2000) Gating of store-operated channels by conformational coupling to ryanodine receptors. Mol Cell 6(2):421–431. https://doi.org/10.1016/s1097-2765(00)00041-1
142. Eltit JM, Ding X, Pessah IN, Allen PD, Lopez JR (2013) Nonspecific sarcolemmal cation channels are critical for the pathogenesis of malignant hyperthermia. FASEB J 27(3):991–1000. https://doi.org/10.1096/fj.12-218354
143. Lopez JR, Kaura V, Hopkins P, Liu X, Uryach A, Adams J et al (2020) Transient receptor potential cation channels and calcium dyshomeostasis in a mouse model relevant to malignant hyperthermia. Anesthesiology 133(2):364–376. https://doi.org/10.1097/ALN.0000000000003387
144. Iyer KA, Hu Y, Nayak AR, Kurebayashi N, Murayama T, Samso M (2020) Structural mechanism of two gain-of-function cardiac and skeletal RyR mutations at an equivalent site by cryo-EM. Sci Adv 6(31):eabb2964. https://doi.org/10.1126/sciadv.abb2964
145. Durham WJ, Aracena-Parks P, Long C, Rossi AE, Goonasekera SA, Boncompagni S et al (2008) RyR1 S-nitrosylation underlies environmental heat stroke and sudden death in Y522S RyR1 knockin mice. Cell 133(1):53–65. https://doi.org/10.1016/j.cell.2008.02.042
146. Lopez JR, Uryash A, Adams J, Hopkins PM, Allen PD (2021) Molecular modification of transient receptor potential canonical 6 channels modulates calcium dyshomeostasis in a mouse model relevant to malignant hyperthermia. Anesthesiology 134(2):234–247. https://doi.org/10.1097/ALN.0000000000003635
147. Lopez JR, Kaura V, Diggle CP, Hopkins PM, Allen PD (2018) Malignant hyperthermia, environmental heat stress, and intracellular calcium dysregulation in a mouse model expressing the p.G2435R variant of RYR1. Br J Anaesth 121(4):953–961. https://doi.org/10.1016/j.bja.2018.07.008
148. Altamirano F, Eltit JM, Robin G, Linares N, Ding X, Pessah IN et al (2014) Ca^{2+} influx via the Na^{+}/Ca^{2+} exchanger is enhanced in malignant hyperthermia

skeletal muscle. J Biol Chem 289(27):19180–19190. https://doi.org/10.1074/jbc.M114.550764

149. Cully TR, Choi RH, Bjorksten AR, Stephenson DG, Murphy RM, Launikonis BS (2018) Junctional membrane Ca(2+) dynamics in human muscle fibers are altered by malignant hyperthermia causative RyR mutation. Proc Natl Acad Sci U S A 115 (32):8215–8220. https://doi.org/10.1073/pnas.1800490115
150. Duke AM, Hopkins PM, Calaghan SC, Halsall JP, Steele DS (2010) Store-operated Ca^{2+} entry in malignant hyperthermia-susceptible human skeletal muscle. J Biol Chem 285(33):25645–25653. https://doi.org/10.1074/jbc.M110.104976
151. Altamirano F, Riazi S, Ibarra Moreno CA, Kraeva N, Uryash A, Allen PD et al (2019) Is malignant hyperthermia associated with hyperglycaemia? Br J Anaesth 122(1):e3–e5. https://doi.org/10.1016/j.bja.2018.09.014
152. Tammineni ER, Kraeva N, Figueroa L, Manno C, Ibarra CA, Klip A et al (2020) Intracellular calcium leak lowers glucose storage in human muscle, promoting hyperglycemia and diabetes. Elife 9:e53999. https://doi.org/10.7554/eLife.53999
153. Allen DG, Whitehead NP, Froehner SC (2016) Absence of dystrophin disrupts skeletal muscle signaling: roles of Ca^{2+}, reactive oxygen species, and nitric oxide in the development of muscular dystrophy. Physiol Rev 96(1):253–305. https://doi.org/10.1152/physrev.00007.2015
154. Turner PR, Westwood T, Regen CM, Steinhardt RA (1988) Increased protein degradation results from elevated free calcium levels found in muscle from mdx mice. Nature 335(6192):735–738. https://doi.org/10.1038/335735a0
155. Burr AR, Millay DP, Goonasekera SA, Park KH, Sargent MA, Collins J et al (2014) Na^{+} dysregulation coupled with Ca^{2+} entry through NCX1 promotes muscular dystrophy in mice. Mol Cell Biol 34 (11):1991–2002. https://doi.org/10.1128/MCB.00339-14
156. Altamirano F, Perez CF, Liu M, Widrick J, Barton ER, Allen PD et al (2014) Whole body periodic acceleration is an effective therapy to ameliorate muscular dystrophy in mdx mice. PLoS One 9(9): e106590. https://doi.org/10.1371/journal.pone.0106590
157. Goonasekera SA, Davis J, Kwong JQ, Accornero F, Wei-LaPierre L, Sargent MA et al (2014) Enhanced Ca(2)(+) influx from STIM1-Orai1 induces muscle pathology in mouse models of muscular dystrophy. Hum Mol Genet 23(14):3706–3715. https://doi.org/10.1093/hmg/ddu079
158. Uehara A, Yasukochi M, Imanaga I, Nishi M, Takeshima H (2002) Store-operated Ca^{2+} entry uncoupled with ryanodine receptor and junctional membrane complex in heart muscle cells. Cell Calcium 31(2):89–96. https://doi.org/10.1054/ceca.2001.0257
159. Luo X, Hojayev B, Jiang N, Wang ZV, Tandan S, Rakalin A et al (2012) STIM1-dependent store-operated Ca(2)(+) entry is required for pathological cardiac hypertrophy. J Mol Cell Cardiol 52 (1):136–147. https://doi.org/10.1016/j.yjmcc.2011.11.003
160. Hulot JS, Fauconnier J, Ramanujam D, Chaanine A, Aubart F, Sassi Y et al (2011) Critical role for stromal interaction molecule 1 in cardiac hypertrophy. Circulation 124(7):796–805. https://doi.org/10.1161/CIRCULATIONAHA.111.031229
161. Correll RN, Goonasekera SA, van Berlo JH, Burr AR, Accornero F, Zhang H et al (2015) STIM1 elevation in the heart results in aberrant Ca(2)(+) handling and cardiomyopathy. J Mol Cell Cardiol 87:38–47. https://doi.org/10.1016/j.yjmcc.2015.07.032
162. Benard L, Oh JG, Cacheux M, Lee A, Nonnenmacher M, Matasic DS et al (2016) Cardiac Stim1 silencing impairs adaptive hypertrophy and promotes heart failure through inactivation of mTORC2/Akt signaling. Circulation 133 (15):1458–1471.; discussion 71. https://doi.org/10.1161/CIRCULATIONAHA.115.020678
163. Bartoli F, Bailey MA, Rode B, Mateo P, Antigny F, Bedouet K et al (2020) Orai1 channel inhibition preserves left ventricular systolic function and normal Ca(2+) handling after pressure overload. Circulation 141(3):199–216. https://doi.org/10.1161/CIRCULATIONAHA.118.038891
164. Cacheux M, Strauss B, Raad N, Ilkan Z, Hu J, Benard L et al (2019) Cardiomyocyte-specific STIM1 (stromal interaction molecule 1) depletion in the adult heart promotes the development of arrhythmogenic discordant alternans. Circ Arrhythm Electrophysiol 12(11):e007382. https://doi.org/10.1161/CIRCEP.119.007382
165. Zhao G, Li T, Brochet DX, Rosenberg PB, Lederer WJ (2015) STIM1 enhances SR Ca^{2+} content through binding phospholamban in rat ventricular myocytes. Proc Natl Acad Sci U S A 112(34):E4792–E4801. https://doi.org/10.1073/pnas.1423295112
166. Molkentin JD, Lu JR, Antos CL, Markham B, Richardson J, Robbins J et al (1998) A calcineurin-dependent transcriptional pathway for cardiac hypertrophy. Cell 93(2):215–228. https://doi.org/10.1016/s0092-8674(00)81573-1
167. Flanagan WM, Corthesy B, Bram RJ, Crabtree GR (1991) Nuclear association of a T-cell transcription factor blocked by FK-506 and cyclosporin A. Nature 352(6338):803–807. https://doi.org/10.1038/352803a0
168. Loh C, Shaw KT, Carew J, Viola JP, Luo C, Perrino BA et al (1996) Calcineurin binds the transcription factor NFAT1 and reversibly regulates its activity. J Biol Chem 271(18):10884–10891. https://doi.org/10.1074/jbc.271.18.10884
169. Rao A, Luo C, Hogan PG (1997) Transcription factors of the NFAT family: regulation and function.

Annu Rev Immunol 15:707–747. https://doi.org/10.1146/annurev.immunol.15.1.707

170. Berry CT, May MJ, Freedman BD (2018) STIM- and Orai-mediated calcium entry controls NF-kappaB activity and function in lymphocytes. Cell Calcium 74:131–143. https://doi.org/10.1016/j.ceca.2018.07.003
171. Eder P, Molkentin JD (2011) TRPC channels as effectors of cardiac hypertrophy. Circ Res 108(2):265–272. https://doi.org/10.1161/CIRCRESAHA.110.225888
172. Dietrich A, Kalwa H, Rost BR, Gudermann T (2005) The diacylgylcerol-sensitive TRPC3/6/7 subfamily of cation channels: functional characterization and physiological relevance. Pflugers Arch 451(1):72–80. https://doi.org/10.1007/s00424-005-1460-0
173. Hofmann T, Obukhov AG, Schaefer M, Harteneck C, Gudermann T, Schultz G (1999) Direct activation of human TRPC6 and TRPC3 channels by diacylglycerol. Nature 397(6716):259–263. https://doi.org/10.1038/16711
174. Ohba T, Watanabe H, Murakami M, Takahashi Y, Iino K, Kuromitsu S et al (2007) Upregulation of TRPC1 in the development of cardiac hypertrophy. J Mol Cell Cardiol 42(3):498–507. https://doi.org/10.1016/j.yjmcc.2006.10.020
175. Seth M, Zhang ZS, Mao L, Graham V, Burch J, Stiber J et al (2009) TRPC1 channels are critical for hypertrophic signaling in the heart. Circ Res 105(10):1023–1030. https://doi.org/10.1161/CIRCRESAHA.109.206581
176. Kuwahara K, Wang Y, McAnally J, Richardson JA, Bassel-Duby R, Hill JA et al (2006) TRPC6 fulfills a calcineurin signaling circuit during pathologic cardiac remodeling. J Clin Invest 116(12):3114–3126. https://doi.org/10.1172/JCI27702
177. Nakayama H, Wilkin BJ, Bodi I, Molkentin JD (2006) Calcineurin-dependent cardiomyopathy is activated by TRPC in the adult mouse heart. FASEB J 20(10):1660–1670. https://doi.org/10.1096/fj.05-5560com
178. Wu X, Eder P, Chang B, Molkentin JD (2010) TRPC channels are necessary mediators of pathologic cardiac hypertrophy. Proc Natl Acad Sci U S A 107(15):7000–7005. https://doi.org/10.1073/pnas.1001825107
179. Treves S, Jungbluth H, Voermans N, Muntoni F, Zorzato F (2017) Ca(2+) handling abnormalities in early-onset muscle diseases: novel concepts and perspectives. Semin Cell Dev Biol 64:201–212. https://doi.org/10.1016/j.semcdb.2016.07.017

TRPV1 in Pain and Itch

12

Fengxian Li and Fang Wang

Abstract

Transient receptor potential vanilloid type 1 (TRPV1) is a nonselective cation channel that is intensively expressed in the peripheral nerve system and involved in a variety of physiological and pathophysiological processes in mammals. Its activity is of great significance in transmitting pain or itch signals from peripheral sensory neurons to the central nervous system. The alteration or hypersensitivity of TRPV1 channel is well evidenced under various pathological conditions. Moreover, accumulative studies have revealed that TRPV1-expressing ($TRPV1^+$) sensory neurons mediate the neuroimmune crosstalk by releasing neuropeptides to innervated tissues as well as immune cells. In the central projection, $TRPV1^+$ terminals synapse with the secondary neurons for the transmission of pain and itch signalling. The intense involvement of TRPV1 and $TRPV1^+$ neurons in pain and itch makes it a potential pharmaceutical target. Over decades, the basis of TRPV1 channel structure, the nature of its activity, and its modulation in pathological processes have been broadly studied and well documented. Herein, we highlight the role of TRPV1 and its associated neurons in sensing pain and itch. The fundamental understandings of TRPV1-involved nociception, pruriception, neurogenic inflammation, and cell-specific modulation will help bring out more effective strategies of TRPV1 modulation in treating pain- and itch-related diseases.

Keywords

TRPV1 · Itch · Pain · Sensory neurons

12.1 Introduction

The transient receptor potential (TRP) channels are a superfamily that is made up of 28 members in mice and 27 members in humans [1]. As non-selective cation-permeable channels, the TRP family is capable to respond to multiple external and internal stimuli, including changes in thermal, pH, chemical irritants, as well as mechanical and osmotic cues. Therefore, they are implied in numerous physiological and pathological conditions and have become an increasing focus in basic science, translational research, and drug development.

Generally, the TRP superfamily is divided into six subfamilies: canonical (TRPC), vanilloid (TRPV), ankyrin (TRPA), melastatin (TRPM), polycystin (TRPP), and mucolipin (TRPML).

F. Li (✉)
Department of Anesthesiology, Zhujiang Hospital of Southern Medical University, Guangzhou, China

F. Wang
Department of Dermatology, The First Affiliated Hospital of Sun Yat-sen University, Guangzhou, China
e-mail: wangf78@mail.sysu.edu.cn

L. Zhou (ed.), *Ion Channels in Biophysics and Physiology*, Advances in Experimental Medicine and Biology 1349, https://doi.org/10.1007/978-981-16-4254-8_12

Among them, the TRPV subfamily, which includes four members TRPV1 to TRPV4, are broadly involved in nature responses like pain and itch across mammals. The TPRV1 channel is also known as vanilloid receptor 1 and was cloned and identified as the receptor for heat and capsaicin in 1997 by David Julius lab [2]. It consists of six transmembrane domains with a pore formed by transmembrane segments 5–6 (S5–S6) and the intervening pore loop, which is flanked by S1–S4 voltage-sensor-like domains [3]. Various stimuli such as noxious heat (>43 °C), capsaicin, and pH acting on TRPV1 can directly lead to the channel opening and the consequent cation entry [2, 4]. TRPV1 is highly expressed in the peripheral nerve system (PNS) particularly C-fiber [5]. Ever since its discovery, TRPV1 has attracted broad attention in multiple studies that focus on pain or itch [6–11]. In this chapter, we will start with the introduction of basic biology in neuroscience and then interpret how TRPV1 participates in pain and itch perception. Furthermore, based on its biology, we will discuss the potential of TRPV1 modulation as drug interventions in pathologic sensory conditions.

12.2 TRPV1 Biology in Pain and Itch

12.2.1 The Basics of Pain and Itch

In mammals, the skin, mucous membrane, and muscles are highly innervated by primary somatosensory fibers. Their cell bodies are located in trigeminal ganglia which innervate the head and neck and the dorsal root ganglia (DRG) which receive signals from the rest of the body. These afferent neurons are pseudo-unipolar neurons, which project long axons to the skin and deeper body structures, and transmit impulse signaling through the other axonal branches synapse with neurons in the brain stem nuclei or spinal cord dorsal horn. The information is then relayed by higher order neurons towards the cortex where various senses such as warmth, coldness, touch, pressure, pain, and itch are ultimately perceived. While pain is defined as "an unpleasant sensory and emotional experience associated with actual or potential tissue damage" [12], the itch was introduced as "An unpleasant sensation that provokes the desire to scratch" [13]. Generally, sensory neurons that are specialized to sense pain and itch are referred to as nociceptors and pruriceptors, respectively. Due to the protective behaviors subsequentially aroused by pain and itch, hosts can be aware of or protected from environmental dangers.

Historically, sensory neurons have been classified according to the amount of myelination which dictates conduction velocity (fast or slow) and diameter (large or small). There are three main types of sensory neurons: large diameter (heavily and moderately myelinated Aβ fibers), medium diameter (thinly myelinated Aδ fibers), and small diameter (unmyelinated C fibers). While various mechanosensations such as touch and pressure are mainly elicited by Aδ and Aβ fibers, the vast majority of pain and itch is mediated by Aδ fibers and unmyelinated C fibers with free nerve endings.

12.2.2 TRPV1 and TRPV1$^+$ Sensory Neurons

TRPV1 is the most well-characterized TRP channel and is the specific receptor of capsaicin (the spicy ingredient of chili peppers). However, a wide range of endogenous and exogenous stimuli such as noxious temperature (~43 °C), acidic or basic pH, vanilloid compounds, can also activate this ion channel [2, 4].

It is well established that TRPV1 activation and consequential cation ion flow are required for transduction of a number of pain or itch signaling. Meanwhile, nociceptors rely on molecular sensors like TRPV1 to detect noxious stimuli. It has been demonstrated in numerous studies that ~40–50% of sensory neurons express TRPV1 [14, 15]. More recently, emerging studies using unbiased single-cell RNA sequencing (scRNA seq) of murine DRG neurons updated the sensory neuron classification system according to their gene expression [16–19]. In 2015, Usoskin et al. identified that a subpopulation of Aδ fibers expresses the gene of calcitonin-gene related peptide (CGRP, *Calca*) and nerve growth factor receptor (TrkA; *Ntrk1*). They classified this population as peptidergic (PEP) 2. Those C fibers that express the above two genes plus neuropeptides substance P (SP; *Tac1*), are classified into PEP1. Other C fibers that were previously classified as non-peptidergic (NP) neurons (P2rx3 expression and predicted IB4 binding), were classified into three subclusters NP1, NP2, and NP3. Interestingly, some NP neurons like NP2 also contain markers for peptidergic neurons (i.e., CGRP gene *Calca*) [18, 19]. Therefore, NP neurons still have the capability to release neuropeptides. It has been revealed by sc-RNA seq data that *Trpv1* is abundantly expressed in the PEP1, NP2, and NP3 [18, 19], suggesting the critical role played by these three subpopulations in transmitting noxious and pruriceptive stimuli (Figs. 12.1 and 12.2).

12.2.3 TRPV1 in Pain Sensation

12.2.3.1 Pain Classification

Pain can be simply divided into acute pain and chronic pain. Acute pain refers to the new onset of tissue damage or injuries, usually featured as self-limited healing and relief within hours or days. From this direction, acute pain provides warning signals and provokes avoidance from dangers. However, when the pain pathway is altered or pain sensation tends to last longer or recurs, chronic pain can be defined. In order to better identify chronic pain, given the first priority to pain etiology, the Task Force group of IASP (International Association for the Study of Pain) has defined chronic pain as "pain that lasts or recurs for longer than 3 months" in the year 2015 [20], regardless the specific pain phenotypes. Chronic pain is further classified as: (1) chronic primary pain; (2) chronic cancer-related pain; (3) chronic postsurgical or posttraumatic pain; (4) chronic neuropathic pain; (5) chronic secondary headache or orofacial pain; (6) chronic secondary visceral pain; and (7) chronic secondary musculoskeletal pain. More details regarding ICD-11 (the 11th version of International Classification of Diseases) codes for chronic pain were also extensively defined in the year 2019, which benefits to identify patients with chronic pain [21, 22].

The clear-cut definition and ICD-11 codes for chronic pain are attributed to the up-to-date efforts of the majority body of clinical and basic science research. Chronic pain has several debilitating features in which hypersensitivity is the most dominant one. Hypersensitivity refers to the "tissue or nerve damage elicits hyperactivity to promote guarding of the injury area" [23]. As a result, innocuous stimuli such as light touch could be perceived as pain (allodynia), and painful stimuli will induce greater intensity (hyperalgesia). This phenomenon is driven by various sensitization pathways and corresponding mechanisms. The mechanism for sensing and transmitting pain sensation is complex and involves multiple factors, leading to difficulties in pain treatment [23]. As the predominated nociceptor in the peripheral sensory system, TRPV1$^+$ neurons possess numerous receptors for pain mediators such as adenosine triphosphate (ATP), serotonin, and bradykinin. Of the utmost importance, the TRPV1 channel is also a potent sensor for noxious stimuli and can be sensitized in a broad spectrum of chronic pain environments.

12.2.3.2 TRPV1 Serves as the Sensor for Pain Sensation

In general, direct activation of the TRPV1 channel by capsaicin and noxious heat (above 43 °C) can induce a burning sensation in the applied region. Capsaicin is known as the pain mediator long before its receptor was identified in 1997 [2]. By binding to the residue of Y511 located at the transmembrane spanning segment 4 (S4), capsaicin induces the opening of the TRPV1 channel and introduces calcium influx, the first sign for pain sensation singling in the peripheral. Similarly, noxious heat enables TRPV1 channel opening and triggers intracellular calcium increase. The intracellular calcium enhancement due to the receptor potential of TRPV1 activation could further trigger the voltage-gated ion channels to generate action potentials and result in the peripheral transmission of burning pain sensation. The

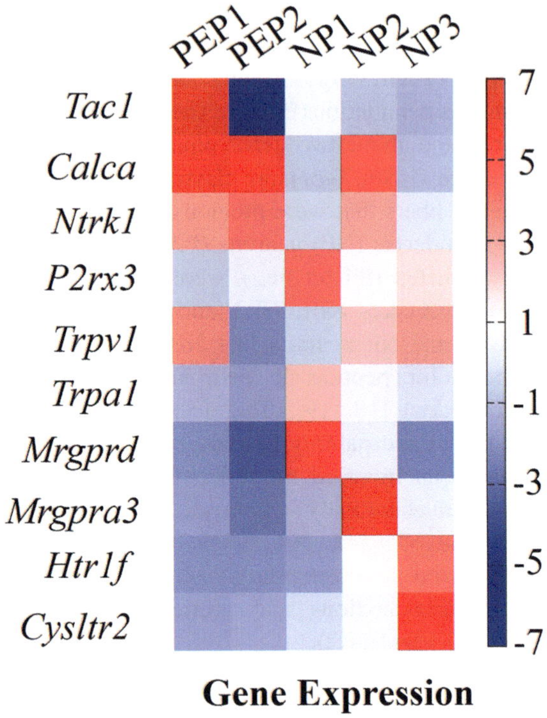

Fig. 12.1 Maximum likelihood estimated expression of selected genes in mouse dorsal root ganglia neuron subpopulations. Nociceptors are considered to primarily be in PEP1 and PEP2 populations, while pruriceptor populations are considered to be NP1, NP2, and NP3. (Full database is available in Usoskin et al. [18])

property of sensing heat and capsaicin to produce pain sensation makes TRPV1 the sensor for these two fundamental noxious as well as thermal and chemical stimuli in the sensory nervous system.

Acidosis is common in the context of inflammation, tissue damage, and ischemia, in which protons are the important pain mediators. Proton (also described as hydrogen ion, acid, or low pH in different research papers) can directly activate the TRPV1 channel when the pH value hits lower than 6 at room temperature, by binding the E648 site of the extracellular loop and triggering cation entry for pain transmission [24]. Interestingly, in

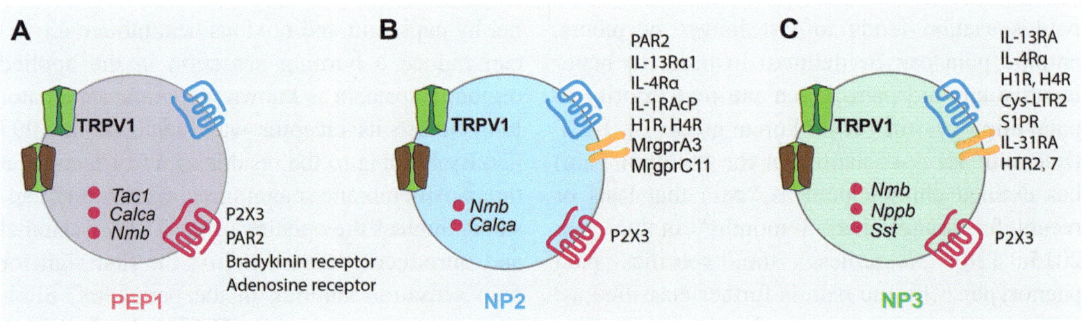

Fig. 12.2 Pain- and itch-sensing receptors are broadly expressed in TRPV1^{+} subpopulations. (**a**) Peptidergic TRPV1^{+} neurons express the majority body of pain-sensing receptors and release the corresponding neuropeptides such as SP (coded by gene *Tac1*), CGRP (coded by *Calca*), and NMB (coded by *Nmb*). (**b**, **c**) Itch-sensing neurons are mainly expressed in NP2 and NP3 TRPV1^{+} subpopulations. Neuropeptides releasing from these two subpopulations including NMB, CGRP, and SST (coded by gene *Sst*), Nppb (coded by gene *Nppb*). Of note, MrgprA3 and Nppb are the two well-accepted makers for itch populations

addition to protons, TRPV1 can also detect basic deviations from homeostatic pH. Alkaline pH such as ammonia can also induce irritant and pain sensation, which is common in the environment of artificial fertilizers and industrial pollutants. Via exposure to the external NH_4Cl solutions to get the diffusion-free NH_3 (ammonia), cultured DRG neurons and TRPV1-coexpressing human embryonic kidney 293 (HEK293) cells showed robust calcium responses [4]. In TRPV1-expressing HEK293 cells, the cell activity induced by intracellular pH of 9.5 was nearly abolished when TRPV1-antagonist was introduced. These results indicated the responsible role of TRPV1 in sensing alkaline chemicals and intracellular basic pH. Distinct from the proton binding site of E648, the base-sensing residue is detected at the site of H378 located in TRPV1 N-terminals [4]. Although TRPA1 (transient receptor potential ankyrin 1) can also respond to ammonia and intracellular base, so far, TRPV1 is the only identified ion channel that senses both acid and alkaline pH [4].

Some toxin peptides like venoms (which contains three types of inhibitor cysteine knot peptides) from spiders, snakes, cone snails, or scorpions, functioning as vanillotoxins, can also directly activate TRPV1 and induce inflammatory pain [25]. One such classical peptide toxin named as double-knot toxin (DkTx) from the Earth Tiger tarantula, can specifically bind to the residues within the S5-P-S6 pore region, and exhibit antibody-like bivalency to produce pain sensation in a virtually irreversible manner [26].

While sensing the noxious stimuli, TRPV1 activity is also magnified via sensitization in the context of inflammatory pain. Studies have revealed that TRPV1 responses to pain-producing chemicals and thermal stimuli, subsequently triggers heat-evoked pain or hypersensitivity in injured tissues [27]. Genetic TRPV1 knockout mice lack the thermal hypersensitivity in the context of inflammation, indicating that TRPV1 serves as the pain sensor upon activation by internal and external stimuli [11].

12.2.3.3 TRPV1$^+$ Sensory Neuron in Pain Sensation

Besides the direct activation by heat and chemical irritants, TRPV1 protein actually labels the nociceptive PEP1 population based on the classification of sensory neuron types by the unbiased large-scale scRNA-seq [18]. Among the aforementioned types of pain classified by IASP, TRPV1 plays an extremely important role in inflammatory pain. Injuries or damaged tissues can release the classical "inflammatory soup", which comprises the major pain-inducing chemicals such as ATP, bradykinin, prostaglandins, 5-HT (5-hydro-xytryptamine, also known as serotonin), and endothelin-1. From experimental data and the unbiased sequencing analysis, TRPV1$^+$ neurons possess the dominant nociceptive populations sensing the above pain mediators. That is, TRPV1$^+$ PEP1 neurons express purinergic receptor P2rX3, bradykinin B2 receptor, protease-activated receptor 2 (PAR2), and adenosine 2 receptor and so on for sensing pain mediators (Summarized in Fig. 12.2a) [1, 18]. In the field of acute pain, postoperative pain is the most common one that requires acute anti-pain treatment otherwise would potentiate chronic pain. Postoperative pain also attributes to inflammation in wounds. TRPV1 is necessary for heat (but not mechanical) hyperalgesia after incision, which has been evidenced by genetic knockout and pharmacologic antagonism in tested mice [28]. Genetic ablation of TRPV1-lineage neurons results in the overall deficit to either thermal hyperalgesia or neurogenic inflammation, although this is partially due to other thermal TRPs co-expressed in the TRPV1-lineage population [9]. Also, ablation of TRPV1$^+$ neurons and fibers by the potent agonist resiniferatoxin (RTX) treatment has been shown to impair thermal nociception without affecting mechanical nociception in adult rats. Furthermore, during the projection from DRG to the dorsal horn, the loss of TRPV1$^+$ neurons significantly eliminates the expression of presynaptic mu opioid receptors in DRG, and also potentiates the analgesic effect of opioid, indicating the inhibition effect of the intact mechano-nociception afferents [29].

Most of the pain mediator receptors are not only located in the TRPV1$^+$ neurons but also employ TRPV1 as the downstream ion channel for nociception sensitization. Activation of the bradykinin B2 receptor by bradykinin during tissue inflammation mainly elicits pain sensation by sensitizing TRPV1 via phospholipase C (PLC) and cyclooxygenase-1 activation [30]. Meanwhile, TRPV1 is also required for a number of G-protein coupled receptors (GPCRs), such as the 5-HT2 receptor and PAR2 receptor when sensing pain mediators [23]. Notably, TRPV1 may have some alterations upon chemical stimulation to augment the pain sensation. For example, during pain sensitization, multiple intracellular signals like protein kinase A (PKA) and protein kinase C (PKC) pathways can phosphorylate the TRPV1 channel and produce pain sensitization, which provides a unique strategy to modulate TRPV1 activity for analgesia [31].

12.2.3.4 TRPV1-TRPA1 Complex in Pain Sensation

TRPV1$^+$ neurons co-express several TRP channels that are also nociception sensors. TRPA1 is the sole member of TRPA subfamily and serves a broad spectrum of biophysiological functions in different species. TRPA1 is largely co-expressed with TRPV1 in sensory neurons, where they act closely in modulating pain [11, 32]. TRPA1 senses exogenous irritant chemicals such as pungent products isothiocyanates (AITC) [33, 34] as well as endogenous metabolites such as reactive carbonyl species like 4-hydroxynonenal (4-HNE) and 4-oxoonenal (4-ONE) [35, 36]. Hence TRPA1 is viewed as the "gate-keeper" for inflammation within sensory nervous system [37]. The role of TRPA1 involved in pain sensation has been well-documented [38, 39]. The most intriguing inhibition effect of TRPV1 on TRPA1 has been observed in pain conditions [40]. Until recently, the finding of Tmen100 protein as the modulator for TRPA1-mediated hyperalgesia in a TRPV1-dependent manner [41], has shed light on a new strategy for pain treatment. In general, Tmem100 is a 134-amino-acid transmembrane protein highly conserved in vertebrates [42]. In the tested rodent DRGs, Tmem100 is exclusively expressed in the peptidergic population. Importantly, Tmem100 has the capacity to release the inhibition of TRPA1 by TRPV1, resulting in inflammatory-mediated hyperalgesia. Mutation of Tmem100 with Q-Q-Q in the charged K-R-R sequence in the C terminal enhances the binding with TRPV1, which causes disconnection or low affinity with TRPA1. The consequence of this structure change contributes to the relief of TRPA1-mediated hyperalgesia [41]. Therefore, the regulation effect of TRPV1-TRPA1 correlation is also considered as a new target in neurogenic inflammation [43].

12.2.4 TRPV1 in Itch Sensation

12.2.4.1 Itch Is a Distinct Neural Process from Pain

Although it has been long considered that an itch is a mild form of pain, recent innovations in neuroscience and neuroimmunology have identified how itch pathways are distinct from pain.

In 2007, Sun and Chen discovered the first itch-specific pathway in the nervous system defined by gastrin-releasing peptide (GRP) and its receptor (GRPR), indicating that itch is a trackable and distinct sensory process from pain [44]. Later in 2009, Liu et al. identified that murine mas-related G-protein-coupled receptor (Mrgpr) A3 is the receptor of chloroquine (CQ), an anti-malaria drug that can induce serve itch sensation [45]. Mrgpr family are GPCRs and comprise ~8 genes and pseudogenes in humans and 27 in murine, most of which are expressed by primary sensory neurons [46]. CQ can cause robust itch in both humans and mice [45]. Interestingly, deletion of a cluster of Mrgpr genes significantly reduced mice scratching behavior induced by CQ but not histamine. They also identified that human MRGPRX1 has a similar primary sequence to MrgprA3 and can be specifically activated by CQ [45]. In the year 2013, the same group further found that

MrgprA3-expressing neurons were the itch selective population. To demonstrate this, they adopted a unique approach by inserting the TRPV1 cation channel only in $MrgprA3^+$ sensory neurons in TRPV1-deficient mice. Following capsaicin intradermal injection, they observed itch-induced scratching behavior rather than pain-related behaviors in those animals [47]. Other Mrgpr family expressing in neurons mediates itch sensation include MrgprC11 and MrgprD, which can be activated by bovine adrenal medulla peptide 8–22 [45] and β-alanine [48], respectively. Collectively, the discoveries of specific Mrgprs on sensory neurons and their itch function provide landmark advances in itch biology and neurology.

12.2.4.2 $TRPV1^+$ Sensory Neuron in Itch Sensation

The scRNA-seq studies classified MrgprA3 (overlapped with MrgprC11)-expressing neurons into NP2 itch subpopulation (Figs. 12.1 and 12.2b). In addition to the NP2 subpopulation, NP3 is the other itch sensory neuron with significant TRPV1 expression (Figs. 12.1 and 12.2c). However, during the sensory nervous system development, TRPV1-lineage neurons in mice differentiate into DRG neuron subsets with diverse TRP markers that include TRPV1, TRPA1, and TRPM8. Therefore, in addition to $TRPV1^+$ neurons, the animal tool of TRPV1-linage deficiency may also have impaired TRPA1-, TRPM8-, and even Mrg-expressing neurons [9]. However, administration of RTX, a potent agonist of TRPV1, provides an effective study approach to ablate only $TRPV1^+$ neurons in the fully developed sensory nervous system [49].

NP3 is characterized by high enrichment of gene *Nppb* for brain natriuretic peptide (BNP), a neurotransmitter that activates itch in the spinal cord, along with genes that code the pruritogen receptors: interleukin (IL)-31RA (*Il-31ra*) and CysLTR2 (*Cysltr2*) (Fig. 12.2c). Stimuli that act on sensory neurons and directly induce itch are defined as pruritogens. IL-31 is a type 2 immune cytokine and is predominantly released by T helper (Th) 2 cells in the context of atopic dermatitis (AD), a common itchy and inflammatory skin disease. IL-31 is the first cytokine that has been identified as a pruritogen. In 2014, Cevikbas et al. identified that IL31RA is expressed by both human and mouse DRG neurons [50]. Either intradermal or intrathecal injection of IL-31 was sufficient to evoke robust itch behavior in naive wild-type (WT) mice. Notably, those IL-31-responsive sensory neurons also largely co-express with TRPV1. Later, the connection between IL-31 and $Nppb^+$ neurons was further confirmed by the findings in which IL-31 is sufficient to upregulate *Nppb* in both skin and DRG [51]. More recently, we identified that the potent pruritogen leukotriene (LT) C4 and its receptor CysLTR2 neural pathway is the key mechanism in AD-associated itch flares [52]. However, although CysLTR2 siRNA knockdown in trigeminal ganglia was sufficient to alleviate acute itch flares in this context, it remains undefined whether this phenotype is specifically dependent on $TRPV1^+$ neurons. Notwithstanding this, the Hoon group directly demonstrated that $Nppb^+$ neurons are sensors of mast cell-derived LTC4 and TRPV1-linage neurons are required for *N*-methyl LTC4 (N-met LTC4)-induced itch. Moreover, they also revealed that itch elicited by serotonin and sphingosine-1 phosphate (S1p) is dependent on the $Nppb^+$ subpopulation and these itch pathways further rely on the canonical GRPR-spinal cord circuit. Taken together, these results highlight that the Nppb-labeled sensory neurons are specifically responsible for IL-31-, LTC4-, serotonin-, and S1p-evoked itch. However, whether the cellular sources of these pruritogens play different roles towards itch neuron populations is still an interesting question for further research.

Histamine is a classical pruritogen that can be released from mast cells and basophils. So far, four distinct GPCRs (H1R, H2R, H3R, and H4R) for histamine have been found. Three of these receptors are expressed by neurons. However, only H1R and H4R are expressed by pruriceptive DRG and mediate histaminergic itch [53, 54]. Both NP2 and NP3 are found to have histamine receptor RNA expressed in sc-RNA seq studies (Fig. 12.2). Consistent with this, TRPV1-DTA animals exhibited a remarkable reduction of scratch responses to histamine intradermal injection [55].

IL-4 and IL-13 are the other two canonical types 2 effector cytokines. They have been shown to directly activate sensory neurons and promote itch in the context of AD-like disease [56, 57]. Compared with the receptors for IL-31 (*Il31ra* and *Osmrβ*), the gene of IL-4 and IL-13 shared receptor subunit (*Il4ra*) is broadly expressed across NP1, NP2, and NP3 itch sensory neurons [18, 57]. More importantly, the humanized anti-IL-4Rα mAb (dupilumab) has demonstrated outstanding anti-itch effects in patients with AD or other chronic pruritic diseases such as chronic pruritus of unknown origin [58–64]. Taken together, these advances highlight how cytokines dramatically promote itch. However, whether these cytokine itch pathways will be inhibited by targeting TRPV1$^+$ neurons is unclear. Further studies may unveil the role and mechanisms of specific sensory subpopulations in cytokine-induced itch.

12.2.4.3 Role of TRPV1 Ion Channel for Itch Signaling

In addition to investigating the function of TRPV1$^+$ sensory neurons in itch signaling, numerous basic studies have explored the molecular mechanisms of itch sensation.

It is well established that histaminergic itch is TRPV1-dependent. TRPV1-deficient mice exhibited less histamine-induced scratch compared with WT controls, whereas α-5HT- or endothelin 1-elicited itch was unaffected. The TRPV1 channel opening is critically required for histamine to activate sensory neurons via H1R [2, 65]. TRPV1 opening consequently causes increases in intracellular calcium (Ca^{2+}) concentrations that prepare for neuronal action potentials. This process likely involves phospholipase (PL) A_2 or lipoxygenases (LO) as a PLA2 inhibitor or a LO inhibitor is sufficient to block the histamine-induced Ca^{2+} influx in sensory neurons [66]. In addition, other analyses indicated that phospholipase (PL) Cbeta3 has overlapped expression with H1R in a subpopulation of C fibers and PLCbeta3 specifically mediates histamine-induced calcium responses through the H1R in cultured sensory neurons. The PLCbeta3-deficient murine strain showed significant defects in histamine-induced scratching behavior [67]. Notwithstanding this, the precise molecular mechanisms underlying histamine receptor-TRPV1 pathways will be an exciting area for further investigation. For the newly discovered H4R, recent studies suggest that TRPV1 and PLC are also required for H4R-mediated itch signaling in vitro [68].

In addition to histaminergic itch, multiple non-histaminergic itch mechanisms that involve TRPV1 include IL-31, IL-4/IL-13, LTB4, and LTC4. To explore which TRP channels are involved in IL-31-mediated itch, Cevikbas et al. employed TRPV1- and TRPA1-deficient mouse strain, respectively, and found that either TRPV1 or TRPA1 deletion can significantly decrease IL-31-evoked itch [50]. Using calcium imaging in vitro, they confirmed that the percentage of IL-31-responsive neurons was significantly reduced in TRPV1-or TRPA1-deficient DRG. Taken together, these results indicate that IL-31 employs both TRPV1 and TRPA1 to fully mediate itch. To investigate the downstream of TRP channels in the IL-31 itch pathway, in the same study, authors first observed that IL-31-stimulated murine DRG neurons induced ERK1/2 phosphorylation in vitro. Then they demonstrated that administration of U0126 which can completely prevent ERK1/2 phosphorylation was sufficient to reduce scratching bouts in WT mice that received IL-31 intradermal injections. Thus, ERK1/2 might provide a future direction to study the TRP downstream phosphorylation in the IL-31-induced itch pathway.

Although it appears that cultured DRG neurons from either *Trpa1*$^{-/-}$ or *Trpv1*$^{-/-}$ mice exhibited fewer responses to both IL-4 and IL-13, whether TRPV1 or TRPA1 is required for IL-4 and IL-13 enhanced scratching behavior remains undefined. However, Oetjen et al. demonstrated that IL-4/IL-13 needs Janus kinase (JAK) to transmit their signals into neurons [57]. The

JAK family has four members: JAK1, JAK2, JAK3, and TYK2 [69]. The conditional deletion of JAK1 in sensory neurons led to a marked reduction in scratching behavior in the setting of AD-like disease. Moreover, in humans, JAK inhibitors are shown to significantly reduce itch severity in patients suffering from chronic pruritus [70]. Therefore, whether there is any connection between TRPV1 and JAKs would be interesting for future investigation.

LTs are a family of eicosanoid inflammatory mediators and are produced in leukocytes, particularly mast cells and basophils [71, 72]. LTs are generated by the oxidation of arachidonic acid and the essential fatty acid eicosapentaenoic acid via the 5-LO pathway. LTs are divided into two classes: the chemoattractant LTB4 and the cysteinyl LTs (CysLTs: LTC4, LTD4, and LTE4) [73]. It has been shown that intradermal injections of LTB4 resulted in scratching behavior in mice [74, 75]. Similar to IL-31-induced itch, either TRPV1 or TRPA1 antagonists are sufficient to inhibit itch behavior proved by LTB4 [75]. For CysLTs, while the LTC4-CysLTR2 itch pathway has been confirmed by multiple research groups [52, 76, 77], the role of LTD4 and LTE4 as pruritogens remains controversial. Interestingly, although LTC4-activated sensory neurons showed large coexpression with TRPV1 and TRPA1, either TRPV1 or TRPA1 knockout could not reduce the percentage of LTC4-responsive DRG neurons. Only TRPV1 and TRPA1 compound deletion is capable to ameliorate the neuronal responsiveness towards LTC4. In line with these in vitro results, mice lacking either TRPV1 or TRPA1 exhibited unaltered scratching behavior following *N*-met LTC4 injection compared with their littermate controls. However, TRPV1 and TRPA1 double knockout attenuated LTC4-induced scratching responses. Taken together, we can conclude that CysLTR2 is equally reliant on either TRPV1 or TRPA1 to fully mediate LTC4-elicited itch. In other words, both canonical TRPV1 and TRPA1 signaling must be impaired to limit LTC4-mediated itch.

12.3 TRPV1 Activity Modulation in Pain and Itch

12.3.1 TRPV1 Upregulation in the Context of Pain

In many pathological conditions of chronic pain, the functional upregulation of TRPV1 is likely induced by inflammatory mediators such as nerve growth factors released from adjacent non-neural cells. The elevation of TRPV1 protein, but not mRNA levels, indicates the cytoplasm assembly upon activation, rather than transcription-independent overexpression [78]. Interestingly, the increased TRPV1 protein then is transported to the peripheral membrane of a cell body, but not the central C-fiber terminals for synapse transmission. The TRPV1 upregulation is implied in tissue inflammation and thermal hypersensitivity. Further studies revealed that p38 MAPK activation is required for TRPV1 overexpression [78], indicating MAPK pathways in induction and maintenance of peripheral sensitization and persistent chronic pain [79].

Under pathological situations, TRPV1 has been detected in previous TRPV1-absent sensory neurons. To a large extent, hyperalgesia could be ascribed to this mechanism. Take TRPM8-dependent cold allodynia as an example, we have demonstrated that pleasant cool temperature that only causes cool sensation to the skin surprisingly induces cold allodynia to the cornea [80]. The significance of this phenomenon is that such cold allodynia becomes more severe in multiple pathological conditions such as dry eyes. We further identified that TRPV1 is highly expressed in $TRPM8^+$ neuron-innervated cornea rather than other tissues. The expression of TRPV1 in $TRPM8^+$ neurons was further elevated under dry eye conditions. Via electrophysiological approaches, we found that TRVP1 enables $TRPM8^+$ cold-sensing neurons to depolarize and fire action potentials more easily. In line with in vitro results, TRPV1 deficiency and pharmacological antagonist significantly alleviated corneal cold allodynia. Meanwhile, TRPV1 upregulation is sufficient to cause cold allodynia as genetical overexpression of TRPV1 in TRPM8-expressing sensory neurons led to cold allodynia in both cornea and skin even without any pathological changes [80]. Taken together, it is reasonable that the upregulated TRPV1 ion channel initiates neural hypersensitivity in the TRPV1$^+$ neuron population and results in phenomena of hyperalgesia.

12.3.2 TRPV1 Upregulation in the Context of Chronic Itch

TRPV1 upregulation has also been seen in itchy conditions. In a pure chronic itch model, dry skin conditioning was performed by the acetone–ether–water procedure for sequential days to induce robust itch behavior, known as dry skin model. In the mouse model, we evidenced that the TRPV1$^+$ fiber largely expanded their territory in the skin compared with the control treatment. This is confirmed by histochemistry staining of placental alkaline phosphatase in the genetic TRPV1-reporter mouse line [81]. In this study, the upregulation of TRPV1 was confirmed on both gene and protein levels and the upregulated TRPV1 was sufficient to provoke enhanced calcium responses to the low dose of capsaicin, indicating its intact function. Importantly, in this dry skin model, a low dose of capsaicin injection strikingly induced robust scratching behavior rather than pain behavior, suggesting that TRPV1 might have broader functions in some specific pathologic conditions.

12.3.3 TRPV1 Structure Modulation

TRPV1 channel possesses highly modulable pockets for cation of calcium, magnesium, and sodium ion. A three-dimensional study revealed that TRPV1 has an ion permeation pathway via S5 and S6 and an intervening pore loop region (S5-P-S6) [3], which serves as dominant modulation sites. Based on the high allosteric coupling between upper and lower gates during activation, inflammatory agents are capable to modulate TRPV1 and such modulation is found to contribute to acute and persistent pain [23]. TRPV1 has two activated structures that have been detected by pharmacological probes (a peptide toxin, DkTx, and a small vanilloid agonist, RTX). Such diversity highlights the modulable property of TRPV1 and implies its potentials in regulating neuron activities [82]. In 2007, the high-resolution structure of the TRPV1 domain was described as cytosolic ankyrin repeat domain (ARD). This advanced finding helps better understand protein interactions folding in TRPV1 and indicates that the calcium-dependent regulation may involve competitive interactions between ATP and calmodulin at the TRPV1-ARD-binding site [83].

12.3.4 TRPV1 Phosphorylation

12.3.4.1 PKC Pathway

As the dominant downstream TRP channels for GPCRs-mediated pain and itch, the TRPV1 channel can be modulated by three major pathways. The process of TRPV1 phosphorylation is the most important pathway for the transmission of pain and itch signaling. The PKC-dependent pathway is the most common way for TRPV1 sensitization. Generally, TRPV1 phosphorylation mediated by PKC involves the Gαq/11-PLC-phosphatidylinositol (4,5) bisphosphate (PIP2)-diacylglycerol (DAG)-PKC intracellular pathway [84]. This pathway is crucial for various GPCRs such as bradykinin B2 receptor, 5-HT receptors, purinergic receptors, protease-activated receptors, and Mrgprs. Many chemokines can also directly sensitize TRPV1 and contribute to hyperalgesia during inflammation. Pro-inflammatory chemokines such as CCL3 can activate CCR1 (co-expressing with TRPV1 in more than 85% of small-diameter neurons) and induce calcium influx and PKC activation, hence are responsible for sensitization in a receptor crosstalk manner [85].

PIP2 is one of the first noticeable intracellular molecules for phosphorylation in sensory neurons. PLC is capable to hydrolyze PIP2 to produce DAG and isositol triphosphate (IP3). IP3 further promotes the release of calcium ions in cells and acts synergistically with DAG in the process of activating PKC. PKC activation phosphorylates the TRPV1 channel, hence triggers the sensitization effect from the upstream signaling. In fact, TRPV1 interaction with PIP2 and calmodulin is the classical pathway for the activity modulation within neurons. Cation permeability induced by TRPV1 agonists is dynamic and could be intensively modulated in a PKC-dependent pathway. The permeability of cation led by TRPV1 opening is time- and agonist concentration-dependent for large cations and calcium entry. This phenomenon has been evidenced in native or recombinant TRPV1 channels in rats and attributes to the cation selectivity led by different agonists that phosphorylate site Ser800 or Ser502 in a PKC-dependent manner. This property may further change the TRPV1 activity in pain sensation, as well as in neurotransmitter release or agonist-related cytotoxicity [86]. The putative binding sites for TRPV1 activity have been indicated to be Lys571 and Arg575 in the linker between the S4 and S5 of one TRPV1 subunit and Lys694, and the predicted PIP2–TRPV1 interaction region has relied on Leu777-Ser820 [87].

However, controversial results have revealed the negative regulatory role of PIP2 on the TRPV1 channel [88]. Researchers reconstituted purified TRPV1 into artificial liposomes and tested the precise effect of various phosphoinositides by introducing them into the TRPV1-expression system, respectively. Through this way, they found that membrane lipids including PIP2, PI4P, and phosphatidylinositol, can actually inhibit TRPV1-mediated sensitivity [88]. They also claimed that TRPV1 is fully functional without phosphoinositides. The ongoing exploration found that phospholipids, specifically phosphoinositides are important for heat-induced channel open status [89]. With regard to the controversial results for PIP2 involvement in TRPV1 sensitization, as the TPRV1 channel possesses multiple binding sites for activity modulation, further evidence could be applied to confirm the role of phosphoinositide in TRPV1 modulation based on the binding site regulation [90].

12.3.4.2 PKA Pathway

The sensitization and phosphorylation of TRPV1 via Gαs-activation are relied on Gαs- adenylyl cyclase (AC)-cAMP-PKA pathway. Briefly, activation of Gαs by pain and itch mediators could activate AC to generate cAMP, leading to the activation of PKA. PKA activation further sensitizes TRPV1 channels and sequentially triggers the hyperactivity of cation entry. Hence, any disturbance of this pathway will impair the overall sensitivity of TRPV1 phosphorylation. The GPCR-related Gαs-activation is very important for pain and itch mediator transmission. While Gαs-coupled receptors (such as 5-HT4 and 5-HT7) have positive regulation effects on the cAMP-dependent modulation of TRPV1, the Gαi/o-coupled receptors (e.g. μ-opioid receptor, cannabinoid receptors 1 and 2) play a negative regulation role [84].

Prostaglandin E2 is a classical TRPV1-mediated noxious stimulus that could act on Gαs. In this pathway, the downstream desensitization of TRPV1 can be achieved by altering Ser116 and Thr370 at the PKA binding site. On the contrary, pretreatment with forskolin to activate AC is ineffective to reduce TRPV1 desensitization. These results indicate the involvement of PKA-dependent reduction of desensitization of capsaicin-activated currents [91]. Nevertheless, other binding sites for PKA-dependent phosphorylation at the site of Thr144 or Ser502 have also been discovered [87].

Either opioids or cannabis is effective to reduce nociception provoked by noxious stimuli. Interestingly, both of their receptors, particularly at the postsynaptic membrane, are largely co-expressed with TRPV1 in peripheral sensory neurons. They both negatively regulate TRPV1 activity by acting on Gαi/o-coupled receptors. μ-opioid receptor (MOR)-mediated activation of Gαi/o-coupled receptors is sufficient to reduce AC activity and cAMP levels, leading to the decreased neural activity and thus anti-pain effects [92]. However, opioid withdrawal can significantly increase the cAMP level and increase the PKA activity. It has been demonstrated in transfected HEK293 cells and dissociated DRG neurons that capsaicin-induced TRPV1 activity plays a vital role in hyperalgesia resulted from opioid withdrawal [93]. Cannabis has two receptors like cannabinoid receptor subtype 1 and 2, both of which are Gαi/o-coupled receptors and are expressed in many types of neurons including TRPV1 positive ones. Cannabis has been prescribed to treat many refractory pain disorders including headache, arthritis, and post-operation pain, indicating the importance of cannabinoid receptors activation in pain relief [94]. Endogenous cannabinoids are bioactive lipids known as endocannabinoids and are derived from dietary omega-3 and omega-6 polyunsaturated fatty acids [95]. The endocannabinoid-TRPV1 axis represents an intrinsic prologetic pathway. Recently, a study identified epoNADA (epoxidation of N-arachidonic acid-dopamine) and epoNA5HT (epoxidation of *N*-arachidonic acid-serotonin), which are formed by CYP peroxygenases under inflammatory conditions, are bifunctional rheostat modulators of the endocannabinoid-TRPV1 axis. They demonstrated that epoNADA and epoNA5HT are more potent than their precursors (NADA and NA5HT respectively) in modulating TRPV1 activity. Their capability as strong antagonists is shown by suppressing intracellular calcium response and membrane currents provoked by capsaicin in sensory neurons. In addition, epoNA5HT is also a complete cannabinoid receptor subtype 1 agonist. These molecules can further effectively reduce pro-inflammatory biomarkers such as IL-6, IL-1β, and TNF-α. Collectively, these results indicate that cannabinoids are potential candidates in the development of anti-pain and anti-inflammation therapeutics [96].

12.3.4.3 CaMKII-Dependent Phosphorylation

CaMKII, refers to Ca^{2+}-calmodulin-dependent kinase II, is essential for synaptic plasticity in peripheral $TRPV1^{+}$ fibers. CaMKII phosphorylates TRPV1, thereby regulates vanilloid agonist binding to the receptor. Studies have shown that the dephosphorylation of TRPV1 by the protein phosphatase 2B, also known as calcineurin, leads to a desensitization of the receptor [97]. Moreover, point mutations in TRPV1 at two putative consensus sites (Ser502 and Thr704) that are required for CaMKII binding lead to the failure of capsaicin-stimulate currents and cause a concomitant reduction in TRPV1 phosphorylation. In a word, CaMKII and calcineurin could control the balance between activation and desensitization in nociceptors by regulating TRPV1 binding [98]. A further study also revealed an interesting phenomenon named potentiation related to CaMKII. It is shown that activation of CaMKII and extracellular signal-regulated kinase (ERK)1/2 contribute to a time-dependent potentiation of Ca^{2+} responses elicited by repeated application of capsaicin during a 40-min interval. However, pretreatment with deltamethrin to block calcineurin and tachyphylaxis can enhance the potentiation effect. Therefore, potentiation may be an important peripheral auto sensitization mechanism that needs further investigation [99].

12.3.5 Other Modulators

12.3.5.1 Protons

Acidosis is the feature of many pathological microenvironments, hence acid-sensing ion channels represent the ion channel families that detect noxious chemical stimuli. It has been validated that the proton-sensing channels named ASIC1a and ASIC3, are preferentially expressed in $TRPV1^{+}$ sensory neurons and function as nociceptive sensors [100]. In addition to binding its specific acid channels, protons could act as TRPV1 agonists by acting on the S6 domain or modulators by combining the S5 domain. Thermal hypersensitivity mainly attributes to TRPV1 sensitization by acidosis in pathological conditions [31]. By lowering pH value, TRPV1 capability responsive to heat and capsaicin is potentiated in a dose-dependent manner. When the extracellular pH decreases to 6.4, spontaneous openings of the TRPV1 channel are observed even at body temperature. The TRPV1 binding site of E600, but not E648, is responsible for this type of regulation and serves as a regulator in the acid environment rather than heat- or capsaicin-evoked responses [24]. The potent effect of protons to lower the nociceptive threshold for temperature directly drives cation entry via TRPV1 opening at body temperature and contributes to the mechanisms of thermal pain (or hypersensitivity) in the context of tissue injury. This phenomenon can explain the clinical observation in which thermal hypersensitivity is usually exhibited in the context of infection, inflammation, and ischemia situations.

12.3.5.2 Pirt

Pirt is a membrane protein that is expressed in all peripheral neural tissues. It can bind both PIP2 and TRPV1 and is a positive regulator for TRPV1 [101]. Studies have shown that Pirt-deficient mice showed impaired responsiveness to noxious heat and capsaicin and responsible currents were significantly attenuated. Gain-of-function approaches indicated that TRPV1-mediated currents are enhanced by Pirt co-expression in a heterologous expression system. To achieve the modulation mechanism, the C terminus of Pirt directly binds to TRPV1 and PIP2 to strengthen TRPV1 activity [101].

Pirt regulation for TRPV1-dependent neuropathic pain has been observed in a chronic constriction injury model, which is a neuropathic pain model widely utilized in murine. In this context, pain behavior and TRPV1 overexpression were attenuated by knocking out Pirt gene [102]. In a visceral pain model that mimics uterine contraction-induced pain, Pirt expression and capsaicin-mediated activity were enhanced, leading to distinct pain behavior featured as writhing responses. However, this pain behavior was significantly attenuated in Pirt deficient-mice, indicating that Pirt might be involved in visceral pain that is mediated by TRPV1 [103].

In addition to pain, Pirt is also involved in itch modulation. Pirt deficit attenuates cellular and behavioral responses to various pruritogens such as histamine and CQ in animals. It has been shown that both TRPV1-dependent and TRPV1-independent itch likely need Pirt, suggesting its broad function in itch transmission [104].

12.3.5.3 GABA-Autocrine Feedback

Interestingly, TRPV1 activation can also trigger GABA release from peripheral nerve endings. GABA release serves as an autocrine feedback mechanism exclusively limiting TRPV1 sensitization in the setting of pathological inflammation. This feedback effect did not affect the TRPV1 response to capsaicin under physiological status. Also, this effect is independent of canonical G protein signaling but relies on the close juxtaposition of the GABAB1 receptor subunit and TRPV1 [105]. These modulation effects are essential for signaling transmission, helping expand neuroimmune interactions, and demonstrating self-protection by autocrine feedback mechanisms.

12.3.6 Pain-to-Itch Switch

In order to distinguish pain and itch phenotype in experimental mice, researchers have identified that upon intradermal injection of aversive chemicals into the cheek, pruritogens induce mice scratching behavior using hind paws, while pain signaling induces wiping behavior with forepaws [106]. Via behavior studies, substances like histamine, 5-HT, and agonists of protease-activated receptors PAR-2 and PAR-4, have been identified as itch-inducers. In contrast, capsaicin, AITC, and bradykinin that usually elicit dose-related forelimb wiping contribute to pain behavior [107]. However, under pathological conditions like allergic contact dermatitis in which bovine adrenal medulla 8–22 (a peptide that elicits histamine-independent itch via MrgprC11)-related scratching is enhanced, bradykinin that is supposed to cause pain behavior evokes both scratching behavior and wiping behavior, indicating a phenotype named as pain-to-itch switch [108]. This phenomenon has been confirmed in a dry skin mouse model, in which the increased innervation of TRPV1$^+$ fibers in dry skin was demonstrated [81]. Although the precise mechanisms remain undetermined, this phenomenon implies TRPV1 may contribute to neural function plasticity in pathological conditions.

A research group has carried out studies to test human sensation evoked by capsaicin and histamine spicules in volunteer subjects [109, 110]. They observed that while single spicule of capsaicin, histamine, or cowhage induced dose-dependent pruritic sensations that were usually accompanied by pricking/stinging and burning, a wider and deeper application by capsaicin intradermal injection only induced pain with hyperalgesia but not itch or alloknesis [109, 110]. Although why capsaicin could induce different sensations by different application methods (punctate v.s injection) remains unknown, it provides clues to further study the phenomenon of pain-to-itch switch.

12.4 TRPV1$^+$ Neurons as the Center of Neuroimmune Interactions

While pain and itch allow the host to sense and expel environmental stimuli, the immune system provides a following protective mechanism by fighting pathogens. Emerging evidences have revealed that immunity and nociception have crosstalk particularly in the skin and mucous barriers. On one hand, cytokines derived from immune cells or even environmental materials alone could act on sensory neurons. On the other hand, sensory neurons particularly nociceptors in turn release neuropeptides and cause immunity remodeling. In the sections above, we talk about mechanisms of pain and itch that are mediated by inflammatory mediators. In this section, we will focus on how the peripheral sensory nervous system regulates immune responses via neuropeptide releasing (Fig. 12.3).

Neuropeptides are signaling peptides that are mainly made by and released from neurons. Neuropeptides can affect both neurons and non-neuronal cells and are considered to be key mediators in the communication between neurons (in particular sensory neurons) and effector cells. A number of candidate neurotransmitters/neuropeptides released from peripheral sensory neurons towards the spinal cord include GRP, glutamate, SP, BNP, and neuromedin B (NMB), etc. [44, 55, 111–115]. Meanwhile, activated nociceptors are capable of releasing neuropeptides to act on non-neural cells. These processes are particularly significant in pathologic conditions such as infectious immunity and allergic reactions.

12.4.1 Mast Cells: A Classic Neuroimmune Paradigm

The axon reflex represents a classic neuroimmune paradigm. The activation of nociceptors not only leads to the transmission of action potentials toward the central nervous system but also causes neuropeptide releasing along its fibers back to skin endings. These released peptides mainly include CGRP and SP. Both CGRP and SP have been shown to act on the vasculature and mast cells to induce vasodilatation, edema, degranulation, and consequent immune cell recruitment [116]. Vasodilatation caused by CGRP is mediated by receptors at precapillary arterioles, whereas SP induces plasma extravasation from venoles. Unlike neurokinin-1 receptor (NK1R) which is the receptor for SP in the spinal cord, recent studies have shown that MrgprB2 in mice (MRGPRX2 in humans) is the key receptor that induces SP-driven activation of mast cells [117, 118]. Therefore, mast cells appear to be the key cellular mechanism that bridges immunity and somatosensation. On the one hand, they release inflammatory mediators to promote pain and itch upon stimuli. On the other hand, they accept neuropeptide signaling from sensory nerves to further exemplify the noxious responses. Thus, the reaction of axon reflex is also called “neurogenic inflammation.”

More interestingly, a recent study uncovered that common environmental allergens that have cysteine protease activity, e.g., house dust mites, can directly activate TRPV1$^+$ neurons to release SP. Further, SP acts on MrgprB2 on mast cells to promote type 2 inflammation in the skin. These findings elucidate how environmental stimuli directly activate nociceptors and regulate inflammation via the sensory system [119].

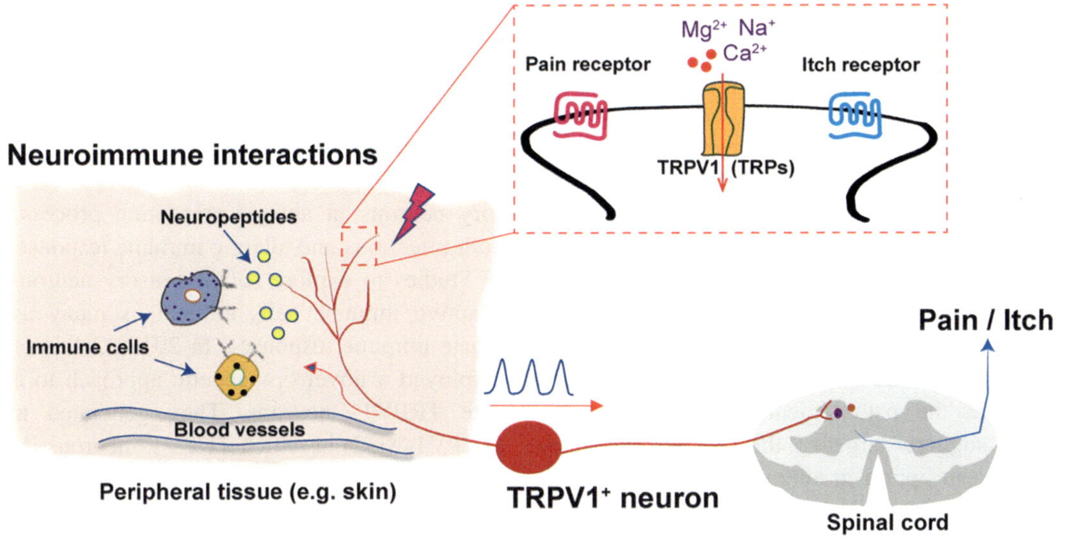

Fig. 12.3 Neuroimmune interactions mediated by TRPV1$^+$ neurons in the context of pain and itch. TRPV1$^+$ fibers in the peripheral tissues directly sense pain and itch mediators, evoking action potential to transmit the nociceptive signaling to feed into the central circuitry. On the other hand, TRPV1$^+$ peripheral axons release abundant neuropeptides to provoke neuroimmune interactions

12.4.2 Beyond Mast Cells: Other Immune Cells Regulated by Nociceptors

In addition to mast cells, more innate immune cells have been found capable to respond to neuropeptides. Neutrophils are an innate immune cell population that is well known for rapidly responding to infections, helping resolve infections, and healing the damaged tissue. A number of bacteria infections provoke host pain sensation, but the precise mechanisms are not fully understood. In 2013, Chiu et al. firstly found that bacterial infections cause pain by directly activating nociceptors [120]. Later in 2018, the Chiu group identified that TRPV1$^+$ nociceptors suppressed protective immunity against lethal *Staphylococcus aureus* pneumonia through suppression of the recruitment and surveillance of neutrophils and alteration of lung γδ T cell numbers [121]. Further, in the same year, the same research team reported that on the basis of nociceptor activation stimulated by *Streptococcus pyogenes*-released streptolysin S to produce pain, nociceptors, in turn, secrete neuropeptide CGRP to inhibit the recruitment of neutrophils and the killing of bacteria [122]. In the above studies, they employed the approach of RTX administration and TRPV1-DTR strain to demonstrate the role of TRPV1$^+$ nociceptors in the context of bacterial infections. Taken together, it appears that nociceptors and their related neuropeptide CGRP are likely negative factors to defense bacterial pathogens across barriers from the skin to the lung.

Group 2 innate lymphoid cells (ILC2) belong to the family of innate lymphocytes and are a prominent source of type 2 cytokines [123]. ILC2s are found constitutively at the skin and mucosal barriers and in the spleen, fat-associated lymphoid clusters, and lymph nodes. The significant role of ILC2 plays in type 2 immunity like AD, metabolic homeostasis, and chronic pathologies like fibrosis has been demonstrated in both patients and animal models [124]. So far, the identified neuropeptides that can regulate ILC2s include vasoactive intestinal peptide (VIP), neuromedin U (NMU), and NMB [125–129]. However, the most relevant neuropeptide released from nociceptors is VIP. VIP was first shown to stimulate gut ILC2s to release IL-5 through the VPAC2 receptor and cause eosinophil accumulation [130]. Another study conducted by Talbot et al. identified that IL-5 directly activates nociceptors to induce the secretion of VIP. Then VIP further activates resident ILC2s and effector $CD4^+$ T cells via VPAC2. This positive feedback loop formed by cytokine, nociceptors, and immune cells significantly amplifies the pathological allergic conditions, which suggests targeting the nervous system as a treatment option in allergies. NMU and NMB have been, respectively, found to promote lung ILC2 and helminth-induced ILC2 responses [127–129]. However, while the main cellular source of NMU is cholinergic neurons, the major source of NMB in the setting of helminth infection remains unclear. Notwithstanding this, these studies highlight how significant the nervous system is in innate immunity regulation and imply that nervous systems might also play a role in the link between innate immunity and adaptive immune responses.

Dendritic cells (DCs) are important antigen-presenting cells and are resident in multiple barriers. They are responsible to process and present antigens or allergens to adaptive immune cells like T cells and promote their differentiation. *Candida albicans* is the most common pathogen that causes cutaneous candidiasis. In a murine model of cutaneous candidiasis, Kashem et al. discovered that IL-23-derived from $CD301^+$ dermal DCs could augment γδ T cells to release IL-17A and inhibit cutaneous *Candida albicans* infection. Interestingly, CGRP released from $TRPV1^+$ neurons that are directly activated by *Candida albicans* is sufficient to promote IL-23 releasing from dermal DCs [131]. The role of the other neuropeptide SP released from $TRPV1^+$ neurons play in DCs and antigen-presenting processes were reported recently. In the study published by Perner et al. [132], authors identified that environmental allergens like papain and Alternaria extract were sufficient to provoke scratching behavior in mice and this phenotype was dependent on $TRPV1^+$ sensory neurons. Intriguingly, using TRPV1-DTR mouse strain, they surprisingly found that allergen-induced $CD301b^+$ DC migration required the $TRPV1^+$ neuron subset as well as SP. Collectively, these studies highlight the significance of $TRPV1^+$ sensory neurons in antigen-presenting process for both infectious and allergic immune responses.

Studies to explore somatosensory neurons in adaptive immune cells are not as many as in innate immune responses. In 2019, Cohen et al. employed a novel optogenetic approach to activate $TRPV1^+$ neurons. They generated mice (TRPV1-Ai32) in which $TRPV1^+$ neurons in the skin could express the light-gated cation channel channelrhodopsin-2 and be activated by cutaneous light stimulation [133]. Strikingly, this optogenetic stimulation is capable of initiating skin inflammation which is predominated by Th17 (type 3) immune responses. The biological effect of such skin immunity was further shown to enhance host local defense to *Candida albicans* and *Staphylococcus aureus*. More importantly, this defense effect can be propagated to adjacent, unstimulated skin areas through a nerve reflex arc. Thus, although the concept of "neurogenic inflammation" was established a century ago, this study demonstrated that activation of sensory neurons alone is sufficient to results in inflammation and might be a future target to augment local immune responses in fighting pathogens.

12.5 Therapeutic Strategy and Perspectives

12.5.1 Agonist

The basic concept for analgesic effects of capsaicin refers to the selective excitation and subsequent desensitization of $TRPV1^+$ fibers (pain and itch sensing fibers). This is widely used in relieving pain and itch related to many sensory disorders. This acute desensitization occurs rapidly within 20 s upon activation and is dependent on the calcium influx, which represents a feedback mechanism protecting nociceptors from toxic calcium overload [87]. This effect of "activation triggers sequential desensitization" has been repeatedly verified in capsaicin topical application to alleviate pain and itch. In pain clinic, 8% capsaicin topical patch has been used in the painful disorders, such as diabetic neuropathy, chemotherapy-induced peripheral neuropathy, and peripheral neuropathy associated with hypereosinophilic syndrome [134–138]. Furthermore, direct capsaicin injection into the joint has been used to relieve osteoarthritis-related pain [139]. Unlike "activation triggers sequential desensitization," tachyphylaxis is another type of desensitization and refers to the reduced responses due to repeated activation. Further studies are needed to verify its anti-itch or anti-pain effects in clinical practice.

Although topical capsaicin has been approved by Food and Drug Administration (FDA) for neuropathic pain treatment, its advantages are significant as it not only desensitizes the TRPV1 channel-related pain sensation, but also disrupts TRPV1$^+$ peripheral terminals to abolish the potential nociception transmission via other co-expressed channels or receptors. However, one major concern is the disruption of physiological functions mediated by neuropeptides released TRPV1$^+$ peripheral nerves. Disrupting or desensitizing the TRPV1$^+$ fibers in a period of time will inevitably lead to the absence of neuropeptide releasing in various noxious conditions such as axon reflex. Furthermore, long-term use of the capsaicin patch might null the somatosensation of the treated skin, which might attenuate the response upon harmful stimulation, such as noxious heat. Finally, activation of TRPV1 in skin lesions causes painful sensation before desensitization occurs, which is not acceptable to some fatigue patients or children.

12.5.2 Antagonist

TRPV1 antagonists are expected to treat pain and itch sensation by inhibiting receptor potential and preventing convey of nociceptive signaling from peripheral sites to the central nervous system. A series of novel drugs have been tested in clinical trials. GlaxoSmithKline SB-705498 is a small molecular and a potent, selective, orally bioavailable antagonist for TRPV1 [140]. The anti-pain effect of SB-705498 has been tested in a cohort of 19 healthy volunteers who had pain symptom that was evoked by noxious heat, capsaicin or ultraviolet radiation B (UVB) irradiation [141]. The results appear positive as SB-705498 was revealed safe and well-tolerated and exhibited pharmacodynamic activities in alleviating pain and hyperalgesia.

However, some tested antagonists were found to potentiate hyperthermia or hypothermia. Given hyperthermia/hypothermia as the most common side effects, the pharmaceutical value of TRPV1 agonists was questioned. Thermoregulatory effects of TRPV1 antagonists in humans have been intensively studied. Controversial views are emerging as the tested compounds in vitro may have different binding potency in vivo and thus, the safety and efficacy of new TRPV1 antagonists should be reexamined. Accordingly, clinical trials revealed that polymodal TRPV1 antagonists (ABT-102, AZD1386, and V116517) cause an increase of body temperature, which was not shown in patients treated with mode-selective blocker NEO6860 [142]. In a Phase I clinical trial in which AMG517 is examined, subjects exhibited marked dose-dependent hyperthermia that might be related to the vasoconstriction and increasing thermogenesis [143]. Another potent and selective TRPV1 channel antagonist, JNJ-38893777, was tested in a single-center, double-blind, placebo-controlled, sequential group, single-ascending-dose phase 1 study. Results turned out promising as increases in body temperature or changes in Fridericia-corrected QT interval (QTcF) were not observed [144].

While systemic application of TRPV1 antagonist may face unexpected thermogenesis problems, topical administration could be a safer strategy. To treat itch-related skin problem, TRPV1-inhibitor 4-t-butylcyclohexanol based skin care cream has been tested to treat perioral dermatitis [145]. Perioral dermatitis is a skin problem and characterized by impaired skin barrier function that results in redness, dryness, burning, and pruritus. After TRPV1-inhibitor treatment, skin barrier function was surprisingly restored, featured as the decreased transepidermal water loss values and increased stratum corneum hydration [145]. These results support that topical application of TRPV1 antagonists is better tolerated.

More profound TRPV1 antagonists in pain- and itch-associated phenotypes have been tested in laboratory animals. For example, pharmacological blockade of TRPV1 using AMG9810 (a potent and selective TRPV1 antagonist [146, 147]) significantly decreased calcium responses in sensitized neurons, suggesting pharmacological effects in hyperalgesia conditions. Indeed, in an allergic conjunctivitis mouse model, antagonizing TRPV1 by AMG9810 could effectively disrupt histamine-dependent itch behavior [148], indicating the topical application of AMG9810 might be effective in dealing with ocular discomforts.

The most common concern about the overall antagonization of the TRPV1 channel is that the normal physiological function is likely disrupted. Thus, more studies are still needed to explore how to reserve TRPV1 physiological function when targeting pathologic pain and itch [149].

12.5.3 TRPV1 Activity-Dependent Silencing by QX-314

In addition to molecular antagonists, cell-specific design and synthesis may open a new avenue to save the protective role of pain and itch. The newly proof-of-concept to silence a specific activated nociceptive population has become available when QX-314 was introduced into the field. QX-314 is a charged, membrane-impermeant lidocaine derivative and can only penetrate into the neurons via a hydrophilic pore in the cell membrane. After entering cells, QX-314 sequentially blocks sodium channels to inhibit the overall neural activity and thus functions as local anesthetics. This specific characteristic enables the selective effect in which nociceptors are silenced via an activity-dependent manner. In 2007, Binshtok, Woolf, and colleagues generated a novel strategy to inhibit pain sensation [150]. They simultaneously applied QX-314 and capsaicin into rat hind paws or sciatic nerve and found that pain sensitivity was suppressed and the effect lasted for greater than 2 h without motor or tactile function impact [150]. This strategy opens a door to inhibit pain-responsible firing that is mediated by a broad range of factors. Since the opening of TRPV1 or/and TRPA1 channels is widely exhibited upon stimulation of pain mediators, which provides the critical condition of QX-314 entry. Besides for inflammatory mediators, pain induced by bacterial infections can also be blocked by QX-314. Three classes of PFTs-alpha-hemolysin (Hla), phenol-soluble modulins (PSMs), and the leukocidin HlgAB associated with methicillin-resistant *Staphylococcus aureus* (MRSA) have been identified to induce pain during the infection. Strikingly, QX-314 application revealed immediate and long-lasting blockade of pain sensation in a murine model infected by MRSA. Moreover, this approach is even more effective than analgesic reagents like lidocaine or ibuprofen [151].

Similarly, the effect of QX-314 was also confirmed in itch-related study. By applying QX-314 and histamine or chloroquine together, itch behavior was significantly inhibited [152]. Importantly, blocking itch-transmitting fibers did not reduce pain-associated behavior. Overall, this strategy not only helps identify the distinct populations for pain and itch mediators, but also provides a clinical anti-pruritic therapeutic approach for both histaminergic and non-histaminergic pruritus [152]. A specific case is the distinguishment of ocular pain and itch sensation. Although ocular discomforts can present as pain or itch, whether these signals are transmitted by the same or different sensory neurons remains not clear. Via tracing approaches, we found that pain and itch sensing sensory fibers have different anatomic territories. To be specific, while ocular itch is mediated by a subset of conjunctival-selective sensory fibers marked by MrgprA3, pain-sensing fibers were only observed in corneal innervations. By selectively silencing MrgprA3-expressing conjunctiva sensory fibers by using QX-314 and CQ (the agonist of MrgprA3), we observed that the itching behavior was markedly reduced in both pruritogen-stimulated and allergic-related conditions [153]. Taken together, these results indicate that TRPV1 activity-dependent silencing by QX-314 may have great significance in research and development for anti-pruritic medications.

12.6 Summary

The broad integration of physiological and pathophysiological of TRPV1 has been well studied and documented in pain- and itch-related fields. Over time, advanced research progresses emerge and indicate TRPV1 as a potential pharmaceutical target for treating pathologic pain and itch. The resolution of TRPV1 structure in the steady-state and the observation and mechanisms of dynamic change in TRPV1 upon stimulation makes it possible to modulate TRPV1 in a variety of conditions [3, 82]. Of note, although adjustment of expression level and the channel activity have revealed prosperity in pain and itch relieving, TRPV1 antagonists in clinical trials have encountered difficulties due to their adverse effects such as hyperthermia or hypothermia [142, 143]. Notwithstanding this, how to modulate TRPV1 activity in specific tissues and avoid affecting its basic physiological function remains an active research field to achieve the anti-pain and anti-itch effects. Excitingly, the novel concept of TRPV1 activity-dependent silencing by QX-314 has provided a promising approach to block the selective pain sensation. Over time, the consistent studies of TRPV1 in pain and itch will lead a new era to solve debilitating sensory disorders.

References

1. Venkatachalam K, Montell C (2007) TRP channels. Annu Rev Biochem 76:387–417
2. Caterina MJ, Schumacher MA, Tominaga M, Rosen TA, Levine JD, Julius D (1997) The capsaicin receptor: a heat-activated ion channel in the pain pathway. Nature 389:816–824
3. Liao M, Cao E, Julius D, Cheng Y (2013) Structure of the TRPV1 ion channel determined by electron cryo-microscopy. Nature 504:107–112
4. Dhaka A, Uzzell V, Dubin AE, Mathur J, Petrus M, Bandell M, Patapoutian A (2009) TRPV1 is activated by both acidic and basic pH. J Neurosci 29:153–158
5. Kobayashi K, Fukuoka T, Obata K, Yamanaka H, Dai Y, Tokunaga A, Noguchi K (2005) Distinct expression of TRPM8, TRPA1, and TRPV1 mRNAs in rat primary afferent neurons with adelta/c-fibers and colocalization with trk receptors. J Comp Neurol 493:596–606
6. Kuruvilla M, Kalangara J, Lee FEE (2019) Neuropathic pain and itch mechanisms underlying allergic conjunctivitis. J Investig Allergol Clin Immunol 29: 349–356
7. Jara-Oseguera A, Simon SA, Rosenbaum T (2008) TRPV1: on the road to pain relief. Curr Mol Pharmacol 1:255–269
8. McKemy DD (2011) A spicy family tree: TRPV1 and its thermoceptive and nociceptive lineage. EMBO J 30:453–455
9. Mishra SK, Tisel SM, Orestes P, Bhangoo SK, Hoon MA (2011) TRPV1-lineage neurons are required for thermal sensation. EMBO J 30:582–593
10. Kim YS, Chu Y, Han L, Li M, Li Z, LaVinka PC, Sun S, Tang Z, Park K, Caterina MJ, Ren K, Dubner R, Wei F, Dong X (2014) Central terminal sensitization of TRPV1 by descending serotonergic facilitation modulates chronic pain. Neuron 81:873–887
11. Caterina MJ, Leffler A, Malmberg AB, Martin WJ, Trafton J, Petersen-Zeitz KR, Koltzenburg M, Basbaum AI, Julius D (2000) Impaired nociception and pain sensation in mice lacking the capsaicin receptor. Science 288:306–313
12. Raja SN, Carr DB, Cohen M, Finnerup NB, Flor H, Gibson S, Keefe FJ, Mogil JS, Ringkamp M, Sluka KA, Song XJ, Stevens B, Sullivan MD, Tutelman PR, Ushida T, Vader K (2020) The revised International Association for the Study of Pain definition of pain: concepts, challenges, and compromises. Pain 161:1976–1982
13. Ikoma A, Steinhoff M, Stander S, Yosipovitch G, Schmelz M (2006) The neurobiology of itch. Nat Rev Neurosci 7:535–547
14. Saloman JL, Chung MK, Ro JY (2013) P2X(3) and TRPV1 functionally interact and mediate sensitization of trigeminal sensory neurons. Neuroscience 232:226–238
15. Barabas ME, Stucky CL (2013) TRPV1, but not TRPA1, in primary sensory neurons contributes to cutaneous incision-mediated hypersensitivity. Mol Pain 9:9
16. Li CL, Li KC, Wu D, Chen Y, Luo H, Zhao JR, Wang SS, Sun MM, Lu YJ, Zhong YQ, Hu XY, Hou R, Zhou BB, Bao L, Xiao HS, Zhang X (2016) Somatosensory neuron types identified by high-coverage single-cell RNA-sequencing and functional heterogeneity. Cell Res 26:967

17. Li C, Wang S, Chen Y, Zhang X (2018) Somatosensory neuron typing with high-coverage single-cell RNA sequencing and functional analysis. Neurosci Bull 34:200–207
18. Usoskin D, Furlan A, Islam S, Abdo H, Lonnerberg P, Lou D, Hjerling-Leffler J, Haeggstrom J, Kharchenko O, Kharchenko PV, Linnarsson S, Ernfors P (2015) Unbiased classification of sensory neuron types by large-scale single-cell RNA sequencing. Nat Neurosci 18:145–153
19. Zeisel A, Hochgerner H, Lonnerberg P, Johnsson A, Memic F, van der Zwan J, Haring M, Braun E, Borm LE, La Manno G, Codeluppi S, Furlan A, Lee K, Skene N, Harris KD, Hjerling-Leffler J, Arenas E, Ernfors P, Marklund U, Linnarsson S (2018) Molecular architecture of the mouse nervous system. Cell 174:999–1014.e22
20. Treede RD, Rief W, Barke A, Aziz Q, Bennett MI, Benoliel R, Cohen M, Evers S, Finnerup NB, First MB, Giamberardino MA, Kaasa S, Kosek E, Lavand'homme P, Nicholas M, Perrot S, Scholz J, Schug S, Smith BH, Svensson P, Vlaeyen JWS, Wang SJ (2015) A classification of chronic pain for ICD-11. Pain 156:1003–1007
21. Treede RD, Rief W, Barke A, Aziz Q, Bennett MI, Benoliel R, Cohen M, Evers S, Finnerup NB, First MB, Giamberardino MA, Kaasa S, Korwisi B, Kosek E, Lavand'homme P, Nicholas M, Perrot S, Scholz J, Schug S, Smith BH, Svensson P, Vlaeyen JWS, Wang SJ (2019) Chronic pain as a symptom or a disease: the IASP Classification of Chronic Pain for the International Classification of Diseases (ICD-11). Pain 160:19–27
22. Nugraha B, Gutenbrunner C, Barke A, Karst M, Schiller J, Schafer P, Falter S, Korwisi B, Rief W, Treede RD (2019) Pain ITftCoC: the IASP classification of chronic pain for ICD-11: functioning properties of chronic pain. Pain 160:88–94
23. Basbaum AI, Bautista DM, Scherrer G, Julius D (2009) Cellular and molecular mechanisms of pain. Cell 139:267–284
24. Jordt SE, Tominaga M, Julius D (2000) Acid potentiation of the capsaicin receptor determined by a key extracellular site. Proc Natl Acad Sci U S A 97:8134–8139
25. Siemens J, Zhou S, Piskorowski R, Nikai T, Lumpkin EA, Basbaum AI, King D, Julius D (2006) Spider toxins activate the capsaicin receptor to produce inflammatory pain. Nature 444:208–212
26. Bohlen CJ, Priel A, Zhou S, King D, Siemens J, Julius D (2010) A bivalent tarantula toxin activates the capsaicin receptor, TRPV1, by targeting the outer pore domain. Cell 141:834–845
27. Caterina MJ, Julius D (2001) The vanilloid receptor: a molecular gateway to the pain pathway. Annu Rev Neurosci 24:487–517
28. Pogatzki-Zahn EM, Shimizu I, Caterina M, Raja SN (2005) Heat hyperalgesia after incision requires TRPV1 and is distinct from pure inflammatory pain. Pain 115:296–307
29. Chen SR, Pan HL (2006) Loss of TRPV1-expressing sensory neurons reduces spinal mu opioid receptors but paradoxically potentiates opioid analgesia. J Neurophysiol 95:3086–3096
30. Tang HB, Inoue A, Oshita K, Nakata Y (2004) Sensitization of vanilloid receptor 1 induced by bradykinin via the activation of second messenger signaling cascades in rat primary afferent neurons. Eur J Pharmacol 498:37–43
31. Abdelhamid RE, Sluka KA (2015) ASICs mediate pain and inflammation in musculoskeletal diseases. Physiology (Bethesda) 30:449–459
32. Bautista DM, Jordt SE, Nikai T, Tsuruda PR, Read AJ, Poblete J, Yamoah EN, Basbaum AI, Julius D (2006) TRPA1 mediates the inflammatory actions of environmental irritants and proalgesic agents. Cell 124:1269–1282
33. Bautista DM, Movahed P, Hinman A, Axelsson HE, Sterner O, Hogestatt ED, Julius D, Jordt SE, Zygmunt PM (2005) Pungent products from garlic activate the sensory ion channel TRPA1. Proc Natl Acad Sci U S A 102:12248–12252
34. Macpherson LJ, Dubin AE, Evans MJ, Marr F, Schultz PG, Cravatt BF, Patapoutian A (2007) Noxious compounds activate TRPA1 ion channels through covalent modification of cysteines. Nature 445:541–545
35. Trevisani M, Siemens J, Materazzi S, Bautista DM, Nassini R, Campi B, Imamachi N, Andre E, Patacchini R, Cottrell GS, Gatti R, Basbaum AI, Bunnett NW, Julius D, Geppetti P (2007) 4-Hydroxynonenal, an endogenous aldehyde, causes pain and neurogenic inflammation through activation of the irritant receptor TRPA1. Proc Natl Acad Sci U S A 104:13519–13524
36. Taylor-Clark TE, McAlexander MA, Nassenstein C, Sheardown SA, Wilson S, Thornton J, Carr MJ, Undem BJ (2008) Relative contributions of TRPA1 and TRPV1 channels in the activation of vagal bronchopulmonary C-fibres by the endogenous autacoid 4-oxononenal. J Physiol 586:3447–3459
37. Bautista DM, Pellegrino M, Tsunozaki M (2013) TRPA1: a gatekeeper for inflammation. Annu Rev Physiol 75:181–200
38. Taylor-Clark TE, Undem BJ, Macglashan DW Jr, Ghatta S, Carr MJ, McAlexander MA (2008) Prostaglandin-induced activation of nociceptive neurons via direct interaction with transient receptor potential A1 (TRPA1). Mol Pharmacol 73:274–281
39. McMahon SB, Wood JN (2006) Increasingly irritable and close to tears: TRPA1 in inflammatory pain. Cell 124:1123–1125
40. Akopian AN, Ruparel NB, Jeske NA, Hargreaves KM (2007) Transient receptor potential TRPA1 channel desensitization in sensory neurons is agonist dependent and regulated by TRPV1-directed internalization. J Physiol 583:175–193

41. Weng HJ, Patel KN, Jeske NA, Bierbower SM, Zou W, Tiwari V, Zheng Q, Tang Z, Mo GC, Wang Y, Geng Y, Zhang J, Guan Y, Akopian AN, Dong X (2015) Tmem100 is a regulator of TRPA1-TRPV1 complex and contributes to persistent pain. Neuron 85:833–846
42. Moon EH, Kim MJ, Ko KS, Kim YS, Seo J, Oh SP, Lee YJ (2010) Generation of mice with a conditional and reporter allele for Tmem100. Genesis 48:673–678
43. Weyer AD, Stucky CL (2015) Loosening pain's grip by tightening TRPV1-TRPA1 interactions. Neuron 85:661–663
44. Sun YG, Chen ZF (2007) A gastrin-releasing peptide receptor mediates the itch sensation in the spinal cord. Nature 448:700–703
45. Liu Q, Tang Z, Surdenikova L, Kim S, Patel KN, Kim A, Ru F, Guan Y, Weng HJ, Geng Y, Undem BJ, Kollarik M, Chen ZF, Anderson DJ, Dong X (2009) Sensory neuron-specific GPCR Mrgprs are itch receptors mediating chloroquine-induced pruritus. Cell 139:1353–1365
46. Dong X, Han S, Zylka MJ, Simon MI, Anderson DJ (2001) A diverse family of GPCRs expressed in specific subsets of nociceptive sensory neurons. Cell 106:619–632
47. Han L, Ma C, Liu Q, Weng HJ, Cui Y, Tang Z, Kim Y, Nie H, Qu L, Patel KN, Li Z, McNeil B, He S, Guan Y, Xiao B, Lamotte RH, Dong X (2013) A subpopulation of nociceptors specifically linked to itch. Nat Neurosci 16:174–182
48. Liu Q, Sikand P, Ma C, Tang Z, Han L, Li Z, Sun S, LaMotte RH, Dong X (2012) Mechanisms of itch evoked by β-alanine. J Neurosci 32:14532–14537
49. Mishra SK, Hoon MA (2010) Ablation of TrpV1 neurons reveals their selective role in thermal pain sensation. Mol Cell Neurosci 43:157–163
50. Cevikbas F, Wang X, Akiyama T, Kempkes C, Savinko T, Antal A, Kukova G, Buhl T, Ikoma A, Buddenkotte J, Soumelis V, Feld M, Alenius H, Dillon SR, Carstens E, Homey B, Basbaum A, Steinhoff M (2014) A sensory neuron-expressed IL-31 receptor mediates T helper cell-dependent itch: involvement of TRPV1 and TRPA1. J Allergy Clin Immunol 133:448–460
51. Meng J, Moriyama M, Feld M, Buddenkotte J, Buhl T, Szollosi A, Zhang J, Miller P, Ghetti A, Fischer M, Reeh PW, Shan C, Wang J, Steinhoff M (2018) New mechanism underlying IL-31-induced atopic dermatitis. J Allergy Clin Immunol 141:1677–1689.e8
52. Wang F, Trier AM, Li F, Kim S, Chen Z, Chai JN, Mack MR, Morrison SA, Hamilton JD, Baek J, Yang TB, Ver Heul AM, Xu AZ, Xie Z, Dong X, Kubo M, Hu H, Hsieh CS, Dong X, Liu Q, Margolis DJ, Ardeleanu M, Miller MJ, Kim BS (2021) A basophil-neuronal axis promotes itch. Cell 184:422–440.e17
53. Dimitriadou V, Rouleau A, Dam Trung Tuong M, Newlands GJ, Miller HR, Luffau G, Schwartz JC, Garbarg M (1994) Functional relationship between mast cells and C-sensitive nerve fibres evidenced by histamine H3-receptor modulation in rat lung and spleen. Clin Sci (Lond) 87:151–163
54. Shim WS, Tak MH, Lee MH, Kim M, Kim M, Koo JY, Lee CH, Kim M, Oh U (2007) TRPV1 mediates histamine-induced itching via the activation of phospholipase A2 and 12-lipoxygenase. J Neurosci 27:2331–2337
55. Mishra SK, Hoon MA (2013) The cells and circuitry for itch responses in mice. Science 340:968–971
56. Campion M, Smith L, Gatault S, Metais C, Buddenkotte J, Steinhoff M (2019) Interleukin-4 and interleukin-13 evoke scratching behaviour in mice. Exp Dermatol 28:1501–1504
57. Oetjen LK, Mack MR, Feng J, Whelan TM, Niu H, Guo CJ, Chen S, Trier AM, Xu AZ, Tripathi SV, Luo J, Gao X, Yang L, Hamilton SL, Wang PL, Brestoff JR, Council ML, Brasington R, Schaffer A, Brombacher F, Hsieh CS, Gereau RW, Miller MJ, Chen ZF, Hu H, Davidson S, Liu Q, Kim BS (2017) Sensory neurons co-opt classical immune signaling pathways to mediate chronic itch. Cell 171:217–228.e13
58. Stanger R, Rivera-Oyola R, Lebwohl M (2020) Dupilumab as a treatment for generalized idiopathic pruritus: a report of two cases. Br J Dermatol 182:1494–1495
59. Zhai LL, Savage KT, Qiu CC, Jin A, Valdes-Rodriguez R, Mollanazar NK (2019) Chronic pruritus responding to dupilumab-a case series. Medicines (Basel) 6(3):72
60. Calugareanu A, Jachiet M, Tauber M, Nosbaum A, Aubin F, Misery L, Droitcourt C, Barbarot S, Debarbieux S, Saussine A, Bagot M, de Masson A, Seneschal J, Staumont-Salle D, Bouaziz JD, French Group of Research and Study in Atopic Dermatitis from the French Society of Dermatology (2020) Effectiveness and safety of dupilumab for the treatment of prurigo nodularis in a French multicenter adult cohort of 16 patients. J Eur Acad Dermatol Venereol 34:e74–e76
61. Giura MT, Viola R, Fierro MT, Ribero S, Ortoncelli M (2020) Efficacy of dupilumab in prurigo nodularis in elderly patient. Dermatol Ther 33:e13201
62. Napolitano M, Fabbrocini G, Scalvenzi M, Nistico SP, Dastoli S, Patruno C (2020) Effectiveness of dupilumab for the treatment of generalized prurigo nodularis phenotype of adult atopic dermatitis. Dermatitis 31:81–84
63. Tanis R, Ferenczi K, Payette M (2019) Dupilumab treatment for prurigo nodularis and pruritis. J Drugs Dermatol 18:940–942
64. Jeon J, Wang F, Badic A, Kim BS (2021) Treatment of patients with chronic pruritus of unknown origin with dupilumab. J Dermatolog Treat:1–4

65. Imamachi N, Park GH, Lee H, Anderson DJ, Simon MI, Basbaum AI, Han SK (2009) TRPV1-expressing primary afferents generate behavioral responses to pruritogens via multiple mechanisms. Proc Natl Acad Sci U S A 106:11330–11335
66. Kim BM, Lee SH, Shim WS, Oh U (2004) Histamine-induced Ca(2+) influx via the PLA(2)/lipoxygenase/TRPV1 pathway in rat sensory neurons. Neurosci Lett 361:159–162
67. Han SK, Mancino V, Simon MI (2006) Phospholipase Cbeta 3 mediates the scratching response activated by the histamine H1 receptor on C-fiber nociceptive neurons. Neuron 52:691–703
68. Jian T, Yang N, Yang Y, Zhu C, Yuan X, Yu G, Wang C, Wang Z, Shi H, Tang M, He Q, Lan L, Wu G, Tang Z (2016) TRPV1 and PLC participate in histamine H4 receptor-induced itch. Neural Plast 2016:1682972
69. Villarino AV, Kanno Y, O'Shea JJ (2017) Mechanisms and consequences of Jak-STAT signaling in the immune system. Nat Immunol 18:374–384
70. Wang F, Morris C, Bodet ND, Kim BS (2019) Treatment of refractory chronic pruritus of unknown origin with tofacitinib in patients with rheumatoid arthritis. JAMA Dermatol 155(12):1426–1428
71. Salmon JA, Higgs GA (1987) Prostaglandins and leukotrienes as inflammatory mediators. Br Med Bull 43:285–296
72. O'Byrne PM, Israel E, Drazen JM (1997) Antileukotrienes in the treatment of asthma. Ann Intern Med 127:472–480
73. Luster AD, Tager AM (2004) T-cell trafficking in asthma: lipid mediators grease the way. Nat Rev Immunol 4:711–724
74. Andoh T, Kuraishi Y (1998) Intradermal leukotriene B4, but not prostaglandin E2, induces itch-associated responses in mice. Eur J Pharmacol 353:93–96
75. Fernandes ES, Vong CT, Quek S, Cheong J, Awal S, Gentry C, Aubdool AA, Liang L, Bodkin JV, Bevan S, Heads R, Brain SD (2013) Superoxide generation and leukocyte accumulation: key elements in the mediation of leukotriene B(4)-induced itch by transient receptor potential ankyrin 1 and transient receptor potential vanilloid 1. FASEB J 27:1664–1673
76. Solinski HJ, Kriegbaum MC, Tseng PY, Earnest TW, Gu X, Barik A, Chesler AT, Hoon MA (2019) Nppb neurons are sensors of mast cell-induced itch. Cell Rep 26:3561–3573.e4
77. Voisin T, Perner C, Messou MA, Shiers S, Ualiyeva S, Kanaoka Y, Price TJ, Sokol CL, Bankova LG, Austen KF, Chiu IM (2021) The CysLT2R receptor mediates leukotriene C4-driven acute and chronic itch. Proc Natl Acad Sci U S A 118(13):e2022087118
78. Ji RR, Samad TA, Jin SX, Schmoll R, Woolf CJ (2002) p38 MAPK activation by NGF in primary sensory neurons after inflammation increases TRPV1 levels and maintains heat hyperalgesia. Neuron 36:57–68
79. Cheng JK, Ji RR (2008) Intracellular signaling in primary sensory neurons and persistent pain. Neurochem Res 33:1970–1978
80. Li F, Yang W, Jiang H, Guo C, Huang AJW, Hu H, Liu Q (2019) TRPV1 activity and substance P release are required for corneal cold nociception. Nat Commun 10:5678
81. Yu G, Yang N, Li F, Chen M, Guo CJ, Wang C, Hu D, Yang Y, Zhu C, Wang Z, Shi H, Gegen T, Tang M, He Q, Liu Q, Tang Z (2016) Enhanced itch elicited by capsaicin in a chronic itch model. Mol Pain 12:1744806916645349
82. Cao E, Liao M, Cheng Y, Julius D (2013) TRPV1 structures in distinct conformations reveal activation mechanisms. Nature 504:113–118
83. Lishko PV, Procko E, Jin X, Phelps CB, Gaudet R (2007) The ankyrin repeats of TRPV1 bind multiple ligands and modulate channel sensitivity. Neuron 54:905–918
84. Salzer I, Ray S, Schicker K, Boehm S (2019) Nociceptor signalling through ion channel regulation via GPCRs. Int J Mol Sci 20(10):2488
85. Zhang N, Inan S, Cowan A, Sun R, Wang JM, Rogers TJ, Caterina M, Oppenheim JJ (2005) A proinflammatory chemokine, CCL3, sensitizes the heat- and capsaicin-gated ion channel TRPV1. Proc Natl Acad Sci U S A 102:4536–4541
86. Chung MK, Guler AD, Caterina MJ (2008) TRPV1 shows dynamic ionic selectivity during agonist stimulation. Nat Neurosci 11:555–564
87. Touska F, Marsakova L, Teisinger J, Vlachova V (2011) A "cute" desensitization of TRPV1. Curr Pharm Biotechnol 12:122–129
88. Cao E, Cordero-Morales JF, Liu B, Qin F, Julius D (2013) TRPV1 channels are intrinsically heat sensitive and negatively regulated by phosphoinositide lipids. Neuron 77:667–679
89. Sun X, Zakharian E (2015) Regulation of the temperature-dependent activation of transient receptor potential vanilloid 1 (TRPV1) by phospholipids in planar lipid bilayers. J Biol Chem 290:4741–4747
90. Xiao R, Xu XZS (2021) Temperature sensation: from molecular thermosensors to neural circuits and coding principles. Annu Rev Physiol 83:205–230
91. Mohapatra DP, Nau C (2003) Desensitization of capsaicin-activated currents in the vanilloid receptor TRPV1 is decreased by the cyclic AMP-dependent protein kinase pathway. J Biol Chem 278:50080–50090
92. Vetter I, Wyse BD, Monteith GR, Roberts-Thomson SJ, Cabot PJ (2006) The mu opioid agonist morphine modulates potentiation of capsaicin-evoked TRPV1 responses through a cyclic AMP-dependent protein kinase A pathway. Mol Pain 2:22
93. Spahn V, Fischer O, Endres-Becker J, Schafer M, Stein C, Zollner C (2013) Opioid withdrawal increases transient receptor potential vanilloid

1 activity in a protein kinase A-dependent manner. Pain 154:598–608
94. Baron EP, Lucas P, Eades J, Hogue O (2018) Patterns of medicinal cannabis use, strain analysis, and substitution effect among patients with migraine, headache, arthritis, and chronic pain in a medicinal cannabis cohort. J Headache Pain 19:37
95. Freitas HR, Isaac AR, Malcher-Lopes R, Diaz BL, Trevenzoli IH, De Melo Reis RA (2018) Polyunsaturated fatty acids and endocannabinoids in health and disease. Nutr Neurosci 21:695–714
96. Arnold WR, Carnevale LN, Xie Z, Baylon JL, Tajkhorshid E, Hu H, Das A (2021) Anti-inflammatory dopamine- and serotonin-based endocannabinoid epoxides reciprocally regulate cannabinoid receptors and the TRPV1 channel. Nat Commun 12:926
97. Por ED, Samelson BK, Belugin S, Akopian AN, Scott JD, Jeske NA (2010) PP2B/calcineurin-mediated desensitization of TRPV1 does not require AKAP150. Biochem J 432:549–556
98. Jung J, Shin JS, Lee SY, Hwang SW, Koo J, Cho H, Oh U (2004) Phosphorylation of vanilloid receptor 1 by Ca^{2+}/calmodulin-dependent kinase II regulates its vanilloid binding. J Biol Chem 279:7048–7054
99. Zhang X, Daugherty SL, de Groat WC (2011) Activation of CaMKII and ERK1/2 contributes to the time-dependent potentiation of Ca^{2+} response elicited by repeated application of capsaicin in rat DRG neurons. Am J Physiol Regul Integr Comp Physiol 300:R644–R654
100. Ugawa S, Ueda T, Yamamura H, Shimada S (2005) In situ hybridization evidence for the coexistence of ASIC and TRPV1 within rat single sensory neurons. Brain Res Mol Brain Res 136:125–133
101. Kim AY, Tang Z, Liu Q, Patel KN, Maag D, Geng Y, Dong X (2008) Pirt, a phosphoinositide-binding protein, functions as a regulatory subunit of TRPV1. Cell 133:475–485
102. Wang C, Gu L, Ruan Y, Gegen T, Yu L, Zhu C, Yang Y, Zhou Y, Yu G, Tang Z (2018) Pirt together with TRPV1 is involved in the regulation of neuropathic pain. Neural Plast 2018:4861491
103. Wang C, Wang Z, Yang Y, Zhu C, Wu G, Yu G, Jian T, Yang N, Shi H, Tang M, He Q, Lan L, Liu Q, Guan Y, Dong X, Duan J, Tang Z (2015) Pirt contributes to uterine contraction-induced pain in mice. Mol Pain 11:57
104. Patel KN, Liu Q, Meeker S, Undem BJ, Dong X (2011) Pirt, a TRPV1 modulator, is required for histamine-dependent and -independent itch. PLoS One 6:e20559
105. Hanack C, Moroni M, Lima WC, Wende H, Kirchner M, Adelfinger L, Schrenk-Siemens K, Tappe-Theodor A, Wetzel C, Kuich PH, Gassmann M, Roggenkamp D, Bettler B, Lewin GR, Selbach M, Siemens J (2015) GABA blocks pathological but not acute TRPV1 pain signals. Cell 160:759–770
106. Shimada SG, LaMotte RH (2008) Behavioral differentiation between itch and pain in mouse. Pain 139: 681–687
107. Akiyama T, Carstens MI, Carstens E (2010) Differential itch- and pain-related behavioral responses and micro-opoid modulation in mice. Acta Derm Venereol 90:575–581
108. Fu K, Qu L, Shimada SG, Nie H, LaMotte RH (2014) Enhanced scratching elicited by a pruritogen and an algogen in a mouse model of contact hypersensitivity. Neurosci Lett 579:190–194
109. Sikand P, Shimada SG, Green BG, LaMotte RH (2009) Similar itch and nociceptive sensations evoked by punctate cutaneous application of capsaicin, histamine and cowhage. Pain 144:66–75
110. Sikand P, Shimada SG, Green BG, LaMotte RH (2011) Sensory responses to injection and punctate application of capsaicin and histamine to the skin. Pain 152:2485–2494
111. Akiyama T, Tominaga M, Takamori K, Carstens MI, Carstens E (2014) Roles of glutamate, substance P, and gastrin-releasing peptide as spinal neurotransmitters of histaminergic and nonhistaminergic itch. Pain 155:80–92
112. Brumovsky P, Watanabe M, Hokfelt T (2007) Expression of the vesicular glutamate transporters-1 and -2 in adult mouse dorsal root ganglia and spinal cord and their regulation by nerve injury. Neuroscience 147:469–490
113. Sun YG, Zhao ZQ, Meng XL, Yin J, Liu XY, Chen ZF (2009) Cellular basis of itch sensation. Science 325:1531–1534
114. Carstens EE, Carstens MI, Simons CT, Jinks SL (2010) Dorsal horn neurons expressing NK-1 receptors mediate scratching in rats. Neuroreport 21: 303–308
115. Liu Y, Abdel Samad O, Zhang L, Duan B, Tong Q, Lopes C, Ji RR, Lowell BB, Ma Q (2010) VGLUT2-dependent glutamate release from nociceptors is required to sense pain and suppress itch. Neuron 68: 543–556
116. Achanta S, Chintagari NR, Brackmann M, Balakrishna S, Jordt SE (2018) TRPA1 and CGRP antagonists counteract vesicant-induced skin injury and inflammation. Toxicol Lett 293:140–148
117. McNeil BD, Pundir P, Meeker S, Han L, Undem BJ, Kulka M, Dong X (2015) Identification of a mast-cell-specific receptor crucial for pseudo-allergic drug reactions. Nature 519:237–241
118. Meixiong J, Anderson M, Limjunyawong N, Sabbagh MF, Hu E, Mack MR, Oetjen LK, Wang F, Kim BS, Dong X (2019) Activation of mast-cell-expressed Mas-related G-protein-coupled receptors drives non-histaminergic itch. Immunity 50:1163–1171.e5
119. Serhan N, Basso L, Sibilano R, Petitfils C, Meixiong J, Bonnart C, Reber LL, Marichal T, Starkl P, Cenac N, Dong X, Tsai M, Galli SJ, Gaudenzio N (2019) House dust mites activate

nociceptor-mast cell clusters to drive type 2 skin inflammation. Nat Immunol 20:1435–1443
120. Chiu IM, Heesters BA, Ghasemlou N, Von Hehn CA, Zhao F, Tran J, Wainger B, Strominger A, Muralidharan S, Horswill AR, Bubeck Wardenburg J, Hwang SW, Carroll MC, Woolf CJ (2013) Bacteria activate sensory neurons that modulate pain and inflammation. Nature 501:52–57
121. Baral P, Umans BD, Li L, Wallrapp A, Bist M, Kirschbaum T, Wei Y, Zhou Y, Kuchroo VK, Burkett PR, Yipp BG, Liberles SD, Chiu IM (2018) Nociceptor sensory neurons suppress neutrophil and gammadelta T cell responses in bacterial lung infections and lethal pneumonia. Nat Med 24:417–426
122. Pinho-Ribeiro FA, Baddal B, Haarsma R, O'Seaghdha M, Yang NJ, Blake KJ, Portley M, Verri WA, Dale JB, Wessels MR, Chiu IM (2018) Blocking neuronal signaling to immune cells treats streptococcal invasive infection. Cell 173:1083–1097.e22
123. Spits H, Di Santo JP (2011) The expanding family of innate lymphoid cells: regulators and effectors of immunity and tissue remodeling. Nat Immunol 12: 21–27
124. McKenzie AN (2014) Type-2 innate lymphoid cells in asthma and allergy. Ann Am Thorac Soc 11(Suppl 5):S263–S270
125. Talbot S, Abdulnour RE, Burkett PR, Lee S, Cronin SJ, Pascal MA, Laedermann C, Foster SL, Tran JV, Lai N, Chiu IM, Ghasemlou N, DiBiase M, Roberson D, Von Hehn C, Agac B, Haworth O, Seki H, Penninger JM, Kuchroo VK, Bean BP, Levy BD, Woolf CJ (2015) Silencing nociceptor neurons reduces allergic airway inflammation. Neuron 87:341–354
126. Cardoso V, Chesne J, Ribeiro H, Garcia-Cassani B, Carvalho T, Bouchery T, Shah K, Barbosa-Morais NL, Harris N, Veiga-Fernandes H (2017) Neuronal regulation of type 2 innate lymphoid cells via neuromedin U. Nature 549:277–281
127. Wallrapp A, Riesenfeld SJ, Burkett PR, Abdulnour RE, Nyman J, Dionne D, Hofree M, Cuoco MS, Rodman C, Farouq D, Haas BJ, Tickle TL, Trombetta JJ, Baral P, Klose CSN, Mahlakoiv T, Artis D, Rozenblatt-Rosen O, Chiu IM, Levy BD, Kowalczyk MS, Regev A, Kuchroo VK (2017) The neuropeptide NMU amplifies ILC2-driven allergic lung inflammation. Nature 549:351–356
128. Klose CSN, Mahlakoiv T, Moeller JB, Rankin LC, Flamar AL, Kabata H, Monticelli LA, Moriyama S, Putzel GG, Rakhilin N, Shen X, Kostenis E, Konig GM, Senda T, Carpenter D, Farber DL, Artis D (2017) The neuropeptide neuromedin U stimulates innate lymphoid cells and type 2 inflammation. Nature 549:282–286
129. Inclan-Rico JM, Ponessa JJ, Valero-Pacheco N, Hernandez CM, Sy CB, Lemenze AD, Beaulieu AM, Siracusa MC (2020) Basophils prime group 2 innate lymphoid cells for neuropeptide-mediated inhibition. Nat Immunol 21:1181–1193
130. Nussbaum JC, Van Dyken SJ, von Moltke J, Cheng LE, Mohapatra A, Molofsky AB, Thornton EE, Krummel MF, Chawla A, Liang HE, Locksley RM (2013) Type 2 innate lymphoid cells control eosinophil homeostasis. Nature 502:245–248
131. Kashem SW, Riedl MS, Yao C, Honda CN, Vulchanova L, Kaplan DH (2015) Nociceptive sensory fibers drive interleukin-23 production from CD301b+ dermal dendritic cells and drive protective cutaneous immunity. Immunity 43:515–526
132. Perner C, Flayer CH, Zhu X, Aderhold PA, Dewan ZNA, Voisin T, Camire RB, Chow OA, Chiu IM, Sokol CL (2020) Substance P release by sensory neurons triggers dendritic cell migration and initiates the type-2 immune response to allergens. Immunity 53:1063–1077.e7
133. Cohen JA, Edwards TN, Liu AW, Hirai T, Jones MR, Wu J, Li Y, Zhang S, Ho J, Davis BM, Albers KM, Kaplan DH (2019) Cutaneous TRPV1(+) neurons trigger protective innate type 17 anticipatory immunity. Cell 178:919–932.e14
134. Anand P, Elsafa E, Privitera R, Naidoo K, Yiangou Y, Donatien P, Gabra H, Wasan H, Kenny L, Rahemtulla A, Misra P (2019) Rational treatment of chemotherapy-induced peripheral neuropathy with capsaicin 8% patch: from pain relief towards disease modification. J Pain Res 12:2039–2052
135. Cabezon-Gutierrez L, Custodio-Cabello S, Palka-Kotlowska M, Khosravi-Shahi P (2020) High-dose 8% capsaicin patch in treatment of chemotherapy-induced peripheral neuropathy. A systematic review. J Pain Symptom Manage 60:1047–1054.e1
136. Abrams RMC, Pedowitz EJ, Simpson DM (2021) A critical review of the capsaicin 8% patch for the treatment of neuropathic pain associated with diabetic peripheral neuropathy of the feet in adults. Expert Rev Neurother 21:259–266
137. Galvez R, Navez ML, Moyle G, Maihofner C, Stoker M, Ernault E, Nurmikko TJ, Attal N (2017) Capsaicin 8% patch repeat treatment in nondiabetic peripheral neuropathic pain: a 52-week, open-label, single-arm, safety study. Clin J Pain 33:921–931
138. Ruivo EFM, Gestosa SVS, Mulenas NME, Lares AMG (2020) Peripheral neuropathy associated with hypereosinophilic syndrome: a clinical therapeutic success with capsaicin 8% patch. J Pain Palliat Care Pharmacother 34:155–158
139. Campbell JN, Stevens R, Hanson P, Connolly J, Meske DS, Chung MK, Lascelles BDX (2021) Injectable capsaicin for the management of pain due to osteoarthritis. Molecules 26(4):778
140. Rami HK, Thompson M, Stemp G, Fell S, Jerman JC, Stevens AJ, Smart D, Sargent B, Sanderson D, Randall AD, Gunthorpe MJ, Davis JB (2006) Discovery of SB-705498: a potent, selective and orally

bioavailable TRPV1 antagonist suitable for clinical development. Bioorg Med Chem Lett 16:3287–3291
141. Chizh BA, O'Donnell MB, Napolitano A, Wang J, Brooke AC, Aylott MC, Bullman JN, Gray EJ, Lai RY, Williams PM, Appleby JM (2007) The effects of the TRPV1 antagonist SB-705498 on TRPV1 receptor-mediated activity and inflammatory hyperalgesia in humans. Pain 132:132–141
142. Garami A, Shimansky YP, Rumbus Z, Vizin RCL, Farkas N, Hegyi J, Szakacs Z, Solymar M, Csenkey A, Chiche DA, Kapil R, Kyle DJ, Van Horn WD, Hegyi P, Romanovsky AA (2020) Hyperthermia induced by transient receptor potential vanilloid-1 (TRPV1) antagonists in human clinical trials: insights from mathematical modeling and meta-analysis. Pharmacol Ther 208:107474
143. Gavva NR, Treanor JJ, Garami A, Fang L, Surapaneni S, Akrami A, Alvarez F, Bak A, Darling M, Gore A, Jang GR, Kesslak JP, Ni L, Norman MH, Palluconi G, Rose MJ, Salfi M, Tan E, Romanovsky AA, Banfield C, Davar G (2008) Pharmacological blockade of the vanilloid receptor TRPV1 elicits marked hyperthermia in humans. Pain 136:202–210
144. Manitpisitkul P, Mayorga A, Shalayda K, De Meulder M, Romano G, Jun C, Moyer JA (2015) Safety, tolerability and pharmacokinetic and pharmacodynamic learnings from a double-blind, randomized, placebo-controlled, sequential group first-in-human study of the TRPV1 antagonist, JNJ-38893777, in healthy men. Clin Drug Investig 35:353–363
145. Srour J, Bengel J, Linden T, Jovanovic Z, Roggenkamp D, Reinholz M, Rothenberger C, Neufang G, Wollenberg A (2020) Efficacy of a skin care cream with TRPV1 inhibitor 4-t-butylcyclohexanol in the topical therapy of perioral dermatitis. J Cosmet Dermatol 19:1409–1414
146. Gavva NR, Tamir R, Qu Y, Klionsky L, Zhang TJ, Immke D, Wang J, Zhu D, Vanderah TW, Porreca F, Doherty EM, Norman MH, Wild KD, Bannon AW, Louis JC, Treanor JJ (2005) AMG 9810 [(E)-3-(4-t-butylphenyl)-N-(2,3-dihydrobenzo[b][1,4] dioxin-6-yl)acrylamide], a novel vanilloid receptor 1 (TRPV1) antagonist with antihyperalgesic properties. J Pharmacol Exp Ther 313:474–484
147. Norman MH, Zhu J, Fotsch C, Bo Y, Chen N, Chakrabarti P, Doherty EM, Gavva NR, Nishimura N, Nixey T, Ognyanov VI, Rzasa RM, Stec M, Surapaneni S, Tamir R, Viswanadhan VN, Treanor JJ (2007) Novel vanilloid receptor-1 antagonists: 1. Conformationally restricted analogues of trans-cinnamides. J Med Chem 50:3497–3514
148. Huang CC, Kim YS, Olson WP, Li F, Guo C, Luo W, Huang AJW, Liu Q (2016) A histamine-independent itch pathway is required for allergic ocular itch. J Allergy Clin Immunol 137:1267–1270.e6
149. Moore C, Gupta R, Jordt SE, Chen Y, Liedtke WB (2018) Regulation of pain and itch by TRP channels. Neurosci Bull 34:120–142
150. Binshtok AM, Bean BP, Woolf CJ (2007) Inhibition of nociceptors by TRPV1-mediated entry of impermeant sodium channel blockers. Nature 449: 607–610
151. Blake KJ, Baral P, Voisin T, Lubkin A, Pinho-Ribeiro FA, Adams KL, Roberson DP, Ma YC, Otto M, Woolf CJ, Torres VJ, Chiu IM (2018) Staphylococcus aureus produces pain through pore-forming toxins and neuronal TRPV1 that is silenced by QX-314. Nat Commun 9:37
152. Roberson DP, Gudes S, Sprague JM, Patoski HA, Robson VK, Blasl F, Duan B, Oh SB, Bean BP, Ma Q, Binshtok AM, Woolf CJ (2013) Activity-dependent silencing reveals functionally distinct itch-generating sensory neurons. Nat Neurosci 16: 910–918
153. Huang CC, Yang W, Guo C, Jiang H, Li F, Xiao M, Davidson S, Yu G, Duan B, Huang T, Huang AJW, Liu Q (2018) Anatomical and functional dichotomy of ocular itch and pain. Nat Med 24:1268–1276

Lysosomal TRPML1 Channel: Implications in Cardiovascular and Kidney Diseases

13

Guangbi Li and Pin-Lan Li

Abstract

Lysosomal ion channels mediate ion flux from lysosomes and regulate membrane potential across the lysosomal membrane, which are essential for lysosome biogenesis, nutrient sensing, lysosome trafficking, lysosome enzyme activity, and cell membrane repair. As a cation channel, the transient receptor potential mucolipin 1 (TRPML1) channel is mainly expressed on lysosomes and late endosomes. Recently, the normal function of TRPML1 channels has been demonstrated to be important for the maintenance of cardiovascular and renal glomerular homeostasis and thereby involved in the pathogenesis of some cardiovascular and kidney diseases. In arterial myocytes, it has been found that Nicotinic Acid Adenine Dinucleotide Phosphate (NAADP), an intracellular second messenger, can induce Ca^{2+} release through the lysosomal TRPML1 channel, leading to a global Ca^{2+} release response from the sarcoplasmic reticulum (SR). In podocytes, it has been demonstrated that lysosomal TRPML1 channels control lysosome trafficking and exosome release, which contribute to the maintenance of podocyte functional integrity. The defect or functional deficiency of lysosomal TRPML1 channels has been shown to critically contribute to the initiation and development of some chronic degeneration or diseases in the cardiovascular system or kidneys. Here we briefly summarize the current evidence demonstrating the regulation of lysosomal TRPML1 channel activity and related signaling mechanisms. We also provide some insights into the canonical and non-canonical roles of TRPML1 channel dysfunction as a potential pathogenic mechanism for certain cardiovascular and kidney diseases and associated therapeutic strategies.

Keywords

Lysosome · TRPML1 channel · NAADP · Autophagy · Exosomes · Atherosclerosis · Chronic kidney disease

13.1 Introduction

Transient receptor potential mucolipin 1 (TRPML1) channel is a cation channel expressed in late endosomes and lysosomes [1–4]. As a member of the TRPML family of ion channels, TRPML1 is coded by the gene MCOLN1. TRPML1 channel was identified in a search for genesis of mucolipidosis type IV (MLIV), one of lysosomal storage diseases

G. Li · P.-L. Li (✉)
Department of Pharmacology and Toxicology, School of Medicine, Virginia Commonwealth University, Richmond, VA, USA
e-mail: guangbi.li@vcuhealth.org; pin-lan.li@vcuhealth.org

L. Zhou (ed.), *Ion Channels in Biophysics and Physiology*, Advances in Experimental Medicine and Biology 1349, https://doi.org/10.1007/978-981-16-4254-8_13

[5, 6]. The major symptoms of MLIV include psychomotor retardation and visual impairment [7], which are associated with the defect of lysosome trafficking and lysosomal accumulation of lipofuscin and other macromolecules including phospholipids, gangliosides, and mucopolysaccharides [8, 9]. In this chapter, we will describe the common characteristics of lysosomal TRPML1 channel and summarize the regulatory mechanisms of TRPML1 channel and related cellular activities mediated by TRPML1 channels and their regulation. We will also present some recent findings regarding the pathogenic role of lysosomal TRPML1 channel deficiency or dysregulation in cardiovascular and glomerular diseases.

13.2 Characteristics of Lysosomal TRPML1 Channels

13.2.1 Subcellular Localization of Mammalian TRPML1 Channels

As a lysosomal transmembrane protein, TRPML1 channel has two di-leucine motifs for its trafficking to lysosome [2–4]. Deletion of these di-leucine motifs may induce the accumulation of TRPML1 channel in the plasma membrane [2–4]. Also, the expression of a dominant-negative variant of dynamin was found to induce the accumulation of TRPML1 channel in the plasma membrane through abrogation of clathrin-mediated endocytosis [3]. These findings indicate that TRPML1 channel may traffic from the plasma membrane to the lysosome. However, there are evidences showing the direct delivery of TRPML1 channel to the lysosome from the trans-Golgi without intermediate translocation to the plasma membrane [4, 10]. Interestingly, overactivation of TRPML1 channel may also result in its traffic from lysosomes to the plasma membrane. For example, elevated Ca^{2+} release through TRPML1 channel due to gene mutation enhanced lysosomal exocytosis and consequent translocation of TRPML1 channel to plasma membrane [11]. In Drosophila cells, activation of the mechanistic target of rapamycin complex-1 (mTORC1) also led to the elevated level of TRPML1 channel in the plasma membrane [12]. Moreover, three mammalian TRPMLs including TRPML1, 2, and 3 are capable of forming heteromultimers [3, 13–15], and the difference in biophysical properties and regulatory mechanisms between homomultimers and heteromultimers may enhance the functional diversity of these proteins [13].

13.2.2 Biophysical Properties of TRPML1 Channels

Various sensations and ligands participate in the regulation of TRP channel activity [16]. TRPMLs are relatively small proteins consisting of <600 amino acids, and the expectation of their molecular sizes is approximately 56–65 kDa. The different sizes of TRPML1 channel may range from 36 to 75 kDa due to cleavage and other modifications in various native cells [6, 17]. A similar topology to other TRP channels was found in TRPML1 channel. The ion channel is formed at the center of the canonical homotetrameric assembly which is formed by TRPML1 channels. Each subunit is composed of six transmembrane helices (S1–S6), two-pore helices (PH1 and PH2), and a luminal domain [18]. The characteristics of TRPML1 channel include a TRP channel-homologous region and an internal pore region for the passage of Ca^{2+} and Na^+. Also, TRPML1 channel is characterized with the location of both NH_2 and COOH terminal chains in cytosol. The activity of TRPML1 as a nonspecific cation channel is permeant to Ca^{2+} and H^+ and thereby controlled by lysosomal luminal and cytosolic Ca^{2+} and H^+ concentrations [19, 20].

The activity of TRPML1 reaches the maximum at pH 4.6, which is nearby the mean value of lysosomal lumen pH of mammalian cells [21, 22]. In Drosophila, the optimal pH for maximal activation of TRPML1 channel is pH 5.2, consistent with the mean luminal pH of insect lysosomes [23].

As channels insensitive to voltage but gated by ligands, TRPMLs have been extensively studied regarding their ion selectivity. For example, it has been demonstrated that human TRPML1 ion selectivity is $Ba^{2+} > Mn^{2+} > Fe^{2+} = Ca^{2+} = Mg^{2+}$- $> Ni^{2+} = Co^{2+} = Cd^{2+} > Zn^{2+} \gg Cu^{2+}$. In the presence of 30–105 mM Fe^{2+}, its pS is 32–40 at pH 4.6 [21, 22]. Nevertheless, most of previous studies characterized lysosome TRPML1 channel in cell lines transfected with TRPML1 genes. Recently, by lipid bilayer reconstitution and single lysosome patch clamping, we have demonstrated the normal function of the TRPML1 channel in lysosomes isolated from rat hepatocytes, coronary arterial myocytes, human fibroblasts, and murine podocytes [24–27]. The native TRPML1 channels can be activated by Nicotinic acid adenine dinucleotide phosphate (NAADP). In addition, the opening of TRPML channel can be blocked by its specific TRPML1 channel blockers, TRPML1 siRNA, and neutralized antibody against TRPML1 channel.

Many studies have elucidated the physiological relevance of the Ca^{2+} conductivity of TRPML1 channel to the regulation of lysosome maturation, autophagic flux, lysosomal exocytosis, and Ca^{2+} signaling-related cellular activities [20, 28]. However, the functional significance of the conductance of TRPML1 channel to other ions beyond Ca^{2+} has not yet been well clarified. Based on the conductance of the TRMPL1 channel to Fe^{2+}, it has been assumed that TRPML1 may contribute to the hematological and degenerative abnormality in MLIV via regulation of the cellular homeostasis of Fe^{2+} [11, 21]. Clearly, more studies are needed to explore the physiological relevance of the conductance of TRPML1 channel to other ions beyond Ca^{2+}.

13.3 Agonists and Blockers of TRPML1 Channel

13.3.1 NAADP

As a metabolite of $NADP^+$, NAADP has been reported to induce Ca^{2+} release from sea urchin egg microsomes [29–31]. Since then, it was demonstrated that NAADP is produced by the soluble protein Aplysia ADP-ribosylcyclase and its membrane-bound homologs, CD38 and CD157 [32–35]. These enzymes can exchange the terminal nicotinamide group of the $NADP^+$ with nicotinic acid to produce NAADP through a base-exchange reaction, which has been found in various cells and tissues, such as smooth muscle cells, pancreatic acinar cells, human T lymphocytes, sea urchin eggs, and rat brain [36–39]. In our recent studies, we have demonstrated that CD38 and its cytosolic isoforms have been found to produce NAADP in mouse coronary artery in response to death receptor activation, endothelin, and oxidant stimulation [40, 41].

Our previous studies on redox signaling and lipid raft-associated transmembrane signaling mechanisms [40, 42, 43] indicate that some agonists or stimuli may generate an acid microenvironment in vascular cells. For example, both FasL and ET-1 have been found to preferably stimulate NAADP production in vascular smooth muscle cells (VSMCs) [41, 44], which may be associated with their ability to generate a local acidic microenvironment at the cell membrane, facilitating a base exchange reaction via CD38. The formation of this local acidic environment might be attributed to lipid raft clustering if agonist receptors are linked to lipid rafts. Previous studies have shown that the clustering of membrane lipid rafts may induce the activation of CD38 [42, 45–47]. The agonist-induced membrane raft clustering is usually initiated by lysosome fusion to the cell membrane, which translocates critical proteins to the cell membrane for raft clustering such as acid sphingomyelinase (ASM) and lysosomal vacuolar H^+-ATPase. A local acidic environment is provided by lysosomal vacuolar H^+-ATPase, which maintains the normal function of translocated ASM to produce ceramide for the construction of lipid raft platforms [43]. Such membrane microenvironment of acidic pH generated and maintained by vacuolar H^+-ATPase is critical for the production of NAADP by CD38 within lipid raft platforms.

When we searched for the molecular target of NAADP action in lysosomes, reconstitution, and characterization of NAADP-sensitive Ca^{2+}

channel from lysosome of rat liver was first done in our laboratory using lipid bilayer. In 2007, we have demonstrated a reconstituted Ca^{2+} channel with all the biophysical and pharmacological features of TRPML1 in liver lysosomes [26]. In bovine coronary arterial muscle cells, we further characterized this TRPML1 channel using lysosome preparations by the constitution in the lipid bilayer. In arterial muscle preparation, we have demonstrated that the reconstituted lysosomal channel is also a voltage-dependent Ca^{2+} channel [25]. It was found that NAADP activates TRPML1 channel reconstituted from the liver and coronary arterial muscle lysosome preparations in a concentration-dependent manner. In addition, high concentrations of NAADP may result in self-desensitization of TRPML1 channel. In the bilayer preparations containing coronary arterial muscle lysosomal channels, pretreatment with a subthreshold concentration of NAADP substantially attenuated the activity of these channels in response to higher concentrations of NAADP, which indicated the self-desensitization property of these lysosomal channels. This self-desensitization property of lysosomal ion channels has also been shown in the actions of NAADP as a Ca^{2+}-releasing second messenger in other cells such as sea urchin egg fractions or intact egg cells [48–50].

Pharmacologically, these reconstituted lysosomal channels in both liver and coronary arterial lysosomes were blocked by commonly used antagonists of TRPML1 channels such as NAADP receptor antagonist PPADS, sodium channel antagonist amiloride, and dihydropyridine derivatives nifedipine and verapamil [26, 51]. Using gene silencing, deprivation of TRPML1 protein, and interference of its channel pore formation, NAADP was shown to induce Ca^{2+} release from the lysosomal store via TRPML1 channel-mediated Ca^{2+} release [25]. More recently, lysosomal TRPML1 channels have been reported to regulate local compartmental Ca^{2+} which importantly controls lysosome function such as trafficking or intracellular signaling [27].

13.3.2 Phosphoinositides

As a low-abundance lysosome-specific phosphoinositide [52–55], phosphatidylinositol-3,5-bisphosphate (PI(3,5)P2) can be generated from PI(3)P through PIKfyve/Fab1, a PI 5-kinase that locates in the lysosome of both yeast and mammalian cells [54, 56–58]. Several associated proteins, such as Fig4, Vac14, and Vac7, can enhance the activity of PIKfyve/Fab1 [53–55, 59]. Also, the myotubularin (MTM/MTMR)-family of PI-3 phosphatase can metabolize PI (3,5)P2 into PI(5)P [54, 58, 60]. A variety of neurodegenerative diseases can be induced by gene mutations of PI(3,5)P2-metabolizing enzymes and their regulators, such as Charcot–Marie–Tooth disease and amyotrophic lateral sclerosis [53, 54, 61]. Recently, it has been found that PI(3,5)P2 activates lysosome-localized TRPML1 channels specifically and potently [22, 62–64]. Enlarged lysosomes and vacuoles and deficient lysosome trafficking were found in both PI(3,5)P2-deficient cells and TRPML1-deficient cells [53–55, 59]. In PI(3,5)P2-deficient mouse fibroblasts, the enlargement of vacuole can be inhibited by overexpression of TRPML1 [62].

On the contrary, PI(4,5)P2 has been found to potently inhibit the opening of TRPML1 channel [64]. Overexpression and plasma membrane translocation of 5-phosphatase can potentiate whole-cell TRPML1 currents through degradation of PI(4,5)P2. Thus, inhibition by PI(4,5)P2 may serve as a mechanism to inactivate lysosome-operating TRPML1 channels upon plasma membrane insertion. It has been known that different PIP2 products may have different effects on TRPML1 channel activity, such as the inhibitory effects of the plasma membrane-specific phosphoinositides PI(4,5)P2, PI(3,4,5) P3, and PI(3,4)P2 and the stimulatory effect of lysosome-specific PI(3,5)P2. This suggests that the activity of TRPML1 channel is economically and efficiently regulated by these compartment-specific phosphoinositides.

In a recent study, the structure of mouse TRPML1 channel has been uncovered by single-particle cryo-electron microscopy after its embedding in nanodiscs [65]. Its structure indicates that the coupling of ligand binding to pore opening may depend on the binding of PIP2 to the N-terminus of the channel and the helix-turn-helix extension between S2 and S3. Under the closed condition, two equally distributed conformations of the S4–S5 linker have been observed, which implies that S4–S5 linker may mediate the gating of TRPML1 channel by PIP2.

13.3.3 Synthetic Agonists and Blockers

Xu and his associates developed an elegant approach using patch-clamp techniques and recorded Fe^{2+} and Ca^{2+} channel activity directly from the lysosome membrane in different cell types with TRPML1 transgene. They have discovered that TRPML1 is a proton-impermeable and inwardly rectifying cation channel, which is dually permeable to Ca^{2+} and Fe^{2+}/Mn^{2+} [11, 21, 62]. More recently, the same group used a genetically encoded Ca^{2+} indicator (GCaMP3) attached directly to TRPML1 to directly measure Ca^{2+} release from lysosomes via TRP-ML1 [66], which further confirms the nature of TRPML1 as a lysosomal Ca^{2+} release channel. These studies have demonstrated that all three members of the mammalian TRPML subfamily, TRPML1–3, are activated by mucolipin synthetic agonist 1 (ML-SA1). More recently, it has been revealed that ML-SA1 may have therapeutic potential against some lysosome storage diseases, in which TRPML1 expression and function may be diminished. For example, the defect in lysosomal function was found in cells from patients with Niemann–Pick (NP) disease, one of the lysosome storage diseases that resulted in alterations in Ca^{2+} and Fe^{2+} homoeostasis, defects in lipid trafficking, and impairments in autophagosome–lysosome fusion and/or lysosome reformation [67, 68]. Similar pathological changes can also be observed in MLIV, which led to a broad examination of TRPML1 channel function in cells with NP disease type A and type C (NPA and NPC) mutations and cells lack of NPC1 gene ($NPC1^{-/-}$ cells). It was found that the activity of TRPML1 channel was remarkably inhibited by NPA and NPC mutations or NPC1 gene deletion, suggesting that reduction of TRPML1 channel activity is a consequence of NP disease [67]. Mechanically, the activity of the TRPML1 channel was directly inhibited by sphingomyelin, which is abnormally accumulated in lysosomes in all NP disease cells. Moreover, the lysosomal function in $NPC1^{-/-}$ cells was recovered by enhancement of TRPML1 expression or activation of TRPML1 by ML-SA1 [67]. In a recent study, it has been reported that Duchenne muscular dystrophy in mouse models can be ameliorated by ML-SA5 through activation of TRPML1 channel [69]. Therefore, the pharmacological intervention of the TRPML1 channel by synthetic agonist may be an effective strategy to rescue lysosomal function in various lysosomal storage disorders. However, the mechanism mediating the action of ML-SA1 on TRPML1 remains unclear. Before the discovery of ML-SAs, previous studies showed that SF-22 and SF-51 have stimulatory effects on TRPML1 channel [67, 70]. It has been found that ML-SA1 induced more potent activation of TRPML1 channel compared to SF-51 [67]. As an mTOR inhibitor, rapamycin has been reported to have direct activating action on TRPML1 channels [71]. All these activators can be used to study TRPML1 channel activity and its physiological and pathological relevance in different cells or tissues.

It was found that amiloride, ruthenium red, nifedipine, 2APB, and SKF 96365 as commonly used cation channel and/or Ca^{2+} channel blockers failed to inhibit TRPML1 or TRPML3. Verapamil and La^{3+}, however, have been demonstrated to block both TRPML1 channel and TRPML3 channel [25, 72, 73]. ML-SI1, a novel selective TRPML1 inhibitor, has also been found to block the opening of TRPML1 channel, leading to inhibition of particle uptake and lysosomal exocytosis in bone marrow-derived macrophages [74]. More recently, ML-SI3 and ML-SI6 have been reported to inhibit TRPML1 channel activity [69] in a more specific manner, which can be used as tool drugs for studies on these lysosomal channels.

13.4 Associated Proteins of TRPML1 Channel

13.4.1 ALG-2

A recent study has shown that TRPML1 channel interacts with apoptosis-linked gene 2 (ALG-2) in a Ca^{2+}-dependent manner, which impacts endolysosomal vesicle trafficking [75]. As a Ca^{2+}-binding protein, ALG-2 belongs to the penta-EF-hand protein family [76]. The conformation of ALG-2 is changed in response to Ca^{2+}, which enhances its affinities to its binding partners [76–78]. It has been found that ALG-2 binds to the amino-terminal tail of TRPML1 channel in a Ca^{2+}-dependent manner. Furthermore, mutation of the ALG-2-binding domain in TRPML1 channel may result in the disarrangement of its distribution and function, suggesting that TRPML1 channel activity is regulated by ALG-2. More recently, there is evidence showing that lysosomal distribution is controlled by ALG-2 as a Ca^{2+} effector of the TRPML1 channel [79]. The overexpression of ALG-2 in fibroblasts resulted in a dramatic perinuclear distribution of lysosomes, which was abolished by inhibition of TRPML1 channel. On the other hand, activation of TRPML1 channel with ML-SA1 increased the amount of ALG-2 on the lysosomal membrane. Importantly, ALG-2 pulled down dynamitin, a dynactin complex component, in coimmunoprecipitation assays in a Ca^{2+}-independent manner [79]. We have also found that NAADP induces Ca^{2+} release through TRPML1 channels via ALG-2 (data not published). In mouse coronary arterial myocytes, ALG-2 was found abundantly expressed in the cytosol. Using confocal microscopy, FasL and ET-1, two well-known NAADP stimulators were found to increase the colocalization coefficient of ALG-2-GFP with TRPML1-RFP in these arterial myocytes. Immunoprecipitation with anti-ALG-2 antibody, the interaction of ALG-2 with TRPML1 channel in the lysosome, was confirmed, which was enhanced upon NAADP stimulation. In addition, recombinant ALG-2 protein binding with NAADP as receptors was significantly reduced by NAADP, and NAADP-induced lysosomal Ca^{2+} release was blocked by ALG-2 siRNA. These results demonstrate that ALG-2 indeed directly binds to NAADP and thereby may mediate the action of NAADP-induced lysosomal Ca^{2+} release via lysosomal TRPML1 channels.

13.4.2 Hsc70

The chaperone-mediated autophagy (CMA) selectively determines the recycle of soluble cytosolic proteins [80]. The selectivity of CMA depends on the recognition of a targeting motif in soluble cytosolic proteins by a chaperone, which is essential for the transportation of these proteins to the lysosomal surface [80]. As a constitutively expressed member of the 70-kDa family of chaperones, the heat shock cognate protein of 70 kDa (Hsc70) recognizes the substrate of CMA in the cytoplasm [81]. Recently, it has been demonstrated that Hsc70 interacts with TRPML1 channel in yeast two-hybrid and co-immunoprecipitation experiments [82]. In starvation studies, a working model has been suggested that lysosomal TRPML1 activity upon starvation may release Ca^{2+} to promote lysosomal Hsc70 recruitment. This TRMPL1–Hsc70 interaction enhances CMA and shifts metabolism from an anabolic toward a catabolic, thereby exerting nutrient-liberating action [79]. In macrophages of MLIV patients, the defect of CMA has been found in response to serum withdrawal, indicating that the interaction between TRPML1 channel and Hsc70 is essential for CMA. Furthermore, the reduction of lysosome-associated membrane protein type 2A is associated with the elevation of oxidized proteins in MLIV fibroblasts. All these results tell us that the delivery of substrates to the lysosomal surface depends on TRPML1 channels as a docking site for Hsc70 during CMA [82].

13.4.3 LAPTM

More recently, a novel interaction between the TRPML1 channel and the members of the lysosome-associated protein transmembrane (LAPTM) family has been identified [83]. In ARPE-19 cells, LAPTM4a, LAPTM4b, and LAPTM5 showed very high degrees of colocalization with the TRPML1 channel in late endosomes and lysosomes [83]. Functionally, overexpression of LAPTM4b caused enlargement of late endosomes and lysosomes, which can be prevented by overexpression of TRPML1 channels [83]. In HeLa cells, dysfunction of LAPTM4 resulted in a phenotype similar to lysosomal storage disease found in MLIV patients [83]. Correspondingly, a previous study has shown that expressions of LAPTM4a and LAPTM5 in MLIV cells increased threefold and sevenfold, respectively [84]. This changes the compensation attempted by MLIV cells due to the lack of LAPTM activity. In summary, LAPTM–TRPML1 interaction is important for the normal function of lysosomes and that the defect of this interaction is implicated in the pathogenesis of MLIV.

13.5 Regulatory Mechanisms of TRPML1 Channel Activity

13.5.1 Cathepsin B

The lysosomal proteases contribute to antimicrobial host defense against infection. As lysosomal proteases, cathepsins provide antimicrobial host defense against infection through degradation of intracellular bacteria within lysosomes. The reduction of capacity to kill bacteria has been reported in macrophages lacking cathepsin D during pneumococcal infection [85]. Also, the susceptibility to infections by *Porphyromonas gingivalis* and *Staphylococcus aureus* was amplified by the knockout of the cathepsin E gene [86]. Moreover, cathepsin L has been reported to inhibit the infection of *Mycoplasma pulmonis* [87]. On the contrary, there is evidence indicating that the activity of cathepsin B may contribute to bacterial infection.

TRPML1 channel may be inactivated after cleavage by lysosomal proteases. Recently, it has been reported that cathepsin B mediates the critical or final cleavage of TRPML1 channel [88]. The cleavage of the TRPML1 channel constitutes a regulatory mechanism to limit the duration of TRPML1 channel activity, which is important to the maintenance of lysosomal ionic homeostasis. In a recent study, the elevation of resistance to Francisella novicida, a cytosolic bacterial pathogen, has been found in mice and macrophages lacking cathepsin B activity [89]. Mechanically, defect in cathepsin B function inhibited mTOR activity and prevented cleavage of TRPML1 channel. In response to these changes, transcription of lysosomal and autophagy genes increased lysosomal biogenesis and enhanced the activity of autophagy initiation kinase ULK1 for bacteria clearance, which depended on transcription factor EB (TFEB). Interestingly, it has been found that inhibition of TRPML1 channel expression leads to the leak of lysosomal protease cathepsin B into the cytoplasm [90]. The leak of cathepsin B is associated with apoptosis, which can be prevented by pharmacological inhibition of cathepsin B activity. These findings indicate that modulation of cathepsin B activity may be a potential therapeutic strategy to enhance host immunity against certain bacterial infections. Also, cathepsin B might be a therapeutic target for the treatment of debilitating lysosome storage diseases.

13.5.2 TOR–TFEB Signaling Pathway

In autophagy, unwanted cellular components are digested for the generation of catabolites that are used for housekeeping biosynthesis processes in response to nutrient starvation. A recent study has reported that the activity of TRPML1 is potently and rapidly increased upon nutrient starvation [91]. In this study, starvation-induced boost of lysosomal degradation capability was completely abolished by pharmacological inhibition or genetic deletion of TRPML1, indicating the essential role of TRPML1 channel in the lysosomal adaptation to nutrient starvation [91]. Similar to nutrient starvation, the complete suppression of mTORC1 function remarkably increased TRPML1 channel-mediated Ca^{2+} release [91]. Mechanically, TFEB nuclear translocation is triggered after nutrient starvation or complete inhibition of mTORC1 [91, 92]. The expression of TRPML1 channel is upregulated by nuclear translocation or overexpression of TFEB [91, 93]. Furthermore, TOR as a nutrient-sensitive protein kinase has been found to directly target and inactivate the TRPML1 channel through phosphorylation [94]. The mutations of phosphorylation sites of the TRPML1 channel blocked the inhibition of TRPML1 channel activity by TOR [94].

13.5.3 Phosphorylation

The function of protein is importantly regulated by post-translational modifications. Phosphorylation and dephosphorylation events play vital roles in the regulation of the activity of many TRP channels. Serine, threonine, and tyrosine residues are targets of TRP phosphorylation, which is catalyzed by various kinases, including tyrosine kinase, PKA (protein kinase A), PKC (protein kinase C), PKG (protein kinase G), and CaMKII (Ca^{2+}/calmodulin-dependent protein kinase II) [95]. In a recent study, it has been demonstrated that the activity of TRPML1 channel is modulated by phosphorylation [96]. FSK-induced activation of PKA enhanced phosphorylation of TRPML1 channel, leading to inhibition of TRPML1 channel activity. On the contrary, the activity of the TRPML1 channel was remarkably amplified after the inhibition of PKA by H89. These results indicate that PKA may be involved in the regulation of lysosomal function through modulation of TRPML1 channel activity.

13.5.4 Regulation of TRPML1 Channel Activity by Sphingolipids

Ceramide is the central core in sphingolipid metabolism and plays an essential role in normal cell and tissue homeostasis and in the development of numerous diseases [97, 98]. The hydrolysis of sphingomyelin by acid sphingomyelinase (ASM) leads to the production of ceramide, which preferentially occurs in lysosomes or other acidic vesicles. Lysosomal acid ceramidase (AC) metabolizes ceramide into sphingosine which can be phosphorylated by sphingosine kinase for the generation of sphingosine-1-phosphate (S1P) [99, 100]. Recent studies have shown that sphingomyelin accumulation inhibits lysosomal TRPML1 channel-mediated Ca^{2+} release and thereby leads to impaired lysosome trafficking and lysosomal storage disease as shown in Niemann–Pick disease [67, 101]. On the contrary, lysosomal Ca^{2+} release through the TRPML1 channel and associated lysosome trafficking are enhanced by sphingosine. More recently, we have demonstrated that sphingolipid metabolism by AC regulates TRPML1 channel activity [24]. The defect in AC function may block TRPML1 channel, leading to lysosome dysfunction and increased exosome release from podocytes [24]. These findings suggest that sphingolipid signaling plays a pivotal role in the regulation of lysosomal TRPML1 channel activity and thereby determines Ca^{2+}-dependent lysosome trafficking and fusion to multivesicular bodies (MVBs) that govern exosome release. However, it remains unknown how sphingolipids alter the activity of TRPML1 channel. Further study on this regulatory mechanism would be beneficial to our understanding of the regulation of TRPML1 channel under both physiological and pathological conditions.

13.5.5 Redox Regulation of TRPML1 Channel Activity

A variety of active oxygen-containing compounds are generally called reactive oxygen species (ROS), including hydrogen peroxide (H_2O_2), superoxide anion (O_2^-), and free radical (superoxide and hydroxyl radicals) [102]. Biomacromolecules involved in different cellular activities can be damaged by ROS [103]. On the contrary, ROS is essential for the redox signaling cascade in many crucial cellular processes [104]. A simultaneous balance of ROS is maintained by cellular oxidation and antioxidation. Cellular damage can be induced by oxidative stress when oxidation exceeds antioxidation [105]. There is increasing evidence showing that the boost of autophagy may be attributed to the production of ROS, which is a vital defensive mechanism against cellular stress [106, 107]. Under conditions such as ischemia, hypoxia, and nutrient starvation, autophagy may be induced by mitochondrial ROS [108–110]. Nevertheless, autophagosome accumulation was observed after induction of autophagy by ROS [111–113], which indicated that lysosome-dependent autophagosome degradation may be interfered. Therefore, the actions of ROS on the molecules controlling lysosomal functions, such as TRPML1 channel, need to be studied.

A recent study has focused on the regulation of TRPML1 channel by ROS or oxidants in some cell lines [113]. The actions of a variety of commonly used oxidants on the TRPML1 channel have been tested in this study. It was found that chloramine-T, a nonselective strong oxidant, activated TRPML1 channel with a potency comparable to those of PI(3,5)P2 and ML-SA1. Furthermore, several other commonly used oxidants, including NaOCl, H_2O_2, *N*-chlorosuccinimide, t-butyl hydroperoxide (TBHP), and thimerosal, have been found to activate TRPML1 channel less potently. Cysteine-modifying oxidants such as DTNP and DTNB24, however, failed to induce the opening of TRPML1 channel. Similarly, SNAP, a NO-donor, and 4-HNE25, a reactive lipid peroxidation intermediate, did not affect the activity of TRPML1 channel. However, among these different oxidants, only H_2O_2 mimics the mitochondrial ROS while the dose of H_2O_2 in this study was too high (10 mM). Furthermore, various forms of ROS or oxidants that produced effects on TRPML1 channel activity vary a lot in this study, which made the conclusion of this study uncertain. At least, ROS from different resources and ROS at different levels may play diverse roles in the regulation of TRPML1 channel activity. Functionally, carbonyl cyanide *m*-chlorophenylhydrazone (CCCP), a mitochondrial respiration inhibitor commonly used to induce ROS production, was found to induce autophagic clearance of damaged mitochondria through activation of TRPML1 channel [113]. Conversely, another previous study has shown that CCCP induced autophagosome accumulation and apoptosis through activation of TRPML1 channel [114]. Recently, we have tested whether ROS affects TRPML1 channel activity in murine podocytes (data not published). It was found that pretreatment of H_2O_2 (100 μM) totally blocked the opening of TRPML1 channel induced by ML-SA1. Moreover, pretreatment with homocysteine (Hcy), an inducer of endogenous ROS production [115], remarkably attenuated Ca^{2+} release through the TRPML1 channel induced by ML-SA1. These results indicate that ROS may act as a danger factor by inhibiting TRPML1 channel activity under pathological conditions, such as hyperhomocysteinemia (hHcy). In this regard, it has been found that many compounds, including *N*-acetylcysteine, Ferulic acid, and Trehalose, reduce ROS and enhance autophagy simultaneously [116–118]. These findings may support the possibility that ROS inhibits lysosome function through blockade of TRPML1 channels.

Interestingly, the TRPML1 channel has been found to mediate the removal of ROS [113, 119]. In RPE1 (retinal pigmented epithelial 1) cells, inhibition of TRPML1 channel expression remarkably increased lipid peroxidation, which indicated the elevation of ROS production [119]. Also, depolarization and morphological changes in mitochondria were induced by siRNA of TRPML1 channel in these cells [119]. In another study, repolarization of mitochondrial membrane potential after CCCP treatment was much slower in MLIV cells compared with wild-type cells [113]. The basal level of ROS in MLIV fibroblasts was much more than the basal level of ROS in wild-type cells. Constitutive elevation of ROS and chronic inhibition of TRPML1 were observed in ML-IV cells and NPC cells [67]. Furthermore, blockade of TRPML1 channel by ML-SI4 markedly enhanced CCCP-induced ROS production in HeLa cells [113]. Collectively, these findings unveil that inhibition of TRPML1 channels by ROS and clearance of ROS by TRPML1 channel may form a balance under normal conditions, and this balance may be broken by pathological stimuli.

In addition to the controversy of the role of ROS in TRPML1 channel regulation, the mechanisms by which ROS control TRPML1 channel activity are under exploration. For example, previous studies have demonstrated that ROS enhanced the activity of Cathepsin B while having no effects on the expression of Cathepsin B [120, 121]. The enhancement of Cathepsin B activity by ROS may lead to inhibition of TRPML1 channel activity through increased cleavage of TRPML1 channels. Furthermore, it has been found that sanguinarine (SNG), a naturally occurring benzophenanthridine alkaloid, can induce ROS production and apoptotic cell death in human leukemic cells [122]. In addition, ceramide accumulation due to activation of ASM and inhibition of AC may be the mechanism mediating SNG-induced ROS production and apoptosis in human leukemic cells [122]. In MDA-MB-231 cells, the combination of C6-ceramide and DM-102, an AC inhibitor, was reported to enhance ROS production remarkably [123]. These findings indicate the complex interactions in the regulatory system of TRPML1 channel activity. For our better understanding of the impact of ROS on TRPML1 channels, further studies are imperative on this regulatory pathway. Regulatory mechanisms of TRPML1 channel activity are summarized in Fig. 13.1.

13.6 Functions of TRPML1 Channels in Health and Diseases

13.6.1 Lysosomal pH Control

The regulatory effect of the TRPML1 channel on lysosomal pH remains debatable. It has been reported that TRPML1 channel may mediate the release of H^+ from the lysosomal lumen [88, 124–126]. The dysfunction of TRPML1 channel may lower the lysosomal pH. Also, it was found that elevation of the lysosomal pH by Nigericin or Chloroquine attenuated the lysosomal storage in MLIV cells [125]. However, another study found that treatment with Nigericin or Chloroquine had no significant effect on the lysosomal storage in MLIV cells [127]. Moreover, there is evidence showing that TRPML1 channel is not permeable to H^+ [21, 73]. Taken together, the exact role of TRPML1 channel in the regulation of lysosomal pH remains unclear and further studies are necessary.

13.6.2 Fusion and Fission of Cell Membrane

Previous studies have reported that TRPML1 channel activity is essential to cellular activities associated with vesicular trafficking, such as autophagy, lysosomal exocytosis, and the formation of early endosomes [20]. All these cellular functions require the fusion and fission of the membrane. In this regard, defect in TRPML1 channel or PI(3,5)P2 has been found to block lysosome-to-Golgi retrograde trafficking, a process requiring membrane fission [2, 8, 54–58, 128–131]. Also, it has been reported that membrane-fusion processes including exocytosis and fusion of lysosome and autophagosome are dependent on the TRPML1 channel and PI(3,5)P2-metabolizing enzymes [11, 54, 58, 132, 133]. Protein complexes are recruited by the local elevation of PI(3,5)P2, leading to the generation of membrane curvature and consequent membrane fusion and fission [58]. Also, the activity of TRPML1 can be enhanced by the local

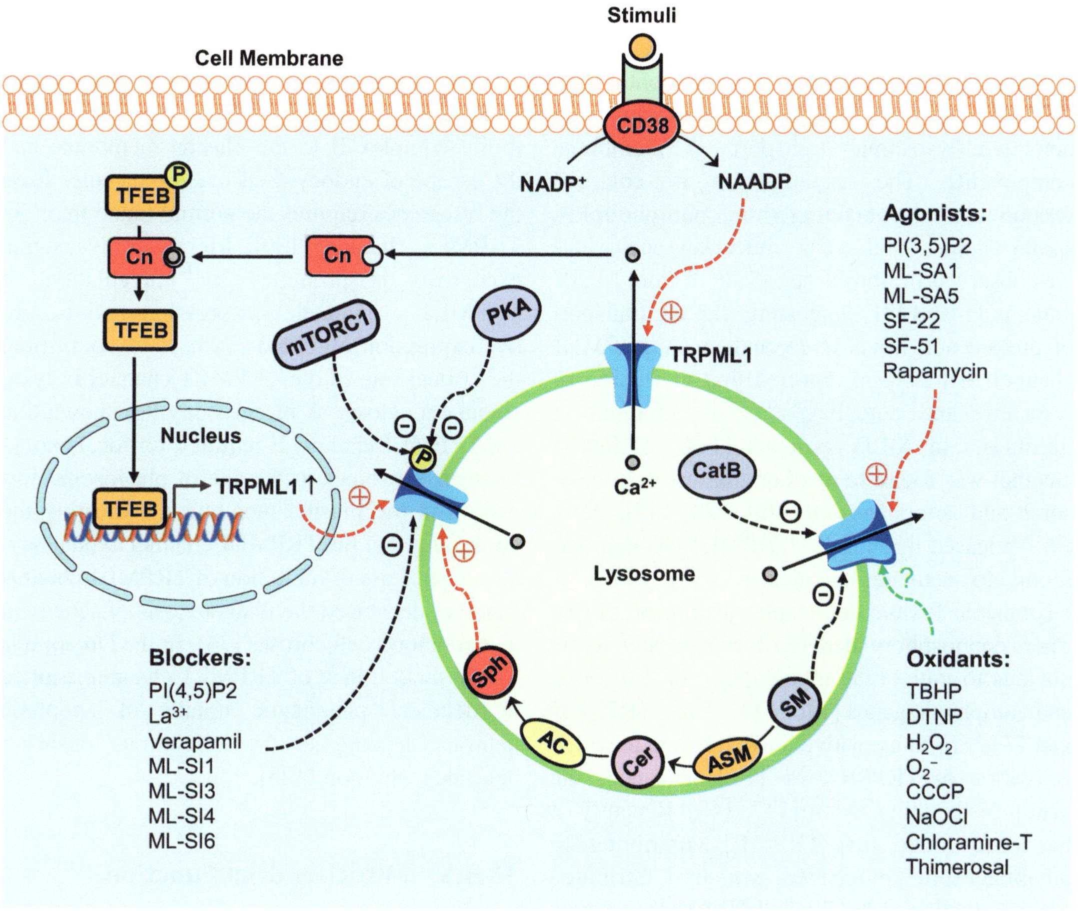

Fig. 13.1 Regulatory mechanisms of TRPML1 channel activity. TRPML1 channel agonists such as PI(3,5)P2 and ML-SA1 may induce the opening of this channel. TRPML1 channel blockers such as PI(4,5)P2 or ML-SI1 may block this channel. In response to stimuli such as FasL and ET-1, CD38 converts $NADP^+$ to NAADP that may activate the TRPML1 channel. PKA and mTORC1 are responsible for the phosphorylation of the TRPML1 channel which may result in dysfunction of this channel. Ca^{2+} released through the TRPML1 channel activates calcineurin (Cn) which is responsible for the dephosphorylation of TFEB. Dephosphorylated TFEB translocates to the nucleus to initiate the transcription of lysosomal and autophagic genes. Sphingolipids have different effects on TRPML1 channel activity, with inhibition by sphingomyelin (SM), no effect from ceramide (Cer), and enhancement by sphingosine (Sph). Cathepsin B mediates the critical or final cleavage of TRPML1 channel which results in dysfunction of this channel. The effects of oxidants on the TRPML1 channel remain controversial

elevation of PI(3,5)P2, leading to the increase in juxtaorganellar Ca^{2+} that affects SNARE proteins and lipid bilayer fusion [58, 134] via binding to putative Ca^{2+} sensor proteins such as ALG-2 [75] and Synaptotagmin/CaM [135]. These findings have shown that the normal function of the TRPML1 channel is essential for membrane fission and fusion.

13.6.3 Autophagy

Lysosome as a primary digestive organelle is responsible for the degradation of membrane proteins, membrane polysaccharides, complex lipids, endocytosed membranes, and autophagocytosed organelles. In recent studies, it has been found that a number of neurodegenerative disorders may be attributed to impaired lysosomal function and consequent autophagic deficiency [136–139]. As a lysosomal storage disorder, MLIV is featured by severe ophthalmological and neurological abnormalities caused by defective lysosomal transport of membrane components. The accumulation of enlarged vacuoles containing phospholipids, sphingolipids, and acid mucopolysaccharides was observed in fibroblasts obtained from MLIV patients [140–143], suggesting that the transport of protein and lipids is dependent on TRPML1 channel. It has been reported that the fusion of lysosome and autophagosome is inhibited in fibroblasts of MLIV patients [144]. TRPML1 channel was found to mediate the fusion of lysosome and autophagosome [20, 144, 145]. Also, Ca^{2+} released through the TRPML1 channel was found to activate calcineurin (Cn) which is responsible for the dephosphorylation of TFEB. Then, dephosphorylated TFEB translocated to the nucleus to initiate the transcription of lysosomal and autophagic genes [146, 147]. Thus, TRPML1 and TFEB form a positive feedback loop in which activation of TRPML1 or TFEB enhances the action of the other factor [147, 148]. Recently, it has been found that TRPML1 contributes to autophagosome biogenesis, which is attributed to the induction of the Beclin1/VPS34 autophagic complex, the activation of calcium/calmodulin-dependent protein kinase kinase β (CaMKKβ), the boost of AMP-activated protein kinase (AMPK), and the generation of phosphatidylinositol 3-phosphate (PI3P) [149]. These findings suggest that the TRPML1 channel plays an important role in the regulation of autophagy and the control of cellular metabolism.

13.6.4 Lysosomal Exocytosis

Previous studies have shown that the TRPML1 channel mediates the exocytosis of lysosomes [150–152]. Also, the elevation of TRPML1 channel activity due to gene mutations may induce constitutive lysosomal exocytosis, leading to remarkable translocation of TRPML1 channels to the plasma membrane [11]. Furthermore, it has been found that the TRPML1 channel contributes to the construction of tubular extensions on the cell surface as a consequence of lysosomal exocytosis [153]. In mouse macrophages, the transport of major histocompatibility complex II to the plasma membrane and the escape of endocytosed macromolecules from the lysosomes requires the normal function of the TRPML1 channel [130]. Moreover, lysosomal exocytosis is promoted by upregulation of TRPML1 channel expression due to overexpression of TFEB [148, 154], confirming the critical role of the TRPML1 channel in lysosomal exocytosis. A recent study has shown that the TRPML1 channel is required for focal exocytosis of lysosomes to the site of phagosome formation at the plasma membrane, indicating the contribution of the TRPML1 channel to phagocytosis [74]. Loss or inhibition of TRPML1 channel function decreased the phagocytosis of senescent and apoptotic cell corpses [74]. In the Drosophila MLIV model, lack of TRPML1 channel resulted in deficient phagocytic uptake of apoptotic neurons, leading to the precipitous onset of neurodegeneration [126].

13.6.5 Mitochondrial Function

There is increasing evidence suggesting the functional crosstalk between lysosome and mitochondrion [155, 156]. In unbiased protein interaction studies, the physical interaction between the TRPML1 channel and several mitochondrial proteins was observed [157]. Also, aberrations in mitochondrial function, morphology, and Ca^{2+} buffering capacity were found in cells lacking the TRPML1 channel [158]. Furthermore, genetic deletion of TRPML1 channel induced neuronal cell loss in the brains of Drosophila adults, which was attributed to dysfunction of the mitochondrion, accumulation of mitochondria with dissipated electrochemical membrane potentials, diminished autophagic clearance of damaged mitochondria, and elevation of ROS [126]. In a recent study, loss of mitochondrial membrane potential and the enhancement of ROS production were observed in mammalian cells lacking the TRPML1 channel [119].

13.6.6 Triggering of Large Ca^{2+} Release from Sarcoplasmic Reticulum

As a ubiquitous Ca^{2+} messenger, NAADP acts as one of the most potent intracellular Ca^{2+} mobilizing molecules [35, 159–161]. The Ca^{2+} mobilizing action of NAADP is even stronger than that induced by commonly known Ca^{2+} mobilizing second messengers IP3 and cADPR [30, 31]. It has been reported that the ryanodine receptor (RyR) is a possible target for the action of NAADP in mobilizing Ca^{2+} from intracellular stores [162–167]. However, other studies agreed more with the findings that in sea urchin eggs, there is an acidic compartment related to lysosomes [168]. In recent studies, NAADP has been demonstrated to first activate Ca^{2+} bursts as a triggering mechanism and then lead to global Ca^{2+} mobilization through IP3Rs and RyRs in the sarcoplasmic reticulum (SR), a so-called two-pool mechanism. It is assumed that the NAADP-sensitive Ca^{2+} store or acidic Ca^{2+} store is responsible for a localized signal, where the latter triggers Ca^{2+}-induced Ca^{2+} release (CICR) to cause global Ca^{2+} increases through RyRs or IP3Rs on the SR [41, 169, 170]. Recent studies in our laboratory have shown that NAADP may not directly activate RyRs on the SR in VSMCs [25, 26, 44]. Instead, we have demonstrated that NAADP induces Ca^{2+} release through the lysosomal TRPML1 channel, which leads to large Ca^{2+} release through RyRs on the SR [25].

Although the two-pool mechanism is attractive in its interpretation of a two-phase Ca^{2+} release induced by NAADP, some issues remain to be addressed. For example, a long delay of the second phase Ca^{2+} release (seconds to minutes) has been confirmed in NAADP-induced CICR [41, 171, 172], which is very different from the classical CICR reported previously [173–177]. Between lysosomal clusters and a subpopulation of SR, the lysosome–SR junction has been found to form a trigger zone, where NAADP-induced lysosomal Ca^{2+} release through the TRPML1 channel activates a global Ca^{2+} response via CICR in pulmonary VSMCs and some other cells [41, 44, 169]. In collaboration with Dr. van Breemen, who has extensive experience in characterizing plasma membrane–SR and mitochondria–SR junctions in smooth muscle cells under resting and contracting condition [178, 179], we performed electron microscopy and found that there are lysosome–SR junctions around 30–80 nm in VSMCs, which are relatively large compared to plasma membrane–SR and mitochondria–SR junctions. However, these Lysosome–SR junctions are rather heterogeneous within arterial myocytes (Fig. 13.2a). Based on the dimensional characteristics of the lysosome–SR junction obtained by electron microscopy, we built a three-dimensional reconstruction of a typical lysosome–SR junction, including lysosome (green), SR (yellow), TRPML1 channel (blue), SERCA2 pumps (red), and Ca^{2+} (white) (Fig. 13.2b). Given the great mobility of lysosomes in cytosol, lysosome–SR junction possibly depends on lysosomal movement toward SR [25]. It has been proposed that lysosomal aggregation around SR contributes to the global Ca^{2+} release following small Ca^{2+} bursts from lysosomes. Although lysosomal Ca^{2+} release through TRPML1 channel may not be enough to induce global Ca^{2+} release from the SR, it may be enough to drive lysosomal movement toward SR. When these clustered lysosomes work together, global Ca^{2+} release from the SR is activated [25]. This interaction of lysosome and SR was also proposed later by Zhu et al. in pulmonary VSMCs [172]. Furthermore, we have demonstrated that NAADP stimulates lysosome trafficking through its Ca^{2+} mobilizing action [27].

Recently, the two-pool mechanism has been found to regulate vasoconstriction. In this regard, Evans and associates first reported that intracellular dialysis of NAADP induced spatially restricted "bursts" of Ca^{2+} release, leading to global Ca^{2+} wave and contraction in pulmonary artery smooth muscle cells [180]. Depletion of SR Ca^{2+} stores with thapsigargin and inhibition of RyRs with ryanodine both blocked the global Ca^{2+} waves by NAADP [180]. In VSMCs, this two-pool mechanism has been demonstrated to

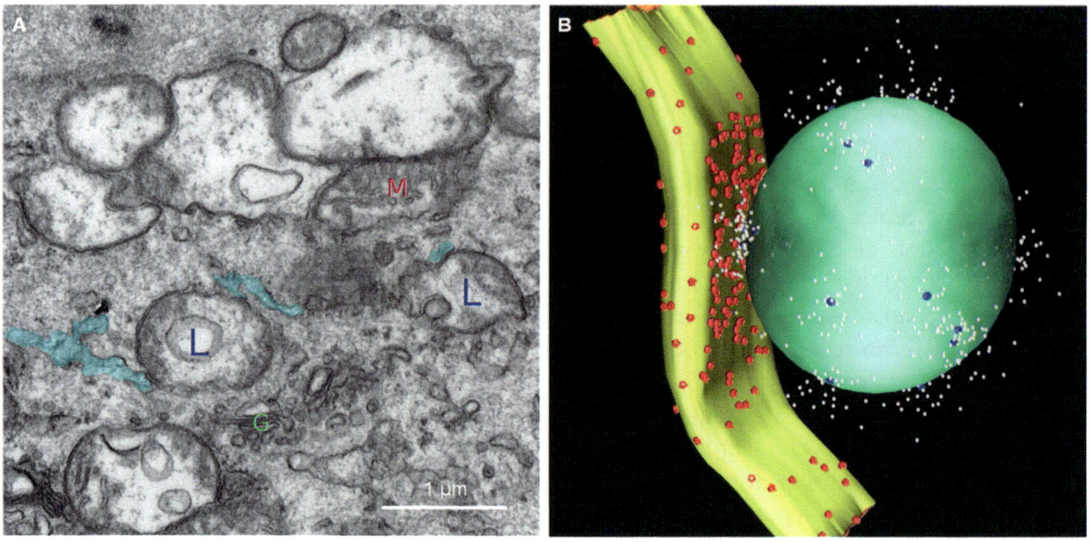

Fig. 13.2 Lysosome–SR junction. The image obtained by electron microscopy shows the structure of the lysosome–SR junction. Lysosomes are labeled with L; SRs are labeled with blue color; the mitochondrion is labeled with M; the Golgi is labeled with G. Based on the dimensional characteristics of the lysosome–SR junction obtained by electron microscopy, we built a three-dimensional reconstruction of a typical lysosome–SR junction, including lysosome (green), SR (yellow), TRPML1 channel (blue), SERCA2 pumps (red), and Ca^{2+} (white)

function in response to different agonists such as FasL and ET-1 or by delivery of NAADP into the cells [41, 44]. In coronary arteries, we also found that ET-1-induced NAADP production mobilized intracellular Ca^{2+}, which depends on the normal function of the lysosome [44]. The lysosome function inhibitor bafilomycin A1 and NAADP antagonist PPADS substantially blocked ET-1-induced maximal coronary arterial constriction. These results indicate that NAADP-induced Ca^{2+} release through lysosomal TRPML1 channel may contribute to ET-1-induced Ca^{2+} mobilization in CASMCs and consequent vasoconstriction of coronary arteries [44]. More recently, we further demonstrated that FasL increased the production of NAADP but failed to induce vasoconstriction in coronary arterial preparation. However, IP3-producing agonist U46619-induced coronary arterial contraction was significantly enhanced by FasL, suggesting that arterial contraction may be sensitized by elevation of NAADP production [41]. CICR in VSMCs is shown in Fig. 13.3.

13.6.7 Podocyte Differentiation and Podocytopathy

According to previous studies, the differentiation and maturation of podocytes are highly dependent on normal autophagy [181–183]. Our recent studies have demonstrated that autophagic flux or autophagy maturation is importantly attributed to lysosomal function in mouse podocytes and that lysosome dysfunction may result in autophagic deficiency and consequent podocyte dedifferentiation [183, 184]. Recently, we have demonstrated that CD38 controls lysosome function and thereby regulates autophagic flux in podocytes [183]. It was found that inhibition of CD38 attenuated the fusion of lysosome and autophagosome, leading to autophagosome accumulation in podocytes. Moreover, NAADP, the product of CD38, has been shown to play an essential role in the autophagic flux in podocytes. PPADS, an antagonist of the NAADP receptor, decreased GPN-induced lysosomal Ca^{2+} release and inhibited autophagic flux in podocytes. These results indicate that NAADP-induced lysosomal

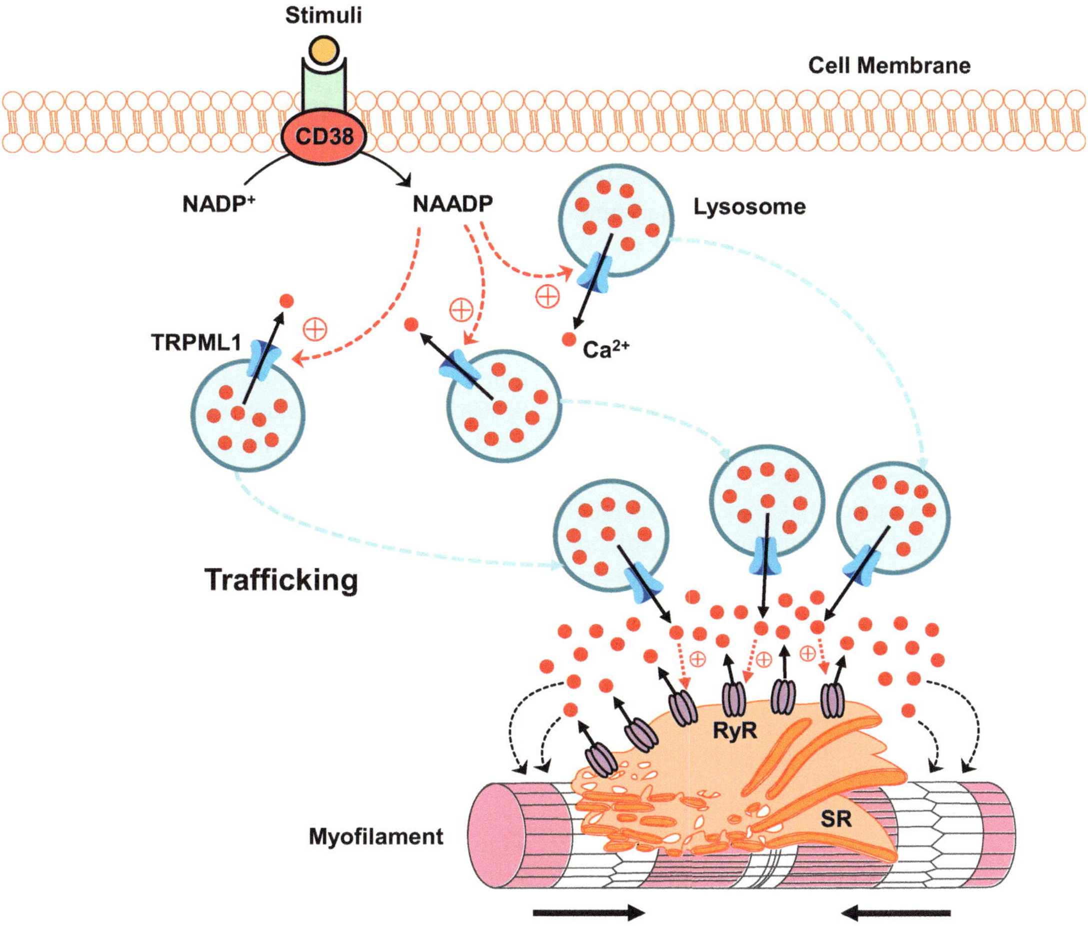

Fig. 13.3 CICR in VSMCs. In response to stimuli such as FasL and ET-1, CD38 converts $NADP^+$ to NAADP which may activate TRPML1 channel. Ca^{2+} released through the TRPML1 channel drives lysosome trafficking toward the SR, where aggregated lysosomes further release Ca^{2+} to activate RyRs to produce large Ca^{2+} release from the SR, leading to the contraction of myofilament

Ca^{2+} release through the TRPML1 channel may contribute to autophagic flux in podocytes. Furthermore, we have demonstrated that dysfunction of CD38 enhances podocyte dedifferentiation, leading to glomerular injury and sclerosis [184]. These findings indicate that CD38–NAADP–TRPML1 signaling pathway may be a therapeutic target for podocyte dedifferentiation and glomerular diseases due to autophagic deficiency.

Previous studies have disclosed that obesity is a risk factor for chronic kidney disease (CKD) and end-stage renal disease (ESRD) [185, 186]. In adolescents with severe obesity, an early glomerular injury may be detected using urinary sphingolipid excretion as a parameter, which is proposed according to a recent clinical study reporting that urinary sphingolipid excretion occurs in adolescents with severe obesity despite the absence of microalbuminuria [187]. The elevated urinary sphingolipids include ceramides, sphingomyelin, and glycosphingolipids. In the development of obesity-induced glomerular diseases, exosomes may play an important role, given the recent discovery that ceramide-enriched exosome stimulates further damage in response to danger signals [188–190]. As one of the extracellular vesicles (EVs), the exosome is released after the fusion of MVB to the plasma membrane [191]. The involvement of exosomes in cell communication and the pathogenesis of different diseases has been extensively studied [192–194]. There is evidence that lysosomal function determines the fate of various intracellular vesicles, including phagosomes, autophagosomes, and MVBs [27, 183, 195, 196]. Recently, we have reported that lysosome trafficking and lysosome–MVB interaction in podocytes are controlled by TRPML1 channel-mediated Ca^{2+} release [72]. Dysfunction of lysosomal AC may block the TRPML1 channel. The lack of Ca^{2+} release through the TRPML1 channel may decrease lysosome–MVB interaction, leading to enhancement of exosome secretion from podocytes. Under pathological conditions, enhancement of podocyte-derived exosome release due to blockade of TRPML1 channel may contribute to podocyte injury and glomerular damage. Regulations of autophagic flux and exosome release by TRPML1 channels in podocytes are summarized in Fig. 13.4.

13.6.8 Exosome Release and Arterial Medial Calcification

The process of apatite calcium salts accumulate and deposit within the vascular wall is known as arterial calcification which has been reported to be associated with atherosclerosis, diabetes mellitus, aging, and CKD [197]. Arterial calcification is anatomically classified into intimal and medial calcification [198]. In arterial intimal calcification, artery occlusion often occurs due to lipid accumulation, inflammation, and fibrosis, which is featured by irregularly scattered deposits in the atherosclerotic plaques [199]. In arterial medial calcification (AMC), the elevation of arterial stiffness is attributed to continuous deposition of hydroxyapatite in the absence of inflammatory cells along the internal elastic lamina [200, 201].

In recent studies, arterial SMC-derived exosomes in the vascular interstitial space have been reported to contribute to the development of arterial medial calcification [202–204]. Given the important role of the lysosomal TRPML1 channel in the regulation of exosome release [72], we hypothesized that dysfunction of the lysosomal TRPML1 channel may block lysosome trafficking and fusion to MVBs in arterial SMCs, leading to enhanced exosome release from arterial SMCs. A recent study in our laboratory has demonstrated that the normal function of lysosomal AC may be essential for the maintenance of contractile phenotype and lysosome-dependent degradation of MVBs in arterial SMCs [205]. The SMC-specific deletion of the Asah1 gene was found to result in enhanced exosome secretion, SMC phenotypic transition, and AMC. The regulatory role of AC is attributed to the action of AC-associated sphingolipids on the lysosomal TRPML1-mediated Ca^{2+} release. In arterial SMCs lack of Asah1 gene, the dysfunction of the TRPML1 channel led to reduced lysosome–MVB interaction and enhanced exosome release, which contributed to the development of AMC [205]. Furthermore, a recent study has demonstrated that phenotypic transition of arterial medial SMCs can be enhanced by deletion of Mcoln1 gene in a mouse model of AMC [206]. The deletion of the Mcoln1 gene induced an abnormality of lysosome positioning and elevation of exosome release, which contributed to the development of AMC [206]. These findings have further confirmed that the normal function of the lysosomal TRPML1 channel is essential for the maintenance of contractile phenotype and lysosome-dependent degradation of MVBs in arterial SMCs.

13.6.9 Lysosome-Mediated Autophagic Flux and Atherogenesis

Acute experiments in cells and isolated vessels have indicated that NAADP is involved in the development of hypertension and pulmonary hypertension [159]. However, so far there is no direct evidence showing that this CD38-derived second messenger is implicated in any vascular diseases. Given the diversity of cell types in the artery wall and the involvement of autophagy in different vascular cell functions, the role of

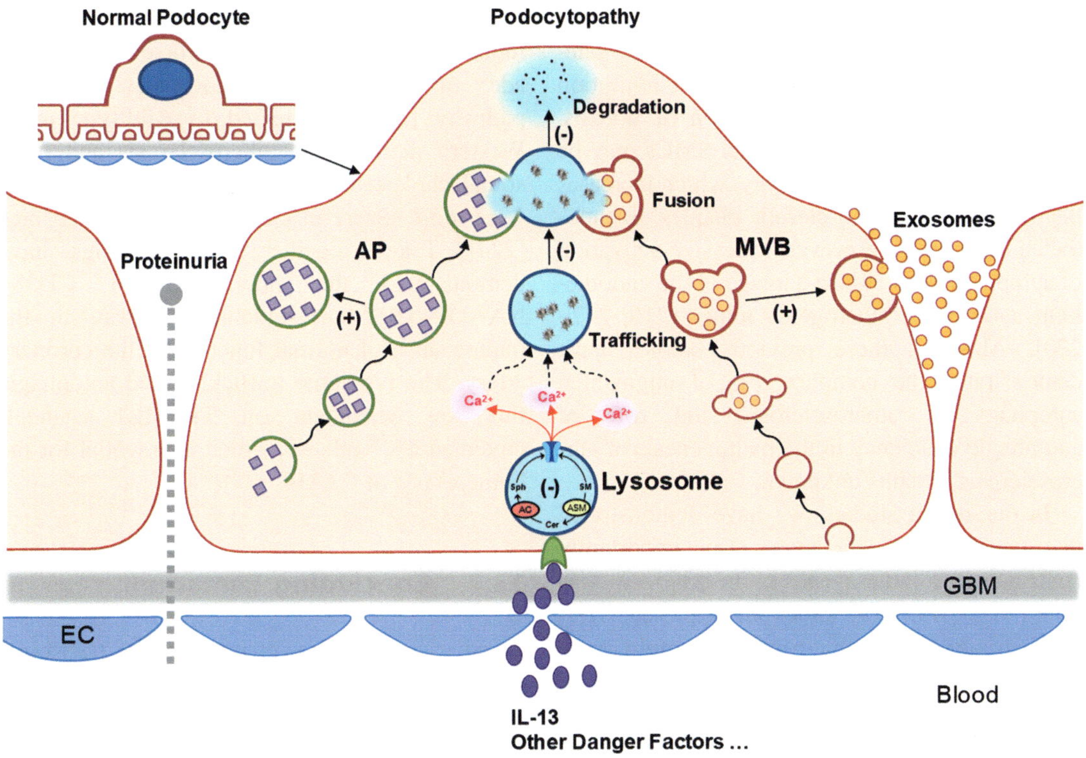

Fig. 13.4 Regulations of autophagic flux and exosome release by TRPML1 channels in podocytes. Autophagosomes (APs) and multivesicular bodies (MVBs) can fuse with and deliver content to lysosomes for degradation, which is regulated by lysosome trafficking. This lysosome-mediated regulatory mechanism constitutively controls the fate of APs and MVBs. Sphingolipids have different effects on TRPML1 channel activity, with inhibition by sphingomyelin (SM), no effect from ceramide (Cer), and enhancement by sphingosine (Sph). Ca^{2+} released through TRPML1 channel drives lysosome trafficking to and fusion with APs or MVBs. AC deficiency or inhibition may undermine lysosome–AP and lysosome–MVB interactions, leading to accumulation of APs and increased release of exosomes, which in turn triggers or promotes podocyte phenotypic transition and podocytopathy

autophagy in the development of atherosclerosis is complex [207–209]. It has been demonstrated that autophagy can generate either protective or detrimental effect on atherosclerosis, which is determined by the status of autophagy and the stage of atherosclerosis [210, 211]. Damaged components in the arterial wall can be cleaned up by autophagy, leading to the recovery of cells from the damage in response to atherosclerotic stimuli. Moreover, autophagosomes can engulf defective or damaged mitochondria for autophagic degradation, which inhibits proapoptotic protein release and cell apoptosis [212–214]. This autophagic flux may protect arterial cells from atherogenic injury. However, damaged lysosomes may release hydrolases, engage as part of oxidative stress, and enhance cellular damages in response to acute or persistent oxidative stress during atherosclerosis [211, 215–217]. Enhanced or reduced autophagy plays different roles in the development of atherosclerosis depending on the different cells involved. For example, cholesterol transport out of macrophages is increased by autophagy, which may reduce foam cell formation by prevention of lipid droplet formation. Enhanced autophagic death of macrophages also attenuates foam cell formation, which may inhibit atherosclerotic injury. Autophagy contributes to the maintenance of differentiated, quiescent, and contractile phenotype of arterial SMCs, leading to the inhibition of cell proliferation and prevention of fibrosis. Nevertheless, the death of arterial SMCs may be induced by excessive autophagy, which increases the instability of atherosclerotic plaques. In endothelial cells (ECs); however, excessive activation of autophagy may induce damage of the endothelium and initiate atherogenic injury [215, 217–220]. Although these previous studies have demonstrated the complex role of augmented autophagy in atherosclerosis, the role of autophagic deficiency in the pathogenesis of atherosclerosis remains unknown.

In our recent studies, we have demonstrated that autophagic deficiency also importantly contributes to atherogenesis. In vivo, it was found that CD38 gene knockout amplified Western diet-induced autophagosome accumulation in coronary arterial media [221]. In vitro, both deletion of CD38 gene and inhibition of TRPML1 channel expression remarkably attenuated fusion of lysosome and autophagosome, leading to autophagosome accumulation in CAMs. Defects in the CD38–NAADP–TRPML1 signaling pathway promoted CAM dedifferentiation and stimulated the production of the extracellular matrix, which contributed to the development of atherosclerosis [221]. Another study has shown that CD38 is essential for the normal function of nuclear factor E2-related factor 2 (Nrf2) in CAMs [222]. The dedifferentiation of CAMs is attributed to the downregulation of Nrf2 in CD38 gene knockout CAMs [222]. As an event observed in the early stage of atherosclerosis, increased collagen I deposition in the extracellular matrix may be caused by excess formation or decreased degradation or both of collagen I in CAMs [223]. In this regard, we have demonstrated that CD38–NAADP–TRPML1 signaling pathway plays an important role in the regulation of autophagy and thereby controls collagen metabolism [195]. The collagen I deposition and arterial wall thickening in mice fed with Western diet were enhanced by CD38 gene knockout [195]. As a hallmark in the development of atherosclerosis, cholesterol accumulation in macrophages was worsened by blockade of CD38–NAADP–TRPML1 signaling pathway [224]. Compared to wild-type mice, Western diet-induced atherosclerosis and lysosomal cholesterol sequestration in macrophages were more severe in CD38 gene knockout mice [224]. Taken together, these findings have demonstrated the importance of CD38–NAADP–TRPML1 signaling pathway in the maintenance of normal function of the coronary artery. The lysosome trafficking and autophagic flux are dependent on TRPML1 channel-mediated Ca^{2+} release, which is essential for the homeostasis of CAMs.

13.7 Concluding Remarks

There is increasing evidence that the lysosomal TRPML1 channel is implicated in the regulation of cardiovascular and glomerular functions and may be involved in the development of cardiovascular and glomerular diseases. TRPML1 channel-mediated lysosome Ca^{2+} bursts activate CICR that is an important regulatory mechanism of vasoconstriction. Furthermore, TRPML1 channel regulates various cellular activities, including autophagy, lysosomal exocytosis, apoptosis, lipid transportation, and exosome release. All these studies have provided innovative insights into the physiology and Pathobiology of TRPML1 channels, which may help develop therapeutic strategies preventing the development of vascular diseases such as arterial calcification and atherosclerosis. In addition, dysfunction of TRPML1 channels has also been implicated in the development of a variety of glomerular diseases due to their induction of podocyte dysfunction and injury, resulting in glomerular sclerosis and ultimate ESRD. A deeper mechanistic investigation is of the utmost importance to understand how pathological stimuli regulate lysosomal TRPML1 channel activity, which may further promote the development of more effective therapies for the prevention or treatment of chronic degenerative cardiovascular and glomerular diseases.

References

1. Manzoni M, Monti E, Bresciani R, Bozzato A, Barlati S, Bassi MT, Borsani G (2004) Overexpression of wild-type and mutant mucolipin proteins in mammalian cells: effects on the late endocytic compartment organization. FEBS Lett 567(2–3):219–224
2. Pryor PR, Reimann F, Gribble FM, Luzio JP (2006) Mucolipin-1 is a lysosomal membrane protein required for intracellular lactosylceramide traffic. Traffic 7(10):1388–1398
3. Venkatachalam K, Hofmann T, Montell C (2006) Lysosomal localization of TRPML3 depends on TRPML2 and the mucolipidosis-associated protein TRPML1. J Biol Chem 281(25):17517–17527
4. Vergarajauregui S, Puertollano R (2006) Two di-leucine motifs regulate trafficking of mucolipin-1 to lysosomes. Traffic 7(3):337–353
5. Bassi MT, Manzoni M, Monti E, Pizzo MT, Ballabio A, Borsani G (2000) Cloning of the gene encoding a novel integral membrane protein, mucolipidin- and identification of the two major founder mutations causing mucolipidosis type IV. Am J Hum Genet 67(5):1110–1120
6. Sun M, Goldin E, Stahl S, Falardeau JL, Kennedy JC, Acierno JS Jr, Bove C, Kaneski CR, Nagle J, Bromley MC, Colman M, Schiffmann R, Slaugenhaupt SA (2000) Mucolipidosis type IV is caused by mutations in a gene encoding a novel transient receptor potential channel. Hum Mol Genet 9(17):2471–2478
7. Amir N, Zlotogora J, Bach G (1987) Mucolipidosis type IV: clinical spectrum and natural history. Pediatrics 79(6):953–959
8. Chen CS, Bach G, Pagano RE (1998) Abnormal transport along the lysosomal pathway in mucolipidosis, type IV disease. Proc Natl Acad Sci U S A 95(11):6373–6378
9. Goldin E, Blanchette-Mackie EJ, Dwyer NK, Pentchev PG, Brady RO (1995) Cultured skin fibroblasts derived from patients with mucolipidosis 4 are auto-fluorescent. Pediatr Res 37(6):687–692
10. Miedel MT, Weixel KM, Bruns JR, Traub LM, Weisz OA (2006) Posttranslational cleavage and adaptor protein complex-dependent trafficking of mucolipin-1. J Biol Chem 281(18):12751–12759
11. Dong XP, Wang X, Shen D, Chen S, Liu M, Wang Y, Mills E, Cheng X, Delling M, Xu H (2009) Activating mutations of the TRPML1 channel revealed by proline-scanning mutagenesis. J Biol Chem 284(46):32040–32052
12. Wong CO, Li R, Montell C, Venkatachalam K (2012) Drosophila TRPML is required for TORC1 activation. Curr Biol 22(17):1616–1621
13. Curcio-Morelli C, Zhang P, Venugopal B, Charles FA, Browning MF, Cantiello HF, Slaugenhaupt SA (2010) Functional multimerization of mucolipin channel proteins. J Cell Physiol 222(2):328–335
14. Zeevi DA, Frumkin A, Offen-Glasner V, Kogot-Levin A, Bach G (2009) A potentially dynamic lysosomal role for the endogenous TRPML proteins. J Pathol 219(2):153–162
15. Zeevi DA, Lev S, Frumkin A, Minke B, Bach G (2010) Heteromultimeric TRPML channel assemblies play a crucial role in the regulation of cell viability models and starvation-induced autophagy. J Cell Sci 123(Pt 18):3112–3124
16. Venkatachalam K, Montell C (2007) TRP channels. Annu Rev Biochem 76:387–417
17. Yamaguchi S, Jha A, Li Q, Soyombo AA, Dickinson GD, Churamani D, Brailoiu E, Patel S, Muallem S (2011) Transient receptor potential mucolipin 1 (TRPML1) and two-pore channels are functionally independent organellar ion channels. J Biol Chem 286(26):22934–22942
18. Schmiege P, Fine M, Blobel G, Li X (2017) Human TRPML1 channel structures in open and closed conformations. Nature 550(7676):366–370
19. Dong XP, Wang X, Xu H (2010) TRP channels of intracellular membranes. J Neurochem 113 (2):313–328
20. Venkatachalam K, Wong CO, Zhu MX (2015) The role of TRPMLs in endolysosomal trafficking and function. Cell Calcium 58(1):48–56
21. Dong XP, Cheng X, Mills E, Delling M, Wang F, Kurz T, Xu H (2008) The type IV mucolipidosis-associated protein TRPML1 is an endolysosomal iron release channel. Nature 455(7215):992–996
22. Feng X, Huang Y, Lu Y, Xiong J, Wong CO, Yang P, Xia J, Chen D, Du G, Venkatachalam K, Xia X, Zhu MX (2014) Drosophila TRPML forms PI(3,5)P2-activated cation channels in both endolysosomes and plasma membrane. J Biol Chem 289 (7):4262–4272
23. Swetha MG, Sriram V, Krishnan KS, Oorschot VM, ten Brink C, Klumperman J, Mayor S (2011) Lysosomal membrane protein composition, acidic pH and sterol content are regulated via a light-dependent pathway in metazoan cells. Traffic 12(8):1037–1055
24. Yuan X, Bhat OM, Lohner H, Zhang Y, Li PL (2019) Endothelial acid ceramidase in exosome-mediated release of NLRP3 inflammasome products during hyperglycemia: evidence from endothelium-specific deletion of Asah1 gene. Biochim Biophys Acta Mol Cell Biol Lipids 1864(12):158532
25. Zhang F, Jin S, Yi F, Li PL (2009) TRP-ML1 functions as a lysosomal NAADP-sensitive Ca^{2+} release channel in coronary arterial myocytes. J Cell Mol Med 13(9B):3174–3185
26. Zhang F, Li PL (2007) Reconstitution and characterization of a nicotinic acid adenine dinucleotide phosphate (NAADP)-sensitive Ca^{2+} release channel from liver lysosomes of rats. J Biol Chem 282 (35):25259–25269
27. Zhang F, Xu M, Han WQ, Li PL (2011) Reconstitution of lysosomal NAADP-TRP-ML1 signaling

pathway and its function in TRP-ML1(−/−) cells. Am J Physiol Cell Physiol 301(2):C421–C430
28. Li PL, Zhang Y, Abais JM, Ritter JK, Zhang F (2013) Cyclic ADP-ribose and NAADP in vascular regulation and diseases. Messenger (Los Angel) 2(2):63–85
29. Chini EN, Beers KW, Dousa TP (1995) Nicotinate adenine dinucleotide phosphate (NAADP) triggers a specific calcium release system in sea urchin eggs. J Biol Chem 270(7):3216–3223
30. Clapper DL, Walseth TF, Dargie PJ, Lee HC (1987) Pyridine nucleotide metabolites stimulate calcium release from sea urchin egg microsomes desensitized to inositol trisphosphate. J Biol Chem 262 (20):9561–9568
31. Lee HC, Aarhus R (1995) A derivative of NADP mobilizes calcium stores insensitive to inositol trisphosphate and cyclic ADP-ribose. J Biol Chem 270(5):2152–2157
32. Aarhus R, Graeff RM, Dickey DM, Walseth TF, Lee HC (1995) ADP-ribosyl cyclase and CD38 catalyze the synthesis of a calcium-mobilizing metabolite from NADP. J Biol Chem 270(51):30327–30333
33. Galione A (1993) Cyclic ADP-ribose: a new way to control calcium. Science 259(5093):325–326
34. Lee HC (1997) Mechanisms of calcium signaling by cyclic ADP-ribose and NAADP. Physiol Rev 77 (4):1133–1164
35. Lee HC (2005) Nicotinic acid adenine dinucleotide phosphate (NAADP)-mediated calcium signaling. J Biol Chem 280(40):33693–33696
36. Ge ZD, Li PL, Chen YF, Gross GJ, Zou AP (2002) Myocardial ischemia and reperfusion reduce the levels of cyclic ADP-ribose in rat myocardium. Basic Res Cardiol 97(4):312–319
37. Ge ZD, Zhang DX, Chen YF, Yi FX, Zou AP, Campbell WB, Li PL (2003) Cyclic ADP-ribose contributes to contraction and Ca^{2+} release by M1 muscarinic receptor activation in coronary arterial smooth muscle. J Vasc Res 40(1):28–36
38. Lee HC, Aarhus R (2000) Functional visualization of the separate but interacting calcium stores sensitive to NAADP and cyclic ADP-ribose. J Cell Sci 113 (Pt 24):4413–4420
39. Li PL, Tang WX, Valdivia HH, Zou AP, Campbell WB (2001) cADP-ribose activates reconstituted ryanodine receptors from coronary arterial smooth muscle. Am J Physiol Heart Circ Physiol 280(1): H208–H215
40. Xu M, Zhang Y, Xia M, Li XX, Ritter JK, Zhang F, Li PL (2012) NAD(P)H oxidase-dependent intracellular and extracellular O2*-production in coronary arterial myocytes from CD38 knockout mice. Free Radic Biol Med 52(2):357–365
41. Zhang F, Xia M, Li PL (2010) Lysosome-dependent Ca(2+) release response to Fas activation in coronary arterial myocytes through NAADP: evidence from CD38 gene knockouts. Am J Physiol Cell Physiol 298(5):C1209–C1216
42. Jia SJ, Jin S, Zhang F, Yi F, Dewey WL, Li PL (2008) Formation and function of ceramide-enriched membrane platforms with CD38 during M1-receptor stimulation in bovine coronary arterial myocytes. Am J Physiol Heart Circ Physiol 295(4):H1743–H1752
43. Xu M, Xia M, Li XX, Han WQ, Boini KM, Zhang F, Zhang Y, Ritter JK, Li PL (2012) Requirement of translocated lysosomal V1 H(+)-ATPase for activation of membrane acid sphingomyelinase and raft clustering in coronary endothelial cells. Mol Biol Cell 23(8):1546–1557
44. Zhang F, Zhang G, Zhang AY, Koeberl MJ, Wallander E, Li PL (2006) Production of NAADP and its role in Ca^{2+} mobilization associated with lysosomes in coronary arterial myocytes. Am J Physiol Heart Circ Physiol 291(1):H274–H282
45. Deaglio S, Vaisitti T, Billington R, Bergui L, Omede P, Genazzani AA, Malavasi F (2007) CD38/CD19: a lipid raft-dependent signaling complex in human B cells. Blood 109(12):5390–5398
46. Munoz P, Navarro MD, Pavon EJ, Salmeron J, Malavasi F, Sancho J, Zubiaur M (2003) CD38 signaling in T cells is initiated within a subset of membrane rafts containing Lck and the CD3-zeta subunit of the T cell antigen receptor. J Biol Chem 278 (50):50791–50802
47. Zilber MT, Setterblad N, Vasselon T, Doliger C, Charron D, Mooney N, Gelin C (2005) MHC class II/CD38/CD9: a lipid-raft-dependent signaling complex in human monocytes. Blood 106(9):3074–3081
48. Aarhus R, Dickey DM, Graeff RM, Gee KR, Walseth TF, Lee HC (1996) Activation and inactivation of Ca^{2+} release by NAADP+. J Biol Chem 271 (15):8513–8516
49. Bach G (2005) Mucolipin 1: endocytosis and cation channel—a review. Pflugers Arch 451(1):313–317
50. Genazzani AA, Empson RM, Galione A (1996) Unique inactivation properties of NAADP-sensitive Ca^{2+} release. J Biol Chem 271(20):11599–11602
51. Yusufi AN, Cheng J, Thompson MA, Burnett JC, Grande JP (2002) Differential mechanisms of Ca (2+) release from vascular smooth muscle cell microsomes. Exp Biol Med (Maywood) 227 (1):36–44
52. Bonangelino CJ, Nau JJ, Duex JE, Brinkman M, Wurmser AE, Gary JD, Emr SD, Weisman LS (2002) Osmotic stress-induced increase of phosphatidylinositol 3,5-bisphosphate requires Vac14p, an activator of the lipid kinase Fab1p. J Cell Biol 156 (6):1015–1028
53. Chow CY, Zhang Y, Dowling JJ, Jin N, Adamska M, Shiga K, Szigeti K, Shy ME, Li J, Zhang X, Lupski JR, Weisman LS, Meisler MH (2007) Mutation of FIG4 causes neurodegeneration in the pale tremor mouse and patients with CMT4J. Nature 448 (7149):68–72
54. Dove SK, Dong K, Kobayashi T, Williams FK, Michell RH (2009) Phosphatidylinositol

3,5-bisphosphate and Fab1p/PIKfyve underPPIn endo-lysosome function. Biochem J 419(1):1–13
55. Zhang Y, Zolov SN, Chow CY, Slutsky SG, Richardson SC, Piper RC, Yang B, Nau JJ, Westrick RJ, Morrison SJ, Meisler MH, Weisman LS (2007) Loss of Vac14, a regulator of the signaling lipid phosphatidylinositol 3,5-bisphosphate, results in neurodegeneration in mice. Proc Natl Acad Sci U S A 104(44):17518–17523
56. Botelho RJ, Efe JA, Teis D, Emr SD (2008) Assembly of a Fab1 phosphoinositide kinase signaling complex requires the Fig4 phosphoinositide phosphatase. Mol Biol Cell 19(10):4273–4286
57. Duex JE, Nau JJ, Kauffman EJ, Weisman LS (2006) Phosphoinositide 5-phosphatase Fig 4p is required for both acute rise and subsequent fall in stress-induced phosphatidylinositol 3,5-bisphosphate levels. Eukaryot Cell 5(4):723–731
58. Poccia D, Larijani B (2009) Phosphatidylinositol metabolism and membrane fusion. Biochem J 418 (2):233–246
59. Jin N, Chow CY, Liu L, Zolov SN, Bronson R, Davisson M, Petersen JL, Zhang Y, Park S, Duex JE, Goldowitz D, Meisler MH, Weisman LS (2008) VAC14 nucleates a protein complex essential for the acute interconversion of PI3P and PI(3,5)P(2) in yeast and mouse. EMBO J 27(24):3221–3234
60. Shen J, Yu WM, Brotto M, Scherman JA, Guo C, Stoddard C, Nosek TM, Valdivia HH, Qu CK (2009) Deficiency of MIP/MTMR14 phosphatase induces a muscle disorder by disrupting Ca(2+) homeostasis. Nat Cell Biol 11(6):769–776
61. Chow CY, Landers JE, Bergren SK, Sapp PC, Grant AE, Jones JM, Everett L, Lenk GM, McKenna-Yasek DM, Weisman LS, Figlewicz D, Brown RH, Meisler MH (2009) Deleterious variants of FIG4, a phosphoinositide phosphatase, in patients with ALS. Am J Hum Genet 84(1):85–88
62. Dong XP, Shen D, Wang X, Dawson T, Li X, Zhang Q, Cheng X, Zhang Y, Weisman LS, Delling M, Xu H (2010) PI(3,5)P(2) controls membrane trafficking by direct activation of mucolipin Ca (2+) release channels in the endolysosome. Nat Commun 1:38
63. Feng X, Xiong J, Lu Y, Xia X, Zhu MX (2014) Differential mechanisms of action of the mucolipin synthetic agonist, ML-SA1, on insect TRPML and mammalian TRPML1. Cell Calcium 56(6):446–456
64. Zhang X, Li X, Xu H (2012) Phosphoinositide isoforms determine compartment-specific ion channel activity. Proc Natl Acad Sci U S A 109 (28):11384–11389
65. Chen Q, She J, Zeng W, Guo J, Xu H, Bai XC, Jiang Y (2017) Structure of mammalian endolysosomal TRPML1 channel in nanodiscs. Nature 550 (7676):415–418
66. Martelli AM, Chiarini F, Evangelisti C, Cappellini A, Buontempo F, Bressanin D, Fini M, McCubrey JA (2012) Two hits are better than one: targeting both phosphatidylinositol 3-kinase and mammalian target of rapamycin as a therapeutic strategy for acute leukemia treatment. Oncotarget 3(4):371–394
67. Shen D, Wang X, Li X, Zhang X, Yao Z, Dibble S, Dong XP, Yu T, Lieberman AP, Showalter HD, Xu H (2012) Lipid storage disorders block lysosomal trafficking by inhibiting a TRP channel and lysosomal calcium release. Nat Commun 3:731
68. Weiss N (2012) Cross-talk between TRPML1 channel, lipids and lysosomal storage diseases. Commun Integr Biol 5(2):111–113
69. Yu L, Zhang X, Yang Y, Li D, Tang K, Zhao Z, He W, Wang C, Sahoo N, Converso-Baran K, Davis CS, Brooks SV, Bigot A, Calvo R, Martinez NJ, Southall N, Hu X, Marugan J, Ferrer M, Xu H (2020) Small-molecule activation of lysosomal TRP channels ameliorates Duchenne muscular dystrophy in mouse models. Sci Adv 6(6):eaaz2736
70. Chen CC, Keller M, Hess M, Schiffmann R, Urban N, Wolfgardt A, Schaefer M, Bracher F, Biel M, Wahl-Schott C, Grimm C (2014) A small molecule restores function to TRPML1 mutant isoforms responsible for mucolipidosis type IV. Nat Commun 5:4681
71. Zhang X, Chen W, Gao Q, Yang J, Yan X, Zhao H, Su L, Yang M, Gao C, Yao Y, Inoki K, Li D, Shao R, Wang S, Sahoo N, Kudo F, Eguchi T, Ruan B, Xu H (2019) Rapamycin directly activates lysosomal mucolipin TRP channels independent of mTOR. PLoS Biol 17(5):e3000252
72. Li G, Huang D, Hong J, Bhat OM, Yuan X, Li PL (2019) Control of lysosomal TRPML1 channel activity and exosome release by acid ceramidase in mouse podocytes. Am J Physiol Cell Physiol 317(3):C481–C491
73. Xu H, Delling M, Li L, Dong X, Clapham DE (2007) Activating mutation in a mucolipin transient receptor potential channel leads to melanocyte loss in varitint-waddler mice. Proc Natl Acad Sci U S A 104 (46):18321–18326
74. Samie M, Wang X, Zhang X, Goschka A, Li X, Cheng X, Gregg E, Azar M, Zhuo Y, Garrity AG, Gao Q, Slaugenhaupt S, Pickel J, Zolov SN, Weisman LS, Lenk GM, Titus S, Bryant-Genevier M, Southall N, Juan M, Ferrer M, Xu H (2013) A TRP channel in the lysosome regulates large particle phagocytosis via focal exocytosis. Dev Cell 26 (5):511–524
75. Vergarajauregui S, Martina JA, Puertollano R (2009) Identification of the penta-EF-hand protein ALG-2 as a Ca^{2+}-dependent interactor of mucolipin-1. J Biol Chem 284(52):36357–36366
76. Lo KW, Zhang Q, Li M, Zhang M (1999) Apoptosis-linked gene product ALG-2 is a new member of the calpain small subunit subfamily of Ca^{2+}-binding proteins. Biochemistry 38(23):7498–7508
77. Maki M, Kitaura Y, Satoh H, Ohkouchi S, Shibata H (2002) Structures, functions and molecular evolution

of the penta-EF-hand Ca^{2+}-binding proteins. Biochim Biophys Acta 1600(1–2):51–60
78. Tarabykina S, Mollerup J, Winding P, Berchtold MW (2004) ALG-2, a multifunctional calcium binding protein? Front Biosci 9:1817–1832
79. Li X, Rydzewski N, Hider A, Zhang X, Yang J, Wang W, Gao Q, Cheng X, Xu H (2016) A molecular mechanism to regulate lysosome motility for lysosome positioning and tubulation. Nat Cell Biol 18 (4):404–417
80. Bejarano E, Cuervo AM (2010) Chaperone-mediated autophagy. Proc Am Thorac Soc 7(1):29–39
81. Chiang HL, Terlecky SR, Plant CP, Dice JF (1989) A role for a 70-kilodalton heat shock protein in lysosomal degradation of intracellular proteins. Science 246(4928):382–385
82. Venugopal B, Mesires NT, Kennedy JC, Curcio-Morelli C, Laplante JM, Dice JF, Slaugenhaupt SA (2009) Chaperone-mediated autophagy is defective in mucolipidosis type IV. J Cell Physiol 219 (2):344–353
83. Vergarajauregui S, Martina JA, Puertollano R (2011) LAPTMs regulate lysosomal function and interact with mucolipin 1: new clues for understanding mucolipidosis type IV. J Cell Sci 124(Pt 3):459–468
84. Bozzato A, Barlati S, Borsani G (2008) Gene expression profiling of mucolipidosis type IV fibroblasts reveals deregulation of genes with relevant functions in lysosome physiology. Biochim Biophys Acta 1782 (4):250–258
85. Bewley MA, Marriott HM, Tulone C, Francis SE, Mitchell TJ, Read RC, Chain B, Kroemer G, Whyte MK, Dockrell DH (2011) A cardinal role for cathepsin d in co-ordinating the host-mediated apoptosis of macrophages and killing of pneumococci. PLoS Pathog 7(1):e1001262
86. Tsukuba T, Yamamoto S, Yanagawa M, Okamoto K, Okamoto Y, Nakayama KI, Kadowaki T, Yamamoto K (2006) Cathepsin E-deficient mice show increased susceptibility to bacterial infection associated with the decreased expression of multiple cell surface Toll-like receptors. J Biochem 140(1):57–66
87. Xu X, Greenland J, Baluk P, Adams A, Bose O, McDonald DM, Caughey GH (2013) Cathepsin L protects mice from mycoplasmal infection and is essential for airway lymphangiogenesis. Am J Respir Cell Mol Biol 49(3):437–444
88. Kiselyov K, Chen J, Rbaibi Y, Oberdick D, Tjon-Kon-Sang S, Shcheynikov N, Muallem S, Soyombo A (2005) TRP-ML1 is a lysosomal monovalent cation channel that undergoes proteolytic cleavage. J Biol Chem 280(52):43218–43223
89. Qi X, Man SM, Malireddi RK, Karki R, Lupfer C, Gurung P, Neale G, Guy CS, Lamkanfi M, Kanneganti TD (2016) Cathepsin B modulates lysosomal biogenesis and host defense against Francisella novicida infection. J Exp Med 213(10):2081–2097
90. Colletti GA, Miedel MT, Quinn J, Andharia N, Weisz OA, Kiselyov K (2012) Loss of lysosomal ion channel transient receptor potential channel mucolipin-1 (TRPML1) leads to cathepsin B-dependent apoptosis. J Biol Chem 287 (11):8082–8091
91. Wang W, Gao Q, Yang M, Zhang X, Yu L, Lawas M, Li X, Bryant-Genevier M, Southall NT, Marugan J, Ferrer M, Xu H (2015) Up-regulation of lysosomal TRPML1 channels is essential for lysosomal adaptation to nutrient starvation. Proc Natl Acad Sci U S A 112(11):E1373–E1381
92. Settembre C, Zoncu R, Medina DL, Vetrini F, Erdin S, Erdin S, Huynh T, Ferron M, Karsenty G, Vellard MC, Facchinetti V, Sabatini DM, Ballabio A (2012) A lysosome-to-nucleus signalling mechanism senses and regulates the lysosome via mTOR and TFEB. EMBO J 31(5):1095–1108
93. Sardiello M, Palmieri M, di Ronza A, Medina DL, Valenza M, Gennarino VA, Di Malta C, Donaudy F, Embrione V, Polishchuk RS, Banfi S, Parenti G, Cattaneo E, Ballabio A (2009) A gene network regulating lysosomal biogenesis and function. Science 325(5939):473–477
94. Onyenwoke RU, Sexton JZ, Yan F, Diaz MC, Forsberg LJ, Major MB, Brenman JE (2015) The mucolipidosis IV Ca^{2+} channel TRPML1 (MCOLN1) is regulated by the TOR kinase. Biochem J 470(3):331–342
95. Yao X, Kwan HY, Huang Y (2005) Regulation of TRP channels by phosphorylation. Neurosignals 14 (6):273–280
96. Vergarajauregui S, Oberdick R, Kiselyov K, Puertollano R (2008) Mucolipin 1 channel activity is regulated by protein kinase A-mediated phosphorylation. Biochem J 410(2):417–425
97. Gault CR, Obeid LM, Hannun YA (2010) An overview of sphingolipid metabolism: from synthesis to breakdown. Adv Exp Med Biol 688:1–23
98. Gulbins E, Li PL (2006) Physiological and pathophysiological aspects of ceramide. Am J Physiol Regul Integr Comp Physiol 290(1):R11–R26
99. Bhat OM, Yuan X, Li G, Lee R, Li PL (2018) Sphingolipids and redox signaling in renal regulation and chronic kidney diseases. Antioxid Redox Signal 28(10):1008–1026
100. Futerman AH, Hannun YA (2004) The complex life of simple sphingolipids. EMBO Rep 5(8):777–782
101. Piccoli E, Nadai M, Caretta CM, Bergonzini V, Del Vecchio C, Ha HR, Bigler L, Dal Zoppo D, Faggin E, Pettenazzo A, Orlando R, Salata C, Calistri A, Palu G, Baritussio A (2011) Amiodarone impairs trafficking through late endosomes inducing a Niemann-Pick C-like phenotype. Biochem Pharmacol 82(9):1234–1249
102. Siti HN, Kamisah Y, Kamsiah J (2015) The role of oxidative stress, antioxidants and vascular inflammation in cardiovascular disease (a review). Vasc Pharmacol 71:40–56

103. Huang J, Lam GY, Brumell JH (2011) Autophagy signaling through reactive oxygen species. Antioxid Redox Signal 14(11):2215–2231
104. Droge W (2002) Free radicals in the physiological control of cell function. Physiol Rev 82(1):47–95
105. Skowronska M, Albrecht J (2013) Oxidative and nitrosative stress in ammonia neurotoxicity. Neurochem Int 62(5):731–737
106. Scherz-Shouval R, Elazar Z (2011) Regulation of autophagy by ROS: physiology and pathology. Trends Biochem Sci 36(1):30–38
107. Scherz-Shouval R, Shvets E, Fass E, Shorer H, Gil L, Elazar Z (2007) Reactive oxygen species are essential for autophagy and specifically regulate the activity of Atg4. EMBO J 26(7):1749–1760
108. Filomeni G, De Zio D, Cecconi F (2015) Oxidative stress and autophagy: the clash between damage and metabolic needs. Cell Death Differ 22(3):377–388
109. Mi Y, Xiao C, Du Q, Wu W, Qi G, Liu X (2016) Momordin Ic couples apoptosis with autophagy in human hepatoblastoma cancer cells by reactive oxygen species (ROS)-mediated PI3K/Akt and MAPK signaling pathways. Free Radic Biol Med 90:230–242
110. Scherz-Shouval R, Shvets E, Elazar Z (2007) Oxidation as a post-translational modification that regulates autophagy. Autophagy 3(4):371–373
111. Kim RJ, Hah YS, Sung CM, Kang JR, Park HB (2014) Do antioxidants inhibit oxidative-stress-induced autophagy of tenofibroblasts? J Orthop Res 32(7):937–943
112. Liu GY, Jiang XX, Zhu X, He WY, Kuang YL, Ren K, Lin Y, Gou X (2015) ROS activates JNK-mediated autophagy to counteract apoptosis in mouse mesenchymal stem cells in vitro. Acta Pharmacol Sin 36(12):1473–1479
113. Zhang X, Cheng X, Yu L, Yang J, Calvo R, Patnaik S, Hu X, Gao Q, Yang M, Lawas M, Delling M, Marugan J, Ferrer M, Xu H (2016) MCOLN1 is a ROS sensor in lysosomes that regulates autophagy. Nat Commun 7:12109
114. Morelli MB, Amantini C, Tomassoni D, Nabissi M, Arcella A, Santoni G (2019) Transient receptor potential mucolipin-1 channels in glioblastoma: role in patient's survival. Cancers (Basel) 11(4):525
115. Abais JM, Xia M, Li G, Gehr TW, Boini KM, Li PL (2014) Contribution of endogenously produced reactive oxygen species to the activation of podocyte NLRP3 inflammasomes in hyperhomocysteinemia. Free Radic Biol Med 67:211–220
116. Chowdhury S, Ghosh S, Das AK, Sil PC (2019) Ferulic acid protects hyperglycemia-induced kidney damage by regulating oxidative insult, inflammation and autophagy. Front Pharmacol 10:27
117. Cui J, Tang L, Hong Q, Lin S, Sun X, Cai G, Bai XY, Chen X (2019) N-Acetylcysteine ameliorates gentamicin-induced nephrotoxicity by enhancing autophagy and reducing oxidative damage in miniature pigs. Shock 52(6):622–630
118. Wang XY, Yang H, Wang MG, Yang DB, Wang ZY, Wang L (2017) Trehalose protects against cadmium-induced cytotoxicity in primary rat proximal tubular cells via inhibiting apoptosis and restoring autophagic flux. Cell Death Dis 8(10):e3099
119. Coblentz J, St Croix C, Kiselyov K (2014) Loss of TRPML1 promotes production of reactive oxygen species: is oxidative damage a factor in mucolipidosis type IV? Biochem J 457(2):361–368
120. Rybicka JM, Balce DR, Chaudhuri S, Allan ER, Yates RM (2012) Phagosomal proteolysis in dendritic cells is modulated by NADPH oxidase in a pH-independent manner. EMBO J 31(4):932–944
121. Rybicka JM, Balce DR, Khan MF, Krohn RM, Yates RM (2010) NADPH oxidase activity controls phagosomal proteolysis in macrophages through modulation of the lumenal redox environment of phagosomes. Proc Natl Acad Sci U S A 107 (23):10496–10501
122. Rahman A, Thayyullathil F, Pallichankandy S, Galadari S (2016) Hydrogen peroxide/ceramide/Akt signaling axis play a critical role in the antileukemic potential of sanguinarine. Free Radic Biol Med 96:273–289
123. Flowers M, Fabrias G, Delgado A, Casas J, Abad JL, Cabot MC (2012) C6-ceramide and targeted inhibition of acid ceramidase induce synergistic decreases in breast cancer cell growth. Breast Cancer Res Treat 133(2):447–458
124. Miedel MT, Rbaibi Y, Guerriero CJ, Colletti G, Weixel KM, Weisz OA, Kiselyov K (2008) Membrane traffic and turnover in TRP-ML1-deficient cells: a revised model for mucolipidosis type IV pathogenesis. J Exp Med 205(6):1477–1490
125. Soyombo AA, Tjon-Kon-Sang S, Rbaibi Y, Bashllari E, Bisceglia J, Muallem S, Kiselyov K (2006) TRP-ML1 regulates lysosomal pH and acidic lysosomal lipid hydrolytic activity. J Biol Chem 281 (11):7294–7301
126. Venkatachalam K, Long AA, Elsaesser R, Nikolaeva D, Broadie K, Montell C (2008) Motor deficit in a Drosophila model of mucolipidosis type IV due to defective clearance of apoptotic cells. Cell 135(5):838–851
127. Kogot-Levin A, Zeigler M, Ornoy A, Bach G (2009) Mucolipidosis type IV: the effect of increased lysosomal pH on the abnormal lysosomal storage. Pediatr Res 65(6):686–690
128. Cheng X, Shen D, Samie M, Xu H (2010) Mucolipins: intracellular TRPML1-3 channels. FEBS Lett 584(10):2013–2021
129. Duex JE, Tang F, Weisman LS (2006) The Vac14p-Fig4p complex acts independently of Vac7p and couples PI3,5P2 synthesis and turnover. J Cell Biol 172(5):693–704
130. Thompson EG, Schaheen L, Dang H, Fares H (2007) Lysosomal trafficking functions of mucolipin-1 in murine macrophages. BMC Cell Biol 8:54

131. Treusch S, Knuth S, Slaugenhaupt SA, Goldin E, Grant BD, Fares H (2004) Caenorhabditis elegans functional orthologue of human protein h-mucolipin-1 is required for lysosome biogenesis. Proc Natl Acad Sci U S A 101(13):4483–4488
132. Ferguson CJ, Lenk GM, Meisler MH (2009) Defective autophagy in neurons and astrocytes from mice deficient in PI(3,5)P2. Hum Mol Genet 18 (24):4868–4878
133. Puertollano R, Kiselyov K (2009) TRPMLs: in sickness and in health. Am J Physiol Renal Physiol 296 (6):F1245–F1254
134. Roth MG (2004) Phosphoinositides in constitutive membrane traffic. Physiol Rev 84(3):699–730
135. Luzio JP, Pryor PR, Bright NA (2007) Lysosomes: fusion and function. Nat Rev Mol Cell Biol 8 (8):622–632
136. Filimonenko M, Stuffers S, Raiborg C, Yamamoto A, Malerod L, Fisher EM, Isaacs A, Brech A, Stenmark H, Simonsen A (2007) Functional multivesicular bodies are required for autophagic clearance of protein aggregates associated with neurodegenerative disease. J Cell Biol 179(3):485–500
137. Parkinson N, Ince PG, Smith MO, Highley R, Skibinski G, Andersen PM, Morrison KE, Pall HS, Hardiman O, Collinge J, Shaw PJ, Fisher EM, MRC Proteomics in ALS Study; FReJA Consortium (2006) ALS phenotypes with mutations in CHMP2B (charged multivesicular body protein 2B). Neurology 67(6):1074–1077
138. Reid E, Connell J, Edwards TL, Duley S, Brown SE, Sanderson CM (2005) The hereditary spastic paraplegia protein spastin interacts with the ESCRT-III complex-associated endosomal protein CHMP1B. Hum Mol Genet 14(1):19–38
139. Skibinski G, Parkinson NJ, Brown JM, Chakrabarti L, Lloyd SL, Hummerich H, Nielsen JE, Hodges JR, Spillantini MG, Thusgaard T, Brandner S, Brun A, Rossor MN, Gade A, Johannsen P, Sorensen SA, Gydesen S, Fisher EM, Collinge J (2005) Mutations in the endosomal ESCRTIII-complex subunit CHMP2B in frontotemporal dementia. Nat Genet 37(8):806–808
140. Berman ER, Livni N, Shapira E, Merin S, Levij IS (1974) Congenital corneal clouding with abnormal systemic storage bodies: a new variant of mucolipidosis. J Pediatr 84(4):519–526
141. Goldin E, Cooney A, Kaneski CR, Brady RO, Schiffmann R (1999) Mucolipidosis IV consists of one complementation group. Proc Natl Acad Sci U S A 96(15):8562–8566
142. Riedel KG, Zwaan J, Kenyon KR, Kolodny EH, Hanninen L, Albert DM (1985) Ocular abnormalities in mucolipidosis IV. Am J Ophthalmol 99 (2):125–136
143. Slaugenhaupt SA, Acierno JS Jr, Helbling LA, Bove C, Goldin E, Bach G, Schiffmann R, Gusella JF (1999) Mapping of the mucolipidosis type IV gene to chromosome 19p and definition of founder haplotypes. Am J Hum Genet 65(3):773–778
144. Vergarajauregui S, Connelly PS, Daniels MP, Puertollano R (2008) Autophagic dysfunction in mucolipidosis type IV patients. Hum Mol Genet 17 (17):2723–2737
145. Ahuja M, Park S, Shin DM, Muallem S (2016) TRPML1 as lysosomal fusion guard. Channels (Austin) 10(4):261–263
146. Medina DL, Ballabio A (2015) Lysosomal calcium regulates autophagy. Autophagy 11(6):970–971
147. Medina DL, Di Paola S, Peluso I, Armani A, De Stefani D, Venditti R, Montefusco S, Scotto-Rosato-A, Prezioso C, Forrester A, Settembre C, Wang W, Gao Q, Xu H, Sandri M, Rizzuto R, De Matteis MA, Ballabio A (2015) Lysosomal calcium signalling regulates autophagy through calcineurin and TFEB. Nat Cell Biol 17(3):288–299
148. Settembre C, Di Malta C, Polito VA, Garcia Arencibia M, Vetrini F, Erdin S, Erdin SU, Huynh T, Medina D, Colella P, Sardiello M, Rubinsztein DC, Ballabio A (2011) TFEB links autophagy to lysosomal biogenesis. Science 332 (6036):1429–1433
149. Scotto Rosato A, Montefusco S, Soldati C, Di Paola S, Capuozzo A, Monfregola J, Polishchuk E, Amabile A, Grimm C, Lombardo A, De Matteis MA, Ballabio A, Medina DL (2019) TRPML1 links lysosomal calcium to autophagosome biogenesis through the activation of the CaMKKbeta/VPS34 pathway. Nat Commun 10(1):5630
150. LaPlante JM, Falardeau J, Sun M, Kanazirska M, Brown EM, Slaugenhaupt SA, Vassilev PM (2002) Identification and characterization of the single channel function of human mucolipin-1 implicated in mucolipidosis type IV, a disorder affecting the lysosomal pathway. FEBS Lett 532(1–2):183–187
151. LaPlante JM, Sun M, Falardeau J, Dai D, Brown EM, Slaugenhaupt SA, Vassilev PM (2006) Lysosomal exocytosis is impaired in mucolipidosis type IV. Mol Genet Metab 89(4):339–348
152. LaPlante JM, Ye CP, Quinn SJ, Goldin E, Brown EM, Slaugenhaupt SA, Vassilev PM (2004) Functional links between mucolipin-1 and Ca^{2+}-dependent membrane trafficking in mucolipidosis IV. Biochem Biophys Res Commun 322 (4):1384–1391
153. LaPlante JM, Falardeau JL, Brown EM, Slaugenhaupt SA, Vassilev PM (2011) The cation channel mucolipin-1 is a bifunctional protein that facilitates membrane remodeling via its serine lipase domain. Exp Cell Res 317(6):691–705
154. Medina DL, Fraldi A, Bouche V, Annunziata F, Mansueto G, Spampanato C, Puri C, Pignata A, Martina JA, Sardiello M, Palmieri M, Polishchuk R, Puertollano R, Ballabio A (2011) Transcriptional activation of lysosomal exocytosis promotes cellular clearance. Dev Cell 21(3):421–430

155. Terman A, Gustafsson B, Brunk UT (2006) The lysosomal-mitochondrial axis theory of postmitotic aging and cell death. Chem Biol Interact 163 (1–2):29–37
156. Terman A, Gustafsson B, Brunk UT (2006) Mitochondrial damage and intralysosomal degradation in cellular aging. Mol Asp Med 27(5–6):471–482
157. Spooner E, McLaughlin BM, Lepow T, Durns TA, Randall J, Upchurch C, Miller K, Campbell EM, Fares H (2013) Systematic screens for proteins that interact with the mucolipidosis type IV protein TRPML1. PLoS One 8(2):e56780
158. Jennings JJ Jr, Zhu JH, Rbaibi Y, Luo X, Chu CT, Kiselyov K (2006) Mitochondrial aberrations in mucolipidosis Type IV. J Biol Chem 281 (51):39041–39050
159. Evans AM, Wyatt CN, Kinnear NP, Clark JH, Blanco EA (2005) Pyridine nucleotides and calcium signalling in arterial smooth muscle: from cell physiology to pharmacology. Pharmacol Ther 107(3):286–313
160. Galione A (2006) NAADP, a new intracellular messenger that mobilizes Ca^{2+} from acidic stores. Biochem Soc Trans 34(Pt 5):922–926
161. Yamasaki M, Churchill GC, Galione A (2005) Calcium signalling by nicotinic acid adenine dinucleotide phosphate (NAADP). FEBS J 272 (18):4598–4606
162. Dammermann W, Guse AH (2005) Functional ryanodine receptor expression is required for NAADP-mediated local Ca^{2+} signaling in T-lymphocytes. J Biol Chem 280(22):21394–21399
163. Gerasimenko J, Maruyama Y, Tepikin A, Petersen OH, Gerasimenko O (2003) Calcium signalling in and around the nuclear envelope. Biochem Soc Trans 31(Pt 1):76–78
164. Gerasimenko JV, Maruyama Y, Yano K, Dolman NJ, Tepikin AV, Petersen OH, Gerasimenko OV (2003) NAADP mobilizes Ca^{2+} from a thapsigargin-sensitive store in the nuclear envelope by activating ryanodine receptors. J Cell Biol 163(2):271–282
165. Hohenegger M, Suko J, Gscheidlinger R, Drobny H, Zidar A (2002) Nicotinic acid-adenine dinucleotide phosphate activates the skeletal muscle ryanodine receptor. Biochem J 367(Pt 2):423–431
166. Langhorst MF, Schwarzmann N, Guse AH (2004) Ca^{2+} release via ryanodine receptors and Ca^{2+} entry: major mechanisms in NAADP-mediated Ca^{2+} signaling in T-lymphocytes. Cell Signal 16(11):1283–1289
167. Mojzisova A, Krizanova O, Zacikova L, Kominkova V, Ondrias K (2001) Effect of nicotinic acid adenine dinucleotide phosphate on ryanodine calcium release channel in heart. Pflugers Arch 441 (5):674–677
168. Churchill GC, Okada Y, Thomas JM, Genazzani AA, Patel S, Galione A (2002) NAADP mobilizes Ca(2+) from reserve granules, lysosome-related organelles, in sea urchin eggs. Cell 111(5):703–708
169. Kinnear NP, Boittin FX, Thomas JM, Galione A, Evans AM (2004) Lysosome-sarcoplasmic reticulum junctions. A trigger zone for calcium signaling by nicotinic acid adenine dinucleotide phosphate and endothelin-1. J Biol Chem 279(52):54319–54326
170. Kinnear NP, Wyatt CN, Clark JH, Calcraft PJ, Fleischer S, Jeyakumar LH, Nixon GF, Evans AM (2008) Lysosomes co-localize with ryanodine receptor subtype 3 to form a trigger zone for calcium signalling by NAADP in rat pulmonary arterial smooth muscle. Cell Calcium 44(2):190–201
171. Evans AM, Cannell MB (1997) The role of L-type Ca^{2+} current and Na^{+} current-stimulated Na/Ca exchange in triggering SR calcium release in guinea-pig cardiac ventricular myocytes. Cardiovasc Res 35(2):294–302
172. Zhu MX, Ma J, Parrington J, Calcraft PJ, Galione A, Evans AM (2010) Calcium signaling via two-pore channels: local or global, that is the question. Am J Physiol Cell Physiol 298(3):C430–C441
173. Fleischer S, Inui M (1989) Biochemistry and biophysics of excitation-contraction coupling. Annu Rev Biophys Biophys Chem 18:333–364
174. Franco L, Zocchi E, Calder L, Guida L, Benatti U, De Flora A (1994) Self-aggregation of the transmembrane glycoprotein CD38 purified from human erythrocytes. Biochem Biophys Res Commun 202 (3):1710–1715
175. Galione A, Lee HC, Busa WB (1991) Ca(2+)-induced Ca^{2+} release in sea urchin egg homogenates: modulation by cyclic ADP-ribose. Science 253 (5024):1143–1146
176. Galione A, White A, Willmott N, Turner M, Potter BV, Watson SP (1993) cGMP mobilizes intracellular Ca^{2+} in sea urchin eggs by stimulating cyclic ADP-ribose synthesis. Nature 365(6445):456–459
177. Hirst DG, Kennovin GD, Flitney FW (1994) The radiosensitizer nicotinamide inhibits arterial vasoconstriction. Br J Radiol 67(800):795–799
178. Dai J, Kuo KH, Leo JM, van Breemen C, Lee CH (2005) Rearrangement of the close contact between the mitochondria and the sarcoplasmic reticulum in airway smooth muscle. Cell Calcium 37(4):333–340
179. Poburko D, Kuo KH, Dai J, Lee CH, van Breemen C (2004) Organellar junctions promote targeted Ca^{2+} signaling in smooth muscle: why two membranes are better than one. Trends Pharmacol Sci 25(1):8–15
180. Boittin FX, Galione A, Evans AM (2002) Nicotinic acid adenine dinucleotide phosphate mediates Ca^{2+} signals and contraction in arterial smooth muscle via a two-pool mechanism. Circ Res 91(12):1168–1175
181. Asanuma K, Mundel P (2003) The role of podocytes in glomerular pathobiology. Clin Exp Nephrol 7 (4):255–259
182. Li G, Li CX, Xia M, Ritter JK, Gehr TW, Boini K, Li PL (2015) Enhanced epithelial-to-mesenchymal transition associated with lysosome dysfunction in podocytes: role of p62/Sequestosome 1 as a signaling hub. Cell Physiol Biochem 35(5):1773–1786
183. Xiong J, Xia M, Xu M, Zhang Y, Abais JM, Li G, Riebling CR, Ritter JK, Boini KM, Li PL (2013)

Autophagy maturation associated with CD38-mediated regulation of lysosome function in mouse glomerular podocytes. J Cell Mol Med 17 (12):1598–1607
184. Boini KM, Xia M, Xiong J, Li C, Payne LP, Li PL (2012) Implication of CD38 gene in podocyte epithelial-to-mesenchymal transition and glomerular sclerosis. J Cell Mol Med 16(8):1674–1685
185. Boini KM, Zhang C, Xia M, Han WQ, Brimson C, Poklis JL, Li PL (2010) Visfatin-induced lipid raft redox signaling platforms and dysfunction in glomerular endothelial cells. Biochim Biophys Acta 1801 (12):1294–1304
186. Hall JE, Henegar JR, Dwyer TM, Liu J, Da Silva AA, Kuo JJ, Tallam L (2004) Is obesity a major cause of chronic kidney disease? Adv Ren Replace Ther 11 (1):41–54
187. Davis S, Nehus E, Inge T, Zhang W, Setchell K, Mitsnefes M (2018) Effect of bariatric surgery on urinary sphingolipids in adolescents with severe obesity. Surg Obes Relat Dis 14(4):446–451
188. Kakazu E, Mauer AS, Yin M, Malhi H (2016) Hepatocytes release ceramide-enriched pro-inflammatory extracellular vesicles in an IRE1alpha-dependent manner. J Lipid Res 57 (2):233–245
189. Podbielska M, Szulc ZM, Kurowska E, Hogan EL, Bielawski J, Bielawska A, Bhat NR (2016) Cytokine-induced release of ceramide-enriched exosomes as a mediator of cell death signaling in an oligodendroglioma cell line. J Lipid Res 57(11):2028–2039
190. Wang G, Dinkins M, He Q, Zhu G, Poirier C, Campbell A, Mayer-Proschel M, Bieberich E (2012) Astrocytes secrete exosomes enriched with proapoptotic ceramide and prostate apoptosis response 4 (PAR-4): potential mechanism of apoptosis induction in Alzheimer disease (AD). J Biol Chem 287(25):21384–21395
191. Hessvik NP, Llorente A (2018) Current knowledge on exosome biogenesis and release. Cell Mol Life Sci 75(2):193–208
192. Bruno S, Porta S, Bussolati B (2016) Extracellular vesicles in renal tissue damage and regeneration. Eur J Pharmacol 790:83–91
193. Erdbrugger U, Le TH (2016) Extracellular vesicles in renal diseases: more than novel biomarkers? J Am Soc Nephrol 27(1):12–26
194. Pomatto MAC, Gai C, Bussolati B, Camussi G (2017) Extracellular vesicles in renal pathophysiology. Front Mol Biosci 4:37
195. Bao JX, Zhang QF, Wang M, Xia M, Boini KM, Gulbins E, Zhang Y, Li PL (2017) Implication of CD38 gene in autophagic degradation of collagen I in mouse coronary arterial myocytes. Front Biosci (Landmark Ed) 22:558–569
196. Xu M, Li X, Walsh SW, Zhang Y, Abais JM, Boini KM, Li PL (2013) Intracellular two-phase Ca^{2+} release and apoptosis controlled by TRP-ML1 channel activity in coronary arterial myocytes. Am J Physiol Cell Physiol 304(5):C458–C466
197. Shanahan CM, Crouthamel MH, Kapustin A, Giachelli CM (2011) Arterial calcification in chronic kidney disease: key roles for calcium and phosphate. Circ Res 109(6):697–711
198. Demer LL, Tintut Y (2008) Vascular calcification: pathobiology of a multifaceted disease. Circulation 117(22):2938–2948
199. Ehara S, Kobayashi Y, Yoshiyama M, Shimada K, Shimada Y, Fukuda D, Nakamura Y, Yamashita H, Yamagishi H, Takeuchi K, Naruko T, Haze K, Becker AE, Yoshikawa J, Ueda M (2004) Spotty calcification typifies the culprit plaque in patients with acute myocardial infarction: an intravascular ultrasound study. Circulation 110(22):3424–3429
200. Mackey RH, Venkitachalam L, Sutton-Tyrrell K (2007) Calcifications, arterial stiffness and atherosclerosis. Adv Cardiol 44:234–244
201. Shroff RC, Shanahan CM (2007) The vascular biology of calcification. Semin Dial 20(2):103–109
202. Kapustin AN, Chatrou ML, Drozdov I, Zheng Y, Davidson SM, Soong D, Furmanik M, Sanchis P, De Rosales RT, Alvarez-Hernandez D, Shroff R, Yin X, Muller K, Skepper JN, Mayr M, Reutelingsperger CP, Chester A, Bertazzo S, Schurgers LJ, Shanahan CM (2015) Vascular smooth muscle cell calcification is mediated by regulated exosome secretion. Circ Res 116(8):1312–1323
203. Kapustin AN, Schoppet M, Schurgers LJ, Reynolds JL, McNair R, Heiss A, Jahnen-Dechent W, Hackeng TM, Schlieper G, Harrison P, Shanahan CM (2017) Prothrombin loading of vascular smooth muscle cell-derived exosomes regulates coagulation and calcification. Arterioscler Thromb Vasc Biol 37(3):e22–e32
204. Kapustin AN, Shanahan CM (2016) Emerging roles for vascular smooth muscle cell exosomes in calcification and coagulation. J Physiol 594(11):2905–2914
205. Bhat OM, Li G, Yuan X, Huang D, Gulbins E, Kukreja RC, Li PL (2020) Arterial medial calcification through enhanced small extracellular vesicle release in smooth muscle-specific Asah1 gene knockout mice. Sci Rep 10(1):1645
206. Bhat OM, Yuan X, Camus S, Salloum FN, Li PL (2020) Abnormal lysosomal positioning and small extracellular vesicle secretion in arterial stiffening and calcification of mice lacking mucolipin 1 gene. Int J Mol Sci 21(5):1713
207. Fasano T, Pisciotta L, Bocchi L, Guardamagna O, Assandro P, Rabacchi C, Zanoni P, Filocamo M, Bertolini S, Calandra S (2012) Lysosomal lipase deficiency: molecular characterization of eleven patients with Wolman or cholesteryl ester storage disease. Mol Genet Metab 105(3):450–456
208. Ryter SW, Lee SJ, Smith A, Choi AM (2010) Autophagy in vascular disease. Proc Am Thorac Soc 7(1):40–47
209. Seedorf U, Wiebusch H, Muntoni S, Christensen NC, Skovby F, Nickel V, Roskos M, Funke H, Ose L,

Assmann G (1995) A novel variant of lysosomal acid lipase (Leu336-->Pro) associated with acid lipase deficiency and cholesterol ester storage disease. Arterioscler Thromb Vasc Biol 15(6):773–778
210. Bampton ET, Goemans CG, Niranjan D, Mizushima N, Tolkovsky AM (2005) The dynamics of autophagy visualized in live cells: from autophagosome formation to fusion with endo/lysosomes. Autophagy 1(1):23–36
211. Martinet W, De Meyer GR (2009) Autophagy in atherosclerosis: a cell survival and death phenomenon with therapeutic potential. Circ Res 104 (3):304–317
212. Gutierrez MG, Master SS, Singh SB, Taylor GA, Colombo MI, Deretic V (2004) Autophagy is a defense mechanism inhibiting BCG and Mycobacterium tuberculosis survival in infected macrophages. Cell 119(6):753–766
213. Kim JS, Nitta T, Mohuczy D, O'Malley KA, Moldawer LL, Dunn WA Jr, Behrns KE (2008) Impaired autophagy: a mechanism of mitochondrial dysfunction in anoxic rat hepatocytes. Hepatology 47 (5):1725–1736
214. Zhu H, Tannous P, Johnstone JL, Kong Y, Shelton JM, Richardson JA, Le V, Levine B, Rothermel BA, Hill JA (2007) Cardiac autophagy is a maladaptive response to hemodynamic stress. J Clin Invest 117 (7):1782–1793
215. Jia G, Cheng G, Gangahar DM, Agrawal DK (2006) Insulin-like growth factor-1 and TNF-alpha regulate autophagy through c-jun N-terminal kinase and Akt pathways in human atherosclerotic vascular smooth cells. Immunol Cell Biol 84(5):448–454
216. Martinet W, De Meyer GR (2008) Autophagy in atherosclerosis. Curr Atheroscler Rep 10(3):216–223
217. Xu K, Yang Y, Yan M, Zhan J, Fu X, Zheng X (2010) Autophagy plays a protective role in free cholesterol overload-induced death of smooth muscle cells. J Lipid Res 51(9):2581–2590
218. Jia G, Cheng G, Agrawal DK (2007) Autophagy of vascular smooth muscle cells in atherosclerotic lesions. Autophagy 3(1):63–64
219. Schrijvers DM, De Meyer GR, Herman AG, Martinet W (2007) Phagocytosis in atherosclerosis: molecular mechanisms and implications for plaque progression and stability. Cardiovasc Res 73(3):470–480
220. Verheye S, Martinet W, Kockx MM, Knaapen MW, Salu K, Timmermans JP, Ellis JT, Kilpatrick DL, De Meyer GR (2007) Selective clearance of macrophages in atherosclerotic plaques by autophagy. J Am Coll Cardiol 49(6):706–715
221. Zhang Y, Xu M, Xia M, Li X, Boini KM, Wang M, Gulbins E, Ratz PH, Li PL (2014) Defective autophagosome trafficking contributes to impaired autophagic flux in coronary arterial myocytes lacking CD38 gene. Cardiovasc Res 102(1):68–78
222. Xu M, Li XX, Wang L, Wang M, Zhang Y, Li PL (2015) Contribution of Nrf2 to atherogenic phenotype switching of coronary arterial smooth muscle cells lacking CD38 gene. Cell Physiol Biochem 37 (2):432–444
223. Adiguzel E, Ahmad PJ, Franco C, Bendeck MP (2009) Collagens in the progression and complications of atherosclerosis. Vasc Med 14 (1):73–89
224. Xu X, Yuan X, Li N, Dewey WL, Li PL, Zhang F (2016) Lysosomal cholesterol accumulation in macrophages leading to coronary atherosclerosis in CD38(−/−) mice. J Cell Mol Med 20(6):1001–1013

Store-Operated Calcium Entry in the Cardiovascular System

14

Xian Liu and Zui Pan

Abstract

Calcium (Ca^{2+}) is a critical regulator of cardiovascular function. The Ca^{2+} channels, pumps, and exchangers contributing to cytosolic Ca^{2+} signals governing cardiac contraction and vascular tone are well known. In addition to these Ca^{2+} components, store-operated calcium entry (SOCE) is a ubiquitous mechanism recently recognized underlying cardiovascular function maintenance and disease development and progression. With this review article, we hope to highlight the accumulated knowledge about the SOCE machinery and its potential contribution to cardiac and vascular function and its roles in cardiovascular pathogenesis and pathology.

Keywords

SOCE · Orai · STIM · Calcium signaling · Cardiovascular disease

X. Liu · Z. Pan (✉)
Department of Kinesiology, The University of Texas at Arlington, Arlington, TX, USA

Department of Graduate Nursing, College of Nursing and Health Innovation, The University of Texas at Arlington, Arlington, TX, USA
e-mail: zui.pan@uta.edu

Abbreviations

2-APB	2-Aminoethoxydiphenyl borate
AP	Action potential
apoE	Apolipoprotein E
Ca^{2+}	Calcium
CaMK	Calmodulin-dependent protein kinase
CICR	Calcium-induced calcium release
CRAC	Calcium release-activated channel
EC	Endothelial cell
EPC	Endothelial precursor cell
I_{CRAC}	Calcium-release activated current
NFAT	Nuclear factor of activated T-cells
NO	Nitric oxide
PASMC	Pulmonary artery smooth muscle cell
PDGF	Plate-derived growth factor
PM	Plasma membrane
RyR	Ryanodine receptor
SAM	Sterile alpha motif
SOAR	STIM1 Orai-activating region
SOCE	Store-operated calcium entry
SR	Sarcoplasmic reticulum
STIM1	Stromal-interacting molecule 1
TAC	Transverse aortic constriction
TM	Transmembrane domain
TRP	Transient receptor potential
t-tubule	Transverse tubule
VEGF	Vascular endothelial growth factor
VGCC	Voltage-gated calcium channel
VSMC	Vascular smooth muscle cell

L. Zhou (ed.), *Ion Channels in Biophysics and Physiology*, Advances in Experimental Medicine and Biology 1349, https://doi.org/10.1007/978-981-16-4254-8_14

14.1 Introduction

Store-operated calcium entry (SOCE) was originally identified in non-excitable cells about three decades ago, and its roles in skeletal muscle and cardiomyocytes have been recognized much later. It consists of two main components—Orai1 and Stromal-Interacting Molecule 1 (STIM1). Orai1 is located on the plasma membrane (PM) and STIM1 is located on the SR membrane. When SR Ca^{2+} stores are reduced, STIM1 is activated and forms patches and induces the aggregation of Orai1, which further triggers the activation of SOCE, leading to the flow of extracellular Ca^{2+} into the cell. This inward current is named calcium-release activated current (I_{CRAC}). Besides I_{CRAC}, another type of store-operated currents (I_{SOC}) not selective for Ca^{2+} has also been identified, which requires the presence and interaction among Orai, STIM, and transient receptor potential (TRP) cation channels (Fig. 14.1). This review will first introduce these channels and their roles in the maintenance of cardiac function and the development and progression of cardiac diseases in detail. Cardiovascular diseases are associated with Ca^{2+} dysregulation not only in the cardiomyocytes but also in the vascular cells and tissues. Since the evidence of dysregulated SOCE in vascular tissues is emerging, the second part of this review is focused on revealing the role of SOCE and its components in vascular function and their contribution to the vascular aspects in cardiovascular diseases.

14.2 Cardiac Excitation–Contraction Coupling

Ca^{2+} is a ubiquitous cellular second messenger and controls cardiac excitation–contraction coupling, which is the process from the myocyte electrical excitation to the heart contraction [1]. The heart contracts generating heart beat that involves the synchronized atrial and ventricular contraction in a process called "cardiac cycle" [2]. The cardiac cycle is initiated with the sinoatrial node formed with a special group of pacemaking myocytes in the right atrial wall. The SA node generates action potentials (APs) that propagate down through the atrial chambers and ventricular chambers, thereby leading to their contraction and generation of the pumping force pushing blood into the circulation system throughout the body [1].

On the sarcolemma, a region penetrating deep into the cell forms the transverse tubule (t-tubule). On the t-tubule locates the L-type voltage-gated calcium channels (VGCCs) (also known as dihydropyridine receptors), which can be activated by the cardiac AP. During a cardiac AP, Ca^{2+} enters the cardiomyocyte through depolarization-activated dihydropyridine receptor as inward Ca^{2+} current, contributing to the AP plateau [3]. This inward Ca^{2+} current then binds and activates another receptor called type 2 ryanodine receptor (RyR2) located in the terminal cisternae of SR, which is a calcium store located close to the t-tubule [4]. Activation of RyR2 causes SR to open and release even more calcium into the cell (calcium-induced calcium release, CICR) [5], leading to a calcium spark [6]. Both the Ca^{2+} influx and the CICR contribute to the increased concentration of free intracellular Ca^{2+}, which binds to the myofilament protein troponin C and switch on the contractile machinery [1], including the interaction between actin and myosin, thereby causing cardiac contraction [7].

14.3 Expression of SOCE Components in the Heart

14.3.1 STIM

Mammals have two STIM homologs, STIM1 and STIM2 [8].

STIM 1 is a single-pass transmembrane protein. Human STIM1 is a 90-kDa phosphoprotein made of 685 amino acids [9]. Despite the majority ubiquitously expressed in the intracellular organelles (e.g. endoplasmic reticulum-ER/SR), about 10–20% of STIM1 is also found localized in the PM, where its functions are unrelated to

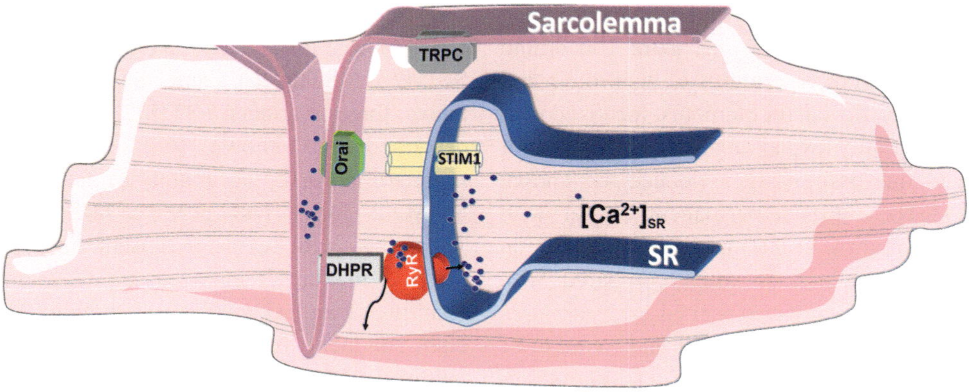

Fig. 14.1 Molecular machinery in the cardiomyocytes. On the sarcolemma of a cardiomyocyte, a region penetrating deep into the cell forms the transverse tubule (t-tubule). L-type voltage-gated calcium channels (VGCCs) (also known as dihydropyridine receptors, DHPR) are located on the t-tubule. During a cardiac action potential, Ca^{2+} enters the cardiomyocyte through depolarization-activated DHPR as inward Ca^{2+} current. These Ca^{2+} further bind and activate the ryanodine receptor (RyR), the major SR Ca^{2+} release channel located in the terminal cisternae of the SR and carry out excitation–contraction coupling. Another Ca^{2+} influx pathway is store-operated Ca^{2+} entry mediated by STIM1 as Ca^{2+} sensor located at SR and Orai or TRPC proteins as Ca^{2+} channel located at the sarcolemma and/or t-tubule

SOCE [10]. The carboxyl-terminus of STIM1 projects into the cytosol wherever its location (ER/SR or PM), whereas the amino-terminus is located either in the ER/SR lumen (for ER/SR membrane-resident STIM1) or the extracellular medium (for PM-resident STIM1) [11, 12]. The structure of functional domains of STIM1 has been studied and revealed. The amino-terminal regions include an ER/SR signal peptide, a canonical EF-hand Ca^{2+}-binding motif, a hidden EF-hand, and a sterile alpha motif (SAM) (aa 1–22, 63–96, 97–128, 132–200, respectively) [13, 14], followed by the transmembrane domain (aa 214–234). The ezrin/radixin/moesin domain-containing carboxyl (cytosol) region is highly conserved among the STIM proteins. This region includes three coiled-coil domains, CC1, CC2, and CC3, a calcium release-activated channel (CRAC)-modulatory domain, a proline/serine-rich region, and a polybasic lysine-rich region (aa 238–342, 364–389, 399–423, 470–491, 600–629, 671–685, respectively) [15–17].

Four groups almost simultaneously identified the key region of STIM1 for interaction with store-operated channels. This region overlapping with the coiled-coil domains of the cytosol region of STIM1 was given different names: STIM1 Orai-activating region (SOAR), Orai-activating small fragment, CRAC-activating domain, coiled-coil domain-containing region, with aa 344–442 [15], 233–450 [18], 342–448 [19], 339–444 [20].

STIM1 can go through a variety of post-translational modifications, including phosphorylation on serine [21] and tyrosine residues [22] and N-linked glycosylation on the SAM domain N^{131} and N^{171} [23].

Despite the many reports of STIM1 expressed in cardiomyocytes [24, 25], other cell types might have a higher expression level of STIM1. For example, in a study generating a novel cardiomyocyte-restricted STIM1 knockout mouse, the whole heart tissue demonstrated approximately a 20% reduction in total STIM1 protein expression in the knockout group compared to control, indicating the remaining 80% expressed in cell types other than cardiomyocytes in the heart [26].

STIM1L is a long splice variant of STIM1 and was found to be expressed in human skeletal muscle [27], neonatal rat cardiomyocytes [28], and differentiated myotubes [27]. STIM1L is reportedly expressed in myotubes during human myogenesis. This isoform has extra 106 amino acids on the carboxyl-terminus of the protein and is bound to cortical actin filaments, localized near the PM, partially explaining the rapid onset of SOCE in this tissue [27]. Evidence has shown that STIM1L forms a permanent cluster with Orai1 contributing to the rapid SOCE activation in skeletal muscle cells when compared with other cells [29].

STIM2 was identified in 2001 and found to be a transmembrane protein sharing a similar structure to that of STIM1. Unlike STIM1, so far there has been no report showing the presence of STIM2 in PM [12, 30]. However, STIM2 has been observed to be present in the ER membrane and acidic intracellular stores [8, 31, 32]. The amino-terminal regions of STIM2 include the EF-hand Ca^{2+}-binding motif and SAM domain (aa 67–100, 136–204, respectively). The ezrin/radixin/moesin domain-containing region at the carboxyl terminus is highly conserved with three coiled-coil domains just as STIM1 [32]. Overlapping with these coiled-coil domains is SOAR [33], playing roles in the activation of SOCE [34]. The function of the proline- and histidine-rich region next to the SOAR remains to be further investigated [21, 35]. There is a calmodulin-binding region and a polybasic lysine-rich region close to the end of the carboxyl-terminal region [35, 36]. Both STIM1 and STIM2 can activate Orai1, but STIM2 is a weaker activator likely as a result of the difference in the SOAR [37, 38] and SAM domains [39]. However, the lower affinity for Ca^{2+} by the EF-hand of STIM2 allows it to activate Orai1 at a lower agonist-induced stimulation level and subsequent ER store depletion [40, 41].

Both human and mouse tissues express STIM2 [21, 36], and it is the dominant homolog expressed in the mouse brain, pancreas, placenta, and heart but found almost absent in skeletal muscle, kidney, liver, and lung. The co-expression of STIM1 and STIM2 in many human cell lines [21] as well as cell types [42, 43] demonstrates the co-existence of these two isoforms and potential interaction.

Researchers have found three splice variants of STIM2 to this date: STIM2.1 (STIM2β), STIM2.2 (STIM2α), and STIM2.3. Among these three variants, STIM2.2 is the best described and known [44], and previously reported STIM2 is mainly referred to as STIM2.2 [45]. The gene *STIM2* comprises 13 exons with exon 9 absent in STIM2.2 mRNA, making it encoded by 12 exons and resulting in an 833-amino acid protein [34]. The eight-residue insert (383-VAASYLIQ-392) within the SOAR domain encoded by exon 9 in STIM2.1 was initially considered responsible for impairing the interaction of Orai1 and its activation [34]. Yet, new evidence has shown the opposite. The heterodimer comprising the SOAR regions of STIM1 and STIM2.1 has been shown capable of fully activating the Orai1 channel while preventing its crosslinking and clustering [46]. STIM2.1 expressed ubiquitously and might form a heterodimer with STIM1 and STIM2.2, attenuating SOCE-mediated Ca^{2+} influx [34]. STIM2.3 expresses an alternative exon 13, leading to translation at the upstream end, encoding a protein 17 kDa smaller [44]. The expression of STIM2.3 is limited and its function remains to be elucidated [44].

14.3.2 ORAI

The name of Orai is from Greek mythology, who is heaven's gatekeeper [47]. The human Orai family consists of three homologs, Orai1, Orai2, and Orai3, and cardiomyocytes express all these three homologs. In 1992, an intracellular Ca^{2+} stored discharge-activated current was identified and termed as the I_{CRAC} which is highly Ca^{2+}-selective and inwardly rectifying [48]. Orai1, the canonical isoform, was first identified as the channel conducting I_{CRAC} in 2006 through whole-genome screening of *Drosophila* S2 cells and gene mapping in I_{CRAC} deficiency-induced hereditary severe combined immune deficiency patients [49–51]. Orai1 is a 33-kDa protein of 301 amino acids that do not share homology with other known ion channels. It has four transmembrane domains (TM1–TM4) with both the amino and the carboxyl terminal tails located in the cytosol [52–55], which are both proved to be essential for STIM1 interaction and regulation [15, 18, 19, 56–58].

Initial studies have demonstrated that a tetramer is the most likely Orai1 subunit stoichiometry of the mammalian CRAC channels [59–61]. However, the crystallization of *Drosophila* Orai has identified the hexameric assembly of Orai subunits forming the channel [62]. Despite the unavailability of the mammalian crystal structure of Orai1 so far, a study analyzing the biophysical properties of hexameric and tetrameric human Orai1 has revealed the difference between these two stoichiometry hypotheses. A tetrameric structure is shown to display highly Ca^{2+}-selective conductance characteristics of I_{CRAC}, whereas the hexameric architecture forms a nonselective cation channel [63]. In the hexameric CRAC channel, six of the TM1 domains of Orai1 subunits form a hexamer with a pore including residues 74–90 (ETON region) of amino terminus at the center, with the ETON essential for STIM1 interaction [56]. Negatively charged residues (D110, D112, and D114) form an external vestibule working as a funnel (the pore) and attract Ca^{2+} to the pore, which is followed by the selectivity filter (aa E106), contributing to the CRAC channel high Ca^{2+} selectivity [64, 65], a hydrophobic region (aa V102, F99, and L95), and a basic region (aa R91, K87, and R83). The other three transmembrane domains TM2–TM4 surround the pore [62] with residues of these domains essential for regulating CRAC channel function.

Two splice variants of Orai1 (the long Orai1α and the short Orai1β) have been discovered through alternative translation-initiation sites at Methionine 1 and Methionine 64 from the same messenger RNA. Both variants mediate I_{CRAC} [66, 67], but Orai1β shows a much smaller Ca^{2+}-dependent inactivation [67].

All three Orai homologs are able to mediate I_{CRAC} with STIM1 overexpression [68, 69], but Orai2 and Orai3 mediate smaller currents than Orai1 [52], with a few specific exceptional studies [70, 71] and how they mediate native Ca^{2+} entry pathways remains to be explored [72].

14.3.3 TRPC

Besides I_{CRAC}, another type of nonselective Ca^{2+} store-operated currents (I_{SOC}) has been identified. They have different biophysical properties and exhibit a greater conductance than I_{CRAC} [73]. The interaction among STIM1, Orai1, and TRPC1 is required for the activation of I_{SOC} [67, 74–76]. The coiled-coil domains of both the amino and the carboxyl termini of TRP channels are relevant with the interaction with STIM1 [77]. TRPC subfamily includes seven members, that is, TRPC1–TRPC7 [78]. The first mammalian TRP protein was TRPC1, and it was identified in both humans [79, 80] and mice [81].

The involvement of TRP channels in SOCE has been widely investigated and heated debates have been present, with the TRPC subfamily in particular. Gated by store Ca^{2+} depletion, the TRPC channel's roles have been examined with a variety of tools including endogenously knocking down TRPs, generating TRPC knockout models, as well as overexpression of specific TRPC proteins [82–84]. Researchers have reached a consensus that TRPC1, STIM1, and Orai1 form a complex underlying the less selectiveI_{SOC} current. However, two hypotheses about the molecular basis of I_{SOC} still co-exist. The first hypothesis states that the I_{SOC} equals to I_{CRAC} plus a less selective TRPC-mediated current [85]. The second hypothesis insists there is still an undiscovered component contributing to I_{SOC} besides TRPC1 and Orai1 subunits [86].

Several TRPC subfamily proteins have been found in the heart [24, 87, 88]. It appears that TRPC proteins together with L-type VGCCs form a complex, responsible for developing heartbeat initiation [88]. Correlation has been found between the increased expression of TRPC channels and enhanced SOCE and spontaneous Ca^{2+} waves underlying arrhythmia [87]. During chronic cardiac diseases in both human and animal models, increased expression levels of almost all TRPC family members (TRPC1, TRPC3–7) have been identified [89–91]. In addition, both the diacylglycerol-sensitive (TRPC3/6/7) and the IP_3R-sensitive (TRPC1/4/5) subfamily members have been found responsible for store-dependent activation [90, 92, 93]. More details regarding the roles of TRPC channels in cardiomyocytes can be found in two recent excellent reviews [87, 88].

14.4 SOCE During Cardiac Development

At different stages of cardiac maturation, SOCE seems to be playing differentiated roles. In embryonic cardiomyocytes, SOCE appears to be more prominent, whereas with postnatal development it tends to decline. This is manifested by the relatively higher expression of STIM1 in early cardiomyocytes than later in adult cells [28]. Furthermore, the morphology, organization, and signaling properties are different in embryonic and neonatal cardiomyocytes compared to those in adult cells [94]. Many of the studies stating SOCE to be functional in cardiomyocytes are done within neonatal cells, and the results do not necessarily correspond to the situation in adult muscle.

However, numerous evidences have demonstrated the expression of molecular components of SOCE (such as STIM, Orai, and TRPC proteins) from embryo to adult.

In fact, STIM1 and Orai1 proteins have been identified by several groups simultaneously in embryonic, neonatal, and adult rat ventricular cardiomyocytes [95–99] and demonstrated them to be essential in SOCE. Moreover, a significant amount of a splice variant of STIM1 (STIM1L), whose expression decreases with cardiomyocyte maturation, was also found to be expressed in neonatal hearts [28]. Sabourin et al. reported a physical interaction between STIM1 and Orai1 induced by Ca^{2+} store depletion in neonatal cardiomyocytes [92]. It has also been reported in neonatal cardiomyocytes the SR Ca^{2+} depletion causes STIM1 to form puncta [97] and interact with Orai1 [99], which recapitulates studies in non-excitable cells. The co-localization of STIM1 and SERCA, phospholamban, and RyR has also been reported in neonatal and adult rat hearts [24, 99, 100].

TRPC1, STIM1, and Orai1 were found to be involved in the formation of SOCE channels in human cardiac c-kit^{+} progenitor cells and can regulate cell cycling and migration [101]. However, in a cardiomyocyte-restricted STIM1 knockout mouse model, the cardiac function of the mice did not decline before they reached 20 weeks of age [26]. By 36 weeks, the STIM1-deficient mice demonstrated marked left ventricular dilation. The presence of an inflammatory infiltrate and cardiac fibrosis also appeared from 20 weeks and progressively worsened by 36 weeks of age. These data demonstrate that deletion of STIM1 does not affect the early development of the heart but precipitates deleterious heart remodeling in postnatal mice. What's more, despite the exhibition of immunodeficiencies, muscular hypotonia, ectodermal dysplasia, autoimmunity, and lymphoproliferative diseases, Orai1-/STIM1-deficient patients or mice don't show overt cardiac muscle-related phenotypes and no prejudice to the cardiovascular function [102]. This is consistent with the idea that the SOCE pathway is not indispensable during cardiac muscle development and/or contractility under physiological circumstances. SOCE-associated cardiac dysfunction might be seen with aging, stress stimuli, or prolonged exercise but couldn't be examined due to the death in utero or in the early life of patients or deficient mice. Interestingly, the overexpression of STIM1 in cardiomyocytes is also detrimental. Mice with cardiomyocyte-specific overexpression showed no phenotype until 10 weeks of age when they started demonstrating declined cardiac function [24]. These studies indicate the significance of STIM1 in the cardiomyocyte structure and function; either too little or too much STIM1 can contribute to pathological changes in cardiomyocytes. However, it is still elusive whether these changes caused by reducing or enhancing STIM1 are due to changes in SOCE.

14.5 SOCE in the Vascular System

The blood vessel walls mainly consist of two cell types, the vascular smooth muscle cells (VSMCs) and endothelial cells (ECs). The majority of VSMCs are in the tunica, especially for muscular arteries and arterioles. Despite VSMCs being maintained in a partially constricted state, the dynamic regulation of intracellular Ca^{2+} concentrations can rapidly adjust to the vascular tone and maintain hemodynamic stability [103]. Normally, VSMCs are quiescent and contractile and infrequently divide; however, they can switch to a proliferative and migratory state in certain circumstances, such as arterial injury or inflammation [104–106]. This change is reckoned as a critical step in the pathogenesis of multiple vascular diseases, including arterial stenosis in atherosclerosis, neointimal hyperplasia following angioplasty or stent placement, and arteriolar remodeling in hypertension. ECs are most abundant in the intimal layer of the blood vessel and are involved in various functions, such as selective barrier formation for cellular and nutrient trafficking, local blood flow, and vascular tone modulation as well as oxidative stress and inflammation, thrombosis, hemostasis, and VSMC proliferation resistance [107]. Notably, ECs demonstrate distinctive diversity indicated by their significantly heterogeneous structural and functional phenotypes throughout the cardiovascular system [108]. Similar to VSMCs, ECs display proliferative and migratory phenotypes under pathological conditions [109]. Understanding the signaling pathways underlying the switch of phenotypes in VSMCs and ECs is essential for the development of targeted therapies against vascular diseases.

Thrombin is a physiological agonist for proteinase-activated receptors located on both VSMCs and ECs. Despite the limited expression of the proteinase-activated receptors on healthy artery VSMCs, an increased expression of these receptors is seen in vascular lesions [110]. The activation of thrombin is crucial to mediate VSMC contraction, proliferation, migration, hypertrophy, and extracellular matrix production [111]; therefore, its signaling in VSMCs contributes importantly to the pathogenesis of vascular occlusive diseases. Synthetic VSMCs stimulated with thrombin in vitro are reported to enable the activation of an innovative non-SOCE/CRAC but STIM1/Orai1/Orai3-requiring and highly selective Ca^{2+} entry pathway [112], similar to the arachidonic acid-modulated Ca^{2+} channels [113, 114]. Thrombin activates store-independent leukotrieneC4-regulated Ca^{2+} entry channels in VSMCs [112]. The involvement of either ER or plasma membrane STIM1 pools, the oligomeric state of functional STIM1, and the nature of the interactions between STIM1 and Orai1/3 LRC channels remains to be further elucidated. The LRC channels are the first store-independent Orai channels reported to be stimulated with a physiological agonist in the vasculature. On the other hand, thrombin in ECs has been reported to activate store-dependent CRAC channels [115]. In conclusion, not only can we see distinct effects of different agonists in the same cell type, but the same agonist can result in diverse Ca^{2+} entry pathways in different cell types.

14.5.1 SOCE in Vascular Smooth Muscle Cells

The contribution of SOCE to vascular smooth muscle cell (VSMC) growth, proliferation, and migration has been demonstrated in studies using vasoactive agonists and growth factors. The growth and proliferation of aortic VSMCs have been reported to be mediated by STIM1 and Orai1-dependent SOCE when stimulated by angiotensin-II [116, 117]. In addition, treatment of urotensin-II promotes VSMCs proliferation and Ca^{2+}/cAMP response element-binding protein activation. These require a complex signaling pathway involving on the one hand SOCE mediated by STIM1, Orai1, and TRPC1, and on the other hand epithelium growth factor receptor, extracellular signal-regulated kinase, and calmodulin-dependent protein kinase (CaMK) activation [118]. Meanwhile, the paracrine release of plate-derived growth factor (PDGF) by VSMCs, ECs, and macrophages promotes VSMC migration, while induction of intimal proliferation is considered as a secondary phenomenon [119]. STIM1- and Orai1-(but not TRPC1/4/6- nor Orai2/3-) mediated SOCE specifically contributes to the PDGF-induced Ca^{2+} entry [120], with STIM1 and Orai1 being the essential components for VSMCs and airway smooth muscle migration when stimulated with PDGF in vitro [120, 121]. On the contrary, in smooth muscle-specific STIM1 knockout mice, PDGF-induced VSMC proliferation is strongly reduced as a result of SOCE-dependent nuclear factor of activated T-cell (NFAT) activation reduction [122]. Similarly, studies using PDGF receptor inhibitors demonstrate neointima formation attenuation in animal models of restenosis [123–125].

Although the contribution of SOCE to VSMC growth, proliferation, and migration has been well established, as shown above, its role in VSMC contractility remains contentious. A number of studies have used non-specific SOCE inhibitors like 2-Aminoethoxydiphenyl borate (2-APB) and lanthanides to examine the role of SOCE in VSMC contractility [126–128]. In freshly isolated contractile VSMCs compared with cultured synthetic proliferative VSMCs, minimal SOCE/CRAC activity and STIM1/Orai1 protein expression levels have been reported [112, 129–131]. However, the contribution of CRAC channels to vascular tone can be achieved by their vasorelaxation roles mediated by ECs, delivered by Ca^{2+}-modulated nitric oxide (NO) production [132]. Vasoactive agonists increase cytosolic endothelial Ca^{2+} concentrations, activate endothelial No synthase, and initiate NO-dependent vasodilation [133]. In porcine aortic ECs, STIM1 inhibition blocked SOCE and was associated with thrombin-induced partial NO production decrease [134]. Meanwhile, in mouse aortic ECs, the reduced SOCE activity is associated with impaired acetylcholine-induced vasorelaxation [135]. In addition, in diabetic mice, decreased STIM1 protein expression caused impaired ER Ca^{2+} refilling, attenuates endothelium-dependent vasorelaxation in coronary arteries, while STIM1 overexpression has a beneficial and therapeutic effect on coronary endothelial dysfunction in diabetes [136].

Interestingly, although detectable in cultured human aortic VSMCs, Orai1 is much less expressed than the abundant Orai2 and Orai3 when normalized to expression levels in lymphocytes [131]. Orai2 and Orai3 expression levels are also higher in cultured synthetic aortic VSMCs than in contractile cells [112, 128]. Despite the high expression levels, Orai2 and Orai3 being silenced in vitro did not change the SOCE amplitude in synthetic VSMCs, indicating only Orai1 as the component of CRAC channels in these cells [130]. However, after carotid injury, the expression level of Orai3 was increased and in vivo knockdown of Orai3 significantly inhibited neointima formation [112], indicating the contribution of Orai3 in channels or signaling pathways other than SOCE (namely store-independent leukotrieneC4-regulated Ca^{2+} entry channels), thus promoting VSMC synthetic phenotypes.

In addition to the abovementioned in vitro data, the important contrition of SOCE to VSMC proliferation and migration on the luminal side of injured vessels to form neointima has also been demonstrated in vivo, taking advantage of the animal model of angioplasty, accomplished by mechanical injury procedure of rat carotid arteries [137]. In fact, in the medial and neointimal layer, the lentivirus-mediated STIM1 and Orai1 inhibition reduces the formation of neointima and prevents the increase of these two proteins 14 days after injury. Furthermore, knockdown of STIM1 and Orai1 blocks VSMC proliferation reduction-associated NFAT activation [129]. Similarly, angiotensin-II significantly upregulates STIM1 and Orai1 expression levels in the neointimal layer [117, 138]. When STIM1 and Orai1 are silenced, the angiotensin-II-induced VSMCs proliferation and accelerated neointimal growth are suppressed [117, 138]. In line with this, after carotid artery ligation, the smooth muscle-specific STIM1 knockout mice demonstrate significantly reduced neointimal formation compared with control mice [122]. A recent study identifies the scaffolding protein Homer1 (which can bind to several Ca^{2+}-signaling molecules [139]) as a binding partner for Orai1 and TRPC channels forming SOC complexes in the neointima and that Homer1 can regulate the migration and proliferation of VSMCs [140].

14.5.2 SOCE in Endothelial Cells

Endothelial cells have the ability to produce both potent vasoconstrictor endothelin-1 and vasodilator NO. When there is damage in the vessels (for example caused by tissue hypoxia [109]), they need to be replaced by newly grown ones either from preexisting blood vessels or from bone marrow-generated endothelial precursor cells (EPCs). However, the growth of new vessels is also a contributor to the development of tumor growth and metastasis-promoting vascular networks [109]. Upon stimulation of various growth factors, vascular endothelial growth factor (VEGF) in particular, signaling pathways are activated and induce the proliferation and migration of ECs [109]. The activation of the VEGF receptor triggers IP_3-mediated Ca^{2+} release and induces a low Ca^{2+} conductance across the plasma membrane, which is essential for EC proliferation [141, 142]. This conductance was initially difficult to measure due to its relatively low and variable amplitude in endothelial cells. The first evidence demonstrates that ECs can activate store-operated small inward rectifying CRAC-like current induced by IP_3, thapsigargin, or Ca^{2+} ionophores [143]. Following that, in the human umbilical vein, ECs STIM1 and Orai1 are shown to be required in mediating CRAC currents post passive store depletion, and CRAC channels are shown to contribute to EC proliferation [115]. Genetic inhibition of Orai1 using siRNA or a dominant-negative Orai1 mutant, as well as pharmacological inhibition of Orai1 using the CRAC channel blocker S66, are both able to inhibit and prevent the endothelial tube from forming in vitro [144]. Conversely, in primary human umbilical vein ECs, STIM1, TRPC1, and TRPC4, but not Orai1, are associated with the endothelial tube formation [145], indicating a potential dissociation of STIM1 from Orai1 in tubulogenesis. Meanwhile, knockdown of STIM1 or STIM2 only shows mild inhibitory effects compared to that from knockdown of Orai1 [115]. Consistently, when STIM1 is knocked down specifically in EC in the mice, no abnormal endothelial migration or vasculogenesis is observed [146].

EPCs exist in postnatal circulation and they proliferate, migrate, and develop into mature EC phenotype [147]. The expression of SOCE in EPC was first demonstrated in 2010 [148]. The SOCE in EPC was further characterized by evidence showing enhanced STIM1-mediated SOCE upon stimulation of hepatocyte grow factor [149], with STIM1 being essential for EPC proliferation and migration [138, 149, 150]. SOCE-mediated IP_3-dependent calcium oscillations are seen when EPCs are treated with VEGF [151]. All these data provide evidence demonstrating the involvement of CRAC channels in EPC Ca^{2+} signaling pathways regulating proliferation and tubulogenesis.

SOCE is identified as a key mechanism in tissue vascularization, vascular repair, and vasculogenesis, which is the de novo formation of new blood vessels.

Angiogenesis is the expansion of the existing vascular system, which first takes place during embryonic development, and also continues in the mature animals most commonly seen in either wound healing [152] or tumor metastasis [153]. In angiogenesis, the sprouting of endothelial cells is mainly controlled by two categories of tyrosine receptors, the VEGF receptors and the tie and tek kinases [154]. Cells surrounding the vessels synthesize and secrete angiopoietin-1 and angiopoietin-2, whose endothelial receptors are tie-2 [155]. The tie-2/angiopoietin signaling pathway is able to induce the production of growth factors like PDGF and VEGF, which trigger the mesenchymal cells to differentiate into pericytes or smooth muscle cells to form vessel wall [156]. Interestingly, angiopoietin-1 inhibits VEGF-triggered SOCE, thus negatively regulating and protecting the endothelial cells from growth factor-induced hyperpermeability [157].

Vasculogenesis is the de novo formation of the vascular system in the early embryo, while in the mature organism, a similar process is responsible for vascular repair. At the initial stage, both vascular and hematopoietic tissue development occur simultaneously, involving angioblasts and hematopoietic stem cells. Then endodermal cells secrete growth factors to induce vasculogenesis in the embryo [158–160]. Angioblasts and endothelial progenitor cells have also been identified in the adult [147], suggesting the vasculogenesis does not only occur during embryonic development. Endothelial repair involves a heterogeneous population of endothelial progenitors existing in both peripheral blood and within specific niches of the vessel wall. When transplanted into immunodeficient mice, a subtype of endothelial progenitor cells, cells that can form an endothelial colony, display great ability to proliferate and form de novo blood vessels [161]. These cells may play roles in neointima hyperplasia and vascular repair after injury. Endothelial repair accomplished by progenitor cells is tightly modulated by specific growth factors, with the underlying signaling pathways likely associated with SOCE. Dysfunctional SOCE has been found to contribute to the impaired proliferation of endothelial progenitor cells in an atherosclerosis mouse model [162]. Similar to the mature endothelium, the CRAC channels' contribution to growth hormone and mediator signaling pathways are also dependable on the developmental stage and differentiation. Notably, at certain phenotypical stages during ex vivo proliferation of progenitor cell clusters, TRPC3 expression level has been shown to increase [163].

The involvement of this classical SOCE and STIM1/Orai1 as the core of Ca^{2+} entry pathway in particular progenitor populations and stages of vasculogenesis remains to be further validated. The proliferation and differentiation of endothelial precursors seem to be regulated by multiple Ca^{2+} entry signaling pathways in a highly cooperative manner. In addition, these pathways appear to have differential influences dependent on different vasculogenesis or vascular endothelial repair stages.

14.5.3 SOCE and Vascular Diseases

Vascular diseases include systemic and pulmonary arterial hypertension and a number of vascular occlusive diseases such as restenosis, thrombosis, atherosclerosis. Evidences have strongly supported that dysregulated SOCE being a crucial contributor to the development of vascular diseases [104, 120, 131, 164–168] (see Table 14.1).

14.5.3.1 Thrombosis

Thrombosis is a pathological process of thrombus formation within the blood vessels. It is a defense mechanism against post-traumatic blood loss at sites of vascular injury when platelets and fibrin are drawn to the injured location, forming clots to prevent the bleeding. However, under severe circumstances, a process called embolism occurs. The thrombus formed gets released into the circulation system, traveling freely in the blood vessels, leading to infarction and stroke. In addition, thrombosis causes the narrowing of blood vessels, thus obstructing blood flow. The obstruction of coronary arteries leads to myocardial infarction and further to heart attack, while the obstruction of cerebral arteries causes ischemic brain infarction and ultimately leads to stroke.

During thrombus formation, a key step is the platelet activation achieved through the binding of agonists such as thrombin, thromboxane A_2, and ADP to PLC-coupled receptors, which leads to increased cytosolic Ca^{2+} concentration inside the platelets [169, 170]. This elevation of cytosolic Ca^{2+} concentration is essential for platelet aggregation at the thrombus formation [169].

The identification of Orai1 as the platelet SOC channels was reported to be essential for thrombus formation [171]. The authors showed strongly expressed Orai1 in human and mouse platelets. To examine the role of Orai1 in blood clotting, Orai1-deficient ($Orai1^{-/-}$) mice were generated and their platelets demonstrated severely defective SOCE, agonist (thrombin, ADP, collagen-related peptide-CRP, collagen)-induced Ca^{2+} responses, along with impaired activation and thrombus formation compared with wild-type mice under flow in vitro. Further in vivo results showed that $Orai1^{-/-}$ chimeras (wild-type mice infused with the bone marrow of an Orai1 knockout mouse) were protected from cerebral ischemia without displaying major

Table 14.1 Summary of the roles of Orai, STIM, and TRPC in cardiovascular pathophysiology

Disease	Cell/tissue type	Species	Main finding	References
Thrombosis				
	Platelet	Mouse	Strong Orai1 expression present. Orai1$^{-/-}$ mice show defective SOCE, agonist-induced Ca^{2+} responses, and thrombus activation and formation. Orai1$^{-/-}$ chimeras protected from cerebral ischemia	[171]
	Platelet	Mouse	Orai1^{R93W} mice demonstrate integrin activation reduction and degranulation impairment. Orai1^{R93W} platelets aggregation or adherence to collagen unaffected. Orai1^{R93W} platelets defective in surface phosphatidylserine exposure	[172]
	Platelet	Mouse	The platelets in the mouse model constitutively expressing STIM1 demonstrate premature activation and these mice demonstrate symptoms of macrothrombocytopenia	[173]
	Platelet	Mouse	Gain-of-function mouse model with mutant STIM1^{R304W} shows impaired platelet activation	[174, 175]
	Platelet	Mouse	Platelets of mice with specific conditional STIM1 knockout less likely to generate fibrin at laser-induced injury site. STIM1 knockout lowers thrombus stability	[176, 177]
		Human	No prolonged bleeding time in patients with Orai1 and STIM1 mutations and abrogated SOCE compared with control patients	[178, 179]
	Platelet	Mouse	TRPC1 not a contributor to platelet SOCE not its activation and function	[181]
	Platelet	Human	Both Orai1 and TRPC1 are involved in platelet SOCE and aggregation	[182]
	Platelet	Mouse	STIM2 not involved in thrombus formation	[177]
Atherosclerosis				
	VSMCs in endothelial denuded aortic rings	mouse	Apolipoprotein E (apoE) knockout mice more prone to hyperlipidemia and atherosclerosis, higher ATP-induced SOCE occurring before the significant development of atherosclerosis plaques	[164]
	Isolated aortic tissue	Mouse	ApoE knockout mice fed with high-fat diet demonstrate increased Orai1 mRNA and protein levels. Orai1 knockdown demonstrate decreased atherosclerosis plaque size	[164, 185]
	Isolated VSMCs	Pig	Higher SOCE, STIM1, and Orai1 (not significantly) expression levels were found in pigs fed with pro-atherosclerosis high-calorie diet. SOCE and STIM1 expression level decreased with exercise	[186]

(continued)

Table 14.1 (continued)

Disease	Cell/tissue type	Species	Main finding	References
	Aortic VSMCs and Aortic tissue	Mouse	Ox-LDL promotes proliferation, migration, and invasion of mouse aortic VSMCs, increases STIM1 expression and decreases miR-185 expression. 3′-UTR of STIM1 contains miR-185 binding site. miR-185 silencing or STIM1 overexpression promotes ox-LDL-induced mouse aortic VSMC viability, migration, and invasion, while miR-185 overexpression or STIM1 silencing shows opposite effect. miR-185 silencing increases VEGF and MMP-9 *levels* in vitro, and increases the lesions of arterial wall tissues and STIM1 positive rate *in vivo*. STIM1 silencing reverses these effects	[187]
	EPC	Mouse	ApoE knockout mice demonstrate attenuated EPC proliferation and migration. EPCs of these mice show reduced SOCE, Orai1, and STIM1 protein levels	[189]
	EPC	Mouse	SOCE showing protective effect from ox-LDL-induced EPC proliferation decrease	[190]
	Polymorphonuclear leukocytes (PMN)	Human	Orai1 shown necessary for neutrophil and monocytes migration into inflammatory vascular endothelium	[192]
	Macrophages	Human Mouse	Macrophage Ca^{2+} entry mediated by Orai1 when treated with ox-LDL. ox-LDL-triggered Ca^{2+} entry activates calcineurin, which activates c-Jun N-terminal kinase and p38 kinase enhancing the scavenger receptor A expression, and further induces LDL uptake, thus promoting the transition from macrophage to foam cells	[185]
Systemic arterial hypertension				
	Endothelial smooth muscle	Mouse	Smooth muscle STIM1 specific knockout mice show partial protection against endothelial dysfunction and hypertension development after angiotensin II infusion	[196, 197]
	VSMCs	Rat Human	Increased Ca^{2+} found in hypertensive rat models and human patients	[198, 199]
	Endothelium-denuded aortic rings	Rat	STIM1/Orai1 involved in elevated SOCE-mediated Ca^{2+} entry contributes to potentiating vascular activity	[200, 201]
	Basilar artery	Rat	STIM1/Orai1 involved in elevated SOCE-mediated Ca^{2+} entry contributes to vascular tone and force generation	[194, 195, 202]
	VSMCs	Rat	The SOCE inhibitors Gd^{3+} and SKF96365 were reported to inhibit systemic hypertension in rats	[206]
		Rat	Rats subjected to chronic ethanol consumption for 30 days demonstrated	[207]

(continued)

Table 14.1 (continued)

Disease	Cell/tissue type	Species	Main finding	References
			higher systemic blood pressure, enhanced SOCE, and increased STIM1 expression	
	Aortic rings	Rat	Male spontaneous hypertensive rats show more contractile force, higher Orai1 and STIM1 mRNA and protein levels than female	[200]
Pulmonary arterial hypertension				
	PASMCs	Mouse	STMI1 activation leads to Orai1-mediated SOCE	[210]
	Proliferative PASMCs and Contractile PASMCs isolated from pulmonary artery rings with denuded endothelium	Rat	STIM2, Orai2, and TRPC6 expression levels as well as SOCE higher in proliferative cultured PASMCs	[211]
	PASMCs	Human	STIM2 is higher in idiopathic PAH patient PASMCs and contributes to SOCE enhancement	[212]
	Distal pulmonary arteries and PASMCs	Rat	Increased SOCE and Orai1 and Orai2 expression levels	[213, 214]
	PASMCs	Rat	PASMCs STIM1 knockdown leads to inhibition of SOCE and decrease of hypoxia-induced proliferation and cell cycle progression	[215]
	Heart and vessels	Mouse	Wild type mice infused with Ang II develop hypertension and cardiovascular dysfunction with enhanced expression of STIM1 in both heart and vessels, whereas this development is absent in mice with STIM1 specifically knocked out in smooth muscle	[197]
Cardiac hypertrophy and heart failure				
	Neonatal ventricular myocytes	Rat	Cells treated with phenylephrine and angiotensin II showed increased intracellular Ca^{2+} and cell area and NFAT activation, which were all prevented by SOC inhibitor SKF-96365 and to a lesser extent by LRCC inhibitor	[231]
	Neonatal cardiomyocytes	Rat	Endothelin-1 treatment for 48 h enhanced TRPC1 expression, SOCE, and NFAT activation without upregulating STIM1. However, STIM1 KD suppressed these effects	[95]
	Neonatal ventricular myocytes	Rat	KD of both STIM1 and Orai1 completely abolished phenylephrine-induced hypertrophic growth in neonatal cardiomyocytes by inhibiting CaMKII and ERK1/2 signaling pathway, whereas merely Orai1 KD prevented phenylephrine-mediated signaling in a calcineurin-dependent manner	[96]
	Neonatal cardiomyocytes	Rat	Neonatal cardiomyocytes with STIM1 overexpression showed significantly larger size and increased NFAT activity, and both were prevented by SKF-96365	[97]

(continued)

Table 14.1 (continued)

Disease	Cell/tissue type	Species	Main finding	References
	Adult heart	Rat	Up-regulation of STIM1 protein and enhanced SOC current in the pressure overload-induced left ventricular hypertrophy. Silencing STIM1 gene expression reduces SOCE and protects the heart from hypertrophy development through decreasing the CnA/NFAT4 signaling pathway	[97]
	Heart	Mouse	Deletion of STIM1 protects the heart from pressure overload-induced cardiac hypertrophy	[232]
	Heart	Mouse	STIM1 transgenic (overexpression) mice exhibited sudden cardiac death, while surviving mice developed heart failure with hypertrophy, induction of the fetal gene program, histopathology and mitochondrial structural alterations, loss of ventricular functional performance, and pulmonary edema	[17]
	Adult cardiomyocytes	Mouse	Cardiac myocytes isolated from STIM1 transgenic mice displayed spontaneous Ca^{2+} transients that were prevented by SOCE blocker SKF-96365, increased LTCC current, and enhanced Ca^{2+} spark frequency	[17]
		Mouse	Mice undergoing TAC presented increased mRNA and protein STIM1L levels and enhanced SOCE compared with sham animals	[21]
	Adult cardiomyocytes	Mouse	Phenylephrine induced of STIM1L expression	[21]
	Embryonic stem cell-derived cardiomyocytes	Human	Phenylephrine treatment for 48 h induced a marked hypertrophy along with increased Orai1 protein expression level	[234]
		Zebrafish	Inactivation of Orai1 resulted in heart failure, reduced ventricular systolic function, bradycardia, and skeletal muscle weakness	[98]
		Mouse	After 8 weeks of TAC, Orai1 deficient mice showed a significantly reduced survival rate, a much earlier loss of cardiac function, and an earlier, greater dilation of the left ventricle, and significantly higher expression levels of apoptotic markers, indicating that Orai1 deficiency seems to accelerate or exacerbate the progression of the decease, rapidly leading to dilated cardiomyopathy, heart failure, and earlier death	[18]
		Mouse	Overexpression of SOCE-associated regulatory factors in the heart prevents cardiac hypertrophy by suppressing the up-regulation of STIM1 and Orai1	[235]

(continued)

Table 14.1 (continued)

Disease	Cell/tissue type	Species	Main finding	References
		Mouse	In a novel genetically-modified mouse model that specifically disrupts Orai1 in cardiomyocytes, Orai1 functional inhibition preserves alterations of Ca^{2+} homeostasis, fibrosis, and systolic function without affecting hypertrophy during pressure overload	[236]
	Heart	Mouse	TRPC6 functions as a positive regulator of calcineurin-NFAT signaling and a key component of a calcium-dependent regulatory loop driving pathologic cardiac remodeling	[241]
	Cardiomyocytes	Rat	DAG-induced Ca^{2+} signaling pathway through TRPC3 and TRPC6 is essential for angiotensin II-induced NFAT activation and cardiac hypertrophy	[242]
		Rat	TRPC3 expression was upregulated in the spontaneous hypertensive heart failure (SHHF) rat model through activation of calcineurin and its downstream effector NFAT	[243]
	Ventricular cardiac tissue	Human	TRPC5 expression is induced in failing human heart	[243]
		Mouse	Maladaptive hypertrophy induced by pressure overload was suppressed by deletion of either *Trpc3* or *Trpc6* in mice	[244]
		Mouse	TRPC1 KO mice are shown to have the calcineurin-NFAT signaling pathway inhibited, which reduces TAC-induced hypertrophic response and is related to a better survival rate	[246]
	Cultured adult feline myocytes	Cat	KD of *Trpc4* decreased TAC-induced hypertrophy and contractile dysfunction in response to myocardial infarction	[83]
		Mouse	TRPC1/4 double KO prevents cardiac hypertrophy and fibrotic infiltration after TAC and chronic neurohumoral stimulation	[240]
		Mouse	Overexpression of dominant-negative gene variants of TRPC3, TRPC4, and TRPC6 is confirmed to have protective effects against TAC-induced hypertrophy	[86]
	Adult cardiomyocyte	Rat	The mineralocorticoid pathway specifically promotes TRPC1/TRPC5-mediated SOCE in adult rat cardiomyocytes	[247]
Arrhythmias				
	Isolated sinoatrial nodes (SAN)	Mouse	Mouse SAN exhibits SOC activity which may be attributable to TRPC expression, and SOCCs may be involved in regulating pacemaker firing rate	[252]

(continued)

Table 14.1 (continued)

Disease	Cell/tissue type	Species	Main finding	References
		Mouse	STIM1 is demonstrated to be crucial in maintaining the Ca^{2+} content of intracellular Ca^{2+} stores, thus contributing to maintaining the regular sinus rhythm of the heart in mouse	[253]
	HL-1 cells	Mouse	STIM1 KD could perturb cell contraction rate and induce irregular spontaneous Ca^{2+} oscillations, thus presenting proarrhythmogenic activities, including early or delayed after depolarizations	[255]
		Mouse	Early mortality in STIM1-KD (inducible and myocyte-specific) mice was reported likely related to enhanced susceptibility to ventricular tachycardia/ ventricular fibrillation secondary to the pathogenesis of spatially discordant action potential duration alternans	[256]
	Ventricular myocytes	Rat	The overexpression of STIM1 generates spontaneous Ca^{2+} transients, thus causing arrhythmogenic Ca^{2+} waves and cytosolic and SR Ca^{2+} overload, potentially triggering sudden cardiac death	[100]
	Heart	Rat	Application of 2-APB induced a period of tachycardic ectopy and progressed to spontaneous ventricular depolarization in Langendorff perfused rat heart and sinus rhythm and heart mechanical output was restored upon SKF-96365 application	[257]
		Mouse	TRPC3, although unlikely to function as the primary SOCE channel in pacemakers, was reported to have implication in both sinoatrial and atrial arrhythmias	[262]
	Ventricular myocytes	Mouse	TRPC channels and SOCE mechanism are involved in cardiac arrhythmogenesis via the promotion of spontaneous Ca^{2+} waves and triggered activities under hyperactivated conditions	[263]

bleeding [171]. The involvement of Orai1 in platelet SOCE was also supported by another study using chimeric mice expressing a mutated inactive form of Orai1 in blood cells only ($Orai1^{R93W}$, a naturally occurring mutation found in patients with severe combined immunodeficiency) [172]. Reduced integrin activation and impaired degranulation were shown in these $Orai1^{R93W}$ platelets when stimulated with low agonist (collagen) concentrations under static conditions. However, the $Orai1^{R93W}$ platelets' ability to aggregate or adhere to collagen was not significantly affected under arterial flow conditions ex vivo. In contrast, these adherent $Orai1^{R93W}$ platelets were defective in surface phosphatidylserine exposure, indicating Orai1 as a crucial component for the platelets' pro-coagulant response rather than for other Ca^{2+}-dependent cellular responses [172].

Accumulating evidences supporting the role of STIM1 in platelet Ca^{2+} concentration increase and thrombosis occurrence are strong. STIM1 is implicated to be required for platelet activation in a study using a gain-of-function STIM1 mutant mouse model ($STIM1^{D84G}$), which constitutively expresses active STIM1 [173]. The platelets in these mice with STIM1 mutation showed premature activation, and these mice demonstrated symptoms of macrothrombocytopenia. Furthermore, another gain-of-function mouse model with mutant $STIM1^{R304W}$ showed impaired platelet activation, likely as a result of decreased expression of STIM1 in platelets [174, 175]. STIM1 is also shown to be a critical participant in thrombus stability. A platelet-specific conditional STIM1 knockout mouse model was generated to display thrombosis induced by laser injury [176]. The platelets in these mice showed reduced ability to generate fibrin at the injury site, indicating less stable thrombi in these mice compared to wild-type mice. The ability to express phosphatidylserine at the plasma membrane of the platelets was reduced when STIM1 was knocked out, lowering the thrombus stability [176, 177]. Collectively, these data have established the crucial role of STIM1 in thrombus formation and stability. In addition, human patients with Orai1 and STIM1 mutations and abrogated SOCE did not demonstrate prolonged bleeding time compared to control patients [178, 179]. Cyclophilin-A was identified as an important regulator of SOCE by regulating STIM1 phosphorylation [180].

Although there is consensus about the involvement of Orai1 and STIM1 in thrombosis, whether TRPCs are functioning similarly is a contentious issue. Varga-Szabo group revealed that TRPC1 does not contribute to platelet SOCE nor to their activation and function [181]. The authors used mice lacking TRPC1 and platelets from these mice display normal SOCE compared to wild-type mice. Furthermore, platelet function both in vitro and in vivo did not change in the absence of TRPC1. In addition, human platelets SOCE was not altered when treated with presumably inhibitory anti-TRPC1 antibodies [181]. $TRPC1^{-/-}$ mice platelets SOCE mediated by thapsigargin, thrombin, and CRP was shown to be independent of TRPC1 [171]. Taken together, TRPC1 neither contributes to the physiological function of platelets nor to the pathological thrombosis. Contrary to this, Galan et al. reported both Orai1 and TRPC1 are involved in platelet SOCE and aggregation [182]. Interestingly, it has been hypothesized that platelets contain two mechanisms of Ca^{2+} entry and phosphatidylserine exposure, with only one depending on STIM1-Orai1 interaction, and the other being receptor-operated Ca^{2+} entry pathway [177]. Interestingly, STIM2 is shown not to play significant roles in thrombus formation as $STIM2^{-/-}$ reacted normally to collagen treatment [177].

14.5.3.2 Restenosis

Restenosis is the pathological remodeling and reoccurrence of arteries (most often coronary arteries) causing restricted blood flow. This is oftentimes following percutaneous angioplasty or stenting. Restenosis happens in 1–3% of patients and often leads to acute myocardial infarction or even acute cardiac arrest [183, 184]. The two most important contributors to restenosis are thrombosis (see the above section) and neointima formation, which is due to VSMC proliferation and migration into the lumen of vessels (see Sect. 14.5.1).

14.5.3.3 Atherosclerosis

The pathogenesis of atherosclerosis is the narrowing of the blood vessel lumen caused by the subendothelial deposition of lipids on the blood vessel wall. The stimulation of the lipid accumulation leads to chronic inflammation of the arterial wall, which is a hallmark of atherosclerosis. Monocytes and macrophages migrate into the plaques between ECs and VSMCs upon lipid stimulation. These immune cells engulf the excessive fatty materials and turn into foam cells, promoting a chronic inflammatory environment in the vasculature. This chronic inflammation triggers VSMC remodeling and further causes endothelial dysfunction, thus leading to the final blood vessel narrowing. Diet is also a key factor during atherosclerosis development and progression.

Emerging evidences have implicated the involvement of Orai1 and STIM1 in atherosclerosis. A study using apolipoprotein E (apoE) knockout mouse model reported these mice are prone to hyperlipidemia and atherosclerosis and demonstrate higher SOCE induced by ATP in endothelial denuded aortic rings. Notably, the enhancement of SOCE in VSMCs occurred in apoE knockout mice before the significant development of atherosclerosis plaques, indicating that SOCE and VSMC remodeling might be early events in atherosclerosis development [164]. Another study also using the apoE knockout mouse model demonstrated when these mice are fed with a high-fat diet, the Orai1 expression both at mRNA and protein levels are increased in isolated aortic tissue [164]. The atherosclerosis plaque size was decreased when Orai1 was knocked down by either siRNA or SOCE pharmacological inhibitor SKF96365 application [185]. This study did not identify a specific cell type, and thus further studies are needed for clarification. Similarly, in pigs fed with pro-atherosclerosis high-calorie diet, enhanced SOCE, STIM1, and Orai1 expression levels are observed in isolated VSMCs compared to those in pigs fed with normal chow, although Orai1 increase was not significant [186]. Interestingly, exercise decreased the expression level of STIM1 as well as SOCE [186], indicating that SOCE and its related protein expression level together with coronary atherosclerosis can be attenuated with proper diet and exercise. A recent study identified miR-185 as a modulator of STIM1 in atherosclerosis models both in vitro and in vivo. The authors demonstrated SIMT1 was a potential target gene of miR-185 in atherosclerosis, the progression of which was promoted by knocking down of miR-185 through enhancing cell proliferation, migration, and invasion via targeting STIM1. This provides insight into miR-185/STIM1 axis function in atherosclerosis development [187].

Endothelial dysfunction is an early step promoting vascular inflammation which further contributes to the formation of atherosclerosis plaques. There is an excellent review about the endothelial cell dysfunction and the pathobiology of atherosclerosis [188]. EPCs also play vital roles in the regeneration of healthy intima, and the proliferation and migration of EPCs are attenuated when apoE is knocked out in mice [189]. The EPCs in these mice also displayed a reduction in SOCE, Orai1, and STIM1 protein expression levels [189]. The same group also demonstrated protective effects of SOCE from ox-LDL-induced cellular proliferation decrease in EPCs [190].

The recruitment of monocytes and neutrophils by the endothelium at injury sites is one key step in atherosclerosis plaque formation. In other occlusive diseases, the recruitment of these cells has been reported at the early stage of disease development [191]. Orai1 is shown as a necessary component for neutrophil and monocyte migration into inflammatory vascular endothelium [192]. Orai1 has also been demonstrated as important for the formation of foam cells from macrophages upon stimulation of lipids accumulation [185]. Macrophage Ca^{2+} entry was mediated by Orai1 when treated with ox-LDL. The authors showed this ox-LDL-triggered Ca^{2+} entry activates calcineurin, which activates c-Jun N-terminal kinase and p38 kinase enhancing the scavenger receptor A expression and further induces LDL uptake, thus promoting the transition from macrophage to foam cells [185].

14.5.3.4 Systemic Arterial Hypertension

The association between arterial hypertension and elevated intracellular Ca^{2+} levels as well as abnormal expression of Ca^{2+}-handling proteins has been well characterized [193–195]. In a mouse model with STIM1 specifically knocked out in the smooth muscle, partial protection against endothelial dysfunction and hypertension development after infusion of angiotensin II has been reported [196, 197]. Increased Ca^{2+} has also been demonstrated in hypertensive rat models and human patients [198, 199]. Elevated SOCE-mediated Ca^{2+} influx with STIM1/Orai1 as the molecule involved has been reported to contribute to potentiating vascular activity [200, 201], vascular tone, and force generation [194, 195,

202]. Male spontaneous hypertensive rats treated with high concentrations of SOCE blockers (either 2-APB and Gd^{3+}, or STIM1 and Orai1 neutralizing antibodies) in their aortic rings lead to reduced spontaneous tone and force generation to levels close to those in normotensive rats [201]. These spontaneous hypertensive rats also displayed higher expression levels of Orai1 and STIM1 at both mRNA and protein levels [201]. However, these data should be interpreted with caution due to the neutralizing antibodies' nonspecific effects. In addition, the high concentration of SOCE inhibitors has been reported to influence other ion channels, including ER Ca^{2+} release channels and pumps such as IP_3 receptor-mediated channels and various TRP channels [203, 204], which can further inhibit SOCE [201, 205]. The SOCE inhibitors Gd^{3+} and SKF96365 have also been reported to inhibit systemic hypertension in rats [206]. Similarly, rats subjected to chronic ethanol consumption for 30 days demonstrated higher systemic blood pressure, enhanced SOCE, and increased STIM1 expression [207].

The gender difference in hypertension susceptibility has been supported, showing males having a higher rate of incidence than females [208]. More contractile force and higher mRNA and protein expression levels of Orai1 and STIM1 are found in the aortic rings of male spontaneous hypertensive rats compared to female rats [200]. The differences in Orai and STIM isoform expression in male and female patients with hypertension warrant further investigation. Studies examining the influence of sex hormones on regulating SOCE-related molecular components have been done. Estrogen has been shown to increase Orai3 but not Orai1 in estrogen-positive breast cancer cells [70]. Since Orai3 is an essential component of heteromeric ARC channels, it might be an important triggering downstream signaling pathways distinct from SOCE [113].

14.5.3.5 Pulmonary Arterial Hypertension

Despite being a relatively rare disease, pulmonary arterial hypertension (PAH) can lead to major complications such as dyspnea, heart failure, and even death. Similar to systemic arterial hypertension, PAH also has idiopathic pathogenesis. Yet the etiologies of PAH and systemic arterial hypertension are different, with the pathological hallmark of PAH being smooth muscle and endothelial proliferation and migration together with thrombosis [209]. The dysfunction of endothelium is proposed to drive PAH, although in systemic arterial hypertension, endothelium only plays a supporting role. Orai1-mediated SOCE has been found as a result of STIM1 activation in mouse pulmonary artery smooth muscle cells (PASMCs) [210]. Fernandez et al. have shown that STIM2, Orai2, and TRPC6 expression levels, as well as SOCE, are upregulated in the proliferative cultured PASMCs compared to the contractile PASMSC isolated from rat pulmonary artery rings with denuded endothelium [211]. The upregulation of STIM2 is also found by the same group in PASMCs from patients with idiopathic PAH and contributes to enhanced SOCE [212]. Increased SOCE and expression levels of Orai1 and Orai2 (but not Orai3 and STIM1) were observed in both rat distal pulmonary arteries and PASMCs under chronic hypoxia [213, 214]. This upregulation is selectively in pulmonary only and not in coronary under hypoxia [213]. The expression level of Orai3 has been measured, yet its function in PAH warrants further investigation [214]. STIM1 has been shown to play a key role in hypoxia-induced PAH [215]. When STIM1 was knocked down in PASMCs, SOCE was inhibited; NFAT nuclear translocation was reduced and hypoxia-induced proliferation and cell cycle progression of PASMCs were decreased [215]. Another group used smooth muscle-specific STIM1 knockout mouse model to demonstrate the essential role of STIM1 in hypertension [197]. The authors show that wild-type mice infused with angiotensin II develop hypertension and cardiovascular dysfunction with enhanced expression of STIM1 in both heart and vessels. All these pathologies were significantly blunted in mice lacking STIM1 specifically in smooth muscle. Angiotensin II-induced hypertension was found associated with enhanced ER stress through pathways mediated by TGF-β and NADPH oxidase [197].

14.6 SOCE and Cardiac Diseases (see Table 14.1)

14.6.1 Cardiac Hypertrophy and Heart Failure

The heart functions to pump blood into and perfuse the peripheral organs, satisfying the demand under both normal and stress conditions. To achieve this goal, the heart and individual cardiomyocytes undergo enlargement (termed as hypertrophy) during increased preload or afterload [216]. Initially, cardiac hypertrophy increases contractility by adding sarcomere units in parallel [216]. Then following Laplace's law, the increased left ventricular wall thickness decreases the left ventricular wall stress, thus maintaining cardiac efficiency [216]. Accompanied by cardiac hypertrophy are also qualitative changes including those in gene expression, which induce changes in metabolism, contractility, and cardiomyocyte survival [216]. Cardiac hypertrophy can be divided into two categories: physiological and pathological, both of which develop as an adaptive response to cardiac stress but their underlying molecular mechanisms, cardiac phenotype, and prognosis are quite different. Physiological hypertrophy keeps normal cardiac function over time, while pathological hypertrophy often undergoes deleterious remodeling of cardiomyocytes further progressing to adverse cardiovascular events including but not limited to heart failure, arrhythmias, and even sudden cardiac death [89, 217–219]. It is the nature of upstream stimuli and downstream signaling pathways that determine the development of physiological or pathological hypertrophy and not the duration of cardiac stress per se [220–223]. Ca^{2+}-related genes are only changed during pathological hypertrophy but not in physiological hypertrophy [216]. We will mainly focus on pathological hypertrophy in our review.

To match a greater hemodynamic demand or some stress conditions induced by myocardial infarction or hypertension-induced pressure overload, cardiac hypertrophy occurs in the preservation of the pump function [224]. Several lines of evidence have demonstrated the involvement of a number of signaling pathways in the hypertrophic growth of cardiomyocytes, and the Ca^{2+} signals have been established as the triggering events. However, not all Ca^{2+} signals in the cardiomyocytes can initiate hypertrophic response. For example, the global cytosol Ca^{2+} signals accompanying every contraction are not necessary activators for hypertrophy [225]. It is rather the local signaling action-triggering Ca^{2+} signals from discrete sources that can initiate hypertrophic growth [226]. The IP_3 receptors near the nucleus [226] or on the nuclear envelope generates increased cytosolic Ca^{2+} signals, activating phosphatase calcineurin, which subsequently dephosphorylates NFAT causing it to translocate into the nucleus and activate gene transcription; or activating CaMKII, which subsequently phosphorylates histone deacetylase [227–229], both of which switch on the fetal genes leading to cardiac hypertrophy [89].

STIM1/Orai-mediates SOCE may serve as the necessary source for local Ca^{2+} elevation to promote hypertrophic growth [230]. The first study in neonatal cardiomyocytes showed that 48 h administration of IP_3-activating agonists such as phenylephrine and angiotensin II increased intracellular Ca^{2+} and cell area and lead to NFAT activation, which were all prevented by the treatment of a nonselective SOC inhibitor SKF-96365 and to a lesser extent by the treatment of LRCC inhibitor [231]. Following this study, several lines of evidence found similar results in STIM1 and Orai1 knockdown neonatal cardiomyocytes, where 48 h treatment of endothelin-1 or phenylephrine induced SOCE enhancement, NFAT activation, and cell size increase were suppressed [95–97]. Another study demonstrated that knockdown of both STIM1 and Orai1 completely abolished phenylephrine-induced hypertrophic growth in neonatal cardiomyocytes by inhibiting CaMKII and extracellular signal-regulated kinase 1/2 (extracellular signal-regulated kinases 1/2) signaling pathway, whereas merely Orai1 knockdown prevented phenylephrine-mediated signaling in a calcineurin-dependent manner [96]. On the other hand, when STIM1 was overexpressed in neonatal cardiomyocytes the cells showed significantly larger size and enhanced NFAT activity, both of which were prevented by SKF-96365 treatment [97].

These in vitro data clearly identify the STIM1/Orai1-mediated Ca^{2+} entry as the fundamental mechanism underlying cardiac hypertrophy development. However, how it contributes to adults and in vivo models is still relatively limited. The first in vivo study to demonstrate upregulation of STIM1 protein and enhanced SOC current in the pressure overload-induced left ventricular hypertrophy was carried out in rats [97]. On the contrary, silencing STIM1 gene expression reduces SOCE and protects the heart from hypertrophy development through decreasing the calcineurin/NFAT4 signaling pathway [97]. Another study in mice demonstrated that deletion of STIM1 protects the heart from pressure overload-induced cardiac hypertrophy [232]. Other studies have also confirmed that transverse aortic constriction (TAC)-induced STIM1 upregulation activates the NFAT and CaMKII signaling pathway through enhancing SOCE, thereby promoting cardiac hypertrophy and arrhythmias [24, 233].

STIM1L has been found predominant at neonatal stages and its expression is decreased in adults and reappears upon hypertrophic agonist application or afterload-induced cardiac stress. Mice undergoing TAC presented increased mRNA and protein expression levels of STIM1L and enhanced SOCE compared to sham animals, in line with the evidence showing reactivation of STIM1L expression enhances SOCE during the hypertrophic process [28]. In isolated adult cardiomyocytes, with the application of phenylephrine induction of STIM1L was also observed [28].

Studies examining the role of Orai1 in cardiac hypertrophy and heart failure remain limited. Human embryonic stem cell-derived cardiomyocytes treated with phenylephrine for 48 h induced marked hypertrophy along with increased Orai1 protein expression level [234]. Hypertrophy was inhibited by suppression of Orai1 expression/activity using siRNAs or a dominant-negative construct Orai1 (G98A), or inhibited by NO and cyclic guanosine monophosphate via activating PKG. Importantly, when Orai1 is mutated on serine 34, the anti-hypertrophic responses were abolished, indicating that hat NO, cyclic guanosine monophosphate, and PKG inhibit the hypertrophy of human embryonic stem cell-derived cardiomyocytes via PKG-mediated phosphorylation on Orai1-Ser-34 [234]. As for in vivo studies, in a zebra model using reverse genetics, the inactivation of the highly conserved zebrafish orthologue of Orai1 resulted in heart failure, reduced ventricular systolic function, bradycardia, and skeletal muscle weakness [98]. This is the first study showing Orai1 deficiency in zebrafish causes heart failure. Loss of Orai1 is found to lead to defective signal transduction at the cardiac z-disc [98]. Their findings link Orai1-mediated calcium signaling to sarcomere physiology, in part by affecting z-disc composition and function mediated by the calcineurin-calsarcin-NFAT signaling pathway, shedding light upon the potential role of SOCE in serving as the pharmacological target to modulate Ca^{2+}-dependent pathways to improve the outcome of cardiac hypertrophy. Horton et al. applied a TAC pressure overload model to mice deficient in Orai1 (global heterozygous) and compared their response to wild-type mice [25]. After 8 weeks of TAC, the Orai1-deficient mice show a significantly reduced survival rate, a much earlier loss of cardiac function, and an earlier, greater dilation of the left ventricle, and significantly higher expression levels of apoptotic markers, indicating that Orai1 deficiency seems to accelerate or exacerbate the progression of the decease, rapidly leading to dilated cardiomyopathy, heart failure, and earlier death [25]. Notably, contrary to the above-mentioned phenotypes, the authors did not observe any change in the heart weight or rate of increase in heart weight early on nor in cellular hypertrophy [25]. They attribute this discrepancy to the theory that a maximum rate of hypertrophy has been achieved in both Orai1-deficient and wild-type mice, but the Orai1-deficient mice are unable to compensate for the overload (to yield equivalent functional compensation to wild-type mice) [25]. A recent in vivo study shows that overexpression of SOCE-associated regulatory factors in the heart prevents cardiac hypertrophy probably through suppressing the upregulation of STIM1 and Orai1 [235]. In an even more recent study, using a novel genetically-modified mouse that specifically disrupts the Orai1 channel in cardiomyocytes, the authors show that even if Orai1 is not instrumental in regulating normal EC coupling and cardiac function, its functional inhibition preserves alterations of Ca^{2+} homeostasis, fibrosis and systolic function without affecting hypertrophy during pressure overload [236].

TRPC family members have also been demonstrated upregulated in several studies of cardiac hypertrophy and heart failure [237–239]. Seven isoforms (TRPC1–TRPC7) are found to control pathological hypertrophy through signaling effectors including calcineurin and NFAT [89, 237, 239, 240]. TRPC3 and TRPC6 are particularly vital for the development of hypertrophy through calcineurin-dependent signaling pathways [241–243]. Maladaptive hypertrophy induced by pressure overload was suppressed by deletion of either *Trpc3* or *Trpc6* in mice [244]. TRPC6 is phosphorylated by protein kinase G (PKG), which reduces channel conductance, therefore negatively regulating TRPC-mediated hypertrophy [245]. TRPC1 knockout mice are also shown to have the calcineurin-NFAT signaling pathway inhibited, which reduces the TAC-induced hypertrophic response and is related to a better survival rate [246]. Knockdown of *Trpc4* also decreased TAC-induced hypertrophy and contractile dysfunction in response to myocardial infarction [90]. Another study further shows TRPC1/4 double knockout prevents cardiac hypertrophy and fibrotic infiltration after TAC and chronic neurohumoral stimulation [240]. Overexpression of dominant-negative gene variants of certain TRPCs (TRPC3, TRPC4, and TRPC6) is confirmed to have protective effects against TAC-induced hypertrophy [93]. A recent study demonstrates that the mineralocorticoid pathway specifically promotes TRPC1/TRPC5-mediated SOCE in adult rat cardiomyocytes, which might be the underlying mechanism of abnormally enhanced SOCE during cardiac hypertrophy and heart failure [247].

14.6.2 Arrhythmias

As stated earlier in the EC coupling section, the calcium concentration within the cardiomyocytes experiences dynamic changes following an action potential. However, for tens to hundreds of milliseconds after AP, Ca^{2+} remains refractory to the electrical stimuli by returning to diastolic levels mediated by the sarcolemma Ca^{2+}-ATPase and NCX as well as the SR NCX [248].

In humans, normal cardiac sinus rhythm is typically around 60 beats per minute, and this relies on a pacemaker mechanism called "coupled-clock" mechanism which is a coordinated crosstalk between sarcolemma and SR and involves multiple Ca^{2+}-dependent ion transport processes [249, 250]. This mechanism integrates an electric oscillator located on the sarcolemma and an intracellular SR Ca^{2+} cycling mechanism. Cellular Ca^{2+} oscillation requires a Ca^{2+} entry mechanism to compensate for the loss of Ca^{2+} that goes extracellular during cytosolic Ca^{2+} transients and to replenish the Ca^{2+} stores. The vital role of store-replenishing mechanism in cardiac pacemaking has been clearly reported, and thus the important role of SOCE as one key mechanism for store-filling has been recognized [251]. The search for pacemaker activity regulating SOCE channels was first done in the TRPC family [252] and later in STIM/Orai complexes [253, 254]. The TRPC, STIM, and Orai proteins have been found expressed in sinoatrial node cells and involved in SOCE pacemaker activity [252, 254]. In addition, STIM1 has been demonstrated to be crucial in maintaining the Ca^{2+} content of intracellular Ca^{2+} stores, thus contributing to maintaining the regular sinus rhythm of the heart in mouse [253]. The knockdown of STIM1 was found to perturb cell contraction rate and induce irregular spontaneous Ca^{2+} oscillations, thus presenting proarrhythmogenic activities, including early or delayed after depolarizations [255]. In an adult murine model with inducible and myocyte-specific STIM1 depletion, STIM1 was demonstrated to regulate spatially discordant alternans [256]. Early mortality in STIM1-knockdown mice was reported likely related to enhanced susceptibility to ventricular tachycardia/ventricular fibrillation secondary to the pathogenesis of spatially discordant action potential duration alternans [256]. The overexpression of STIM1 in vivo (in the adult heart) and in vitro (in ventricular myocytes) generates spontaneous Ca^{2+} transients, thus causing arrhythmogenic Ca^{2+} waves and cytosolic and SR Ca^{2+} overload, potentially triggering sudden cardiac death [24, 100]. The role of Orai1 and Orai3-activating 2-APB (when applied at high concentration) in the initiation of atrial and ventricular arrhythmias has been reported in cardiomyocytes [257]. Application of 2-APB induced a period of tachycardic ectopy and progressed to spontaneous ventricular depolarization in Langendorff perfused rat heart and sinus rhythm and heart mechanical output was restored upon SKF-96365 application, indicating that activating myocardial voltage-independent calcium channels, possibly the Orais, may be a novel cause of ventricular arrhythmia [257].

In some disease conditions related to arrhythmias, abnormal calcium signals have been reported. During atrial fibrillation, spurious electrical signals can lead to atrial tachycardic with the atrial chamber displaying more than 300 beats per minute [258]. Spontaneous Ca^{2+} signals have been recognized as the cause of such arrhythmic activity [2, 259, 260]. In cardiomyocytes, many mechanisms can contribute to the generation of Ca^{2+} spontaneous signals, including increased SR Ca^{2+} content, increased RyR or IP_3R activity, and CICR-triggering Ca^{2+} source introduction. Although generating a more modest current than VGCC, Ca^{2+} entry pathways such as SOCE could also lead to arrhythmogenic spontaneous Ca^{2+} signals generation [261]. Notably, TRPC3, although unlikely to function as the primary SOCE channel in pacemakers, was reported to have implication in both sinoatrial and atrial arrhythmias [262]. It has also been demonstrated that SOCE, which is at least partially mediated by TRPC channels, exists in adult mouse ventricular myocytes. TRPC channels and SOCE mechanism are also reported to be involved in cardiac arrhythmogenesis via the promotion of spontaneous Ca^{2+} waves and triggered activities under hyperactivated conditions [263]. A recent study demonstrated local transient Ca^{2+} (LoCE) events that comprise cardiac SOCE [264]. These LoCEs were found concentrated at the myocyte periphery from a genetic murine model of arrhythmic disease (catecholaminergic ventricular tachycardia, CPVT), particularly at the intercalated disk, close to intercellular mechanical junctions. Furthermore, SOCE proteins and LoCEs were found upregulated at the intercalated disk in CPVT and myocytes showed characteristic arrhythmogenic spontaneous Ca^{2+} waves under cholinergic stress, which was effectively prevented by SOCE inhibition, further indicating the role of cardiac SOCE-mediated signaling in arrhythmias and providing more details with the fundamental cardiac SOCE properties.

References

1. Bers DM (2002) Cardiac excitation-contraction coupling. Nature 415(6868):198–205
2. Bers DM (2008) Calcium cycling and signaling in cardiac myocytes. Annu Rev Physiol 70:23–49
3. Pan Z, Brotto M, Ma J (2014) Store-operated Ca^{2+} entry in muscle physiology and diseases. BMB Rep 47(2):69–79
4. MacLeod KT (2016) Recent advances in understanding cardiac contractility in health and disease. F1000Res 5:F1000 Faculty Rev-1770
5. Roderick HL, Berridge MJ, Bootman MD (2003) Calcium-induced calcium release. Curr Biol 13(11): R425
6. Cheng H et al (1996) Calcium sparks and $[Ca^{2+}]i$ waves in cardiac myocytes. Am J Phys 270(1 Pt 1): C148–C159
7. Berridge MJ (2003) Cardiac calcium signalling. Biochem Soc Trans 31(Pt 5):930–933
8. Liou J et al (2005) STIM is a Ca^{2+} sensor essential for Ca^{2+}-store-depletion-triggered Ca^{2+} influx. Curr Biol 15(13):1235–1241
9. Manji SS et al (2000) STIM1: a novel phosphoprotein located at the cell surface. Biochim Biophys Acta 1481(1):147–155
10. Thompson JL, Shuttleworth TJ (2012) A plasma membrane-targeted cytosolic domain of STIM1 selectively activates ARC channels, an

arachidonate-regulated store-independent Orai channel. Channels (Austin) 6(5):370–378
11. Lopez JJ et al (2006) Interaction of STIM1 with endogenously expressed human canonical TRP1 upon depletion of intracellular Ca^{2+} stores. J Biol Chem 281(38):28254–28264
12. Spassova MA et al (2006) STIM1 has a plasma membrane role in the activation of store-operated Ca(2+) channels. Proc Natl Acad Sci U S A 103 (11):4040–4045
13. Stathopulos PB et al (2008) Structural and mechanistic insights into STIM1-mediated initiation of store-operated calcium entry. Cell 135(1):110–122
14. Baba Y et al (2006) Coupling of STIM1 to store-operated Ca^{2+} entry through its constitutive and inducible movement in the endoplasmic reticulum. Proc Natl Acad Sci U S A 103(45):16704–16709
15. Yuan JP et al (2009) SOAR and the polybasic STIM1 domains gate and regulate Orai channels. Nat Cell Biol 11(3):337–343
16. Li Z et al (2007) Mapping the interacting domains of STIM1 and Orai1 in Ca^{2+} release-activated Ca^{2+} channel activation. J Biol Chem 282 (40):29448–29456
17. Muik M et al (2008) Dynamic coupling of the putative coiled-coil domain of ORAI1 with STIM1 mediates ORAI1 channel activation. J Biol Chem 283(12):8014–8022
18. Muik M et al (2009) A cytosolic homomerization and a modulatory domain within STIM1 C terminus determine coupling to ORAI1 channels. J Biol Chem 284(13):8421–8426
19. Park CY et al (2009) STIM1 clusters and activates CRAC channels via direct binding of a cytosolic domain to Orai1. Cell 136(5):876–890
20. Kawasaki T, Lange I, Feske S (2009) A minimal regulatory domain in the C terminus of STIM1 binds to and activates ORAI1 CRAC channels. Biochem Biophys Res Commun 385(1):49–54
21. Williams RT et al (2001) Identification and characterization of the STIM (stromal interaction molecule) gene family: coding for a novel class of transmembrane proteins. Biochem J 357(Pt 3):673–685
22. Lopez E et al (2012) STIM1 tyrosine-phosphorylation is required for STIM1-Orai1 association in human platelets. Cell Signal 24 (6):1315–1322
23. Williams RT et al (2002) Stromal interaction molecule 1 (STIM1), a transmembrane protein with growth suppressor activity, contains an extracellular SAM domain modified by N-linked glycosylation. Biochim Biophys Acta 1596(1):131–137
24. Correll RN et al (2015) STIM1 elevation in the heart results in aberrant Ca(2)(+) handling and cardiomyopathy. J Mol Cell Cardiol 87:38–47
25. Horton JS et al (2014) The calcium release-activated calcium channel Orai1 represents a crucial component in hypertrophic compensation and the development of dilated cardiomyopathy. Channels (Austin) 8(1):35–48
26. Collins HE et al (2014) Stromal interaction molecule 1 is essential for normal cardiac homeostasis through modulation of ER and mitochondrial function. Am J Physiol Heart Circ Physiol 306(8):H1231–H1239
27. Darbellay B et al (2011) STIM1L is a new actin-binding splice variant involved in fast repetitive Ca^{2+} release. J Cell Biol 194(2):335–346
28. Luo X et al (2012) STIM1-dependent store-operated Ca(2)(+) entry is required for pathological cardiac hypertrophy. J Mol Cell Cardiol 52(1):136–147
29. Rosado JA et al (2015) STIM and Orai1 variants in store-operated calcium entry. Front Pharmacol 6:325
30. Jardin I et al (2013) The polybasic lysine-rich domain of plasma membrane-resident STIM1 is essential for the modulation of store-operated divalent cation entry by extracellular calcium. Cell Signal 25 (5):1328–1337
31. Zbidi H et al (2011) STIM1 and STIM2 are located in the acidic Ca^{2+} stores and associates with Orai1 upon depletion of the acidic stores in human platelets. J Biol Chem 286(14):12257–12270
32. Soboloff J et al (2006) STIM2 is an inhibitor of STIM1-mediated store-operated Ca^{2+} entry. Curr Biol 16(14):1465–1470
33. Wang JY et al (2015) STIM1 overexpression promotes colorectal cancer progression, cell motility and COX-2 expression. Oncogene 34 (33):4358–4367
34. Rana A et al (2015) Alternative splicing converts STIM2 from an activator to an inhibitor of store-operated calcium channels. J Cell Biol 209 (5):653–669
35. Ercan E et al (2012) Di-arginine signals and the K-rich domain retain the Ca(2)(+) sensor STIM1 in the endoplasmic reticulum. Traffic 13(7):992–1003
36. Bauer MC et al (2008) Calmodulin binding to the polybasic C-termini of STIM proteins involved in store-operated calcium entry. Biochemistry 47 (23):6089–6091
37. Wang X et al (2014) Distinct Orai-coupling domains in STIM1 and STIM2 define the Orai-activating site. Nat Commun 5:3183
38. Zheng S et al (2018) Identification of molecular determinants that govern distinct STIM2 activation dynamics. PLoS Biol 16(11):e2006898
39. Zheng L et al (2011) Auto-inhibitory role of the EF-SAM domain of STIM proteins in store-operated calcium entry. Proc Natl Acad Sci U S A 108 (4):1337–1342
40. Brandman O et al (2007) STIM2 is a feedback regulator that stabilizes basal cytosolic and endoplasmic reticulum Ca^{2+} levels. Cell 131(7):1327–1339
41. Subedi KP et al (2018) STIM2 induces activated conformation of STIM1 to control Orai1 function in ER-PM junctions. Cell Rep 23(2):522–534
42. Oh-Hora M et al (2008) Dual functions for the endoplasmic reticulum calcium sensors STIM1 and

STIM2 in T cell activation and tolerance. Nat Immunol 9(4):432–443
43. Darbellay B et al (2010) Human muscle economy myoblast differentiation and excitation-contraction coupling use the same molecular partners, STIM1 and STIM2. J Biol Chem 285(29):22437–22447
44. Miederer AM et al (2015) A STIM2 splice variant negatively regulates store-operated calcium entry. Nat Commun 6:6899
45. Berna-Erro A et al (2017) Role of STIM2 in cell function and physiopathology. J Physiol 595 (10):3111–3128
46. Zhou Y et al (2018) Cross-linking of Orai1 channels by STIM proteins. Proc Natl Acad Sci U S A 115 (15):E3398–E3407
47. Guo RW, Huang L (2008) New insights into the activation mechanism of store-operated calcium channels: roles of STIM and Orai. J Zhejiang Univ Sci B 9(8):591–601
48. Hoth M, Penner R (1992) Depletion of intracellular calcium stores activates a calcium current in mast cells. Nature 355(6358):353–356
49. Feske S et al (2006) A mutation in Orai1 causes immune deficiency by abrogating CRAC channel function. Nature 441(7090):179–185
50. Vig M et al (2006) CRACM1 is a plasma membrane protein essential for store-operated Ca^{2+} entry. Science 312(5777):1220–1223
51. Zhang SL et al (2006) Genome-wide RNAi screen of Ca(2+) influx identifies genes that regulate Ca(2+) release-activated Ca(2+) channel activity. Proc Natl Acad Sci U S A 103(24):9357–9362
52. Mercer JC et al (2006) Large store-operated calcium selective currents due to co-expression of Orai1 or Orai2 with the intracellular calcium sensor, Stim1. J Biol Chem 281(34):24979–24990
53. Peinelt C et al (2006) Amplification of CRAC current by STIM1 and CRACM1 (Orai1). Nat Cell Biol 8 (7):771–773
54. Prakriya M et al (2006) Orai1 is an essential pore subunit of the CRAC channel. Nature 443 (7108):230–233
55. Soboloff J et al (2006) Orai1 and STIM reconstitute store-operated calcium channel function. J Biol Chem 281(30):20661–20665
56. Derler I et al (2013) The extended transmembrane Orai1 N-terminal (ETON) region combines binding interface and gate for Orai1 activation by STIM1. J Biol Chem 288(40):29025–29034
57. Palty R, Isacoff EY (2016) Cooperative binding of stromal interaction molecule 1 (STIM1) to the N and C termini of calcium release-activated calcium modulator 1 (Orai1). J Biol Chem 291(1):334–341
58. Palty R, Stanley C, Isacoff EY (2015) Critical role for Orai1 C-terminal domain and TM4 in CRAC channel gating. Cell Res 25(8):963–980
59. Penna A et al (2008) The CRAC channel consists of a tetramer formed by Stim-induced dimerization of Orai dimers. Nature 456(7218):116–120
60. Maruyama Y et al (2009) Tetrameric Orai1 is a teardrop-shaped molecule with a long, tapered cytoplasmic domain. J Biol Chem 284(20):13676–13685
61. Mignen O, Thompson JL, Shuttleworth TJ (2008) Orai1 subunit stoichiometry of the mammalian CRAC channel pore. J Physiol 586(2):419–425
62. Hou X et al (2012) Crystal structure of the calcium release-activated calcium channel Orai. Science 338 (6112):1308–1313
63. Thompson JL, Shuttleworth TJ (2013) How many Orai's does it take to make a CRAC channel? Sci Rep 3:1961
64. Peinelt C et al (2008) 2-Aminoethoxydiphenyl borate directly facilitates and indirectly inhibits STIM1-dependent gating of CRAC channels. J Physiol 586 (13):3061–3073
65. Yamashita M et al (2007) Orai1 mutations alter ion permeation and Ca^{2+}-dependent fast inactivation of CRAC channels: evidence for coupling of permeation and gating. J Gen Physiol 130(5):525–540
66. Fukushima M et al (2012) Alternative translation initiation gives rise to two isoforms of Orai1 with distinct plasma membrane mobilities. J Cell Sci 125 (Pt 18):4354–4361
67. Desai PN et al (2015) Multiple types of calcium channels arising from alternative translation initiation of the Orai1 message. Sci Signal 8(387):ra74
68. Frischauf I et al (2009) Molecular determinants of the coupling between STIM1 and Orai channels: differential activation of Orai1-3 channels by a STIM1 coiled-coil mutant. J Biol Chem 284 (32):21696–21706
69. Lis A et al (2007) CRACM1, CRACM2, and CRACM3 are store-operated Ca^{2+} channels with distinct functional properties. Curr Biol 17(9):794–800
70. Motiani RK et al (2013) Orai3 is an estrogen receptor alpha-regulated Ca(2)(+) channel that promotes tumorigenesis. FASEB J 27(1):63–75
71. Vaeth M et al (2017) ORAI2 modulates store-operated calcium entry and T cell-mediated immunity. Nat Commun 8:14714
72. Trebak M, Kinet JP (2019) Calcium signalling in T cells. Nat Rev Immunol 19(3):154–169
73. Parekh AB, Putney JW Jr (2005) Store-operated calcium channels. Physiol Rev 85(2):757–810
74. Ambudkar IS, de Souza LB, Ong HL (2017) TRPC1, Orai1, and STIM1 in SOCE: friends in tight spaces. Cell Calcium 63:33–39
75. Jardin I et al (2008) Orai1 mediates the interaction between STIM1 and hTRPC1 and regulates the mode of activation of hTRPC1-forming Ca^{2+} channels. J Biol Chem 283(37):25296–25304
76. Sabourin J et al (2015) Store-operated Ca^{2+} entry mediated by Orai1 and TRPC1 participates to insulin secretion in rat beta-cells. J Biol Chem 290 (51):30530–30539
77. Lee KP et al (2014) Molecular determinants mediating gating of Transient Receptor Potential Canonical (TRPC) channels by stromal interaction

molecule 1 (STIM1). J Biol Chem 289 (10):6372–6382
78. Flockerzi V, Nilius B (2014) TRPs: truly remarkable proteins. Handb Exp Pharmacol 222:1–12
79. Wes PD et al (1995) TRPC1, a human homolog of a Drosophila store-operated channel. Proc Natl Acad Sci U S A 92(21):9652–9656
80. Zhu X et al (1995) Molecular cloning of a widely expressed human homologue for the Drosophila trp gene. FEBS Lett 373(3):193–198
81. Petersen CC et al (1995) Putative capacitative calcium entry channels: expression of Drosophila trp and evidence for the existence of vertebrate homologues. Biochem J 311(Pt 1):41–44
82. Huang GN et al (2006) STIM1 carboxyl-terminus activates native SOC, I(crac) and TRPC1 channels. Nat Cell Biol 8(9):1003–1010
83. Pani B et al (2013) Impairment of TRPC1-STIM1 channel assembly and AQP5 translocation compromise agonist-stimulated fluid secretion in mice lacking caveolin1. J Cell Sci 126(Pt 2):667–675
84. Jardin I et al (2008) Functional relevance of the de novo coupling between hTRPC1 and type II IP3 receptor in store-operated Ca^{2+} entry in human platelets. Cell Signal 20(4):737–747
85. Cheng KT et al (2011) Local Ca(2)+ entry via Orai1 regulates plasma membrane recruitment of TRPC1 and controls cytosolic Ca(2)+ signals required for specific cell functions. PLoS Biol 9(3):e1001025
86. Ong EC et al (2013) A TRPC1 protein-dependent pathway regulates osteoclast formation and function. J Biol Chem 288(31):22219–22232
87. Dominguez-Rodriguez A et al (2015) Proarrhythmic effect of sustained EPAC activation on TRPC3/4 in rat ventricular cardiomyocytes. J Mol Cell Cardiol 87:74–78
88. Sabourin J, Robin E, Raddatz E (2011) A key role of TRPC channels in the regulation of electromechanical activity of the developing heart. Cardiovasc Res 92(2):226–236
89. Eder P, Molkentin JD (2011) TRPC channels as effectors of cardiac hypertrophy. Circ Res 108 (2):265–272
90. Makarewich CA et al (2014) Transient receptor potential channels contribute to pathological structural and functional remodeling after myocardial infarction. Circ Res 115(6):567–580
91. Kirschmer N et al (2016) TRPC4alpha and TRPC4beta similarly affect neonatal cardiomyocyte survival during chronic GPCR stimulation. PLoS One 11(12):e0168446
92. Sabourin J et al (2016) Transient receptor potential canonical (TRPC)/Orai1-dependent store-operated Ca^{2+} channels: new targets of aldosterone in cardiomyocytes. J Biol Chem 291(25):13394–13409
93. Wu X et al (2010) TRPC channels are necessary mediators of pathologic cardiac hypertrophy. Proc Natl Acad Sci U S A 107(15):7000–7005
94. Bootman MD et al (2011) Atrial cardiomyocyte calcium signalling. Biochim Biophys Acta 1813 (5):922–934
95. Ohba T et al (2009) Essential role of STIM1 in the development of cardiomyocyte hypertrophy. Biochem Biophys Res Commun 389(1):172–176
96. Voelkers M et al (2010) Orai1 and Stim1 regulate normal and hypertrophic growth in cardiomyocytes. J Mol Cell Cardiol 48(6):1329–1334
97. Hulot JS et al (2011) Critical role for stromal interaction molecule 1 in cardiac hypertrophy. Circulation 124(7):796–805
98. Volkers M et al (2012) Orai1 deficiency leads to heart failure and skeletal myopathy in zebrafish. J Cell Sci 125(Pt 2):287–294
99. Zhu-Mauldin X et al (2012) Modification of STIM1 by O-linked N-acetylglucosamine (O-GlcNAc) attenuates store-operated calcium entry in neonatal cardiomyocytes. J Biol Chem 287(46):39094–39106
100. Zhao G et al (2015) STIM1 enhances SR Ca^{2+} content through binding phospholamban in rat ventricular myocytes. Proc Natl Acad Sci U S A 112(34): E4792–E4801
101. Che H et al (2015) Roles of store-operated Ca^{2+} channels in regulating cell cycling and migration of human cardiac c-kit+ progenitor cells. Am J Physiol Heart Circ Physiol 309(10):H1772–H1781
102. Feske S (2010) CRAC channelopathies. Pflugers Arch 460(2):417–435
103. Bolton TB et al (2004) Smooth muscle cells and interstitial cells of blood vessels. Cell Calcium 35 (6):643–657
104. House SJ et al (2008) The non-excitable smooth muscle: calcium signaling and phenotypic switching during vascular disease. Pflugers Arch 456 (5):769–785
105. Lompre AM et al (2013) STIM1 and Orai in cardiac hypertrophy and vascular proliferative diseases. Front Biosci (Schol Ed) 5:766–773
106. Gomez D, Owens GK (2012) Smooth muscle cell phenotypic switching in atherosclerosis. Cardiovasc Res 95(2):156–164
107. Aird WC (2005) Spatial and temporal dynamics of the endothelium. J Thromb Haemost 3(7):1392–1406
108. Aird WC (2007) Phenotypic heterogeneity of the endothelium: I. Structure, function, and mechanisms. Circ Res 100(2):158–173
109. Carmeliet P, Jain RK (2011) Molecular mechanisms and clinical applications of angiogenesis. Nature 473 (7347):298–307
110. Nelken NA et al (1992) Thrombin receptor expression in normal and atherosclerotic human arteries. J Clin Invest 90(4):1614–1621
111. Hirano K (2007) The roles of proteinase-activated receptors in the vascular physiology and pathophysiology. Arterioscler Thromb Vasc Biol 27(1):27–36
112. Gonzalez-Cobos JC et al (2013) Store-independent Orai1/3 channels activated by intracrine leukotriene

C4: role in neointimal hyperplasia. Circ Res 112 (7):1013–1025
113. Mignen O, Thompson JL, Shuttleworth TJ (2008) Both Orai1 and Orai3 are essential components of the arachidonate-regulated Ca^{2+}-selective (ARC) channels. J Physiol 586(1):185–195
114. Mignen O, Thompson JL, Shuttleworth TJ (2007) STIM1 regulates Ca^{2+} entry via arachidonate-regulated Ca^{2+}-selective (ARC) channels without store depletion or translocation to the plasma membrane. J Physiol 579(Pt 3):703–715
115. Abdullaev IF et al (2008) Stim1 and Orai1 mediate CRAC currents and store-operated calcium entry important for endothelial cell proliferation. Circ Res 103(11):1289–1299
116. Simo-Cheyou ER et al (2017) STIM-1 and ORAI-1 channel mediate angiotensin-II-induced expression of Egr-1 in vascular smooth muscle cells. J Cell Physiol 232(12):3496–3509
117. Guo RW et al (2012) Stim1- and Orai1-mediated store-operated calcium entry is critical for angiotensin II-induced vascular smooth muscle cell proliferation. Cardiovasc Res 93(2):360–370
118. Rodriguez-Moyano M et al (2013) Urotensin-II promotes vascular smooth muscle cell proliferation through store-operated calcium entry and EGFR transactivation. Cardiovasc Res 100(2):297–306
119. Myllarniemi M et al (1997) Inhibition of platelet-derived growth factor receptor tyrosine kinase inhibits vascular smooth muscle cell migration and proliferation. FASEB J 11(13):1119–1126
120. Bisaillon JM et al (2010) Essential role for STIM1/Orai1-mediated calcium influx in PDGF-induced smooth muscle migration. Am J Physiol Cell Physiol 298(5):C993–C1005
121. Spinelli AM et al (2012) Airway smooth muscle STIM1 and Orai1 are upregulated in asthmatic mice and mediate PDGF-activated SOCE, CRAC currents, proliferation, and migration. Pflugers Arch 464 (5):481–492
122. Mancarella S et al (2013) Targeted STIM deletion impairs calcium homeostasis, NFAT activation, and growth of smooth muscle. FASEB J 27(3):893–906
123. Levitzki A (2005) PDGF receptor kinase inhibitors for the treatment of restenosis. Cardiovasc Res 65 (3):581–586
124. Jandt E et al (2010) Stent-based release of a selective PDGF-receptor blocker from the bis-indolylmethanon class inhibits restenosis in the rabbit animal model. Vasc Pharmacol 52(1–2):55–62
125. Makiyama Y et al (2008) Imatinib mesilate inhibits neointimal hyperplasia via growth inhibition of vascular smooth muscle cells in a rat model of balloon injury. Tohoku J Exp Med 215(4):299–306
126. Trepakova ES et al (2001) Properties of a native cation channel activated by Ca^{2+} store depletion in vascular smooth muscle cells. J Biol Chem 276 (11):7782–7790
127. Smani T et al (2007) Role of Ca^{2+}-independent phospholipase A2 and store-operated pathway in urocortin-induced vasodilatation of rat coronary artery. Circ Res 101(11):1194–1203
128. Hopson KP et al (2011) S1P activates store-operated calcium entry via receptor- and non-receptor-mediated pathways in vascular smooth muscle cells. Am J Physiol Cell Physiol 300(4):C919–C926
129. Zhang W et al (2011) Orai1-mediated I (CRAC) is essential for neointima formation after vascular injury. Circ Res 109(5):534–542
130. Potier M et al (2009) Evidence for STIM1- and Orai1-dependent store-operated calcium influx through ICRAC in vascular smooth muscle cells: role in proliferation and migration. FASEB J 23 (8):2425–2437
131. Berra-Romani R et al (2008) Ca^{2+} handling is altered when arterial myocytes progress from a contractile to a proliferative phenotype in culture. Am J Physiol Cell Physiol 295(3):C779–C790
132. Taniguchi H et al (1999) Possible involvement of Ca^{2+} entry and its pharmacological characteristics responsible for endothelium-dependent, NO-mediated relaxation induced by thapsigargin in guinea-pig aorta. J Pharm Pharmacol 51(7):831–840
133. Shaul PW (2002) Regulation of endothelial nitric oxide synthase: location, location, location. Annu Rev Physiol 64:749–774
134. Hirano K, Hirano M, Hanada A (2009) Involvement of STIM1 in the proteinase-activated receptor 1-mediated Ca^{2+} influx in vascular endothelial cells. J Cell Biochem 108(2):499–507
135. Boittin FX et al (2008) Ca^{2+}-independent PLA2 controls endothelial store-operated Ca^{2+} entry and vascular tone in intact aorta. Am J Physiol Heart Circ Physiol 295(6):H2466–H2474
136. Estrada IA et al (2012) STIM1 restores coronary endothelial function in type 1 diabetic mice. Circ Res 111(9):1166–1175
137. Zhang W, Trebak M (2014) Vascular balloon injury and intraluminal administration in rat carotid artery. J Vis Exp (94):52045
138. Guo RW et al (2009) An essential role for stromal interaction molecule 1 in neointima formation following arterial injury. Cardiovasc Res 81(4):660–668
139. Jardin I et al (2012) Homers regulate calcium entry and aggregation in human platelets: a role for Homers in the association between STIM1 and Orai1. Biochem J 445(1):29–38
140. Jia S et al (2017) Homer binds to Orai1 and TRPC channels in the neointima and regulates vascular smooth muscle cell migration and proliferation. Sci Rep 7(1):5075
141. Garnier-Raveaud S et al (2001) Identification of membrane calcium channels essential for cytoplasmic and nuclear calcium elevations induced by vascular endothelial growth factor in human endothelial cells. Growth Factors 19(1):35–48

142. Faehling M et al (2002) Essential role of calcium in vascular endothelial growth factor A-induced signaling: mechanism of the antiangiogenic effect of carboxyamidotriazole. FASEB J 16(13):1805–1807
143. Fasolato C, Nilius B (1998) Store depletion triggers the calcium release-activated calcium current (ICRAC) in macrovascular endothelial cells: a comparison with Jurkat and embryonic kidney cell lines. Pflugers Arch 436(1):69–74
144. Li J et al (2011) Orai1 and CRAC channel dependence of VEGF-activated Ca^{2+} entry and endothelial tube formation. Circ Res 108(10):1190–1198
145. Antigny F, Girardin N, Frieden M (2012) Transient receptor potential canonical channels are required for in vitro endothelial tube formation. J Biol Chem 287 (8):5917–5927
146. Gandhirajan RK et al (2013) Blockade of NOX2 and STIM1 signaling limits lipopolysaccharide-induced vascular inflammation. J Clin Invest 123(2):887–902
147. Asahara T et al (1997) Isolation of putative progenitor endothelial cells for angiogenesis. Science 275 (5302):964–967
148. Sanchez-Hernandez Y et al (2010) Store-operated Ca (2+) entry is expressed in human endothelial progenitor cells. Stem Cells Dev 19(12):1967–1981
149. Shi Y et al (2010) Knockdown of stromal interaction molecule 1 attenuates hepatocyte growth factor-induced endothelial progenitor cell proliferation. Exp Biol Med (Maywood) 235(3):317–325
150. Kuang CY et al (2010) Silencing stromal interaction molecule 1 by RNA interference inhibits the proliferation and migration of endothelial progenitor cells. Biochem Biophys Res Commun 398(2):315–320
151. Dragoni S et al (2011) Vascular endothelial growth factor stimulates endothelial colony forming cells proliferation and tubulogenesis by inducing oscillations in intracellular Ca^{2+} concentration. Stem Cells 29(11):1898–1907
152. Bao P et al (2009) The role of vascular endothelial growth factor in wound healing. J Surg Res 153 (2):347–358
153. Hanahan D, Folkman J (1996) Patterns and emerging mechanisms of the angiogenic switch during tumorigenesis. Cell 86(3):353–364
154. Sato TN et al (1995) Distinct roles of the receptor tyrosine kinases Tie-1 and Tie-2 in blood vessel formation. Nature 376(6535):70–74
155. Suri C et al (1996) Requisite role of angiopoietin-1, a ligand for the TIE2 receptor, during embryonic angiogenesis. Cell 87(7):1171–1180
156. Folkman J, D'Amore PA (1996) Blood vessel formation: what is its molecular basis? Cell 87 (7):1153–1155
157. Jho D et al (2005) Angiopoietin-1 opposes VEGF-induced increase in endothelial permeability by inhibiting TRPC1-dependent Ca2 influx. Circ Res 96(12):1282–1290
158. Pardanaud L, Yassine F, Dieterlen-Lievre F (1989) Relationship between vasculogenesis, angiogenesis and haemopoiesis during avian ontogeny. Development 105(3):473–485
159. Ribatti D, Nico B, Crivellato E (2009) Morphological and molecular aspects of physiological vascular morphogenesis. Angiogenesis 12(2):101–111
160. Cines DB et al (1998) Endothelial cells in physiology and in the pathophysiology of vascular disorders. Blood 91(10):3527–3561
161. Mead LE, et al (2008) Isolation and characterization of endothelial progenitor cells from human blood. Curr Protoc Stem Cell Biol. Chapter 2: p. Unit 2C 1
162. Wang QC et al (2016) TMCO1 is an ER Ca(2+) load-activated Ca(2+) channel. Cell 165(6):1454–1466
163. Poteser M et al (2008) Identification of a rare subset of adipose tissue-resident progenitor cells, which express CD133 and TRPC3 as a VEGF-regulated Ca^{2+} entry channel. FEBS Lett 582(18):2696–2702
164. Van Assche T et al (2007) Altered Ca^{2+} handling of smooth muscle cells in aorta of apolipoprotein E-deficient mice before development of atherosclerotic lesions. Cell Calcium 41(3):295–302
165. Leung FP et al (2008) Store-operated calcium entry in vascular smooth muscle. Br J Pharmacol 153 (5):846–857
166. Zhang W, Trebak M (2011) STIM1 and Orai1: novel targets for vascular diseases? Sci China Life Sci 54 (8):780–785
167. Ruhle B, Trebak M (2013) Emerging roles for native Orai Ca^{2+} channels in cardiovascular disease. Curr Top Membr 71:209–235
168. Spinelli AM, Trebak M (2016) Orai channel-mediated Ca^{2+} signals in vascular and airway smooth muscle. Am J Physiol Cell Physiol 310(6):C402–C413
169. Mazzucato M et al (2002) Sequential cytoplasmic calcium signals in a 2-stage platelet activation process induced by the glycoprotein Ibalpha mechanoreceptor. Blood 100(8):2793–2800
170. Rink TJ, Sage SO (1990) Calcium signaling in human platelets. Annu Rev Physiol 52:431–449
171. Braun A et al (2009) Orai1 (CRACM1) is the platelet SOC channel and essential for pathological thrombus formation. Blood 113(9):2056–2063
172. Bergmeier W et al (2009) R93W mutation in Orai1 causes impaired calcium influx in platelets. Blood 113(3):675–678
173. Grosse J et al (2007) An EF hand mutation in Stim1 causes premature platelet activation and bleeding in mice. J Clin Invest 117(11):3540–3550
174. Varga-Szabo D et al (2008) The calcium sensor STIM1 is an essential mediator of arterial thrombosis and ischemic brain infarction. J Exp Med 205 (7):1583–1591
175. Gamage TH et al (2018) STIM1 R304W causes muscle degeneration and impaired platelet activation in mice. Cell Calcium 76:87–100
176. Ahmad F et al (2011) Relative contributions of stromal interaction molecule 1 and CalDAG-GEFI to

calcium-dependent platelet activation and thrombosis. J Thromb Haemost 9(10):2077–2086
177. Gilio K et al (2010) Roles of platelet STIM1 and Orai1 in glycoprotein VI- and thrombin-dependent procoagulant activity and thrombus formation. J Biol Chem 285(31):23629–23638
178. McCarl CA et al (2009) ORAI1 deficiency and lack of store-operated Ca^{2+} entry cause immunodeficiency, myopathy, and ectodermal dysplasia. J Allergy Clin Immunol 124(6):1311–1318.e7
179. Picard C et al (2009) STIM1 mutation associated with a syndrome of immunodeficiency and autoimmunity. N Engl J Med 360(19):1971–1980
180. Elvers M et al (2012) Intracellular cyclophilin A is an important Ca(2+) regulator in platelets and critically involved in arterial thrombus formation. Blood 120 (6):1317–1326
181. Varga-Szabo D et al (2008) Store-operated Ca(2+) entry in platelets occurs independently of transient receptor potential (TRP) C1. Pflugers Arch 457 (2):377–387
182. Galan C et al (2009) STIM1, Orai1 and hTRPC1 are important for thrombin- and ADP-induced aggregation in human platelets. Arch Biochem Biophys 490 (2):137–144
183. Daemen J et al (2007) Early and late coronary stent thrombosis of sirolimus-eluting and paclitaxel-eluting stents in routine clinical practice: data from a large two-institutional cohort study. Lancet 369 (9562):667–678
184. Iakovou I et al (2005) Incidence, predictors, and outcome of thrombosis after successful implantation of drug-eluting stents. JAMA 293(17):2126–2130
185. Liang SJ et al (2016) Inhibition of Orai1 store-operated calcium channel prevents foam cell formation and atherosclerosis. Arterioscler Thromb Vasc Biol 36(4):618–628
186. Edwards JM et al (2010) Exercise training decreases store-operated Ca^{2+} entry associated with metabolic syndrome and coronary atherosclerosis. Cardiovasc Res 85(3):631–640
187. Fang M et al (2019) miR-185 silencing promotes the progression of atherosclerosis via targeting stromal interaction molecule 1. Cell Cycle 18(6–7):682–695
188. Gimbrone MA Jr, Garcia-Cardena G (2016) Endothelial cell dysfunction and the pathobiology of atherosclerosis. Circ Res 118(4):620–636
189. Wang LY et al (2015) Reduction of store-operated Ca (2+) entry correlates with endothelial progenitor cell dysfunction in atherosclerotic mice. Stem Cells Dev 24(13):1582–1590
190. Yang J et al (2017) Store-operated calcium entry-activated autophagy protects EPC proliferation via the CAMKK2-MTOR pathway in ox-LDL exposure. Autophagy 13(1):82–98
191. Welt FG et al (2000) Neutrophil, not macrophage, infiltration precedes neointimal thickening in balloon-injured arteries. Arterioscler Thromb Vasc Biol 20(12):2553–2558
192. Schaff UY et al (2010) Orai1 regulates intracellular calcium, arrest, and shape polarization during neutrophil recruitment in shear flow. Blood 115(3):657–666
193. Wellman GC et al (2001) Membrane depolarization, elevated Ca(2+) entry, and gene expression in cerebral arteries of hypertensive rats. Am J Physiol Heart Circ Physiol 281(6):H2559–H2567
194. Kitazono T et al (2002) Increased activity of calcium channels and Rho-associated kinase in the basilar artery during chronic hypertension in vivo. J Hypertens 20(5):879–884
195. Goulopoulou S, Webb RC (2014) Symphony of vascular contraction: how smooth muscle cells lose harmony to signal increased vascular resistance in hypertension. Hypertension 63(3):e33–e39
196. Kassan M et al (2015) Differential role for stromal interacting molecule 1 in the regulation of vascular function. Pflugers Arch 467(6):1195–1202
197. Kassan M et al (2016) Essential role of smooth muscle STIM1 in hypertension and cardiovascular dysfunction. Arterioscler Thromb Vasc Biol 36 (9):1900–1909
198. Bendhack LM, Sharma RV, Bhalla RC (1992) Altered signal transduction in vascular smooth muscle cells of spontaneously hypertensive rats. Hypertension 19(2 Suppl):II142–II148
199. Bohr DF, Webb RC (1988) Vascular smooth muscle membrane in hypertension. Annu Rev Pharmacol Toxicol 28:389–409
200. Giachini FR et al (2012) STIM1/Orai1 contributes to sex differences in vascular responses to calcium in spontaneously hypertensive rats. Clin Sci (Lond) 122 (5):215–226
201. Giachini FR et al (2009) Increased activation of stromal interaction molecule-1/Orai-1 in aorta from hypertensive rats: a novel insight into vascular dysfunction. Hypertension 53(2):409–416
202. Tanwar J, Trebak M, Motiani RK (2017) Cardiovascular and hemostatic disorders: role of STIM and Orai proteins in vascular disorders. Adv Exp Med Biol 993:425–452
203. Bootman MD et al (2002) 2-Aminoethoxydiphenyl borate (2-APB) is a reliable blocker of store-operated Ca^{2+} entry but an inconsistent inhibitor of InsP3-induced Ca^{2+} release. FASEB J 16(10):1145–1150
204. Trebak M et al (2002) Comparison of human TRPC3 channels in receptor-activated and store-operated modes. Differential sensitivity to channel blockers suggests fundamental differences in channel composition. J Biol Chem 277(24):21617–21623
205. Cortes SF, Lemos VS, Stoclet JC (1997) Alterations in calcium stores in aortic myocytes from spontaneously hypertensive rats. Hypertension 29 (6):1322–1328
206. Xu YJ, Elimban V, Dhalla NS (2015) Reduction of blood pressure by store-operated calcium channel blockers. J Cell Mol Med 19(12):2763–2770
207. Souza Bomfim GH et al (2017) Functional upregulation of STIM-1/Orai-1-mediated

store-operated Ca^{2+} contributing to the hypertension development elicited by chronic EtOH consumption. Curr Vasc Pharmacol 15(3):265–281
208. Gillis EE, Sullivan JC (2016) Sex differences in hypertension: recent advances. Hypertension 68 (6):1322–1327
209. Farber HW, Loscalzo J (2004) Pulmonary arterial hypertension. N Engl J Med 351(16):1655–1665
210. Ng LC et al (2010) Orai1 interacts with STIM1 and mediates capacitative Ca^{2+} entry in mouse pulmonary arterial smooth muscle cells. Am J Physiol Cell Physiol 299(5):C1079–C1090
211. Fernandez RA et al (2015) Upregulated expression of STIM2, TRPC6, and Orai2 contributes to the transition of pulmonary arterial smooth muscle cells from a contractile to proliferative phenotype. Am J Physiol Cell Physiol 308(8):C581–C593
212. Song MY, Makino A, Yuan JX (2011) STIM2 contributes to enhanced store-operated Ca entry in pulmonary artery smooth muscle cells from patients with idiopathic pulmonary arterial hypertension. Pulm Circ 1(1):84–94
213. He X et al (2018) Hypoxia selectively upregulates cation channels and increases cytosolic [Ca(2+)] in pulmonary, but not coronary, arterial smooth muscle cells. Am J Physiol Cell Physiol 314(4):C504–C517
214. Wang J et al (2017) Orai1, 2, 3 and STIM1 promote store-operated calcium entry in pulmonary arterial smooth muscle cells. Cell Death Discov 3:17074
215. Hou X et al (2013) Silencing of STIM1 attenuates hypoxia-induced PASMCs proliferation via inhibition of the SOC/Ca^{2+}/NFAT pathway. Respir Res 14:2
216. Nakamura M, Sadoshima J (2018) Mechanisms of physiological and pathological cardiac hypertrophy. Nat Rev Cardiol 15(7):387–407
217. Ljubojevic S et al (2014) Early remodeling of perinuclear Ca^{2+} stores and nucleoplasmic Ca^{2+} signaling during the development of hypertrophy and heart failure. Circulation 130(3):244–255
218. Iemitsu M et al (2001) Physiological and pathological cardiac hypertrophy induce different molecular phenotypes in the rat. Am J Physiol Regul Integr Comp Physiol 281(6):R2029–R2036
219. Umar S et al (2012) Cardiac structural and hemodynamic changes associated with physiological heart hypertrophy of pregnancy are reversed postpartum. J Appl Physiol (1985) 113(8):1253–1259
220. Maillet M, van Berlo JH, Molkentin JD (2013) Molecular basis of physiological heart growth: fundamental concepts and new players. Nat Rev Mol Cell Biol 14(1):38–48
221. Shimizu I, Minamino T (2016) Physiological and pathological cardiac hypertrophy. J Mol Cell Cardiol 97:245–262
222. Tham YK et al (2015) Pathophysiology of cardiac hypertrophy and heart failure: signaling pathways and novel therapeutic targets. Arch Toxicol 89 (9):1401–1438
223. Bernardo BC et al (2010) Molecular distinction between physiological and pathological cardiac hypertrophy: experimental findings and therapeutic strategies. Pharmacol Ther 128(1):191–227
224. Samak M et al (2016) Cardiac hypertrophy: an introduction to molecular and cellular basis. Med Sci Monit Basic Res 22:75–79
225. Molkentin JD (2006) Dichotomy of Ca^{2+} in the heart: contraction versus intracellular signaling. J Clin Invest 116(3):623–626
226. Lipp P et al (2000) Functional InsP3 receptors that may modulate excitation-contraction coupling in the heart. Curr Biol 10(15):939–942
227. Nakayama H et al (2010) The IP3 receptor regulates cardiac hypertrophy in response to select stimuli. Circ Res 107(5):659–666
228. Hohendanner F et al (2014) Calcium and IP3 dynamics in cardiac myocytes: experimental and computational perspectives and approaches. Front Pharmacol 5:35
229. Higazi DR et al (2009) Endothelin-1-stimulated InsP3-induced Ca^{2+} release is a nexus for hypertrophic signaling in cardiac myocytes. Mol Cell 33 (4):472–482
230. Collins HE et al (2013) STIM1/Orai1-mediated SOCE: current perspectives and potential roles in cardiac function and pathology. Am J Physiol Heart Circ Physiol 305(4):H446–H458
231. Hunton DL et al (2002) Capacitative calcium entry contributes to nuclear factor of activated T-cells nuclear translocation and hypertrophy in cardiomyocytes. J Biol Chem 277(16):14266–14273
232. Parks C et al (2016) STIM1-dependent Ca(2+) microdomains are required for myofilament remodeling and signaling in the heart. Sci Rep 6:25372
233. Troupes CD et al (2017) Role of STIM1 (Stromal Interaction Molecule 1) in hypertrophy-related contractile dysfunction. Circ Res 121(2):125–136
234. Wang Y et al (2015) Nitric oxide-cGMP-PKG pathway acts on Orai1 to inhibit the hypertrophy of human embryonic stem cell-derived cardiomyocytes. Stem Cells 33(10):2973–2984
235. Dai F et al (2018) Overexpression of SARAF ameliorates pressure overload-induced cardiac hypertrophy through suppressing STIM1-Orai1 in Mice. Cell Physiol Biochem 47(2):817–826
236. Bartoli F et al (2020) Orai1 channel inhibition preserves left ventricular systolic function and normal Ca(2+) handling after pressure overload. Circulation 141(3):199–216
237. Smani T et al (2015) Functional and physiopathological implications of TRP channels. Biochim Biophys Acta 1853(8):1772–1782
238. Bartoli F, Sabourin J (2017) Cardiac remodeling and disease: current understanding of STIM1/Orai1-mediated store-operated Ca(2+) entry in cardiac function and pathology. Adv Exp Med Biol 993:523–534

239. Yue Z et al (2015) Role of TRP channels in the cardiovascular system. Am J Physiol Heart Circ Physiol 308(3):H157–H182
240. Camacho Londono JE et al (2015) A background Ca^{2+} entry pathway mediated by TRPC1/TRPC4 is critical for development of pathological cardiac remodelling. Eur Heart J 36(33):2257–2266
241. Kuwahara K et al (2006) TRPC6 fulfills a calcineurin signaling circuit during pathologic cardiac remodeling. J Clin Invest 116(12):3114–3126
242. Onohara N et al (2006) TRPC3 and TRPC6 are essential for angiotensin II-induced cardiac hypertrophy. EMBO J 25(22):5305–5316
243. Bush EW et al (2006) Canonical transient receptor potential channels promote cardiomyocyte hypertrophy through activation of calcineurin signaling. J Biol Chem 281(44):33487–33496
244. Seo K et al (2014) Combined TRPC3 and TRPC6 blockade by selective small-molecule or genetic deletion inhibits pathological cardiac hypertrophy. Proc Natl Acad Sci U S A 111(4):1551–1556
245. Takahashi S et al (2008) Nitric oxide-cGMP-protein kinase G pathway negatively regulates vascular transient receptor potential channel TRPC6. J Physiol 586(17):4209–4223
246. Seth M et al (2009) TRPC1 channels are critical for hypertrophic signaling in the heart. Circ Res 105 (10):1023–1030
247. Bartoli F et al (2019) Specific upregulation of TRPC1 and TRPC5 channels by mineralocorticoid pathway in adult rat ventricular cardiomyocytes. Cell 9(1):47
248. Niggli E (2011) Ryanodine receptors: waking up from refractoriness. Cardiovasc Res 91(4):563–564
249. Lakatta EG et al (2006) The integration of spontaneous intracellular Ca^{2+} cycling and surface membrane ion channel activation entrains normal automaticity in cells of the heart's pacemaker. Ann N Y Acad Sci 1080:178–206
250. Yaniv Y, Lakatta EG, Maltsev VA (2015) From two competing oscillators to one coupled-clock pacemaker cell system. Front Physiol 6:28
251. Imtiaz MS et al (2010) SR Ca^{2+} store refill—a key factor in cardiac pacemaking. J Mol Cell Cardiol 49 (3):412–426
252. Ju YK et al (2007) Store-operated Ca^{2+} influx and expression of TRPC genes in mouse sinoatrial node. Circ Res 100(11):1605–1614
253. Zhang H et al (2015) STIM1-Ca^{2+} signaling modulates automaticity of the mouse sinoatrial node. Proc Natl Acad Sci U S A 112(41):E5618–E5627
254. Liu J et al (2015) Store-operated calcium entry and the localization of STIM1 and Orai1 proteins in isolated mouse sinoatrial node cells. Front Physiol 6:69
255. Nguyen N et al (2013) STIM1 participates in the contractile rhythmicity of HL-1 cells by moderating T-type Ca(2+) channel activity. Biochim Biophys Acta 1833(6):1294–1303
256. Cacheux M et al (2019) Cardiomyocyte-specific STIM1 (Stromal Interaction Molecule 1) depletion in the adult heart promotes the development of arrhythmogenic discordant alternans. Circ Arrhythm Electrophysiol 12(11):e007382
257. Wang P et al (2012) Evidence that 2-aminoethoxydiphenyl borate provokes fibrillation in perfused rat hearts via voltage-independent calcium channels. Eur J Pharmacol 681(1–3):60–67
258. Nattel S (2003) Atrial electrophysiology and mechanisms of atrial fibrillation. J Cardiovasc Pharmacol Ther 8(Suppl 1):S5–S11
259. Heijman J et al (2012) Calcium handling and atrial fibrillation. Wien Med Wochenschr 162 (13–14):287–291
260. Voigt N, Nattel S, Dobrev D (2012) Proarrhythmic atrial calcium cycling in the diseased heart. Adv Exp Med Biol 740:1175–1191
261. Bootman MD, Rietdorf K (2017) Tissue specificity: store-operated Ca(2+) entry in cardiac myocytes. Adv Exp Med Biol 993:363–387
262. Ju YK et al (2015) The involvement of TRPC3 channels in sinoatrial arrhythmias. Front Physiol 6:86
263. Wen H et al (2018) Potential arrhythmogenic role of TRPC channels and store-operated calcium entry mechanism in mouse ventricular myocytes. Front Physiol 9:1785
264. Bonilla IM et al (2019) Enhancement of cardiac store operated calcium entry (SOCE) within novel intercalated disk microdomains in arrhythmic disease. Sci Rep 9(1):10179

15 Physiological Functions, Biophysical Properties, and Regulation of KCNQ1 (K_V7.1) Potassium Channels

Michael C. Sanguinetti and Guiscard Seebohm

Abstract

KCNQ1 (K_V7.1) K^+ channels are expressed in multiple tissues, including the heart, pancreas, colon, and inner ear. The gene encoding the KCNQ1 protein was discovered by a positional cloning effort to determine the genetic basis of long QT syndrome, an inherited ventricular arrhythmia that can cause sudden death. Mutations in *KCNQ1* can also cause other types of arrhythmia (i.e., short QT syndrome, atrial fibrillation) and the gene may also have a role in diabetes and certain cancers. KCNQ1 α-subunits can partner with accessory β-subunits (KCNE1–KCNE5) to form K^+-selective channels that have divergent biophysical properties. In the heart, KCNQ1 α-subunits coassemble with KCNE1 β-subunits to form channels that conduct I_{Ks}, a very slowly activating delayed rectifier K^+ current. K_V7.1 channels are highly regulated by PIP_2, calmodulin, and phosphorylation, and rich pharmacology includes blockers and gating modulators. Recent biophysical studies and a cryo-EM structure of the KCNQ1-calmodulin complex have provided new insights into K_V7.1 channel function, and how interactions between KCNQ1 and KCNE subunits alter the gating properties of heteromultimeric channels.

Keywords

Potassium channel · KCNQ1 · KCNE1 · K_V7.1 · Biophysics · Pharmacology · Long QT syndrome

15.1 Introduction

The *KCNQ1* channel gene was discovered by a positional cloning effort to determine the genetic basis of an inherited ventricular arrhythmia. However, before the gene was cloned, the cardiac current conducted by KCNQ1 (K_V7.1) channels had been described in biophysical terms. In the heart, delayed rectifier K^+ currents mediate repolarization of myocellular action potentials. The outward currents that mediate repolarization of cardiomyocytes were first characterized using voltage-clamp techniques and Purkinje fibers isolated from sheep heart [1]. Repolarization from the plateau phase of action potentials in these fibers was found to be mediated by two current components with widely differing rates of activation. Based on ionic current reversal potentials, it was concluded that the fast

M. C. Sanguinetti (✉)
Division of Cardiovascular Medicine, University of Utah, Salt Lake City, UT, USA
e-mail: m.sanguinetti@utah.edu

G. Seebohm
Cellular Electrophysiology and Molecular Biology, Institute for Genetics of Heart Diseases, University Hospital Münster, Münster, Germany
e-mail: Guiscard.Seebohm@ukmuenster.de

L. Zhou (ed.), *Ion Channels in Biophysics and Physiology*, Advances in Experimental Medicine and Biology 1349, https://doi.org/10.1007/978-981-16-4254-8_15

activating component ("i_{x1}") was K^+ ion-selective, whereas the very slow activating component ("i_{x2}") was conducted by both K^+ and Na^+ ions. Later attempts to replicate these findings in Purkinje fibers were unsuccessful [2] and it was suggested that the slow i_{x2} component may have been an experimental artifact related to poor voltage control and extracellular K^+ accumulation during long depolarizing voltage-clamp pulses of a multicellular preparation. However, further studies confirmed the existence of a very slowly activating and K^+-selective outward current (I_K) in isolated single bullfrog atrial myocytes [3], and both a rapid (i_{x1} or I_{Kr}) and slow (i_{x2} or I_{Ks}) components of I_K in aggregates of neonatal chick myocytes [4] and single guinea pig ventricular [5] and atrial [6] myocytes. In both the chick and guinea pig cardiomyocytes, the fully-activated current-voltage (*I–V*) relationship of the fast component is inwardly rectified due to the fast inactivation of channels at positive potentials, whereas the *I–V* relationship of the slow component is linear because the underlying channels do not inactivate. It is now known that in the heart, hERG channels [7] conduct I_{Kr} [8, 9] and that I_{Ks} channels are conducted by heteromultimeric channels formed by coassembly of KCNQ1 pore-forming α-subunits and KCNE1 β-subunits [10, 11].

In 1988, it was reported that a very small (130 amino acids) protein called "IsK" cloned from rat kidney induced an extremely slowly activating outward K^+ current when heterologously expressed in *Xenopus* oocytes [12]. IsK was subsequently cloned from neonatal rat heart and uterus [13] and an antisense oligonucleotide to IsK specifically inhibited the expression of the current observed in oocytes injected with total mRNAs isolated from rat kidney, heart, or uterus. IsK, encoded by the gene *KCNE1* was subsequently renamed "minK" (*min*imal *K*) because it was the smallest protein known to induce a K^+ current. Although the slow kinetics and voltage dependence for activation of IsK in oocytes resemble cardiac I_{Ks}, based on what was already know about the α-subunit proteins that formed other K^+ channels such as Shaker, it seemed unlikely that proteins as small as minK could coassemble to form functional K^+-selective channels. The α-subunit protein partner of minK that was predicted to be required for the formation of functional channels was discovered by positional cloning of a gene (*KVLQT1*, human chromosomal location 11p15.5) associated with congenital long QT syndrome, a disease characterized by a prolonged QT interval on the electrocardiogram, ventricular arrhythmia, and sudden cardiac death [14]. The *KVLQT1* gene was later renamed *KCNQ1*, and other closely related potassium channel genes were soon discovered (*KCNQ2*, *KCNQ3*, *KCNQ5*) and linked to a rare form of congenital epilepsy [15–17] or to deafness (*KCNQ4*) [18]. The biophysical properties of the K^+-selective current conducted by heterologously expressed *KCNQ1* alone did not match any previously described cardiac K^+ current. This finding suggested that KCNQ1 α-subunits were likely to coassemble with another subunit to form channels that conducted one of the many well-characterized cardiac K^+ currents. A prime candidate for this presumed subunit partner of KCNQ1 was KCNE1 and indeed the K^+ current produced by co-expression of the two subunits [10, 11] was strikingly similar to the well-described slowly activating delayed rectifier K^+ current (i_{X2} or I_{Ks}) previously described in cardiomyocytes [1, 3–5]. KCNQ1 channel proteins are also called "$K_V7.1$" (IUPHAR/BPS Guide to Pharmacology database on ion channels; https://www.guidetopharmacology.org/). In this review, $K_V7.1$ refers to homotetrameric KCNQ1 channels, whereas heteromultimeric channels are named according to their α-subunit/β-subunit partner (e.g., KCNQ1/KCNE1 channels that conduct I_{Ks}).

15.2 Physiological Roles for $K_V7.1$ and I_{Ks} Channels

KCNQ1 is expressed in the heart, pancreas, and epithelial tissues of several tissues including the inner ear, airways, intestine, colon, and stomach (Fig. 15.1). In the heart, KCNQ1 subunits partner with KCNE1 subunits to form channels that

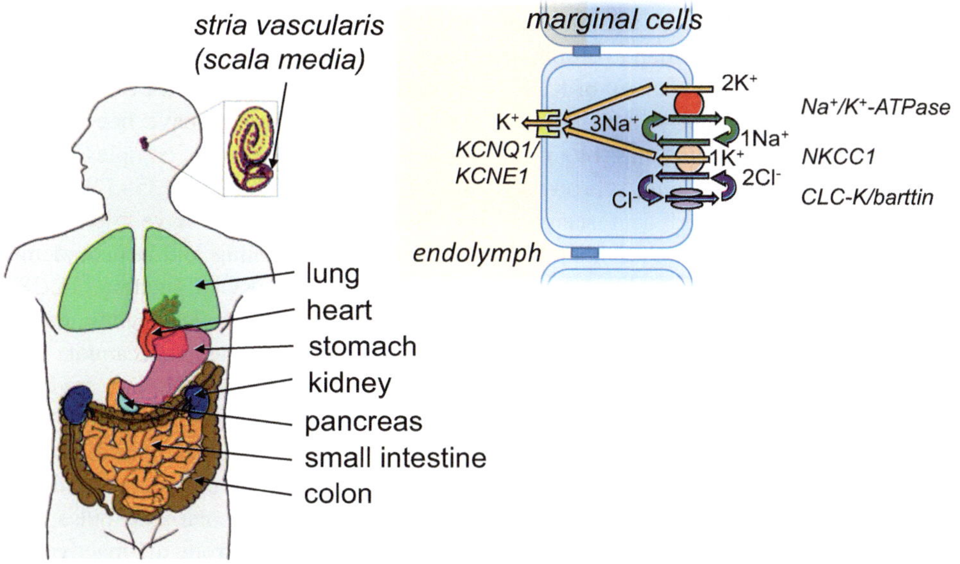

Fig. 15.1 KCNQ1/KCNEx channels are expressed in multiple human organs. KCNQ1/KCNE1 channels in the marginal cells of the *stria vascularis* aid in the generation of the potassium-rich endolymph by allowing the efflux of K^+

conduct I_{Ks}, the current that together with I_{Kr} mediates repolarization of ventricular cardiomyocyte action potentials. KCNQ1 subunits also partner with KCNE1 subunits in epithelial cells of the inner ear, airways, and exocrine pancreas to conduct I_{Ks} [19]. Immunohistochemical analysis was used to first demonstrate that the KCNE1 protein is expressed on the endolymphatic surface of the marginal cells in the *stria vascularis* of the inner ear [20]. In some cells, *KCNQ1* is expressed in the absence of *KCNE1*. For example, in mice, mRNA for *KCNQ1*, but not *KCNE1* was found in epithelial tissues of the small intestine, lung, liver, and thymus [21].

In epithelial cells of many tissues, KCNQ1 partners with KCNE3 to form a time- and voltage-independent outward K^+ current [22]. Electrogenic transport of Cl^- ions is required for the secretion of fluid across the small and large intestine. In these epithelial cells, KCNQ1 partners with KCNE3 (also called MiRP2, "minK related protein 2") subunits to form a time- and voltage-independent outward K^+ current [22] that counters the flux of Cl^- ions. For example, CFTR Cl^- channels, activated by cAMP, are responsible for anion and fluid secretion across the intestinal epithelium, and this anion transport is largely counterbalanced by the flux of K^+ ions mediated by KCNQ1-KCNE3 channels located on the basolateral membrane of the epithelial cells. However, with knockout of either *Kcnq1* or *Kcne3* in mice, a significant amount of cAMP-activated anion secretion persists in colon epithelia [23, 24], indicating an important role for another K^+ conductance. A recent study suggests that TASK-2 channels are responsible for this alternative source of counter-balancing K^+ flux [25]. In the stomach, gastric secretion of protons (H^+) is accomplished by H^+/K^+-ATPase and is coupled to the influx of K^+ at the luminal surface. In the apical membrane of parietal cells of the stomach, KCNQ1, KCNE2, and KCNE3 [26] and Kir1.1 (ROMK) [27] colocalizes with the ATPase. When heterologously expressed in COS cells KCNQ1-KCNE3 channel current was H^+-insensitive, whereas KCNQ1-KCNE2 channel current was activated by low extracellular pH [26, 28]. The relative roles of KCNQ1 and Kir1.1 in regulating gastric secretion are uncertain as inhibition of either channel type has been reported to eliminate gastric secretion. Specifically, knockout of Kir1.1 in mice eliminated secretagogue-stimulated gastric acid secretion [27], while the block of KCNQ1-containing channels with chromanol 293B also completely inhibited acid secretion [26]. Together these findings suggest that both Kir1.1 and KCNQ1-containing channels are essential for K^+ recycling and H^+ secretion across the apical membrane of parietal cells in the stomach.

15.3 $K_V7.1$ and KCNQ1/KCNEx Channel Biophysics

$K_V7.1$ channels activate rapidly with a time constant in the range of 80–100 ms at +50 mV. Rapid activation is followed by a delayed slow inactivation. The kinetic behavior of $K_V7.1$ can be approximated by a linear gating scheme [29, 30] of the form:

$$C1 \leftrightarrow C2 \leftrightarrow C3 \leftrightarrow O1 \leftrightarrow O2 \leftrightarrow I$$

In this model, C1, C2, C3 represent closed states; and O1 and O2 represent open states; the transitions between these states are voltage-dependent. Alternative, circular gating schemes have been used to capture special features of $K_V7.1$ channel gating. One of these models [31] accounts for an open component that is determined by the allosteric coupling of residues in the S4–S5 linker/pore domain (PD). The model introduces an alternative deactivation path connecting O2 with C1 and describes voltage-dependent gating for steps C1 $\leftrightarrow$ C2 and O1 $\leftrightarrow$ O2. However, this model does not include an explicit inactivation step. Rapid activation is followed by a delayed slow and incomplete inactivation, with about 50–60% of the channels being inactivated at +60 mV at steady-state [29, 30]. Cui et al. [32] introduced two linear-branched gating schemes for the KCNQ1/KCNE1 channel. Interestingly, Tzounopoulos et al. reported a cross-over gating after hyperpolarizing prepulsing which cannot be explained by a classical Cole–Moore shift in a linear gating model [33], but can be accounted for by a branched model [34]. Several other mathematical gating schemes have been proposed to describe the different complex gating of KCNQ1/KCNE1 channels [35–37]. In addition, several research groups have tried to combine mathematical modeling and structural modeling to address KCNQ1/KCNE1 gating [34, 38–41]. It is likely that the complex gating behavior of KCNQ1/KCNE1 channels cannot be fully recapitulated with simple Markov models.

Slow $K_V7.1$ inactivation involves the central pore as evident in the appearance of characteristic "hooked" tail currents upon repolarization of the membrane from a depolarizing pulse, reflecting channels recovering from an inactivated to an open state [30, 42, 43]. Interestingly, especially prominent inactivation of macroscopic outward currents is seen in several mutant $K_V7.1$ channels [44–46]. The slow-developing C-type inactivation mechanism in other Kv channels involves peripheral protein regions and an allosteric modification of a hydrogen network that alters the conformation of the selectivity filter [47, 48]. While C-type inactivation has been proposed to be caused by a constriction followed by a collapse of the selectivity filter conduction pathway [49, 50], the precise molecular mechanisms of such inactivation are not well understood. This holds even more true for $K_V7.1$ channels which inactivate by a "modified" C-type inactivation mechanism [44]. Recently, Hou et al. [51] suggested that $K_V7.1$ inactivation derives from a different mechanism of voltage sensor domain (VSD)-pore coupling. They propose that the activation occurs via an intermediate open (IO) state to the activated-open (AO) state. According to Hou et al., this AO state is less efficient in VSD-pore coupling, which produces inactivation [51]. Rate-limiting gating steps along with two open states had previously been revealed by intracellular Na^+ block of $K_V7.1$ channels [52]. Similar to classical C-type inactivation, the probable role of the selectivity filter in $K_V7.1$ channel inactivation is exemplified by the presence of extracellular Ba^{2+}-induced pore block and the correlation between inward Rb^+ vs. K^+ flux and degree of inactivation of mutant channels [44, 53]. Besides inactivation, $K_V7.1$ channel currents are suppressed by a fast flickery block and a small single-channel conductance of about 1.8 pS [29, 54]. The structural (molecular) basis for low ion flux and flicker activity remains elusive.

The diversity in biophysical properties and physiological roles of channels containing KCNQ1 subunits results from its tissue-dependent coassembly with different KCNE β-subunits. In the heart and kidney, KCNQ1 associates with KCNE1 to conduct I_{Ks} [10, 11, 55]. Compared to $K_V7.1$, KCNQ1/KCNE1 channels activate at higher membrane potentials, have a larger single-channel conductance (3.2 pS), exhibit a pronounced delay in the onset and a much slower rate of activation (time constants measured in seconds), and do not inactivate [10, 11, 29, 30, 37, 43]. A detailed analysis of single KCNQ1/KCNE1 channel activity revealed up to five long-lived sub-conductance states, with single-channel currents ranging from 0.13 to 0.66 pA at 60 mV [37]. These sub-conductance states may result from discrete movements of the four different voltage sensors. Application of cAMP reduced first latency to channel opening and increased higher sub-conductance level occupancy, linking cAMP modulation to single-channel events [56, 57]. Channels formed by coassembly of KCNQ1 with KCNE2 subunits are characterized by reduced current amplitude, instantaneous activation with fast partial deactivation, and a linear *I–V* relationship [58]. KCNQ1-KCNE3 channel currents also have a linear *I–V* relationship, a larger conductance than $K_V7.1$ channels, and exhibit only a minimal time dependency of gating such that current onset is nearly instantaneous [22, 59, 60]. By contrast, KCNE4 strongly suppresses KCNQ1 currents [61] and KCNE5 right-shifts the voltage dependence of $K_V7.1$ activation therewith suppressing currents at physiological voltages [62, 63].

Similar to other Kv channels, functional $K_V7.1$ channels are tetramers. KCNE1 β-subunits coassemble to the outside of these tetrameric channels and molecular interactions that take place at various sites within the channel complex. The stoichiometry of these subunit interactions has been a matter of intense debate. Whereas some research groups have provided data in favor of a fixed KCNQ1-KCNE1 subunit stoichiometry of 4–2, others have argued for a flexible stoichiometry with regard to the number (1–4) of KCNE1 subunits. Experimental findings in favor of a fixed KCNQ1-KCNE1 stoichiometry of 4–2 were provided by Wang and Goldstein [64] based on current suppression upon mutant coexpression, and by Chen et al. [65] analyzing ^{3}H-charybdotoxin and an antibody binding to heteromeric channels. Morin and Kobertz [66] covalently modified individual KCNE-subunits with a derivatized charybdotoxin. By repeating this step after removal of free charybdotoxin they could count the available KCNE1 subunits in KCNQ1/KCNE1 complexes, supporting a fixed KCNQ1-KCNE1 stoichiometry of 4–2. Finally, Kang et al. [67] suggested that in silico binding of more than two KCNE1 subunits to the $K_V7.1$ tetramer might be sterically hindered. However, this notion is not supported by other findings [34], or if the recent KCNQ1/calmodulin single particle-based structure [68] is considered. On the contrary, Cui et al. [32] and Morokuma et al. [69] showed that current amplitude, activation kinetics, and voltage dependence of KCNQ1/KCNE1 channels vary with the amount of coexpressed KCNE1. Nakajo and Kubo [70] used a single-molecule fluorescent bleaching approach to provide data in support of variable KCNE1 stoichiometry. By application of free KCNE1 C-terminal peptides to expressed KCNQ1/KCNE1 channels and analyzing voltage dependence of activating currents, Zheng et al. also provided experimental evidence supporting variable stoichiometry [71]. Recently, we could show that KCNE1 induced a lateral crevice in the pore domain constituting a high-affinity adamantine compound binding site in the $K_V7.1$ channel. KCNQ1/KCNE1 coexpression at variable stoichiometry resulted in a phenotype of channel kinetics that paralleled adamantine compound sensitivity [72]. Although controversial, a variable KCNQ1-KCNE1 stoichiometry appears to be most likely. This notion is remarkable as a variable stoichiometry of protein-protein assembly is a rare.

15.4 K_V7.1 Channel Structure

Like other Kv channels, each KCNQ1 subunit has six α-helical transmembrane segments (S1–S6), intracellular N- and C-termini, a voltage sensing domain formed by S1–S4 segments, and a pore domain formed by the S5 and S6 segments. Functional K_V7.1 channels are tetramers, formed by coassembly of four identical KCNQ1 α-subunits. Based on the extensive physiology of native channels, it is clear that in most cells K_V7.1 subunits form a macromolecular complex with PIP_2, calmodulin (CaM), and most often with regulatory subunits such as KCNE1, KCNE2, or KCNE3. In addition, the regulatory proteins protein kinase A, protein phosphatase 1, adenylyl cyclase 9, phosphodiesterase 4D3, and yotiao (AKAP9) can associate with K_V7.1 channels to form a signaling complex in the plasma membrane [73–75].

The single-particle cryo-electron microscopy (cryo-EM) structure of K_V7.1 channels was published in 2017 [68]. Specifically, the structure of amino acids 67–610 of *Xenopus laevis* (African clawed frog) K_V7.1 in association with CaM was solved at an overall resolution of 3.7 Å. The region of the frog K_V7.1 used for structural determination is 78% homologous to the human channel subunit, and this construct had biophysical properties very similar, but not identical to the full-length frog channel. The association of the construct with CaM is relevant because CaM is likely to be always associated with KCNQ1 subunits in the mature channel complex and their interaction is required for normal channel assembly and gating function [76]. The published K_V7.1/CaM structure was in a PIP_2-free state, with the voltage sensors in their activated configuration (i.e., outwardly displaced as would usually occur at depolarized transmembrane voltage), but the ionic conductance pore was in a closed conformation (smallest pore radius of 0.8 Å). In the absence of PIP_2, the VSD appeared uncoupled from the "activation gate" (pore domain).

The K_V7.1 tetrameric channel was found to have several unusual features compared to other Kv channels. The S2–S3 linker has nine more amino acids compared to most other Kv channels, the extracellular loop that connects the S5 segment to the pore helix forms a negatively charged cap element that surrounds the extracellular facing channel entrance, and the S6 helix does not have the "glycine hinge" or Pro-X-Pro motif proposed to serve as hinges that can narrow (close) the channel pore in other Kv channels. In place of the Pro-Val-Pro hinge motif, K_V7.1 has a Pro-Ala-Gly motif. The structure also shows a likely region near the S4–S5 linker responsible for PIP_2 binding that serves to strengthen the coupling between the voltage sensor and the pore domain. The channel exhibits a "domain-swapped" transmembrane topology, where the VSD interacts with the pore domain of an adjacent subunit rather than forming an intra-subunit interaction between these two domains. This domain-swapped arrangement is similar to Kv1 channels but unlike that observed for the structures of many other Kv channels. Finally, the S4–S5 linker of K_V7.1 is partly a loop, and perhaps a PIP_2-binding site, instead of the helical structure found in most other Kv channels, a feature that may allow the uncoupling between the VSD and pore domain that is observed in the absence of PIP_2. The amplitude of the K_V7.1 channel current is decreased when PIP_2 is depleted from the inner membrane leaflet and this effect is proposed to occur because of decoupling of the VSD activation from channel opening [77, 78]. The interacting interface between the VSD and the pore domain includes specific residues in the S2–S3 loop, the S4–S5 linker, and the S6 terminus. Mutation of these key residues impaired VSD to pore coupling by decreasing the apparent binding affinity of PIP_2. Mutagenesis of two residues (K183 in the S2–S3 loop, Arg249 in S4–S5 linker) had the opposite effect, increasing channel current as a result of enhanced VSD to pore coupling [77]. Insights gained from mutagenesis experiments and molecular dynamics simulations provide a model for PIP_2 regulation of channel activation: PIP_2 binds to positively charged residues of KCNQ1 to form intra-subunit salt bridges between key regions of the VSD and the S6 to enhance channel open probability [79]. Future cryo-EM studies will hopefully provide insights into specific residues responsible for interaction with KCNE subunits and how coassembly alters the structure of $K_V7.1$.

The functional properties of KCNQ1/KCNE1 channels were recently simulated by using an artificial intelligence machine learning approach [41]. Mechanistic insights predicted from simulations of KCNQ1/KCNE1 channel gating suggests that the subconductance states of the channel were dependent on the entire channel, not just the energy profile of the pore domain, that KCNQ1-KCNE1 interactions result in the two stepped nature of voltage sensor displacement and slow onset of current activation.

15.5 KCNQ1 Channel Pharmacology

Several inhibitors and activators of $K_V7.1$ and KCNQ1/KCNE1 channels have been described (Fig. 15.2). The first relatively specific blocker to be described was chromanol 293B (trans-6-cyano-4-(*N*-ethylsulfonyl-*N*-methylamino)-3-hydroxy-2,2-dimethyl-chroman) that was reported to inhibit a K^+ conductance in the basolateral membrane of colonic crypt cells [80], later identified as a channel containing KCNQ1 subunits [81]. The IC_{50} for chromanol 293B is enantiomer-dependent and varies from 10 to 30 μM for $K_V7.1$ and KCNQ1/KCNE1 expressed in *Xenopus* oocytes [82]. An analog of this compound (IKs124) was later found to be far more potent, with a potency that varied depending upon the specific KCNE subunit partner examined. IKs124 blocked $K_V7.1$ in CHO cells with an IC_{50} of 8, 370, and 440 nM when coexpressed with KCNE2, KCNE1, and KCNE3, respectively [28].

The putative binding sites for chromanol 293B and the benzodiazepine L-735821 ("L7", *N*-[(3*R*)-2,3-dihydro-1-methyl-2-oxo-5-phenyl-1*H*-1,4-benzodiazepin-3-yl]-3-(2,4-dichlorophenyl)-2-propenamide) were studied by site-directed mutagenesis and by analyzing effects of the compound on $K_V7.1$-$K_V7.2$ chimeras ($K_V7.2$ channels are insensitive to L7) [83, 84]. Chimeric channel results indicated that both compounds interact with a binding site associated with the selectivity filter and the S6 transmembrane segment. For both compounds, mutation of Thr312 of the pore domain and Ile337, Phe339 (for L7), and Phe340 of the S6 domain most affected channel block. Simulated docking of the compounds suggested that they inhibit ion flux by preventing occupancy of K^+ ions of a pore helix-coordinated site within the central hydrated cavity or by electrostatic interaction with the innermost K^+ ion in the selectivity filter. The same binding site has been proposed for amitriptyline that blocks $K_V7.1$ and I_{Ks} channels with IC_{50} values of 8.8 μM and 2.5 μM, respectively [85]. Other synthetic blockers of I_{Ks} include HMR 1556 [86] with an IC_{50} of 84 nM in HEK293 cells [87], and azimilide with an IC_{50} of 5.6 μM in *Xenopus* oocytes [88]. Drug effects on $K_V7.1$ can vary when the channel is partnered with KCNE1. For example, the potency for the block of $K_V7.1$ channels by chromanol 293B, azimilide, and 17-beta-oestradiol is 6- to 100-fold lower than for KCNQ1/KCNE1 channels [88], whereas channel block by XE991, a non-specific KCNQ channel inhibitor, is more potent for $K_V7.1$ (IC_{50} of 0.8 μM) than for KCNQ1/KCNE1 (IC_{50} of 11.1 μM) [89]. KCNE1 was recently reported to affect the activity of adamantine compounds in a previously unexpected manner. These compounds, such as JNJ303 (2-(4-chlorophenoxy)-2-methyl-*N*-[5-[(methylsulfonyl)amino]tricycle[3.3.1.13,7]dec-2-yl]-propanamide) modify KCNQ1/KCNE1 channel gating by binding to fenestrations in $K_V7.1$ that only become accessible when KCNE1 accessory subunits are bound to $K_V7.1$ channels [72].

Two peptide toxins were recently discovered that inhibit $K_V7.1$ and KCNQ1/KCNE1 channels. SSD609 is a 47-residue polypeptide isolated from the venom of the centipede *Scolopendra subspinipes* that inhibits guinea pig myocyte I_{Ks} with an IC_{50} of 209 nM [90]. Based on experiments with heterologous expressed

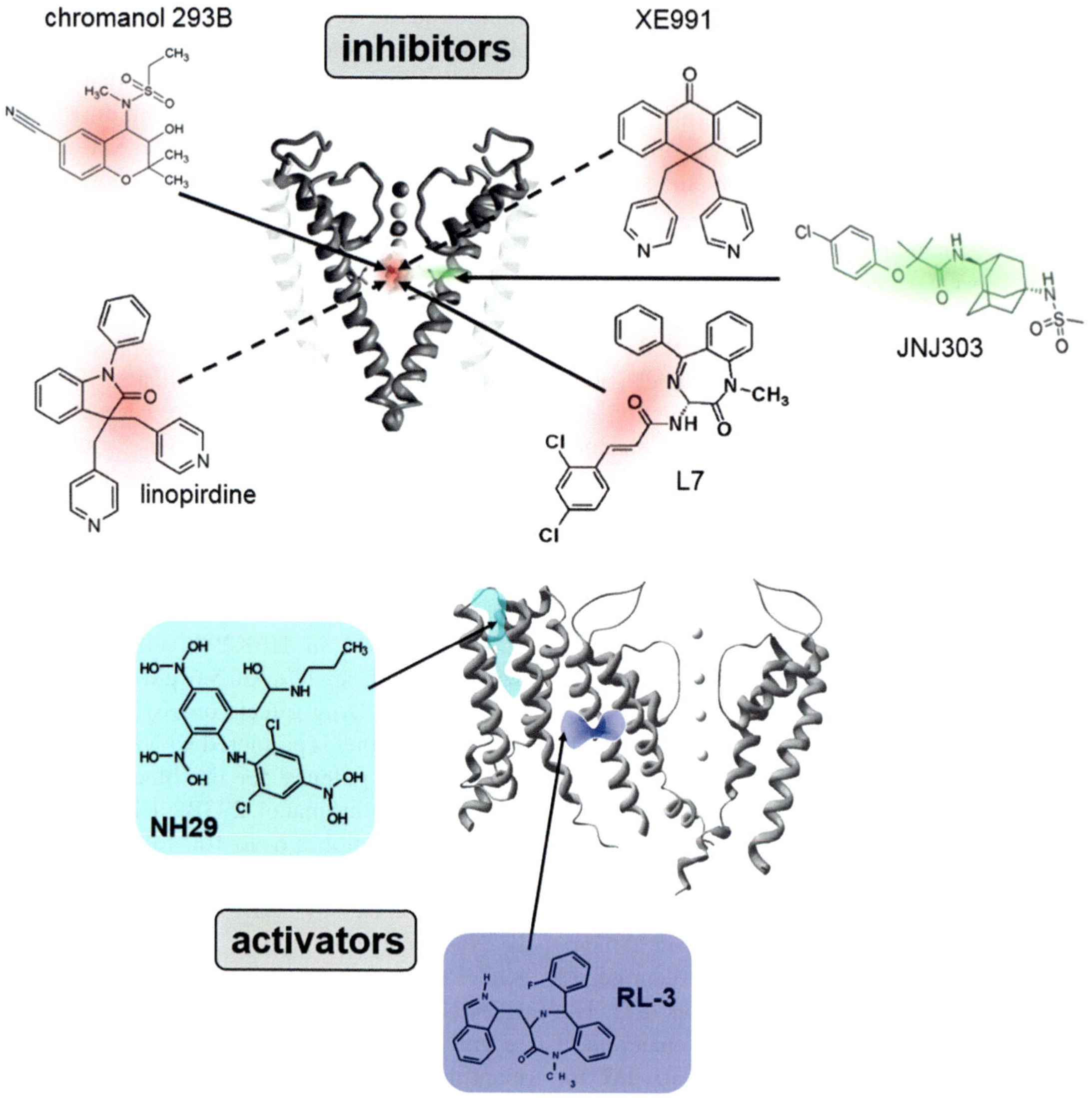

Fig. 15.2 Putative binding sites of KCNQ1/KCNE1 inhibitors and Kv7.1 activators

channels, this toxin interacted most prominently with Glu19 in the extracellular helix of KCNE1 of the KCNQ1/KCNE1 channel complex but does not inhibit K_V7.1 channels or those formed by coassembly of K_V7.1 with KCNE2 or KCNE4 [90]. The scorpion *Scorpiops jendeki* polypeptide (62 amino acids) toxin SjAPI-2 inhibits K_V7.1 channels with an IC_{50} of 771 nM [91]; however, SjAPI peptides are also enzyme inhibitors (K_i 97 nM for α-chymotrypsin) [92].

The first compound discovered to activate cardiac I_{Ks} was another benzodiazepine (1,3-dihydro-5-(2-fluorophenyl)-3-(1*H*-indol-3-ylmethyl)-1-methyl-2*H*-1,4-benzodiazepin-2-one) [93]. The compound has stereospecific activity: the R-enantiomer (R-L3, L364,373) is an agonist, whereas the S-enantiomer inhibits I_{Ks}. R-L3 is a gating modifier that increases I_{Ks} in cardiomyocytes by shifting the voltage dependence of channel activation to more negative potentials (e.g., $V_{0.5}$ shifted by −24 mV at 1 μM) and slowing the rate of channel deactivation. By contrast, KCNQ1/KCNE1 channels formed by excessive overexpression of KCNE1 with KCNQ1 in *Xenopus laevis* oocytes were insensitive to R-L3, suggesting that the $K_V7.1$ binding site for R-L3 may overlap with the interaction domain for KCNE1 [93]. Presumably, R-L3 cannot bind when $K_V7.1$ channels are partnered with 4 KCNE1 subunits but can bind to and affect the gating of $K_V7.1$ channels when partnered with <4 KCNE1 subunits as presumably must occur in vivo since this drug enhances I_{Ks} in cardiomyocytes. Scanning mutagenesis, plus molecular, and kinetic modeling suggests that R-L3 binds to a specific pocket between S5 and S6 transmembrane domains of $K_V7.1$ and accelerates the forward transition rate between two open states (O_1, O_2) in the simple channel gating scheme: $C_1 \leftrightarrow C_2 \leftrightarrow C_3 \leftrightarrow O_1 \leftrightarrow O_2 \leftrightarrow I$ [94]. The natural compound rottlerin (mallotoxin) isolated from the tree *Mallotus phillippinensis* activates I_{Ks} and both $K_V7.1$ and $K_V7.4$ channels, but is without effect on $K_V7.2$/$K_V7.3$ or $K_V7.5$ channels [95]. Simulated docking identified the pocket located between adjacent VSDs and the outer regions of the S5 and S6 segments of the pore domain as the preferred binding site for rottlerin, suggesting that it too may overlap that of R-L3.

Another activator (ML277, (R)-*N*-(4-(4-methoxyphenyl)thiazol-2-yl)-1-tosylpiperidine-2-carboxamide) was discovered by screening a compound library for effects on thallium influx rate of cells expressing $K_V7.1$ channels [96]. This compound activates $K_V7.1$ with an EC_{50} of 260 nM and >100-fold selectivity compared to $K_V7.2$ or $K_V7.4$ channel types. Similar to R-L3, ML277 is without effect on $K_V7.1$ channels that are saturated with KCNE1 accessory subunits (i.e., channel complex consisting of 4 KCNQ1 and 4 KCNE1 subunits) [97]. The mechanisms of action and putative binding sites for ML277 have been characterized by extensive mutagenesis and molecular dynamics simulations [98]. The authors suggest that the compound binds to two sites, one located at the intracellular boundary of the S2–S3 loop and S4–S5 linker, and another pocket formed between the S5 and S6 helices (and between adjacent VSDs) that is only accessible when not precluded by KCNE1. Given the many similarities between the gating effects and KCNE1-dependent activity of ML277 and R-L3, it seems likely these two compounds bind to the same, or overlapping site on the $K_V7.1$ channel. Further studies incorporating concatenated wild-type and mutant $K_V7.1$ subunits are needed to determine unambiguously whether all four pore domain binding sites are required for maximal activity of R-L3 and ML277 as has recently been accomplished to quantify the activity of activators of $K_V7.2$ [99] and hERG [100] channels.

$K_V7.1$/I_{Ks} activators such as ML277 may have antiarrhythmic activity in the long QT syndrome [101]. Activation of I_{Ks} by ML277 shortens action potential duration of atrial myocytes and eliminated alternans, the beat-to-beat alternation in contraction, action potential duration, and intracellular Ca^{2+} transient amplitude that can induce arrhythmia [102].

15.6 KCNQ1 Channel Regulation

Differential regulation of $K_V7.1$ channels by multiple pathways constitute the basis of different native currents in various organs and allow for specific functions in their physiological settings (Fig. 15.3). A large functional variety also derives from the coassembly of KCNQ1 α-subunits with multiple types of KCNE β-subunits. However, KCNEs do not merely transfer only one specific feature to the KCNQ1/KCNEx channel complex but couple the channel complexes to different regulatory mechanisms within the cells. First of all, is the coupling to phosphorylation events. For example, acute stress is associated with β-adrenergic stimulation leading to an increase in cAMP, which activates protein kinase A (PKA) that in turn activates I_{Ks} in native tissues [103]. Phosphorylation at KCNQ1-residue Ser27 causes a prominent leftward-shift in the *I–V* curve of KCNQ1/KCNE1 channel currents, an effect that is much less prominent when Kv7.1 is expressed without KCNE1 [104]. PKA forms a macromolecular complex together with protein phosphatase 1 and the KCNQ1/KCNE1 channel via Yotiao (also named AKAP79), supporting very fast and reliable local kinase signaling

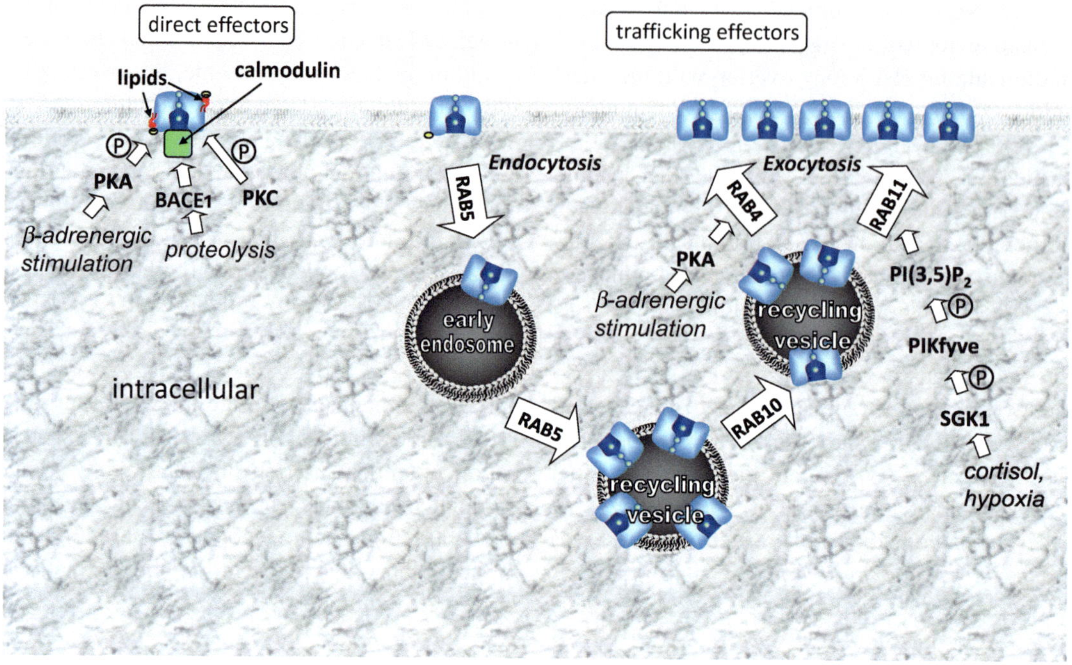

Fig. 15.3 Regulation of KCNQ1/KCNE1 channels

[73]. Besides direct phosphorylation of Kv7.1, cytoskeletal components are modulated by PKA to cause I_{Ks} stimulation [105]. PKA sensitivity critically depends on the KCNE variant that is coassembled with KCNQ1 subunits, whereas the intracellular C-terminus of KCNEs apparently determines PKA-sensitivity [106]. However, PKA does not only alter the biophysical properties of KCNQ1/KCNE1 channels, it also increases the number of channels at the plasma membrane by modulation of channel trafficking. Exocytosis of KCNQ1/KCNE1 channels via RAB4 and RAB11 increases channel density at the plasma membrane and is stimulated by kinases PKA and SGK1 (Fig. 15.3) [107, 108]. Stimulation of plasma membrane expression of KCNQ1/KCNE1 channels can also be mediated by Klotho, β-catenin proteins, or other modulators of channel trafficking, although the exact molecular mechanisms remain elusive [109, 110]. KCNQ1/KCNE1 channels are endocytosed via a RAB5 pathway which in turn is under control of protein kinase C [107, 111]. Similar to increased exocytosis, reduced channel endocytosis by PKC increases the density of KCNQ1/KCNE1 channels at the plasma membrane. Although modulation of I_{Ks} by PKC had been known for a long time [112], analysis of the underlying mechanisms has been complicated because channel activation via modulation of trafficking overlaps with a PKC-mediated phosphatidylinositol-4,5-bisphosphate (PI(4,5)P_2) depletion that reduces I_{Ks}, resulting in a biphasic regulation of I_{Ks} by PKC [113]. The mechanisms of I_{Ks} activation and direct interaction sites between KCNQ1/KCNE1 channels and PI(4,5)P_2 have been intensively studied [77, 114–117]. The situation is even more complex as PI(4,5)P_2 also interacts with calmodulin, an additional protein component of the large macromolecular I_{Ks} channel assembly complex [73]. PI(4,5)P_2-calmodulin interactions converge to KCNQ1 helix B to modulate channel gating and stabilize channel open states [118, 119]. PI(4,5)P_2 represents one active membrane component; however, similar membrane lipids like PI(3,5)P_2 have been described as modulators of I_{Ks} as well, whereas precise interaction sites with this rare analog have not been studied [107]. Other membrane lipids with

charged head groups modulate Kv7.1 channels, whereas electrostatic interactions at the inner membrane surface seem to be central for these functional effects [120]. Lipids interact with the channel at both the inner leaflet and the core of the cell membrane. Polyunsaturated fatty acids may interact at the membrane borders via its negatively charged head groups and via long polyunsaturated acyl chains along with the KCNQ1/KCNE1 channel protein within the membrane core [121, 122]. Interestingly, the protease β-site APP-cleaving enzyme 1 (BACE1) modulates channel gating by physical interaction with the channel complex in a manner that is dependent on the coexpressed KCNE subtype in native tissue [123]. Proteolysis of $K_V7.1$ may introduce a novel line of research as caspases have been recently mentioned as new regulatory components of KCNQ1/KCNE1 channels [124]. Summarizing, $K_V7.1$ channels are in the center of a macromolecular complex that is subject to a multitude of modulatory factors, resulting in tissue-specific channel functions that are adapted to specific physiological settings.

15.7 KCNQ1 and Disease

15.7.1 Cardiac Arrhythmias

The first KCNQ channel gene was discovered using a positional cloning approach to identify the human gene responsible for the most common type of long QT syndrome (LQTS) known as Romano-Ward syndrome (RWS). This disorder is inherited in an autosomal dominant manner. RWS is readily diagnosed by an excessively long QT interval (>440 ms) on the body surface electrocardiogram (ECG), indicative of a delayed rate of ventricular repolarization, and often accompanied by syncope. LQTS is associated with an increased risk of a ventricular arrhythmia called *torsades de pointes* (TdP), observed as a rapid twisting of the QRS axis around the isoelectric line of the ECG that can degenerate into ventricular fibrillation and cause sudden cardiac death. Initial studies had linked the dominantly inherited RWS in affected members of 16 families to the chromosomal location 11p15.5 and further study identified several different mutations in a novel K^+ channel gene *KVLQT1* [14]. The gene name *KVLQT1* was later changed to *KCNQ1* in order to conform to the channel gene nomenclature proposed by the International Union of Basic and Clinical Pharmacology (IUPHAR) that is based on functional characteristics and structural protein motifs. The overall prevalence of LQTS is not known with accuracy, but based on screening of ~44,000 infants in Italy, it is commonly estimated to be 1 in 2000 [125]. *KCNQ1* was the first gene associated with LQTS and therefore RWS caused by any of the several hundred *KCNQ1* mutations identified to date is called "LQT1", with an estimated prevalence of 1 in 5000 and accounting for 35–40% of all LQTS cases [126]. To date, 17 different LQTS genes have been identified [127–129], although strong evidence for disease causality has not been well established for many of these genes. LQT1-associated defects in *KCNQ1* include intragenic deletions, small insertions, frame shifts, and most commonly single nucleotide substitutions [14, 130]. Missense mutant $K_V7.1$ proteins can form heteromultimers with wild-type $K_V7.1$ channel proteins, resulting in altered I_{Ks} channel gating and/or a slight to drastic reduction (in the case of dominant-negative subunit interactions) in the number of fully-functional channels. Some mutations in *KCNQ1* reduce I_{Ks} channel function by affecting $K_V7.1$ interaction with PIP_2 [131], protein kinase A [132], or Yotiao [73]. Once it was discovered that $K_V7.1$ partners with KCNE1 to form I_{Ks} channels, it was predicted and then quickly verified that mutations in *KCNE1* also cause RWS [133]. Since mutations in *KCNE1* was the fifth gene discovered to be associated with congenital LQTS, this form of RWS is commonly called LQT5. Mutations in *KCNE1* can suppress I_{Ks} by several mechanisms, including reducing channel trafficking to the cell membrane [134], shifting the voltage dependence of channel activation to more positive potentials and accelerating channel deactivation [133], or by disrupting physical interaction between KCNE1 and KCNQ1 subunits or its regulation by PKA and PIP_2 [135]. LQT5 also includes alterations that cause the defective assembly of KCNE1 subunits with $K_V7.1$ [136]. Genetic testing for LQT1 is complicated by the large number of missense variants in *KCNQ1*. Amino acid conservation analysis of $K_V7.1$ and a large case-control study was recently used to generate topology-based estimative predictive values [137] to aid in the assigning of potential pathological likelihood of variants of unknown significance that can range from benign to disease causing [138]. Although the human *KCNQ1* gene exhibits genomic imprinting (preferential expression of a specific parental allele in somatic cells of children), this does not occur in the heart and explains the lack of parent-of-origin effect in LQT1 [139].

LQT1-linked arrhythmias are often associated with prolonged exercise such as swimming [140]. The heightened sympathetic nerve activity that occurs during intense exercise can trigger early after depolarizations, premature depolarizations that occur following the plateau phase of the ventricular action potential that can, in turn, initiate polymorphic ventricular tachycardia. The cellular mechanisms of ventricular arrhythmia were recently examined using a transgenic rabbit model of LQT1 (overexpression of KCNQ1-Y315S in the heart). Reduced I_{Ks} leads to prolonged action potential durations in the left ventricle and early after depolarizations in the right ventricle of these rabbits. The resulting dispersion of ventricular refractoriness between the two chambers facilitates conduction blocks that initiate arrhythmia [141].

Autosomal recessive mutations in *KCNQ1* [142] or *KCNE1* can cause Jervell and Lange-Nielsen syndrome, JLNS [143, 144], characterized by congenital bilateral deafness in addition to severe QT prolongation, TdP ventricular arrhythmias, and a high risk of sudden cardiac death. The finding that *KCNQ1* was expressed in the cochlear *stria vascularis* of the inner ear [142] suggested that deafness in JLNS could be caused by a disruption in endolymph homeostasis. This hypothesis was proven by the finding that knockout of either subunit of the I_{Ks} channel results in collapse of the endolymphatic space and deafness in mice [145, 146]. The endocochlear potential of +80 mV is formed by the activity of Na^+, K^+-ATPase, and Na^+, K^+, $2Cl^-$-cotransporter in the basolateral membrane of the marginal cell layer and KCNQ1/KCNE1 channels located in the apical membrane (Fig. 15.1) [147].

Not surprisingly, given that loss of function mutations cause LQTS, gain of function mutations in *KCNQ1* can cause short QT syndrome [148], a disorder characterized by abnormally reduced QT intervals (<320 ms), syncope, atrial fibrillation, ventricular arrhythmia, and sudden death [149]. The KCNQ1 missense mutation V141M was found to be associated with atrial fibrillation and short QT syndrome in utero [150], whereas KCNQ1 missense mutation S140G was associated with familial atrial fibrillation [150, 151]. When currents conducted by S140G or V141M KCNQ1/KCNE1 channels are characterized using repetitive pulsing, the current appears to be instantaneous as though mutant channels are constitutively open (i.e., incapable of deactivating/closing). However, when voltage-clamp pulses were applied at an extremely slow rate, it became apparent that these mutant channels exhibit an extremely slow rate of deactivation. The slow rate component of I_{Ks} deactivation was decreased 62-fold by S140G and 140-fold by the V141M mutation. In addition, the half-point for activation of these mutant channels was shifted by about −50 mV. Altered charge-pair interactions between E160 in S2 and R237 in S4 within the VSD of KCNQ1 subunits may account for slowed channel deactivation by both of these SQT-associated point mutations [152]. Using voltage-clamp fluorometry to measure the gating kinetics of $K_V7.1$ homomeric channels [153], it was found that S140G, but not V141M slows VSD displacement. When KCNQ1 and KCNE1 were coexpressed, both mutations alter voltage sensor-pore coupling to slow channel deactivation. Simulating the gain of channel function induced by the V141M KCNQ1 using cellular and tissue models of human atria suggests that atrial fibrillation results from a shortened tissue excitation wavelength that initiates maintained spiral waves [154].

15.7.2 Diabetes and Cancer

A genome-wide association study of common variants in a large Japanese population first identified *KCNQ1* as a potential susceptibility gene in type 2 diabetes [155]. A subsequent fine-mapping study in Europeans identified several non-coding index variants in the region flanking *KCNQ1* that were associated with type 2 diabetes [156], and DNA methylation at the *KCNQ1* locus is inversely associated with insulin sensitivity [157]. KCNE2 accessory subunits may associate with $K_V7.1$ in the pancreas. *Kcne2* deletion in mice impairs glucose tolerance at an early age and later causes diabetes. These mice also exhibit a down-regulation in the skeletal muscle expression of insulin receptor β and insulin receptor substrate 1 and a greatly reduced secretion of β-cell insulin [158].

Recent reports strongly suggest that *KCNQ1* acts as a tumor suppressor in several different cancers. *Kcnq1* was identified as a tumor suppressor gene and targeted deletion results in more intestinal tumors, including some aggressive adenocarcinomas in mice [159], and T-cell lymphomas, plasma cell tumors, and hemangiosarcomas in hamsters [160]. *KCNQ1* may have a role in human colorectal cancer as low expression of this gene was significantly associated with poor overall survival [159]. DNA hypermethylation of the *KCNQ1* promoter can cause reduced expression of *KCNQ1* in human hepatocellular carcinoma tissues [161]. Loss of *KCNQ1* expression is strongly associated with disease recurrence in stage II and III colon cancer [162]. KCNQ1 regulates Wnt/beta-catenin signaling and epithelial-to-mesenchymal transition in human colorectal cancer cell lines, and its suppression leads to tumor cell proliferation [163]. The potential importance of these findings is suggested by the positive correlation between KCNQ1/KCNE3 channel complex expression and disease-free survival in colorectal carcinoma patients [163].

KCNQ1 is an imprinted gene and part of an imprinting control region (ICR). *KCNQ1* is a maternally expressed allele. The *KCNQ1* antisense transcript *KCNQ1 overlapping transcript 1* (*KCNQ1OT1*), is a long non-coding RNA gene and is a paternally expressed allele [164]. *KCNQ1OT1* modulates transcriptional silencing of the *KCNQ1* locus via histone methylation. Both type 2 diabetes and some cancers are tightly liked to alterations in DNA methylation. *KCNQ1OT1* long non-coding RNA rather than the expressed Kv7.1 channel protein may have a role in these diseases.

References

1. Noble D, Tsien RW (1969) Outward membrane currents activated in the plateau range of potentials in cardiac Purkinje fibres. J Physiol 200:205–231
2. Jaeger JM, Gibbons WR (1985) A re-examination of late outward plateau currents of cardiac Purkinje fibers. Am J Phys 249:H108–H121
3. Hume JR, Giles W, Robinson K, Shibata EF, Nathan RD, Kanai K et al (1986) A time- and voltage-dependent K^+ current in single cardiac cells from bullfrog atrium. J Gen Physiol 88(6):777–798
4. Shrier A, Clay JR (1986) Repolarization currents in embryonic chick atrial heart cell aggregates. Biophys J 50:861–874
5. Sanguinetti MC, Jurkiewicz NK (1990) Two components of cardiac delayed rectifier K^+ current: differential sensitivity to block by class III antiarrhythmic agents. J Gen Physiol 96:195–215
6. Sanguinetti MC, Jurkiewicz NK (1991) I_K is comprised of two components in Guinea pig atrial cells. Am J Phys 260:H393–H399
7. Warmke JW, Ganetzky B (1994) A family of potassium channel genes related to *eag* in *Drosophila* and mammals. Proc Natl Acad Sci U S A 91:3438–3442
8. Sanguinetti MC, Jiang C, Curran ME, Keating MT (1995) A mechanistic link between an inherited and an acquired cardiac arrhythmia: *HERG* encodes the I_{Kr} potassium channel. Cell 81:299–307
9. Trudeau M, Warmke JW, Ganetzky B, Robertson GA (1995) HERG, a human inward rectifier in the voltage-gated potassium channel family. Science 269:92–95
10. Barhanin J, Lesage F, Guillemare E, Fink M, Lazdunski M, Romey G (1996) KvLQT1 and IsK (minK) proteins associate to form the I_{Ks} cardiac potassium channel. Nature 384:78–80

11. Sanguinetti MC, Curran ME, Zou A, Shen J, Spector PS, Atkinson DL et al (1996) Coassembly of KvLQT1 and minK (IsK) proteins to form cardiac I_{Ks} potassium channel. Nature 384:80–83
12. Takumi T, Ohkubo H, Nakanishi S (1988) Cloning of a membrane protein that induces a slow voltage-gated potassium current. Science 242:1042–1045
13. Folander K, Smith JS, Antanavage J, Bennett C, Stein RB, Swanson R (1990) Cloning and expression of the delayed-rectifier I_{sK} channel from neonatal rat heart and diethylstilbestrol-primed rat uterus. Proc Natl Acad Sci U S A 87:2975–2979
14. Wang Q, Curran ME, Splawski I, Burn TC, Millholland JM, VanRaay TJ et al (1996) Positional cloning of a novel potassium channel gene: KVLQT1 mutations cause cardiac arrhythmias. Nat Genet 12: 17–23
15. Singh NA, Charlier C, Stauffer D, DuPont BR, Leach RJ, Melis R et al (1998) A novel potassium channel gene, KCNQ2, is mutated in an inherited epilepsy of newborns. Nat Genet 18:25–29
16. Charlier C, Singh NA, Ryan SG, Lewis TB, Reus BE, Leach RJ et al (1998) A pore mutation in a novel KQT-like potassium channel gene in an idiopathic epilepsy family. Nat Genet 18:53–55
17. Kananura C, Biervert C, Hechenberger M, Engels H, Steinlein OK (2000) The new voltage gated potassium channel KCNQ5 and neonatal convulsions. Neuroreport 11:2063–2067
18. Kubisch C, Schroeder BC, Friedrich T, Lutjohann B, El-Amraoui A, Marlin S et al (1999) KCNQ4, a novel potassium channel expressed in sensory outer hair cells, is mutated in dominant deafness. Cell 96:437–446
19. Bleich M, Warth R (2000) The very small-conductance K^+ channel KvLQT1 and epithelial function. Pflugers Arch 440:202–206
20. Sakagami M, Fukazawa K, Matsunaga T, Fujita H, Mori N, Takumi T et al (1991) Cellular localization of rat I_{sK} protein in the stria vascularis by immunohistochemical observation. Hear Res 56:168–172
21. Demolombe S, Franco D, de Boer P, Kuperschmidt S, Roden D, Pereon Y et al (2001) Differential expression of KvLQT1 and its regulator IsK in mouse epithelia. Am J Physiol Cell Physiol 280:C359–C372
22. Schroeder BC, Waldegger S, Fehr S, Bleich M, Warth R, Greger R et al (2000) A constitutively open potassium channel formed by KCNQ1 and KCNE3. Nature 403:196–199
23. Vallon V, Grahammer F, Volkl H, Sandu CD, Richter K, Rexhepaj R et al (2005) KCNQ1-dependent transport in renal and gastrointestinal epithelia. Proc Natl Acad Sci U S A 102:17864–17869
24. Preston P, Wartosch L, Gunzel D, Fromm M, Kongsuphol P, Ousingsawat J et al (2010) Disruption of the K^+ channel beta-subunit KCNE3 reveals an important role in intestinal and tracheal Cl^- transport. J Biol Chem 285:7165–7175
25. Julio-Kalajzic F, Villanueva S, Burgos J, Ojeda M, Cid LP, Jentsch TJ et al (2018) K2P TASK-2 and KCNQ1-KCNE3 K(+) channels are major players contributing to intestinal anion and fluid secretion. J Physiol 596:393–407
26. Grahammer F, Herling AW, Lang HJ, Schmitt-Graff-A, Wittekindt OH, Nitschke R et al (2001) The cardiac K^+ channel KCNQ1 is essential for gastric acid secretion. Gastroenterology 120:1363–1371
27. Vucic E, Alfadda T, MacGregor GG, Dong K, Wang T, Geibel JP (2015) Kir1.1 (ROMK) and Kv7.1 (KCNQ1/KvLQT1) are essential for normal gastric acid secretion: importance of functional Kir1.1. Pflugers Arch 467:1457–1468
28. Heitzmann D, Grahammer F, von Hahn T, Schmitt-Graff A, Romeo E, Nitschke R et al (2004) Heteromeric KCNE2/KCNQ1 potassium channels in the luminal membrane of gastric parietal cells. J Physiol 561:547–557
29. Pusch M (1998) Increase of the single-channel conductance of KvLQT1 potassium channels induced by the association with minK. Pflugers Arch 437:172–174
30. Tristani-Firouzi M, Sanguinetti MC (1998) Voltage-dependent inactivation of the human K^+ channel KvLQT1 is eliminated by association with minimal K^+ channel (minK) subunits. J Physiol 510(Pt 1): 37–45
31. Ma LJ, Ohmert I, Vardanyan V (2011) Allosteric features of KCNQ1 gating revealed by alanine scanning mutagenesis. Biophys J 100:885–894
32. Cui J, Kline RP, Pennefather P, Cohen IS (1994) Gating of I-Sk expressed in Xenopus-oocytes depends on the amount of messenger-Rna injected. J Gen Physiol 104:87–105
33. Tzounopoulous T, Maylie J, Adelman JP (1998) Gating of I_{sK} channels expressed in Xenopus oocytes. Biophys J 74:2299–2305
34. Strutz-Seebohm N, Pusch M, Wolf S, Stoll R, Tapken D, Gerwert K et al (2011) Structural basis of slow activation gating in the cardiac I-Ks channel complex. Cell Physiol Biochem 27:443–452
35. Silva J, Rudy Y (2005) Subunit interaction determines I-Ks participation in cardiac repolarization and repolarization reserve. Circulation 112: 1384–1391
36. Osteen JD, Sampson KJ, Kass RS (2010) The cardiac I-Ks channel, complex indeed. Proc Natl Acad Sci U S A 107:18751–18752
37. Werry D, Eldstrom J, Wang ZR, Fedida D (2013) Single-channel basis for the slow activation of the repolarizing cardiac potassium current, I-Ks. Proc Natl Acad Sci U S A 110:E996–E1005
38. Xu Y, Wang Y, Meng XY, Zhang M, Jiang M, Cui M et al (2013) Building KCNQ1/KCNE1 channel models and probing their interactions by molecular-dynamics simulations. Biophys J 105:2461–2473

39. Barro-Soria R, Rebolledo S, Liin SI, Perez ME, Sampson KJ, Kass RS et al (2014) KCNE1 divides the voltage sensor movement in KCNQ1/KCNE1 channels into two steps. Nat Commun 5:3750
40. Xu JJ, Rudy Y (2018) Effects of beta-subunit on gating of a potassium ion channel: molecular simulations of cardiac IKs activation. J Mol Cell Cardiol 124:35–44
41. Ramasubramanian S, Rudy Y (2018) The structural basis of IKs ion-channel activation: mechanistic insights from molecular simulations. Biophys J 114: 2584–2594
42. Abitbol I, Peretz A, Lerche C, Busch AE, Attali B (1999) Stilbenes and fenamates rescue the loss of I-KS channel function induced by an LQT5 mutation and other IsK mutants. EMBO J 18:4137–4148
43. Pusch M, Magrassi R, Wollnik B, Conti F (1998) Activation and inactivation of homomeric KvLQT1 potassium channels. Biophys J 75:785–792
44. Gibor G, Yakubovich D, Rosenhouse-Dantsker A, Peretz A, Schottelndreier H, Seebohm G et al (2007) An inactivation gate in the selectivity filter of KCNQ1 potassium channels. Biophys J 93:4159–4172
45. Seebohm G, Scherer CR, Busch AE, Lerche C (2001) Identification of specific pore residues mediating KCNQ1 inactivation. A novel mechanism for long QT syndrome. J Biol Chem 276:13600–13605
46. Seebohm G, Westenskow P, Lang F, Sanguinetti MC (2005) Mutation of colocalized residues of the pore helix and transmembrane segments S5 and S6 disrupt deactivation and modify inactivation of KCNQ1 K^+ channels. J Physiol 563:359–368
47. Pless SA, Galpin JD, Niciforovic AP, Kurata HT, Ahern CA (2013) Hydrogen bonds as molecular timers for slow inactivation in voltage-gated potassium channels. Elife 2:e01289
48. Hoshi T, Armstrong CM (2013) C-type inactivation of voltage-gated K^+ channels: pore constriction or dilation? J Gen Physiol 141:151–160
49. Kurata HT, Fedida D (2006) A structural interpretation of voltage-gated potassium channel inactivation. Prog Biophys Mol Biol 92:185–208
50. Starkus JG, Kuschel L, Rayner MD, Heinemann SH (1997) Ion conduction through C-type inactivated shaker channels. J Gen Physiol 110:539–550
51. Hou PP, Eldstrom J, Shi JY, Zhong L, McFarland K, Gao Y et al (2017) Inactivation of KCNQ1 potassium channels reveals dynamic coupling between voltage sensing and pore opening. Nat Commun 8(1):1730
52. Pusch M, Ferrera L, Friedrich T (2001) Two open states and rate-limiting gating steps revealed by intracellular Na^+ block of human KCNQ1 and KCNQ1/KCNE1 K^+ channels. J Physiol 533:135–143
53. Seebohm G, Sanguinetti MC, Pusch M (2003) Tight coupling of rubidium conductance and inactivation in human KCNQ1 potassium channels. J Physiol 552: 369–378
54. Pusch M, Bertorello L, Conti F (2000) Gating and flickery block differentially affected by rubidium in homomeric KCNQ1 and heteromeric KCNQ1/KCNE1 potassium channels. Biophys J 78:211–226
55. Vallon V, Grahammer F, Richter K, Bleich M, Lang F, Barhanin J et al (2001) Role of KCNE1-dependent K(+) fluxes in mouse proximal tubule. J Am Soc Nephrol 12:2003–2011
56. Westhoff M, Eldstrom J, Murray CI, Thompson E, Fedida D (2019) I-Ks ion-channel pore conductance can result from individual voltage sensor movements. Proc Natl Acad Sci U S A 116:7879–7888
57. Thompson E, Eldstrom J, Westhoff M, McAfee D, Balse E, Fedida D (2017) cAMP-dependent regulation of IKs single-channel kinetics. J Gen Physiol 149:781–798
58. Tinel N, Diochot S, Lauritzen I, Barhanin J, Lazdunski M, Borsotto M (2000) M-type KCNQ2-KCNQ3 potassium channels are modulated by the KCNE2 subunit. FEBS Lett 480:137–141
59. Melman YF, Domenech A, de la Luna S, McDonald TV (2001) Structural determinants of KvLQT1 control by the KCNE family of proteins. J Biol Chem 276:6439–6444
60. Mazhari R, Nuss HB, Armoundas AA, Winslow RL, Marban E (2002) Ectopic expression of KCNE3 accelerates cardiac repolarization and abbreviates the QT interval. J Clin Invest 109:1083–1090
61. Grunnet M, Jespersen T, Rasmussen HB, Ljungstrom T, Jorgensen NK, Olesen SP et al (2002) KCNE4 is an inhibitory subunit to the KCNQ1 channel. J Physiol 542:119–130
62. Angelo K, Jespersen T, Grunnet M, Nielsen MS, Klaerke DA, Olesen SP (2002) KCNE5 induces time- and voltage-dependent modulation of the KCNQ1 current. Biophys J 83:1997–2006
63. Seebohm G, Sanguinetti MC, Pusch M (2003) Tight coupling of rubidium conductance and inactivation in human KCNQ1 potassium channels. J Physiol Lond 552:369–378
64. Wang KW, Goldstein SAN (1995) Subunit composition of minK potassium channels. Neuron 14:1303–1309
65. Chen H, Sesti F, Goldstein SA (2003) Pore- and state-dependent cadmium block of I(Ks) channels formed with MinK-55C and wild-type KCNQ1 subunits. Biophys J 84:3679–3689
66. Morin TJ, Kobertz WR (2008) Counting membrane-embedded KCNE beta-subunits in functioning K^+ channel complexes. Proc Natl Acad Sci U S A 105: 1478–1482
67. Kang C, Tian C, Sonnichsen FD, Smith JA, Meiler J, George AL Jr et al (2008) Structure of KCNE1 and implications for how it modulates the KCNQ1 potassium channel. Biochemistry 47:7999–8006
68. Sun J, MacKinnon R (2017) Cryo-EM structure of a KCNQ1/CaM complex reveals insights into congenital long QT syndrome. Cell 169:1042–1050.e9

69. Morokuma J, Blackiston D, Levin M (2008) KCNQ1 and KCNE1 K^+ channel components are involved in early left-right patterning in Xenopus laevis embryos. Cell Physiol Biochem 21:357–372
70. Nakajo K, Kubo Y (2010) A role of the voltage-sensor domain in the modulation of KCNQ1 channel by KCNE subunits. J Physiol Sci 60:S7
71. Zheng RJ, Thompson K, Obeng-Gyimah E, Alessi D, Chen J, Cheng HY et al (2010) Analysis of the interactions between the C-terminal cytoplasmic domains of KCNQ1 and KCNE1 channel subunits. Biochem J 428:75–84
72. Wrobel E, Rothenberg I, Krisp C, Hundt F, Fraenzel B, Eckey K et al (2016) KCNE1 induces fenestration in the Kv7.1/KCNE1 channel complex that allows for highly specific pharmacological targeting. Nat Commun 7:12795
73. Marx SO, Kurokawa J, Reiken S, Motoike H, D'Armiento J, Marks AR et al (2002) Requirement of a macromolecular signaling complex for beta adrenergic receptor modulation of the KCNQ1-KCNE1 potassium channel. Science 295:496–499
74. Terrenoire C, Houslay MD, Baillie GS, Kass RS (2009) The cardiac IKs potassium channel macromolecular complex includes the phosphodiesterase PDE4D3. J Biol Chem 284:9140–9146
75. Li Y, Chen L, Kass RS, Dessauer CW (2012) The A-kinase anchoring protein Yotiao facilitates complex formation between adenylyl cyclase type 9 and the IKs potassium channel in heart. J Biol Chem 287: 29815–29824
76. Ghosh S, Nunziato DA, Pitt GS (2006) KCNQ1 assembly and function is blocked by long-QT syndrome mutations that disrupt interaction with calmodulin. Circ Res 98:1048–1054
77. Zaydman MA, Silva JR, Delaloye K, Li Y, Liang H, Larsson HP et al (2013) Kv7.1 ion channels require a lipid to couple voltage sensing to pore opening. Proc Natl Acad Sci U S A 110:13180–13185
78. Zaydman MA, Cui J (2014) PIP2 regulation of KCNQ channels: biophysical and molecular mechanisms for lipid modulation of voltage-dependent gating. Front Physiol 5:195
79. Kasimova MA, Zaydman MA, Cui J, Tarek M (2015) PIP(2)-dependent coupling is prominent in Kv7.1 due to weakened interactions between S4-S5 and S6. Sci Rep 5:7474
80. Warth R, Riedemann N, Bleich M, Van Driessche W, Busch AE, Greger R (1996) The cAMP-regulated and 293B-inhibited K^+ conductance of rat colonic crypt base cells. Pflugers Arch 432:81–88
81. Kunzelmann K, Hubner M, Schreiber R, Levy-Holzman R, Garty H, Bleich M et al (2001) Cloning and function of the rat colonic epithelial K^+ channel KVLQT1. J Membr Biol 179:155–164
82. Seebohm G, Lerche C, Pusch M, Steinmeyer K, Bruggemann A, Busch AE (2001) A kinetic study on the stereospecific inhibition of KCNQ1 and I (Ks) by the chromanol 293B. Br J Pharmacol 134: 1647–1654
83. Seebohm G, Chen J, Strutz N, Culberson C, Lerche C, Sanguinetti MC (2003) Molecular determinants of KCNQ1 channel block by a benzodiazepine. Mol Pharmacol 64:70–77
84. Lerche C, Bruhova I, Lerche H, Steinmeyer K, Wei AD, Strutz-Seebohm N et al (2007) Chromanol 293B binding in KCNQ1 (Kv7.1) channels involves electrostatic interactions with a potassium ion in the selectivity filter. Mol Pharmacol 71:1503–1511
85. Villatoro-Gomez K, Pacheco-Rojas DO, Moreno-Galindo EG, Navarro-Polanco RA, Tristani-Firouzi M, Gazgalis D et al (2018) Molecular determinants of Kv7.1/KCNE1 channel inhibition by amitriptyline. Biochem Pharmacol 152:264–271
86. Gogelein H, Bruggemann A, Gerlach U, Brendel J, Busch AE (2000) Inhibition of IKs channels by HMR 1556. Naunyn Schmiedebergs Arch Pharmacol 362: 480–488
87. Dong MQ, Lau CP, Gao Z, Tseng GN, Li GR (2006) Characterization of recombinant human cardiac KCNQ1/KCNE1 channels (I (Ks)) stably expressed in HEK 293 cells. J Membr Biol 210:183–192
88. Busch AE, Busch GL, Ford E, Suessbrich H, Lang H-J, Greger R et al (1997) The role of the I_{sK} protein in the specific pharmacological properties of the I_{Ks} channel complex. Br J Pharmacol 122:187–189
89. Wang HS, Brown BS, McKinnon D, Cohen IS (2000) Molecular basis for differential sensitivity of KCNQ and I(Ks) channels to the cognitive enhancer XE991. Mol Pharmacol 57:1218–1223
90. Sun P, Wu F, Wen M, Yang X, Wang C, Li Y et al (2015) A distinct three-helix centipede toxin SSD609 inhibits I(ks) channels by interacting with the KCNE1 auxiliary subunit. Sci Rep 5:13399
91. Chen J, Zhang C, Yang W, Cao Z, Li W, Chen Z et al (2015) SjAPI-2 is the first member of a new neurotoxin family with Ascaris-type fold and KCNQ1 inhibitory activity. Int J Biol Macromol 79:504–510
92. Chen Z, Wang B, Hu J, Yang W, Cao Z, Zhuo R et al (2013) SjAPI, the first functionally characterized Ascaris-type protease inhibitor from animal venoms. PLoS One 8:e57529
93. Salata JJ, Jurkiewicz NK, Wang J, Evans BE, Orme HT, Sanguinetti MC (1998) A novel benzodiazepine that activates cardiac slow delayed rectifier K^+ channels. Mol Pharmacol 53:220–230
94. Seebohm G, Pusch M, Chen J, Sanguinetti MC (2003) Pharmacological activation of normal and arrhythmia-associated mutant KCNQ1 potassium channels. Circ Res 93:941–947
95. Matschke V, Piccini I, Schubert J, Wrobel E, Lang F, Matschke J et al (2016) The natural plant product Rottlerin activates Kv7.1/KCNE1 channels. Cell Physiol Biochem 40:1549–1558
96. Mattmann ME, Yu H, Lin Z, Xu K, Huang X, Long S et al (2012) Identification of (R)-N-(4-(4-methoxyphenyl)thiazol-2-yl)-1-

tosylpiperidine-2-carboxamide, ML277, as a novel, potent and selective K(v)7.1 (KCNQ1) potassium channel activator. Bioorg Med Chem Lett 22:5936–5941
97. Yu H, Lin Z, Mattmann ME, Zou B, Terrenoire C, Zhang H et al (2013) Dynamic subunit stoichiometry confers a progressive continuum of pharmacological sensitivity by KCNQ potassium channels. Proc Natl Acad Sci U S A 110:8732–8737
98. Xu Y, Wang Y, Zhang M, Jiang M, Rosenhouse-Dantsker A, Wassenaar T et al (2015) Probing binding sites and mechanisms of action of an I (Ks) activator by computations and experiments. Biophys J 108:62–75
99. Wang AW, Yau MC, Wang CK, Sharmin N, Yang RY, Pless SA et al (2018) Four drug-sensitive subunits are required for maximal effect of a voltage sensor-targeted KCNQ opener. J Gen Physiol 150:1432–1443
100. Wu W, Gardner A, Sanguinetti MC (2015) Concatenated hERG1 tetramers reveal stoichiometry of altered channel gating by RPR-260243. Mol Pharmacol 87:401–409
101. Ma D, Wei H, Lu J, Huang D, Liu Z, Loh LJ et al (2015) Characterization of a novel KCNQ1 mutation for type 1 long QT syndrome and assessment of the therapeutic potential of a novel IKs activator using patient-specific induced pluripotent stem cell-derived cardiomyocytes. Stem Cell Res Ther 6:39
102. Kanaporis G, Kalik ZM, Blatter LA (2019) Action potential shortening rescues atrial calcium alternans. J Physiol 597:723–740
103. Walsh KB, Kass RS (1988) Regulation of a heart potassium channel by protein kinase A and kinase C. Science 242:67–69
104. Kurokawa J, Motoike HK, Rao J, Kass RS (2004) Regulatory actions of the A-kinase anchoring protein Yotiao on a heart potassium channel downstream of PKA phosphorylation. Proc Natl Acad Sci U S A 101:17884–17884
105. Nicolas CS, Park KH, Harchi AE, Camonis J, Kass RS, Escande D et al (2008) I(Ks)response to protein kinase A-dependent KCNQ1 phosphorylation requires direct interaction with microtubules. Cardiovasc Res 79:427–435
106. Kurokawa J, Bankston JR, Kaihara A, Chen L, Furukawa T, Kass RS (2009) KCNE variants reveal a critical role of the beta subunit carboxyl terminus in PKA-dependent regulation of the I-Ks potassium channel. Channels 3:16–24
107. Seebohm G, Strutz-Seebohm N, Birkin R, Dell G, Bucci C, Spinosa MR et al (2007) Regulation of endocytic recycling of KCNQ1/KCNE1 potassium channels. Circ Res 100:686–692
108. Piccini I, Fehrmann E, Frank S, Muller FU, Greber B, Seebohm G (2017) Adrenergic stress protection of human iPS cell-derived cardiomyocytes by fast K(v) 7.1 recycling. Front Physiol 8:705
109. Almilaji A, Pakladok T, Munoz C, Elvira B, Sopjani M, Lang F (2014) Upregulation of KCNQ1/KCNE1 K^+ channels by Klotho. Channels 8:222–229
110. Wilmes J, Haddad-Tovolli R, Alesutan I, Munoz C, Sopjani M, Pelzl L et al (2012) Regulation of KCNQ1/KCNE1 by beta-catenin. Mol Membr Biol 29:87–94
111. Parks XX, Ronzier E, O-Uchi J, Lopes CM (2019) Fluvastatin inhibits Rab5-mediated IKs internalization caused by chronic Ca^{2+}-dependent PKC activation. J Mol Cell Cardiol 129:314–325
112. Walsh KB, Begenisich TB, Kass RS (1988) β-Adrenergic modulation in the heart: independent regulation of K and Ca channels. Pflugers Arch 411:232–234
113. Matavel A, Lopes CMB (2009) PKC activation and PIP2 depletion underlie biphasic regulation of IKs by Gq-coupled receptors. J Mol Cell Cardiol 46:704–712
114. Loussouarn G, Park KH, Bellocq C, Baro I, Charpentier F, Escande D (2003) Phosphatidylinositol-4,5-bisphosphate, PIP2, controls KCNQ1/KCNE1 voltage-gated potassium channels: a functional homology between voltage-gated and inward rectifier K^+ channels. EMBO J 22:5412–5421
115. Rodriguez N, Amarouch MY, Montnach J, Piron J, Labro AJ, Charpentier F et al (2010) Phosphatidylinositol-4,5-bisphosphate (PIP2) stabilizes the open pore conformation of the Kv11.1 (hERG) channel. Biophys J 99:1110–1118
116. Li Y, Zaydman MA, Wu D, Shi JY, Guan M, Virgin-Downey B et al (2011) KCNE1 enhances phosphatidylinositol 4,5-bisphosphate (PIP2) sensitivity of I-Ks to modulate channel activity. Proc Natl Acad Sci U S A 108:9095–9100
117. Eckey K, Wrobel E, Strutz-Seebohm N, Pott L, Schmitt N, Seebohm G (2014) Novel K(v)7.1-phosphatidylinositol 4,5-bisphosphate interaction sites uncovered by charge neutralization scanning. J Biol Chem 289:22749–22758
118. Tobelaim WS, Dvir M, Lebel G, Cui M, Buki T, Peretz A et al (2017) Ca^{2+}-calmodulin and PIP2 interactions at the proximal C-terminus of Kv7 channels. Channels 11:686–695
119. Sachyani D, Dvir M, Strulovich R, Tria G, Tobelaim W, Peretz A et al (2014) Structural basis of a Kv7.1 potassium channel gating module: studies of the intracellular C-terminal domain in complex with calmodulin. Structure 22:1582–1594
120. Taylor KC, Sanders CR (2017) Regulation of KCNQ/Kv7 family voltage-gated K^+ channels by lipids. BBA-Biomembranes 1859:586–597
121. Liin SI, Ejneby MS, Barro-Soria R, Skarsfeldt MA, Larsson JE, Harlin FS et al (2015) Polyunsaturated fatty acid analogs act antiarrhythmically on the cardiac I-Ks channel. Proc Natl Acad Sci U S A 112:5714–5719
122. Larsson JE, Larsson HP, Liin SI (2018) KCNE1 tunes the sensitivity of K(v)7.1 to polyunsaturated

fatty acids by moving turret residues close to the binding site. Elife 7:e37257
123. Agsten M, Hessler S, Lehnert S, Volk T, Rittger A, Hartmann S et al (2015) BACE1 modulates gating of KCNQ1 (Kv7.1) and cardiac delayed rectifier KCNQ1/KCNE1 (I-Ks). J Mol Cell Cardiol 89:335–348
124. Strigli A, Raab C, Hessler S, Huth T, Schuldt AJT, Alzheimer C et al (2018) Doxorubicin induces caspase-mediated proteolysis of KV7.1. Commun Biol 1:155
125. Schwartz PJ, Stramba-Badiale M, Crotti L, Pedrazzini M, Besana A, Bosi G et al (2009) Prevalence of the congenital long-QT syndrome. Circulation 120:1761–1767
126. Perrin MJ, Gollob MH (2013) Genetics of cardiac electrical disease. Can J Cardiol 29:89–99
127. Wu J, Ding WG, Horie M (2016) Molecular pathogenesis of long QT syndrome type 1. J Arrhythm 32: 381–388
128. Bohnen MS, Peng G, Robey SH, Terrenoire C, Iyer V, Sampson KJ et al (2017) Molecular pathophysiology of congenital long QT syndrome. Physiol Rev 97:89–134
129. Giudicessi JR, Kullo IJ, Ackerman MJ (2017) Precision cardiovascular medicine: state of genetic testing. Mayo Clin Proc 92:642–662
130. Russell MW, Dick M II, Collins FS, Brody LC (1996) KVLQT1 mutations in three families with familial or sporadic long QT syndrome. Hum Mol Genet 5:1319–1324
131. Park KH, Piron J, Dahimene S, Merot J, Baro I, Escande D et al (2005) Impaired KCNQ1-KCNE1 and phosphatidylinositol-4,5-bisphosphate interaction underlies the long QT syndrome. Circ Res 96: 730–739
132. Wu J, Naiki N, Ding WG, Ohno S, Kato K, Zang WJ et al (2014) A molecular mechanism for adrenergic-induced long QT syndrome. J Am Coll Cardiol 63: 819–827
133. Splawski I, Tristani-Firouzi M, Lehmann MH, Sanguinetti MC, Keating MT (1997) Mutations in the *hminK* gene cause long QT syndrome and suppress I_{Ks} function. Nat Genet 17:338–340
134. Bianchi L, Shen Z, Dennis AT, Priori SG, Napolitano C, Ronchetti E et al (1999) Cellular dysfunction of LQT5-minK mutants: abnormalities of I_{Ks}, I_{Kr} and trafficking in long QT syndrome. Hum Mol Genet 8:1499–1507
135. Dvir M, Strulovich R, Sachyani D, Ben-Tal Cohen I, Haitin Y, Dessauer C et al (2014) Long QT mutations at the interface between KCNQ1 helix C and KCNE1 disrupt I(KS) regulation by PKA and PIP(2). J Cell Sci 127:3943–3955
136. Harmer SC, Wilson AJ, Aldridge R, Tinker A (2010) Mechanisms of disease pathogenesis in long QT syndrome type 5. Am J Physiol Cell Physiol 298:C263–C273
137. Kapplinger JD, Tseng AS, Salisbury BA, Tester DJ, Callis TE, Alders M et al (2015) Enhancing the predictive power of mutations in the C-terminus of the KCNQ1-encoded Kv7.1 voltage-gated potassium channel. J Cardiovasc Transl Res 8:187–197
138. Kapplinger JD, Tester DJ, Salisbury BA, Carr JL, Harris-Kerr C, Pollevick GD et al (2009) Spectrum and prevalence of mutations from the first 2,500 consecutive unrelated patients referred for the FAMILION long QT syndrome genetic test. Heart Rhythm 6:1297–1303
139. Lee MP, Hu R-J, Johnson LA, Feinberg AP (1997) Human *KVLQT1* gene shows tissue-specific imprinting and encompasses Beckwith-Wiedemann syndrome chromosomal rearrangements. Nat Genet 15: 181–185
140. Moss AJ, Robinson JL, Gessman L, Gillespie R, Zareba W, Schwartz PJ et al (1999) Comparison of clinical and genetic variables of cardiac events associated with loud noise versus swimming among subjects with the long QT syndrome. Am J Cardiol 84:876–879
141. Choi BR, Li W, Terentyev D, Kabakov AY, Zhong M, Rees CM et al (2018) Transient outward K(+) current (Ito) underlies the right ventricular initiation of polymorphic ventricular tachycardia in a transgenic rabbit model of long-QT syndrome type 1. Circ Arrhythm Electrophysiol 11:e005414
142. Neyroud N, Tesson F, Denjoy I, Leibovici M, Donger C, Barhanin J et al (1997) A novel mutation in the potassium channel gene *KVLQT1* causes the Jervell and Lange-Nielsen cardioauditory syndrome. Nat Genet 15:186–189
143. Tyson J, Traneebjaerg L, Bellman S, Wren C, Taylor JF, Bathen J et al (1997) IsK and KvLQT1: mutation in either of the two subunits of the slow component of the delayed rectifier potassium channel can cause Jervell and Lange-Nielsen syndrome. Hum Mol Genet 6:2179–2185
144. Splawski I, Timothy KW, Vincent GM, Atkinson DL, Keating MT (1997) Molecular basis of the long-QT syndrome associated with deafness. N Engl J Med 336:1562–1567
145. Vetter DE, Mann JR, Wangemann P, Liu J, McLaughlin KJ, Lesage F et al (1996) Inner ear defects induced by null mutation of the *isk* gene. Neuron 17:1251–1264
146. Casimiro MC, Knollmann BC, Ebert SN, Vary JC Jr, Greene AE, Franz MR et al (2001) Targeted disruption of the Kcnq1 gene produces a mouse model of Jervell and Lange-Nielsen Syndrome. Proc Natl Acad Sci U S A 98:2526–2531
147. Nin F, Hibino H, Doi K, Suzuki T, Hisa Y, Kurachi Y (2008) The endocochlear potential depends on two K^+ diffusion potentials and an electrical barrier in the stria vascularis of the inner ear. Proc Natl Acad Sci U S A 105:1751–1756
148. Bellocq C, van Ginneken AC, Bezzina CR, Alders M, Escande D, Mannens MM et al (2004)

Mutation in the KCNQ1 gene leading to the short QT-interval syndrome. Circulation 109:2394–2397
149. Borggrefe M, Wolpert C, Antzelevitch C, Veltmann C, Giustetto C, Gaita F et al (2005) Short QT syndrome. Genotype-phenotype correlations. J Electrocardiol 38:75–80
150. Hong K, Piper DR, Diaz-Valdecantos A, Brugada J, Oliva A, Burashnikov E et al (2005) De novo KCNQ1 mutation responsible for atrial fibrillation and short QT syndrome in utero. Cardiovasc Res 68:433–440
151. Chen YH, Xu SJ, Bendahhou S, Wang XL, Wang Y, Xu WY et al (2003) KCNQ1 gain-of-function mutation in familial atrial fibrillation. Science 299:251–254
152. Restier L, Cheng L, Sanguinetti MC (2008) Mechanisms by which atrial fibrillation-associated mutations in the S1 domain of KCNQ1 slow deactivation of IKs channels. J Physiol 586:4179–4191
153. Peng G, Barro-Soria R, Sampson KJ, Larsson HP, Kass RS (2017) Gating mechanisms underlying deactivation slowing by two KCNQ1 atrial fibrillation mutations. Sci Rep 7:45911
154. Whittaker DG, Colman MA, Ni H, Hancox JC, Zhang H (2018) Human atrial arrhythmogenesis and sinus bradycardia in KCNQ1-linked short QT syndrome: insights from computational modelling. Front Physiol 9:1402
155. Unoki H, Takahashi A, Kawaguchi T, Hara K, Horikoshi M, Andersen G et al (2008) SNPs in KCNQ1 are associated with susceptibility to type 2 diabetes in East Asian and European populations. Nat Genet 40:1098–1102
156. Gaulton KJ, Ferreira T, Lee Y, Raimondo A, Magi R, Reschen ME et al (2015) Genetic fine mapping and genomic annotation defines causal mechanisms at type 2 diabetes susceptibility loci. Nat Genet 47: 1415–1425
157. Shah UJ, Xie W, Flyvbjerg A, Nolan JJ, Hojlund K, Walker M et al (2019) Differential methylation of the type 2 diabetes susceptibility locus KCNQ1 is associated with insulin sensitivity and is predicted by CpG site specific genetic variation. Diabetes Res Clin Pract 148:189–199
158. Lee SM, Baik J, Nguyen D, Nguyen V, Liu S, Hu Z et al (2017) Kcne2 deletion impairs insulin secretion and causes type 2 diabetes mellitus. FASEB J 31: 2674–2685
159. Than BL, Goos JA, Sarver AL, O'Sullivan MG, Rod A, Starr TK et al (2014) The role of KCNQ1 in mouse and human gastrointestinal cancers. Oncogene 33:3861–3868
160. Li R, Miao J, Tabaran AF, O'Sullivan MG, Anderson KJ, Scott PM et al (2018) A novel cancer syndrome caused by KCNQ1-deficiency in the golden Syrian hamster. J Carcinog 17:6
161. Fan H, Zhang M, Liu W (2018) Hypermethylated KCNQ1 acts as a tumor suppressor in hepatocellular carcinoma. Biochem Biophys Res Commun 503: 3100–3107
162. den Uil SH, Coupe VM, Linnekamp JF, van den Broek E, Goos JA, Delis-van Diemen PM et al (2016) Loss of KCNQ1 expression in stage II and stage III colon cancer is a strong prognostic factor for disease recurrence. Br J Cancer 115:1565–1574
163. Rapetti-Mauss R, Bustos V, Thomas W, McBryan J, Harvey H, Lajczak N et al (2017) Bidirectional KCNQ1:beta-catenin interaction drives colorectal cancer cell differentiation. Proc Natl Acad Sci U S A 114:4159–4164
164. Kanduri C (2011) Kcnq1ot1: a chromatin regulatory RNA. Semin Cell Dev Biol 22:343–350

The Role of Thermosensitive Ion Channels in Mammalian Thermoregulation

16

Yawen Chen and Kun Song

Abstract

Ambient temperature detection and core body temperature maintenance are critical for the environment adaptability of mammals, requiring an elaborate neural network that converts the temperature information sensed by thermoreceptors into physiological and behavioral thermoregulatory responses. The molecular basis of thermosensation lies in the activation of various thermosensitive ion channels with distinct temperature thresholds expressed on the cell membrane of sensory neurons. These channels are able to convert thermal stimuli into electrical activities by gating ions into and out of the cell. In this chapter, we briefly introduce the physiological functions of the main thermosensitive ion channels involved in the core body temperature homeostasis orchestrated by the neural circuits in the peripheral and central nerve systems.

Keywords

Core body temperature · Thermosensation · Thermoregulation · Thermosensitive ion channels · Neural circuitry · Hypothalamus

Y. Chen · K. Song (✉)
Brain research Center and Department of Biology, School of Life Sciences, Southern University of Science and Technology, Shenzhen, Guangdong, China
e-mail: chenyw3@sustech.edu.cn; songk@sustech.edu.cn

16.1 Introduction

The core body temperature (T_{core}) of most living organisms falls into a range from around 0 to 45 °C, since that temperature below 0 °C leads to water crystallization that sharply reduces enzymatic activity, while proteins start denaturing when the temperature exceeds 45 °C [1]. Animals can be categorized as either homeothermic or poikilothermic according to their capability of maintaining the internal/core body temperature—T_{core}. Most mammals are homeothermic, except for some rare ones, such as the naked mole-rat and sloth. Although the temperature of their body shell (skin and subcutaneous tissues) fluctuates mainly along with the environmental temperature, homeothermic mammals usually keep their internal temperature of the brain and viscera to be relatively constant within a narrow range of 37 °C (Some bats have higher T_{core} but are not discussed here.). This specific body temperature has been regarded as a mechanism to optimize enzymatic reactions and cellular functions to generate higher metabolic rates, which is necessary for sustained physiological activities independent of the ambient temperature [2]. Therefore, the homeothermic species often obtain a larger area of habitats and potentially more preys since they can adapt to a broader range of environmental temperatures.

To maintain a stable core body temperature in mammals, various thermal afferent pathways located throughout the body can detect and convey the information of absolute temperatures or

L. Zhou (ed.), *Ion Channels in Biophysics and Physiology*, Advances in Experimental Medicine and Biology 1349, https://doi.org/10.1007/978-981-16-4254-8_16

temperature changes in real time and trigger autonomic or behavioral responses via neural circuits to defend against the ambient temperature changes. The molecular basis of thermosensation relies on the thermosensitive ion channels that detect the peripheral and central temperature information and convert it into the neuronal impulse. Such a signal is further transmitted through different neural circuits for thermoregulation if it is necessary.

Here, we provide an outline of the thermosensitive ion channels and their physiological roles in peripheral and central temperature sensation and regulation in the context of the mammalian thermoregulatory circuits from thermal afferents to thermoeffectors.

16.2 The Organization of the Thermoregulatory Circuits

16.2.1 Thermal Afferent Pathways

Thermal stimuli, broadly categorized into noxious heat (mean detection thresholds (MDT): >42–47 °C), innocuous warmth (MDT: ~34–42 °C), innocuous coolness (MDT: ~14–30 °C), and noxious cold (MDT: <7.3–18.4 °C) [3, 4], can activate corresponding sensory neurons that measure the temperature of the body. Most of these sensory neurons are pseudounipolar, having their cell bodies located in trigeminal ganglia (TG, that innervates the head and face) or dorsal root ganglia (DRG, that innervates the rest of the body). Their neurites (primarily unmyelinated C fibers and thinly myelinated Aδ fibers) split into two branches: one innervates into the skin or viscera to receive temperature information of different thermal stimuli; the other projects and transmits the signal to the superficial lamina (laminae I and II (LI/II)) of the spinal dorsal horn (DH) or the spinal trigeminal nucleus. Another group of thermosensitive neurons is located in the preoptic area (POA) of the hypothalamus within the deep brain, which is regarded as the thermoregulatory center controlling the core body temperature homeostasis [5–8]. Although acute external temperature challenges have little impact on brain temperature [9, 10], heat generated by exercise or fever, or decrease in metabolic rate when animals enter a state of hibernation or torpor, may change brain temperature by several degrees Celsius, which may be detected by warm-sensitive preoptic neurons. Together, all these sensory neurons comprise the primary input into the thermoregulatory circuits.

Thermal information received by the DRG or TG sensory neurons is then relayed to the LI/II DH neurons [11–14]. In vivo calcium imaging in spinal cord revealed that heat-responsive DH neurons encode the absolute skin temperature without adaptation, whereas cold-responsive DH neurons encode temperature changes and are rapidly adapting to ambient temperature, suggesting that periphery thermal inputs that converge into spinal cord will be recoded before being transmitted to the brain [14].

Temperature-responsive DH neurons are supposed to send projections from LI/II to the contralateral somatosensory nuclei of the thalamus as well as the lateral parabrachial nucleus (LPB) of the brainstem [10, 15–18]. Although it has been proposed that projections from the thalamus to the somatosensory cortex mediate the perception of body temperature [19, 20], which should induce thermoregulatory behaviors, the evidence that thalamic lesions do not block behavioral or autonomic responses to the temperature changes in the skin suggests that the spinothalamocortical pathway is not indispensable for thermal perception and regulation [10, 21]. In contrast, projections from LI/II to the LPB and further to the POA may play a critical role in thermoregulation, as the lesion of the LPB abolishes the autonomic responses as well as temperature preference behavior, suggesting that LPB is necessary for both thermal perception and regulation [21].

POA serves as the key integratory hub for thermoregulation, since it can converge the periphery thermal information relayed from LPB with the intrinsic brain temperature detected by thermosensitive POA neurons and send outputs to thermoeffectors via descending nerves to initiate thermoregulatory responses [4, 20]. Studies have shown that about 30% of the recorded POA neurons are warm-sensitive, some of which can also be activated by incoming warm stimuli from the periphery; while less than 10% of the neurons are cold responsive, the remaining ones are temperature insensitive [22, 23]. Local POA warming or cooling elicits heat- or cold-defense responses both autonomically and behaviorally, which mimic the responses to environmental temperature challenges [5, 8, 24–26]. Besides, POA lesion in animals abolishes their ability to maintain their T_{core} in either a hot or cold environment [27–30] but retains most of the thermoregulatory behaviors [20, 27]. These findings imply that POA is sufficient but not necessary for thermal perception. Similar to the case of LPB, there might be circuits communicating thermal information from POA to the cortex, yet no present evidence suggests that (Fig. 16.1).

16.2.2 Efferent Pathways Controlling Thermoeffectors

Integrated thermal information in the brain, particularly in the POA, will then trigger autonomic and behavioral effectors via efferent nerves to generate heat- or cold-defense responses.

As for autonomic thermoregulation, heat-defense responses (e.g. vasodilation and evaporative cooling) and cold-defense responses (e.g. vasoconstriction, thermogenesis of the brown adipose tissue (BAT), and shivering of the skeletal muscle) are produced by controlling the contraction/relaxation of vascular smooth muscle, the mitochondrial leak in the BAT, the contraction/relaxation of skeletal muscle, and the secretion of sweat or salivary gland [4, 20]. Each of these effectors is innervated by distinct efferent pathways, while these pathways have a similar organization in terms of descending relays from POA to peripheral nerves via the brain stem.

Canonical pathways controlling BAT and skeletal muscle involve a tonic, inhibitory output, which blocks the activity of thermogenesis-promoting neurons located in the dorsomedial hypothalamus (DMH), sent from the POA to the DMH. Once the inhibitory effect is removed, these DMH neurons can excite premotor neurons in the raphe pallidus (RPA), which in turn relay the excitatory output to sympathetic nerves triggering non-shivering thermogenesis of the BAT [31, 32], or to somatic motor neurons triggering repetitive contraction of the skeletal muscle (shivering thermogenesis) [33–35].

The pathway controlling cutaneous vasomotion is similar to the one above, with the exception that a relay in the DMH is not required; instead, the POA sends inhibitory output directly to the cutaneous vasoconstrictor premotor neurons located in the RPA and adjacent rostral ventrolateral medulla (RVLM). Once the inhibition from the POA is gone, these premotor neurons will trigger vasoconstriction via the sympathetic nerves to reduce the cutaneous blood flow and thus the heat exchanging rate between the skin and environment [4, 20].

Evaporative cooling in rodents is achieved primarily by spreading saliva on their fur, though they may sweat in their footpad [36]. Control of salivation is involved with an inhibitory output from POA to hypothalamic neurons, probably located in the lateral hypothalamus (LH) [20]. Without preoptic inhibition, these LH neurons will send the excitatory drive to parasympathetic preganglionic neurons in the superior salivatory nucleus (SSN) [37]. The latter will activate the downstream ganglion cells for salivation. By contrast, pathways connecting POA and premotor neurons located in the rostral ventromedial medulla (RVMM), the latter of which innervates sympathetic nerves for sweating [38, 39], are unknown, despite that sweating correlates with POA activation in humans [40].

In the face of thermal challenges, mammals also engage in motivated, voluntary behaviors to help their bodies dissipate or preserve heat. These include warmth or cold seeking, postural extension or huddling, nest or burrow making, and various means used by humans [41]. Rats can even be trained to turn on a heat lamp or a cooling fan by pushing a lever when exposed to cold or heat, respectively [42–44]. However, the neural circuits motivating these behaviors remain, to a wide extent, unclear [4, 20]. Voluntary behaviors

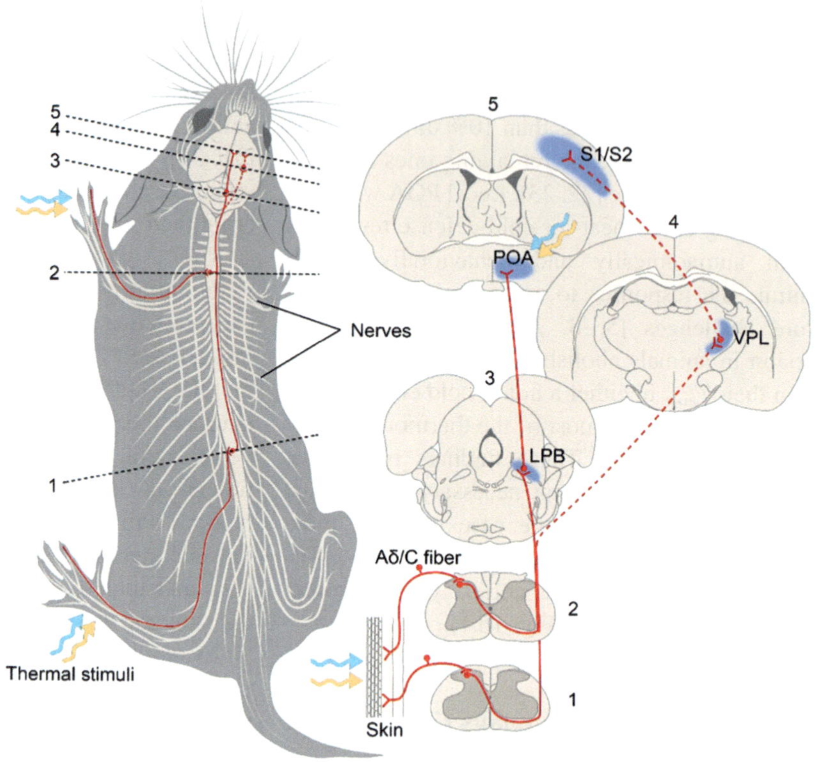

Fig. 16.1 The model for the thermal afferents from the periphery to the POA in rodents. Dashed curves indicate the putative pathways deemed essential to generate thermal perceptions. *LPB* lateral parabrachial nucleus, *VPL* ventral posterolateral thalamic nuclei, *POA* preoptic area, *S1* primary somatosensory cortex, *S2* secondary somatosensory cortex

are thought to be motivated by cortical perceptions. As we mentioned earlier, both the thalamocortical pathway and POA seem to be dispensable for thermal perceptions, yet LPB does, suggesting that thermal perceptions might be generated by both the thalamocortical pathway and the POA, while LPB provides input to both of these pathways (Fig. 16.2).

16.3 Thermosensitive Ion Channels and Their Functions in Thermoregulation

16.3.1 Thermosensitive Properties of the Ion Channels

The molecular mechanisms underlying the detection of ambient and internal temperature by thermosensitive neurons (thermoreceptors) depend on the activation of temperature-gated ion channels expressed on the cell membrane, which are able to convert thermal stimuli into electrical activities by gating ions into and out of the cell. The definition of channel thermosensitivity is based on the temperature dependence of channel open probability, which is quantified by the Q_{10}, a temperature coefficient value representing the increase in the channel activity when the temperature varies by 10° [45]. In fact, temperature increases in a certain range could accelerate the activation of most ion channels, with a Q_{10} of ~3. Only channels with $Q_{10} > 5$ are generally considered to be thermosensitive [3]. Different thermosensitive ion channels are activated by distinct temperature thresholds, leading to membrane depolarization or hyperpolarization according to the sorts and directions of ion currents. Most temperature-gated channels that are known to contribute to mammalian thermoregulation belong to the transient receptor potential (TRP) ion channel superfamily, which is comprised of several subfamilies including vanilloid (TRPV), canonical or classic (TRPC),

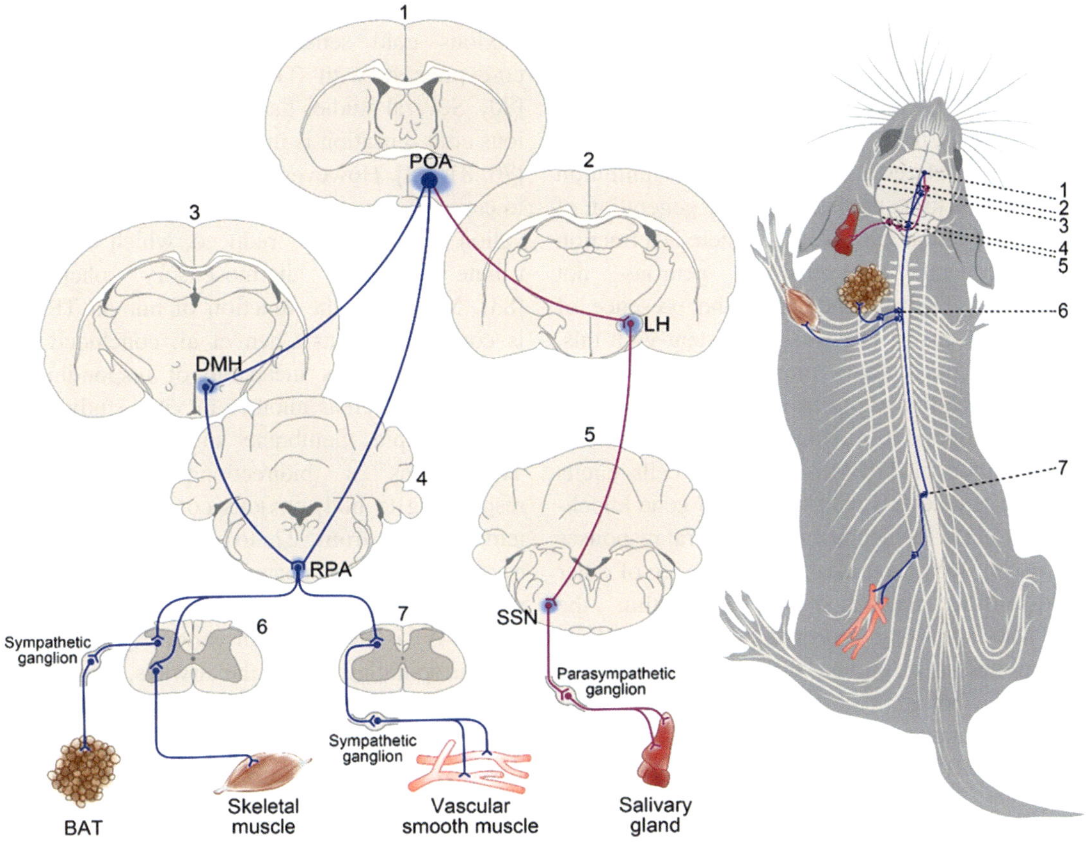

Fig. 16.2 The model for the thermal efferent pathways from the POA to the thermoeffectors in rodents. *POA* preoptic area, *LH* lateral hypothalamus, *DMH* dorsomedial hypothalamus, *RPA* raphe pallidus, *SSN* superior salivatory nucleus

ankyrin (TRPA), and melastatin (TRPM) [46]. Some non-TRP thermosensitive ion channels, including two-pore domain potassium (K2P) channels (TREK-1/2 and TRAAK) [47, 48], chloride channel anoctamin 1 (ANO1 also known as TMEM16A) [49], and STIM1-ORAI1 channel complex [50], may also participate in thermosensation. Here, we focus on the expression pattern and physiological functions of these thermosensitive ion channels.

16.3.1.1 TRP Channels

Thermosensitive TRP channels are cationic channels, most of which are permeable to both monovalent and divalent cations, with a few exceptions [3]. Different TRP channel subtypes are sensitive to either cold/cool or heat/warm temperatures. Besides temperature, many of these TRP channels can also be activated by other physical and chemical stimuli such as ligands, mechanical stress, or osmolarity, thus becoming polymodal receptors [51]. Their activation results in inward cationic currents that evoke the depolarization of neuronal membrane potential and excite the thermoreceptors [48].

TRPM8

TRPM8 is activated by temperature <26–28 °C, having a Q_{10} of 24 [52]. It could alternatively be activated by PIP_2 [53–55] or by cooling compounds such as menthol or its derivatives, eucalyptol, and the supercooling agent icilin [52, 56]. TRPM8 is highly expressed in a subpopulation of C-fiber DRG and TG sensory neurons, rendering these neurons and their fibers cold-sensitive and capable of transducing the cold sensation [57]. It is recently found that in mice, a portion of these TRPM8-expressing fibers fire continuously at physiological skin temperatures (around 27 °C) and increase or shut down their firing activity upon cooling or warming, respectively [58]. Interestingly, such a warm-inhibition effect seems to be necessary for the generation of warm perception [58]. Notably, there is a fraction of peripheral cold-sensitive neurons not expressing TRPM8, indicating the presence of other cold sensors [59–61]. Consistent with this, *Trpm8*$^{-/-}$ mice show impaired yet not fully abrogated ability to discriminate innocuous cold and still respond to noxious cold [62–64]. Selective ablation of TRPM8-expressing cells elicits more severe defects than the *Trpm8* gene knockout manner in the neural and behavioral responses to cooling [65], suggesting the existence of other cold-sensing mechanisms even within the TRPM8-expressing neuronal population. While no studies have shown that T_{core} changed in *Trpm8*$^{-/-}$ mice, treatment in wild-type mice with subcutaneous icilin led to hyperthermia, whereas TRPM8 antagonists led to hypothermia [66, 67]. These changes in T_{core} caused by TRPM8 agonist or antagonists were abolished in *Trpm8*$^{-/-}$ mice [66, 67]. These results suggest that TRPM8 is sufficient but not necessary for controlling the activity of TRPM8-expressing cells that take part in the circuits of T_{core} regulation.

TRPM8 is also expressed in many other tissues such as prostate, vascular tissue, lung, and testis, where the physiological role of TRPM8 is poorly understood [68].

TRPA1

In rodents, TRPA1 is found activated by temperature <17 °C with a Q_{10} of 10, which is close to the painful cold threshold [69–71], although it is a heat receptor in non-mammalian species [72–75]. TRPA1 is also a promiscuous chemical sensor that can be activated by a large number of compounds, including pungent compounds, cannabinol, and cytosolic Ca^{2+} [76]. It may also play a role in mechanosensation [77, 78]. In mice, TRPA1 is present along the axon of a subpopulation of Aδ- and C-fiber DRG and TG nociceptive sensory neurons and was deduced to mediate noxious cold sensation [71, 79]. It is not co-expressed with TRPM8 in DRG neurons [80]. Several studies have shown that acute noxious cold sensation is impaired in *Trpa1*$^{-/-}$ mice [70, 81, 82]. However, *Trpa1*-knockout is found to cause the secondary effect that the skin innervation is substantially reduced, which may contribute to the thermosensory phenotype [83]. Moreover, the function of human TRPA1 is controversial, as Chen et al. concluded that human TRPA1 is incapable of responding to cold [84], whereas another in vitro study using artificial lipid membrane suggests the human TRPA1 as a bidirectional thermosensor, responding to both noxious cold and increasing temperatures from 22 to 40 °C ($Q_{10} \approx 6$) according to the channel conformations and redox states [85].

TRPA1 is also found in various types of cells such as epithelial cells in the airways, gastrointestinal tract, bladder, and skin; mast cells, fibroblasts, β-cells of the Langerhans islets, as well as astrocytes in the rodent hippocampus, where TRPA1 is potentially involved in different regulatory processes [86].

TRPC5

When expressed heterologously in HEK293 cells, TRPC5 monomeric channels can be activated in the temperature range of 25–37 °C with a Q_{10} around 10 [87]. It can also be activated by a multiplicity of other factors, including nitric oxide, lysophospholipids, reduced thioredoxin, protons, lanthanides, and calcium [88]. In terms of thermoregulation, TRPC5 expressed in murine DRG sensory neurons, especially the nerve terminals of these neurons, may play a role in innocuous cold detection as a complement to TRPM8 [87]. Of note, deletion of TRPC5 in 129S1/SvImJ mice resulted in no temperature-sensitive behavioral changes [87].

TRPC5 has the highest expressional level in the brain [88], particularly in the frontal cortex, pyramidal cell layer of the hippocampus, dentate gyrus, hypothalamus, and amygdala [89, 90], where TRPC5 exerts multiple functions including regulation of neurite extension [91], dendrite patterning [92], and motor neuron axon pathfinding [93]. Whether TRPC5 works in temperature sensation in the brain has not yet been explored, as decrease in brain temperature usually doesn't occur. But it may be of interest to investigate this issue when mice are into torpor, a fasting- and/or cold-induced state of decreased metabolic rate in which T_{core} can fall below 31 °C [4].

TRPV1

TRPV1 is activated by temperature >42 °C, having a Q_{10} around 26 [94]. It could alternatively be activated by some vanilloid chemicals such as capsaicin and piperine, extracellular protons (pH < 6.5), as well as by toxins such as resiniferatoxin [95]. TRPV1 is predominantly expressed in Aδ nociceptive- and C-fiber DRG and TG sensory neurons, both in the cell body and along the axon, conferring heat sensitivity to these neurons [96, 97]. A subset of TRPV1-positive sensory neurons is found co-expressed with TRPA1, suggesting that noxious cold and heat stimuli might elicit the same kind of nociception [71]. Mice lacking TRPV1 showed no responses to capsaicin and diminished responses to acute heat [98, 99]. DRG neurons from these mice were also deficient in their responses to these stimuli [98, 99], but a small fraction of the cells showed responses to the temperature over 55 °C [98], suggesting the existence of other receptors of noxious heat. Although TRPV1 antagonists or agonists induce hyperthermia or hypothermia, respectively [100, 101], *Trpv1*$^{-/-}$ mice have normal T_{core} [102], suggesting that TRPV1 is sufficient but not necessary for the activity of TRPV1-expressing cells that participate in controlling core body temperature homeostasis. Besides, whether TRPV1 is required for the response to brain warming or central capsaicin is dubious, as TRPV1 expression is extremely sparse in the brain [103].

TRPV1 is also expressed in arteriolar smooth muscle, which is involved in controls of blood flow in certain thermoregulatory tissues, such as skeletal muscle, skin, and trachea [95]. But the detailed functions of TRPV1 in these tissues are kept elusive.

TRPV2

TRPV2 is activated by temperature >52 °C, having a Q_{10} higher than 100 [94]. Mechanical stresses, 2-Aminoethoxydiphenyl borate (2-APB) (species dependent), cannabinoids, and probenecid also activate TRPV2 [104]. TRPV2 is expressed in DRG and TG sensory neurons, but unlike TRPV1, TRPV2 is abundant in medium- to large-diameter neurons, the axons of which are predominantly myelinated nociceptive Aδ fibers [105–107]. It seems reasonable to consider TRPV2-expressing sensory neurons as sensors that detect harmful heat stimuli; however, in *Trpv2* knockout mice, no abnormality is detected in thermosensation [108]. Moreover, *Trpv1/Trpv2* double knockout mice show no differences from *Trpv1*$^{-/-}$ mice in a behavioral study [108], suggesting that TRPV2 does little to serve as a thermosensor in vivo.

Unlike TRPV1, TRPV2 is highly expressed in various regions of the brain. In the hypothalamus, TRPV2 is significantly expressed in the supraoptic nucleus and paraventricular nucleus, which is critical for osmoregulation [104]. TRPV2 is also abundantly expressed in the immune system and various tissues that are unlikely to be exposed to temperatures above 50 °C [104].

TRPV3

TRPV3 can be activated by innocuous warmth (31–39 °C) as well as noxious heat, having a Q_{10} of 17 [94, 109]. TRPV3 can also be activated by 2-APB, menthol, and plant extracts such as camphor, carvacrol, eugenol, and thymol [110]. TRPV3 is predominantly expressed in keratinocytes of the skin but not in neurons [111]. It has been proposed that heat-activated TRPV3 evokes ATP or prostaglandin E_2 (PGE_2) release from keratinocytes to excite sensory nerves that innervate the skin [112, 113]. Supportively, *Trpv3* knockout mice of intercrossed C57BL6/129J background showed deficits in responses to innocuous and noxious heat stimuli [114]. However, a later study showed that the deletion of TRPV3 in homogeneous C57BL6 background mice resulted in no obvious changes in thermal preference behavior [115]. Meanwhile, *Trpv3* knockout mice of either C57BL6 or a 129S6 background showed no deficits in responses to acute heat [115], suggesting that the thermosensation role of TRPV3 is strain dependent.

TRPV4

TRPV4 can be activated by temperature >27–34 °C, having a Q_{10} of 19 [116]. It has been proved by whole-cell and excised patches that heat activation of TRPV4 requires the participation of other endogenous molecules [117]. TRPV4 is also a polymodal receptor that can be activated by hypoosmolarity [118] and bisandrographolide from Chinese herbal plant *Andrographis paniculate* [119]. TRPV4 is broadly found in various tissues and is strongly expressed in keratinocytes, hippocampal neurons, and choroid plexus epithelial cells [120]. TRPV4 is found to express in the peripheral nerve endings of DRG sensory neurons [120]. The thermosensory function of TRPV4 is controversial. Early studies showed that *Trpv4* knockout mice and *Trpv3*/*Trpv4* double knockout mice have normal thermoregulation [115, 121], whereas a later study showed that a peripheral TRPV4 antagonist increases body temperature, while agonist leads to hypothermia [122]. Nevertheless, temperature can act directly and locally on tissues via TRPV4 activation. For instance, Ca^{2+} influx in keratinocytes mediated by warm-activated TRPV4 leads to the increase of E-cadherin expression and reorganization of actin cytoskeletons, which enhances the barrier function of these cells [123]. This may decipher why skin dehydration happens only in the winter season is the reduced TRPV4 activities in cold temperatures [124]. Besides, it has been proposed that TRPV4 is constitutively active in the hippocampal neurons through its activation by brain temperature, which is of great importance to enhance neuronal excitability in the brain and to regulate social behaviors in mammals [125–127].

TRPM2

TRPM2 can be activated by temperature >35 °C, having a Q_{10} of 44 [128]. In addition to heat, TRPM2 can be activated or sensitized by reactive oxygen species (ROS) such as H_2O_2 [129, 130], which can reduce the temperature threshold of TRPM2 activation. TRPM2 is widely expressed in the brain and peripheral nervous system, including DRG sensory neurons and both sympathetic and parasympathetic neurons [131–133]. In the periphery, TRPM2 is responsible for the detection of non-noxious warmth, as it has been shown that wild-type mice avoided the non-noxious warm temperature of 38 °C, while *Trpm2* knockout mice showed no avoidance of 38 °C [133]. In the brain, TRPM2 is present in a subset of preoptic neurons found to be activated by a temperature of 38 °C or above [134]. Injecting a high dose of PGE_2, a compound thought to inhibit the activity of POA warm-sensitive neurons, in the POA caused higher fever temperature in *Trpm2*$^{-/-}$ mice than in *Trpm2*$^{+/+}$ mice [134], suggesting that TRPM2 acts as a "brake" that limits the magnitude of fever temperature. Under normal conditions, *Trpm2*$^{-/-}$ mice have normal T_{core}, while chemogenetic activation and inhibition of TRPM2-expressing POA neurons in vivo caused hypothermia and hyperthermia, respectively [134], suggesting that TRPM2-expressing neurons are part of the circuits involved in controlling T_{core}, while TRPM2 is sufficient but not necessary for the activity of these neurons.

The presence of TRPM2 in sympathetic and parasympathetic neurons [133] is surprising because, as we mentioned earlier, these neurons belong to efferent pathways of thermoregulation circuits. TRPM2-expressing sympathetic neurons showed a significant calcium increase in response to heat [133], of which the in vivo physiological function is remained to be elucidated. Besides, TRPM2 is widely expressed in the immune system and is thought to be involved in inflammatory responses [129, 135, 136].

TRPM3

TRPM3 can be activated by temperature >40 °C, having a Q_{10} of 7.2 [135]. It can also be activated by pregnenolone sulphate (PS), nifedipine, and β-cyclodextrin [137, 138]. TRPM3 is broadly present in a variety of tissues [139] and is abundantly expressed in DRG and TG neurons and functions as a sensor of noxious heat [140]. It is also suggested that a subset of TRPM3-expressing sensory neurons express TRPV1 as well [140]. Since that a subpopulation of TRPV1-positive neurons express TRPA1 as mentioned above, although lacking direct evidence, it may be possible that all three ion channels are co-expressed in a subset of the nociceptors. *Trpm3*$^{-/-}$ mice have normal T_{core} but a deficit in sensing noxious heat in a similar way to *Trpv1*$^{-/-}$ mice [140]. Double knockout of any two of the three (TRPM3, TRPV1, and TRPA1) channels also results in defect yet not fully abrogated heat avoidance responses [141]. By contrast, combined loss of all these three TRP channels (triple knockout) completely abrogate the heat-induced noxious responses both at cellular and behavioral levels [141]. Furthermore, reintroducing any one of the channels into triple-knockout neurons by transient transfection restored the heat sensitivity of the cells [141], indicating that the triad of these TRP channels is sufficient and necessary for noxious heat transduction in mice. Unlike robust changes in body temperature due to manipulations with antagonists and agonists of TRPV1, neither the activation of TRPM3 channels with PS nor pharmacological blockade of TRPM3 alters the body temperature of mice [140, 142]. Thus, TRPM3-expressing cells appear to be less involved in the homeostasis of the body temperature than TRPV1-expressing cells.

16.3.1.2 TREK Channels

TREK channels (TWIK-related potassium channels), comprised of TREK1, TREK2, and TRAAK (TWIK-related arachidonic acid-activated potassium channel), belong to the two-pore domain K^+ (K2P) channel family. These channels are also called background K^+ channels because at physiological body temperature (37 °C), they are active and contribute to the background K^+ conductance [47, 143]. In addition to their sensitivity to temperature, they can also be modulated by other physical and chemical stimuli such as membrane stretch and free fatty acids [144, 145]. Their activation, in contrast to that of TRP channels, results in outward K^+ currents that evoke hyperpolarization of neuronal membrane potential and reduce the excitability of thermoreceptors [48].

TREK1/2 can be effectively activated by temperature >25 °C, while TRAAK is effectively activated by temperature >31 °C [47], although these channels are not thoroughly silent until the temperature drops to 14 °C [143, 146]. The Q_{10} of these channels are all around 10 [145]. It has been proved by whole-cell and excised patches that heat activation of all of the three channels requires the integrity of the cell and maybe the participation of other endogenous molecules [47, 143]. They are often co-expressed with TRP channels [147, 148] and are all present in DRG and TG sensory neurons as well as in the hypothalamus [47, 48, 143, 146, 148–153], where their activation may counteract the stimulatory effect of TRP channels and increase the temperature thresholds for thermoreceptor firing [48]. Indeed, it has been reported that compared with control mice, *Trek1*$^{-/-}$, *Trek2*$^{-/-}$, *Traak*$^{-/-}$, *Trek1/Traak* double knockout, and *Trek1/Trek2/Traak* triple knockout mice exhibited hypersensitivity in response to a heating ramp (30–50 °C) and suffered from hyperalgesia at temperatures around 45 °C; while *Trek1*$^{-/-}$, *Traak*$^{-/-}$, the double and triple knockout mice but not *Trek2*$^{-/-}$ mice showed hyperalgesia at higher temperatures (46–50 °C) [48, 154–156]. Similar results were obtained by using a skin-nerve preparation to examine the proportion and activity of C-fibers responding to the same temperature range [48, 154–156], indicating that TREK2 regulates C-fiber responses to moderate warmth, while other TREK channels mediate responses to noxious heat. When tested with cooling and cold temperatures, the *Trek2*$^{-/-}$ and the triple knockout mice showed enhanced sensitivity to 20–25 °C; the double and triple knockout mice showed enhanced sensitivity to 10–20 °C; while only triple knockout mice were hypersensitive to 5–15 °C [48, 155, 156]. Thus, TREK channels play a dual role in controlling the temperature thresholds for both the cold and the heat thermoreceptors.

16.3.1.3 ANO1 (TMEM16A)

ANO1 is a Ca^{2+}-activated chloride channel which is found to be activated by temperatures over 44 °C independently from intracellular Ca^{2+}, having a Q_{10} around 19 [49]. ANO1 channels were found to be primarily expressed in DRG neurons, most of which also expressed TRPV1 [49, 157]. Because ANO1 is an anion-selective channel, whether its activation results in inward or outward Cl^- currents, such that evoking hyperpolarization or depolarization of the cell membrane, respectively, depends on intracellular Cl^- concentration [49].

It has been shown that DRG neurons have elevated expression and activity of the sodium–potassium–chloride co-transporter 1 (NKCC1) that accumulates Cl^- within the cells, resulting in a relatively higher intracellular Cl^- concentration in DRG neurons than in other neurons of the central nervous system [157–159], which further leads to a Cl^- equilibrium potential (about −30 mV) more positive than the resting membrane potential (about −60 to −55 mV) [160, 161] and thus depolarization of DRG neurons upon heat activation of ANO1. Consistent with this inference, DRG/TG-specific *Ano1* knockout mice showed a pronounced reduction in response to noxious heat in tail-withdrawal tests, indicating the role of ANO1 as a heat sensor [49].

16.3.1.4 STIM1-ORAI1 Channel Complex

STIM1 is an endoplasmic reticulum (ER) Ca^{2+} sensor [162, 163], while ORAI1 is a Ca^{2+}-selective pore-forming subunit located at the plasma membrane [164, 165]. ER Ca^{2+} depletion will cause conformational changes and oligomerization of STIM1; oligomerized STIM1 then translocates to and clusters at ER–plasma membrane junctions to bind with ORAI1 [163, 166, 167], and ultimately the STIM1-ORAI1 complex mediates inward Ca^{2+} currents [168, 169]. It was found by the heterologous expression that STIM1 clustering could also be induced by temperatures over 37–43 °C (Q_{10} is around 7) without store depletion of ER Ca^{2+} [50]. However, functional coupling of STIM1 and ORAI1 that generates Ca^{2+} influx is blocked by heating but is reversed by cooling the temperature below 37 °C [50]. Thus, STIM1 and ORAI1 mediate a unique response to cooling after a heating step (heat off-response) [50].

STIM1 and ORAI1 are ubiquitously expressed in human and mouse tissues and are predominantly present in lymphoid organs (lymphocytes) and skeletal muscle, while they are very lowly expressed in neuronal tissue [170]. Matching with this expression pattern, STIM1- and ORAI1-deficiencies mainly lead to immunodeficiency (such as impaired T- and B-cell activation) and myopathy [171–174], while retaining normal cognitive and neuronal functions [170, 171, 174–176]. It indicates a minor role of STIM1–ORAI1 channel complex in neural circuits of thermoregulation. Nevertheless, the observation that temperature-induced Ca^{2+} influx mediated by STIM1–ORAI1 activation modifies gene expression in Jurkat T cells suggests that the thermosensory ability of non-neurocyte might be of physiological importance [50], as immune cells can experience large temperature changes while they circulate from central organs to peripheral tissues [177], and almost all cells will be faced with T_{core} changes in cases such as fever, torpor, and hibernation [4].

16.4 Summary

Temperature sensation and thermoregulation are one of the cornerstones for mammals to adapt to the environment, relying on various temperature-gated ion channels to switch on or off the neural circuits or local cellular mechanisms in response to the temperature changes. Several aspects about thermosensitive ion channels are worthy of note in order to comprehensively understand the thermoregulatory mechanisms.

First, many channels are not solely activated by temperature changes but display multimodal activation by various chemical and physical stimuli. It is common that different stimuli synergistically act on the channel and reduce each other's thresholds detected by the channel. Thus, the thermosensory functions may vary with the cellular context, and it is of importance to ascertain the role of these channels under pathophysiologic conditions.

Second, although temperature sensation is mainly carried out by sensory neurons that provide input for thermal perceptions and responses of thermo-effectors, TRPM2 is found to be expressed in efferent neurons [133], and many temperature-gated channels are also present in cells outside the presently identified thermoregulatory circuits. Are these channels still function as thermosensors? If so, the physiological significance underlying the functions remains to be elucidated.

Third, the activation of thermoreceptors could be controlled by the collaboration of different thermosensitive channels, since that some types of thermosensitive channels are reported to be co-expressed in the same neurons, which may complicate the interpretation of the relative importance of each channel as well as the phenotypes. Therefore, profiling the co-expression of thermosensitive channels in the same cell may help to understand the full picture of thermosensation, in which the single-cell RNA sequencing is becoming an effective strategy [178].

Fourth, although the temperature ranges where thermosensitive channels change their open probability have been extensively revealed, the physicochemical and structural bases underlying the temperature-gating processes still remain elusive, since that the channel structures of the temperature-activated states are hard to be obtained and could be different from the ligand-bound states [46]. A recent study tactfully clarified the key residues that determine the cold sensitivity of TRPM8 by comparing the TRPM8 orthologs in vertebrate species inhabiting distinct ambient temperatures such as African elephant, human, and emperor penguin [179]. This finding not only showed that the cold sensitivity of TRPM8 is tuned by side-chain hydrophobicity within the pore domain, leading to different capacities of cold hardiness in different vertebrate species [179], but also provided an idea for studying the temperature-sensing regions of other channels. It could be expected that thermosensitive ion channels will be widely applied to the field of synthetic biology once the molecular nature of these "thermo-switch" is fully disclosed.

Prospectively, the interrogation of thermosensitive ion channels will provide insight on thermoregulatory mechanisms at the molecular level. Besides, to fully understand the principle of core body temperature homeostasis, one also needs to elucidate the neurons that coordinate this pivotal homeostatic function by applying optogenetic or chemogenetic tools in vivo. In this way, we can build up our knowledge on the mechanisms of thermoregulation at both molecular and neural circuit levels.

Acknowledgments This work was supported by the grant from Shenzhen-Hong Kong Institute of Brain Science-Shenzhen Fundamental Research Institutions (2021SHIBS0002).

References

1. Romanovsky AA (2018) The thermoregulation system and how it works. Handb Clin Neurol 156:3–43
2. Clarke A, Portner HO (2010) Temperature, metabolic power and the evolution of endothermy. Biol Rev Camb Philos Soc 85(4):703–727
3. Tan CH, McNaughton PA (2018) TRPM2 and warmth sensation. Pflugers Arch 470(5):787–798
4. Tan CL, Knight ZA (2018) Regulation of body temperature by the nervous system. Neuron 98(1):31–48
5. Baldwin BA, Ingram DL (1967) Effect of heating & cooling the hypothalamus on behavioral thermoregulation in the pig. J Physiol 191(2):375–392
6. Carlisle HJ (1966) Behavioural significance of hypothalamic temperature-sensitive cells. Nature 209 (5030):1324–1325
7. Folkow B et al (1949) Cutaneous vasodilatation elicited by local heating of the anterior hypothalamus in cats and dogs. Acta Physiol Scand 17(4):317–326
8. Hammel HT et al (1960) Thermoregulatory responses to hypothalamic cooling in unanesthetized dogs. Am J Physiol 198(3):481–486
9. Bratincsak A, Palkovits M (2005) Evidence that peripheral rather than intracranial thermal signals induce thermoregulation. Neuroscience 135 (2):525–532
10. Nakamura K, Morrison SF (2008) A thermosensory pathway that controls body temperature. Nat Neurosci 11(1):62–71
11. Andrew D, Craig AD (2001) Spinothalamic lamina I neurones selectively responsive to cutaneous warming in cats. J Physiol Lond 537(2):489–495

12. Christensen BN, Perl ER (1970) Spinal neurons specifically excited by noxious or thermal stimuli: marginal zone of the dorsal horn. J Neurophysiol 33 (2):293–307
13. Craig AD et al (2001) Quantitative response characteristics of thermoreceptive and nociceptive lamina I spinothalamic neurons in the cat. J Neurophysiol 86(3):1459–1480
14. Ran C et al (2016) The coding of cutaneous temperature in the spinal cord. Nat Neurosci 19 (9):1201–1209
15. Cechetto DF et al (1985) Spinal and trigeminal dorsal horn projections to the parabrachial nucleus in the rat. J Comp Neurol 240(2):153–160
16. Hylden JL et al (1986) Physiology and morphology of the lamina I spinomesencephalic projection. J Comp Neurol 247(4):505–515
17. Hylden JLK et al (1989) Spinal lamina-I projection neurons in the rat - collateral innervation of parabrachial area and thalamus. Neuroscience 28 (1):27–37
18. Li JL et al (2006) Medullary dorsal horn neurons providing axons to both the parabrachial nucleus and thalamus. J Comp Neurol 498(4):539–551
19. Craig AD (2002) How do you feel? Interoception: the sense of the physiological condition of the body. Nat Rev Neurosci 3(8):655–666
20. Madden CJ, Morrison SF (2019) Central nervous system circuits that control body temperature. Neurosci Lett 696:225–232
21. Yahiro T et al (2017) The lateral parabrachial nucleus, but not the thalamus, mediates thermosensory pathways for behavioural thermoregulation. Sci Rep 7(1):5031
22. Hori T et al (1980) Thermo-sensitive neurons in hypothalamic tissue slices in vitro. Brain Res 186 (1):203–207
23. Kelso SR et al (1982) Thermosensitive single-unit activity of in vitro hypothalamic slices. Am J Physiol 242(1):R77–R84
24. Banet M et al (1978) The central control of shivering and non-shivering thermogenesis in the rat. J Physiol 283:569–584
25. Carlisle HJ, Laudenslager ML (1979) Observations on the thermoregulatory effects of preoptic warming in rats. Physiol Behav 23(4):723–732
26. Mohammed M et al (2018) Preoptic area cooling increases the sympathetic outflow to brown adipose tissue and brown adipose tissue thermogenesis. Am J Physiol Regul Integr Comp Physiol 315(4):R609–R618
27. Almeida MC et al (2006) Neural substrate of cold-seeking behavior in endotoxin shock. PLoS One 1:e1
28. Lipton JM et al (1974) Effects of brainstem lesions on temperature regulation in hot and cold environments. Am J Physiol 226(6):1356–1365
29. Satinoff E et al (1976) Thermoregulatory cold-defense deficits in rats with pre-optic-anterior hypothalamic-lesions. Brain Res Bull 1(6):553–565
30. Srividya R et al (2006) Differences in the effects of medial and lateral preoptic lesions on thermoregulation and sleep in rats. Neuroscience 139(3):853–864
31. Conceicao EPS et al (2019) Neurons in the rat ventral lateral preoptic area are essential for the warm-evoked inhibition of brown adipose tissue and shivering thermogenesis. Acta Physiol (Oxf) 225(4): e13213
32. Nakamura K, Morrison SF (2010) A thermosensory pathway mediating heat-defense responses. Proc Natl Acad Sci U S A 107(19):8848–8853
33. Nakamura K, Morrison SF (2011) Central efferent pathways for cold-defensive and febrile shivering. J Physiol 589(Pt 14):3641–3658
34. Schafer SS, Schafer S (1973) The behavior of the proprioceptors of the muscle and the innervation of the fusimotor system during cold shivering. Exp Brain Res 17(4):364–380
35. Tanaka M et al (2006) Reflex activation of rat fusimotor neurons by body surface cooling, and its dependence on the medullary raphe. J Physiol 572 (Pt 2):569–583
36. Jessen C (1985) Thermal afferents in the control of body temperature. Pharmacol Ther 28(1):107–134
37. Saper CB, Loewy AD (1980) Efferent connections of the parabrachial nucleus in the rat. Brain Res 197 (2):291–317
38. Farrell MJ et al (2013) Brain stem representation of thermal and psychogenic sweating in humans. Am J Phys Regul Integr Comp Phys 304(10):R810–R817
39. Shafton AD, McAllen RM (2013) Location of cat brain stem neurons that drive sweating. Am J Physiol Regul Integr Comp Physiol 304(10):R804–R809
40. Farrell MJ et al (2014) Preoptic activation and connectivity during thermal sweating in humans. Temperature (Austin) 1(2):135–141
41. Terrien J et al (2011) Behavioral thermoregulation in mammals: a review. Front Biosci (Landmark Ed) 16:1428–1444
42. Carlton PL, Marks RA (1958) Cold exposure and heat reinforced operant behavior. Science 128 (3335):1344
43. Lipton JM (1968) Effects of preoptic lesions on heat-escape responding and colonic temperature in rat. Physiol Behav 3(1):165
44. Weiss B, Laties VG (1961) Behavioral thermoregulation. Science 133(3461):1338–1344
45. Ito E et al (2015) Thermodynamic implications of high Q 10 of thermo-TRP channels in living cells. Biophysics (Nagoya-shi) 11:33–38
46. Baez D et al (2014) Gating of thermally activated channels. Curr Top Membr 74:51–87
47. Kang D et al (2005) Thermosensitivity of the two-pore domain K^+ channels TREK-2 and TRAAK. J Physiol 564(Pt 1):103–116
48. Lamas JA et al (2019) Ion channels and thermosensitivity: TRP, TREK, or both? Int J Mol Sci 20 (10):2371

49. Cho H et al (2012) The calcium-activated chloride channel anoctamin 1 acts as a heat sensor in nociceptive neurons. Nat Neurosci 15(7):1015–1021
50. Xiao B et al (2011) Temperature-dependent STIM1 activation induces Ca(2)+ influx and modulates gene expression. Nat Chem Biol 7(6):351–358
51. Clapham DE (2003) TRP channels as cellular sensors. Nature 426(6966):517–524
52. Brauchi S et al (2004) Clues to understanding cold sensation: thermodynamics and electrophysiological analysis of the cold receptor TRPM8. Proc Natl Acad Sci U S A 101(43):15494–15499
53. Liu B, Qin F (2005) Functional control of cold- and menthol-sensitive TRPM8 ion channels by phosphatidylinositol 4,5-bisphosphate. J Neurosci 25(7):1674–1681
54. Phelps CB, Gaudet R (2007) The role of the N terminus and transmembrane domain of TRPM8 in channel localization and tetramerization. J Biol Chem 282 (50):36474–36480
55. Rohacs T et al (2005) PI(4,5)P2 regulates the activation and desensitization of TRPM8 channels through the TRP domain. Nat Neurosci 8(5):626–634
56. Viana F (2011) Chemosensory properties of the trigeminal system. ACS Chem Neurosci 2(1):38–50
57. McKemy DD et al (2002) Identification of a cold receptor reveals a general role for TRP channels in thermosensation. Nature 416(6876):52–58
58. Paricio-Montesinos R et al (2020) The sensory coding of warm perception. Neuron 106(5):830–841.e3
59. Babes A et al (2006) A novel type of cold-sensitive neuron in rat dorsal root ganglia with rapid adaptation to cooling stimuli. Eur J Neurosci 24(3):691–698
60. Fajardo O et al (2008) TRPA1 channels mediate cold temperature sensing in mammalian vagal sensory neurons: pharmacological and genetic evidence. J Neurosci 28(31):7863–7875
61. Thut PD et al (2003) Cold transduction in rat trigeminal ganglia neurons in vitro. Neuroscience 119 (4):1071–1083
62. Bautista DM et al (2007) The menthol receptor TRPM8 is the principal detector of environmental cold. Nature 448(7150):204–208
63. Colburn RW et al (2007) Attenuated cold sensitivity in TRPM8 null mice. Neuron 54(3):379–386
64. Dhaka A et al (2007) TRPM8 is required for cold sensation in mice. Neuron 54(3):371–378
65. Pogorzala LA et al (2013) The cellular code for mammalian thermosensation. J Neurosci 33 (13):5533–5541
66. Almeida MC et al (2012) Pharmacological blockade of the cold receptor TRPM8 attenuates autonomic and behavioral cold defenses and decreases deep body temperature. J Neurosci 32(6):2086–2099
67. Knowlton WM et al (2011) Pharmacological blockade of TRPM8 ion channels alters cold and cold pain responses in mice. PLoS One 6(9):e25894
68. Almaraz L et al (2014) Trpm8. Handb Exp Pharmacol 222:547–579
69. Dhaka A et al (2006) Trp ion channels and temperature sensation. Annu Rev Neurosci 29:135–161
70. Karashima Y et al (2009) TRPA1 acts as a cold sensor in vitro and in vivo. Proc Natl Acad Sci U S A 106(4):1273–1278
71. Story GM et al (2003) ANKTM1, a TRP-like channel expressed in nociceptive neurons, is activated by cold temperatures. Cell 112(6):819–829
72. Castillo K et al (2018) Thermally activated TRP channels: molecular sensors for temperature detection. Phys Biol 15(2):021001
73. Gracheva EO et al (2010) Molecular basis of infrared detection by snakes. Nature 464(7291):1006–1011
74. Saito S et al (2014) Heat and noxious chemical sensor, chicken TRPA1, as a target of bird repellents and identification of its structural determinants by multispecies functional comparison. Mol Biol Evol 31 (3):708–722
75. Saito S et al (2012) Analysis of transient receptor potential ankyrin 1 (TRPA1) in frogs and lizards illuminates both nociceptive heat and chemical sensitivities and coexpression with TRP vanilloid 1 (TRPV1) in ancestral vertebrates. J Biol Chem 287(36):30743–30754
76. Jordt SE et al (2004) Mustard oils and cannabinoids excite sensory nerve fibres through the TRP channel ANKTM1. Nature 427(6971):260–265
77. da Costa DS et al (2010) The involvement of the transient receptor potential A1 (TRPA1) in the maintenance of mechanical and cold hyperalgesia in persistent inflammation. Pain 148(3):431–437
78. Kerstein PC et al (2009) Pharmacological blockade of TRPA1 inhibits mechanical firing in nociceptors. Mol Pain 5:19
79. Kim YS et al (2010) Expression of transient receptor potential ankyrin 1 (TRPA1) in the rat trigeminal sensory afferents and spinal dorsal horn. J Comp Neurol 518(5):687–698
80. Bandell M et al (2004) Noxious cold ion channel TRPA1 is activated by pungent compounds and bradykinin. Neuron 41(6):849–857
81. Gentry C et al (2010) The roles of iPLA2, TRPM8 and TRPA1 in chemically induced cold hypersensitivity. Mol Pain 6:4
82. Kwan KY et al (2006) TRPA1 contributes to cold, mechanical, and chemical nociception but is not essential for hair-cell transduction. Neuron 50 (2):277–289
83. Andersson DA et al (2013) Methylglyoxal evokes pain by stimulating TRPA1. PLoS One 8(10):e77986
84. Chen J et al (2013) Species differences and molecular determinant of TRPA1 cold sensitivity. Nat Commun 4:2501
85. Moparthi L et al (2016) Human TRPA1 is a heat sensor displaying intrinsic U-shaped thermosensitivity. Sci Rep 6:28763
86. Zygmunt PM, Hogestatt ED (2014) Trpa1. Handb Exp Pharmacol 222:583–630

87. Zimmermann K et al (2011) Transient receptor potential cation channel, subfamily C, member 5 (TRPC5) is a cold-transducer in the peripheral nervous system. Proc Natl Acad Sci U S A 108(44):18114–18119
88. Zholos AV (2014) Trpc5. Handb Exp Pharmacol 222:129–156
89. Fowler MA et al (2007) Corticolimbic expression of TRPC4 and TRPC5 channels in the rodent brain. PLoS One 2(6):e573
90. Riccio A et al (2009) Essential role for TRPC5 in amygdala function and fear-related behavior. Cell 137(4):761–772
91. Greka A et al (2003) TRPC5 is a regulator of hippocampal neurite length and growth cone morphology. Nat Neurosci 6(8):837–845
92. Puram SV et al (2011) A TRPC5-regulated calcium signaling pathway controls dendrite patterning in the mammalian brain. Genes Dev 25(24):2659–2673
93. Henley J, Poo MM (2004) Guiding neuronal growth cones using Ca^{2+} signals. Trends Cell Biol 14 (6):320–330
94. Yao J et al (2011) Modular thermal sensors in temperature-gated transient receptor potential (TRP) channels. Proc Natl Acad Sci U S A 108 (27):11109–11114
95. Bevan S et al (2014) Trpv1. Handb Exp Pharmacol 222:207–245
96. Caterina MJ et al (1997) The capsaicin receptor: a heat-activated ion channel in the pain pathway. Nature 389(6653):816–824
97. Helliwell RJ et al (1998) Capsaicin sensitivity is associated with the expression of the vanilloid (capsaicin) receptor (VR1) mRNA in adult rat sensory ganglia. Neurosci Lett 250(3):177–180
98. Caterina MJ et al (2000) Impaired nociception and pain sensation in mice lacking the capsaicin receptor. Science 288(5464):306–313
99. Davis JB et al (2000) Vanilloid receptor-1 is essential for inflammatory thermal hyperalgesia. Nature 405 (6783):183–187
100. Gavva NR (2008) Body-temperature maintenance as the predominant function of the vanilloid receptor TRPV1. Trends Pharmacol Sci 29(11):550–557
101. Hori T (1984) Capsaicin and central control of thermoregulation. Pharmacol Ther 26(3):389–416
102. Toth DM et al (2011) Nociception, neurogenic inflammation and thermoregulation in TRPV1 knockdown transgenic mice. Cell Mol Life Sci 68 (15):2589–2601
103. Cavanaugh DJ et al (2011) Trpv1 reporter mice reveal highly restricted brain distribution and functional expression in arteriolar smooth muscle cells. J Neurosci 31(13):5067–5077
104. Kojima I, Nagasawa M (2014) Trpv2. Handb Exp Pharmacol 222:247–272
105. Koike S et al (2004) Distribution of vanilloid receptors in the rat laryngeal innervation. Acta Otolaryngol 124(4):515–519
106. Lewinter RD et al (2004) Immunoreactive TRPV-2 (VRL-1), a capsaicin receptor homolog, in the spinal cord of the rat. J Comp Neurol 470(4):400–408
107. Ma QP (2001) Vanilloid receptor homologue, VRL1, is expressed by both A- and C-fiber sensory neurons. Neuroreport 12(17):3693–3695
108. Park U et al (2011) TRP vanilloid 2 knock-out mice are susceptible to perinatal lethality but display normal thermal and mechanical nociception. J Neurosci 31(32):11425–11436
109. Xu H et al (2002) TRPV3 is a calcium-permeable temperature-sensitive cation channel. Nature 418 (6894):181–186
110. Yang P, Zhu MX (2014) Trpv3. Handb Exp Pharmacol 222:273–291
111. Peier AM et al (2002) A heat-sensitive TRP channel expressed in keratinocytes. Science 296 (5575):2046–2049
112. Huang SM et al (2008) Overexpressed transient receptor potential vanilloid 3 ion channels in skin keratinocytes modulate pain sensitivity via prostaglandin E2. J Neurosci 28(51):13727–13737
113. Mandadi S et al (2009) TRPV3 in keratinocytes transmits temperature information to sensory neurons via ATP. Pflugers Arch 458(6):1093–1102
114. Moqrich A et al (2005) Impaired thermosensation in mice lacking TRPV3, a heat and camphor sensor in the skin. Science 307(5714):1468–1472
115. Huang SM et al (2011) TRPV3 and TRPV4 ion channels are not major contributors to mouse heat sensation. Mol Pain 7:37
116. Guler AD et al (2002) Heat-evoked activation of the ion channel, TRPV4. J Neurosci 22(15):6408–6414
117. Watanabe H et al (2002) Heat-evoked activation of TRPV4 channels in a HEK293 cell expression system and in native mouse aorta endothelial cells. J Biol Chem 277(49):47044–47051
118. Nilius B et al (2001) Differential activation of the volume-sensitive cation channel TRP12 (OTRPC4) and volume-regulated anion currents in HEK-293 cells. Pflugers Arch 443(2):227–233
119. Smith PL et al (2006) Bisandrographolide from Andrographis paniculata activates TRPV4 channels. J Biol Chem 281(40):29897–29904
120. Shibasaki K (2016) TRPV4 ion channel as important cell sensors. J Anesth 30(6):1014–1019
121. Liedtke W, Friedman JM (2003) Abnormal osmotic regulation in trpv4−/− mice. Proc Natl Acad Sci U S A 100(23):13698–13703
122. Vizin RC et al (2015) TRPV4 activates autonomic and behavioural warmth-defence responses in Wistar rats. Acta Physiol (Oxf) 214(2):275–289
123. Sokabe T et al (2010) The TRPV4 channel contributes to intercellular junction formation in keratinocytes. J Biol Chem 285(24):18749–18758
124. Kida N et al (2012) Importance of transient receptor potential vanilloid 4 (TRPV4) in epidermal barrier function in human skin keratinocytes. Pflugers Arch 463(5):715–725

125. Shibasaki K et al (2015) TRPV4 activation at the physiological temperature is a critical determinant of neuronal excitability and behavior. Pflugers Arch 467(12):2495–2507
126. Shibasaki K et al (2007) Effects of body temperature on neural activity in the hippocampus: regulation of resting membrane potentials by transient receptor potential vanilloid 4. J Neurosci 27(7):1566–1575
127. Shibasaki K et al (2015) Hippocampal neuronal maturation triggers post-synaptic clustering of brain temperature-sensor TRPV4. Biochem Biophys Res Commun 458(1):168–173
128. Togashi K et al (2006) TRPM2 activation by cyclic ADP-ribose at body temperature is involved in insulin secretion. EMBO J 25(9):1804–1815
129. Hara Y et al (2002) LTRPC2 Ca^{2+}-permeable channel activated by changes in redox status confers susceptibility to cell death. Mol Cell 9(1):163–173
130. Kashio M et al (2012) Redox signal-mediated sensitization of transient receptor potential melastatin 2 (TRPM2) to temperature affects macrophage functions. Proc Natl Acad Sci U S A 109 (17):6745–6750
131. Chung KK et al (2011) Expression and functional properties of TRPM2 channels in dopaminergic neurons of the substantia nigra of the rat. J Neurophysiol 106(6):2865–2875
132. Fonfria E et al (2006) Tissue distribution profiles of the human TRPM cation channel family. J Recept Signal Transduct Res 26(3):159–178
133. Tan CH, McNaughton PA (2016) The TRPM2 ion channel is required for sensitivity to warmth. Nature 536(7617):460–463
134. Song K et al (2016) The TRPM2 channel is a hypothalamic heat sensor that limits fever and can drive hypothermia. Science 353(6306):1393–1398
135. Sano Y et al (2001) Immunocyte Ca^{2+} influx system mediated by LTRPC2. Science 293 (5533):1327–1330
136. Wehage E et al (2002) Activation of the cation channel long transient receptor potential channel 2 (LTRPC2) by hydrogen peroxide. A splice variant reveals a mode of activation independent of ADP-ribose. J Biol Chem 277(26):23150–23156
137. Naylor J et al (2010) Pregnenolone sulphate- and cholesterol-regulated TRPM3 channels coupled to vascular smooth muscle secretion and contraction. Circ Res 106(9):1507–1515
138. Wagner TF et al (2008) Transient receptor potential M3 channels are ionotropic steroid receptors in pancreatic beta cells. Nat Cell Biol 10(12):1421–1430
139. Oberwinkler J, Philipp SE (2014) Trpm3. Handb Exp Pharmacol 222:427–459
140. Vriens J et al (2011) TRPM3 is a nociceptor channel involved in the detection of noxious heat. Neuron 70 (3):482–494
141. Vandewauw I et al (2018) A TRP channel trio mediates acute noxious heat sensing. Nature 555 (7698):662
142. Straub I et al (2013) Flavanones that selectively inhibit TRPM3 attenuate thermal nociception in vivo. Mol Pharmacol 84(5):736–750
143. Maingret F et al (2000) TREK-1 is a heat-activated background K^+ channel. EMBO J 19(11):2483–2491
144. Patel AJ et al (1998) A mammalian two pore domain mechano-gated S-like K^+ channel. EMBO J 17 (15):4283–4290
145. Schneider ER et al (2014) Temperature sensitivity of two-pore (K2P) potassium channels. Curr Top Membr 74:113–133
146. Kang D, Kim D (2006) TREK-2 (K2P10.1) and TRESK (K2P18.1) are major background K^+ channels in dorsal root ganglion neurons. Am J Physiol Cell Physiol 291(1):C138–C146
147. La JH et al (2011) Differences in the expression of transient receptor potential channel V1, transient receptor potential channel A1 and mechanosensitive two pore-domain K^+ channels between the lumbar splanchnic and pelvic nerve innervations of mouse urinary bladder and colon. Neuroscience 186:179–187
148. Yamamoto Y et al (2009) Immunohistochemical colocalization of TREK-1, TREK-2 and TRAAK with TRP channels in the trigeminal ganglion cells. Neurosci Lett 454(2):129–133
149. Medhurst AD et al (2001) Distribution analysis of human two pore domain potassium channels in tissues of the central nervous system and periphery. Brain Res Mol Brain Res 86(1–2):101–114
150. Talley EM et al (2003) Two-pore-domain (KCNK) potassium channels: dynamic roles in neuronal function. Neuroscientist 9(1):46–56
151. Talley EM et al (2001) Cns distribution of members of the two-pore-domain (KCNK) potassium channel family. J Neurosci 21(19):7491–7505
152. Viatchenko-Karpinski V et al (2018) Characterization of temperature-sensitive leak K(+) currents and expression of TRAAK, TREK-1, and TREK2 channels in dorsal root ganglion neurons of rats. Mol Brain 11(1):40
153. Wechselberger M et al (2006) Ionic channels and conductance-based models for hypothalamic neuronal thermosensitivity. Am J Physiol Regul Integr Comp Physiol 291(3):R518–R529
154. Alloui A et al (2006) TREK-1, a K^+ channel involved in polymodal pain perception. EMBO J 25 (11):2368–2376
155. Noel J et al (2009) The mechano-activated K^+ channels TRAAK and TREK-1 control both warm and cold perception. EMBO J 28(9):1308–1318
156. Pereira V et al (2014) Role of the TREK2 potassium channel in cold and warm thermosensation and in pain perception. Pain 155(12):2534–2544
157. Yang YD et al (2008) TMEM16A confers receptor-activated calcium-dependent chloride conductance. Nature 455(7217):1210–1215
158. Deschenes M et al (1976) A model for an estimate in vivo of the ionic basis of presynaptic inhibition: an

intracellular analysis of the GABA-induced depolarization in rat dorsal root ganglia. Brain Res 118 (3):486–493
159. Price TJ et al (2009) Chloride regulation in the pain pathway. Brain Res Rev 60(1):149–170
160. Kaneko H et al (2002) Determination of intracellular chloride concentration in dorsal root ganglion neurons by fluorescence lifetime imaging. Curr Topics Membr 53:167
161. Rocha-Gonzalez HI et al (2008) Na^{+},K^{+},$2Cl^{-}$ cotransport and intracellular chloride regulation in rat primary sensory neurons: thermodynamic and kinetic aspects. J Neurophysiol 100(1):169–184
162. Roos J et al (2005) STIM1, an essential and conserved component of store-operated Ca^{2+} channel function. J Cell Biol 169(3):435–445
163. Zhang SL et al (2005) STIM1 is a Ca^{2+} sensor that activates CRAC channels and migrates from the Ca^{2+} store to the plasma membrane. Nature 437 (7060):902–905
164. Prakriya M et al (2006) Orai1 is an essential pore subunit of the CRAC channel. Nature 443 (7108):230–233
165. Vig M et al (2006) CRACM1 is a plasma membrane protein essential for store-operated Ca^{2+} entry. Science 312(5777):1220–1223
166. Luik RM et al (2008) Oligomerization of STIM1 couples ER calcium depletion to CRAC channel activation. Nature 454(7203):538–U11
167. Stathopulos PB et al (2008) Structural and mechanistic insights into STIM1-mediated initiation of store-operated calcium entry. Cell 135(1):110–122
168. Park CY et al (2009) STIM1 clusters and activates CRAC channels via direct binding of a cytosolic domain to Orai1. Cell 136(5):876–890
169. Yuan JP et al (2009) SOAR and the polybasic STIM1 domains gate and regulate Orai channels. Nat Cell Biol 11(3):337–343
170. Berna-Erro A et al (2012) Orais and STIMs: physiological mechanisms and disease. J Cell Mol Med 16 (3):407–424
171. Feske S et al (2006) A mutation in Orai1 causes immune deficiency by abrogating CRAC channel function. Nature 441(7090):179–185
172. Le Deist F et al (1995) A primary T-cell immunodeficiency associated with defective transmembrane calcium influx. Blood 85(4):1053–1062
173. Partiseti M et al (1994) The calcium current activated by T cell receptor and store depletion in human lymphocytes is absent in a primary immunodeficiency. J Biol Chem 269(51):32327–32335
174. Picard C et al (2009) STIM1 mutation associated with a syndrome of immunodeficiency and autoimmunity. N Engl J Med 360(19):1971–1980
175. Feske S (2009) ORAI1 and STIM1 deficiency in human and mice: roles of store-operated Ca^{2+} entry in the immune system and beyond. Immunol Rev 231 (1):189–209
176. Feske S et al (2010) Immunodeficiency due to mutations in ORAI1 and STIM1. Clin Immunol 135 (2):169–182
177. Hanson DF (1997) Fever, temperature, and the immune response. Thermoregulation 813:453–464
178. Usoskin D et al (2015) Unbiased classification of sensory neuron types by large-scale single-cell RNA sequencing. Nat Neurosci 18(1):145–153
179. Yang S et al (2020) A paradigm of thermal adaptation in penguins and elephants by tuning cold activation in TRPM8. Proc Natl Acad Sci U S A 117 (15):8633–8638

Mechanotransduction Ion Channels in Hearing and Touch

17

Songling Li and Zhiqiang Yan

Abstract

The ability of living organisms to detect mechanical force originates from mechanotransduction ion channels, which convert membrane tension into electrical or chemical signals that are transmitted to the brain. A variety of studies on touch and sound perception in both vertebrates and invertebrates have broadened our understanding of mechanotransduction and identified promising candidates for mechanotransduction ion channels. Here, we discussed the physiological properties of mechanotransduction ion channels in hearing and touch, the identification of their molecular entities, and recent structural studies providing insights to their gating mechanisms in force sensing. We present an updated review of the evidence supporting several candidates, including NOMPC, Brv1, and TMC channels, as mechanotransduction ion channels and highlight their qualifications satisfying the specific criteria proposed for a mechanotransducer.

S. Li · Z. Yan (✉)
Institute of Molecular Physiology, Shenzhen Bay Laboratory, Shenzhen, China

State Key Laboratory of Medical Neurobiology and MOE Frontiers Center for Brain Science, School of Life Sciences, Fudan University, Shanghai, China
e-mail: 15110700001@fudan.edu.cn; zqyan@szbl.ac.cn

Keywords

Mechanotransduction · Mechanogating · Sensory transduction · NOMPC · TMC · Force sensation

17.1 Introduction

Hearing and touch, together with vision, olfaction, and taste constitute the five primal senses defined by Aristotle. Each of these senses adopts morphologically and functionally distinct sensory cell types in which unique integral membrane proteins are tuned to transduce external stimuli into electrical or biochemical signals that activate our sensory system. G protein-coupled receptors (GPCR) called rhodopsin, which is activated by photons of light, is abundant in rod cells in the vision system. Olfactory epithelial cells also contain multiple GPCRs that are responsive to volatile odorant molecules and enable us to detect smells. Similarly, GPCRs expressed in the taste buds sense chemicals in the food [1]. Unlike these three sensory modalities, the molecular mechanism of hearing and touch remains less understood for several reasons. It has been proposed long ago that sensing touch and hearing sound rely on specific ion channels that are gated by force [2]. The problem, however, is that ion channels differ greatly in their sequence and function, which is quite different from the structural similarities of GPCRs. While rhodopsin is

L. Zhou (ed.), *Ion Channels in Biophysics and Physiology*, Advances in Experimental Medicine and Biology 1349, https://doi.org/10.1007/978-981-16-4254-8_17

abundant in photoreceptor cells, the ion channels show a relatively low expression level in the sensory system. Moreover, ion channels generally form a complex with other auxiliary subunits that may help in their assembly or trafficking [3]. These challenges have hampered the elucidation of the molecules necessary for force transduction over decades and thus impede our understanding of the molecular mechanism underlying hearing and touch.

Mechanotransduction refers to a process in which molecular receptors converts external forces into electrical signals, including the detection of vibration, indentation, gravity, and sound waves. It constitutes a key step in the sense of hearing, touch, balance, and proprioception. Moreover, it also partakes in the internal sensation of blood pressure and osmolality [4]. Although these sensory modalities appear quite different, they each encode the form and magnitude of mechanical force. Mechanotransduction occurs through the opening of an ion channel pore in response to a mechanical stimulus. Activation of mechanosensitive ion channel at the plasma membrane is directly driven by force, the time course in the microsecond range excludes the involvement of a second messenger [3, 5, 6]. In other words, the transformation of mechanical stress into an intracellular electrical signal is fulfilled by the opening of mechanotransduction ion channels. Thus, mechanotransduction channels respond to a variety of stimuli in both internal and external environments, but molecular entities that sense mechanical forces have been difficult to identify. In both vertebrates and invertebrates, the identification of most mechanosensitive channels was finally fulfilled by reverse genetic screening but not conventional biochemical methods or homology approaches [1].

Four criteria have been previously proposed that must be satisfied for a candidate to qualify as a bona fide mechanotransduction channel [3, 6]. First, the channel encoding gene must show an appropriate expression level in mechanoreceptor cells. Disrupting the gene function should eliminate mechanosensitive response of the cell. Moreover, ectopic expression of this gene should suffice to endow a naive cell with mechanosensitivity. If auxiliary subunits are not associated with mechanogating, mechanosensitive currents should be recorded when incorporating purified ion channel proteins into artificial liposome bilayers. Only a few ion channels met all of the prerequisite qualifications for a mechanotransducer, including PIEZO and NOMPC [4]. PIEZO2 are the touch sensor in mamalian, the related work has been awarded Nobel Prize this year recently, there are numerous outstanding reviews about Piezo so we did not cover this part here.

In this chapter, on the basis of recent molecular, functional, and structural studies, we summarize the significant progress in identifying mechanotransduction ion channels, solving their structures as well as characterizing the gating mechanisms of different mechanotransduction channels in hearing and touch sensation. Then, we discuss main difficulties to define the properties and physiological functions of mechanotransduction ion channels, review emerging cell biological principles underlying force sensing, and highlight questions pending further investigations in the field.

17.2 Mechanosensitive Ion Channels in Touch Sensation

17.2.1 NOMPC in Gentle Touch Sensation

The sense of touch is the physical measurement of forces that contact the body's surface. Detection of surface forces is essential to the development and survival of multicellular organisms [7]. In metazoans, cells are frequently exposed to multiple mechanical stresses such as gravity, pressure, flow, and stretch. After 30 years of investigation, a plenty of molecules required for touch sensation have been uncovered via genetic screens in *C. elegans* and *Drosophila melanogaster*. Recent studies have shed light on the molecular identity of mechanotransduction ion channels and their gating mechanisms implicated in the control of force sensitivity.

Research in non-mammalian model organisms has well documented the function of transient receptor potential (TRP) channels, especially TRPN (NOMPC) orthologs, in hearing and touch [1, 4, 8, 9]. TRPN (NOMPC) is the first TRP channel related to mechanosensation, which was initially found in a genetic screening for fly larvae with deficiency in touch response and mechanoreceptor potentials [10, 11].

TRP-4 is the TRPN ortholog in *C. elegans*, which is highly enriched in the cilium of the cephalic neurons (CEP). These neurons are required for behavioral responses when exposed to light touch stimuli [12]. Loss of *trp-4* gene in CEP neurons eliminates mechanotransduction currents and results in defective animal responses. The phenotype in touch-induced behavioral response and mechanosensitive currents could be rescued by overexpressing *trp-4*, demonstrating that TRP-4 is essential for touch sensation. Additionally, point mutations introduced in the putative pore of TRP-4 shift the reverse potential of the mechanotransduction current [12]. Based on these findings, TRPN orthologs in worms are established as a promising candidate for mechanotransduction channel, whereas technical problems have hindered the identification of worm TRPN ortholog as the inherent mechanogated channel.

Although mammals lack TRPN orthologs, the *Drosophila* TRPN ortholog, NOMPC, acts as a mechanosensor for multiple physiological functions, including proprioception, gentle touch, hearing, visceral sensation, and food texture detection, thus it is a valuable model for exploring mechanotransduction [13–20, 139].

The body wall of *Drosophila* larvae was covered by two distinct group of sensory neurons termed as class III and class IV dendritic arborization (da) neurons tile [21]. Physiological and behavioral experiments show that class III da neurons are responsible for gentle touch sensation, whereas class IV da neurons are responsible for sensing mechanical pain and noxious heat [22–25, 139]. NOMPC displays a high expression level in the cell body and throughout dendrites of class III da neurons. Larvae with *nompC* null mutations are severely defective in the gentle touch responsiveness. Accordingly, when challenged with micrometer displacements in the body wall covered by dendrites of class III da neurons, they fail to respond with action potential firings. In addition, touch-insensitive neurons show light touch sensitivity after ectopically expressing NOMPC [26, 139]. To approach the question of whether NOMPC suffices to produce mechanosensitive channels, the decisive step was made by heterologously expressing NOMPC proteins in cultured *Drosophila* S2 cells. Fortunately, mechanosensitive nonselective cation channels were produced in *Drosophila* S2 cells expressing NOMPC [139]. Mechanical stimuli activate the channel within 2 ms, suggesting that the force directly gates the channel since second messenger signaling was unlikely to happen during this short latency [3, 5, 6]. Moreover, the amplitude of the mechanically activated currents and ion selectivity are altered by mutations in the predicted pore region [139]. These results finally confirm *Drosophila* NOMPC as the pore-forming subunit for a bona fide mechanotransduction channel responsible for touch sensation. Besides, NOMPC also plays a pivotal role in auditory transduction in *Drosophila* chordotonal organ, although there are discrepancies about whether NOMPC or two TRPV proteins, Nanchung and Inactive, form the channel that carries the receptor currents [18, 27, 28].

17.2.2 Mechanogating Mechanism of NOMPC

Mechanosensitive channels have rapid activation kinetics, which seems too short for second messenger cascades [3, 5, 6]. In terms of this point, mechanical forces on the plasma membrane should directly activate these channels. Two nonexclusive models by which membrane forces gates the channels have been proposed [29–32]. In the force-from-lipid model, the force acting on the plasma membrane changes the shape of surface area, leading to opening of the transmembrane channel. Alternatively, the tether model assumes that gating springs transmit force from the cytoskeleton or extracellular matrix to the channel. The mechanogating mechanism of MscL channel in bacteria [33] and two-pore potassium channels in eukaryotes [34–37] coincide with the force-from-lipid model. Although there are several examples of channels that are directly gated by membrane force, direct molecular evidence supporting the tether model has long been pursued.

One of the characteristics of NOMPC is that it includes 29 ankyrin repeats (ARs) in N-terminal region, which are essential for its function [14, 38–40]. Long AR chains are usually assumed to be a coil-like structure. They have been proposed to constitute gating spring component of a tethered mechanosensor [41–44]. While specific antibodies against either the N- or C- terminus of NOMPC-stained cells after permeabilization, an antibody specific to the pore loop region was able to label nonpermeabilized cells [45]. The immunostaining results suggest the trafficking of NOMPC to the plasma membrane, and importantly confirm the ARs at the intracellular side, which means ARs meet a prerequisite of binding the cytoskeleton. Truncated NompC channels are designed by altering numbers or arrangements of ARs, then their functional significance is determined by channel recording. When partially or completely deleting the ARs of NOMPC, the channel could act as mechanosensor; however, the channel functions as a force sensor when the AR region is intact or duplicated [45]. Accordingly, only those channels with the intact AR region restored touch-induced firing of class III da neurons and rescued gentle touch behavioral responses in the *nompC* null mutant background.

In the *Drosophila* halters, membrane-microtubule connectors (MMCs) are highly ordered filamentous structures at the distal tips of mechanoreceptors. These organized structures were considered to represent the tether structure connecting NOMPC to the microtubules [46]. MMCs did not exist in null mutants of *nompC*, but was restored by wild-type transcripts, and elongated during AR domain replication. Immunoprecipitation assays verified that NOMPC binds microtubules. Using pharmacological drugs to break down microtubules also abolished mechanosensitivity of NOMPC [45]. Thus, ARs of NOMPC tethered the channel to the microtubule cytoskeleton and the mechanogating of the channel relies on both the ARs and microtubules. Hence, the molecular entity for the tether model in mechanotransduction eventually falls on NOMPC, of which the gating spring can even been visible as a MMC structure by transmission electron microscopy methods.

Strikingly, the AR region of NOMPC was further shown to confer mechanosensitivity on an evolutionarily related ion channel. This step begins with the construct of a chimera in which the N-terminal region of NOMPC containing the 29 ARs was fused to the structurally similar but functionally distinct voltage-gated potassium channels [47–49]. While expression of this chimeric channel yields mechanogated channels in cultured *Drosophila* S2 cells, the wild-type potassium channels themselves confers no mechanical response [45]. Importantly, the integrity of microtubules was also essential for the mechanical response in those cells, consistent with the notion that ARs that are tethered to the microtubule cytoskeleton endow NOMPC with the responsiveness to mechanical stimuli.

As the cryo-EM structure of NOMPC was recently solved, new insights about the mechanogating mechanism of this channel have emerged [50]. The exciting new finding is that a bundle of closely arranged helices, which are composed of ARs in each subunit of the NOMPC homotetramer, was observed in the structural model. This suggests that the gating spring may literally be a coiled spring that conveys the forces to open the channel. The linker helices and the TRP domain connected each AR helix to the transmembrane helices and the channel pore region. In this way, the AR helices can transmit force to the channel pore. Another important feature is that each two neighboring helices in the AR bundle make several contacts with each other. It is possible that the helices would stretch and compress as a unit upon mechanical stimuli, especially when the contacts are proved to be rigid. Therefore, all the helices in the tetramer would function as a whole unit and produce a coordinated conformational change to switch the pore from its closed state to open state in response to mechanical force [50]. The N-terminal of the AR helix bundle has an appropriate spacing distance which well fit the wall of the 25 nm diameter microtubule. It is likely that a microtubule-binding domain is located in the N-terminal region preceding the first AR. This sequence is not evolutionarily conserved among different species and is predicted to be disordered [51]. Based on above cryo-EM structure, a recent study carried out electrophysiological recording together with molecular dynamics simulations to study NompC gating in atomistic details. The study showed that NompC could be opened by pushing the intracellular ankyrin repeats but not by stretching. Under intracellular compression, the ankyrin repeats from four subunits bundle together to act like a spring, with a spring and a number of hydrogen bonds along the force conveying pathway to enable the mechanosensitivity. The linker helix region works as a bridge between the transient receptor potential domain and the ankyrin repeats, which passes on the compression force to the TRP motif to undergo a clockwise rotation, resulting in the pore opening. This study proposes a universal gating mechanism of similar tethered mechanosensitive TRP channels, which can enable cells to feel compression physiological forces [141].

17.2.3 *Drosophila* Brv1 in Light Touch Sensation

Brv1 belongs to the TRPP (TRP polycystin) subfamily of TRP ion channels [52]. It has long been known that Brv1 is required for cold sensation in the fly antenna, which is put into question by recent studies on *Drosophila* thermosensors [53–55]. Neither electrophysiological recording nor calcium imaging from antennal cold cells reveals an impact of *brv1* null mutation on the cooling response of these cells. However, the avoidance behavioral response to cold temperature is indeed largely reduced in *brv1* mutants, suggesting Brv1 might contribute to thermosensation through other mechanism [55]. After tissue damage, *Drosophila* larvae experience a sensitization in the behavioral responses to cold stimulus. This cold nociceptive sensitization is shown to involve *brv1* gene and mechanosensory neurons on the body wall, implying that Brv1 might regulate cold sensation through mechanosensory pathway [56].

A wide range of sensory functions requires TRP ion channels, an evolutionarily conserved family of nonselective cation channels [40, 57]. *Drosophila* Brv1 shares protein homology with the mammalian PKD2 channels. PKD2 functions as a co-subunit of the channel complex that mediates shear stress responses in kidney primary cilia and forms a Ca^{2+}-permeable cation channel when expressed in heterologous system [58, 59]. These findings suggest a role of PKD2 in mechanosensation; however, PKD2 has not been shown to reconstitute mechanically activated currents in cultured cells [1, 60]. Thus, further investigations of *Drosophila* Brv1 focused on whether Brv1 might participate in mechanosensation and render intrinsic mechanogated currents.

Recently, a study reports that Brv1 produces an inherent mechanically gated channel and plays an important role in fly's tactile sensation [61, 62]. Behavioral analysis indicates that the *brv1* loss-of-function larvae are defective in behavioral response to gentle touch. Immunostaining identifies high level of Brv1 in Class III da neurons, which was previously found to mediate touch sensation via the bona fide mechanotransduction channel NOMPC [139]. Specific knockdown of *brv1* expression in Class III da neurons leads to a decrease in gentle touch responses. Additionally, expressing *brv1* transcript in Class III da neurons with mutant background fully rescued the touch sensitivity and the Ca^{2+} response to mechanical displacements [61, 62]. These observations demonstrate that Brv1 is required for Class III da neurons to mediate touch sensation, likely working in concert with NOMPC.

When expressed in HEK293T cells, Brv1 yields a nonselective cation channel with outwardly rectifying feature [61, 62]. Brv1 channels show dose-dependent current response to membrane stretch in cultured cells. Similar sensitivity to negative pressure was observed in liposomes with reconstituted Brv1 proteins, suggesting Brv1 is an intrinsic mechanosensitive ion channel. Negative pressure in both inside-out and outside-out patch configurations elicits current responses, which implies that mechanogating of the Brv1 channel is insensitive to the direction of membrane deformation. In addition, the mechanogated currents of Brv1 channels do not sensitize to repetitive stimulations [61, 62]. While the identified touch sensors, such as NOMPC and PIEZO, respond to diverse type of mechanical forces, including stretching and poking, Brv1 is insensitive to poking [61–63, 139]. In contrast, cells co-transfected with NOMPC and Brv1 exhibited an enhanced current response to mechanical indentation compared with that of cells transfected with NOMPC alone [61, 62].

The poking-induced currents of NOMPC and Brv1 co-transfected cells present larger amplitude and an increased adaptation time constant [61, 62]. Intriguingly, the prolonged adaptation time constant, which is about 100 ms, closely matches the properties of touch-induced spikes recorded in Class III da neurons [139]. In terms of the adaptation kinetics, there has been a discrepancy between the in vivo and in vitro recordings of touch-induced current responses of the touch sensor NOMPC. In *Drosophila* S2 cells, NOMPC displays an adapting constant ten times smaller than that analyzed from touch-sensitive neurons [139]. The new results from HEK cells co-transfected with NOMPC and Brv1 now perfectly fill this gap, implying that Brv1 might function in the same manner in vivo, modulating the adaptation kinetics of the touch-induced currents. In vascular smooth muscle cells, PKD2 has been shown to exploit a similar mechanism to regulate pressure sensing, by inhibiting the mechanically activated ion channels [64].

A tether gating mechanism of NOMPC has been elucidated, in which the N-terminal ARs act as a gating spring between the channel and the cytoskeletons to transmit force [45, 50]. Purified Brv1 proteins exhibited mechanosensitivity in artificial lipid bilayers, which suggests that the Brv1 channel could be activated directly by mechanical force from lateral membrane tension [61, 62]. When functioning together, Brv1 likely serves as an amplifier to modulate NOMPC channel currents, thus sensitizing class III da neurons to gentle touch stimuli. Consequently, Brv1 plays dual roles in tactile sensation: the primary role as a mechanosensor as well as the auxiliary role as an amplifier for the native touch sensor NOMPC.

17.3 Mammalian TMCs in Hearing

17.3.1 Molecular Components of MET Channels in Hair Cells

Since the discovery of the mechanotransduction (MET) current in hair cells in the late 1970s [2], a central question has been identifying the molecules that make up the conductance of MET channel. Several promising candidates have emerged, including orthologs of the epithelial Na^+ channel and multiple TRP channels, but later prove to be unnecessary for auditory mechanotransduction [65–67]. Most recently, the transmembrane channel-like (TMC) proteins, which belong to a protein family of previously unknown function, were characterized based on the fact that mutations mapped to Tmc genes lead to deafness and lack of the MET current [68–71].

It has been a major challenge over decades to identify the molecular entity of the MET channel in hair cells, due to the small number of hair cells per cochlea [65, 72]. Besides, the expected number of MET channels in each hair cell (~100) has hampered classical methods that were successfully applied in other sensory systems, for instance, the approaches to identify components of the phototransduction complex [66]. Studies on mutations related to various types of hereditary deafness have greatly facilitated isolation of the hair-cell transduction complex. More than 35 human hereditary deafness-related mutations are located in Tmc1 locus, making TMC1 one of the most noteworthy deafness proteins [73].

TMC1 is a member of the family comprising eight transmembrane protein encoding genes. Three subfamilies are included in this family:

TMC1, TMC2, and TMC3; TMC5 and TMC6; and TMC4, TMC7, and TMC8 [74, 75]. In adult cochlear hair cells, TMC1 shows a highly enriched expression pattern, while TMC2 only shows transient presence in the cochlea during early neonatal development and also in vestibular hair cells as previously reported [68, 76, 77].

In addition to TMC1 and TMC2, some other molecules, such as LHFPL5 and TMIE, have also been reported as deafness proteins that are essential for mechanotransduction [78–80]. Expression analysis showed their localization at the tips of the stereocilia which is considered to be the site for the transduction complex. Moreover, both LHFPL5 and TMIE interact with a component of the tip link, the N-terminus of PCDH15 [80, 81], as the same way TMC does [82]. LHFPL5 and TMIE are composed of 219 and 153 amino acid residues, respectively. This appears too short to assemble multimeric channels, in comparison to ion channel proteins such as voltage-gated K^+ channels bearing more than 500 amino acid residues [66]. Thus, it is unlikely that they form the central pore region of an ion channel.

LHFPL5 (also called TMHS) has similar structural properties to the TARP proteins which contain four transmembrane domains but serve distinct function. TARP proteins are essential for the proper targeting of AMPA receptors to the postsynaptic membrane as well as channel conductivity [83]. Similarly, antibody labeling revealed that *Lhfpl5* null mutants abolished TMC1 expression in hair bundles, in contrast to the results in wild-type mice [81]. Thus, proper localization of TMC1 in hair cells requires LHFPL5 and the deafness phenotypes of *Lhfpl5* mutants can be attributed to the loss of TMC1 function in hair bundles.

TMIE is a transmembrane protein involved in deafness in mammals and is predicted to harbor only two transmembrane helices [84, 85]. Yeast two hybrid screens have revealed that TMIE binds to PCDH15 and TMHS, which are directly required for the integrity of the transduction complex [80]. Furthermore, TMIE is localized to the tips of the stereocilia near the MET channel [80]. In cochlear hair cells which lack TMIE, tip links are normal and TMC1/2 showed proper targeting in stereocilia, while no MET currents can be recorded [80]. A recent study reported that TMIE regulates hair cell activity by helping to target and stabilize TMC proteins at the MET channel site in zebrafish [86]. These results raise the possibility that TMCs form the pore of the MET channel.

Two other proteins CIB2 and TOMT are also required for mechanotransduction but will not be discussed here because they unlikely encode multiple transmembrane proteins and therefore are not regarded to be candidate pore-forming subunits of the MET channel [87–90].

17.3.2 Evidence Supporting TMC as Pore-Forming Subunit of the MET Channel

In the last decade, various studies provide evidence supporting that TMC1 and TMC2 are integral parts of the MET channel for hearing. There are several reasons for which TMC1/2 are proposed to be pore-forming subunits of the MET channel. Both in humans and mice, *Tmc1* gene mutations are related to phenotypes of dominant or recessive hearing loss [69, 71]. Moreover, studies on gene-targeted mice have shown that MET in early postnatal cochlear hair cells relies on TMC1 and TMC2. In *Tmc1/2* deficient hair cells, MET current was not detected, which can be rescue by the expression of TMC1 and TMC2 [68, 70]. Antibody immunostaining showed that TMC1/2 protein was located at the lower stereocilia tip of inner hair cells and outer hair cells (IHC and OHC), where auditory transduction occurs [91]. Furthermore, yeast two-hybrid screening and co-immunoprecipitation approaches identified that TMC1/2 combines with a component of the tip link which convey force to the MET channel [81, 82, 92, 93]. Besides, properties of MET channel are altered by TMC1 and TMC2 mutations. Ca^{2+} selectivity in developing IHCs and OHCs lacking TMC2 differs from that of normal hair cells, whereas the MET channel conductance of Tmc2 knockouts remains unchanged [70, 94–

96]. Conversely, *Tmc1* knockouts lead to a change in MET channel conductance and adaptation time constant but not in Ca^{2+} permeation [70, 96]. Additionally, in TMC1-deficient OHCs, the tonotopic gradient of single-channel conductance is decreased. Finally, in IHCs, a point mutation M412K in *Tmc1* affect single-channel conductance and Ca^{2+} permeability of the MET channel [70]. These investigations demonstrate that *Tmc1* or *Tmc2* defines the pore properties of the MET channel.

In addition to humans and mice studies, TMC homologs have also been found in sensory neurons of invertebrates [97–101]. These investigations report that various type of sensory responses evoked by physical stimuli requires TMC proteins. However, it is still under debate whether TMC1 and TMC2 are pore-forming subunits of a channel. A key bottleneck of solving this problem has been the fact that little evidence supports that pore-lining residues of TMCs are located within the MET channel pore. Moreover, there is no direct evidence that TMCs have a structure qualified for forming an ion channel.

Using multiple methods, including chemical crosslinking, nonreducing gel, size exclusion chromatography and low-resolution cryo-EM, a recent piece of work shows that TMC1 assembles as a dimer [102]. It is predicted that TMCs likely fold into a topology comprising 10 transmembrane domains [103], which is similar to the structure of the TMEM16 ion channels or phospholipid scramblases [104–106]. The putative transmembrane residues of TMC1 are fitted in a previously resolved TMEM16A cryo-EM structure, yielding a predicted structure model of TMC1 dimers by the use of I-TASSER server [102].

Strikingly, a central pore is not presented in this TMC1 structure, which is in contrast to classical multimeric channels such as PIEZO or TRPs [50, 107–111]. Each TMC1 subunit possesses an independent ion conduction pathway, surrounded by S4, S5, S6, and S7 transmembrane segments [102]. Taking advantage of molecular dynamics simulations, Pan et al. were allowed to embed the predicted structure of TMC1 into the lipid membrane and assess the ion permeation ability through the putative pore. The groove surrounded by S4 to S7 helices showed distribution of K^+ and water molecules, raising the possibility that this region may form the channel pore for ion permeation [102].

In light of the structural model, Pan et al. mutated several key residues of TMC1 to cysteine to map the pore of TMC1, which has been successfully applied in the functional analysis of adaptation motor in previous studies [112]. Using this strategy, they selected TMC1 residues that are predicted to be located near the pore and substituted these residues to cysteine. Using AAV virus delivery of the *Tmc1* constructs carrying point mutation, they expressed cysteine mutant TMC1 into *Tmc1* and *Tmc2* double knockout hair cells. Most of the TMC1 point mutations rescued the deficiency in the knockout background [102]. Then, cysteine modification reagents MTS, which forms covalent linkage with cysteine, were added to examine the effects on MET currents. When MTS is applied to wild-type control hair cells, no apparent alteration of the MET currents was observed. In contrast, when MTS is applied to some cysteine mutants from the extracellular side, MET currents can be effectively blocked [102], suggesting these mutated residues likely contribute to the ionic pore.

Of all the 17 substitutions tested, 16 of them yielded viable transduction currents [102]. Other effects were also observed when exposed to cysteine modification reagents. Some mutations decreased Ca^{2+} selectivity, while others reduced the MET current or single-channel conductance. The corresponding mutated sites of TMC1 are very likely to be located in the pore of the MET channel, since the effects arising from point mutations all reflect properties of ion channel pore. Furthermore, when deflecting hair cell bundles in negative direction or applying channel blockers, cysteine mutations no longer affect the MET channel function, thus it is unlikely that these mutated sites indirectly contribute to the channel conductivity via conformational changes in other region outside the pore of the channel complex [113]. Moreover, they performed noise analysis on the MET currents data to estimate the amplitude of single-channel currents and numbers of channels recorded. The analysis indicates that cysteine inhibition reduces current amplitude in each channel without altering channel number in hair cells [102]. This result support the idea that cysteine are lining the channel's pore, which may only partially block the channel.

Given that two independent permeation pathways exist in different subunit of TMC channel dimers, the channel conductance with a single pore open should be half that when both pores of the dimer are open. Noise analysis measurements also give rise to a single-channel current of ~13 pA [102], which is in accord with previous data reporting a single-channel conductance of ~150 pS [70, 114]. Interestingly, another study has reported a single-channel current of ~6 pA for hair cell MET channels [94], approximately half those analyzed from noise analysis. It is likely that the smaller value represents the single-pore conductance, while the larger one corresponds to TMC channel dimers with two pore open.

17.3.3 Recent Evidence for TMC as a Mechanosensitive Channel

Although compelling evidence is presented here supporting TMC proteins as the pore-forming component of MET channel, there are still discrepancies among prior studies about the role of TMC in mechanotransduction [115]. It was proposed that the tonotopic gradient of MET channel conductance might originates from variations in the stoichiometry of TMC1/2 proteins [70]. It should be noted that TMC2 expression is not detected in adult hair cells and is not essential for hearing function; moreover, little evidence indicates that TMC1 and TMC2 colocalize in hair cells [68, 91]. Another study even failed to repeat the results that MET single-channel conductance is altered by one TMC1 mutation as originally reported [116]. Strikingly, it is recently reported that all deficiencies of MET current in *Tmc1* and *Tmc2* mutant mice can result from regulation of PIP2 concentration in hair bundles [117], suggesting that channel conductivity might not have a direct role for these changes. The key point is that mammalian and invertebrate TMCs could not be expressed on the surface of cultured cells so far, and it appears that most TMC proteins remained in the ER [80, 118]. Hence, although TMC1/2 are proposed as candidates for the pore-forming subunits of the MET channel, whether TMCs produce a channel permeable to cations and whether this channel is mechanosensitive still need to be determined.

Recent evidence shows that TMC channels also share structural similarity with TMEM63/OSCAs [61, 62, 102, 103, 119, 120], an evolutionarily conserved family of mechanosensitive channels [121]. OSCA channels mediate the hyperosmolarity-evoked physiological response in plants [122, 123]. In *Arabidopsis thaliana*, OSCA1.1 and OSCA1.2 are pore-forming subunits of mechanically activated ion channels [61, 62, 121]. A large number of *Arabidopsis* OSCAs render high-threshold mechanically activated currents in cultured cells. When reconstituted into liposomes, purified AtOSCA1.2 induces strong mechanically activated currents, suggesting intrinsic mechanosensitivity of this protein [121]. Moreover, as shown for other mechanosensitive channels [124, 125], LPC (lyso-Phosphatidylcholine) application increased the mechanosensitive response of OSCA channels [61, 62], indicating that the local membrane curvature caused by transbilayer pressure was sufficient to activate the channel. *Arabidopsis* OSCAs are dimers with each subunit consisting of 11 transmembrane domains surrounding a pore domain. The hydrophobic neck of the channel pore tended to be open upon membrane tension change in molecular dynamics simulation [61, 62]. Considering the structural analogy of TMC to OSCAs, whether a local bilayer tension change also occurs to gate TMC1 or the channel is gated by force from tip links requires further investigations [126]. Consequently, it should not be excluded that the gating mechanism of MET channel in the hair cells follows the force-from-lipid model.

Notably, while TMC proteins are not trafficked to the cell membrane when expressed in cultured cells [80, 98, 118], the purified TMC1 and TMC2 proteins are very recently shown to be embedded into the artificial lipid bilayers [140]. Subsequently, Jia et al. have finished a major advance by characterizing the properties of TMC channels in liposome.

To identify suitable candidates for membrane protein purification, ortholog screening by size-exclusion chromatography has been a commonly adopted strategy, which is successfully applied in numerous studies [127–132]. Expression screening of 21 TMC1/2 homologues in various organisms for high expression level in culture cells helped Jia et al. to identify two TMC1/2 proteins appropriate for purification, which are the green sea turtle TMC1 (CmTMC1) and the budgerigar TMC2 (MuTMC2). Sequence alignment showed both CmTMC1 and MuTMC2 are highly conserved with human TMC1 and TMC2, respectively [140]. Accordingly, like the arrangement of stereocilia in descending heights observed in mammals, birds and reptiles possess hair cells with stereocilia distributed similarly and also tip links to transmit forces [133–138]. Therefore, studying the functional properties of CmTMC1 and MuTMC2 also provides insights into the properties of their mammalian orthologs. Fortunately, they first reconstituted the purified TMC proteins into liposome membrane [140]. When applying a series of voltage steps, spontaneous single-channel currents can be recorded in patches containing either CmTMC1 or MuTMC2 in a voltage-dependent manner. The spontaneous currents are nearly eliminated in a cation-free solution but keep unaffected in Na-gluconate solution, suggesting the currents of CmTMC1 and MuTMC2 are mainly carried by cation channels [140]. These liposome recording data support that both CmTMC1 and MuTMC2 fold into ion channels. Importantly, pressure application evokes stretch-activated currents at −120 mV in patches with CmTMC1 and MuTMC2 proteins. Upon pressure application, the open probability of CmTMC1 and MuTMC2 channels also increased in a pressure-dependent manner [140]. Moreover, when introducing human deafness point mutations into CmTMC1 proteins, ion channel activity and pressure-activated activity were reduced or abolished [140]. Given the high sequence similarity of CmTMC1 and MuTMC2 to human TMC1 and TMC2, the channel activity and force sensitivity of CmTMC1 and MuTMC2 suggest that mammalian TMC1 and TMC2 might also form inherently mechanosensitive ion channels. Of note, the N termini of TMC1 and TMC2 are truncated to facilitate protein purification ([140]). Considering the interaction between the N terminus of TMCs and PCDH15, a tip-link component [82], this interaction might prevent the channel from opening when there is no force stimuli. As a consequence, the truncated TMC1 and TMC2 channels are able to open spontaneously in the absence of mechanical force.

17.4 Conclusions and Perspectives

Mechanotransduction, the conversion of mechanical forces into electrical signals, occurs by the activation of mechanosensitive ion channels. It plays a key role in balance, touch, hearing, and proprioception. Moreover, it also partakes in the regulation of blood pressure and osmolality. Over the past few years, a tremendous amount of advances have been made in understanding mechanotransduction processes in both hearing and touch sensation. Despite considerable progress, fundamental questions still remain.

Through genetic screens, a variety of ion channels have been implicated in mechanotransduction, including MEC/DEG/ASICS and TRPs in invertebrates, as well as PIEZOs, TMCs, and OSCAs across eukaryotes. Current studies are mainly focused on identifying the candidate genes encoding these channels and elucidating the mechanism of force gating. However, scarce evidence has hitherto been provided demonstrating their intrinsic mechanosensitivity. Only PIEZO and NOMPC meet all the prerequisites proposed for a bona fide mechanotransduction channel. Whether the other candidates might be auxiliary subunits or modulators of bona fide mechanotransduction channels require further investigations.

The rich diversity of the molecular architecture of mechanotransduction channels suggests that they have occurred independently in the adaptive evolution. Although structural designs of mechanotransduction channels are quite diverse, most of them utilize lipids to fill fenestrations in the wall of the ion permeation pore [126]. Lipid action through these lateral fenestrations is likely to affect the mechanogating of these channels, further investigations are necessary to answer this question.

Of note, defects in mechanotransduction potentially result in diverse inherited diseases such as cancer, muscular dystrophies, cardiomyopathies, hearing loss, and chronic pain. Therefore, developing potent agonists or blockers for mechanotransduction channels will undoubtedly be very prospective for clinical applications.

Further progress will require answering the major questions listed below. Do mechanotransduction channels function within specialized subcellular compartments, such as organelle or nuclear membrane? The fact that multiple mechanosensitive channels are usually expressed in a common sensory cell, which lead to depolarization or hyperpolarization, raises the question of how cells encode the physical measurement of different form of force? Insight into a general role for mechanosensitive channels with unidentified function such as OSCAs in animal kingdom should also help us to further understand how mechanical forces regulate a series of cell functions and deepen the understanding of mechanotransduction. Moreover, in vertebrates and invertebrates, identification of new candidates for mechanically activated ion channels is still a challenging task. As always, studies on touch and sound perception spanning multiple species will broaden our understanding of mechanotransduction mechanisms.

Acknowledgments This work is supported by grants from the National Key R&D Program of China Project (2017YFA0103900), the National Natural Science Foundation of China (31571083, 31970931), the Program for Professor of Special Appointment (Eastern Scholar of Shanghai, TP2014008), the Shanghai Municipal Science and Technology Major Project (No. 2017SHZDZX01 and No. 2018SHZDZX01) and ZJLab, and the Shanghai Rising-Star Program (14QA1400800).

References

1. Ranade SS, Syeda R, Patapoutian A (2015) Mechanically activated ion channels. Neuron 87 (6):1162–1179
2. Corey DP, Hudspeth AJ (1979) Ionic basis of the receptor potential in a vertebrate hair cell. Nature 281(5733):675–677
3. Arnadottir J, Chalfie M (2010) Eukaryotic mechanosensitive channels. Annu Rev Biophys 39:111–137
4. Delmas P, Coste B (2013) Mechano-gated ion channels in sensory systems. Cell 155(2):278–284
5. Chalfie M (2009) Neurosensory mechanotransduction. Nat Rev Mol Cell Biol 10 (1):44–52
6. Christensen AP, Corey DP (2007) TRP channels in mechanosensation: direct or indirect activation? Nat Rev Neurosci 8(7):510–521
7. Lumpkin EA, Marshall KL, Nelson AM (2010) The cell biology of touch. J Cell Biol 191(2):237–248
8. Albert JT, Gopfert MC (2015) Hearing in Drosophila. Curr Opin Neurobiol 34:79–85
9. Sidi S, Friedrich RW, Nicolson T (2003) NompC TRP channel required for vertebrate sensory hair cell mechanotransduction. Science 301(5629):96–99
10. Kernan M, Cowan D, Zuker C (1994) Genetic dissection of mechanosensory transduction: mechanoreception-defective mutations of Drosophila. Neuron 12(6):1195–1206
11. Walker RG, Willingham AT, Zuker CS (2000) A Drosophila mechanosensory transduction channel. Science 287(5461):2229–2234
12. Kang L, Gao J, Schafer WR et al (2010) C. elegans TRP family protein TRP-4 is a pore-forming subunit of a native mechanotransduction channel. Neuron 67 (3):381–391
13. Chadha A, Kaneko M, Cook B (2015) NOMPC-dependent mechanotransduction shapes the dendrite of proprioceptive neurons. Neurosci Lett 597:111–116
14. Cheng LE, Song W, Looger LL et al (2010) The role of the TRP channel NompC in Drosophila larval and adult locomotion. Neuron 67(3):373–380
15. Effertz T, Wiek R, Gopfert MC (2011) NompC TRP channel is essential for Drosophila sound receptor function. Curr Biol 21(7):592–597
16. Ramdya P, Lichocki P, Cruchet S et al (2015) Mechanosensory interactions drive collective behaviour in Drosophila. Nature 519(7542):233–236
17. Sanchez-Alcaniz JA, Zappia G, Marion-Poll F et al (2017) A mechanosensory receptor required for food texture detection in Drosophila. Nat Commun 8:14192

18. Zhang W, Yan Z, Jan LY et al (2013) Sound response mediated by the TRP channels NOMPC, NANCHUNG, and INACTIVE in chordotonal organs of Drosophila larvae. Proc Natl Acad Sci U S A 110(33):13612–13617
19. Zhang W, Yan Z, Li B et al (2014) Identification of motor neurons and a mechanosensitive sensory neuron in the defecation circuitry of Drosophila larvae. elife 3:e03293
20. Zhou Y, Cao LH, Sui XW et al (2019) Mechanosensory circuits coordinate two opposing motor actions in Drosophila feeding. Sci Adv 5(5): eaaw5141
21. Grueber WB, Jan LY, Jan YN (2002) Tiling of the Drosophila epidermis by multidendritic sensory neurons. Development 129(12):2867–2878
22. Kim SE, Coste B, Chadha A et al (2012) The role of Drosophila Piezo in mechanical nociception. Nature 483(7388):209–212
23. Tracey WD Jr, Wilson RI, Laurent G et al (2003) painless, a Drosophila gene essential for nociception. Cell 113(2):261–273
24. Tsubouchi A, Caldwell JC, Tracey WD (2012) Dendritic filopodia, Ripped Pocket, NOMPC, and NMDARs contribute to the sense of touch in Drosophila larvae. Curr Biol 22(22):2124–2134
25. Zhong L, Hwang RY, Tracey WD (2010) Pickpocket is a DEG/ENaC protein required for mechanical nociception in Drosophila larvae. Curr Biol 20 (5):429–434
26. Gong J, Wang Q, Wang Z (2013) NOMPC is likely a key component of Drosophila mechanotransduction channels. Eur J Neurosci 38(1):2057–2064
27. Effertz T, Nadrowski B, Piepenbrock D et al (2012) Direct gating and mechanical integrity of Drosophila auditory transducers require TRPN1. Nat Neurosci 15(9):1198–1200
28. Lehnert BP, Baker AE, Gaudry Q et al (2013) Distinct roles of TRP channels in auditory transduction and amplification in Drosophila. Neuron 77 (1):115–128
29. Gillespie PG, Walker RG (2001) Molecular basis of mechanosensory transduction. Nature 413 (6852):194–202
30. Kung C (2005) A possible unifying principle for mechanosensation. Nature 436(7051):647–654
31. Lumpkin EA, Caterina MJ (2007) Mechanisms of sensory transduction in the skin. Nature 445 (7130):858–865
32. Orr AW, Helmke BP, Blackman BR et al (2006) Mechanisms of mechanotransduction. Dev Cell 10 (1):11–20
33. Anishkin A, Kung C (2013) Stiffened lipid platforms at molecular force foci. Proc Natl Acad Sci U S A 110 (13):4886–4892
34. Brohawn SG, Campbell EB, MacKinnon R (2014a) Physical mechanism for gating and mechanosensitivity of the human TRAAK K^+ channel. Nature 516(7529):126–130
35. Brohawn SG, del Marmol J, MacKinnon R (2012) Crystal structure of the human K2P TRAAK, a lipid- and mechano-sensitive K^+ ion channel. Science 335 (6067):436–441
36. Brohawn SG, Su Z, MacKinnon R (2014b) Mechanosensitivity is mediated directly by the lipid membrane in TRAAK and TREK1 K^+ channels. Proc Natl Acad Sci U S A 111(9):3614–3619
37. Lolicato M, Riegelhaupt PM, Arrigoni C et al (2014) Transmembrane helix straightening and buckling underlies activation of mechanosensitive and thermosensitive K(2P) channels. Neuron 84 (6):1198–1212
38. Montell C (2004) Molecular genetics of Drosophila TRP channels. Novartis Found Symp 258:3–12
39. Montell C (2005) Drosophila TRP channels. Pflugers Arch 451(1):19–28
40. Venkatachalam K, Montell C (2007) TRP channels. Annu Rev Biochem 76:387–417
41. Gaudet R (2008) A primer on ankyrin repeat function in TRP channels and beyond. Mol Biosyst 4 (5):372–379
42. Howard J, Bechstedt S (2004) Hypothesis: a helix of ankyrin repeats of the NOMPC-TRP ion channel is the gating spring of mechanoreceptors. Curr Biol 14 (6):R224–R226
43. Lee G, Abdi K, Jiang Y et al (2006) Nanospring behaviour of ankyrin repeats. Nature 440 (7081):246–249
44. Sotomayor M, Corey DP, Schulten K (2005) In search of the hair-cell gating spring elastic properties of ankyrin and cadherin repeats. Structure 13 (4):669–682
45. Zhang W, Cheng LE, Kittelmann M et al (2015) Ankyrin repeats convey force to gate the NOMPC mechanotransduction channel. Cell 162 (6):1391–1403
46. Liang X, Madrid J, Gartner R et al (2013) A NOMPC-dependent membrane-microtubule connector is a candidate for the gating spring in fly mechanoreceptors. Curr Biol 23(9):755–763
47. Hao J, Padilla F, Dandonneau M et al (2013) Kv1.1 channels act as mechanical brake in the senses of touch and pain. Neuron 77(5):899–914
48. Kalia J, Swartz KJ (2013) Exploring structure-function relationships between TRP and Kv channels. Sci Rep 3:1523
49. Long SB, Campbell EB, Mackinnon R (2005) Crystal structure of a mammalian voltage-dependent Shaker family K^+ channel. Science 309(5736):897–903
50. Jin P, Bulkley D, Guo Y et al (2017) Electron cryo-microscopy structure of the mechanotransduction channel NOMPC. Nature 547(7661):118–122
51. Liang X, Howard J (2017) Structural biology: a force-sensitive ion channel springs to life. Curr Biol 27(18):R1017–R1020
52. Zanini D, Gopfert MC (2014) TRPs in hearing. Handb Exp Pharmacol 223:899–916

53. Budelli G, Ni L, Berciu C et al (2019) Ionotropic receptors specify the morphogenesis of phasic sensors controlling rapid thermal preference in Drosophila. Neuron 101(4):738–747
54. Gallio M, Ofstad TA, Macpherson LJ et al (2011) The coding of temperature in the Drosophila brain. Cell 144(4):614–624
55. Ni L, Klein M, Svec KV et al (2016) The ionotropic receptors IR21a and IR25a mediate cool sensing in Drosophila. elife 5:e13254
56. Turner HN, Patel AA, Cox DN et al (2018) Injury-induced cold sensitization in Drosophila larvae involves behavioral shifts that require the TRP channel Brv1. PLoS One 13(12):e0209577
57. Dhaka A, Viswanath V, Patapoutian A (2006) Trp ion channels and temperature sensation. Annu Rev Neurosci 29:135–161
58. Cantiello HF (2004) Regulation of calcium signaling by polycystin-2. Am J Physiol Renal Physiol 286(6): F1012–F1029
59. Martinac B (2004) Mechanosensitive ion channels: molecules of mechanotransduction. J Cell Sci 117 (12):2449–2460
60. Peyronnet R, Sharif-Naeini R, Folgering JH et al (2012) Mechanoprotection by polycystins against apoptosis is mediated through the opening of stretch-activated K(2P) channels. Cell Rep 1 (3):241–250
61. Zhang M, Li X, Zheng H et al (2018a) Brv1 is required for Drosophila larvae to sense gentle touch. Cell Rep 23(1):23–31
62. Zhang M, Wang D, Kang Y et al (2018b) Structure of the mechanosensitive OSCA channels. Nat Struct Mol Biol 25(9):850–858
63. Coste B, Mathur J, Schmidt M et al (2010) Piezo1 and Piezo2 are essential components of distinct mechanically activated cation channels. Science 330 (6000):55–60
64. Sharif-Naeini R, Folgering JH, Bichet D et al (2009) Polycystin-1 and -2 dosage regulates pressure sensing. Cell 139(3):587–596
65. Corey DP, Holt JR (2016) Are TMCs the mechanotransduction channels of vertebrate hair cells? J Neurosci 36(43):10921–10926
66. Fettiplace R (2016) Is TMC1 the hair cell mechanotransducer channel? Biophys J 111(1):3–9
67. Fettiplace R, Kim KX (2014) The physiology of mechanoelectrical transduction channels in hearing. Physiol Rev 94(3):951–986
68. Kawashima Y, Geleoc GS, Kurima K et al (2011) Mechanotransduction in mouse inner ear hair cells requires transmembrane channel-like genes. J Clin Invest 121(12):4796–4809
69. Kurima K, Peters LM, Yang Y et al (2002) Dominant and recessive deafness caused by mutations of a novel gene, TMC1, required for cochlear hair-cell function. Nat Genet 30(3):277–284
70. Pan B, Geleoc GS, Asai Y et al (2013) TMC1 and TMC2 are components of the mechanotransduction channel in hair cells of the mammalian inner ear. Neuron 79(3):504–515
71. Vreugde S, Erven A, Kros CJ et al (2002) Beethoven, a mouse model for dominant, progressive hearing loss DFNA36. Nat Genet 30(3):257–258
72. Gillespie PG, Muller U (2009) Mechanotransduction by hair cells: models, molecules, and mechanisms. Cell 139(1):33–44
73. Kawashima Y, Kurima K, Pan B et al (2015) Transmembrane channel-like (TMC) genes are required for auditory and vestibular mechanosensation. Pflugers Arch 467(1):85–94
74. Keresztes G, Mutai H, Heller S (2003) TMC and EVER genes belong to a larger novel family, the TMC gene family encoding transmembrane proteins. BMC Genomics 4(1):24
75. Kurima K, Yang Y, Sorber K et al (2003) Characterization of the transmembrane channel-like (TMC) gene family: functional clues from hearing loss and epidermodysplasia verruciformis. Genomics 82 (3):300–308
76. Liu H, Pecka JL, Zhang Q et al (2014) Characterization of transcriptomes of cochlear inner and outer hair cells. J Neurosci 34(33):11085–11095
77. Scheffer DI, Shen J, Corey DP et al (2015) Gene expression by mouse inner ear hair cells during development. J Neurosci 35(16):6366–6380
78. Gleason MR, Nagiel A, Jamet S et al (2009) The transmembrane inner ear (Tmie) protein is essential for normal hearing and balance in the zebrafish. Proc Natl Acad Sci U S A 106(50):21347–21352
79. Xiong W, Grillet N, Elledge HM et al (2012) TMHS is an integral component of the mechanotransduction machinery of cochlear hair cells. Cell 151 (6):1283–1295
80. Zhao B, Wu Z, Grillet N et al (2014) TMIE is an essential component of the mechanotransduction machinery of cochlear hair cells. Neuron 84 (5):954–967
81. Beurg M, Goldring AC, Fettiplace R (2015a) The effects of Tmc1 Beethoven mutation on mechanotransducer channel function in cochlear hair cells. J Gen Physiol 146(3):233–243
82. Maeda R, Kindt KS, Mo W et al (2014) Tip-link protein protocadherin 15 interacts with transmembrane channel-like proteins TMC1 and TMC2. Proc Natl Acad Sci U S A 111(35):12907–12912
83. Jackson AC, Nicoll RA (2011) The expanding social network of ionotropic glutamate receptors: TARPs and other transmembrane auxiliary subunits. Neuron 70(2):178–199
84. Mitchem KL, Hibbard E, Beyer LA et al (2002) Mutation of the novel gene Tmie results in sensory cell defects in the inner ear of spinner, a mouse model of human hearing loss DFNB6. Hum Mol Genet 11 (16):1887–1898
85. Naz S, Giguere CM, Kohrman DC et al (2002) Mutations in a novel gene, TMIE, are associated

with hearing loss linked to the DFNB6 locus. Am J Hum Genet 71(3):632–636
86. Pacentine IV, Nicolson T (2019) Subunits of the mechano-electrical transduction channel, Tmc1/2b, require Tmie to localize in zebrafish sensory hair cells. PLoS Genet 15(2):e1007635
87. Cunningham CL, Wu Z, Jafari A et al (2017) The murine catecholamine methyltransferase mTOMT is essential for mechanotransduction by cochlear hair cells. elife 6:e24318
88. Erickson T, Morgan CP, Olt J et al (2017) Integration of Tmc1/2 into the mechanotransduction complex in zebrafish hair cells is regulated by transmembrane O-methyltransferase (Tomt). elife 6:e28474
89. Giese APJ, Tang YQ, Sinha GP et al (2017) CIB2 interacts with TMC1 and TMC2 and is essential for mechanotransduction in auditory hair cells. Nat Commun 8(1):43
90. Wang Y, Li J, Yao X et al (2017) Loss of CIB2 causes profound hearing loss and abolishes mechanoelectrical transduction in mice. Front Mol Neurosci 10:401
91. Kurima K, Ebrahim S, Pan B et al (2015) TMC1 and TMC2 localize at the site of mechanotransduction in mammalian inner ear hair cell stereocilia. Cell Rep 12 (10):1606–1617
92. Ahmed ZM, Goodyear R, Riazuddin S et al (2006) The tip-link antigen, a protein associated with the transduction complex of sensory hair cells, is protocadherin-15. J Neurosci 26(26):7022–7034
93. Kazmierczak P, Sakaguchi H, Tokita J et al (2007) Cadherin 23 and protocadherin 15 interact to form tip-link filaments in sensory hair cells. Nature 449 (7158):87–91
94. Beurg M, Kim KX, Fettiplace R (2014) Conductance and block of hair-cell mechanotransducer channels in transmembrane channel-like protein mutants. J Gen Physiol 144(1):55–69
95. Corns LF, Jeng JY, Richardson GP et al (2017) TMC2 modifies permeation properties of the mechanoelectrical transducer channel in early post-natal mouse cochlear outer hair cells. Front Mol Neurosci 10:326
96. Kim KX, Fettiplace R (2013) Developmental changes in the cochlear hair cell mechanotransducer channel and their regulation by transmembrane channel-like proteins. J Gen Physiol 141(1):141–148
97. Chatzigeorgiou M, Bang S, Hwang SW et al (2013) tmc-1 encodes a sodium-sensitive channel required for salt chemosensation in C. elegans. Nature 494 (7435):95–99
98. Guo Y, Wang Y, Zhang W et al (2016) Transmembrane channel-like (tmc) gene regulates Drosophila larval locomotion. Proc Natl Acad Sci U S A 113 (26):7243–7248
99. Wang X, Li G, Liu J et al (2016) TMC-1 mediates alkaline sensation in C. elegans through nociceptive neurons. Neuron 91(1):146–154
100. Yue X, Zhao J, Li X et al (2018) TMC proteins modulate egg laying and membrane excitability through a background leak conductance in C. elegans. Neuron 97(3):571–585.e5
101. Zhang YV, Aikin TJ, Li Z et al (2016) The basis of food texture sensation in Drosophila. Neuron 91 (4):863–877
102. Pan B, Akyuz N, Liu XP et al (2018) TMC1 forms the pore of mechanosensory transduction channels in vertebrate inner ear hair cells. Neuron 99 (4):736–753.e6
103. Ballesteros A, Fenollar-Ferrer C, Swartz KJ (2018) Structural relationship between the putative hair cell mechanotransduction channel TMC1 and TMEM16 proteins. elife 7:e38433
104. Brunner JD, Lim NK, Schenck S et al (2014) X-ray structure of a calcium-activated TMEM16 lipid scramblase. Nature 516(7530):207–212
105. Dang S, Feng S, Tien J et al (2017) Cryo-EM structures of the TMEM16A calcium-activated chloride channel. Nature 552(7685):426–429
106. Paulino C, Kalienkova V, Lam AKM et al (2017) Activation mechanism of the calcium-activated chloride channel TMEM16A revealed by cryo-EM. Nature 552(7685):421–425
107. Liao M, Cao E, Julius D et al (2013) Structure of the TRPV1 ion channel determined by electron cryo-microscopy. Nature 504(7478):107–112
108. McGoldrick LL, Singh AK, Saotome K et al (2018) Opening of the human epithelial calcium channel TRPV6. Nature 553(7687):233–237
109. Saotome K, Murthy SE, Kefauver JM et al (2018) Structure of the mechanically activated ion channel Piezo1. Nature 554(7693):481–486
110. Wang L, Zhou H, Zhang M et al (2019) Structure and mechanogating of the mammalian tactile channel PIEZO2. Nature 573(7773):225–229
111. Zhao Q, Zhou H, Chi S et al (2018) Structure and mechanogating mechanism of the Piezo1 channel. Nature 554(7693):487–492
112. Holt JR, Gillespie SK, Provance DW et al (2002) A chemical-genetic strategy implicates myosin-1c in adaptation by hair cells. Cell 108(3):371–381
113. Corey DP, Akyuz N, Holt JR (2019) Function and dysfunction of TMC channels in inner ear hair cells. Cold Spring Harb Perspect Med 9(10):a033506
114. Beurg M, Evans MG, Hackney CM et al (2006) A large-conductance calcium-selective mechanotransducer channel in mammalian cochlear hair cells. J Neurosci 26(43):10992–11000
115. Qiu X, Muller U (2018) Mechanically gated ion channels in mammalian hair cells. Front Cell Neurosci 12:100
116. Beurg M, Xiong W, Zhao B et al (2015b) Subunit determination of the conductance of hair-cell mechanotransducer channels. Proc Natl Acad Sci U S A 112(5):1589–1594
117. Effertz T, Becker L, Peng AW et al (2017) Phosphoinositol-4,5-bisphosphate regulates auditory

hair-cell mechanotransduction-channel pore properties and fast adaptation. J Neurosci 37 (48):11632–11646
118. Labay V, Weichert RM, Makishima T et al (2010) Topology of transmembrane channel-like gene 1 protein. Biochemistry 49(39):8592–8598
119. Jojoa-Cruz S, Saotome K, Murthy SE et al (2018) Cryo-EM structure of the mechanically activated ion channel OSCA1.2. elife 7:e41845
120. Liu X, Wang J, Sun L (2018) Structure of the hyperosmolality-gated calcium-permeable channel OSCA1.2. Nat Commun 9(1):5060
121. Murthy SE, Dubin AE, Whitwam T et al (2018) OSCA/TMEM63 are an evolutionarily conserved family of mechanically activated ion channels. elife 7:e41844
122. Hou C, Tian W, Kleist T et al (2014) DUF221 proteins are a family of osmosensitive calcium-permeable cation channels conserved across eukaryotes. Cell Res 24(5):632–635
123. Yuan F, Yang H, Xue Y et al (2014) OSCA1 mediates osmotic-stress-evoked Ca^{2+} increases vital for osmosensing in Arabidopsis. Nature 514 (7522):367–371
124. Martinac B, Bavi N, Ridone P et al (2018) Tuning ion channel mechanosensitivity by asymmetry of the transbilayer pressure profile. Biophys Rev 10 (5):1377–1384
125. Perozo E, Kloda A, Cortes DM et al (2002) Physical principles underlying the transduction of bilayer deformation forces during mechanosensitive channel gating. Nat Struct Biol 9(9):696–703
126. Douguet D, Honore E (2019) Mammalian mechanoelectrical transduction: structure and function of force-gated ion channels. Cell 179(2):340–354
127. Coleman JA, Green EM, Gouaux E (2016) X-ray structures and mechanism of the human serotonin transporter. Nature 532(7599):334–339
128. Hibbs RE, Gouaux E (2011) Principles of activation and permeation in an anion-selective Cys-loop receptor. Nature 474(7349):54–60
129. Kawate T, Michel JC, Birdsong WT et al (2009) Crystal structure of the ATP-gated P2X(4) ion channel in the closed state. Nature 460(7255):592–598
130. Lee CH, Lu W, Michel JC et al (2014) NMDA receptor structures reveal subunit arrangement and pore architecture. Nature 511(7508):191–197
131. Penmatsa A, Wang KH, Gouaux E (2013) X-ray structure of dopamine transporter elucidates antidepressant mechanism. Nature 503(7474):85–90
132. Sobolevsky AI, Rosconi MP, Gouaux E (2009) X-ray structure, symmetry and mechanism of an AMPA-subtype glutamate receptor. Nature 462 (7274):745–756
133. Crawford AC, Evans MG, Fettiplace R (1991) The actions of calcium on the mechano-electrical transducer current of turtle hair cells. J Physiol 434:369–398
134. Manley GA, Koppl C (2008) What have lizard ears taught us about auditory physiology? Hear Res 238 (1–2):3–11
135. Ohmori H (1985) Mechano-electrical transduction currents in isolated vestibular hair cells of the chick. J Physiol 359:189–217
136. Pickles JO, Brix J, Comis SD et al (1989) The organization of tip links and stereocilia on hair cells of bird and lizard basilar papillae. Hear Res 41(1):31–41
137. Ricci AJ, Fettiplace R (1997) The effects of calcium buffering and cyclic AMP on mechano-electrical transduction in turtle auditory hair cells. J Physiol 501(Pt 1):111–124
138. Tsuprun V, Goodyear RJ, Richardson GP (2004) The structure of tip links and kinocilial links in avian sensory hair bundles. Biophys J 87(6):4106–4112
139. Yan Z, Zhang W, He Y et al (2013) Drosophila NOMPC is a mechanotransduction channel subunit for gentle-touch sensation. Nature 493(7431):221–225
140. Jia Y, Zhao Y, Kusakizako T et al (2020) TMC1 and TMC2 proteins are pore-forming subunits of mechanosensitive ion channels. Neuron 105(2):310–321
141. Wang Y, Guo Y, Li G, Liu C, Wang L, Zhang A, Yan Z, Song C (2021) The push-to-open mechanism of the tethered mechanosensitive ion channel NompC. eLife. 1010.7554/eLife.58388

The Functional Properties, Physiological Roles, Channelopathy and Pharmacological Characteristics of the Slack (KCNT1) Channel

18

Qi Zhang, Ye Liu, Jie Xu, Yue Teng, and Zhe Zhang

Abstract

The KCNT1 gene encodes the sodium-activated potassium channel that is abundantly expressed in the central nervous system of mammalians and plays an important role in reducing neuronal excitability. Structurally, the KCNT1 channel is absent of voltage sensor but possesses a long C-terminus including RCK1 and RCK2domain, to which the intracellular sodium and chloride bind to activate the channel. Recent publications using electron cryo-microscopy (cryo-EM) revealed the open and closed structural characteristics of the KCNT1 channel and co-assembly of functional domains. The activation of the KCNT1 channel regulates various physiological processes including nociceptive behavior, itch, spatial learning. Meanwhile, malfunction of this channel causes important pathophysiological consequences, including Fragile X syndrome and a wide spectrum of seizure disorders. This review comprehensively describes the structure, expression patterns, physiological functions of the KCNT1 channel and emphasizes the channelopathy of gain-of-function KCNT1 mutations in epilepsy.

Keywords

Slack channel · KCNT1 · Fragile X syndrome · Epilepsy

18.1 The Slack Channel (Slo2.2, KCNT1, KCa4.1)

The α subunit of Slack (refers to "sequence like a Ca^{2+}-activated K^+ channel") channel, which belongs to the BK channel family, is encoded by the KCNT1 gene [1]. The functional Slack channel is composed of four α subunits. Initially, the Slack channel was considered as a Ca^{2+} activated potassium channel which is composed of two K^+ channel subunits, Slo1 and Slack (Slo2.2) [2]. But subsequent research reveals that the Slack channel activates in response to binding to intracellular sodium and chloride with a single conductance of approximately 145 pS [1, 3]. However, the Slack channel probably evolves from a Ca^{2+} activated potassium channel to sodium-activated potassium channel because the *C. elegans* Slo2 channel is a Ca^{2+} and voltage-activated channel, which shows high sequence homology to the mammalian Slack channels [4].

Qi Zhang and Ye Liu have contributed equally to this work.

Q. Zhang · Y. Liu · J. Xu · Y. Teng · Z. Zhang (✉)
Jiangsu Province Key Laboratory of Anesthesiology, Xuzhou Medical University, Xuzhou, Jiangsu Province, China

Jiangsu Province Key Laboratory of Anesthesia and Analgesia Application Technology, Xuzhou Medical University, Xuzhou, Jiangsu Province, China
e-mail: zhangzhe70@xzhmu.edu.cn

L. Zhou (ed.), *Ion Channels in Biophysics and Physiology*, Advances in Experimental Medicine and Biology 1349, https://doi.org/10.1007/978-981-16-4254-8_18

The Slack channel is widely expressed in the nervous systems of mammalians and is involved in various physiological processes including nociceptive sensing [5, 6], itching [7], spatial learning [8], and others. Both the malfunction and gain-of-function of Slack channels are involved in many pathophysiological processes, such as Fragile X syndrome, epilepsy, and heart arrhythmia [9–11].

What are the structural characteristics of the Slack channel that are different from the Slo1 channel result in different biophysical properties of the Slack channel? Furthermore, how does the sodium binding lead to the channel opening? What are the mechanisms of malfunction or gain-of-function of Slack channel that caused neurological disorders? Which drugs can enhance or block the Slack channel as the potential drug targets for the treatment of Slack channel-related disease? Recent studies and the latest findings that provide new insights into the biophysical gating mechanism of the Slack channel, pharmacological properties, and the mechanism that neurological disorders related to the Slack channel will be discussed in this review.

18.2 Structural and Functional Domains of Slack Channels

The Slack gene KCNT1 was firstly reported as the gene that encoded a potassium channel in 1998 [2]. This gene is localized to chromosome 9q34.3 in humans, to chromosome 3 in rats, and chromosome 2 in mice [12]. The KCNT1 gene is transcribed to generate a 4.7 kb mRNA that is translated to a 138 kDa protein, which is abundantly expressed in neurons in the central nervous system of mammalians [12, 13]. Alternative N-terminal splices give rise to more than five distinct transcripts of the rodent Slack channels, which are categorized into two groups: Slack-A and Slack-B. They show different membrane expression levels and expression patterns in the mice brain, suggesting possible distinct modulating roles in different brain regions [12, 14]. In the genebank, many different human KCNT1 gene isoforms are predicted or identified from cDNA libraries (Fig. 18.1a, b). However, the functional expression of these isoforms needs to be confirmed by further research.

The topology and assembly of the Slack channel share similar characteristics with voltage-activated potassium channels (Kv). In contrast with the Slo1 channel, the N terminus of the Slack α subunit is located on the intracellular side. The Slack channel α subunit possesses six hydrophobic transmembrane segments (S1–S6) along with a pore-lining loop between S5 and S6 (Fig. 18.2a) [2]. However, different from the Kv family of channels, the S4 segment of the Slack channel does not possess any charged residues that served as the voltage sensor in the Kv channels. Thus, it is not surprising that the Slack channel is not a voltage-dependent potassium channel. Interestingly, the S4 segment of the C.elegan Slo2 channel also does not include any positively charged residue, but the CSlo2 channel demonstrates Ca^{2+} and voltage-dependent outward current recorded in the inside-out patch configuration [4] (Fig. 18.2b). This phenomenon indicates the Slack channel may evolve from a Ca^{2+} sensitive channel to a Na^+ sensitive channel. But how the voltage dependency of the *C. elegans* Slo2 is generated remains to be addressed.

The large cytoplasmic domain (CTD) of the Slack α subunit contains two RCK (regulator-of-K^+-conductance) domains, in which the sodium-sensitive site and putative NAD binding sites are located [2, 3, 15–17]. Neutralization of the negative charged residue D818 on the RCK2 domain decreases the sodium sensitivity of the rat Slack channel by more than fivefold [3]. However, this site does not completely remove the sodium sensitivity of the Slack channel. Therefore, further study is still needed for the identification of all sodium binding sites.

The cryo-electron microscopy (cryo-EM) structure of the chicken Slack channel is available and provides useful information for understanding the mechanism of Slack channel gating

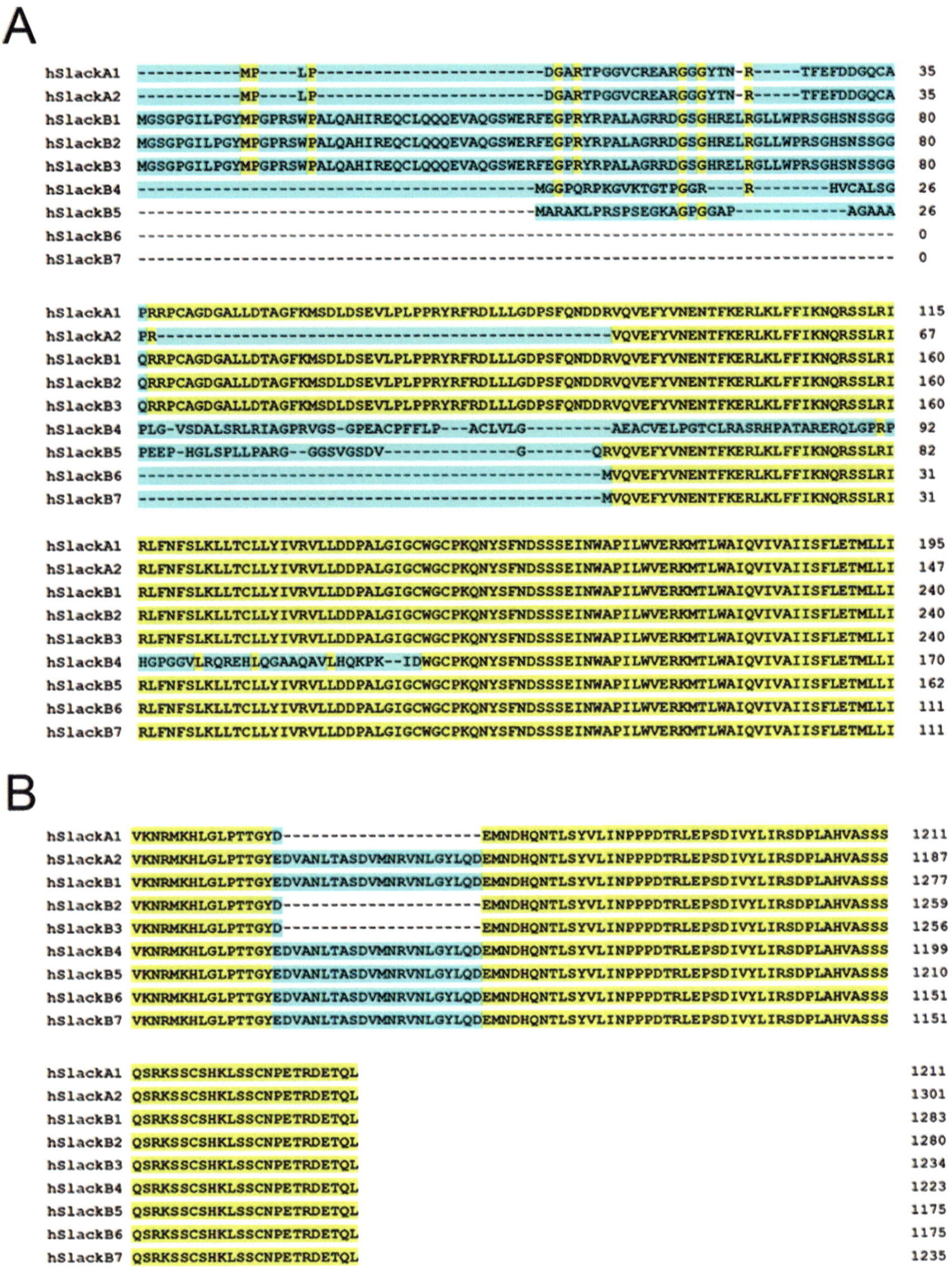

Fig. 18.1 Sequence alignment of human Slack isoforms. The protein abbreviations represent different alternative splicing encoded [45] subunit proteins of the human Slack channel that are shown in the NCBI GenBank: hSlackA1: NP_065873.2; hSlackA2: NP_001258932.1; hSlackB1: XP_011517179.1; hSlackB2: XP_011517180.1; hSlackB3: XP_011517181.1; hSlackB4: XP_016870420.1; hSlackB5: XP_011517182.1; hSlackB6: XP_016870421.1; hSlackB7: XP_024303386.1. Conserved amino residues are shown with a yellow background. Non-conserved residues are shown with a cyan background. (**a**) The N-terminal alternative splicing variants of the Slack channel α subunits. (**b**) The partial C-terminal sequences of splicing variants of the Slack channel α subunits

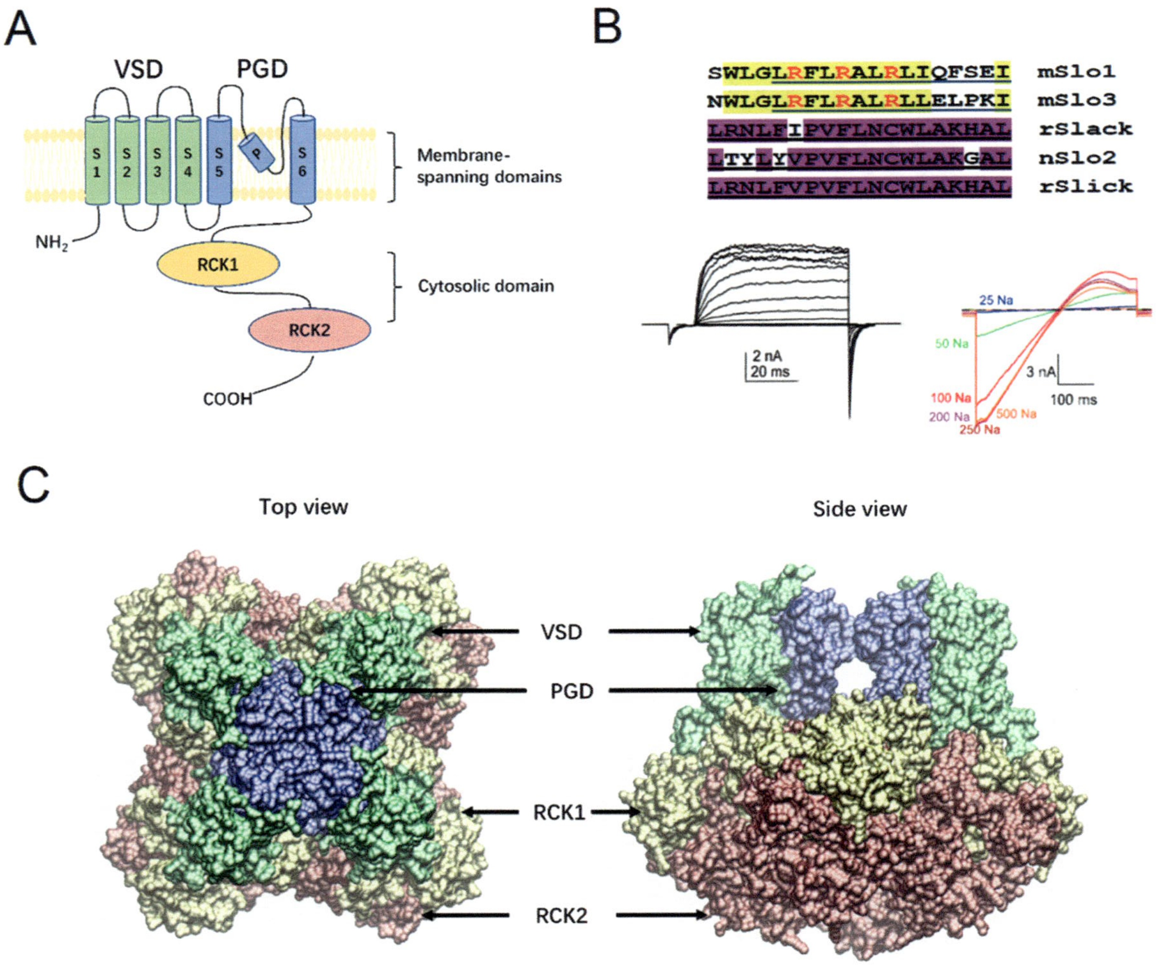

Fig. 18.2 Slack channel structure and current characteristics. (**a**) Membrane Topology of the Slack channels, which has no S0 segment but has S1–S4 transmembrane domain without the function of sensing voltage (VSD), pore-gate domain (PGD; S5, P, and S6 segments), and cytosolic domain (RCK1 and RCK2). (**b**) The S4 segments of Slo1 and Slo3 channels possess positively charged residues, which served as a voltage sensor (shown in Red). But the S4 segments of the rat Slack channel and cSlo2 do not possess positively charged residues whereas the cSlo2 shows voltage dependent current while the Slack channel current has no voltage dependence. The typical currents of the cSlo2 channel and Slack currents were recorded in an inside-out patch configuration with a −140 to 200 mV step protocol and a −100 to +100 mV ramp protocol respectively. (**c**) The Cryo-EM structure (PDB ID:5U70) of the chicken Slack channel. Different structural domains of the Slack channels are depicted in surface representation, with the same color scheme was used to show the same domain of the channel. Left: top view as seen from the extracellular side. Right: side view

[15, 16] (Fig. 18.2c). First of all, the cryo-EM structure reveals an inner helical gate that exists at the end of the S6 segment of the Slack channel. The narrowest segment of the ion permeation pathway below the selectivity filter at Met 333 residue is approximately 6 Å diameter in the closed state, which is less than the diameter of a hydrated K^+ ion (8–20 Å). The mutant M333A exhibits residual current in the absence of Na^+ confirmed the H6 helix served as a gate to control the channel open. The gating open upon sodium binding requires the conformation change of the N lobes of the RCK1 domains. In addition to pulling the S6 linker, the expansion of the N lobes also requires the interaction of the complementary surface of the TMD, and the gating ring involves the N lobes of the RCK1 domain. Second, the two nonidentical RCK domains on the C terminus from the same α subunit interact through a "flexible interface" while the four RCK domains pairs from different α subunits assemble through an "assembly interface". Different from locus Ca^{2+} binding with the Slo1 channel, the sodium activated gating ring in which the sodium ion binds near the flexible interface rather than the assembly interface [15]. Third, there are multiple closed conformations of the Slack channel while a non-conductive state (probably inactivation state) exists when the Slack channel is in the open conformation by titrating with Na^+. The open conformation emerges from an ensemble of closed conformations in a concerted manner, without evidence of Na^+-dependent intermediates [16].

For understanding the large conductance of the Slack channel, the EM-cryo structure provides a picture that a massive funnel-like gating ring with a 40 Å wide top and a gradually narrowed bottom at the pore inside the TMD domain. The inner surface of the funnel is highly electronegative because of the existence of many aspartates and glutamate amino acids. Functioned as a cation attractor, this electrostatically negative funnel contributes to the unusually high conductance. The unitary conductance of the Slack channel exhibits heterogeneity and ranges from 88 to 180 pS when expressed in symmetrical potassium solution [1, 10, 14, 18]. One reasonable explanation of the wide conductance range is the multiple sub-conductance states possessed by both native KNa channels and heterologously expressed Slack channels [1, 17].

18.3 The Phosphorylation Modulation on the Gating and Membrane Expression of the Slack Channel

In addition to being activated by sodium and Cl^- binding, other factors also regulate the gating of the Slack channel. Application of PMA outside of the Xenopus oocytes in which the Slack channel was expressed induced large enhancement of Slack channel current. In inside-out patch configuration, perfusing PKM (a constitutively active form of PKC) to the cytoplasmic face of the Slack channel remained the Slack channel activity in low Na^+ and Cl^- concentration but direct perfusion of PMA did not have any effect. This result indicated that the PKC may enhance the Slack channel activity by phosphorylating the Slack channels [19]. The possible phosphorylation site probably is S407 because the currents of the S407 mutated Slack channel were not suppressed by the Phactr-1 recruited protein phosphatase 1 (PP1) anymore [20, 21]. However, the gating of the Slack channel is not regulated by PKA whereas the internalization probably is affected by PKA [22, 23]. Besides, the p38 mitogen-activated protein kinase phosphorylation and SOD triggered MAP kinase cascade may also be involved in the regulation of membrane expression of the Slack channel [24, 25]. But the mechanism underlying these modulations needs to be further addressed.

18.4 The Expression Patterns and Physiological Function of the Slack Channel

Sodium-activated K^+ channel was first reported existing in guinea pig cardiac cells in 1984. The *K*d of this sodium-sensitive channel is 68 mM, which is approximately twofold higher than the *K*d of sodium dependency of Girk channels

Table 18.1 Distribution and major physiological functions of Slack channel

System	Physiological process	Organ/nucleus	Specie	Slack contribution	References
Nervous system	Auditory	MNTB	Mice	Accuracy of AP time	[18, 30]
	Olfactory	OB	Mice	Interact with Kv1.3	[29]
		OB	Rat	Delayed outward current	[72]
	Vestibular	VANs	Rat	AHP	[28]
	Nociceptive behavior	DRG	Rat	Neuron excitability	[32]
		DRG	Rat	Interact with PKA	[23]
		Dorsal horn	Mice	Synaptic transmission	[5]
		Spinal	Lamprey	sAHP	[34]
	Itch	DRG	Mice	Depolarization-elicited AP	[7]
	Others	Nucleus supraopticus	Rat	sAHP	[73]
Circulatory system		Heart	Guinea pig	Gated by Na^+	[26]

MNTB medial nucleus of the trapezoid body, *OB* olfactory bulb, *AP* action potential, *VANs* vestibular afferent neurons, *DRG* dorsal root ganglion, *PKA* protein kinase A, *sAHP* slow afterhyperpolarization

[26]. Subsequently, a neuronal outward K^+ current with dependence on Na^+ influx was described in cultured avian trigeminal ganglion neurons [27]. More studies were performed to investigate the distribution of the Slack channel in the central nervous system. In situ hybridization and immunohistochemistry with affinity-purified antibody found abundant expression of the Slack channel protein in the olfactory bulb, vestibular system, trigeminal system, red nucleus, deep cerebellar nuclei, thalamus, substantia nigra, amygdala, frontal cortex (the only cortical region), hippocampus, lateral septal nuclei, and some other brain regions [10, 28–31]. Marginal expression of Slack channel was also detected in the mammalian heart [1] (Table 18.1). Consistent with its expression patterns, the function study of the Slack channel is also focused on its roles in the nervous system and circulation system.

18.5 The Role of the Slack Channel in Pain and Itch Sensing

Since the Slack channel is abundantly expressed in dorsal root ganglion, spinal cord dorsal horn, and brain, the role of the Slack channel in the regulation of pain-sensing attracted much attention. The interaction of the Slack channel with chloride TMEM16C channel in pain processing was first reported because the Slack channel is also a chloride-activated channel. The rats with Slack channel knockdown by intrathecal injection of short interfering RNA exhibited increased thermal and mechanical sensitivity. The TMEM16C knockout rats also show enhanced thermal and mechanical sensitivity with an underlying mechanism that can be attributed to TMEM16C enhancing KNa channel activity in $IB4^+$ DRG neurons [32]. However, another paper using the Slack knockout mice demonstrated that global ablation of the Slack channel in mice increased hypersensitivity in models of neuropathic pain, whereas the behavior in models of inflammatory and acute nociceptive pain was normal [6]. But subsequent analysis from another group using another line of Slack knockout mice showed Slack KO mice have enhanced nociceptive responsiveness to localized thermal stimuli compared to wildtype mice [33]. Moreover, the Slack channel deletion altered intrinsic properties and synaptic drive to favor an overall enhanced excitatory tone in the dorsal horn neurons

[34]. Besides, another group also reported that the Slack channel is involved in the itching sense by enhancing the action potential of DRG neurons [7, 25]. Although the inconsistency of these data can partially be explained as the difference between species, further research is still needed to clarify the different roles of Slack channels in dorsal root ganglion and spinal cord dorsal horn.

18.6 The Possible Role of Slack Channel in FMRP Syndrome

Loss of fragile X mental retardation protein (FMRP) causes fragile X syndrome (FXS) in humans. FXS is the most common heritable form of intellectual disability that exhibits mental retardation and autism. The FMRP is an mRNA-binding protein that represses local translation of specific mRNAs, such as mGluR1/5, Kv3.1, KV4.2. Thus, loss of FMRP protein alters synaptic activity by losing the control of expression of the pre-and postsynaptic membrane proteins [35–38]. However, recent studies show the FMRP protein can directly modulate both the BK channel and the Slack channel activity by interacting with the β4 subunit of the BK channel and directly binding to the cytoplasmic carboxy-terminal tail of the Slack channel respectively, which is independent of the mRNA binding activity of FMRP protein [10, 39]. More recently, an observation on Bag cell (BC) neurons in Aplysia indicated perfusion of FMRP enhanced the open probability of the Slack channel and produced narrowing of action potentials. Suppressing the expression of Slack channel through siRNA disturbance failed to alter the ability of BC neurons to undergo a long-lasting discharge induced by synaptic stimulation, but succeeded in diminishing recovery from the inhibitory period following normal discharges [9]. Nevertheless, the latest research using Slack KO mice showed the abnormal social behavior deficits of Slack channel null mice are different from the social behavior deficits of FMRP-KO mice [40]. Thus, the role of the Slack channel in Fragile-X syndrome is still bewildering and requires further investigation.

18.7 The Basic Role of the Potassium Channel in Controlling Neuron Excitability

Following sodium inflow-induced action potential (AP), the slow afterhyperpolarization (sAHP) determines spike frequency regulation. In lamprey spinal neurons, the sodium-dependent sAHP cannot be attributed to the role of quick activation of the Na^+/K^+ pump because it can be produced even with the Na^+/K^+ pump inhibitor ouabain. High immunoreactivity of anti-Slack antibody in those neurons and quinidine sensitivity of sAHP implicated the role of Slack channel in sAHP [34]. The contribution of the Slack channel to spike frequency adaptation and neuronal excitability mediated by sAHP was also observed in the rat thalamic paraventricular nucleus (PVT) neurons and vestibular afferent neurons [28, 41].

18.8 The Potential Roles of Slack Channel in Auditory Signal Transduction

The inside-out patch-clamp recording showed sodium-dependent current existed in the neurons of the medial nucleus of the trapezoid body (MNTB). The characteristic of this sodium-dependent current is consistent with the characteristics of currents generated through the Slack channel [18, 30]. Increasing cytoplasmic Na^+ in MNTB neurons enhances temporal fidelity of action potentials of MNTB neurons' response to a high-frequency stimulus. However, the physiological consequence of loss of the Slack channel in the auditory signal transduction system remains to be investigated.

18.9 The Role of KCNT1 Channel Mutations in Epilepsy

Since the mutations in KCNT1 channel genes had been linked with the epilepsy disease from 2012 [42], 61 mutations have been identified from epileptic patients till now (Table 18.2). The KCNT1

Table 18.2 The list of 61 KCNT1 mutations associated with epilepsy

	hSlackA	hSlackBX5	Rat SlackB	References
1.	R85S	R52S	Q66S	[74]
2.	R133H	R100H	R114H	[75]
3.	R209C	R176C	H190C	[45]
4.	H257D	H224D	H238D(x)	[11]
5.	A259D	A226D	A240D	[57]
6.	A259V	A226V	A240V	[57]
7.	R262Q	R229Q	R243Q	[11]
8.	M267T	M234T	M248T	[76]
9.	Q270E	Q237E	Q251E	[77]
10.	V271F	V238F	V252F©	[48]
11.	L274I	L241I	L255I	[54]
12.	G288S	G255S	G269S©	[78]
13.	V340M	V307M	V321M	[11]
14.	F346L	F313L	F327L	[54]
15.	R356W	R323W	R337W	[76]
16.	C377S	C344S	C358S	[48]
17.	R398Q	R365Q	R379Q©	[42]
18.	R398L	R365L	R379Q	[52]
19.	P409S	P376S	P390S	[77]
20.	R428Q	R395Q	R409Q©	[44]
21.	S435C	S402C	S416C	[50]
22.	L437F	L404F	L418F	[79]
23.	R474C	R441C	R455C	[80]
24.	R474G	R441G	R455C	[64]
25.	R474H	R441H	R455H©	[44]
26.	W476R	W443R	W457R	[65]
27.	A477T	A444T	A458T	[77]
28.	D480N	D447N	D461N	[74]
29.	F502V	F469V	F483V	[54]
30.	M516V	M483V	M497V	[81]
31.	Q550del	Q517del	Q531del	[57]
32.	K629E	K596E	K610E	[82]
33.	K629N	K596N	K610N	[55]
34.	Q651R	Q618R	Q632R	[52]
35.	G652V	G619V	A633V	[59]
36.	I760M	I727M	I739M©	[44]
37.	I760F	I727F	I739F	[60]
38.	Y796H	Y763H	Y775H©	[42]
39.	E893K	E860K	E872K	[52]
40.	M896K	M863K	M875K	[54]
41.	M896I	M863I	M875I©	[42]
42.	Q906H	Q873H	Q885H	[83, 84]
43.	F909L	F876L	F888L	[76]
44.	P924L	P891L	P903L	[80]
45.	R928C	R895C	R907C©	[42]
46.	R929Q	R896Q	R908Q	[52]
47.	F932I	F899I	F911I©	[47]
48.	F932S	F899S	F911S	[53]
49.	F932L	F899L	F899L	[76]

(continued)

Table 18.2 (continued)

	hSlackA	hSlackBX5	Rat SlackB	References
50.	R933G	R900G	R912G	[82]
51.	A934T	A901T	A913T©	[44, 50]
52.	K947E	K914E	K926E	[52]
53.	R950Q	R917Q	R929Q	[11, 57]
54.	R961H	R928H	R940H	[52]
55.	R961S	R928S	R940S	[44, 52]
56.	L962P	L929P	L941P	[74]
57.	A966T	A933T	A945T©	[84]
58.	K985N	K952N	K964N	[62]
59.	R1106Q	R1073Q	R1085Q	[49]
60.	R1107H	R1074H	R1086H	[75]
61.	R1114W	R1081W	R1093W	[57, 76]

The sequence number of the mutations of the Slack associated with epilepsy on alternative splicing variants of the Slack channel

channel became one of the channels that had large numbers of mutations that are associated with epilepsy within 10 years. Those mutations caused diverse syndromes of epilepsy, such as autosomal dominant nocturnal frontal lobe epilepsy (ADNFLE) [42, 43], malignant migrating partial seizures of infancy (MMPSI) [44], Lennox-Gastaut Syndrome [45], Ohtahara syndrome (OS) [46], etc. Most patients who suffered from epilepsy associated with these mutations are infants. In addition to epilepsy, many patients also suffered from comorbidity with mental and cognitive delay, the arrest of psychomotor development, autism, intellectual disability, and hypomyelinating leukodystrophy, which suggested the role of this gene in the development of the nervous system [42, 44, 47]. Moreover, some patients have morbidities with symptoms associated with the circulation systems such as Brugada syndrome, arrhythmia, and massive systemic to pulmonary collateral arteries with life-threatening hemoptysis and heart failure [48, 49]. This phenomenon suggests that arrhythmia and epilepsy may be attributed to the same channelopathy. Many cases are treatment-resistant and have a poor outcome, high mortality, and sudden unexpected death of epilepsy (SUDEP) [50]. As the class I potassium channel inhibitor, quinidine is the most tested drug for treating Slack channel-associated epilepsy [51]. However, the therapeutic effect of quinidine is undesirable. We summarized the treatment effects of quinidine on 63 patients who suffered from MMFSI, ADNFLE, and other KCNT1-related epilepsy syndromes [45, 52–63]. The effective percentage, characterized by over 50% reduction in seizure frequency, occupied only 33.3% of those patients [45, 52–62]. In addition, significant cardiac side effects were also observed even with a low dose of quinidine [56]. In the meantime, other common anti-seizure drugs, such as sodium valproate, clonazepam, gabapentin, are also not effective [53, 64, 65]. Thus, finding an effective and safe blocker of the KCNT1 channel as a treatment drug is an important and emergent task to accomplish.

The distribution of epilepsy-associated mutations on the KCNT1 channel is wide: including the N-terminus, Pore region, RCK1 domain, RCK2 domain, and the far end of the C-terminus. Most epilepsy-associated mutations are conservative among mammalian species, but three of them (R85S, R209C, and G652V) are not conservative among mammalian species (Table 18.2). Based on the mutations that had been characterized, most of the mutations associated with epilepsy are gain-of-function mutations [42, 45]. Those mutants of the KCNT1 channel either increase the maximal Po or enhance the sodium sensitivity of the KCNT1 channel. Some mutants have double effects [66]. However, there are still many mutants that need to be further characterized.

Also, how the structural changing of these mutations alters the biophysical properties remains to be further addressed.

How the gain-of-function channel mutations alter the excitability of neurons is an important question to answer for understanding the mechanism that these mutations cause epilepsy. A previous study indicates that loxapine, like a Slack channel opener, activates native KNa channels and reduces neuronal excitability [67]. Consistent with this study, another study showed enhanced excitability and reduced action potential threshold in DRG neurons of the Slack/Slick double KO mice [6]. Thus, one plausible explanation of epilepsy caused by gain-of-function mutation of the KCNT1 channel is that these mutations lead to disinhibition by decreasing the excitability of interneurons. However, one recent study claimed that an epilepsy-associated KCNT1 mutation enhances the excitability of human iPSC-derived neurons by increasing Slack KNa currents neurons [68]. Thus, the neuronal activity, neural circuits involved in seizures, and mechanism of epilepsy caused by Slack channel mutations still need to be further investigated.

18.10 The Pharmacological Properties of the KCNT1 Channel and Potential Drugs for the Treatment of Epilepsy That Are Associated with Slack Channel Mutations

For any ion channels, finding a specific channel opener or blocker of the channel is an important task for identification the channel in vivo and studying the role of the channel in cells in which they are expressed in. It is a special emergent task for finding a safe and effective blocker targeted the Slack channel as a drug for treating refractory seizures because the numbers of gain-of-function KCNT1 mutations that caused seizure disorders are sharply growing. It has been proposed that antipsychotic drug loxapine and niclosamide worked as effective openers of the Slack channel withEC50 4.4 μM and 2.9 μM respectively [67]. But high specific and efficient blocker of the Slack channel has not been found. As the most used drug for treating epilepsy, quinidine has been reported as an effective blocker of the Slack channel in two studies by whole-cell recording in HEK cells, two groups got the EC50 89.6 μM and 125 μM respectively [51, 69]. As a result, the blockade of quinidine on the Slack is not strong enough as an effective drug for the treatment of Slack mutations associated with epilepsy, especially on the mutations that are closed to the S6 gate, such as F346L [57, 69]. Two studies also reported that the blockade of whole-cell current generated through the Slack channel by another inhibitor bepridil with EC50 approximate 1 μM or 6 μM, respectively [51]. A series of chemical compounds with a similar structure to bepridil were tested for the blockade effect on the Slack channel as a drug candidate for treatment epilepsy [69]. Earlier experiments were performed on cardiac cells that exhibited the sodium-activated potassium current can be inhibited approximately 50% by verapamil (1 μM), bepridil (1 μM), amiodarone (1 μM) [70, 71] in an inside-out patch configuration. However, these studies only tested one dose of these drugs without giving IC50. Thus, a lot of effort and computer-aided design are needed for searching for an effective specific blocker of the KCNT1 channel for treating epilepsy.

18.11 Conclusive Remarks

The studies on the Slack channel have a relatively brief history since the first KCNT1 gene was cloned in 1998 [2]. Recent advancements have identified the gating ions, different alternative splicing, and the functional domain that sodium binding to activate the Slack channel [3, 12]. However, we are still in the initial stage to understand the multiple conformations that exist in the closed state and open state, especially the conformation change coupling the sodium binding with the gating open. The recently published cryo-EM structures have provided information that how the quaternary subunits were assembled to form a functional channel so that we can build up an accurate model to simulate the gating process.

Recent results also indicate how the functional characteristics of the Slack channel are altered by some epilepsy-associated mutations [3]. Dissecting the details of gating mechanisms of the KCNT1 channel will be beneficial for understanding how the epilepsy-related mutations alter the gating properties of the Slack channel but also the development of effective blockers of the Slack channel as treatment drugs.

References

1. Yuan A et al (2003) The sodium-activated potassium channel is encoded by a member of the Slo gene family. Neuron 37:765–773. https://doi.org/10.1016/S0896-6273(03)00096-5
2. Joiner WJ et al (1998) Formation of intermediate-conductance calcium-activated potassium channels by interaction of Slack and Slo subunits. Nat Neurosci 1:462–469. https://doi.org/10.1038/2176
3. Zhang Z, Rosenhouse-Dantsker A, Tang QY, Noskov S, Logothetis DE (2010) The RCK2 domain uses a coordination site present in Kir channels to confer sodium sensitivity to Slo2.2 channels. J Neurosci 30:7554–7562. https://doi.org/10.1523/JNEUROSCI.0525-10.2010
4. Zhang Z et al (2013) SLO-2 isoforms with unique Ca (2+) - and voltage-dependence characteristics confer sensitivity to hypoxia in C. elegans. Channels (Austin) 7:194–205. https://doi.org/10.4161/chan.24492
5. Evely KM et al (2017) Slack KNa channels influence dorsal horn synapses and nociceptive behavior. Mol Pain 13:1744806917714342. https://doi.org/10.1177/1744806917714342
6. Lu R et al (2015) Slack channels expressed in sensory neurons control neuropathic pain in mice. J Neurosci 35:1125–1135. https://doi.org/10.1523/jneurosci.2423-14.2015
7. Martinez-Espinosa PL et al (2015) Knockout of Slo2.2 enhances itch, abolishes KNa current, and increases action potential firing frequency in DRG neurons. Elife 4:e10013. https://doi.org/10.7554/eLife.10013
8. Bausch AE et al (2015) The sodium-activated potassium channel Slack is required for optimal cognitive flexibility in mice. Learn Mem 22:323–335. https://doi.org/10.1101/lm.037820.114
9. Zhang Y et al (2012) Regulation of neuronal excitability by interaction of fragile X mental retardation protein with slack potassium channels. J Neurosci 32:15318–15327. https://doi.org/10.1523/JNEUROSCI.2162-12.2012
10. Brown MR et al (2010) Fragile X mental retardation protein controls gating of the sodium-activated potassium channel Slack. Nat Neurosci 13:819–821. https://doi.org/10.1038/nn.2563
11. Moller RS et al (2015) Mutations in KCNT1 cause a spectrum of focal epilepsies. Epilepsia 56:e114–e120. https://doi.org/10.1111/epi.13071
12. Brown MR et al (2008) Amino-termini isoforms of the Slack K^+ channel, regulated by alternative promoters, differentially modulate rhythmic firing and adaptation. J Physiol 586:5161–5179. https://doi.org/10.1113/jphysiol.2008.160861
13. Nagase T et al (2000) Prediction of the coding sequences of unidentified human genes. XIX. The complete sequences of 100 new cDNA clones from brain which code for large proteins in vitro. DNA Res 7:347–355. https://doi.org/10.1093/dnares/7.6.347
14. Chen H et al (2009) The N-terminal domain of Slack determines the formation and trafficking of Slick/Slack heteromeric sodium-activated potassium channels. J Neurosci 29:5654–5665. https://doi.org/10.1523/JNEUROSCI.5978-08.2009
15. Hite RK et al (2015) Cryo-electron microscopy structure of the Slo2.2 Na(+)-activated K(+) channel. Nature 527:198–203. https://doi.org/10.1038/nature14958
16. Hite RK, MacKinnon R (2017) Structural titration of Slo2.2, a Na(+)-dependent K(+) channel. Cell 168:390–399.e311. https://doi.org/10.1016/j.cell.2016.12.030
17. Tamsett TJ, Picchione KE, Bhattacharjee A (2009) NAD+ activates KNa channels in dorsal root ganglion neurons. J Neurosci 29:5127–5134. https://doi.org/10.1523/JNEUROSCI.0859-09.2009
18. Yang B, Desai R, Kaczmarek LK (2007) Slack and Slick K(Na) channels regulate the accuracy of timing of auditory neurons. J Neurosci 27:2617–2627. https://doi.org/10.1523/JNEUROSCI.5308-06.2007
19. Santi CM et al (2006) Opposite regulation of Slick and Slack K^+ channels by neuromodulators. J Neurosci 26:5059–5068. https://doi.org/10.1523/JNEUROSCI.3372-05.2006
20. Ali SR, Malone TJ, Zhang Y, Prechova M, Kaczmarek LK (2019) Phactr1 regulates Slack (KCNT1) channels via protein phosphatase 1 (PP1). FASEB J 34:1591–1601. https://doi.org/10.1096/fj.201902366R
21. Fleming MR et al (2016) Stimulation of Slack K(+) channels alters mass at the plasma membrane by triggering dissociation of a phosphatase-regulatory complex. Cell Rep 16:2281–2288. https://doi.org/10.1016/j.celrep.2016.07.024
22. Nuwer MO, Picchione KE, Bhattacharjee A (2009) cAMP-dependent kinase does not modulate the Slack sodium-activated potassium channel. Neuropharmacology 57:219–226. https://doi.org/10.1016/j.neuropharm.2009.06.006
23. Nuwer MO, Picchione KE, Bhattacharjee A (2010) PKA-induced internalization of slack KNa channels produces dorsal root ganglion neuron hyperexcitability. J Neurosci 30:14165–14172. https://doi.org/10.1523/JNEUROSCI.3150-10.2010

24. Gururaj S, Fleites J, Bhattacharjee A (2016) Slack sodium-activated potassium channel membrane expression requires p38 mitogen-activated protein kinase phosphorylation. Neuropharmacology 103:279–289. https://doi.org/10.1016/j.neuropharm.2015.12.016
25. Zhang Y, Ni W, Horwich AL, Kaczmarek LK (2017) An ALS-associated mutant SOD1 rapidly suppresses KCNT1 (Slack) Na(+)-activated K(+) channels in aplysia neurons. J Neurosci 37:2258–2265. https://doi.org/10.1523/JNEUROSCI.3102-16.2017
26. Kameyama M et al (1984) Intracellular Na^+ activates a K^+ channel in mammalian cardiac cells. Nature 309:354–356. https://doi.org/10.1038/309354a0
27. Bader CR, Bernheim L, Bertrand D (1985) Sodium-activated potassium current in cultured avian neurones. Nature 317:540–542. https://doi.org/10.1038/317540a0
28. Cervantes B, Vega R, Limon A, Soto E (2013) Identity, expression and functional role of the sodium-activated potassium current in vestibular ganglion afferent neurons. Neuroscience 240:163–175. https://doi.org/10.1016/j.neuroscience.2013.02.052
29. Lu S, Das P, Fadool DA, Kaczmarek LK (2010) The slack sodium-activated potassium channel provides a major outward current in olfactory neurons of Kv1.3−/− super-smeller mice. J Neurophysiol 103:3311–3319. https://doi.org/10.1152/jn.00607.2009
30. Bhattacharjee A, Gan L, Kaczmarek LK (2002) Localization of the Slack potassium channel in the rat central nervous system. J Comp Neurol 454:241–254. https://doi.org/10.1002/cne.10439
31. Rizzi S, Knaus HG, Schwarzer C (2016) Differential distribution of the sodium-activated potassium channels slick and slack in mouse brain. J Comp Neurol 524:2093–2116. https://doi.org/10.1002/cne.23934
32. Huang F et al (2013) TMEM16C facilitates Na(+)-activated K^+ currents in rat sensory neurons and regulates pain processing. Nat Neurosci 16:1284–1290. https://doi.org/10.1038/nn.3468
33. Zhang D et al (2020) Deficiency of SCAMP5 leads to pediatric epilepsy and dysregulation of neurotransmitter release in the brain. Hum Genet 139:545–555. https://doi.org/10.1007/s00439-020-02123-9
34. Wallen P et al (2007) Sodium-dependent potassium channels of a Slack-like subtype contribute to the slow afterhyperpolarization in lamprey spinal neurons. J Physiol 585:75–90. https://doi.org/10.1113/jphysiol.2007.138156
35. Bear MF, Huber KM, Warren ST (2004) The mGluR theory of fragile X mental retardation. Trends Neurosci 27:370–377. https://doi.org/10.1016/j.tins.2004.04.009
36. Strumbos JG, Brown MR, Kronengold J, Polley DB, Kaczmarek LK (2010) Fragile X mental retardation protein is required for rapid experience-dependent regulation of the potassium channel Kv3.1b. J Neurosci 30:10263–10271. https://doi.org/10.1523/JNEUROSCI.1125-10.2010
37. Darnell JC et al (2011) FMRP stalls ribosomal translocation on mRNAs linked to synaptic function and autism. Cell 146:247–261. https://doi.org/10.1016/j.cell.2011.06.013
38. Gross C, Yao X, Pong DL, Jeromin A, Bassell GJ (2011) Fragile X mental retardation protein regulates protein expression and mRNA translation of the potassium channel Kv4.2. J Neurosci 31:5693–5698. https://doi.org/10.1523/JNEUROSCI.6661-10.2011
39. Deng PY et al (2013) FMRP regulates neurotransmitter release and synaptic information transmission by modulating action potential duration via BK channels. Neuron 77:696–711. https://doi.org/10.1016/j.neuron.2012.12.018
40. Bausch AE et al (2018) Loss of sodium-activated potassium channel Slack and FMRP differentially affect social behavior in mice. Neuroscience 384:361–374. https://doi.org/10.1016/j.neuroscience.2018.05.040
41. Zhang L, Kolaj M, Renaud LP (2010) Ca^{2+}-dependent and Na^+-dependent K^+ conductances contribute to a slow AHP in thalamic paraventricular nucleus neurons: a novel target for orexin receptors. J Neurophysiol 104:2052–2062. https://doi.org/10.1152/jn.00320.2010
42. Heron SE et al (2012) Missense mutations in the sodium-gated potassium channel gene KCNT1 cause severe autosomal dominant nocturnal frontal lobe epilepsy. Nat Genet 44:1188–1190. https://doi.org/10.1038/ng.2440
43. Licchetta L et al (2020) Sleep-related hypermotor epilepsy (SHE): contribution of known genes in 103 patients. Seizure 74:60–64. https://doi.org/10.1016/j.seizure.2019.11.009
44. Barcia G et al (2012) De novo gain-of-function KCNT1 channel mutations cause malignant migrating partial seizures of infancy. Nat Genet 44:1255–1259. https://doi.org/10.1038/ng.2441
45. Jia Y et al (2019) Quinidine therapy for Lennox-Gastaut syndrome with KCNT1 mutation. A case report and literature review. Front Neurol 10:64. https://doi.org/10.3389/fneur.2019.00064
46. Evely KM, Pryce KD, Bhattacharjee A (2017) The Phe932Ile mutation in KCNT1 channels associated with severe epilepsy, delayed myelination and leukoencephalopathy produces a loss-of-function channel phenotype. Neuroscience 351:65–70. https://doi.org/10.1016/j.neuroscience.2017.03.035
47. Vanderver A et al (2014) Identification of a novel de novo p.Phe932Ile KCNT1 mutation in a patient with leukoencephalopathy and severe epilepsy. Pediatr Neurol 50:112–114. https://doi.org/10.1016/j.pediatrneurol.2013.06.024
48. Kawasaki Y et al (2017) Three cases of KCNT1 mutations: malignant migrating partial seizures in infancy with massive systemic to pulmonary collateral arteries. J Pediatr 191:270–274. https://doi.org/10.1016/j.jpeds.2017.08.057

49. Juang JM et al (2014) Disease-targeted sequencing of ion channel genes identifies de novo mutations in patients with non-familial Brugada syndrome. Sci Rep 4:6733. https://doi.org/10.1038/srep06733
50. Kuchenbuch M et al (2019) KCNT1 epilepsy with migrating focal seizures shows a temporal sequence with poor outcome, high mortality and SUDEP. Brain 142:2996–3008. https://doi.org/10.1093/brain/awz240
51. Yang B et al (2006) Pharmacological activation and inhibition of Slack (Slo2.2) channels. Neuropharmacology 51:896–906. https://doi.org/10.1016/j.neuropharm.2006.06.003
52. Dilena R et al (2018) Early treatment with quinidine in 2 patients with epilepsy of infancy with migrating focal seizures (EIMFS) due to gain-of-function KCNT1 mutations: functional studies, clinical responses, and critical issues for personalized therapy. Neurotherapeutics 15:1112–1126. https://doi.org/10.1007/s13311-018-0657-9
53. Fitzgerald MP et al (2019) Treatment responsiveness in KCNT1-related epilepsy. Neurotherapeutics 16:848–857. https://doi.org/10.1007/s13311-019-00739-y
54. McTague A et al (2018) Clinical and molecular characterization of KCNT1-related severe early-onset epilepsy. Neurology 90:e55–e66. https://doi.org/10.1212/WNL.0000000000004762
55. Mikati MA et al (2015) Quinidine in the treatment of KCNT1-positive epilepsies. Ann Neurol 78:995–999. https://doi.org/10.1002/ana.24520
56. Mullen SA et al (2018) Precision therapy for epilepsy due to KCNT1 mutations: a randomized trial of oral quinidine. Neurology 90:e67–e72. https://doi.org/10.1212/WNL.0000000000004769
57. Numis AL et al (2018) Lack of response to quinidine in KCNT1-related neonatal epilepsy. Epilepsia 59:1889–1898. https://doi.org/10.1111/epi.14551
58. Bearden D et al (2014) Targeted treatment of migrating partial seizures of infancy with quinidine. Ann Neurol 76:457–461. https://doi.org/10.1002/ana.24229
59. Fukuoka M et al (2017) Quinidine therapy for West syndrome with KCNTI mutation: a case report. Brain Dev 39:80–83. https://doi.org/10.1016/j.braindev.2016.08.002
60. Baumer FM, Sheehan M (2017) Quinidine-associated skin discoloration in KCNT1-associated pediatric epilepsy. Neurology 89:2212. https://doi.org/10.1212/WNL.0000000000004674
61. Milligan CJ et al (2014) KCNT1 gain of function in 2 epilepsy phenotypes is reversed by quinidine. Ann Neurol 75:581–590. https://doi.org/10.1002/ana.24128
62. Abdelnour E et al (2018) Does age affect response to quinidine in patients with KCNT1 mutations? Report of three new cases and review of the literature. Seizure 55:1–3. https://doi.org/10.1016/j.seizure.2017.11.017
63. Hebbar M, Mefford HC et al (2020) F1000Res 9. https://doi.org/10.12688/f1000research.21366.1
64. Datta AN, Michoulas A, Guella I, Study E, Demos M (2019) Two patients with KCNT1-related epilepsy responding to phenobarbital and potassium bromide. J Child Neurol 34:728–734. https://doi.org/10.1177/0883073819854853
65. Poisson K, Wong M, Lee C, Cilio MR (2020) Response to cannabidiol in epilepsy of infancy with migrating focal seizures associated with KCNT1 mutations: an open-label, prospective, interventional study. Eur J Paediatr Neurol 25:77–81. https://doi.org/10.1016/j.ejpn.2019.12.024
66. Tang Q-Y et al (2016) Epilepsy-related Slack channel mutants lead to channel over-activity by two different mechanisms. Cell Rep 14:129–139. https://doi.org/10.1016/j.celrep.2015.12.019
67. Biton B et al (2012) The antipsychotic drug loxapine is an opener of the sodium-activated potassium channel Slack (Slo2.2). J Pharmacol Exp Ther 340:706–715. https://doi.org/10.1124/jpet.111.184622
68. Quraishi IH et al (2019) An epilepsy-associated KCNT1 mutation enhances excitability of human iPSC-derived neurons by increasing Slack KNa currents. J Neurosci 39:7438–7449. https://doi.org/10.1523/JNEUROSCI.1628-18.2019
69. Cole BA et al (2020) Structure-based identification and characterization of inhibitors of the epilepsy-associated KNa1.1 (KCNT1) potassium channel. iScience 23:101100. https://doi.org/10.1016/j.isci.2020.101100
70. Mori K, Kobayashi S, Saito T, Masuda Y, Nakaya H (1998) Inhibitory effects of class I and IV antiarrhythmic drugs on the Na^+-activated K^+ channel current in guinea pig ventricular cells. Naunyn Schmiedeberg Arch Pharmacol 358:641–648. https://doi.org/10.1007/pl00005306
71. Mori K, Saito T, Masuda Y, Nakaya H (1996) Effects of class III antiarrhythmic drugs on the Na(+)-activated K^+ channels in guinea-pig ventricular cells. Br J Pharmacol 119:133–141. https://doi.org/10.1111/j.1476-5381.1996.tb15686.x
72. Budelli G et al (2009) Na^+-activated K^+ channels express a large delayed outward current in neurons during normal physiology. Nat Neurosci 12:745–750. https://doi.org/10.1038/nn.2313
73. Bansal V, Fisher TE (2016) Na(+)-activated K(+) channels in rat supraoptic neurones. J Neuroendocrinol 28. https://doi.org/10.1111/jne.12394
74. Borlot F et al (2020) KCNT1-related epilepsy: an international multicenter cohort of 27 pediatric cases. Epilepsia 61:679–692. https://doi.org/10.1111/epi.16480
75. Hildebrand MS et al (2016) A targeted resequencing gene panel for focal epilepsy. Neurology 86:1605–1612. https://doi.org/10.1212/WNL.0000000000002608
76. Burgess R et al (2019) The genetic landscape of epilepsy of infancy with migrating focal seizures. Ann Neurol 86:821–831. https://doi.org/10.1002/ana.25619

77. Ohba C et al (2015) De novo KCNT1 mutations in early-onset epileptic encephalopathy. Epilepsia 56: e121–e128. https://doi.org/10.1111/epi.13072
78. Ishii A et al (2013) A recurrent KCNT1 mutation in two sporadic cases with malignant migrating partial seizures in infancy. Gene 531:467–471. https://doi.org/10.1016/j.gene.2013.08.096
79. Gertler TS, Thompson CH, Vanoye CG, Millichap JJ, George AL Jr (2019) Functional consequences of a KCNT1 variant associated with status dystonicus and early-onset infantile encephalopathy. Ann Clin Transl Neurol 6:1606–1615. https://doi.org/10.1002/acn3.50847
80. Shimada S et al (2014) A novel KCNT1 mutation in a Japanese patient with epilepsy of infancy with migrating focal seizures. Hum Genome Var 1:14027. https://doi.org/10.1038/hgv.2014.27
81. Rizzo F et al (2016) Characterization of two de novoKCNT1 mutations in children with malignant migrating partial seizures in infancy. Mol Cell Neurosci 72:54–63. https://doi.org/10.1016/j.mcn.2016.01.004
82. Chen Y et al (2018) Genetic and clinical analysis of children with early-onset epilepsy encephalopathy caused by KCNT1 gene mutation. Zhonghua Er Ke Za Zhi 56:824–828. https://doi.org/10.3760/cma.j.issn.0578-1310.2018.11.007
83. Allen NM et al (2016) Unexplained early onset epileptic encephalopathy: exome screening and phenotype expansion. Epilepsia 57:e12–e17. https://doi.org/10.1111/epi.13250
84. Martin HC et al (2014) Clinical whole-genome sequencing in severe early-onset epilepsy reveals new genes and improves molecular diagnosis. Hum Mol Genet 23:3200–3211. https://doi.org/10.1093/hmg/ddu030

Ion Channels in Anesthesia

19

Wei Zhou and Zhonghui Guan

Abstract

Ion channels play a pivotal role in anesthesia, including general and regional anesthesia. Two main classes of general anesthetics (GAs) are inhalational anesthetics, such as isoflurane, sevoflurane, and nitrous oxide; injectable anesthetics, such as propofol, etomidate, and ketamine. Besides hypnotic agents, muscle relaxants for immobility and opioids for analgesia are needed to achieve balanced anesthesia. Although our understanding of anesthesia is far from complete, recent studies have revealed the molecular interactions between anesthetic drugs and ion channels, particularly, the ligand-gated ion channels (LGICs). Ionotropic $GABA_A$ receptors ($GABA_ARs$), the main mediators of the inhibitory signals in the central nervous system (CNS), are the key to hypnosis by general anesthetics. Ionotropic cholinergic receptors (nAChRs), expressed at the neuromuscular junction and the nervous system, are the molecular targets of muscle relaxants. $GABA_ARs$ and nAChRs belong to the same family of pentameric LGICs. With a completely different architecture, ionotropic glutamate receptors (iGluRs) carry the excitatory signals in the CNS and are targeted by inhalational anesthetics and ketamine. Another distinct family of ion channels, two-pore-domain K^+ (K2P) channels, can be activated by inhalational anesthetics and cause neuron hyperpolarization. In this chapter, we will discuss the recent advance in understanding the molecular mechanisms underlying anesthesia through the molecular structures of these ion channels.

Keywords

Anesthetics · $GABA_A$ receptors · Cholinergic receptors · Glutamate receptors · Two-pore-domain K^+ channels · Isoflurane · Sevoflurane · Nitrous oxide · Propofol · Etomidate · Ketamine

Hypnosis, amnesia, immobility, and analgesia are the four main targets anesthesiologists aim to achieve in the operating room to establish an appropriate general anesthesia state [1]. Anesthetics can be categorized into two major families: general anesthetics and local anesthetics. General anesthetics (GAs) are administered through the blood circulation and mainly work in the central nervous system (CNS) to induce and maintain the reversible anesthesia coma state. In contrast, local anesthetics are applied either near the spinal cord or the peripheral nerves to introduce regional analgesia. It is a routine practice to combine general and local

W. Zhou (✉) · Z. Guan
Department of Anesthesia and Perioperative Care, University of California San Francisco, San Francisco, CA, USA
e-mail: wei.zhou@ucsf.edu; Zhonghui.guan@ucsf.edu

L. Zhou (ed.), *Ion Channels in Biophysics and Physiology*, Advances in Experimental Medicine and Biology 1349, https://doi.org/10.1007/978-981-16-4254-8_19

anesthetics to create balanced anesthesia for better safety and efficiency.

Two main ways to deliver GAs are inhalation through the airway into the lungs followed by absorption in the pulmonary circulation, or injection either intravenously or intramuscularly into the systemic circulation. Common inhalational anesthetics include volatile ether-derived halogenated agents such as isoflurane, sevoflurane, desflurane; and gaseous nitrous oxide (Fig. 19.1). Older generations of volatile anesthetics, halothane, enflurane, and methoxyflurane, are rarely used clinically anymore. Common injectable anesthetics are propofol, etomidate, benzodiazepines, and ketamine. Barbiturates such as methohexital and thiopental are less used clinically due to their adverse effects.

Known as hypnotic agents, GAs produce hypnosis and amnesia. Although GAs can produce a certain degree of immobility, it is not sufficient for intubation or other invasive procedures. Therefore, another important category of drugs to achieve the goal of immobility is the muscle relaxant, also known as neuromuscular blocking agents (NMBs). GAs and NMBs are both used for the purpose of immobility, however, through different mechanisms. A commonly used term to gauge the dosing of the inhalational anesthetics is the minimum alveolar concentration (MAC), which indicates the alveolar concentration of inhaled anesthetic that produces immobility in response to a noxious stimulus in 50% of subjects. Although much work is still needed to understand how GAs work, it is generally recognized that they work through the molecular targets in the brain and spinal cord. In contrast, the NMBs are well known to target the neuromuscular junction to either block the nicotinic acetylcholine receptors (nAChRs) as antagonists or over-stimulate the nAChRs to cause the receptor desensitization and muscle paralysis (Fig. 19.1). The nAChRs are non-selective cation channels that are permeable to Na^+, K^+, and Ca^{2+}, with net effects of Na^+ and Ca^{2+} entering and K^+ exiting the muscle cells. One crucial aspect of balanced anesthesia is to maintain immobility with the specific agents—NMBs—in order to spare the dosage of the hypnotic agents for the purpose of faster recovery from anesthesia.

Similarly, the analgesia effect produced by GAs alone is incomplete. An indispensable part of anesthesia management is opioids, whose role in analgesia simply cannot be replaced by hypnotic agents. Distinct from other classes of anesthetic drugs, which target ligand-gated ion channels (LGICs), opioids target the G-protein-coupled receptors (GPCRs). Many studies suggested the involvement of G-protein gated inwardly rectifying K^+ channels and voltage-gated Ca^{2+} channels downstream of the opioid signaling pathway [2–4].

Thus, ion channels and receptors are the key signaling molecules in the body, particularly in the nervous system. Anesthetics execute their functions through interactions with the membrane-bound channels and receptors. Here we discuss the direct interactions between anesthetics and ion channels.

19.1 $GABA_A$ Receptor

γ-Aminobutyric acid (GABA) is the primary inhibitory neurotransmitter in CNS to increase the membrane permeability to Cl^- and HCO_3^-. $GABA_A$ receptors ($GABA_AR$s) belong to a family of pentameric ligand-gated ion channels (pLGICs), also known as Cys-loop receptors. This superfamily of pLGICs, sharing similar overall architecture, includes cation-selective 5-HT_3 receptors (5-HT_3Rs) and nAChRs, and anion-selective $GABA_AR$s and glycine receptors, and invertebrate glutamate-gated chloride channels (GluCl) (Fig. 19.2) [5–7]. Cys-loop receptors play a major role in neural functions and are the targets of many neurological agents such as nicotine, alcohol, benzodiazepines, barbiturates, steroids, local anesthetics, and general anesthetics [8].

$GABA_AR$s are heteropentamers, consisting of subunits from six α subunits, three β subunits, three γ subunits, and others. The most common one in the brain is composed of two α_1, two β_2, and one γ_2. Subunit composition and expression patterns are the two key components affecting the

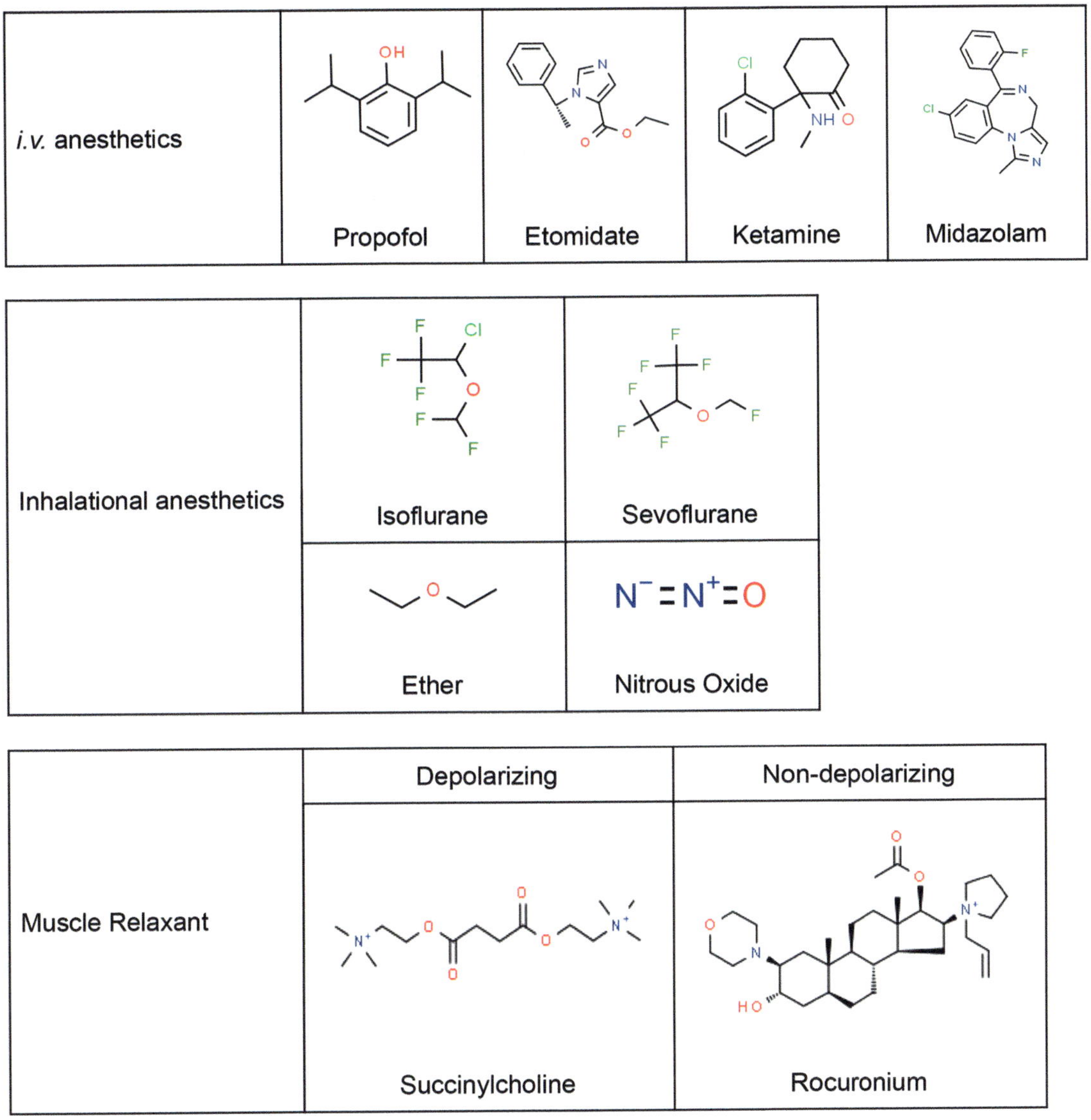

Fig. 19.1 Common anesthetic drugs

roles of the inhibitory currents generated by $GABA_AR$s. $GABA_AR$s, expressed at the postsynaptic membrane, generate fast inhibitory postsynaptic currents (IPSCs) and mediate phasic synaptic inhibition. Most general anesthetics are known to potentiate $GABA_AR$ activity, except Xenon and cyclopropane (non-halogenated alkane), which mainly target NMDA receptors [9, 10].

In addition to the conventional phasic synaptic transmission, a tonic inhibition by low ambient concentrations of GABA targeting the perisomatic and extrasynaptic receptors have been shown throughout the brain and is crucial in regulating the neuron's input conductance and so the excitability of neural circuitry [11, 12]. Anesthetic drugs propofol and midazolam have been shown to potentiate more tonic currents than IPSCs, which represent two different ways how general anesthetics work [13].

Indeed, different $GABA_AR$ isoforms have been shown to be involved in the different aspects

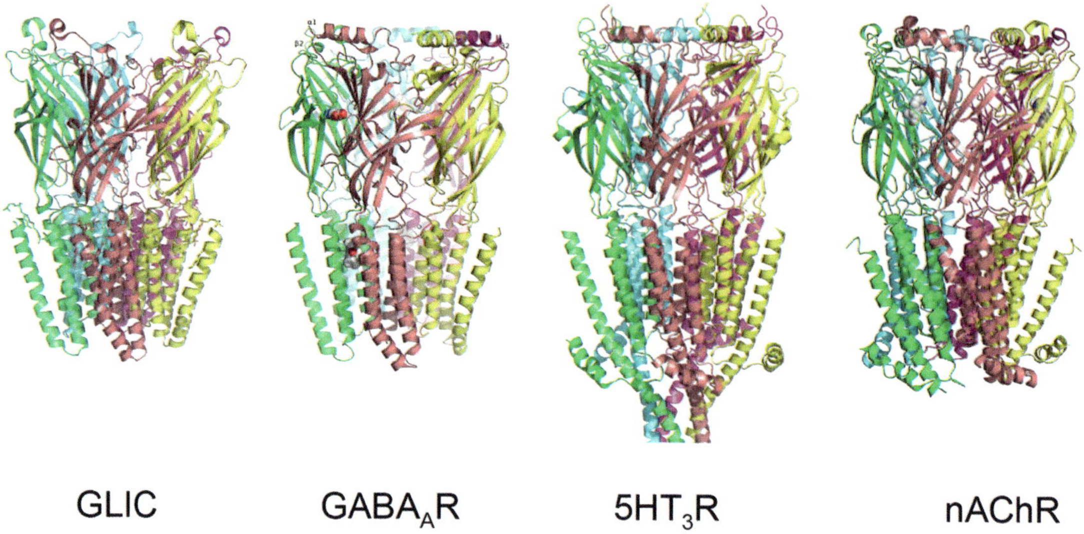

Fig. 19.2 Structures of pentameric ligand-gated ion channels (pLGIC), including GLIC, GABAAR, 5HT3R, nAChR. They share the same overall architecture. The corresponding PDB IDs are 3P50 (GLIC), 6X3T (GABAAR), 4PIR (5HT3R), and 5KXI (nAChR)

of anesthesia. GABA$_A$Rs with $\alpha_1\beta_2\gamma_2$ are diffusely expressed throughout the brain and are thought to be the key for anesthetic sedation [14, 15]. Experiments using brain slices and cultured neurons showed that the tonic currents by α_5GABA$_A$Rs can be potentiated by propofol, isoflurane, and etomidate [16–18], and the α_5 subunit seems to have a restricted expression pattern in the brain and might be involved in the amnesia effect by anesthetics. The mutant mice with β_3(N265M) subunit, a mutation in transmembrane M2 segment, showed failure in inducing the loss of hind-limb withdrawal reflex (LHLR) by intravenous etomidate and propofol [19]. These mice also showed a significantly shorter duration of loss of righting reflex induced by intravenous agents but not volatile ones, suggesting different mechanisms.

Based on the evidence that anesthetics enhance GABA receptor β1 subunits and glycine receptors, but not the GABA ρ1 receptors, by creating chimera channels between Glycine receptor and GABA$_A$ receptor, the authors identified critical amino acids in the transmembrane, M2 and M3 are involved in the allosteric modulation by alcohols and volatile anesthetics [20]. M2 and M3 were also found to be important for the *i.v.* anesthetics such as propofol and etomidate.

In 2011, the crystal structures of a pH-gated bacterial homolog, GLIC, of homo-pentameric ligand-gated ion channels (pLGICs), in complex with general anesthetics propofol or desflurane, revealed a binding pocket in the upper portion of the transmembrane domain (TMD) through van der Waals interactions and hydrogen bonds [21]. The structural data also suggest that endogenous lipids and/or neurosteroids play important roles in the channel activity, which might be targeted by the GAs [22]. Given the nAChRs belong to the same superfamily and can also be blocked by general anesthetics, the GLIC structure also shares a striking similarity with nAChR structure, which suggests that intrasubunit and/or intersubunit binding pockets in the extracellular part of the transmembrane domain mediate the general anesthetic binding. One should note that propofol potentiates GABA$_A$Rs whereas inhibits bacterial GLIC. Interestingly, the crystal structure of the human GABA$_A$R with β_3 homo-pentamer showed large binding pockets in the transmembrane domain in a similar fashion but with distinct structure details [23]. Although the crystal structure of the $GABA_AR$-β_3 didn't have propofol bound, the binding pocket data is consistent with other mutagenesis study results [24].

Further studies on high-resolution structures of bacterial homologs in presumed open and closed conformations showed the differences in the domain organization and shed light on the gating mechanism [25–28]. One prokaryotic pLGIC, ELIC, can be activated by GABA and modulated by benzodiazepines. The structure of ELIC in complex with GABA and benzodiazepines showed that the binding pocket of GABA in the extracellular domain is contributed by the neighboring subunits, while the benzodiazepine binds into two different sites with one intersubunit and the other one intrasubunit [29]. The GABA binding also shares striking similarities with the glutamate binding in the structure of GluCl [5] and the nicotine binding in the structure of the acetylcholine-binding protein [30].

Very recently, the cryo-EM structures of the human $\alpha_1\beta_2\gamma_2$ $GABA_ARs$ (β_2-α_1-β_2-α_1-γ_2), in complex with GABA and the potentiators (diazepam, etomidate, propofol, and barbiturate), or the antagonist (flumazenil), revealed a similar theme of the intersubunit binding pockets for GABA and the modulators (Figs. 19.3 and 19.4) [31]. Etomidate and propofol share two identical binding pockets at the two β_2–α_1 interfaces in TMD. Phenobarbital shares a similar TMD binding pocket, however, located at the other two interfaces, γ_2–β_2 and α_1–β_2. Further comparison with the other pGLIC structures of nicotine-bound nAChR and palonosetron-bound $5\text{-}HT_{3A}R$ showed remarkably similar binding pockets with loop C shared by all these ligands and modulators (Fig. 19.5) [32–34]. Comparing the positive modulator-bound structures with the negative ones suggested that benzodiazepine and anesthetics close the gaps between the subunits and stabilize the channel conformation. In contrast, the binding to the ECD α_1–γ_2 interface by flumazenil, a clinically used anesthesia reversal agent, destabilizes the structure so that the accesses to diazepam, etomidate, and propofol, are blocked by a long-range allosteric mechanism. Therefore, all these structures suggest that the subtle conformational changes triggered by the binding of ligands and modulators can lead to long-range allosteric changes of the transmembrane pore conformations [33, 35, 36].

19.2 Cholinergic Receptor

Cholinergic receptors (AChRs) are divided into two distinct families, ionotropic receptors (a.k.a nicotinic AChR) and metabotropic receptors (a.k. a muscarinic AChR). nAChRs belong to the pLGIC superfamily and can be further divided into Nm, expressed at the neuromuscular junction, and Nn, found in the nervous system. nAChR permeates Na^+, K^+, and Ca^{2+}. The opening of one nAChR channel by two acetylcholine ligands at the neuromuscular junction leads to the depolarizing end-plate potential followed by action potential through voltage-gated Na^+ channels and muscle contraction. The wide distribution of nAChR in the central and peripheral nervous system in addition to the neuromuscular junction suggests the involvement in autonomic regulation, which is an important aspect of anesthesia.

nAChRs share conserved pentameric architecture as $GABA_ARs$. The composition of the subunits also varies depending on the locations of the channels. The Nm receptors consist of 2α and β, δ, and γ or ε, whereas the Nn receptors are more complex and less understood. The knockout mice lacking α_4 or β_2 only showed minimal cognitive deficits [37, 38]. In contrast, mice lacking α_3 in the ganglionic Nn receptors have megacystis and mydriasis, suggesting a crucial role in autonomic regulation [39].

The physiological functions of the CNS Nn receptors are still not clear. It's thought that the activation of Nn receptors in the brain is involved in fine-tuning the balance between excitatory and inhibitory circuits to enhance the induction of long-term potentiation [40, 41]. The Nn receptors in the spinal cord play a certain role in nociception. Alleviation of allodynia in the neuropathic rat model by intrathecal clonidine is thought to involve the nAChRs and mAChRs in the spinal cord [42]. Furthermore, the knockout

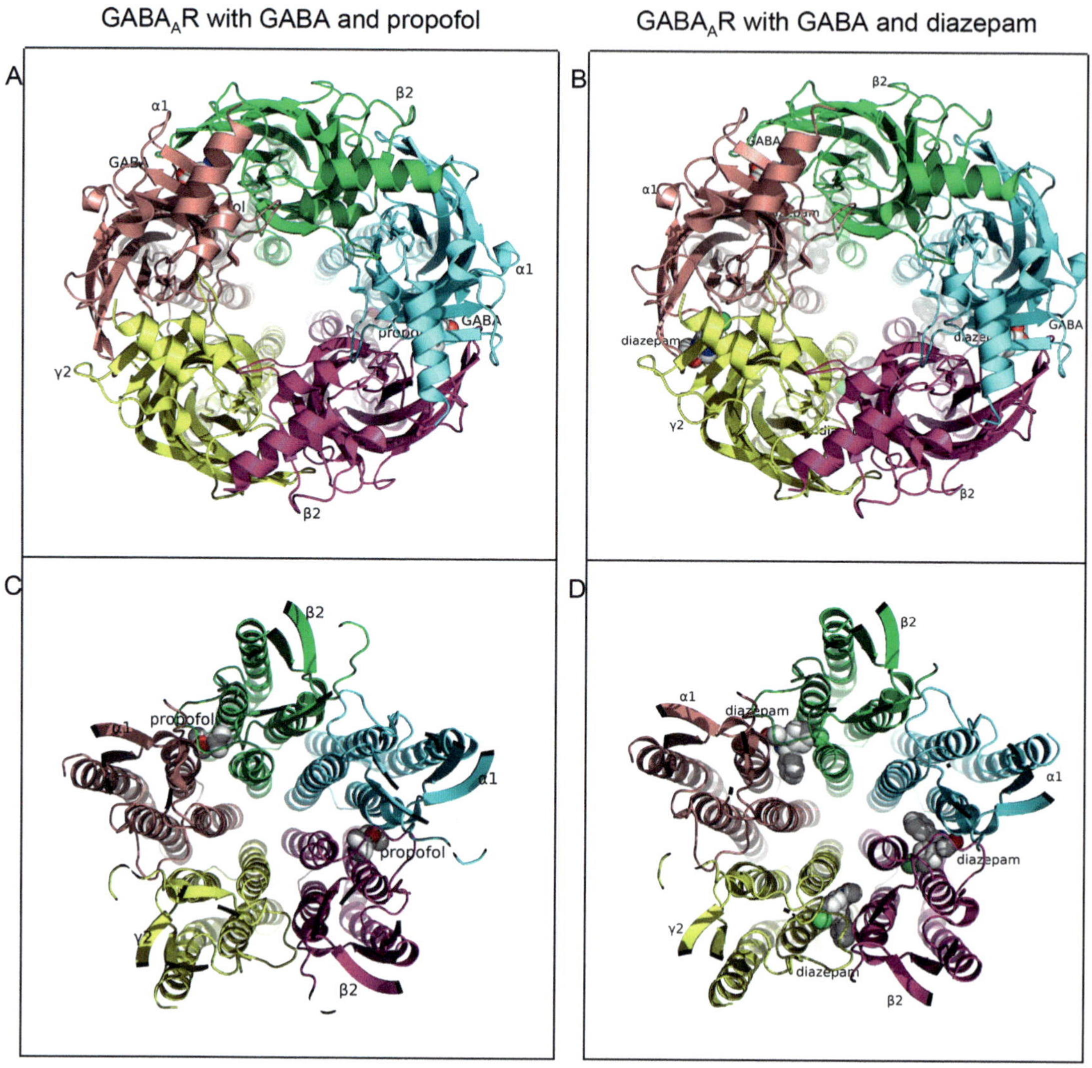

Fig. 19.3 The intersubunit binding pockets for GABA, diazepam, and propofol. The models of GABA$_A$R in complex with GABA and propofol (**a** and **c**), GABA$_A$R in complex with GABA and diazepam (**b** and **c**). The views are from outside into the cell, highlighting the ECD (**a** and **b**) and TMD (**c** and **d**). The GABA, diazepam, and propofol molecules are shown as spheres. PDB ID: 6X3T (**a** and **c**), 6X3X (**b** and **d**)

mice lacking α_4 or β_2 showed less antinociceptive effect by nicotine than the wild-type ones on the hot-plate and tail-flick tests [37].

Two main classes of muscle relaxants, widely used in the operating rooms, are nondepolarizing curarine agents and depolarizing succinylcholine. Rocuronium is the most commonly used nondepolarizing agent and blocks the binding of acetylcholine ligands onto the receptor by mimicking the structure of the acetylcholine molecule. Each succinylcholine molecule is composed of two acetylcholine molecules connected via the acetyl groups. One interesting crystal structure of the ACh binding protein (AChBP) in complex with ACh actually showed the density of two ACh molecules in each of five binding pockets [43]. The succinylcholine molecules can overstimulate the nAChRs to produce muscle twitching throughout the body and eventually lead to desensitization and so muscle paralysis.

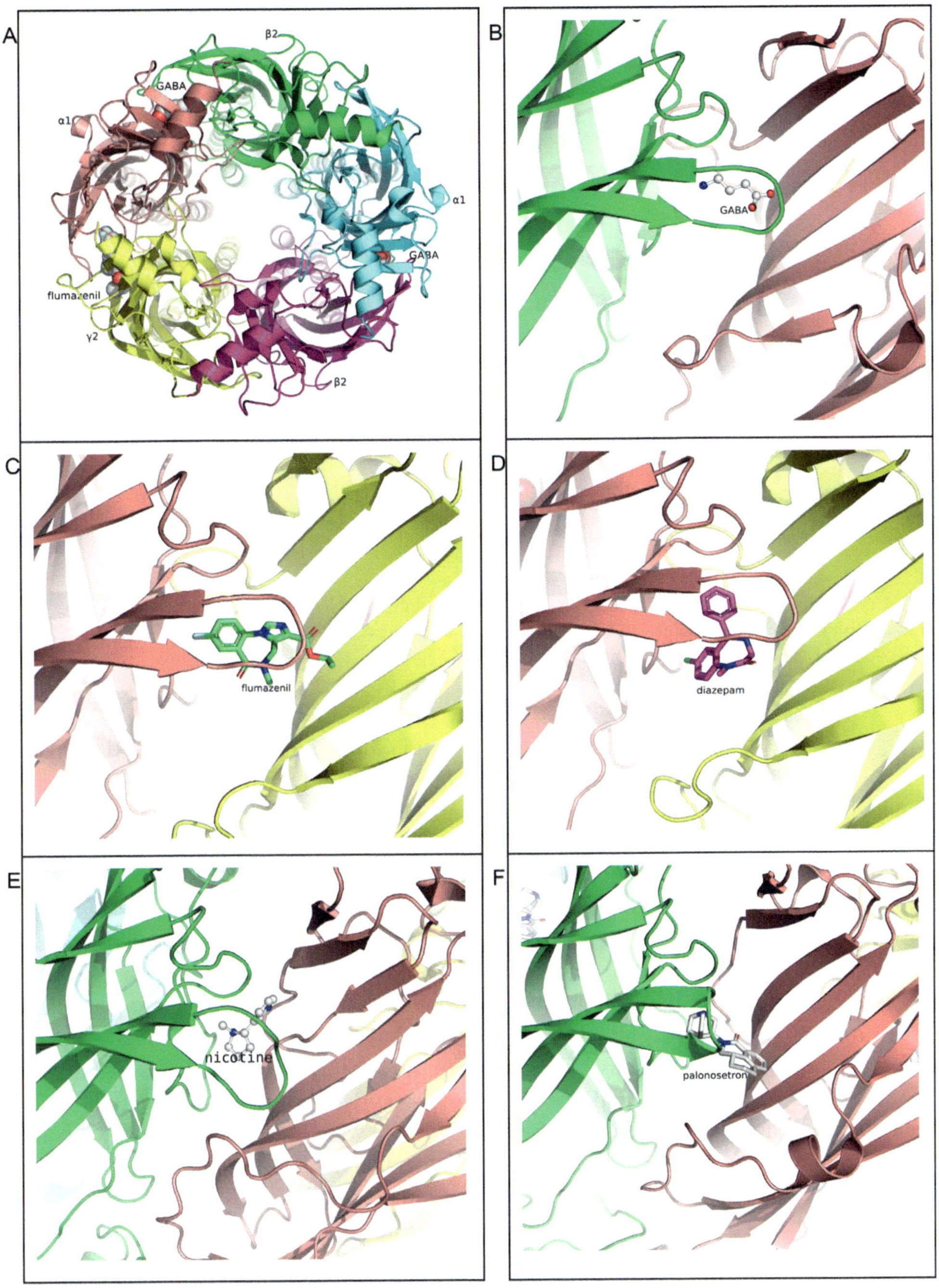

Fig. 19.4 The comparison of the intersubunit binding pockets for GABA (**b**), diazepam (**d**), flumazenil (**c**), nicotine (**e**), and palonosetron (**f**) in the ECDs. The model structures are $GABA_AR$ in complex with GABA and flumazenil (**a–c**), $GABA_AR$ with GABA and diazepam (**d**), nAChR with nicotine (**e**), and 5-HT3AR with palonosetron (**f**). Note the remarkable similarity at the intersubunit packets shared by three different receptors. PDB ID: 6X3U (**a–c**), 6X3X (**d**), 5KXI (**e**), and 6Y1Z (**f**)

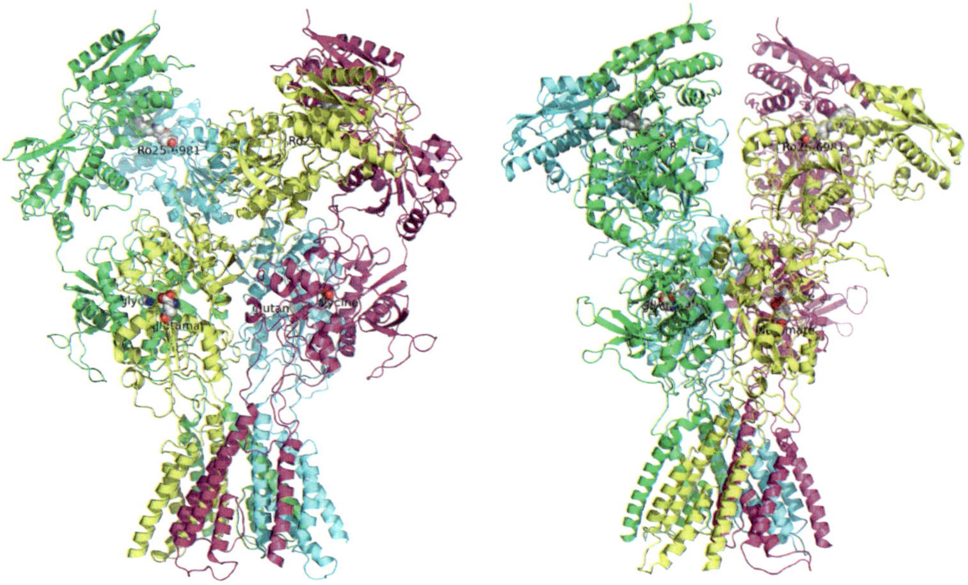

Fig. 19.5 Cryo-EM structure of GluN1/GluN2B NMDA receptor in complex with glutamate, glycine, and allosteric inhibitor Ro25-6981. The ligands and inhibitors are shown as spheres. Note the three-layered structure from top to bottom is ATD, LBD, and TMD. PDB ID: 5IOV

Clinically, muscle twitching indicates succinylcholine is taking effect. When the nondepolarizing muscle relaxants are used during the surgery, it's often needed to restore the normal function of the neuromuscular junction by administering cholinesterase inhibitors to decrease the local degradation of the endogenous acetylcholine in order to increase the ligand concentration.

Similar to the $GABA_A$Rs, most general anesthetics also affect nAChR's activities, however, negatively [44, 45]. Neuronal $\alpha_4\beta_2$ receptors are much more sensitive to general anesthetics (halothane, isoflurane, sevoflurane, and propofol) inhibition than the muscle αβγδ receptors [46]. The single-channel recording also showed the inhibition of nAChR activity in the BC3H1 cells by inhalational anesthetics including isoflurane and N_2O [47]. The IC_{50} of volatile anesthetics on $\alpha_4\beta_2$ receptors is significantly lower than their clinical EC_{50} values, suggesting their important clinical relevance. However, the inhibition by the *i.v.* anesthetics, propofol, has not been found to be clinically relevant.

In 1905, John Newport Langley published his work on the neuromuscular junction as the site of action for nicotine and curare [48]. Hundred years later, the first high-resolution EM structure of nAChR from the *Torpedo* electric ray showed a pentameric channel with a large N-terminal ECD, a transmembrane pore, and a smaller intracellular domain [49–51]. The gate is formed by a hydrophobic girdle in the middle of the lipid bilayer. The structure suggests that the binding of Ach stabilizes the ECD conformation and causes rotational movements which eventually get translated to the conformational changes of the pore-lining helices. Built upon the structural knowledge, studies with mutagenesis, photolabeling, and molecular dynamics simulations have identified multiple binding sites for anesthetics, including pore-lining residues from M2 helices, intersubunit residues, and intrasubunit residues [45, 52–54]. More structural studies are needed to visualize further details on their interactions with anesthetics.

19.3 Glutamate Receptor

Glutamate is the major excitatory neurotransmitter in CNS, targeting metabotropic (mGluRs) or ionotropic glutamate receptors (iGluRs). iGluRs are further divided into three subfamilies, AMPA (GluA), kainate (GluK), and NMDA receptors (GluN, NMDARs). Ca^{2+} influx upon the activation of NMDARs by the concurrent binding of glycine/serine and glutamate, and membrane depolarization, mediates the signaling cascade. The structure of iGluR is composed of the amino-terminal domain (ATD), ligand-binding domain (LBD), transmembrane domain (TMD), and Carboxy-terminal domain (CTD) (Fig. 19.5). The TMD has the voltage sensor In the NMDARs, but not in the non-NMDA receptors. Although AMPA and kainate receptors share sequence similarities with NMDARs, they have distinct structural and functional features. AMPA and kainate receptors are homotetramers and exhibit faster kinetics, whereas NMDARs are obligate heterotetramers and kinetically much slower. NMDARs consist of two GluN1 and two GluN2 subunits. GluN1s bind glycine and GluN2s bind glutamate. NMDARs are considered a key player in learning and memory because the knockout mice lacking NMDAR subunits all displayed strong phenotypes, either lethal or significant impairment in learning and memory [55–57].

In contrast to non-NMDA receptors, NMDARs are major targets of many inhalational anesthetics and ketamine [58–60]. N_2O and Xe have little effect on $GABA_AR$, however, are potent NMDAR inhibitors [61, 62]. Mutagenesis studies suggest that halogenated volatile anesthetics like isoflurane do not share the same binding mechanism as N_2O or ketamine [60]. The residues in the M3 and M4 helices are important for isoflurane and xenon, but not related to the N_2O and ketamine. In contrast, intravenous anesthetics propofol seems to only have an inhibitory effect on the Ca^{2+} influx through the NMDARs at concentrations higher than the clinically relevant ones [63]. Similarly, etomidate does not affect the activity of human NMDARs expressed in *Xenopus* oocytes at the clinically relevant concentrations [64]. The exact role of NMDAR blockade in maintaining anesthesia remains unclear [65]. The NMDAR knockout mice lacking subunit GluN2A showed normal response to sevoflurane and isoflurane but resistance to ketamine and N_2O [66, 67]. This is possibly related to the activity changes of other signaling pathways such as GABAergic, cholinergic, or monoaminergic pathways.

In 2009, the first crystal structure of a homotetrameric AMPA receptor, GluA2, showed a Y-shaped molecular with twofold symmetry in the ATD and LBD, and fourfold symmetry in the TMD [68]. In 2014, the structure of a heterotetrameric NMDAR showed a high degree of overall similar architecture to the GluA2 structure, but with distinct intersubunit and interdomain interfaces [69]. ATDs and LBDs are in a configuration of a dimer of dimers, N1-N2-N1-N2, while the TMDs are in a pseudo-fourfold symmetry. Interestingly, the pore-lining helical structure highly resembles that of an inverted potassium channel. The recent structure of the GluN1/GluN2B NMDA receptor without ATD in complex with two antagonists, MK-801 and memantine, showed that the two drugs bind into the center pore between the M3 bundle crossing and the M2 pore loop [70]. Another study on structures of GluN1/N2B in complex with the agonists, antagonists, or allosteric inhibitors showed different conformations in the LBDs and ATDs [71]. The binding of the allosteric inhibitor Ro25-6981 into the ATD stabilizes the LBD gating structure where glutamate and glycine are bound (Fig. 19.5). Similar to the pLGIC structures, long-range allosteric conformational changes are required from the ATD to LBD, then to the TMD. Residues located in the TMD were identified to be important for the interactions with volatile anesthetics, barbiturates, and ketamine [72–74]. More structure studies in the future are needed to show the antagonism by the anesthetics.

19.4 Two-Pore-Domain Background K^+ Channel

Two-pore-domain K^+ (K2P) channels have two sets of transmembrane domains and re-entry pore helices encoded by a single gene so that only two of such subunits are needed to form a functional

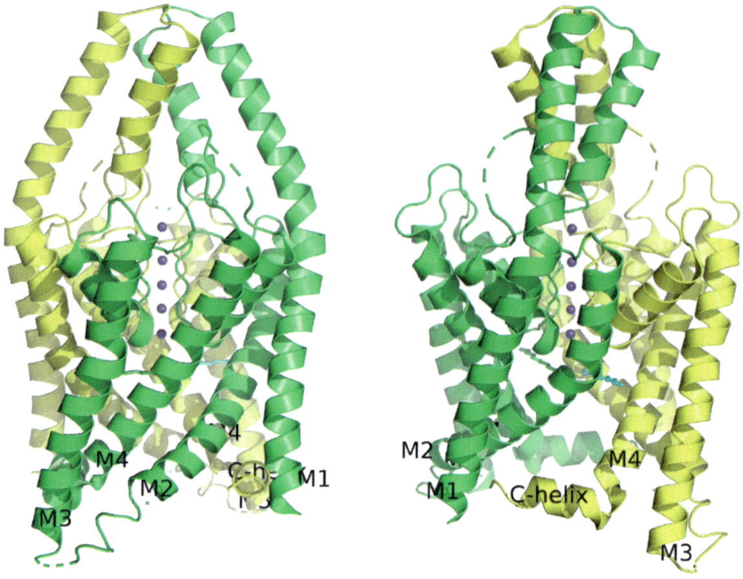

Fig. 19.6 Crystal structure of the human K2P channel (TWIK-1). Two subunits are colored in green and yellow. The extracellular cap domain is dimeric. The overall architecture of the TMD is similar to tetrameric K^+ channels. The C-helix is involved in gating. K^+ ions in the purple line up in the selectivity filter

channel. K2P channels share similar transmembrane domain architecture as other tetrameric K^+ channels. What is different in K2P channels is the extracellular dimeric cap-domain, contributed by the sequence between the M1 and pore-helix from each subunit (Fig. 19.6). K2P channels are thought to regulate resting membrane potential and the excitability of the cells in the nervous system, heart, and muscles. In the mammalian genome, 15 genes of K2P have been identified. K2P channels are activated by a variety of stimuli such as phosphorylation, polyunsaturated fatty acids, inhalational anesthetics, pH, a mechanical stretch of the membrane, and temperature changes.

A novel neuronal K^+ current, identified in one of the largest neurons with spontaneous activities in the great pond snail, *Lymnea stagnalis*, can be activated by inhalational anesthetics: halothane, sevoflurane, isoflurane, enflurane, and chloroform, which lead to neuron hyperpolarization [75, 76]. It was soon found that the mammalian TREK-1 and TASK channels share similar electrophysiological properties with the *Lymnea* channels and can also be activated by volatile anesthetics [77]. Additionally, the CTD after M4 was shown to be critical for activation. The molecular identity of the *Lymnea* K^+ current was later identified as the K2P channel and the cytoplasmic regions of M2 and M3 were also involved in the activation by anesthetics [78]. The importance of K2P channels in anesthesia was further demonstrated by the TREK-1 knockout mice, which showed longer latency to loss of right reflex and higher MAC value to volatile anesthetics, but a normal response to the *i.v.* anesthetics, phenobarbiturate [79]. TASK-3 channels can also be activated by pH and anesthetics, and the TASK-3 knockout mice not only had reduced sensitivity to volatile anesthetics but also displayed fragmented sleep patterns and changes in EEG waveforms [80].

The crystal structures of the human TWIK-1 and human TRAAK channel showed an overall conserved tetrameric K^+ channels core structure in the transmembrane region (Fig. 19.6) [81, 82]. Distinctively, they contain a dimeric extracellular cap-domain which lines on the top of the K^+ permeation pathway so that K^+ ions have to enter from the side portals. The C-terminal helix, as shown in the structure, is located close to the intracellular portion of the neighboring subunit, suggesting the role of the C-terminal helix in the inner gating mechanism of the channel. This structure feature is consistent with the previous findings that the channel activation by the volatile anesthetics is through the involvement of the intracellular end of M2 and

M3, and the C-helix [83–85]. Additionally, local anesthetics bupivacaine was shown to interact with the residues at the pore helix level to allosterically inhibit the gating at the selectivity filter [86].

19.5 Conclusion

Our knowledge of the molecular mechanisms underlying general anesthesia has advanced tremendously in recent years, especially with the help of high-resolution cryo-EM structural studies. In this chapter, we have discussed the current understanding of the interactions between anesthetic agents and the $GABA_A$Rs, nAChRs, NMDARs, and K2P channels, although this list is far from complete. Many other channels and receptors are involved in the anesthesia. A better understanding of the molecular mechanisms will guide us to design more specific drugs targeting particular pathways to achieve the goal of balanced anesthesia with better safety and efficiency.

References

1. Miller RD (2010) Miller's anesthesia. Churchill Livingstone, Philadelphia, PA
2. Al-Hasani R, Bruchas MR (2011) Molecular mechanisms of opioid receptor-dependent signaling and behavior. Anesthesiology 115:1363–1381
3. Wickman K, Clapham DE (1995) Ion channel regulation by G proteins. Physiol Rev 75:865–885
4. Zamponi GW, Snutch TP (1998) Modulation of voltage-dependent calcium channels by G proteins. Curr Opin Neurobiol 8:351–356
5. Hibbs RE, Gouaux E (2011) Principles of activation and permeation in an anion-selective Cys-loop receptor. Nature 474:54–60
6. Corringer P-J et al (2012) Structure and pharmacology of pentameric receptor channels: from bacteria to brain. Structure 20:941–956
7. Ortells MO, Lunt GG (1995) Evolutionary history of the ligand-gated ion-channel superfamily of receptors. Trends Neurosci 18:121–127
8. Brohan J, Goudra BG (2017) The role of GABA receptor agonists in anesthesia and sedation. CNS Drugs 31:845–856
9. de Sousa SLM, Dickinson R, Lieb WR, Franks NP (2000) Contrasting synaptic actions of the inhalational general anesthetics isoflurane and xenon. Anesthesiology 92:1055–1066
10. Raines DE, Claycomb RJ, Scheller M, Forman SA (2001) Nonhalogenated alkane anesthetics fail to potentiate agonist actions on two ligand-gated ion channels. Anesthesiology 95:470–477
11. Bonin RP, Martin LJ, MacDonald JF, Orser BA (2007) α5GABAA receptors regulate the intrinsic excitability of mouse hippocampal pyramidal neurons. J Neurophysiol 98:2244–2254
12. Cavelier P, Hamann M, Rossi D, Mobbs P, Attwell D (2005) Tonic excitation and inhibition of neurons: ambient transmitter sources and computational consequences. Prog Biophys Mol Biol 87:3–16
13. Bai D et al (2001) Distinct functional and pharmacological properties of tonic and quantal inhibitory postsynaptic currents mediated by gamma-aminobutyric acid(A) receptors in hippocampal neurons. Mol Pharmacol 59:814–824
14. Farrant M, Nusser Z (2005) Variations on an inhibitory theme: phasic and tonic activation of GABA A receptors. Nat Rev Neurosci 6:215–229
15. Hentschke H, Schwarz C, Antkowiak B (2005) Neocortex is the major target of sedative concentrations of volatile anaesthetics: strong depression of firing rates and increase of GABAA receptor-mediated inhibition. Eur J Neurosci 21:93–102
16. Cheng VY et al (2006) α5GABAA receptors mediate the amnestic but not sedative-hypnotic effects of the general anesthetic etomidate. J Neurosci 26:3713–3720
17. Bieda MC, MacIver MB (2004) Major role for tonic GABAA conductances in anesthetic suppression of intrinsic neuronal excitability. J Neurophysiol 92:1658–1667
18. Caraiscos VB et al (2004) Selective enhancement of tonic GABAergic inhibition in murine hippocampal neurons by low concentrations of the volatile anesthetic isoflurane. J Neurosci 24:8454–8458
19. Jurd R et al (2003) General anesthetic actions in vivo strongly attenuated by a point mutation in the GABA (A) receptor beta3 subunit. FASEB J 17:250–252
20. Mihic SJ et al (1997) Sites of alcohol and volatile anaesthetic action on GABA A and glycine receptors. Nature 389:385–389
21. Nury H et al (2011) X-ray structures of general anaesthetics bound to a pentameric ligand-gated ion channel. Nature 469:428–431
22. Hosie AM, Wilkins ME, da Silva HMA, Smart TG (2006) Endogenous neurosteroids regulate GABA A receptors through two discrete transmembrane sites. Nature 444:486–489
23. Miller PS, Aricescu AR (2014) Crystal structure of a human GABAA receptor. Nature 512:270–275
24. Yip GMS et al (2013) A propofol binding site on mammalian GABAA receptors identified by photolabeling. Nat Chem Biol 9:715–720
25. Bocquet N et al (2007) A prokaryotic proton-gated ion channel from the nicotinic acetylcholine receptor family. Nature 445:116–119

26. Bocquet N et al (2009) X-ray structure of a pentameric ligand-gated ion channel in an apparently open conformation. Nature 457:111–114
27. Hilf RJC, Dutzler R (2008) X-ray structure of a prokaryotic pentameric ligand-gated ion channel. Nature 452:375–379
28. Hilf RJC, Dutzler R (2009) Structure of a potentially open state of a proton-activated pentameric ligand-gated ion channel. Nature 457:115–118
29. Spurny R et al (2012) Pentameric ligand-gated ion channel ELIC is activated by GABA and modulated by benzodiazepines. Proc Natl Acad Sci U S A 109: E3028–E3034
30. Xiu X, Puskar NL, Shanata JAP, Lester HA, Dougherty DA (2009) Nicotine binding to brain receptors requires a strong cation-π interaction. Nature 458:534–537
31. Kim JJ et al (2020) Shared structural mechanisms of general anaesthetics and benzodiazepines. Nature 585:303–308
32. Morales-Perez CL, Noviello CM, Hibbs RE (2016) X-ray structure of the human α4β2 nicotinic receptor. Nature 538:411–415
33. Zarkadas E et al (2020) The binding of palonosetron and other antiemetic drugs to the serotonin 5-HT3 receptor. Structure 28:1131–1140.e4
34. Purohit P, Auerbach A (2013) Loop C and the mechanism of acetylcholine receptor-channel gating. J Gen Physiol 141:467–478
35. Basak S et al (2020) High-resolution structures of multiple 5-HT3AR-setron complexes reveal a novel mechanism of competitive inhibition. Elife 9:e57870
36. Hassaine G et al (2014) X-ray structure of the mouse serotonin 5-HT3 receptor. Nature 512:276–281
37. Marubio LM et al (1999) Reduced antinociception in mice lacking neuronal nicotinic receptor subunits. Nature 398:805–810
38. Picciotto MR et al (1995) Abnormal avoidance learning in mice lacking functional high-affinity nicotine receptor in the brain. Nature 374:65–67
39. Xu W et al (1999) Megacystis, mydriasis, and ion channel defect in mice lacking the α3 neuronal nicotinic acetylcholine receptor. Proc Natl Acad Sci U S A 96:5746–5751
40. Ji D, Lape R, Dani JA (2001) Timing and location of nicotinic activity enhances or depresses hippocampal synaptic plasticity. Neuron 31:131–141
41. Wu M, Shanabrough M, Leranth C, Alreja M (2000) Cholinergic excitation of septohippocampal GABA but not cholinergic neurons: implications for learning and memory. J Neurosci 20:3900–3908
42. Pan H-L, Chen S-R, Eisenach JC (1999) Intrathecal clonidine alleviates allodynia in neuropathic rats: interaction with spinal muscarinic and nicotinic receptors. Anesthesiology 90:509–514
43. Brams M et al (2011) Crystal structures of a cysteine-modified mutant in loop D of acetylcholine-binding protein. J Biol Chem 286:4420–4428
44. Forman SA, Miller KW (2011) Anesthetic sites and allosteric mechanisms of action on Cys-loop ligand-gated ion channels. Can J Anaesth 58:191–205
45. Forman SA, Chiara DC, Miller KW (2015) Anesthetics target interfacial transmembrane sites in nicotinic acetylcholine receptors. Neuropharmacology 96:169–177
46. Violet JM, Downie DL, Nakisa RC, Lieb WR, Franks NP (1997) Differential sensitivities of mammalian neuronal and muscle nicotinic acetylcholine receptors to general anesthetics. Anesthesiology 86:866–874
47. Wachtel RE (1995) Relative potencies of volatile anesthetics in altering the kinetics of ion channels in BC3H1 cells. J Pharmacol Exp Ther 274:1355–1361
48. Langley JN (1905) On the reaction of cells and of nerve-endings to certain poisons, chiefly as regards the reaction of striated muscle to nicotine and to curari. J Physiol 33:374–413
49. Unwin N (1993) Nicotinic acetylcholine receptor an 9 Å resolution. J Mol Biol 229:1101–1124
50. Miyazawa A, Fujiyoshi Y, Unwin N (2003) Structure and gating mechanism of the acetylcholine receptor pore. Nature 423:949–955
51. Unwin N (2005) Refined structure of the nicotinic acetylcholine receptor at 4 Å resolution. J Mol Biol 346:967–989
52. Brannigan G, LeBard DN, Hénin J, Eckenhoff RG, Klein ML (2010) Multiple binding sites for the general anesthetic isoflurane identified in the nicotinic acetylcholine receptor transmembrane domain. Proc Natl Acad Sci U S A 107:14122–14127
53. Liu LT, Willenbring D, Xu Y, Tang P (2009) General anesthetic binding to neuronal α4β2 nicotinic acetylcholine receptor and its effects on global dynamics. J Phys Chem B 113:12581–12589
54. Yamakura T, Borghese C, Harris RA (2000) A transmembrane site determines sensitivity of neuronal nicotinic acetylcholine receptors to general anesthetics. J Biol Chem 275:40879–40886
55. Forrest D et al (1994) Targeted disruption of NMDA receptor 1 gene abolishes NMDA response and results in neonatal death. Neuron 13:325–338
56. Sakimura K et al (1995) Reduced hippocampal LTP and spatial learning in mice lacking NMDA receptor ε1 subunit. Nature 373:151–155
57. Das S et al (1998) Increased NMDA current and spine density in mice lacking the NMDA receptor subunit NR3A. Nature 393:377–381
58. Liu H-T, Hollmann MW, Liu W-H, Hoenemann CW, Durieux ME (2001) Modulation of NMDA receptor function by ketamine and magnesium: Part I. Anesth Analg 92:1173–1181
59. Hollmann MW, Liu H-T, Hoenemann CW, Liu W-H, Durieux ME (2001) Modulation of NMDA receptor function by ketamine and magnesium. Part II: Interactions with volatile anesthetics. Anesth Analg 92:1182–1191

60. Ogata J et al (2006) Effects of anesthetics on mutant N-methyl-d-aspartate receptors expressed in Xenopus oocytes. J Pharmacol Exp Ther 318:434–443
61. Yamakura T, Harris RA (2000) Effects of gaseous anesthetics nitrous oxide and xenon on ligand-gated ion channels: comparison with isoflurane and ethanol. Anesthesiology 93:1095–1101
62. Franks NP, Dickinson R, de Sousa SLM, Hall AC, Lieb WR (1998) How does xenon produce anaesthesia? Nature 396:324–324
63. Grasshoff C, Gillessen T (2005) Effects of propofol on N-methyl-D-aspartate receptor-mediated calcium increase in cultured rat cerebrocortical neurons. Eur J Anaesthesiol 22:467–470
64. Pierce DW, Pejo E, Raines DE, Forman SA (2012) Carboetomidate inhibits alpha4/beta2 neuronal nicotinic acetylcholine receptors at concentrations affecting animals. Anesth Analg 115:70–72
65. Petrenko AB, Yamakura T, Sakimura K, Baba H (2014) Defining the role of NMDA receptors in anesthesia: are we there yet? Eur J Pharmacol 723:29–37
66. Sato Y, Kobayashi E, Murayama T, Mishina M, Seo N (2005) Effect of N-methyl-d-aspartate receptor ε1 subunit gene disruption of the action of general anesthetic drugs in mice. Anesthesiology 102:557–561
67. Petrenko AB, Yamakura T, Kohno T, Sakimura K, Baba H (2013) Increased brain monoaminergic tone after the NMDA receptor GluN2A subunit gene knockout is responsible for resistance to the hypnotic effect of nitrous oxide. Eur J Pharmacol 698:200–205
68. Sobolevsky AI, Rosconi MP, Gouaux E (2009) X-ray structure of AMPA-subtype glutamate receptor: symmetry and mechanism. Nature 462:745–756
69. Karakas E, Furukawa H (2014) Crystal structure of a heterotetrameric NMDA receptor ion channel. Science 344:992–997
70. Song X et al (2018) Mechanism of NMDA receptor channel block by MK-801 and memantine. Nature 556:515–519
71. Zhu S et al (2016) Mechanism of NMDA receptor inhibition and activation. Cell 165:704–714
72. Minami K et al (1998) Sites of volatile anesthetic action on kainate (glutamate receptor 6) receptors. J Biol Chem 273:8248–8255
73. Yamakura T, Mori H, Masaki H, Shimoji K, Mishina M (1993) Different sensitivities of NMDA receptor channel subtypes to non-competitive antagonists. Neuroreport 4:687–690
74. Birnir B, Tierney ML, Dalziel JE, Cox GB, Gage PW (1997) A structural determinant of desensitization and allosteric regulation by pentobarbitone of the GABAA receptor. J Membr Biol 155:157–166
75. Lopes CMB, Franks NP, Lieb WR (1998) Actions of general anaesthetics and arachidonic acid pathway inhibitors on K^+ currents activated by volatile anaesthetics and FMRFamide in molluscan neurones. Br J Pharmacol 125:309–318
76. Franks NP, Lieb WR (1988) Volatile general anaesthetics activate a novel neuronal K^+ current. Nature 333:662–664
77. Patel AJ et al (1999) Inhalational anesthetics activate two-pore-domain background K^+ channels. Nat Neurosci 2:422–426
78. Andres-Enguix I et al (2007) Determinants of the anesthetic sensitivity of two-pore domain acid-sensitive potassium channels molecular cloning of an anesthetic-activated potassium channel from Lymnaea stagnalis. J Biol Chem 282:20977–20990
79. Heurteaux C et al (2004) TREK-1, a K^+ channel involved in neuroprotection and general anesthesia. EMBO J 23:2684–2695
80. Pang DSJ et al (2009) An unexpected role for TASK-3 potassium channels in network oscillations with implications for sleep mechanisms and anesthetic action. Proc Natl Acad Sci U S A 106:17546–17551
81. Miller AN, Long SB (2012) Crystal structure of the human two-pore domain potassium channel K2P1. Science 335:432–436
82. Brohawn SG, del Marmol J, MacKinnon R (2012) Crystal structure of the human K2P TRAAK, a lipid- and mechano-sensitive K^+ ion channel. Science 335:436–441
83. Mathie A, Veale EL, Cunningham KP, Holden RG, Wright PD (2020) Two-pore domain potassium channels as drug targets: anesthesia and beyond. Annu Rev Pharmacol Toxicol. https://doi.org/10.1146/annurev-pharmtox-030920-111536
84. Luethy A, Boghosian JD, Srikantha R, Cotten JF (2017) Halogenated ether, alcohol, and alkane anesthetics activate TASK-3 tandem pore potassium channels likely through a common mechanism. Mol Pharmacol 91:620–629
85. Bertaccini EJ, Dickinson R, Trudell JR, Franks NP (2014) Molecular modeling of a tandem two pore domain potassium channel reveals a putative binding site for general anesthetics. ACS Chem Nerosci 5:1246–1252
86. Rinné S et al (2019) The molecular basis for an allosteric inhibition of K^+-flux gating in K2P channels. Elife 8:e39476

GPSR Compliance

The European Union's (EU) General Product Safety Regulation (GPSR) is a set of rules that requires consumer products to be safe and our obligations to ensure this.

If you have any concerns about our products, you can contact us on ProductSafety@springernature.com

In case Publisher is established outside the EU, the EU authorized representative is:

Springer Nature Customer Service Center GmbH
Europaplatz 3
69115 Heidelberg, Germany

Batch number: 10607495

Printed by Printforce, the Netherlands